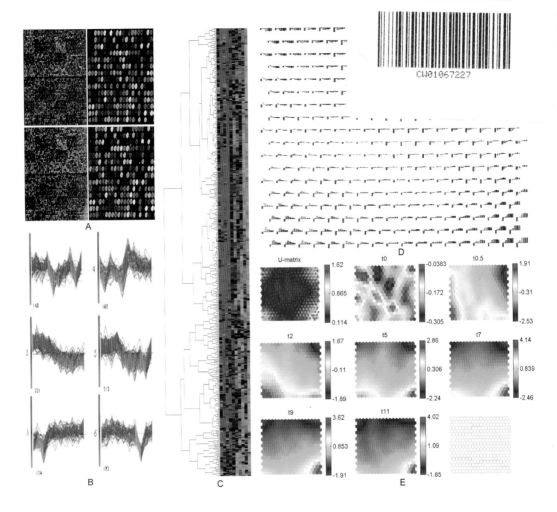

Fig. 1.2. Microarray data visualizations using different algorithms. **A.** Colour images illustrating similarities and dissimilarities between two brain development stages. The upper left is the full image generated by the Cy3 labelled day 11.5 p.c. mouse brain cDNA pool vs. a Cy5 labelled mouse embryonic liver cDNA pool and lower left is the full image generated from the Cy3 labelled day 12.5 p.c. mouse brain cDNA vs. the same control. The right panel depicts partials of the left panels to illustrate details. **B.** Graphic distributions showing representative clusters obtained by K-means clustering. The horizontal scale represents RNA samples obtained from ten different time points of mouse embryonic brain development. The vertical scale weighs changes in expression, from high expression (red) to low expression (green) with units in log ratio, subtracted by median. **C.** A partial tree view obtained through hierarchical clustering of 4608 mouse genes over ten embryonic development samples. Red represents up-regulation and green represents down regulation. **D.** A bar graphic display of SOM illustrating gene clustering and expression patterns of regulated genes during the yeast sporulation process. All genes were organized into 324 (18 x 18) hexagonal map units. Each bar in a given unit illustrates the average expression of genes mapped to that unit. **E.** U-matrix and component plane presentations. The colour coding in U-matrix stands for Euclidean distance. The darker the colour, the smaller the distance. The large dark-blue area that occupies the majority of the display represents unregulated genes, which form some noise clustering. The component plane presentations (t0–t11) illustrate differential displays of regulated genes during sporulation of yeast at the genome-wide scale. The colour coding index stands for the expression values of genes. The brighter the colour, the higher the value. All these differential displays are linked by position: in each display, the hexagon in a certain position corresponds to the same map unit. It is straightforward to compare expression patterns in same positions of different displays. The last label display shows positions of each unit on the map.

Molecular Nutrition

Molecular Nutrition

Edited by

Janos Zempleni

University of Nebraska
Lincoln
USA

and

Hannelore Daniel

Technical University of Munich
Germany

CABI Publishing

CABI Publishing is a division of CAB International

CABI Publishing
CAB International
Wallingford
Oxon OX10 8DE
UK
Tel: +44 (0) 1491 832111
Fax: +44 (0)1491 833508
E-mail: cabi@cabi.org
Web site: www.cabi-publishing.org

CABI Publishing
44 Brattle Street
4th Floor
Cambridge, MA 02138
USA
Tel: +1 617 395 4056
Fax: +1 617 354 6875
E-mail: cabi-nao@cabi.org

©CAB International 2003. All rights reserved. No part of this publication may be reproduced in any form or by any means, electronically, mechanically, by photocopying, recording or otherwise, without the prior permission of the copyright owners.

A catalogue record for this book is available from the British Library, London, UK.

Library of Congress Cataloging-in-Publication Data

Molecular nutrition/edited by Janos Zempleni and Hannelore Daniel.
    p. ; cm.
Includes bibliographical references and index.
    ISBN 0-85199-679-5 (alk. paper)
    1. Nutrient interactions. 2. Nutrition. 3. Molecular biology.
    [DNLM: 1. Nutrition. 2. Molecular Biology. QU 145 M718 2002]
I. Zempleni, Janos. II. Daniel, Hannelore.
    QP143.7 .M656 2002
    517.6--dc21
                                                                                    2002154031

ISBN 0 85199 679 5

Typeset by AMA DataSet Ltd, UK
Printed and bound in the UK by Biddles Ltd, Guildford and King's Lynn

Contents

Contributors ix

Preface xi

PART 1: METHODS IN MOLECULAR NUTRITION RESEARCH

1 **Genomics and Beyond** 1
   Ji Zhang

2 **Perspectives in Post-genomic Nutrition Research** 13
   Hannelore Daniel

PART 2: CELLULAR NUTRIENT HOMEOSTASIS, PROLIFERATION AND APOPTOSIS

3 **Molecular Physiology of Plasma Membrane Transporters for Organic Nutrients** 21
   Hannelore Daniel

4 **Intracellular Trafficking and Compartmentalization of Vitamins and Their Physiologically Active Forms** 43
   Donald B. McCormick

5 **Nutrient Homeostasis in Proliferating Cells** 61
   Janos Zempleni

6 **Nutrients and Apoptosis** 73
   John C. Mathers

PART 3: ROLES FOR NUTRIENTS IN SIGNAL TRANSDUCTION, GENE EXPRESSION AND PROTEOLYSIS

7 **Glucose Regulation of Gene Expression in Mammals** 91
   Fabienne Foufelle and Pascal Ferré

8	**Amino Acid-dependent Control of Transcription in Mammalian Cells** *Michael S. Kilberg, Van Leung-Pineda and Chin Chen*	105
9	**Fatty Acids and Gene Expression** *Ulrike Beisiegel, Joerg Heeren and Frank Schnieders*	121
10	**Role of RARs and RXRs in Mediating the Molecular Mechanism of Action of Vitamin A** *Dianne R. Soprano and Kenneth J. Soprano*	135
11	**Regulation of Gene Expression by Biotin, Vitamin B_6 and Vitamin C** *Krishnamurti Dakshinamurti*	151
12	**Selenium and Vitamin E** *Alexandra Fischer, Josef Pallauf, Jonathan Majewicz, Anne Marie Minihane and Gerald Rimbach*	167
13	**Sphingolipids: a New Strategy for Cancer Treatment and Prevention** *Eva M. Schmelz*	187
14	**The Health Effects of Dietary Isoflavones** *Thomas M. Badger, Martin J.J. Ronis and Nianbai Fang*	201
15	**Mechanisms of Ubiquitination and Proteasome-dependent Proteolysis in Skeletal Muscle** *Didier Attaix, Lydie Combaret, Anthony J. Kee and Daniel Taillandier*	219

PART 4: NUCLEIC ACIDS AND NUCLEIC ACID-BINDING COMPOUNDS

16	**Diet, DNA Methylation and Cancer** *Judith K. Christman*	237
17	**Biotinylation of Histones in Human Cells** *Janos Zempleni*	267
18	**Niacin Status, Poly(ADP-ribose) Metabolism and Genomic Instability** *Jennifer C. Spronck and James B. Kirkland*	277

PART 5: MOLECULAR EVENTS AFFECT PHYSIOLOGY

19	**Assembly of Triglyceride-transporting Plasma Lipoproteins** *Joan A. Higgins*	293
20	**Regulation of Cellular Cholesterol** *Ji-Young Lee, Susan H. Mitmesser and Timothy P. Carr*	309
21	**2002 Assessment of Nutritional Influences on Risk for Cataract** *Allen Taylor and Mark Siegal*	321
22	**Nutrition and Immune Function** *Parveen Yaqoob and Philip C. Calder*	349

PART 6: FOODS

23	**Molecular Mechanisms of Food Allergy**	369
	J. Steven Stanley and Gary A. Bannon	
24	**Safety Assessment of Genetically Modified Foods**	381
	Steve L. Taylor	

Index   395

Contributors

D. Attaix, *Nutrition and Protein Metabolism Unit, INRA de Theix, 63122 Ceyrat, France.*
T.M. Badger, *Department of Pediatrics, Arkansas Children's Hospital Research Institute, Slot 512-20, 1120 Marshall Street, Little Rock, AR 72202, USA.*
G.A. Bannon, *Product Safety Center, Monsanto Company, 800 N. Lindbergh Boulevard, St Louis, MO 63167, USA.*
U. Beisiegel, *Abteilung für Molekulane Zellbiologie, Institut für Biochemie und Molekularbiologie, Universitätsklinikum Hamburg-Eppendorf, Martinistraße 52, D-20246 Hamburg, Germany.*
P.C. Calder, *University of Southampton School of Medicine, Fetal Origin of Adult Disease Division, Institute of Human Nutrition, Highfield, Southampton SO17 1BJ, UK.*
T.P. Carr, *Department of Nutritional Science and Dietetics, University of Nebraska-Lincoln, 316 Ruth Leverton Hall, Lincoln, NE 68583-0806, USA.*
C. Chen, *Department of Biochemistry and Molecular Biology, University of Florida College of Medicine, Box 100245, JHMHC, Gainesville, FL 32610-0245, USA.*
J.K. Christman, *Stokes-Shackleford Professor and Chair, Department of Biochemistry and Molecular Biology and Eppley Cancer Center, University of Nebraska Medical Center, 984525 University Medical Center, Omaha, NE 68198-4525, USA.*
L. Combaret, *Nutrition and Protein Metabolism Unit, INRA de Theix, 63122 Ceyrat, France.*
K. Dakshinamurti, *Department of Biochemistry and Molecular Biology, University of Manitoba, 770 Bannatyne Avenue, Room 305 Basic Sciences Building, Winnipeg, Manitoba R3E 0W3, Canada*
H. Daniel, *Department of Food and Nutrition, Technical University of Munich, Hochfeldweg 2, D-85350 Freising-Weihenstephan, Germany.*
N. Fang, *Department of Pediatrics, Arkansas Children's Hospital Research Institute, Slot 512-20, 1120 Marshall Street, Little Rock, AR 72202, USA.*
P. Ferre, *U465 INSERM, French Institute of Health and Medical Research, Centre de Recherches Biomedicales des Cordeliers, 15 rue de l'Ecole de Medicine, 75270 Paris Cedex 06, France.*
A. Fischer, *Institute of Animal Nutrition and Nutrition Physiology, Justus-Leibig-University, Giessen, Germany.*
F. Foufelle, *U465 INSERM, French Institute of Health and Medical Research, Centre de Recherches Biomedicales des Cordeliers, 15 rue de l'Ecole de Medicine, 75270 Paris Cedex 06, France.*
J. Heeren, *Abteilung für Molekulane Zellbiologie, Institut für Biochemie und Molekularbiologie, Universitätsklinikum Hamburg-Eppendorf, Martinistraße 52, D-20246 Hamburg, Germany.*
J.A. Higgins, *Department of Molecular Biology and Biotechnology, University of Sheffield, Firth Court, Western Bank, Sheffield S10 2TN, UK.*
A.J. Kee, *Muscle Development Unit, Children's Medical Research Institute, Locked Bag 23, Wenworthville, NSW 2145, Australia.*

M.S. Kilberg, *Department of Biochemistry and Molecular Biology, University of Florida College of Medicine, Box 100245, JHMHC, Gainesville, FL 32610-0245, USA.*

J.B. Kirkland, *Department of Human Biology and Nutritional Sciences, University of Guelph, 335 Animal Science and Nutrition, Guelph, Ontario N1G 2W1, Canada.*

J.-Y. Lee, *Department of Nutritional Science and Dietetics, University of Nebraska-Lincoln, 316 Ruth Leverton Hall, Lincoln, NE 68583-0806, USA.*

V. Leung-Pineda, *Department of Biochemistry and Molecular Biology, University of Florida College of Medicine, Box 100245, JHMHC, Gainesville, FL 32610-0245, USA.*

J. Majewicz, *School of Food Biosciences, Hugh Sinclair Human Nutrition Unit, University of Reading, Whiteknights, PO BOX 226, Reading RG6 6AP, UK.*

J.C. Mathers, *Human Nutrition Research Centre, School of Clinical Medical Sciences, Faculty of Agriculture and Biological Sciences, University of Newcastle upon Tyne, Newcastle upon Tyne NE1 7RU, UK.*

D.B. McCormick, *Fuller E. Callaway Professor Emeritus, Emory University School of Medicine, Department of Biochemistry, 4013 Rollins Research Center, Atlanta, GA 30322-3050, USA.*

A.M. Minihane, *School of Food Biosciences, Hugh Sinclair Human Nutrition Unit, University of Reading, Whiteknights, PO Box 226, Reading RG6 6AP, UK.*

S.H. Mitmesser, *Department of Nutritional Science and Dietetics, University of Nebraska-Lincoln, 316 Ruth Leverton Hall, Lincoln, NE 68583-0806, USA.*

J. Pallauf, *Institute of Animal Nutrition and Nutrition Physiology, Justus-Leibig-University, Giessen, Germany.*

G.H. Rimbach, *School of Food Biosciences, Hugh Sinclair Human Nutrition Unit, University of Reading, Whiteknights, PO Box 226, Reading RG6 6AP, UK.*

M.J.J. Ronis, *Department of Pediatrics, Arkansas Children's Hospital Research Institute, Slot 512-20, 1120 Marshall Street, Little Rock, AR 72202, USA.*

E.M. Schmelz, *Karmanos Cancer Institute, Wayne State University, HWCRC 608, 110 East Warren Avenue, Detroit, MI 48201, USA.*

F. Schnieders, *Abteilung für Molekulane Zellbiologie, Institut für Biochemie und Molekularbiologie, Universitätsklinikum Hamburg-Eppendorf, Martinistraße 52, D-20246 Hamburg, Germany.*

M. Siegal, *Laboratory for Nutrition and Vision Research, USDA Human Nutrition Research Center on Aging, Tufts University, 711 Washington St, Boston, MA 02111, USA.*

D.R. Soprano, *Temple University School of Medicine, Room 401 MRB, 3307 North Broad Street, Philadelphia, PA 19140, USA.*

K.J. Soprano, *Temple University School of Medicine, Room 401 MRB, 3307 North Broad Street, Philadelphia, PA 19140, USA.*

J.C. Spronck, *Department of Human Biology and Nutritional Sciences, University of Guelph, 335 Animal Science and Nutrition, Guelph, Ontario N1G 2W1, Canada.*

J.S. Stanley, *Department of Pediatrics, Division of Pediatric Allergy and Immunology, University of Arkansas for Medical Sciences and Arkansas Children's Hospital Research Institute, 4301 W Markham St, Slot 512-13, Little Rock, AR 72205, USA.*

D. Taillandier, *Nutrition and Protein Metabolism Unit, INRA de Theix, 63122 Ceyrat, France.*

A. Taylor, *Laboratory for Nutrition and Vision Research, USDA Human Nutrition Research Center on Aging, Tufts University, 711 Washington St, Boston, MA 02111, USA.*

S.L. Taylor, *Department of Food Science and Technology, 143 Food Industry Building, University of Nebraska-Lincoln, Lincoln, NE 68583-0919, USA.*

P. Yaqoob, *School of Food Biosciences, University of Reading, Whiteknights, PO Box 226, Reading RG6 6AP, UK.*

J. Zempleni, *Department of Nutritional Science and Dietetics, University of Nebraska-Lincoln, 316 Ruth Leverton Hall, Lincoln, NE 68583-0806, USA.*

J. Zhang, *Center for Human Molecular Genetics, Munroe-Meyer Institute, University of Nebraska Medical Center, 985455 Nebraska Medical Center, Omaha, NE 68198-5454, USA.*

Preface

Molecular biology has provided us with many powerful tools and techniques over the past 20–30 years, leading to the emergence of molecular nutrition as a new cornerstone in nutrition research. Molecular nutrition investigates roles for nutrients at the molecular level, such as signal transduction, gene expression and covalent modifications of proteins. Research findings that have been generated by using molecular techniques have guided us into new territory, going way beyond classical nutrition studies such as characterization of clinical signs of nutrient deficiencies. Those nutritionists who investigate effects of nutrients at the molecular level have long recognized what an exciting field this is.

In this book, we sought to capture some of the excitement in the field of molecular nutrition. We were lucky enough to recruit chapter authors that play leading roles within their disciplines. These individuals gladly shared their insights into various aspects of molecular nutrition. A book like this one cannot be comprehensive: the editors had to select from a large number of nutrients and an even larger number of effects of nutrients at the molecular level. For each topic that is included in this book, another equally important topic may have been left out. As editors, we apologize to those nutritionists who do not find 'their' nutrient in this book.

We sought to cover a broad range of research in molecular nutrition. The first two chapters by J. Zhang and H. Daniel cover technical advances that have been made in genomics and post-genomics, and the promise that these technologies hold for future nutrition research. Next, H. Daniel, D.B. McCormick and J. Zempleni tell us how nutrients enter cells, how they are being targeted to their sites of action in cells and how physiological processes such as cell proliferation affect nutrient transport. This section is completed by a chapter by J.C. Mathers in which he reviews the roles for nutrients in apoptosis. A series of fine chapters by F. Foufelle and P. Ferré, M.S. Kilberg *et al.*, U. Beisiegel *et al.*, D.R. Soprano and K.J. Soprano, K. Dakshinamurti, A. Fischer *et al.*, E.M. Schmelz, T. Badger *et al.*, and D. Attaix *et al.* provides examples for roles of macro- and micronutrients in signal transduction, gene expression and proteolysis. J.K. Christman, J. Zempleni, and J. Kirkland and J.C. Spronck contributed reviews of nutrient-dependent modifications of nucleic acids and nucleic acid-binding proteins.

Of course, molecular nutrition cannot afford to be a self-serving science. Chapters on regulation of lipoprotein assembly (J.A. Higgins), cellular cholesterol metabolism (J.-Y. Lee *et al.*), oxidative stress (A. Taylor and M. Siegal) and immune function (P. Yaqoob and P.C. Calder) were selected to demonstrate how events at the molecular level can be integrated into multiorgan and whole-body metabolic pathways. Finally, J.S. Stanley and G. Bannon, and S.L. Taylor share their thoughts with us in the fields of genetically modified foods and molecular mechanisms of food allergy.

Our thanks go to all the scientists who contributed chapters for this book. These individuals created time in their busy schedules to produce chapters of outstanding quality. Also, we thank Rebecca Stubbs from CABI Publishing for her continued support during the preparation of this book.

Hannelore Daniel
Janos Zempleni

1 Genomics and Beyond

Ji Zhang[1,2]

[1]Department of Pathology and Microbiology, University of Nebraska Medical Center;
[2]Center for Human Molecular Genetics, Munroe-Meyer Institute,
University of Nebraska Medical Center, Omaha, Nebraska, USA

Introduction

Dramatic advances in genome research in recent years will facilitate the precise determination of molecular mechanisms underlying human health and disease, and thus offer great potential for promoting health, lowering mortality and morbidity, and preventing disease. Nutritional science can benefit greatly from understanding the molecular mechanisms that cause heterogeneous responses to nutrient intake observed in healthy adults. Therefore, it is of considerable value for nutrition scientists to gain knowledge of essential technologies and resources of genome research. Originally, genomics refer to the scientific discipline of mapping, sequencing and analysis of an organism's genome, the entire set of genes and chromosomes. Now, the emphasis of genomics has undergone a transition from structural analysis of the genome (structural genomics) to functional analysis of the genome (functional genomics). Structural genomics aims to construct high-resolution genetic, physical and transcriptional maps of an organism and ultimately to determine its entire DNA sequence. Functional genomics, however, represents a new phase of genome research, referring to the development of innovative technologies based on the vast amount of structural genomics information. The first section of this chapter will focus on tools and reagents utilized in structural genomics; the second section will focus on DNA microarray technology, a representative of today's functional genomics.

Structural Genomics

Genome, genetic mapping and physical mapping

The human genome is composed of approximately 3 billion nucleotide base pairs, carrying genetic codes for 30,000–100,000 genes. The DNA of the diploid genome is organized into 22 pairs of autosomes and two sex chromosomes. Each chromosome is thought to contain one linear DNA molecule featuring three functional elements required for the successful duplication of the chromosome upon cell division: (i) autonomous replication sequences; (ii) the centromere, to which the mitotic or meiotic spindle attaches; and (iii) the telomere, which ensures the complete replication of the chromatid at its ends. A map of the genome defines the relative position of different loci (genes, regulatory sequences, polymorphic marker sequences, etc.) on the DNA molecules. Two distinct approaches are used to map the loci of the genome: genetic mapping and physical mapping.

When the distance between two loci is measured by the meiotic recombination frequency, a genetic map is constructed. The closer two loci are on the DNA molecule, the more likely it is that they are inherited together. The term synteny refers to loci that reside on the same chromosome; these loci are not necessarily linked. Loci, located on different chromosomes or far apart on the same chromosome, segregate independently with the recombination frequency of 50%. Genetic distances are expressed in percentage recombination or centiMorgans (cM). One cM equals 1% recombination and corresponds to approximately 0.8×10^6 base pairs (bp). The genetic map of the human genome spans roughly 3700 cM.

Genetic maps are constructed by using linkage analysis: the analysis of the segregation of polymorphic markers in pedigrees. The first genetic markers used were phenotypic traits (e.g. colour blindness) and protein polymorphisms. However, polymorphic markers of these types are rare. Detailed genetic maps were made possible by the discovery of highly polymorphic sequences in the genome, typically represented by microsatellite sequences; the latter contain variable numbers of small tandem repeats of di-, tri- or tetranucleotides (Litt and Luty, 1989; Stallings et al., 1991). Linkage analysis is a statistical method and its resolution also depends on the number of informative pedigrees.

Unlike genetic mapping, physical mapping directly measures the distance between two loci on the linear DNA molecule by nucleotide base pairs. Therefore, the complete sequence of the genome is considered as the ultimate physical map. Both genetic and physical maps result in an identical order of genes, but the relative distance between genes can vary widely due to local variations in recombination frequency. Molecular biology provides the instruments needed to construct the physical map. At the most detailed level, the nucleotide sequence of a cloned gene can be determined. At a lower level of resolution, the distance between restriction sites is quantified in units of base pairs. Restriction sites are short sequences (typically 4–8 bp) recognized and cleaved specifically by class II restriction enzymes. The analysis of the presence of restriction sites yields a restriction map that typically comprises 10–100 kilobases (kb). Genomic fragments of this size can be cloned and manipulated using plamids, phages, cosmids, BACs (bacterial artificial chromosomes) (Shizuya et al., 1992), PACs (P1 artificial chromosomes) (Ioannou et al., 1994) or YACs (yeast artificial chromosomes) as vectors (Burke et al., 1987).

Tools and reagents utilized in physical mapping

Mapping using human–rodent hybrid panels

Somatic cell hybrids are tools utilized to map human genes physically on to chromosomes. Fusion of human and rodent cells and the subsequent culture selection results in stable hybrids, which usually contain a complete set of rodent chromosomes and a few human chromosomes. A panel of such hybrids allows for localization of a human gene or gene product on a specific human chromosome (Zhang et al., 1990). This approach can be used to develop synteny maps. To localize a gene in subchromosome regions, however, requires special hybrids containing only part of a human chromosome. This has been achieved by various deletion mapping approaches, including microcell-mediated subchromosome transfer or fusion using donor cells with specific translocations or interstitial deletions (Zhang et al., 1989a). The widely used radiation hybrid approach is also based on this deletion mapping concept (Walter et al., 1994). Here, a single human chromosome is hybridized with rodent chromosomes and cleaved into smaller fragments by using X-rays. Fusion of this radiated cell population with a rodent cell and the subsequent selection of human chromosome materials generates a panel of deletion hybrids containing different fragments of the human chromosome. Collections of radiation-reduced hybrids generated from each of 24 human chromosome-specific rodent hybrids produce a complete human radiation hybrid mapping panel. This facilitates the rapid mapping of genes, ESTs (expressed sequence tags), polymorphic DNA markers and STSs (sequence-tagged sites) (Olson et al., 1989; Weber and May, 1989) to subchromosome regions using polymerase chain reaction (PCR) (Deloukas et al., 1998).

Mapping using chromosome in situ hybridization

Genes can be assigned directly to chromosome regions by in situ hybridization, in which a DNA probe containing the sequence of interest is labelled, denatured and hybridized to its complementary chromosomal DNA from denatured metaphase spreads on a glass slide. Traditionally, the probe is labelled by the incorporation of [^{3}H]nucleotides, and the post-hybridization detection is conducted by autoradiography, in which the chromosome spreads are overlaid with a liquid emulsion, exposed and developed. Silver signals appear near the radioactive probe and therefore highlight the chromosome location of the DNA sequence of interest (Marynen *et al.*, 1989; Zhang *et al.*, 1989b). The major drawback of this radioactive *in situ* hybridization is the time needed for one experiment (typically 2–3 weeks). Also, the detection usually yields a relatively high background, and the assignment of a chromosome locus occurs on a statistical basis.

Fluorescence *in situ* hybridization (FISH) circumvents this problem. The probe is labelled by incorporation of nucleotides labelled with antigen (e.g. biotin) and the detection is performed using fluorescence-conjugated antibody (e.g. avidin). Background noise is low, and most metaphase chromosomes show four specific fluorescent signals (one for each chromatid on the two homologous chromosomes). The sensitivity of this procedure is usually low, and thus it requires the use of large genomic probes (ideally >10 kb). Genomic phage (15 kb), cosmid (40 kb), PACs (80–135 kb), BACs (130 kb) and even YACs (200 kb to 2 Mb) can be utilized as probes for FISH mapping (Hardas *et al.*, 1994).

Genome mapping and disease gene isolation

One of the most prominent applications of genetic mapping is reverse genetics, i.e. mapping of a genetic disease without knowing the underlying biochemical basis. On the basis of genetic and physical mapping, it is possible to map any Mendelian phenotypic trait. Linkage of a disease locus to a mapped genetic marker allows for the diagnosis of disease and identification of carriers if informative pedigrees are available. Once a disease locus has been mapped genetically, physical mapping can be applied to identify the disease gene and its primary defect (Collins, 1995). The identification of the primary defect is also essential for the design of a specific treatment for the disease and may lead to somatic gene therapy in the future. A similar approach can be applied to investigate multiple gene diseases. This further extends the application of gene mapping.

Comparative gene mapping, the mapping of homologous genes in genomes of different species, can also be helpful in the identification of disease genes. Linked gene groups have been conserved to some extent during evolution, and homologous chromosome regions containing different genes have been conserved among species, e.g. human, mouse and rat. The existence of many mouse or rat strains with well-defined, mapped genetic diseases or syndromes can then be used to map similar genetic loci in humans (Zhang *et al.*, 1989b).

Alternatively, association of the disease with cytogenetic lesions such as small chromosome interstitial deletions or translocations allows the direct localization of the disease locus on the physical map and the identification of the locus by means of the same technology.

Mapping the malignant genome in cancer

Cancer, characterized by neoplastic transformations of cells, invasion of tissues and finally metastasis, is a 'disease of genes', caused by alterations of specific genes, which are partially known. Oncogenes play a role in the neoplastic evolution when activated by either mutation or gene amplification. Tumour suppressor genes control growth of cells. Loss or inactivation of these genes also contributes to tumorigenesis. Cytologically, specific cancers often display characteristic chromosome abnormalities, including translocations, deletions, inversions and DNA amplifications. Accordingly, efforts to use molecular tools in order to analyse these recurrent chromosomal abnormalities have led to the identification of numerous genes related to tumour initiation and progression.

Conventional chromosome banding techniques have provided the major basis for karyotypical analysis of malignant cells in tumors. As the interpretation of the chromosome banding pattern is a pure experience-dependent procedure, errors and ambiguous data often are difficult to prevent, especially with respect to detecting minor structural changes when analysing complex karyotypes. FISH has contributed significantly to the characterization of chromosome abnormalities in tumours. Rearrangements involving specific chromosomes or their derivatives in malignant cells can be visualized directly by hybridizing the chromosome-specific painting probes, chromosome-specific repetitive sequence probes or chromosomal region-specific DNA probes to the tumour cells. However, using this approach to characterize those frequently observed unknown marker chromosomes or unknown origin genomic segments would require probes for all 24 human chromosomes. The development of spectrum karyotyping (SKY)

(Schrock et al., 1996) has attempted to circumvent this problem by visualizing 24 human chromosomes individually in different arbitrary colours. The resolution of this technique is limited to the chromosome level. In addition, small marker chromosomes may escape detection.

Comparative genome hybridization (CGH) provides an overview of unbalanced genetic alterations, which is based on a competitive *in situ* hybridization of differentially labelled tumour DNA and normal DNA to a normal human metaphase spread (Kallioniemi et al., 1992). Regions of gain or loss of DNA sequences are seen as an increased or decreased colour ratio of two fluorochromes used to detect the labelled DNAs. The genomic information obtained from this technique, however, is restricted to the area where only gain or loss occurs in malignant genomes. In addition, it does not lead to the generation of DNA from the detected chromosome regions.

Chromosome microdissection provides an approach to isolate DNA directly from any cytologically recognizable regions. The isolated DNA can then be used: (i) with region-specific painting probes for the detection of specific chromosomal disease (Zhang et al., 1993a); (ii) in gene amplification studies (Zhang et al., 1993b); and (iii) with region-specific DNA markers for position cloning (Zhang et al., 1995). Technically, the molecular cytogenetic tools described above are complementary to each other, facilitating rapid scanning of malignant genomes and targeting chromosome abnormalities. This may lead to the identification of genes involved in tumorigenesis.

Genome mapping integration

Genome mapping has entered the final stage of integration, i.e. the integration of genetic and physical resources into more complete and comprehensive maps. One of the milestones in this respect was the construction of a human linkage map containing 5264 genetic markers (Dib et al., 1996). This high-resolution genetic map reaches the 1 cM resolution limit of genetic mapping, with marker spacing being $<1 \times 10^6$ bp of physical distance. This provides one of the most comprehensive instruments in genetic disease studies. Soon after, a high-density physical map with >30,000 gene markers, average spacing about 100 kb, was constructed (Deloukas et al., 1998). Based on these comprehensive maps, the assembly of isolated intact genomic fragments in PAC and BAC vectors into clone maps or comprehensive contigs has been facilitated. Representative PAC or BAC clones were then used as the DNA sources for a shotgun plasmid library, in which the average insert size is about 1 kb. Finally, these shotgun clones were used as DNA templates for high-throughput DNA sequencing using dideoxy-termination biochemistry and automated gel electrophoresis with laser fluorescent detection (Venter et al., 1998). Mapping of the human genome is still incomplete. This is due to the presence of a large number of various interspersed and tandem repeated sequences in chromosomes. In particular, the assembly of repeats in centromeres and paracentric heterochromatin regions is not possible with current technologies. Repeats may contain as little as a few base pairs or as many as 200 bp; these regions may tandem repeat hundreds to thousands of times in a single chromosome region. Although additional efforts are required to complete the human genome sequencing, the vast amount of structural information available to date has facilitated precise determination of molecular mechanisms in human cells (Deloukas et al., 1998; Lander et al., 2001; Venter et al., 2001).

Functional Genomics

The use of microarrays in functional genomics

Functional genomics represents a new phase of genome research: to assess genes functionally on the genome-wide scale. This is represented by the emergence of DNA microarray technology (Schena et al., 1995; DeRisi et al., 1997). *In silico* microarray methodology is where inserts from tens of thousands of cDNA clones (i.e. probes) are arrayed robotically on to a glass slide and subsequently probed with two differentially labelled pools of RNA (i.e. target). Typically, the RNA sample is labelled with a nucleotide conjugated to a fluorescent dye such as Cy3-dUTP or Cy5-dUTP. RNA (target) from at least two treatment groups is compared in order to identify differences in mRNA levels, e.g. normal cells versus diseased cells; wild-type versus a transgenic animal; or general control versus a series of study samples. After hybridization, the slide is excited by appropriate wavelength laser beams to generate two 16-bit TIF images. The pixel number of each spot in each wavelength channel is proportional to the number of fluorescent molecules and hence permits the quantification of the number of target molecules that have hybridized to the cDNA clones (probes). The difference in signal intensities at each wavelength parallels the number of molecules from the two different target sources that have hybridized to the same cDNA probe. A general process of DNA microarray is illustrated in Fig. 1.1. Experimental procedures for this technology have been well established. Thus, this chapter focuses on data analysis.

Microarray data analysis

Figure 1.2A illustrates typical DNA microarray images. The amount of data generated by each microarray experiment is substantial, potentially equivalent to that obtained through tens of thousands of individual nucleotide hybridization experiments done in the manner of traditional molecular biology (e.g. Northern blot). It is extremely challenging to convert such a massive amount of data into meaningful biological networks. Therefore, it is important for life scientists to understand working principles of data mining tools utilized in this field.

Data pre-processing

Various laser-based data acquisition scanners are commercially available now. For data analysis, it usually is necessary first to build up a spreadsheet-like matrix, in which rows represent genes, columns represent RNA samples, and each cell contains a ratio (e.g. pixel number of Cy5 versus pixel number of Cy3) featuring the transcriptional level of the particular gene in the particular sample. This matrix can be studied in two ways: comparing rows in the matrix and comparing columns in the matrix. By looking for similarities in expression patterns of genes in rows, functionally related genes that are co-regulated can be identified. By comparing expression profiles in samples, biologically correlated samples or differentially expressed genes can be determined. Usually, the matrix needs to be filtered to remove genes with missing or erroneous values. Then, numerical values in the matrix are scaled by logarithm with base 2 to normalize data distribution and reduce potential data bias by extreme values. When a series of test samples (e.g. clinical samples) is compared with an unpaired control (reference) sample, the logarithm scaled ratios need to be processed further by mean or median centring to allow for data analysis in test samples that is independent of the gene expression level in the unpaired control sample.

Similarity measurements

Current efforts in understanding microarray data are focused primarily on clustering and visualization. Clustering is intended to catalogue genes or RNA samples into meaningful groups based on their similar behaviours; visualization is intended to depict clustering results in a readily accessible format. For comparisons of similarities, the concept of Euclidean distance and calculation of correlation coefficients are usually utilized to set up the similarity measurement. Euclidean distance is the distance between two n-dimensional points, e.g. X and Y. Corresponding values for X are $X_1, X_2, \ldots, X_N$, and corresponding values for Y are $Y_1, Y_2, \ldots, Y_N$. The Euclidean distance between X and Y is

$$d(x, y) = \sqrt{\sum_{i}^{n}(x_i - y_i)^2}$$

where n is the number of the RNA samples for gene comparison, or the number of genes from sample comparison. For example, comparing any two genes (e.g. X and Y) in a three-dimensional (i.e. three samples) space, the Euclidean distance between X and Y is

$$d(\vec{x}, \vec{y}) = \sqrt{(x_1 - y_1)^2 + (x_2 - y_2)^2 + (x_3 - y_3)^2}$$

The closer the distance between two points, the more similar they are.

The correlation coefficient between any two n-dimensional points is defined as

$$r = \frac{1}{N} \sum_{i=1}^{n} \left(\frac{X_i - \overline{X}}{\delta_X} \right) \left(\frac{Y_i - \overline{Y}}{\delta_Y} \right)$$

where n is the number of the RNA samples for gene comparison, or the number of genes for sample comparison, $\overline{X}$ is the average of values in

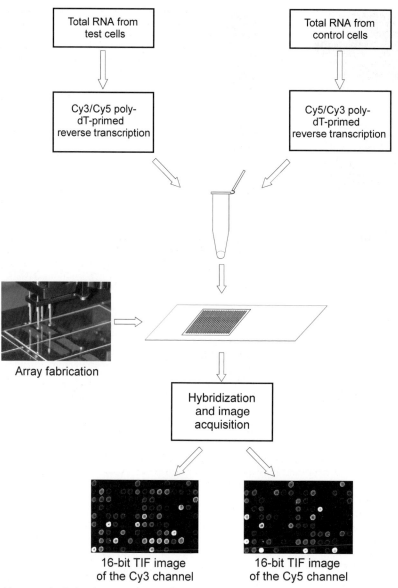

Fig. 1.1. Ideogram depicting the general procedure of *in silico* DNA microarrays. The image on the left illustrates the printing process of microarray fabrication, in which cDNA inserts from individual clones are prepared by PCR and printed on to polylysine-coated glass slides through a GMS 417 arrayer (Affymetrix). After an overnight hybridization with differentially labelled test and control probes, the slide is scanned using a GenePix 4000 scanner (Axon Instruments) to generate two TIF images: green channel and red channel. The pixel ratio between red and green for each spot is used as the numerical value for further data analysis.

point X, and δ_X is the standard deviation of values in point X.

For example, if point X and Y are plotted as curves based on their values in all samples or genes, r will tell how similar the shapes of the two curves are. The correlation coefficient is always between −1 and 1. When r equals 1, the two shapes are identical. When r equals 0, the two shapes are

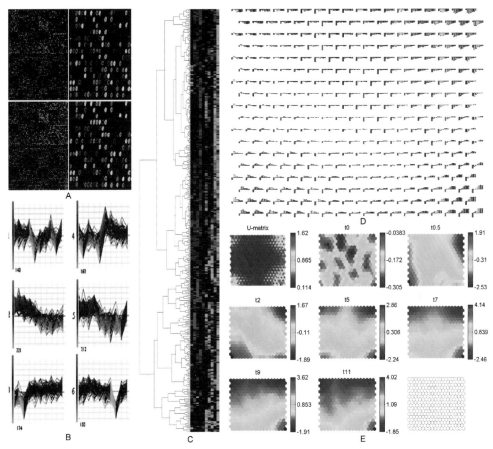

Fig. 1.2. Microarray data visualizations using different algorithms (see colour version in Frontispiece). (A) Colour images illustrating similarities and dissimilarities between two brain development stages. The upper left is the full image generated by the Cy3-labelled day 11.5 post-coitum (p.c.) mouse brain cDNA pool versus a Cy5-labelled mouse embryonic liver cDNA pool, and the lower left is the full image generated from the Cy3-labelled day 12.5 p.c. mouse brain cDNA versus the same control. The right panel depicts partial images of the left panels to illustrate details. (B) Graphic distributions showing representative clusters obtained by K-means clustering. The horizontal scale represents RNA samples obtained from ten different time points of mouse embryonic brain development. The vertical scale weighs changes in expression, from high expression (red) to low expression (green) with units in log ratio, subtracted by median. (C) A partial tree view obtained through hierarchical clustering of 4608 mouse genes over ten embryonic development samples. Red represents up-regulation and green represents down-regulation. (D) A bar graphic display of SOM illustrating gene clustering and expression patterns of regulated genes during the yeast sporulation process. All genes were organized into 324 (18 × 18) hexagonal map units. Each bar in a given unit illustrates the average expression of genes mapped to that unit. (E) U-matrix and component plane presentations. The colour coding in U-matrix stands for Euclidean distance. The darker the colour, the smaller the distance. The large dark blue area that occupies the majority of the display represents unregulated genes, which form some noise clustering. The component plane presentations (t0–t11) illustrate differential displays of regulated genes during sporulation of yeast on the genome-wide scale. The colour coding index stands for the expression values of genes. The brighter the colour, the higher the value. All these differential displays are linked by position: in each display, the hexagon in a certain position corresponds to the same map unit. It is straightforward to compare expression patterns in the same positions of different displays. The last label display shows the positions of each unit on the map.

completely independent. When r equals -1, the two shapes are negatively correlated. Both Euclidean distance and correlation coefficient are used to measure similarities in clustering. There is no clear justification to favour one procedure over the other.

Clustering algorithms

Commonly applied algorithms for gene clustering include hierarchical clustering, K-mean clustering and self-organizing map (SOM). Hierarchical clustering is based primarily on the similarity measure between individuals (genes or samples) using a pairwise average linkage clustering, usually the correlation coefficient (Eisen et al., 1998; White et al., 1999). Through the pairwise comparison, this algorithm eventually clusters individuals into a tree view. The length of the branches of the tree depicts the relationship between individuals, where the shorter the branch is the more similarity there is between individuals (Fig. 1.2C). This algorithm has been used frequently in microarray data analysis, and has proven to be a valuable tool. A major drawback of hierarchical clustering is the polygenetic tree structure of the algorithm, which may be best suited to situations of true hierarchical descent, such as in the evolution of species (Tamayo et al., 1999), rather than situations of multiple distinct pathways in living cells. This may lead to incorrect clustering of genes, especially with large and complex data sets.

K-means clustering allows the partition of individuals into a given number of (K) separated and homogeneous groups based on repeated cycles of computation of the mean vector for all individuals in each cluster and reassignment of individuals to the cluster whose centre is closest to the individual (Fig. 1.2B). Euclidean distance is used commonly as the similarity measurement. A limitation of K-mean clustering is that the arbitrarily determined number of gene clusters may not reflect true situations in living cells. In addition, the relationship between clusters is not defined.

The SOM (Kohonen, 1995; Kohonen et al., 1996), an artificial intelligence algorithm based on unsupervised learning, appears to be particularly promising for microarray data analysis. It is, therefore, of considerable interests to discuss this application in further detail.

Self-organizing map algorithm

This algorithm has properties of both vector quantification and vector projection, and consequently configures output prototype vectors into a topological presentation of original multidimensional input numerical data. SOM consists of a given number of neurons on a usually two-dimensional grid. Each of these neurons is represented by a multidimensional prototype vector. The number of dimensions of prototype vector is equal to that of dimensions (i.e. the number of samples) of input vectors. The number of input vectors is equal to the number of inputs, i.e. the number of genes in the matrix. The neurons are connected to adjacent neurons by a neighbourhood relationship, which dictates the topology, or structure of the map. The prototype vectors are initiated with random numerical values and trained iteratively. Each actual input vector is compared with each prototype vector on the mapping grid based on: $\|\vec{x}-\vec{m}_c\| = \min_i \{\|\vec{x}-\vec{m}_i\|\}$, where $\vec{x}$ stands for input vector and $\vec{m}_c$ for output vector. The best-matching unit (BMU) is defined when the prototype vector of a neurone gives the smallest Euclidean distance to the input vector. Simultaneously, the topological neighbours around the BMU are stretched towards the training input vector so that they are updated as denoted by: $\vec{m}_i(t+1) = \vec{m}_i(t) + \alpha(t)[\vec{x}(t) - \vec{m}_i(t)]$. The SOM training is usually processed in two phases, a first rough training step and then the fine tuning. After iterative trainings, SOM eventually is formed in the format that individuals with similar properties are mapped to the same map unit or nearby neighbouring units, creating a smooth transition of related individuals over the entire map (Kohonen et al., 1996). More importantly, this ordered map provides a convenient platform for various inspections of the numerical data set. Although this algorithm has been utilized in several microarray-based investigations (Tamayo et al., 1999; Toronen et al., 1999; Chen et al., 2001), the full potential of SOM (particularly for visual inspections) has not yet been fully utilized in microarray data analyses. Recently, we have introduced component plane presentations, a more in-depth visualization tool of SOM, for the illustration of microarray data, in order to depict transcriptional changes for genes. By integrating features of this component plane presentation with SOM, microarray analyses go beyond gene clustering to include, for instance, differential displays of regulated genes on a genome-wide scale.

Simultaneous illustrations of gene clusters and genome-wide differential displays using component plane integrated SOM

To demonstrate the advantages of this approach over other analytical methods, we selected a previously analysed yeast sporulation data set with 6400 genes and seven RNA samples over seven time points (Chu *et al.*, 1998). Sporulation in yeast is the process in which diploid cells undergo meiosis to produce haploid germ cells, involving two overlapping steps: meiosis and spore formation; the process can be divided into meiosis I, meiosis II and spore formation. The process of sporulation can be induced using a nitrogen-deficient medium.

For SOM algorithm and its visualizations, we have utilized a SOM toolbox programmed by Vesanto *et al.* (2000). This toolbox, built in the Matlab 5 computation environment, has capacities to pre-process data, train SOM using a range of different kinds of topologies and to visualize SOM in various ways. To maximize the number of neighbourhood contacts topologically, we utilized hexagonal prototype vectors instead of rectangular ones for the SOM training. The algorithm was then conducted using 324 prototype vectors on a two-dimensional lattice (18 × 18 grid). For the visualization, we first utilized a bar graphical display (Fig. 1.2D), similar to previously published displays, to gain a global view of gene clustering and expression patterns of expressed genes. The number of genes mapped to individual map units varied between seven and 62, and the bar chart displayed in each hexagonal unit represented the average expression pattern of genes mapped in the unit. It can be seen that the map has been organized in such a way that related patterns are placed in nearby neighbouring map units, producing a smooth transition of expression patterns over the entire map. Therefore, gene clustering can also be recognized by surrounding neighbouring map units in addition to its core unit.

To illustrate features other than clustering of regulated genes during the sporulation process, we integrated SOM analyses with the powerful visualization tool of component plane presentations.

Component plane presentations provide an in-depth approach to visualize variables that contribute to SOM. Each component plane presentation is considered as a sliced version of SOM, illustrating values of a single vector component in all map units. For example, the first component plane (t0) in Fig. 1.2E shows the SOM slice at time point 0 h and the last component plane (t11) shows the SOM at time 11 h during the sporulation process (Chu *et al.*, 1998). The colours of map units are selected so that the brighter the colour, the greater is the average expression value of the genes mapped to the corresponding unit. Each of these SOM slices can also be considered as a genome-wide differential display of regulated genes, in which all up-regulated units (hexagons in red), down-regulated units (hexagons in blue) and moderately transcribed units (hexagons in green and yellow) are well delineated. By comparing these genome-wide differential displays, we can learn many additional features of regulated genes in cells. For instance, these displays are correlated sequentially with each other, depicting the process of sporulation at the transcriptional level. The sequential inactivation of genes mapped to the two upper corners suggests that the functional group represented by genes on the right is more sensitive to the nitrogen-deficient medium induction than the one on the left, although both of them are suppressed toward the end of the sporulation process. The sequential activation of genes mapped to the two bottom corners gives us a more vivid picture of the process leading to spore formation. Genes in the bottom left corner and left edge are activated at an early stage of the process, indicating that these genes are associated specifically with meiosis I. In contrast, the progressively increased expression of genes in the right corner suggests that these genes are associated with meiosis II and spore formation. This is consistent with the observation that known genes of meiosis II and spore formation have been mapped to these corner units.

The SOM algorithm has great potential, in particular with regard to data visualization. To date, most of the procedures used to visualize microarray data are limited to gene clustering, typically represented by bar graphical displays as depicted in Fig. 1.2D. In contrast, U-matrix (unified distance matrix) as displayed in Fig. 1.2E is a distance matrix method that visualizes the pairwise distance between prototype vectors of neighbouring map units and helps to define the cluster structure of SOM. We have utilized this display successfully to define some core clusters of developmentally related genes expressed during brain development. However, the interpretation of data can be difficult when noise interruption is high. This concern is supported further by the presence of a large number of unregulated genes in the sporulation data set. These genes form clusters in a random manner, producing a visible clustering area in the centre of the SOM (Fig. 1.2D).

Component plane presentations provide an in-depth approach to visualize component variables that contribute to SOM. Thus, SOM can be sliced into multiple sample-specific, genome-wide differential displays. Each of these displays details transcriptional changes of a specific sample on the genome-wide scale. These genome-wide differential displays greatly help to identify the biological meanings of microarray data. As illustrated in this section, we were able to determine directly the functional significance of genes differentially expressed during the process of sporulation at the genome-wide scale. To reach similar conclusions by alternative methods would require a much greater effort (Chu *et al.*, 1998). Component plane presentations are also applicable to microarray data from other organisms. For example, we have applied this approach to microarray data from mouse brain samples using ten time points during early brain development stages. In these studies, we have identified a large number of genes that are related to brain development. These genome-wide differential displays can be used to identify the functional significance of regulated genes. Also, the displays can be used to correlate data from various samples, based on patterns in identical positions of the displays; this is particularly promising for samples from clinical studies. The potential impact of this approach on microarray data analysis can be substantial.

Summary

With advances in genome research, the concept of genomics extends beyond structural analyses of genomes to include functional analysis of the genome. Structural genomics focuses on genetic mapping and physical mapping of the genome by using various tools of molecular biology. A genetic map is based on linkage analysis of the segregation of polymorphic markers in pedigrees. A physical map measures the distance between loci in nucleotide base pairs. The ultimate physical map of a genome is the determination of its complete DNA sequences. One of the most prominent applications of genome mapping is disease gene studies, typically represented by reverse genetics. Cancer genetics is also an important aspect of disease gene studies. Although the completion of the human genome sequencing is approaching, the understanding of tools and reagents involving genome mapping may still be helpful for our current research. This chapter emphasizes functional genomics, represented by DNA microarray technology. This technology allows the measure of tens of thousands of genes in parallel, providing the most comprehensive approach to understanding molecular mechanisms involved in living cells. The most challenging part of DNA microarray analysis is to convert the massive amount of data into biologically meaningful networks. Compared with other data mining tools, we believe that SOMs, in particular if integrated by component plane presentations, is the most powerful tool in this respect. This integrated approach not only allows genes to be clustered but also permits regulated genes to be displayed differentially on the genome-wide scale. This application is particularly appealing for clinical case studies, in which detailed comparison between transcriptional profiles of individual patients often is required. With the great abundance of genomic information and the rapid development of technology, the determination of molecular mechanisms that underlie living human cells has come within reach.

Acknowledgements

The author is grateful to Li Xiao and Yue Teng for their excellent assistance with data calculation and graphics.

References

Burke, D.T., Carle, G.F. and Olson, M.V. (1987) Cloning of large segments of exogenous DNA into yeast by means of artificial chromosome vectors. *Science* 236, 806–812.

Chen, J.J., Peck, K., Hong, T.M., Yang, S.C., Sher, Y.P., Shih, J.Y., Wu, R., Cheng, J.L., Roffler, S.R.,

Wu, C.W. and Yang, P.C. (2001) Global analysis of gene expression in invasion by a lung cancer model. *Cancer Research* 61, 5223–5230.

Chu, S., DeRisi, J., Eisen, M., Mulholland, J., Botstein, D., Brown, P.O. and Herskowitz, I. (1998) The transcriptional program of sporulation in budding yeast. *Science* 282, 699–705.

Collins, F.S. (1995) Positional cloning moves from perditional to traditional. *Nature Genetics* 9, 347–350.

Deloukas, P., Schuler, G.D., Gyapay, G., Beasley, E.M., Soderlund, C., Rodriguez-Tome, P., Hui, L., Matise, T.C., McKusick, K.B., Beckmann, J.S. *et al.* (1998) A physical map of 30,000 human genes. *Science* 282, 744–746.

DeRisi, J.L., Iyer, V.R. and Brown, P.O. (1997) Exploring the metabolic and genetic control of gene expression on a genomic scale. *Science* 278, 680–686.

Dib, C., Faure, S., Fizames, C., Samson, D., Drouot, N., Vignal, A., Millasseau, P., Marc, S., Hazan, J., Seboun, E., Lathrop, M., Gyapay, G., Morissette, J. and Weissenbach, J.A. (1996) Comprehensive genetic map of the human genome based on 5,264 microsatellites. *Nature* 380, 152–154.

Eisen, M.B., Spellman, P.T., Brown, P.O. and Botstein, D. (1998) Cluster analysis and display of genome-wide expression patterns. *Proceedings of the National Academy of Sciences USA* 95, 14863–14868.

Hardas, B.D., Zhang, J., Trent, J.M. and Elder, J. (1994). Direct evidence for homologous sequences on the paracentric regions of human chromosome 1. *Genomics* 21, 359–363.

Ioannou, P.A., Amemiya, C.T., Garnes, J., Kroisel, P.M., Shizuya, H., Chen, C., Batzer, M.A. and de Jong, P.J. (1994) A new bacteriophage P1-derived vector for the propagation of large human DNA fragments. *Nature Genetics* 6, 84–89.

Kallioniemi, A., Kallioniemi, O.P., Sudar, D., Rutovitz, D., Gray, J.W., Waldman, F. and Pinkel, D. (1992) Comparative genomic hybridization for molecular cytogenetic analysis of solid tumors. *Science* 258, 818–821.

Kohonen, T. (1995) *Self-organizing Maps. Springer Series in Information Sciences*, Vol. 30, Springer, Berlin.

Kohonen, T., Oja, E., Simula, O., Visa, A. and Kangas, J. (1996) Engineering applications of the self-organizing map. *Proceedings of the IEFE* 84, 1358–1384.

Lander, E.S., Linton, L.M., Birren, B., Nusbaum, C., Zody, M.C., Baldwin, J., Devon, K., Dewar, K., Doyle, M., FitzHugh, W. *et al.* (2001) Initial sequencing and analysis of the human genome. *Nature* 409, 860–921.

Litt, M. and Luty, J.A. (1989) A hypervariable microsatellite revealed by *in vitro* amplification of a dinucleotide repeat within the cardiac muscle actin gene. *American Journal of Human Genetics* 44, 397–401.

Marynen, P., Zhang, J., Cassiman, J.J., Van den Berghe, H. and David, G. (1989) Partial primary structure of the 48- and 90-kilodalton core proteins of cell surface-associated heparan sulfate proteoglycans of lung fibroblasts. *Journal of Biological Chemistry* 264, 7017–7024.

Olson, M., Hood, L., Cantor, C. and Botstein, D. (1989) A common language for physical mapping of the human genome. *Science* 245, 1434–1435.

Schena, M., Shalon, D., Davis, R.W. and Brown, P.O. (1995) Quantitative monitoring of gene expression patterns with a complementary DNA microarray. *Science* 270, 467–470.

Schrock, E., du Manoir, S., Veldman, T., Schoell, B., Wienberg, J., Ferguson-Smith, M.A., Ning, Y., Ledbetter, D.H., Bar-Am, I., Soenksen, D., Garini, Y. and Ried, T. (1996) Multicolor spectral karyotyping of human chromosomes. *Science* 273, 494–497.

Shizuya, H., Birren, B., Kim, U.J., Mancino, V., Slepak, T., Tachiiri, Y. and Simon, M. (1992) Cloning and stable maintenance of 300-kilobase-pair fragments of human DNA in *Escherichia coli* using an F-factor-based vector. *Proceedings of the National Academy of Sciences USA* 89, 8794–8797.

Stallings, R.L., Ford, A.F., Nelson, D., Torney, D.C., Hildebrand, C.E. and Moyzis, R.K. (1991) Evolution and distribution of (GT)n repetitive sequences in mammalian genomes. *Genomics* 10, 807–815.

Tamayo, P., Slonim, D., Mesiror, J., Zhu, Q., Kitareewan, S., Dmitrovsky, E., Lander, E.S. and Gowb, T.R. (1999) Interpreting patterns of gene expression with self-organizing maps: methods and application to hematopoietic differentiation. *Proceedings of the National Academy of Sciences USA* 96, 2907–2912.

Toronen, P., Kolehmainen, M., Wong, G. and Castren, E. (1999) Analysis of gene expression data using self-organizing maps. *FEBS Letters* 451, 142–146.

Venter, J.C., Adams, M.D., Sutton, G.G., Kerlavage, A.R., Smith, H.O. and Hunkapiller, M. (1998) Shotgun sequencing of the human genome. *Science* 280, 1540–1542.

Venter, J.C., Adams, M.D., Myers, E.W., Li, P.W., Mural, R.J., Sutton, G.G., Smith, H.O., Yandell, M., Evans, C.A., Holt, R.A. *et al.* (2001) The sequence of the human genome. *Science* 291, 1304–1351.

Vesanto, J. (2000) Neural network tool for data mining: SOM toolbox. In: *Proceedings of Symposium on Tool Environments and Development Methods for Intelligent Systems, TOOLMET2000.* Oulun yliopistopaino, Oulu, Finland, pp. 184–196.

Walter, M.A., Spillett, D.J., Thomas, P., Weissenbach, J. and Goodfellow, P.N. (1994) A method for constructing radiation hybrid maps of whole genomes. *Nature Genetics* 7, 22–28.

Weber, J.L. and May, P.E. (1989) Abundant class of human DNA polymorphisms which can be typed using the polymerase chain reaction. *American Journal of Human Genetics* 44, 388–396.

White, K.P., Rifkin, S.A., Hurban, P. and Hogness, D.S. (1999) Microarray analysis of *Drosophila* development during metamorphosis. *Science* 286, 2179–2184.

Zhang, J., Marynen, P., Devriendt, K., Fryns, J.P., Van den Berghe, H. and Cassiman, J.J. (1989a) Molecular analysis of the isochromosome 12P in the Pallister–Killian syndrome. Construction of a mouse–human hybrid cell line containing an i(12p) as the sole human chromosome. *Human Genetics* 83, 359–363.

Zhang, J., Hemschoote, K., Peeters, B., De Clercq, N., Rombauts, W. and Cassiman, J.J. (1989b) Localization of the PRR1 gene coding for rat prostatic proline-rich polypeptides to chromosome 10 by *in situ* hybridization. *Cytogenetics and Cell Genetics* 52, 197–198.

Zhang, J., Devriendt, K., Marynen, P., Van den Berghe, H. and Cassiman, J.J. (1990) Chromosome mapping using polymerase chain reaction on somatic cell hybrids. *Cancer Genetics and Cytogenetics* 45, 217–221.

Zhang, J., Meltzer, P., Jenkins, R., Guan, X.Y. and Trent, J. (1993a) Application of chromosome microdissection probes for elucidation of BCR-ABL fusion and variant Philadelphia chromosome translocations in chronic myelogenous leukemia. *Blood* 81, 3365–3371.

Zhang, J., Trent, J.M. and Meltzer, P.S. (1993b) Rapid isolation and characterization of amplified DNA by chromosome microdissection: identification of IGF1R amplification in malignant melanoma. *Oncogene* 8, 2827–2831.

Zhang, J., Cui, P., Glatfelter, A.A., Cummings, L.M., Meltzer, P.S. and Trent, J.M. (1995) Microdissection based cloning of a translocation breakpoint in a human malignant melanoma. *Cancer Research* 55, 4640–4645.

2 Perspectives in Post-genomic Nutrition Research

Hannelore Daniel
*Molecular Nutrition Unit, Department of Food and Nutrition,
Technical University of Munich, Germany*

Introduction

Every nutritional process relies on the interplay of a large number of proteins encoded by mRNA molecules that are expressed in a given cell. Alterations of mRNA levels and in turn of the corresponding protein levels (although the two variables do not necessarily change in parallel) are critical parameters in controlling the flux of a nutrient or metabolite through a biochemical pathway. Nutrients and non-nutrient components of foods, diets and lifestyle can affect essentially every step in the flow of genetic information, from gene expression to protein synthesis to protein degradation, thereby altering metabolic functions in the most complex ways. There is no doubt that with the genetic information emerging on a daily basis, we are discovering exciting tools that provide us with insights into the molecular basis of human metabolism under normal as well as pathophysiological conditions. There is also no doubt that the interplay of the rather static mammalian genome with its rapidly changing nutritional environment is one of the most attractive and interesting areas in post-genomic research.

From Gene to Function and from Genomics to Phenomics

Although a huge body of information on the number of mammalian genes, on chromosomal localization of individual genes, their genomic structure and in part also on the functions of the encoded proteins has been gathered, we are far from understanding how these individual factors orchestrate metabolism.

The molecular descriptors of metabolism

Genomic data contain only limited information about the dynamic behaviour of integrated cellular processes. Nevertheless, recent technological advances have made it possible to analyse the variability and dynamic changes in the genetic response of a cell or organism by determining the expression level of individual RNA molecules or huge sets of mRNA molecules. Whereas genomics describes large-scale DNA sequencing that provides basic genetic information and insights into sequence heterogeneity (i.e. single nucleotide polymorphisms; SNPs) in coding regions of genes as well as in control elements (i.e. promoters), transcriptomics – also called expression profiling – assesses mRNA levels of a few or up to several thousand open reading frames simultaneously in a biological sample; this is done mainly by DNA hybridization arrays and/or by quantitative polymerase chain reaction (PCR) techniques (Celis, 2000; Lockhart and Winzeler, 2000). Proteomics allows the proteome – as the protein complement of the genome that is expressed in a cell or an organ – to be identified, and changes in protein expression patterns and levels to be determined. Moreover, for individual proteins, post-translational modifications that are crucial for functions, or even amino acid substitutions (polymorphisms) can be detected (Dutt and Lee, 2000; Pandey and Mann, 2000). Applications of these new techniques of genome, transcriptome and proteome analysis are central to the development of nutritional sciences in the next decade and its integration into the rapidly evolving era of functional genomics.

Functional genomics is either based on gene-driven or on phenotype-driven approaches, as shown in Fig. 2.1. The gene-driven approaches use genomic information to identify, clone, express and characterize genes at the molecular level. Phenotype-driven approaches characterize phenotypes from random mutation screens or naturally occurring variants to identify and clone the gene(s) responsible for the particular phenotype, without knowledge of the underlying molecular mechanisms. Of course, the two strategies are highly complementary at virtually all levels of analysis and lead collectively to the correlation of genotypes and phenotypes. Where nature has not provided inborn errors of metabolism that demonstrate the phenotypical consequences of individual gene/protein malfunctions, the role of single genes or groups of genes in the make-up of metabolism can be analysed by gene inactivation ('knock-out') or selective expression ('knock-in') and overexpression models employing experimental animals such as fruit flies (*Drosophila melanogaster*), nematodes (*Caenorhabditis elegans*), mice, rats and human cell lines. These approaches of transgenics have already produced a large number of animal lines lacking one or several genes, or overexpressing others. Although very elegant as new genetic tools for understanding metabolism, these manoeuvres unfortunately quite often do not produce an obvious phenotype. This may tell the investigator that the lack of gene function may be compensated for by other mechanisms. More advanced transgenic technologies in animals such as Cre/lox-controlled cell-, organ- and/or time-dependent gene inactivation or induction of expression allow for analysis of phenotypic consequences in even more elegant ways; these techniques appear to be helpful in particular in those situations where simple gene disruption has been lethal for the developing fetus or newborn (Sauer, 1998).

Although functional genomics starts with the classical one-gene approach, it will rapidly move to more system-based 'holistic' levels by employing the high-throughput technologies (DNA microarrays and proteomics) in combination with targeted gene deletions or selective overexpression to explore the consequences of operational shifts in genetic circuits and cellular systems. This means that we will be facing numerous and huge gene and protein expression databases. To bring meaning and value into these data sets, we require system-based approaches and computational strategies with algorithms that help to describe metabolism (Paton *et al.*, 2000; Tomita, 2001).

Closing the circle: from metabolism to metabolomics

Steady-state levels of human metabolism are determined by the rates of biosynthesis and degradation of proteins (turnover) that function as enzymes, receptors, transporters, channels, hormones and other signalling molecules or that provide structural elements for cells, organs or the skeleton. Between the proteins, there is a variable flow of metabolic intermediates that serve as building blocks for proteins, carbohydrates, lipids or heterooligomers and that provide the fuel for ATP synthesis. Whereas in the past we mainly looked at the phenotypical expression of metabolism by measuring the concentration of a few individual metabolic intermediates, the new molecular tools allow us now to determine every step in the flow of the biological information from DNA to mRNA to proteins and to function. However, biological regulation goes beyond the control of gene expression and protein synthesis: major determinants of metabolism include protein–protein interactions and alterations of protein activity by metabolic intermediates in terms of kinetic effects and/or allosteric modulation of function. So, whatever information is gathered at the levels of mRNA and protein expression will not allow the prediction of metabolic consequences sufficiently (ter Kuile and Westhoff, 2001). The final stage along the line from gene to mRNA, to protein, to function is therefore the analysis of the pattern and the concentrations of the metabolites that flow between proteins, organelles, cells and organs. Thus, we have to end where we started – looking at metabolites. However, this time, analysis of the entire metabolome

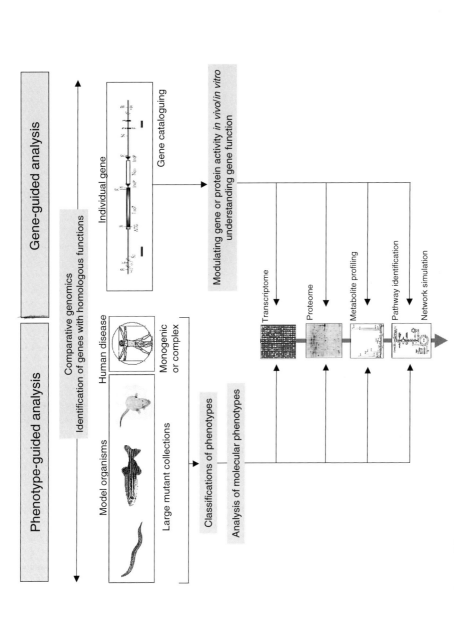

Fig. 2.1. Overview on the two different approaches of functional genomics. One route is guided by the analysis of the phenotype in model organisms or human monogenic diseases to identify the underlying gene and its function. The other route is guided by analysing functions of an individual gene. The main analysis tools and investigative methods are shown with the links between the different levels of information.

as the sum of all detectable low and intermediate molecular weight compounds rather than individual metabolites will be performed by the metabolomics approaches. Different physical methods can be used for the comprehensive analysis of metabolites within a biological sample. In most cases, procedures are coupled to classical chromatographic separation techniques and comprise Fourier transform infrared spectroscopy, electrospray ionization mass spectrometry (ESI-MS) and nuclear magnetic resonance (NMR) spectroscopy (Fiehn et al., 2000; Glassbrook et al., 2000). The potential of comprehensive metabolic analysis coupled to statistical methods of cluster analysis for discriminatory phenotype analysis becomes obvious when inspecting research progress achieved in bacterial systems, yeast and plants employing wild-type as well as transgenic organisms (Delneri et al., 2001; Kose et al., 2001). Phenomics (Schilling et al., 1999) finally takes gene function on such a systemic scale. It utilizes all genomic information, all expression information (at the mRNA and protein level) and all metabolites to describe metabolism in the most comprehensive way on the basis of the networks of biological regulation.

Techniques and Technologies

Whereas genomics employs the first line classical DNA sequence technology, transcriptome analysis utilizes mainly fluorescence-based detection systems to determine message (mRNA) expression levels in a biological sample. Expression profiling is covered in Chapter 1 and thus will not be addressed here under technological aspects.

Proteomics

The term 'proteome' was introduced as the complement to the genome, and comprises all transcribed and translated open reading frames (protein-encoding regions) in a given genome. Proteome analysis is based on the separation of proteins by two-dimensional polyacrylamide gel electrophoresis (2D-PAGE). Although 2D-PAGE may already appear to be an 'ancient' technique, it provides separation with the highest resolution currently available. However, it is still not easy to attempt to obtain standardized and reproducible separation conditions. Similarly, staining of the proteins in the gel is crucial for the subsequent quality of analysis. Both Coomassie blue and silver staining can be performed, with silver-staining being at least ten times more sensitive. Recently, sensitive fluorescent dyes have been developed that improved the sensitivity further (Patton, 2000). 2D-PAGE separates proteins according to their charge (isoelectric point; pI) by isoelectric focusing (IEF) in the first dimension and according to their size (molecular mass) by SDS–PAGE in the second dimension. It therefore has a unique capacity to resolve complex mixtures of proteins, permitting the simultaneous analysis of hundreds or even thousands of gene products (Görg et al., 2000). However, not all proteins are resolved and separated equally well by 2D-PAGE. Analysis of very alkaline, hydrophobic and integral membrane proteins as well as high molecular weight proteins is still a problem. In some cases, a prefractionation according to cellular compartment (membranes/microsomes, cytosol, mitochondria) or according to protein solubility may be necessary (Blackstock and Weir, 1999; Cordwell et al., 2000). In addition, proteins of low cellular abundance, which may be particularly important in view of their cellular functions (e.g. in signalling pathways), are still very difficult to resolve in the presence of large quantities of housekeeping proteins (Corthals et al., 2000). However, new concepts are constantly being developed and include tagging techniques (Adam et al., 2001) and the enrichment of minor proteins prior to separation in 2D gels.

Two developments have been of central importance in the revival of 2D-PAGE protein separation in proteomics: (i) improvement of high-resolution mass spectrometers for ionization and analysis of peptides and proteins based on their molecular mass; and (ii) sequencing of the genomes of various species (including humans) to provide information on coding regions of expressed proteins. Finally, the development of computer algorithms that match uninterpreted (or partially interpreted) mass spectra with the translation of the nucleotide sequence of expressed genes from databases was also important for the advancement of the field of proteomics.

Although the most common procedure for the identification of a protein spot in a gel currently is peptide mapping or 'fingerprint' analysis, there are a number of other techniques and approaches that can also be applied (Chalmers and Gaskell, 2000; Gevaert and Vandekerckhove, 2000). For example, the proteins may be transferred from the gel on to a membrane to make them accessible to specific antibodies that then detect the proteins by an immunoreaction. Proteins can also be excised from the gel and be submitted to classical Edman degradation with N-terminal sequencing or to ESI-MS analysis to obtain an amino acid sequence; this sequence can then be used to identify the protein by comparison with sequences deposited in databases.

To obtain sequence information, the more rapid and advanced techniques employ peptide mass analysis. Here the protein-containing spots are excised from the gel and the gel piece is altered chemically to make the protein accessible to hydrolysis by a protease such as trypsin. Digestion requires a highly purified trypsin that is not contaminated with enzymes that may cleave peptide bonds containing cationic amino acid residues. Based on the site-specific hydrolysis by trypsin, a characteristic pattern of peptide fragments serves as a peptide mass fingerprint to identify the parent protein.

The mixture of peptides produced by digestion with protease is submitted usually to matrix-assisted laser desorption/ionization time-of-flight mass spectrometry (MALDI-TOFMS) analysis to determine the corresponding peptide masses that are characteristic for a given protein. The mass spectrum obtained is analysed by software that interprets the pattern and predicts the protein by various algorithms, based on a comparison with masses predicted by 'virtual digestion' of identified open reading frames (coding amino acid sequences) in a given genome. Nevertheless, the mass spectra require further careful analysis to ensure the accuracy of the observation. For example, if the measured mass exceeded the predicted mass of a given peptide, one could predict that post-translational modifications of the parent protein may have occurred, such as addition of phosphate groups, hydroxylations at lysine or proline residues, glycosylations or addition of fatty acids. Other deviations of measured from predicted masses may be due to polymorphisms in coding sequences with subtle amino acid substitutions, or even more pronounced with deletions or insertions. Such changes in the primary sequence can be resolved, but success rates depend on the type of substitution, and analysis may require internal peptide Edman sequencing or more advanced mass analysis by ESI-MS (Gaskell, 1997).

Proteome analysis is straightforward if homogenous populations of cultured cells or cell lines are analysed. In contrast, tissue samples contain various cell populations that have different expression profiles that may contribute in various ways to the proteome pattern obtained. Analysis is complicated further by variations in biological samples (i.e. different percentages of individual cell types), which may make analysis a particularly difficult task. Such variations may make it necessary to separate the different cell populations by means of cell-specific surface markers and immunoaffinity techniques or by laser-driven microdissection approaches (Banks *et al.*, 1999; Simone *et al.*, 2000).

The future might hold methods of simplified proteome analysis by using antibody libraries that contain specific antibodies against any expressed open reading frame. This would take proteome analysis to the format of high-throughput microplate assays that permit identification and quantification of essentially every known protein (Brody and Gold, 2000).

Metabolomics

The consequences of gene deletions or targeted expression of new genes on the metabolite profiles of organisms are modelled mainly in studies employing bacteria, yeast and plants. Metabolome analysis as a new tool for a comparative display of gene function has the potential not only to provide deeper insights into complex regulatory processes but also to determine phenotypes directly. Automated gas chromatography/mass spectrometry (GC/MS) techniques allow quantification of large numbers of metabolites, the identities of which remain to be determined using chemical procedures. Using these large data sets, various data mining tools for component analysis are employed to assign a given genotype to its characteristic metabolic phenotype (Glassbrook *et al.*, 2000).

Most of the analytical procedures that have been applied to characterize metabolic processes in traditional nutrition studies are based on classical methods of quantitative chemistry. Specific and quantitative analysis is required for focused studies of specific compounds or metabolic pathways. The methods developed have been optimized to produce high-quality data that describe the compounds of interest. However, this type of analysis is poorly suited to gather information simultaneously on the multitude of metabolites that characterize an organism's nutritional processes.

In metabolic profiling, the concept is different. To monitor hundreds or even thousands of metabolites in parallel, high-throughput techniques are required that allow for screening for relative changes rather than absolute concentrations of compounds. Most analytical techniques for profiling of small molecules consist of a high-performance liquid chromatograph (HPLC) or a gas chromatograph (GC) coupled to a mass spectrometer. Mass spectrometers generally are more sensitive and more selective than any other type of detector. When coupled with the appropriate sample introduction and ionization techniques, mass spectrometers can analyse selectively both organic and inorganic compounds. Nevertheless, prior to detection, the metabolites have to be separated by chromatographic techniques that are coupled on-line to the mass detector. GC is used to separate compounds based on their relative vapour pressures and affinities for the material in the chromatography column, but is limited to compounds that are volatile and heat stable. Most biological compounds, such as sugars, amino acids and organic acids, are not sufficiently volatile to be separated by GC in their native state and must therefore be derivatized prior to GC separations. HPLC separations are better suited for the analysis of labile and high molecular weight compounds and for the analysis of non-volatile polar compounds in their natural form. Although GC- and HPLC-based profiling techniques are not truly quantitative, the compounds detected and their relative amounts may be compared between studies by employing the proper standards. The high-throughput screening with GC-MS and HPLC-MS techniques will also generate large volumes of analytical data that require advanced informatics technologies to organize vast amounts of information.

Metabolite profiling technology will allow information to be gathered on the flow of metabolites through biological pathways and the control of these pathways. In particular, high-resolution ^{1}H-NMR spectroscopy (with the advantage of detection of any proton-containing metabolite) appears to have become increasingly important in metabolite profiling. NMR techniques have been used in the past mainly to analyse metabolite changes in mammalian body fluids and tissues; this method may be extended by detecting other nuclides, e.g. ^{31}P or natural isotopes such as ^{13}C.

When metabolomics is applied to studies where substrates enriched in isotopes such as ^{13}C are administered, metabolite analysis can be taken to a dynamic level by quantification of fluxes (Brenna, 2001). Such automated, biochemical profiling techniques will become an important component of multidisciplinary integrated approaches to metabolic and functional genomics studies.

Current limitations of the technologies when applied to studies in humans

Applications of the technologies of genomics, transcriptomics, proteomics and metabolomics in nutritional studies seem unlimited in terms of basic research. In view of applied research interests, these techniques also have a great potential to identify specific markers (biomarkers) that respond to the status of a given nutrient, non-nutrient compound, treatment or diet. Biomarkers of nutrient status may include changes in the levels of individual mRNAs or proteins, but may also include changes in the pattern of a large group of mRNAs or proteins. So far, biomarkers of cellular functions have been identified mainly by rational approaches based on knowledge of metabolism. The new screening approaches are essentially non-logical when analysing several thousands of potentially affected indicator molecules (mRNAs, proteins) simultaneously.

The use of the new technology platforms is essentially unlimited when human cells in culture or model organisms and experimental animals are employed in nutrition research. However, there are limitations with regard to human studies. Expression profiling at the mRNA level is limited by the availability of cells containing sufficient amounts of high-quality RNA for analysis, although the mRNA of a single cell can be used for amplification and quantitation of single genes or sets of genes by PCR techniques. In proteomics, sufficient quantities of human cells for assessing the protein pattern can only be obtained by invasive techniques.

Although certain cells (such as hair follicle cells, skin cells or even exfoliated intestinal cells) may be obtained in sufficient quantities by non-invasive techniques, various types of blood cells are the preferred source of material (RNA and/or proteins) and may be used as 'reporter cells'. They do respond to dietary changes and also have various life expectancies, various gene expression profiles and control systems, and they target various body compartments. In particular, peripheral blood lymphocytes are already used for identification of potential mRNA biomarkers in human studies in response to environmental factors employing DNA arrays (Amundson *et al.*, 2000; Glynne *et al.*, 2000; Marrack *et al.*, 2000). For analysis of SNPs, various sources of DNA can be used; large-scale applications for identification of relevant SNPs are the basis of the rapidly growing field of molecular epidemiology in all areas of biomedical sciences (Haugen, 1999; Tan *et al.*, 1999; Beeley *et al.*, 2000; Perera and Weinstein, 2000; Schork, *et al.*, 2000). Expression profiling, SNP analysis and proteomics are already well established technology platforms in drug discovery studies and are central for pharmacogenomics, a phrase that links genetic heterogeneity (SNPs) and selective expression of drug-metabolizing enzymes to phenotypical differences in the kinetics and dynamics of drug metabolism (Beeley *et al.*, 2000; Rininger *et al.*, 2000; Norton, 2001). To date, some nutrition studies have utilized the new techniques of genomics, proteomics and metabolomics. These studies include application of gene arrays in response to micronutrient status (see Chapter 1) or to caloric restriction (Lee *et al.*, 1999), application of proteomics to identify effects of ligands of peroxisomal proliferator receptors in obese mice (Edvardsson *et al.*, 1999) and metabolic phenograms of plasma components and their diet-induced changes in rats (Vigneau-Callahan *et al.*, 2001).

Summary

Gene–nutrient interactions are the paradigm for the interplay between the genome and the environment. We are just entering the era of post-genomic research, and there is no doubt that molecular nutritional science is going to be of central interest as nutrients and other food components are the key factors in affecting gene and protein activities. The wealth of genetic information and novel techniques with high-throughput capabilities provide exciting tools for nutrition research. Knowledge of the response of mammalian organisms to changes in their nutritional environment may be gathered at the mRNA and/or protein levels by expression arrays, proteome analysis and high-throughput metabolite profiling. As those tools can generate overwhelming data sets, there is clearly an emerging need for bioinformatics in nutritional sciences, but also for nutrition researchers with a good knowledge base in cell biology and biochemistry of metabolism. The ultimate goals in the application of all these techniques are to expand our understanding of metabolism and nutrition and to determine how this relates to health and disease.

References

Adam, G.C., Cravatt, B.F. and Sorensen, E.J. (2001) Profiling the specific reactivity of the proteome with non-directed activity-based probes. *Chemistry and Biology* 8, 81–95.

Amundson, S.A., Do, K.T., Shahab, S., Bittner, M., Meltzer, P., Trent, J. and Fornace, A.J. Jr (2000) Identification of potential mRNA biomarkers in peripheral blood lymphocytes for human exposure to ionizing radiation. *Radiation Research* 154, 342–346.

Banks, R.E., Dunn, M.J., Forbes, M.A., Stanley, A., Pappin, D., Naven, T., Gough, M., Harnden, P. and Selby, P.J. (1999) The potential use of laser capture microdissection to selectively obtain distinct populations of cells for proteomic analysis – preliminary findings. *Electrophoresis* 20, 689–700.

Beeley, L.J., Duckworth, D.M. and Southan, C. (2000) The impact of genomics on drug discovery. *Progress in Medicinal Chemistry* 37, 1–43.

Blackstock, W.P. and Weir, M.P. (1999) Proteomics: quantitative and physical mapping of cellular proteins. *Trends in Biotechnology* 17, 121–127.

Brenna, J.T. (2001) Natural intramolecular isotope measurements in physiology: elements of the case for an effort toward high-precision position-specific isotope analysis. *Rapid Communications in Mass Spectrometry* 5, 1252–1562.

Brody, E.N. and Gold, L. (2000) Aptamers as therapeutic and diagnostic agents. *Journal of Biotechnology* 74, 5–13.

Celis, J.E., Kruhoffer, M., Gromova, I., Frederiksen, C., Ostergaard, M., Thykjaer, T., Gromov, P., Yu, J., Palsdottir, H., Magnusson, N. and Orntoft, T.F. (2000) Gene expression profiling: monitoring transcription and translation products using DNA microarrays and proteomics. *FEBS Letters* 480, 2–16.

Chalmers, M.J. and Gaskell, S.J. (2000) Advances in mass spectrometry for proteome analysis. *Current Opinion in Biotechnology* 11, 384–390.

Cordwell, S.J., Nouwens, A.S., Verrills, N.M., Basseal, D.J. and Walsh, B.J. (2000) Subproteomics based upon protein cellular location and relative solubilities in conjunction with composite two-dimensional electrophoresis gels. *Electrophoresis* 21, 1094–1103.

Corthals, G.L., Wasinger, V.C., Hochstrasser, D.F. and Sanchez, J.C. (2000) The dynamic range of protein expression: a challenge for proteomic research. *Electrophoresis* 21, 1104–1115.

Delneri, D., Brancia, F.L. and Oliver, S.G. (2001) Towards a truly integrative biology through the functional genomics of yeast. *Current Opinion in Biotechnology* 12, 87–91.

Dutt, M.J. and Lee, K.H. (2000) Proteomic analysis. *Current Opinion in Biotechnology* 11, 176–179.

Edvardsson, U., Alexandersson, M., Brockenhuus von Lowenhielm, H., Nystrom, A.C., Ljung, B., Nilsson, F. and Dahllof, B. (1999) A proteome analysis of livers from obese (ob/ob) mice treated with the peroxisome proliferator WY14,643. *Electrophoresis* 20, 935–942.

Fiehn, O., Kopka, J., Dormann, P., Altmann, T., Trethewey, R.N. and Willmitzer, L. (2000) Metabolite profiling for plant functional genomics. *Nature Biotechnology* 18, 1157–1161.

Gaskell, S.J. (1997) Electrospray: principles and practice. *Journal of Mass Spectrometry* 32, 677–688.

Gevaert, K. and Vandekerckhove, J. (2000) Protein identification methods in proteomics. *Electrophoresis* 21, 1145–1154.

Glassbrook, N., Beecher, C. and Ryals, J. (2000) Metabolic profiling on the right path. *Nature Biotechnology* 18, 1142–1143.

Glynne, R., Ghandour, G., Rayner, J., Mack, D.H. and Goodnow, C.C. (2000) B-lymphocyte quiescence, tolerance and activation as viewed by global gene expression profiling on microarrays. *Immunological Reviews* 176, 216–246.

Gorg, A., Obermaier, C., Boguth, G., Harder, A., Scheibe, B., Wildgruber, R. and Weiss, W. (2000) The current state of two-dimensional electrophoresis with immobilized pH gradients. *Electrophoresis* 21, 1037–1053.

Haugen, A. (1999) Progress and potential of genetic susceptibility to environmental toxicants. *Scandinavian Journal of Work Environment and Health* 25, 537–540.

Kose, F., Weckwerth, W., Linke, T. and Fiehn, O. (2001) Visualizing plant metabolomic correlation networks using clique-metabolite matrices. *Bioinformatics* 17, 1198–1208.

Lee, C.K., Klopp, R.G., Weindruch, R. and Prolla, T.A. (1999) Gene expression profile of aging and its retardation by caloric restriction. *Science* 285, 1390–1393.

Lockhart, D.J. and Winzeler, E.A. (2000) Genomics, gene expression and DNA arrays. *Nature* 405, 827–836.

Marrack, P., Mitchell, T., Hildeman, D., Kedl, R., Teague, T.K., Bender, J., Rees, W., Schaefer, B.C. and Kappler, J. (2000) Genomic-scale analysis of gene expression in resting and activated T cells. *Current Opinion in Immunology* 12, 206–209.

Norton, R.M. (2001) Clinical pharmacogenomics: applications in pharmaceutical R&D. *Drug Discovery Today* 6, 180–185.

Pandey, A. and Mann, M. (2000) Proteomics to study genes and genomes. *Nature* 405, 837–846.

Paton, N.W., Khan, S.A., Hayes, A., Moussouni, F., Brass, A., Eilbeck, K., Goble, C.A, Hubbard, S.J. and Oliver, S.G. (2000) Conceptual modelling of genomic information. *Bioinformatics* 16, 548–557.

Patton, W.F. (2000) Making blind robots see: the synergy between fluorescent dyes and imaging devices in automated proteomics. *Biotechniques* 28, 944–948.

Perera, F.P. and Weinstein, I.B. (2000) Molecular epidemiology: recent advances and future directions. *Carcinogenesis* 21, 517–524.

Rininger, J.A., DiPippo, V.A. and Gould-Rothberg, B.E. (2000) Differential gene expression technologies for identifying surrogate markers of drug efficacy and toxicity. *Drug Discovery Today* 5, 560–568.

Sauer, B. (1998) Inducible gene targeting in mice using the Cre/lox system. *Methods* 14, 381–392.

Schilling, C.H., Edwards, J.S. and Palsson, B.O. (1999) Toward metabolic phenomics: analysis of genomic data using flux balances. *Biotechnology Progress* 15, 288–295.

Schork, N.J., Fallin, D. and Lanchbury, J.S. (2000) Single nucleotide polymorphisms and the future of genetic epidemiology. *Clinical Genetics* 58, 250–264.

Simone, N.L., Paweletz, C.P., Charboneau, L., Petricoin, E.F. and Liotta, L.A. (2000) Laser capture microdissection: beyond functional genomics to proteomics. *Molecular Diagnosis* 5, 301–307.

Tan, K.T., Dempsey, A. and Liew, C.C. (1999) Cardiac genes and gene databases for cardiovascular disease genetics. *Current Hypertension Report* 1, 51–58.

ter Kuile, B.H. and Westerhoff, H.V. (2001) Transcriptome meets metabolome: hierarchical and metabolic regulation of the glycolytic pathway. *FEBS Letters* 500, 169–171.

Tomita, M. (2001) Whole-cell simulation: a grand challenge of the 21st century. *Trends in Biotechnology* 19, 205–210.

Vigneau-Callahan, K.E., Shestopalov, A.I., Milbury, P.E., Matson, W.R. and Kristal, B.S. (2001) Characterization of diet-dependent metabolic serotypes: analytical and biological variability issues in rats. *Journal of Nutrition* 131, 924S–932S.

3 Molecular Physiology of Plasma Membrane Transporters for Organic Nutrients

Hannelore Daniel
Molecular Nutrition Unit, Department of Food and Nutrition, Technical University of Munich, Germany

Introduction

Nutrient transport across the plasma membrane of cells is a critical step in metabolism and nutrient homeostasis since the cell membrane is a barrier for permeation and compartmentalizes metabolic processes. The phospholipid bilayer surrounding cells shows an intrinsically low permeability for hydrophilic low molecular weight compounds as well as for larger molecules. However, even lipophilic nutrients such as cholesterol or fatty acids – up until now believed to cross the cell membranes by passive diffusion – appear to require specialized membrane proteins for transport that also allow control of permeation and metabolic flow. The maintenance of an intracellular environment that is distinctly different from the extracellular milieu is essential for life, and therefore a large spectrum of membrane proteins with highly specialized functions has emerged during evolution that control the membrane permeability for ions, water, macro- and micronutrients, and metabolic intermediates, but also that for xenobiotics.

Our knowledge of the molecular architecture and functions of membrane proteins was sparse, as those proteins could in most cases not be isolated by means of classical protein purification techniques. Whereas extracellular and cytoplasmic proteins could be purified in quantities that allowed crystallization and determination of their three-dimensional structure and the structural changes associated with function, membrane transport function could not be related to distinct structural protein elements.

With the advent of molecular biology and new cloning strategies, this changed dramatically. In the 1980s, the first cDNAs encoding mammalian cell membrane proteins were isolated, and allowed predictions to be made about the structure of the proteins and their integration into the cell membrane compartment. The sequences obtained also allowed identification of related sequences and proteins by homology screening. In addition, expression cloning techniques were used to obtain novel sequences encoding nutrient transport proteins (Daniel, 2000). Here, a cDNA library produced from a mRNA pool comprising thousands of individual mRNAs isolated from the tissue of interest is screened for the encoded protein by measuring its function after a cDNA pool has been introduced into a model cell. This was done mainly in oocytes of the South African frog *Xenopus laevis*, as the cells of this organism are large (~ 1 mm) and cDNA pools can be introduced easily by micropipettes. Once the expected nutrient transporter function is obtained by transport measurements in the oocytes, the cDNA pool is divided, and the different pools are reinjected and screened again for the expression of function. Eventually, a single cDNA is identified that encodes the protein with the known function. Sequencing of this cDNA provides the coding region (open reading frame; ORF) and can be translated into the amino acid sequence. A variety of algorithms can then be used to predict the secondary structure of the protein, based on the different polarities of the amino acid residues within the sequence and the assumption that stretches of 10–15 or more consecutive mainly hydrophobic amino acid residues are needed to form an α-helical domain that is inserted into the phosphobilayer of the cell. Accordingly, the more hydrophilic regions are predicted to connect these transmembrane domains and form the intracellular and extracellular linker regions. In many cases, the predicted topologies based on this hydropathy analysis of individual membrane transporters proved to be correct when verified experimentally. Analysis of membrane protein topology is done mainly by introducing flags or special epitopes (small stretches of amino acid sequences) that are accessible to specific antibodies or reagents either from the outside of the cell or after permeabilization of the cells for reaction from the inner side of the membrane. Alternatively, a cysteine scanning approach can be used. Here the protein transport function is used as a reporter system. Codons encoding a cysteine residue are incorporated into the cDNA by site-directed mutagenesis at any interesting position, and the chemical modification of the cysteine residues with special membrane-impermeable or -permeable reagents that eventually will impair protein function allows determination of the location of the cysteine residue as accessible from the inside or outside of the cell by functional analysis in a heterologous expression system.

At the Beginning of Molecular Mammalian Transport Biology: the GLUT Transporters

A landmark for the cloning of human nutrient transporters was the identification of the first GLUT family member from HepG2 cells (Mueckler *et al.*, 1985). At the protein level, the cloned cDNA showed a high similarity and even amino acid identity to the red cell glucose transporter from which a partial sequence had been obtained by classical Edman sequencing. According to the proposed terminology based on HUGO data (human genome sequencing organization) the GLUT transporters are now designated as the SLC2A group with consecutive numbering. The GLUT family (see Table 3.1) currently contains the GLUT-1–GLUT-5 proteins, which are the primary transporters that mediate monosaccharide transport into and out of mammalian cells (Seatter and Gould, 1999). GLUT-6–12 are proteins with as yet not well defined functions, and GLUT-13 represents a proton-dependent myoinositol-transporter mainly expressed in the brain (Uldry *et al.*, 2001). GLUTs generally are comprised of approximately 500 amino acids that are arranged most probably in 12 transmembrane (TM) segments that, by arrangement, form a pore. Both the N- and the C-terminus of the proteins face the intracellular compartment, a large extracellular loop near the N-terminus of the transporter is glycosylated in the mature protein and a large intracellular loop connecting TM segments

Table 3.1. The family of the mammalian GLUT transporters involved in transmembrane translocation of hexoses and related solutes and identified human diseases associated with their malfunction.

SLC2-a family	Major sites of expression	HUGO code	Human diseases
GLUT-1	Erythrocytes, brain (vascular)	SLC2A1	→ GLUT-1 deficiency syndrome
GLUT-2	Liver, pancreas (islets)	SLC2A2	→ Fanconi–Bickel syndrome, NIDDM
GLUT-3	Brain (neuronal)	SLC2A3	→ Fetal glucose toxicity
GLUT-4	Muscle, fat tissue, heart	SLC2A4	→ Iatrogenic diabetes by HIV, NIDDM
GLUT-5	Intestine, kidney, testis	SLC2A5	
GLUT-6	⎫	SLC2A6	
GLUT-7	⎬ Novel isoforms, function not	SLC2A7	
GLUT-8	⎬ yet defined	SLC2A8	
GLUT-9	⎬	SLC2A9	
GLUT-X	⎭		
GLUT-13 (HMIT) (proton-dependent myoinositol transporter)	Brain	SLC2A13	

NIDDM, non-insulin dependent diabetes.

6 and 7 carries consensus sites for phosphorylation (Hruz and Mueckler, 2001). A three-dimensional model of GLUT-1 and its integration into the plasma cell membrane has been reported recently that shows a channel-like central structure that traverses the entire protein and that may form the substrate translocation pore (Zuniga et al., 2001). These models are important in understanding the structure and function as no crystal structures of any of the nutrient transporter proteins are available yet.

In terms of nutrition and glucose homeostasis, the GLUT-2 and GLUT-4 transporters are of central importance. GLUT-2 is responsible for influx and efflux of glucose in hepatocytes and therefore contributes to plasma glucose levels both in the post-prandial state with removal of glucose from portal blood, and in maintaining glucose levels during starvation by contributing to the efflux of glucose provided by gluconeogenesis and/or glycogenolysis. GLUT-2 expression in the basolateral membrane of epithelial cells such as those of the intestine allows glucose, galactose and fructose that enter the cells from the intestinal lumen mainly via the sodium-dependent glucose transporter SGLT-1 (glucose, galactose) or via the GLUT-5 transporter (fructose) to exit into the circulation. As the GLUT transporters (except GLUT-13) mediate monosaccharide transport following concentration gradients, metabolism acts as a modulator of the transmembrane concentration gradients and plays a crucial role in regulating overall transport capacity. GLUT-2 has a particularly important function in the β-cells of the pancreatic islets as the system that provides glucose to the cell that then in turn triggers insulin secretion. GLUT-2 in conjunction with glucokinase is central to the blood glucose-sensing pathway in β-cells. The other most important GLUT family member is GLUT-4 – a key target for modern therapy of non-insulin-dependent type II diabetes (NIDDM). Insulin-dependent incorporation of GLUT-4 transporters into the plasma membrane of adipocytes and muscle cells by translocation from intracellular vesicular stores can serve as a paradigm for the complexity of signalling pathways and the cellular machinery needed for membrane trafficking and membrane fusion. GLUT-4 synthesis and its rapid cycling between the plasma membrane and the storage vesicles is subject to regulation by a large number of factors including a variety of hormones and metabolites, and by exercise (Watson and Pessin, 2001). GLUT-4 is becoming a key player in the therapy of NIDDM with the goal of sensitizing the system for enhanced performance. How critical the function of GLUT-4 is for glucose homeostasis was demonstrated recently in a very elegant study employing mice with a muscle-specific gene inactivation of *glut-4* (cre/loxP gene targeting) that showed dramatic changes in glucose metabolism (Kim et al., 2001). Various types of malfunction of the glucose transporters of the GLUT family have been reported (see Table 3.1) and recently human immunodeficiency virus (HIV) protease inhibitors have been identified as inhibitors of GLUT-4, which may explain some of the metabolic impairments of energy and glucose metabolism in AIDS patients. GLUT-1 deficiency produces a seizure disorder with low glucose concentrations in the cerebrospinal fluid, and GLUT-2 deficiency is the basis of the Fanconi–Bickel syndrome, which resembles type I glycogen storage disease.

The SGLT Family of Electrogenic Transporters

The discovery of the sodium dependency of intestinal glucose transport in the early 1960s fostered vital research to understand the molecular basis of this coupling of glucose influx into enterocytes to sodium ion movement and the energetics of this uphill transport process. The huge daily quantities of glucose absorbed as the end-product of luminal carbohydrate digestion as well as from disaccharide hydrolysis suggested that the transport protein responsible would have a very high transport capacity. The energetics of glucose transport into enterocytes allows glucose to accumulate in the cell and provides a concentration gradient that enables downstream movement of glucose via GLUT-2 at the basolateral side as long as the concentration in the circulation is less than that in the cell. Phlorizin turned out to be a very helpful tool for analysis of glucose transport as it is a high-affinity uptake inhibitor.

The transport protein was identified almost 25 years after the initial proposal of the sodium gradient hypothesis of intestinal glucose transport. In 1987, expression cloning employing *Xenopus* oocytes and selecting mRNA pools that induced sodium-dependent and phlorizin-sensitive glucose transport in oocytes led to the first cDNA encoding an electrogenic glucose transporter named SGLT-1 (Hediger et al., 1987). HUGO data now show the large number of transport proteins that belong to the same family, with individual members not only in transport of hexoses but also, for example, in sodium-dependent transport of iodide, myo-inositol and water-soluble vitamins such as biotin and pantothenate. Proteins homologous to SGLT-1 are also found in prokaryotes and invertebrates, with transport functions for urea, proline, sugars or vitamins.

Various members of the SGLT family, when expressed in heterologous systems, showed remarkable similarities in function. The high-affinity Na^+–glucose co-transporter SGLT-1 from three species (human, rabbit and rat), the low-affinity Na^+–glucose co-transporter SGLT-2, the Na^+-dependent myoinositol co-transporter SMIT1 and the Na^+-dependent iodide co-transporter NIS have all been expressed in *Xenopus* oocytes and analysed by radioactive tracers and electrophysiological techniques (Turk and Wright, 1997). In the absence of substrate, they all show a limited Na^+ permeability but, in the presence of substrates, voltage-dependent positive inward currents indicating Na^+–substrate co-transport are recorded. The transport cycle is ordered, with Na^+ binding to the protein in the first step, followed by substrate binding and the translocation of the loaded complex, and sequential release of the substrate and co-transported ion into the cytoplasm. The rate-limiting step in the process appears to be the reorientation of the unloaded but charged transporter to the outer face of the membrane. More recently, an intrinsic water permeability has been proposed for SGLT-1, with a substantial number of water molecules co-transported in each cycle together with sodium ions and glucose (Loo et al., 1996). Based on the daily amount of glucose transported by SGLT-1, its water transport capacity could account for almost 5 l of fluid absorbed from the upper small intestine solely mediated by SGLT-1.

At the structural level, the SGLT-1 protein has been studied extensively by functional analysis in combination with mutations, epitope tagging, immunological and other cell biological and biophysical techniques. Based on experimental data and computational analyses, a 14 membrane-spanning domain model of human SGLT-1 has been proposed (see Fig. 3.1). All TM domains are most likely to be α-helical, the N-terminus is extracellular and the large highly charged C-terminal domain is cytoplasmic (Turk et al., 1996, 1997). Human SGLT-1 has 664 amino acid residues; other members of the family have between 530 and 735 amino acids. Using truncated and modified SGLT-1-proteins and biophysical techniques, two regions have been identified in the protein that confer the functional domains, with respect to the binding pocket for sodium ions and the region that binds the substrate and thereafter mediates the translocation of sodium and glucose to the cell interior. The conformational change

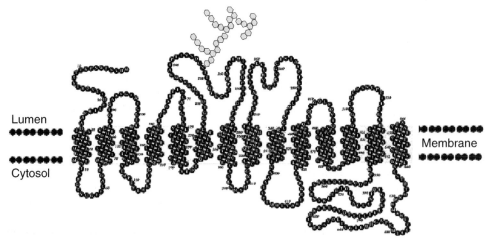

Fig. 3.1. The predicted membrane topology of the electrogenic Na^+-dependent glucose transporter SGLT-1 as found predominantly in the apical membranes of intestinal and renal epithelial cells.

necessary for the movement of sodium ions and substrate through the protein appears to require just the tilting of two TM helices within the plane of the membrane by a movement of 10–15 Å that then generates a pore-like structure for permeation (Eskandari et al., 1998).

Glucose galactose malabsorption syndrome is a hereditary disease in which SGLT-1 does not function properly (Martin et al., 1996). The mutations found in humans have allowed the identification of critical amino acid residues in the SGLT-1 protein that cause the malabsorption of glucose and galactose in the gut leading to life-threatening diarrhoea. Surprisingly, almost all mutations identified in patients cause defects in the biosynthetic pathway of the SGLT-1 protein and a lack of integration into the cell membrane. This addresses one of the important areas in membrane transporter biology. How are proteins guided from the endoplasmic reticulum to the Golgi apparatus and to the specific cell membrane compartment (apical versus basolateral side)?

SGLT-1 expression in the gut is subject to regulation by a variety of hormones and luminal factors. Intestinal glucose transport also shows a significant circadian rhythm: kinetic studies suggested an altered SGLT-1 maximal velocity (V_{max}) associated with changes in gene expression. SGLT-1 may be responsible for the fairly high absorption rate of some of the quercetin glycosides (and perhaps other glycosides) of the flavonoid class (Walgren et al., 2000; Ader et al., 2001). These secondary plant metabolites are good antioxidants and serve as modifiers of proteins in various signalling pathways.

The human kidneys filter approximately 180 g of D-glucose from the plasma each day, and glucose is reabsorbed very efficiently in the proximal tubules. Na^+–glucose co-transport across the renal brush border membrane and facilitated diffusion across the basolateral membrane account for recycling of glucose from primary urine into the circulation. The bulk of glucose is reabsorbed in the convoluted proximal tubule by a low-affinity, high-capacity transporter that might be SGLT-2 (Wright, 2001). Glucose escaping this first reabsorption step is then taken up in the straight proximal tubule by a high-affinity, low-capacity transporter that is most likely to be SGLT-1. Congenital renal defects in glucose reabsorption that cause severe glucosuria appear to be associated mainly with malfunction of the SGLT-2 protein, whereas patients suffering from malfunction of SGLT-1 only have a mild renal glucosuria (Sankarasubbaiyan et al., 2001).

Transport Processes for Nucleosides

Nucleosides are required in the synthesis of DNA and RNA, and as metabolic intermediates; various tissues cannot synthesize nucleosides but depend on an external nucleoside supply. Intertissue exchange of nucleosides requires specific nucleoside transporters. In addition, nucleosides are provided by the diet as the end-products of the gastrointestinal digestion of DNA, RNA and nucleotides. Intestinal epithelial cells express two types of Na^+-dependent nucleoside transporters that couple uphill transport of nucleosides with movement of sodium ions down the electrochemical sodium gradient. These two electrogenic systems are the CNT proteins (concentrative nucleoside transporters). CNT1 transports nucleosides of pyrimidine bases, CNT2 those containing purine bases (Pastor-Anglada et al., 2001). The efflux of the nucleosides from the epithelial cells into the circulation is mediated by the equilibrative nucleoside transporter (the ENT subgroup). The members of this large family of proteins found throughout the human body are uniporters that operate bidirectionally along the transmembrane gradient for nucleosides (Hyde et al., 2001). Members of the ENT family also possess a certain specificity towards pyrimidine or purine bases (Ritzel et al., 2001). The concentrative Na^+-dependent nucleoside transporter proteins have around 650 amino acid residues and most probably contain 13 membrane-spanning domains. Both, ENT and CNT proteins are important not only for metabolism of nucleosides but also for the uptake and delivery of antiviral and cytostatic drugs that structurally resemble nucleosides (Wang et al., 1997).

Monocarboxylate Transporters

The exchange of monocarboxylates among tissues and organs is essential for a variety of metabolic processes including the release of lactate from muscle tissue and other cells followed by its re-uptake into liver for gluconeogenesis (Cori cycle), and the transport of ketone bodies during starvation and ketoacidosis. Other monocarboxylates such as pyruvate, butyrate, propionate and acetate or the keto-derivatives of the branched chain amino acids also show significant interorgan fluxes. Figure 3.2 summarizes some of the cellular functions of the identified transporters for the various monocarboxylates that belong to the MCT family. Monocaboxylate transporters MCT-1–MCT-9 have been identified in humans (HUGO: SLC16A1–SLC16A9), with a large number of related genes in other pro- and eukaryotic organisms (Halestrap and Price, 1999). MCT-1 is expressed ubiquitously in mammals. Very strong expression is found in heart and red muscle, where MCT-1 is up-regulated in response to increased work, suggesting its special role in lactic acid delivery for oxidation (Bonen et al., 1997). It is interesting to note that MCT-1 in striated muscle is also found in mitochondrial membranes based on immunostaining, suggesting that it can be targeted to plasma and mitochondrial membranes (Brooks et al., 1999). MCT-2 is a high-affinity-type transporter expressed in renal proximal tubules, sperm tails and neurons. MCT-3 is expressed uniquely in the retinal pigment epithelium (Halestrap and Price, 1999).

The transporter MCT-4 is most evident in white muscle and other cells with a high rate of glycolysis, including tumour cells and white blood cells, with a particularly high demand for efficient

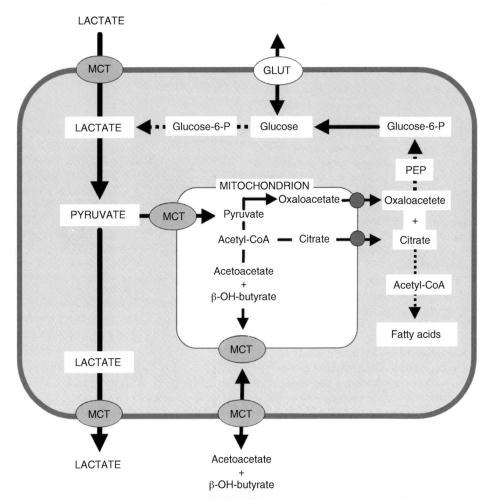

Fig. 3.2. The proposed cellular functions of identified MCT carriers and their role in cellular fluxes of monocarboxylates that link a variety of metabolic pathways.

lactic acid efflux (Bonen, 2000). Functional analysis of heterologously expressed MCT-4 as the major transporter isoform present in white skeletal muscle shows K_m values for L-lactate, D-lactate and pyruvate of 28, 519 and 153 mM, respectively, characterizing MCT-4 as a low-affinity-type transport system. For MCT-1, affinities range from 2.1 mM for pyruvate, to 4.4 mM for L-lactate and to >60 mM for D-lactate (Juel and Halestrap, 1999). Monocarboxylate transport occurs in most cases electroneutrally by symport of the anion with a proton. The kinetics of pyruvate and lactate transport into red blood cells have been characterized thoroughly and show that the characteristics of MCT-1 are consistent with it being the only MCT present in the plasma membrane of red blood cells (Poole and Halestrap, 1993). Transport involves proton binding to the transporter as the initial step, followed by binding of lactate and the translocation of the complex with the sequential release of the solutes into cytoplasm. The transporter can operate in both directions dependent on the concentrations of substrate and co-transported protons; equilibrium is reached when [lactate] in/[lactate]out = $[H^+]$out/$[H^+]$in (Halestrap and Price, 1999).

Regulation of MCT gene expression is an important part of the adaptation of metabolism to various conditions. Advanced physical activity can improve lactate/H^+ transport capacity in muscle by up-regulation of MCT-1 expression. In contrast, in denervated rat muscles, the rate of lactate transport is reduced and expression of MCT-1 and MCT-4 is diminished. Hypoxia is known to increase the expression of lactate dehydrogenase-M and other glycolytic enzymes, as well as GLUT-1. This adaptation involves a variety of transcription factors and response elements including the hypoxia-inducible factor 1, cAMP response element and the erythropoietin hypoxic enhancer. Since the distribution of MCT-4 within muscle fibres is similar to that of lactate dehydrogenase-M, hypoxia could also increase MCT-4 expression by similar molecular mechanisms. Up-regulation of the cardiac monocarboxylate transporter MCT-1 in a rat model of congestive heart failure suggests that lactate may be used as an important respiratory substrate for cardiac metabolism, which becomes more dependent on carbohydrates under these disease conditions (Halestrap and Price, 1999).

Another important tissue in which transport of weak organic acids such as butyrate, lactate, propionate and acetate is mediated by MCT family members is the colonic epithelium. Monocarboxylates are generated in the colon in huge quantities by the microorganisms that metabolize soluble fibres (i.e. inulin and oligofructoses) and non-digested starch. Luminal concentrations of the short chain fatty acids may reach 100–150 mM, and uptake across the brush border membrane provides the fuel for colonocytes (butyrate) and delivers others to the circulation (propionate, lactate and acetate). In addition to its role as an energy substrate, butyrate promotes cell growth in normal colon but inhibits proliferation in established colon adenomas and cancers. The kinetic properties of butyrate transport into colonocytes are consistent with the properties of MCT-1 and therefore this transporter appears to be very important in tissue homeostasis and control of intestinal adaptation (Stein et al., 2000).

The importance of specialized transporters for short chain fatty acids was discussed controversially in view of the high permeability of cell membranes to the non-charged species of these low molecular weight compounds. pH-dependent partition between the two sides of the membrane with unrestricted diffusion of the non-polar species was considered to be the sole mechanism for transport. Of course, this process represents a background route for permeation of monocarboxylates as it is determined by the physicochemical characteristics of the weak acids. However, transport proteins such as MCT-1 and the modulation of the expression level of these carriers provide a level of control for the flux of various organic weak acids into and out of a cell that is mandatory for metabolic homeostasis.

Fatty Acid Transporters

Long chain fatty acids (LCFAs) were also believed to permeate cell membranes without the need for specialized transport proteins as they are highly hydrophobic and insert rapidly into phospholipid bilayers. However, homeostasis demands regulation of membrane transport for a coordinated coupling to metabolism. An increasing volume of evidence suggests that not only the plasma membrane-associated and cytoplasmic fatty acid-binding proteins – known for quite some time – but also specialized transporters for the membrane permeation step are involved in cellular fatty acid uptake. LCFAs not only serve as energy storage substrates and metabolic fuel but are also implicated in the regulation of cell growth and various other cellular functions. Several membrane-associated fatty acid-binding/transport proteins, such as the 43 kDa plasma membrane fatty acid-binding protein (FABPpm), the 88 kDa fatty acid translocase (FAT) and a variety of fatty acid transporter proteins (FATP class), have been identified (Bonen et al., 1998). There are six known proteins of the FATP/SLC27A family that are expressed in tissues and that allow transmembrane movement of LCFAs in mammals (Hirsch et al., 1998). Whereas the saturated fatty acids are very good substrates, polyunsaturated fatty acids are transported very poorly. With the exception of retinoic acid, fat-soluble vitamins are not accepted as substrates. Whereas the coenzyme A (CoA) derivatives of fatty acids are not transported by FATP, overexpression of most FATPs in cells also increases CoA synthase activity 2–5 times, suggesting an intrinsic or associated enzymatic activity of the FATP (Herrmann et al., 2001). The SLC27A family shows a conserved signature motif and a conserved domain for AMP binding, which is essential for transport function. Neither the mechanism nor the exact membrane topology of the FATPs is known yet. However, it may be speculated that LCFAs are co-transported with protons to allow their electroneutral movement into a cell against the inside negative membrane potential.

Expression of FATPs varies with tissue type (Hirsch et al., 1998). Whereas liver expresses various isoforms (designated 2–5) with prominent expression of FATP-5, muscle tissue possesses mainly FATP-1. FATP-2 is found in kidney and FATP-3 predominantly in lung. Heart expresses the FATP-6 isoform. Expression of FATP-4 in the brush border membrane of the intestinal epithelial cells suggests that this protein is involved in intestinal absorption of LCFAs. FAT/CD36 shows a close co-regulation with the expression of FAB. Knock-out mice deficient of CD36 show impaired uptake and use of fatty acids, increased cholesterol levels, decreased body weight, lower than normal plasma glucose and insulin levels, and a reduced metabolic rate (Coburn et al., 2000). In addition, their endurance is reduced. A CD36 deficiency in humans – occurring with an incidence of 0.3–11% in different populations – frequently is associated with defective cardiac fatty acid uptake and hypertension, as well as impaired glucose tolerance. In this context, it should be mentioned that insulin-sensitizing drugs have been shown to regulate CD36 (Yamamoto et al., 1990).

Transmembrane Transport of Amino Acids and Short Chain Peptides

The 20 proteinogenic amino acids and their derivatives are a heterogenous group of compounds that differ in polarity, net charge and molecular mass. They serve as building blocks for protein synthesis, as carriers for nitrogen and carbon units in interorgan metabolism, and as energy substrates, but also as precursors of biologically active compounds such as neurotransmitters. Some are neurotransmitters themselves and others are used for conjugation and excretion of compounds. Amino acid transport proteins mediate and regulate the flow of these nutrients across the plasma membrane for influx and efflux, and thus play a central role in interorgan metabolism. A multitude of membrane transporters with different specificity for the various classes of amino acids mediate transport, due to the different physicochemical characteristics of the amino acids.

The substrate specificity of the different transport pathways was determined in pioneering work by Halvor N. Christensen's group in the 1960s mainly based on analysis of transport in erythrocytes, hepatocytes and fibroblasts (Christensen, 1990). Some general principles were also identified in this early work, such as the stereoselectivity with faster transport of the L-isomers for almost all transport systems and a rather broad substrate specificity of some of the carriers. Some transporters were relatively specific for either acidic or basic amino acids or for amino acids with an aromatic side chain. In addition, differences in the thermodynamic properties of the transport steps were observed with identification of equilibrative systems and systems that were ion-dependent and showed uphill transport capability. Figure 3.3 summarizes the different mechanisms and modes of amino acid transporting systems and representative carrier types.

The first cDNA encoding a mammalian amino acid transporter was identified in 1991 as an ecotropic murine retrovirus receptor that turned

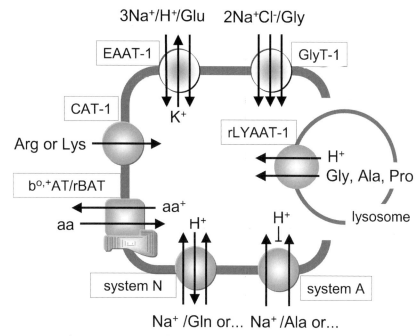

Fig. 3.3. Amino acid transport activities of representative members of the various transporter families and their proposed mode of function, participating ions and flux coupling ratios.

out to transport cationic amino acids (Kim *et al.*, 1991; Wang *et al.*, 1991). Preceding this discovery, the first transporter for the neurotransmitter γ-aminobutyric acid (GABA) was cloned (Guastella *et al.*, 1990). In the last few years, a rapidly growing number of transporters have been isolated by means of expression cloning and homology screening. Currently, about 30 cDNAs have been identified (not including splice variants) that encode proteins with different amino acid transport activities. Table 3.2 summarizes the main transport pathways for amino acids in mammalian cells, subdivided into Na^+-dependent and Na^+-independent pathways and cDNAs that have been cloned that show transport activity when expressed in a target cell.

Transport pathways for zwitterionic amino acids

The zwitterionic amino acid transport systems A, ASC, N and L are present in almost all cell types. Systems A, ASC and N mediate the influx of amino acids with a small side chain by co-transport with sodium ions. System L transporters operate independently of sodium and mediate the flux of amino acids with bulky side chains (i.e. branched and aromatic groups). System L has

Table 3.2. Classified transport activities of the various amino acid-transporting systems found in the plasma membrane of mammalian cells and cDNAs that encode the proteins mediating this activity.

	Isolated cDNA(s) encoding this activity
Na^+ dependent	
A	ATA1–3
N	SN1–3
GLY	GlyT1–2
ASC	ASC1–2
BETA	GAT1–3, BGT-1
IMINO	Not identified yet
B⁰	ATB⁰
B⁰,⁺	ATB⁰,⁺
X⁻$_{AG}$	EAAT1–5
Na^+ independent	
L	4F2hc/LAT(X)
y⁺	CAT1–3
b⁰,⁺	rBAT/b⁰,⁺AT
PAT	LYAAT-1

been postulated to serve in many tissues as an amino acid efflux system rather than for influx. Systems A and N show a preference for glutamine, alanine and serine; system ASC prefers alanine, serine and cysteine. In contrast, amino acids with a small, non-branched side chain are poor substrates for system L, for which the analogue BCH (2-aminoendobycyclo-2,2,1-heptane-2-carboxylic acid) is a model substrate for sodium-independent transport. Zwitterionic amino acids with bulky side chains are, in most cases and in most non-epithelial cells, transported via sodium-independent exchange processes (Palacin et al., 1998). In many epithelia, sodium-dependent transport of alanine can be inhibited by N-methylaminoisobutyric acid (MeAIB), and sodium-dependent transport of MeAIB is used for determining system A-specific transport activity (Reimer et al., 2000). System A is highly pH sensitive and electrogenic, whereas the sodium-dependent system ASC is relatively pH insensitive and operates electroneutrally, suggesting that it may be an antiporter of amino acids associated with the movement of sodium in both directions. An important feature of system A is that its activity in many cell types is highly regulated. This includes an up-regulation during cell cycle progression and cell growth in many cells and tissues, as well as hormonal control by insulin, glucagon, catecholamines, glucocorticoids, various growth factors and mitogens by quite different signalling pathways (McGivan and Pastor-Anglada, 1994). Whereas glucagon and epidermal growth factor induce an immediate increase in system A activity in hepatocytes, insulin up-regulates system A in a gene transcription-dependent manner in hepatocytes and additionally by a rapid pathway – possibly by recruiting preformed transporters to the plasma membrane (also in skeletal muscle). However, insulin deficiency or insulin resistance are also associated with an adaptative up-regulation of system A activity in liver and skeletal muscle, suggesting that the system A transporters are targets of a complex network of regulatory factors for short-term, intermediate and chronic adaptation.

The most interesting amino acid-transporting systems, at least in view of their molecular architecture and the pathophysiology of malfunctions, are the heterodimeric amino acid exchangers that represent the y^+L and $b^{o,+}$ transport activities (see Table 3.2), which are found in a variety of cells (Chillaron et al., 2001; Wagner et al., 2001). The novelty of their structural aspects is that they consist of two separate proteins: one heavy chain (out of two heavy chains that have been identified) and one light chain (out of seven). The two monomers oligomerize via an extracellular disulphate bridge; transport activities and membrane locations of the various complexes depend on the nature of its two subunits. In general terms, the two subunits that form the active amino acid transporters consist of a glycoprotein heavy chain designated as 4F2hc or rBAT that combines with one of the seven light chains (LAT1, LAT2, y^+LAT1, y^+LAT2, ascAT1, xCT and $b^{o,+}$AT) that have been identified (Chillaron et al., 2001). The heavy chain rBAT is expressed mainly in epithelial cells; association with a light chain is followed by translocation of the complex to the apical membrane, resulting in a sodium-independent amino acid exchange process (Palacin et al., 2001a). The other heavy chain 4F2hc associates with various light chains that then mediate amino acid exchange mechanisms in basolateral membranes of epithelial cells as well as in non-epithelial cells. The heavy chains have a glycosidase-like extracellular domain attached to a single TM domain (Verrey et al., 2000). The light chains vary in size, but all possess 12 membrane-spanning domains of a polytopic membrane protein with the N- and C-termini facing the cytoplasm. Figure 3.4 depicts models of how the various light chains associate with either rBAT or 4F2hc to form the heterodimeric complexes that possess different amino acid-transporting selectivities. It appears that the main function of the heavy chain is to serve as an internal carrier that enables the complex to be translocated and integrated into the target cell membrane. The heterodimeric transporters can transport a variety of neutral amino acids in an obligatory exchange mode, which means they mediate influx of certain amino acids at the expense of intracellular amino acids. The $b^{o,+}$ activity of the rBAT-associated complex can, in addition, transport cationic amino acids in exchange for neutral amino acids, which results in transport currents, whereas the y^+LAT1–4F2hc complex exchanges neutral amino acids with sodium co-transport for intracellular cationic amino acids. A variety of mutations in the transporter complexes have been identified that cause malfunctions and severe metabolic disturbances. Defects in $b^{o,+}$AT cause non-type I cysteinuria with clinical manifestations in the kidney (Palacin et al.,

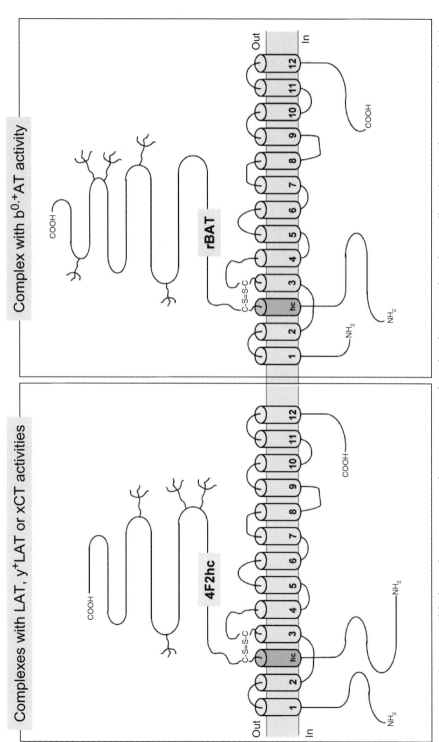

Fig. 3.4. Proposed structure of the heterodimeric amino acid-transporting complexes found in mammalian plasma cell membranes. They consist of either the heavy chain 4F2hc or rBAT linked to one of the various light chains via an extracellular disulphide brigde. The different complexes then constitute transport pathways with different characteristics with regard to substrate specificity and membrane location.

2001b). It represents a reabsorption defect for cysteine in the apical membrane of renal tubules.

Defective function of y$^+$LAT1 causes lysinuric protein intolerance as a transport defect in the basolateral membrane of epithelial cells where it normally mediates the influx of neutral amino acids together with sodium ions in exchange for cationic amino acids from cells. Two recent reviews are recommended for further reading that provide a comprehensive description of the structural and functional aspects of this class of amino acid exchangers (Verrey et al., 2000; Chillaron et al., 2001).

The system N amino acid transporters have been cloned recently (Chaudry et al., 1999; Nakanishi et al., 2001). The most important substrate for this class of amino acid transporters is glutamine. Its central role in interorgan metabolism as a carrier for carbon and nitrogen and its coupling to acid–base metabolism brings glutamine transporters and their regulation into the focus of a variety of metabolic processes. Moreover, as glutamine is the precursor of neuronal glutamate and is taken up efficiently by neurons, system N transporters are also interesting in view of transmitter metabolism. A novelty is the ionic coupling of glutamine influx to cation fluxes. Glutamine uptake mediated by the transporters SN1 and SN2 is sodium dependent but electroneutral. This is caused by the co-transport of a sodium (or lithium) ion together with glutamine and the concomitant efflux of a proton from the cell, that, in turn, causes an intracellular alkalinization. However, glutamine fluxes can also be electrogenic, as SN1 shows an intrinsic proton conductance, allowing proton efflux either coupled to or uncoupled from glutamine uptake (Chaudry et al., 2001). This close functional coupling of glutamine transport to cellular acid–base balance is a key to understanding the metabolic adaptations in glutamine metabolism in states of metabolic acidosis and alkalosis. SN1 and SN2 can transport almost all zwitterionic amino acids (glycine, serine, alanine, asparagine and histidine but not methylaminobutyric acid) in addition to glutamine. The transporter SN1 shows prominent expression in hepatocytes of both the periportal and the perivenous regions, and in brain restricted to astrocytes. SN2 mRNA is present in liver and brain but also in kidney, spleen and testis, and appears to have splice variants (Nakanishi et al., 2001).

Transport of cationic amino acids

Although transport of the cationic amino acids arginine, lysine and ornithine has already been described as an exchange process mediated by the heterodimeric exchangers, the most important transport proteins for these amino acids are the CAT family members initially identified as ecotropic retrovirus receptors (Kim et al., 1991; Wang et al., 1991). In the CAT family, at least three genes encode four different isoforms (CAT-1, CAT-2(A), CAT-2(B) and CAT-3), but there might also be a CAT-4 transporter. They all represent proteins of approximately 70 kDa with 12–14 TM segments. Despite their structural similarity, they differ in tissue distribution, kinetics and regulatory properties. System y+ (CAT transporters) and thus cellular uptake of the cationic amino acids is driven by the inside negative membrane potential in a uniport process. CAT proteins deliver cationic amino acids for growth and development but also for specialized cellular functions such as creatine, carnitine and polyamine synthesis or the delivery of arginine for the different nitric oxide synthases (Closs, 2002).

Transport of anionic amino acids

The anionic amino acids glutamate and aspartate are substrates for the XaG and Xc transport systems (Gadea and Lopez-Colome, 2001a). Four isoforms of the XAG system in the brain belong to the family of Na$^+$-dependent amino acid transporters that can – depending on the isoform – operate in an electrogenic mode with movement of more than one sodium during translocation, or in an electroneutral mode. Some show a potassium efflux during sodium-dependent glutamate influx. These excitatory amino acid transporters (EAATs) have their most important role in the inactivation of glutamatergic neurotransmission in brain by removal of glutamate from the synaptic cleft, but they also seem to be important for the synthesis of glutathione. All EAATs show a very high specificity for the anionic amino acids, with a very high affinity in the micromolar range that is required for efficient and complete removal of glutamate from the synaptic cleft. Although predominantly found in brain, EAAT3 and EAAT4 are also found in a variety of other tissues, and

EAAT5 appears to be expressed ubiquitously. Transporters of the ASC family, namely ASCT1 and ASCT2, also possess the capability for transport of anionic amino acids, although they predominantly take zwitterionic amino acids (alanine, cysteine, serine, glutamine, leucine, isoleucine and valine). These transporters operate in a sodium-dependent manner and are either electrogenic or electroneutral (Utsunomiya-Tate et al., 1996).

Other amino acid transporters include the GLYT carriers that show a preference for glycine with various isoforms and splice variants. They mediate cellular uptake of glycine in co-transport with two Na^+ and one Cl^- ion, and possess 12 putative TM domains (Gadea and Lopez-Colome, 2001b). Very recently, the first proton-dependent electrogenic amino acid transporters have been discovered in mammals (Sagne et al., 2001). They may represent the first lysosomal efflux carriers for amino acids with short side chains (glycine, alanine, serine and proline) by coupling of amino acid flux to the movement of protons. Such proton-dependent transport pathways for small neutral amino acids have also been identified in the apical membrane of intestinal and renal epithelial cells. By their mode of action, one might speculate that these transporters resemble the mammalian roots of archaic proton-coupled amino acid transport pathways as found in prokaryotes, yeast and plants.

Transport of thyroid hormones

Although the two thyroid hormones T4 and T3 are not amino acids but tyrosine derivatives, their transport across the cell membrane involves amino acid transporters and plays a very important role in thyroid hormone metabolism. Based on their high lipophilicity, it was believed originally that thyroid hormones enter cells by passive diffusion. Based on the kinetics of transport and particularly its stereoselectivity, it became clear that specific binding and carrier-mediated processes must be involved. A low-capacity transport mechanism with affinities for T4 and T3 in the nanomolar range and another system characterized by high binding and transport capacity but lower affinities in the micromolar range have been identified. Uptake of T4 and T3 by the high-affinity sites was energy, temperature and often Na^+ dependent, and most probably represents the translocation of thyroid hormone across the plasma membrane by both Na^+-dependent and Na^+-independent mechanisms (Hennemann et al., 2001).

In a variety of tissues and cells (erythrocytes, pituitary cells, astrocytes, the blood–brain barrier and the choroid plexus), uptake of T4 and T3 appears to be mediated by system L or T (aromatic amino acids) transporters. Efflux of T3 – but not of T4 – from some cell types is also saturable (Hennemann et al., 2001). As the liver plays an important role in clearance of thyroid hormones and their conversion, hepatic transport processes are essential for metabolism of thyroid hormones. Hormone uptake in human liver cells is ATP-dependent and appears to be rate limiting for subsequent iodothyronine metabolism. In conditions of starvation, T4 uptake in the liver is decreased probably by inhibition by non-esterified fatty acids, and bilirubin or ATP depletion, resulting in lowered plasma T3 levels as a consequence of reduced conversion. Recently, several organic anion transporters have also been identified as carriers of thyroid hormones (Friesema et al., 1999).

Transport of short chain peptides

Uptake of amino acids in peptide-bound form is a biological phenomenon found throughout nature. Bacteria, yeast and fungi as well as specialized cells of plants, invertebrates and vertebrates express membrane proteins for uptake of dipeptides and tripeptides. Some species (but not mammals) additionally have transporters that also accept larger oligopeptides (≥4 amino acid residues). Based on their molecular and functional characteristics, the membrane peptide transporters have been grouped into the PTR family of proton-dependent oligopeptide transporters (Steiner et al., 1995). In mammals, two genes have been identified that express transport activity for di- and tripeptides: PEPT1 and PEPT2, corresponding to SLC15A1 and SLC15A2 (Fei et al., 1994; Ramamoorthy et al., 1995; Boll et al., 1996). A third protein may be a histidine transporter that also can transport dipeptides (PHT-1) (Yamashita et al., 1997). The human PEPT1 and PEPT2 represent proteins with 708 (PEPT1) and 729 (PEPT2) amino acids residues, 12 TM domains and with N- and

C-terminal ends facing the cytoplasm. The mammalian peptide carriers couple peptide movement across the membrane to movement of protons (H_3O^+) along an inwardly directed electrochemical proton gradient that allows transport of peptides to occur against a substrate gradient. By coupling to proton flux, transport – regardless of the net charge of the substrate – occurs electrogenically and causes intracellular acidification (Amasheh et al., 1997). The required proton gradient for peptide influx is mainly, but not exclusively, provided by electroneutral proton/cation exchangers such as the Na^+/H^+ antiporters that export protons again in exchange for Na^+ ions entering the cells. However, the main driving force for peptide transport is the inside negative membrane potential. Normal dipeptides taken up into the cells are hydrolysed rapidly by a multitude of peptidases present in the cytoplasm that possess a high affinity for di- and tripeptides and a high capacity for hydrolysis. Free amino acids are then delivered into the circulation or are used within the cell for protein synthesis or other purposes.

It is of course tempting to speculate with regards to the advantages for evolutionary conservation of the peptide transporters considering that a large number of amino acid transporters have emerged during evolution. Obviously, transport of a package of amino acids in one transport step requires less cellular energy compared with that required for the individual transport of single amino acids. Moreover, the ability of peptide carriers to transport all possible di- and tripeptides indicates that there is no discrimination either with respect to essential and non-essential amino acids or with respect to the physicochemical characteristics of the substrates, possessing molecular masses of 96.2 Da (di-glycine) to 522.6 Da (tri-tryptophan). Whereas nutritional needs with provision of amino acids for growth, development and metabolism may explain the primary role of peptide transporters, there is evidence that they also may serve special functions in more complex organisms where the transporters are expressed only in specialized cells.

PEPT1 shows highest expression in the apical membrane of intestinal epithelial cells and appears to be responsible for uptake of bulk quantities of di- and tripeptides as end-products of the luminal and membrane-associated digestion of dietary proteins. PEPT1 is also found in renal tubular cells. The predominant peptide transporter expressed in kidney, however, is PEPT2 (Daniel and Herget, 1997). When compared with the intestinal PEPT1 isoform, kidney PEPT2 shows about 50% identity and 70% homology in amino acid sequence, with the highest levels of identity in the TM regions. When in situ hybridization experiments in rabbit kidney were performed with PEPT2-specific antisense probes, high expression of the PEPT2 mRNA was found in proximal tubular cells, with expression levels higher in S1 and S2 than in S3 segments (Smith et al., 1998). Reverse transcription–polymerase chain reaction (RT–PCR), Northern blot analysis and immunodetection identified a variety of extrarenal tissues that express PEPT2, with particularly high expression in brain, choroid plexus, lung epithelium and epithelial cells of the mammary gland (Doring et al., 1998; Berger and Hediger, 1999). The biological role of peptide transport in these tissues has not yet been determined.

The main role of PEPT2 in the apical membrane of renal tubular cells is the reabsorption and conservation of amino acid nitrogen. The amount of peptide-bound amino acids in circulation increases after ingestion of a protein meal, and di- and tripeptides are produced constantly by degradation of endogenous proteins and oligopeptides in circulation and/or are released from various tissues. Although individual di- or tripeptides can be identified in plasma, the total concentration of all circulating short chain peptides in peripheral blood is still not known. Studies in a variety of animals suggest that about 50% of circulating plasma amino acids are peptide bound and that the majority represent di- and tripeptides (Seal and Parker, 1991). In addition to short chain peptides entering the tubular system by filtration, the presence of a variety of renal brush border membrane-bound peptide hydrolases with high catalytic activity could provide considerable quantities of dipeptides and tripeptides by hydrolysis of larger oligopeptides filtered in the glomerulum. Since PEPT2 is a high-affinity transport system (K_m values depending on the peptide >5 to <250 µM), it can remove di- and tripeptides efficiently from the tubular fluids.

Transport by PEPT1 and PEPT2 is stereoselective, with peptides containing L-enantiomers of amino acids possessing a higher affinity for transport than those containing D-enantiomers. Peptides consisting solely of D-amino acids do not show any relevant affinity for transport. The ability of mammalian peptide transporters to transport a variety of peptidomimetics, such as antibiotics of the aminocephalosporin and aminopenicillin classes, or selected peptidase inhibitors, such as bestatin or captopril, makes peptide transporters interesting for drug delivery and pharmacokinetic analyses. The excellent oral bioavailability of these drugs is mediated by PEPT1 in the intestinal epithelium (Adibi, 1997). The renal transporter PEPT2 increases the plasma half-life of these drugs by reabsorption of the compounds after they have been filtered in the glomerulum.

The clinical importance of PEPT1 has been demonstrated in a variety of studies including studies in humans. Here, dipeptides have been shown to be superior to free amino acids for fast intestinal absorption and are also more useful for enteral nutrition as they provide a lower osmolarity of the nutrition solution. Moreover, in a variety of gastrointestinal diseases, the peptide transporter has been found to be less affected by the pathophysiology than the amino acid transporters, and enables delivery of amino acid nitrogen even in states of impaired mucosal function (Adibi, 1997). Figure 3.5 summarizes the different transport pathways for peptides and free amino acids in epithelial cells that mediate, in concert, the transcellular absorption of amino acid nitrogen.

Transport Pathways for Water

Some cells in the body need a very high membrane permeability for water. For example in the kidney, the apical membrane of epithelial cells requires a water permeability that can be adjusted to maintain water homeostasis and that is under hormonal – mainly vasopressin – control. Malfunction of the system demonstrates the quantitative importance of the membrane transport pathways for water. For example, diabetes insipidus, a heriditary disorder, causes decreased capability of renal reabsorption of water that leads to the loss of huge quantities of primary urine (Kwon et al., 2001). Depending on the severity of the disease,

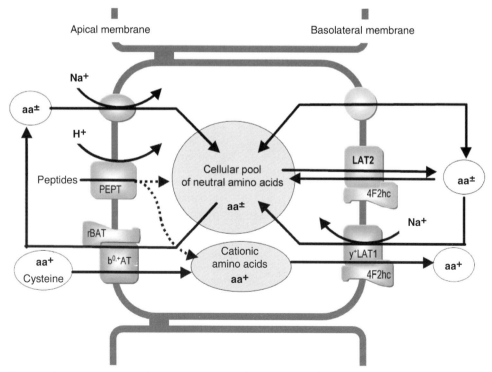

Fig. 3.5. The current view of the concerted action of amino acid and peptide transporters that mediate the transcellular transport of amino acid nitrogen in epithelial cells of the intestine and kidney.

the renal losses of water may account for up to 50 l per day. The underlying defect is in the signalling pathways that control the incorporation and proper function of specialized membrane water channels, the aquaporins (Sansom and Law, 2001). When aquaporins are expressed and inserted into the brush border membrane of tubular cells, a water conductance of the membrane is achieved that allows a large water flow from the lumen to the blood side, that in turn concentrates the urine and conserves water.

Aquaporins (AQPs) are integral membrane proteins that are expressed in a variety of tissues and specialized cells in humans and are also found in bacteria, plants and animals. APQ proteins are subdivided into the AQP proper family, with the human genome containing at least ten related family members, and the aquaglyceroporins, which have a very similar structure but show a permeability for glycerol (Borgnia et al., 1999). These proteins have a channel-like core structure forming a pore. High-resolution X-ray structures of bacterial AQPs and medium-resolution electron microscopy of human AQPs provide detailed insights into the structure and functions of these membrane proteins. They consist of six membrane-spanning helices and two half-TM helices that are only partly inserted into the membrane bilayer. The central pore shows a length of 20 Å and an internal width of 2 Å that account for the water and glycerol permeability (Nollert et al., 2001). In the case of aquaporin proper (AQP1), a single protein molecule allows movement of 10^9 water molecules per second.

Vitamins

Mammalian transport proteins that mediate vitamin uptake or efflux across the plasma cell membranes have – with the exception of vitamin B_{12} – only been identified on a molecular basis very recently. Although the various processes had been analysed in various cell systems by flux measurements and with regard to regulatory processes, the molecular identity of the responsible proteins long remained unknown. As it is not possible to cover all aspects of vitamin transport here, some of the most important recent findings and novel processes will be briefly addressed. Figure 3.6 summarizes the systems involved in transport of vitamins that will be addressed here and the proposed mode of how they mediate TM translocation.

Based on the physicochemical characteristics of the different water-soluble vitamins, different classes of transporters are required to allow neutral,

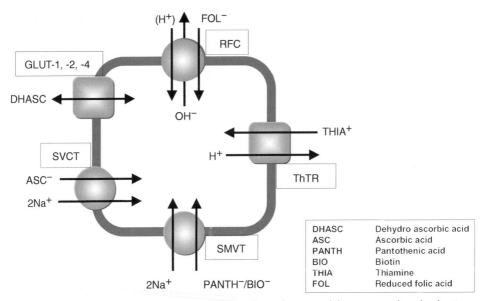

Fig. 3.6. Mammalian vitamin transporters in plasma cell membranes and their proposed mode of action.

anionic or cationic (thiamine) compounds to be transported into a cellular compartment that provides an inside negative membrane potential. In the case of the anionic vitamins such as ascorbic acid, folic acid, pantothenic acid and biotin, cellular uptake requires the co-transport of sodium ions or protons to cause electroneutral transport or a different coupling ratio (excess sodium or protons) to induce an electrogenic process.

The ascorbic acid-transporting proteins SVCT1 and SVCT2 (SLC23A2 and SLC23A1) were identified in 1999 by expression cloning (Tsukaguchi et al., 1999). At that time, it was known that cellular ascorbic acid uptake was saturable and sodium dependent whereas the transport of dehydroascorbic acid had been shown to be mediated by sodium-independent processes involving the GLUT transporters (GLUT-1–GLUT-4), followed by rapid intracellular reduction to ascorbate. The sodium-dependent electrogenic human ascorbate transporter SVCT1 consists of 598 amino acids and has an affinity for ascorbate of approximately 250 µM, whereas SVCT2 has 650 amino acids and represents the high-affinity-type (K_m ~20 µM) transporter. Both are sodium-dependent co-transporters that mediate accumulation of cellular vitamin C but show quite distinct expression patterns (Liang et al., 2001).

A unique transporter that mediates the uptake of pantothenic acid, biotin and lipoate was cloned and designated SMVT for sodium-dependent multivitamin transporter (SLC5A6) (Prasad et al., 1998). The human SMVT represents a 635 amino acid protein with 12 putative TM domains encoded on chromosome 2p23. The SMVT protein transports the substrates with a coupling ratio of two Na^+ ions to one substrate molecule, and is therefore an electrogenic carrier. The protein is expressed in apical membranes of polarized epithelial cells where it mediates uptake of dietary vitamins in intestine and contributes to reabsorption in renal tubular epithelium. Northern blot analysis shows that SMVT transcripts are also present in a variety of other cells and tissues. Protein database comparisons show significant sequence similarity between SMVT and known members of the Na^+-dependent glucose transporter family (Prasad and Ganapathy, 2000).

A number of transporters for reduced folates have also been cloned and characterized recently (Matherly, 2001). Whereas the folic acid-binding proteins α and β were known for some time, the first identified mammalian membrane folate carrier was the RFC-1 (SLC19A1). It mediates a sodium-independent but strongly pH-dependent folic acid uptake process, as already identified functionally in intact tissues prior to cloning. Cellular uptake of reduced folic acid, methyl-tetrahydrofolate and methotrexate increases with decreasing extracellular pH, suggesting a folate–proton symport or an anion-exchange process with folate anions exchanged for intracellular hydroxyl ions. A comparison of the amino acid sequence of human RFC-1 with other protein sequences identifies human ThTr1, the thiamin transporter (SLC19A2), as the closest relative with a sequence identity of 40% and similarity of 55% at the amino acid level (Dutta et al., 1999). Very recently, a third member of this family of proteins was cloned from human and mouse (SLC19A3) and identified as a second thiamine transporter (Rajgopal et al., 2001). The unique feature of thiamine as a water-soluble vitamin is its cationic character. Depending on the pH, it may carry one or two positive charges, and this would suggest that thiamine serves as a substrate of the organic cation transporters that have been identified in large numbers. However, it appears that most of the cation transporters of the OCT family do not transport thiamine. Instead, ThTr1, cloned from human placenta, induces specific thiamine influx in transfected cells that shows the characteristics of a thiamine–proton exchange process. Influx of thiamine into cells accordingly would be coupled to efflux of protons and so thiamine transport shows a pronounced pH dependence with reduced rates of influx at acidic extracellular pH values. A base deletion in the high-affinity thiamine transporter ThTr1 (287delG) was identified recently as the cause of the thiamine-responsive megaloblastic anaemia (TRMA) syndrome that is associated with diabetes and deafness (Diaz et al., 1999). Two patients suffering from TRMA were homozygous for the mutation at nucleotide 287 of the cDNA, relative to the translation start site. The genetic alteration results in a frameshift and a premature stop codon. Fibroblasts of the patients showed almost completely abolished thiamine uptake (Neufeld et al., 2001).

Although not all transport proteins that mediate the influx and efflux of the different water-soluble vitamins have been identified at a molecular level, substantial progress has been made in recent years, and the genes and proteins known so far are already interesting targets for systematic analysis of their associations with inborn errors of vitamin metabolism and screening for genetic heterogeneity in view of functional differences.

References

Ader, P., Block, M., Pietzsch, S. and Wolffram, S. (2001) Interaction of quercetin glucosides with the intestinal sodium/glucose co-transporter (SGLT-1). *Cancer Letters* 162, 175–180.

Adibi, S.A. (1997) The oligopeptide transporter (Pept-1) in human intestine: biology and function. *Gastroenterology* 113, 332–340.

Amasheh, S., Wenzel, U., Boll, M., Dorn, D., Weber, W., Clauss, W. and Daniel, H. (1997) Transport of charged dipeptides by the intestinal H^+/peptide symporter PepT1 expressed in *Xenopus laevis* oocytes. *Journal of Membrane Biology* 155, 247–256.

Berger, U.V. and Hediger, M.A. (1999) Distribution of peptide transporter PEPT2 mRNA in the rat nervous system. *Anatomy and Embryology (Berlin)* 199, 439–449.

Boll, M., Markovich, D., Weber, W.M., Korte, H., Daniel, H. and Murer, H. (1994) Expression cloning of a cDNA from rabbit small intestine related to proton-coupled transport of peptides, beta-lactam antibiotics and ACE-inhibitors. *Pflügers Archiv* 429, 146–149.

Boll, M., Herget, M., Wagener, M., Weber, W.M., Markovich, D., Biber, J., Clauss, W., Murer, H. and Daniel, H. (1996) Expression cloning and functional characterization of the kidney cortex high-affinity proton-coupled peptide transporter. *Proceedings of the National Academy of Sciences USA* 93, 284–289.

Bonen, A. (2000) Lactate transporters (MCT proteins) in heart and skeletal muscles. *Medicine in Science, Sports and Exercise* 32, 778–789.

Bonen, A., Baker, S.K. and Hatta, H. (1997) Lactate transport and lactate transporters in skeletal muscle. *Canadian Journal of Applied Physiology* 22, 531–532.

Bonen, A., Dyck, D.J. and Luiken, J.J. (1998) Skeletal muscle fatty acid transport and transporters. *Advances in Experimental Medicine and Biology* 441, 193–205.

Borgnia, M., Nielsen, S., Engel, A. and Agre, P. (1999) Cellular and molecular biology of the aquaporin water channels. *Annual Review of Biochemistry* 68, 425–458.

Brooks, G.A, Brownk, M.A., Butz, C.E., Sicurello, J.P. and Dubouchaud, H. (1999) Cardiac and skeletal muscle mitochondria have a monocarboxylate transporter MCT1. *Journal of Applied Physiology* 87, 1713–1718.

Chaudhry, F.A., Reimer, R.J., Krizaj, D., Barber, D., Storm-Mathisen, J., Copenhagen, D.R. and Edwards, R.H. (1999) Molecular analysis of system N suggests novel physiological roles in nitrogen metabolism and synaptic transmission. *Cell* 99, 769–780.

Chaudhry, F.A, Krizaj, D., Larsson, P., Reimer, R.J., Wreden, C., Storm-Mathisen, J., Copenhagen, D., Kavanaugh, M. and Edwards, R.H. (2001) Coupled and uncoupled proton movement by amino acid transport system N. *EMBO Journal* 20, 7041–7051.

Chillaron, J., Roca, R., Valencia, A., Zorzano, A. and Palacin, M. (2001) Heteromeric amino acid transporters: biochemistry, genetics, and physiology. *American Journal of Physiology* 281, F995–F1018.

Christensen, H.N. (1990) Role of amino acid transport and countertransport in nutrition and metabolism. *Physiological Reviews* 70, 43–77.

Closs, E.I. (2002) Expression, regulation and function of carrier proteins for cationic amino acids. *Current Opinion in Nephrology and Hypertension* 11, 99–107.

Coburn, C.T., Knapp, F.F. Jr, Febbraio, M., Beets, A.L., Silverstein, R.L. and Abumrad, N.A. (2000) Defective uptake and utilization of long chain fatty acids in muscle and adipose tissues of CD36 knockout mice. *Journal of Biological Chemistry* 275, 32523–32529.

Daniel, H. (2000) Nutrient transporter function studied in heterologous expression systems. *Annals of the New York Academy of Sciences* 915, 184–192.

Daniel, H. and Herget, M. (1997) Cellular and molecular mechanisms of renal peptide transport. *American Journal of Physiology* 273, F1–F88.

Diaz, G.A., Banikazemi, M., Oishi, K., Desnick, R.J. and Gelb, B.D. (1999) Mutations in a new gene encoding a thiamine transporter cause thiamine-responsive megaloblastic anaemia syndrome. *Nature Genetics* 22, 309–312.

Doring, F., Walter, J., Will, J., Focking, M., Boll, M., Amasheh, S., Clauss, W. and Daniel, H. (1998) Delta-aminolevulinic acid transport by intestinal and renal peptide transporters and its physiological and clinical implications. *Journal of Clinical Investigation* 101, 2761–2767.

Dutta, B., Huang, W., Molero, M., Kekuda, R., Leibach, F.H., Devoe, L.D., Ganapathy, V. and Prasad, P.D. (1999) Cloning of the human thiamine transporter, a member of the folate transporter family. *Journal of Biological Chemistry* 274, 31925–31929.

Eskandari, S., Wright, E.M., Kreman, M., Starace, D.M. and Zampighi, G.A. (1998) Structural analysis of cloned plasma membrane proteins by freeze-fracture electron microscopy. *Proceedings of the National Academy of Sciences USA* 95, 11235–11240.

Fei, Y.J., Kanai, Y., Nussberger, S., Ganapathy, V., Leibach, F.H., Romero, M.F., Singh, S.K., Boron, W.F. and Hediger, M.A. (1994) Expression cloning of a mammalian proton-coupled oligopeptide transporter. *Nature* 368, 563–566.

Friesema, E.C., Docter, R., Moerings, E.P., Stieger, B., Hagenbuch, B., Meier, P.J., Krenning, E.P., Hennemann, G. and Visser, T.J. (1999) Identification of thyroid hormone transporters. *Biochemical and Biophysical Research Communications* 254, 497–501.

Gadea, A. and Lopez-Colome, A.M. (2001a) Glial transporters for glutamate, glycine and GABA I. Glutamate transporters. *Journal of Neuroscience Research* 63, 453–460.

Gadea, A. and Lopez-Colome, A.M. (2001b) Glial transporters for glutamate, glycine, and GABA III. Glycine transporters. *Journal of Neuroscience Research* 64, 218–222.

Guastella, J., Nelson, N., Nelson, N., Czyzyk, L., Keynan, S., Miedel, M.C., Davidson, N., Lester, H.A. and Kanner, B.I. (1990) Cloning and expression of a rat brain GABA transporter. *Science* 249, 1303–1306.

Halestrap, A.P. and Price, N.T. (1999) The proton-linked monocarboxylate transporter (MCT) family: structure, function and regulation. *Biochemical Journal* 343, 281–299.

Hediger, M.A., Coady, M.J., Ikeda, T.S. and Wright, E.M. (1987) Expression cloning and cDNA sequencing of the Na^+/glucose co-transporter. *Nature* 330, 379–381.

Hennemann, G., Docter, R., Friesema, E.C., de Jong, M., Krenning, E.P. and Visser, T.J. (2001) Plasma membrane transport of thyroid hormones and its role in thyroid hormone metabolism and bioavailability. *Endocrine Reviews* 22, 451–476.

Herrmann, T., Buchkremer, F., Gosch, I., Hall, A.M., Bernlohr, D.A. and Stremmel, W. (2001) Mouse fatty acid transport protein 4 (FATP4): characterization of the gene and functional assessment as a very long chain acyl-CoA synthetase. *Gene* 270, 31–40.

Hirsch, D., Stahl, A. and Lodish, H.F. (1998) A family of fatty acid transporters conserved from mycobacterium to man. *Proceedings of the National Academy of Sciences USA* 95, 8625–8629.

Hruz, P.W. and Mueckler, M.M. (2001) Structural analysis of the GLUT1 facilitative glucose transporter. *Molecular Membrane Biology* 18, 183–193.

Hyde, R.J., Cass, C.E., Young, J.D. and Baldwin, S.A. (2001) The ENT family of eukaryote nucleoside and nucleobase transporters: recent advances in the investigation of structure/function relationships and the identification of novel isoforms. *Molecular Membrane Biology* 18, 53–63.

Juel, C. and Halestrap, A.P. (1999) Lactate transport in skeletal muscle – role and regulation of the monocarboxylate transporter. *Journal of Physiology* 517, 633–642.

Kim, J.K., Zisman, A., Fillmore, J.J., Peroni, O.D., Kotani, K., Perret, P., Zong, H., Dong, J., Kahn, C.R., Kahn, B.B. and Shulman, G.I. (2001) Glucose toxicity and the development of diabetes in mice with muscle-specific inactivation of GLUT4. *Journal of Clinical Investigation* 108, 153–160.

Kim, J.W., Closs, E.I., Albittron, L.M. and Cunnigham, J.M. (1991) Transport of cationic amino acids by the mouse ecotropic retroviral receptor. *Nature* 352, 725–728.

Kwon, T.H., Hager, H., Nejsum, L.N., Andersen, M.L., Frokiaer, J. and Nielsen, S. (2001) Physiology and pathophysiology of renal aquaporins. *Seminars in Nephrology* 21, 231.

Liang, W.J., Johnson, D. and Jarvis, S.M. (2001) Vitamin C transport systems of mammalian cells. *Molecular Membrane Biology* 18, 87–95.

Loo, D.D., Zeuthen, T., Chandy, G. and Wright, E.M. (1996) Cotransport of water by the Na^+/glucose cotransporter. *Proceedings of the National Academy of Sciences USA* 93, 13367–13370.

Martin, M.G., Turk, E., Lostao, M.P., Kerner, C. and Wright, E.M. (1996) Defects in Na^+/glucose co-transporter (SGLT1) trafficking and function cause glucose–galactose malabsorption. *Nature Genetics* 12, 216–220.

McGivan, J.D. and Pastor-Anglada, M. (1994) Regulatory and molecular aspects of mammalian amino acid transport. *Biochemical Journal* 299, 321–334.

Mueckler, M., Caruso, C., Baldwin, S.A., Panico, M., Blench, I., Morris, H.R., Allard, W.J., Lienhard, G.E. and Lodish, H.F. (1985) Sequence and structure of a human glucose transporter. *Science* 229, 941–945.

Nakanishi, T., Kekuda, R., Fei, Y.J., Hatanaka, T., Sugawara, M., Martindale, R.G., Leibach, F.H., Prasad, P.D. and Ganapathy, V. (2001) Cloning and functional characterization of a new subtype of the amino acid transport system N. *American Journal of Physiology* 281, C1757–C1768.

Neufeld, E.J., Fleming, J.C., Tartaglini, E. and Steinkamp, M.P. (2001) Thiamine-responsive megaloblastic anemia syndrome: a disorder of high-affinity thiamine transport. *Blood Cells, Molecules and Diseases* 27, 135–138.

Nollert, P., Harries, W.E., Fu, D., Miercke, L.J. and Stroud, R.M. (2001) Atomic structure of a glycerol channel and implications for substrate permeation in aqua(glycero)porins. *FEBS Letters* 504, 112–117.

Palacin, M., Estevez, R., Bertran, J. and Zorzano, A. (1998) Molecular biology of mammalian plasma membrane amino acid transporters. *Physiological Reviews* 78, 969–1054.

Palacin, M., Fernandez, E., Chillaron, J. and Zorzano, A. (2001a) The amino acid transport system b(o,+) and cystinuria. *Molecular Membrane Biology* 18, 21–26.

Palacin, M., Borsani, G. and Sebastio, G. (2001b) The molecular bases of cystinuria and lysinuric protein intolerance. *Current Opinion in Genetics and Development* 11, 328–335.

Pastor-Anglada, M., Casado, F.J., Valdes, R., Mata, J., Garcia-Manteiga, J. and Molina, M. (2001) Complex regulation of nucleoside transporter expression in epithelial and immune system cells. *Molecular Membrane Biology* 18, 81–85.

Poole, R.C. and Halestrap, A.P. (1993) Transport of lactate and other monocarboxylates across mammalian plasma membranes. *American Journal of Physiology* 264, C761–C782.

Prasad, P.D. and Ganapathy, V. (2000) Structure and function of mammalian sodium-dependent multivitamin transporter. *Current Opinion in Clinical Nutrition and Metabolic Care* 3, 263–266.

Prasad, P.D., Wang, H., Kekuda, R., Fujita, T., Fei, Y.J., Devoe, L.D., Leibach, F.H. and Ganapathy, V. (1998) Cloning and functional expression of a cDNA encoding a mammalian sodium-dependent vitamin transporter mediating the uptake of pantothenate, biotin, and lipoate. *Journal of Biological Chemistry* 273, 7501–7506.

Rajgopal, A., Edmondnson, A., Goldman, I.D. and Zhao, R. (2001) SLC19A3 encodes a second thiamine transporter ThTr2. *Biochimica et Biophysica Acta* 1537, 175–178.

Ramamoorthy, S., Han, H., Yang-Feng, T.L., Hediger, M.A., Ganapathy, V. and Leibach, F.H. (1995) Human intestinal H^+/peptide cotransporter. Cloning, functional expression, and chromosomal localization. *Journal of Biological Chemistry* 270, 6456–6463.

Reimer, R.J., Chaudhry, F.A., Gray, A.T. and Edwards, R.H. (2000) Amino acid transport system A resembles system N in sequence but differs in mechanism. *Proceedings of the National Academy of Sciences USA* 97, 7715–7720.

Ritzel, M.W., Ng, A.M., Yao, S.Y., Graham, K., Loewen, S.K., Smith, K.M., Hyde, R.J., Karpinski, E., Cass, C.E, Baldwin, S.A. and Young, J.D. (2001) Recent molecular advances in studies of the concentrative Na^+-dependent nucleoside transporter (CNT) family: identification and characterization of novel human and mouse proteins (hCNT3 and mCNT3) broadly selective for purine and pyrimidine nucleosides (system cib). *Molecular Membrane Biology* 18, 65–72.

Sagne, C., Agulhon, C., Ravassard, P., Darmon, M., Hamon, M., El Mestikawy, S., Gasnier, B. and Giros, B. (2001) Identification and characterization of a lysosomal transporter for small neutral amino acids. *Proceedings of the National Academy of Sciences USA* 98, 7206–7211.

Sankarasubbaiyan, S., Cooper, C. and Heilig, C.W. (2001) Identification of a novel form of renal glucosuria with overexcretion of arginine, carnosine, and taurine. *Americal Journal of Kidney Diseases* 37, 1039–1043.

Sansom, M.S. and Law, R.J. (2001) Membrane proteins: aquaporins – channels without ions. *Current Biology* 11, R71–R73.

Seal, C.J. and Parker, D.S. (1991) Isolation and characterization of circulating low molecular weight peptides in steer, sheep and rat portal and peripheral blood. *Comparative Biochemistry and Physiology B* 99, 679–685.

Seatter, M.J. and Gould, G.W. (1999) The mammalian facilitative glucose transporter (GLUT) family. *Pharmaceutical Biotechnology* 12, 201–228.

Smith, D.E., Pavlova, A., Berger, U.V., Hediger, M.A., Yang, T., Huang, Y.G. and Schnermann, J.B. (1998) Tubular localization and tissue distribution of peptide transporters in rat kidney. *Pharmaceutical Research* 15, 1244–1249.

Stein, J., Zores, M. and Schroder, O. (2000) Short-chain fatty acid (SCFA) uptake into Caco-2 cells by a pH-dependent and carrier mediated transport mechanism. *European Journal of Nutrition* 39, 121–125.

Steiner, H.Y., Naider, F. and Becker, J.M. (1995) The PTR family: a new group of peptide transporters. *Molecular Microbiology* 16, 825–834.

Tsukaguchi, H., Tokui, T., Mackenzie, B., Berger, U.V., Chen, X.Z., Wang, Y., Brubaker, R.F. and Hediger, M.A. (1999) A family of mammalian Na^+-dependent L-ascorbic acid transporters. *Nature* 399, 70–75.

Turk, E. and Wright, E.M. (1997) Membrane topology motifs in the SGLT cotransporter family. *Journal of Membrane Biology* 159, 1–20.

Turk, E., Kerner, C.J., Lostao, M.P. and Wright, E.M. (1996) Membrane topology of the human Na^+/glucose cotransporter SGLT1. *Journal of Biological Chemistry* 271, 1925–1934.

Uldry, M., Ibberson, M., Horisberger, J.D., Chatton, J.Y., Riederer, B.M. and Thorens, B. (2001) Identification of a mammalian H(+)-myo-inositol symporter expressed predominantly in the brain. *EMBO Journal* 20, 4467–4477.

Utsunomiya-Tate, N., Endou, H. and Kanai, Y. (1996) Cloning and functional characterization of a system ASC-like Na^+-dependent neutral amino acid transporter. *Journal of Biological Chemistry* 271, 14883–14890.

Verrey, F., Meier, C., Rossier, G. and Kuhn, L.C. (2000) Glycoprotein-associated amino acid exchangers: broadening the range of transport specificity. *Pflügers Archiv* 440, 503–512.

Wagner, C.A., Lang, F. and Broer, S. (2001) Function and structure of heterodimeric amino acid transporters. *American Journal of Physiology* 281, C1077–C1093.

Walgren, R.A., Lin, J.T., Kinne, R.K. and Walle, T. (2000) Cellular uptake of dietary flavonoid quercetin 4′-beta-glucoside by sodium-dependent glucose transporter SGLT1. *Journal of Pharmacology and Experimental Therapeutics* 294, 837–843.

Wang, H., Kavanaugh, M.P., North, R.A. and Kabat D. (1991) Cell-surface receptor for ecotropic murine retroviruses is a basic amino-acid transporter. *Nature* 352, 729–731.

Wang, J., Schaner, M.E., Thomassen, S., Su, S.F., Piquette-Miller, M. and Giacomini, K.M. (1997) Functional and molecular characteristics of Na(+)-dependent nucleoside transporters. *Pharmaceutical Research* 14, 1524–1532.

Watson, R.T. and Pessin, J.E. (2001) Subcellular compartmentalization and trafficking of the insulin-responsive glucose transporter, GLUT4. *Experimental Cell Research* 271, 75–83.

Wright, E.M. (2001) Renal Na(+)–glucose co-transporters. *American Journal of Physiology* 280, F10–F18.

Yamamoto, N., Ikeda, H., Tandon, N.N., Herman, J., Tomiyama, Y., Mitani, T., Sekiguchi, S., Lipsky, R., Kralisz, U. and Jamieson, G.A. (1990) A platelet membrane glycoprotein (GP) deficiency in healthy blood donors: Naka-platelets lack detectable GPIV (CD36). *Blood* 76, 1698–1703.

Yamashita, T., Shimada, S., Guo, W., Sato, K., Kohmura, E., Hayakawa, T., Takagi, T. and Tohyama, M. (1997) Cloning and functional expression of a brain peptide/histidine transporter. *Journal of Biological Chemistry* 272, 10205–10211.

Zuniga, F.A., Shi, G., Haller, J.F., Rubashkin, A., Flynn, D.R., Iserovich, P. and Fischbarg, J. (2001) A three-dimensional model of the human facilitative glucose transporter Glut1. *Journal of Biological Chemistry* 276, 44970–44975.

4 Intracellular Trafficking and Compartmentalization of Vitamins and Their Physiologically Active Forms

Donald B. McCormick
Department of Biochemistry, School of Medicine, Emory University,
Atlanta, Georgia, USA

Overview and Focus

Coverage of what is known about the intracellular trafficking and compartmentalization of the numerous and diverse nutrients that must be supplied to the different cells that comprise many higher organisms, especially such complex mammals as humans, cannot be detailed in a single chapter. The molecular processes and pathways by which macro- and micronutrients are transported and utilized within a cell continue to be summarized in textbooks on biochemistry and nutrition. Suitable examples that treat material that pertains to the human can be found in current editions of a *Textbook of Biochemistry with Clinical Correlations* edited by Devlin (Wiley-Liss, New York) and *Modern Nutrition in Health and Disease* edited by Shils *et al.* (Williams and Wilkins, Baltimore). Chapters updating these subjects are in the *Annual Review of Biochemistry*, *Annual Review of Nutrition* and such compendia as *Present Knowledge in Nutrition* and review chapters in research journals.

It will be the purpose of this chapter to restrict the scope such that at least the main points of what has been learnt about the subject as concerns vitamins can be brought together. This apparently has not been done previously for the particular focus of intracellular trafficking and compartmentalization. Some disconnected statements on this subject can be found in texts and treatises on vitamins, e.g. the third edition of the *Handbook of Vitamins* (Marcel Dekker, New York), which describe many aspects of these essential micronutrients.

Because classically considered vitamins function usually after metabolism to physiologically active forms, i.e. coenzymes or hormones, it is appropriate that aspects of these conversions and subsequent compartmentalizations be included as well. For cohesiveness, it seems reasonable to cover each under the subdivision of the particular vitamin and its functional forms. Events covered are those that occur upon uptake through the plasma membrane of a eukaryotic cell, the subsequent partitioning into cytosol and organelles (mitochondria, endoplasmic reticulum, lysosomes, nucleus, etc.), and those metabolic alterations that are involved.

Vitamin A

Upon intestinal mucosal cell uptake of dietary precursors of vitamin A, namely pro-A carotenoids with a β-ionone ring, they are converted both by the dioxygenase-catalysed endocentric cleavage of the 15,15′ double bond to yield retinal and by eccentric cleavage to β-apocarotenals, which are oxidized further to retinal (Olson, 2001). NAD-dependent dehydrogenases reduce the retinal to its alcohol, which, together with retinol derived from the diet, is largely converted to retinyl esters.

©CAB *International* 2003. *Molecular Nutrition*
(eds J. Zempleni and H. Daniel)

The latter are incorporated into chylomicrons, which are released into the lymph. The uptake of retinyl esters by the liver leads to ester-catalysed hydrolysis followed by complexing of the released retinol with cytoplasmic retinol-binding proteins. Some of the all-*trans*-retinol in liver parenchymal cells binds with a specific protein (RBP) that, together with transthyretin, is secreted into the plasma for distribution to cells in other tissues. Within both the cytosol and the nucleus of the cells, the retinol and a fraction oxidized in the cytosol to retinal and retinoic acid are bound to diverse cytosolic retinoid-binding proteins that are listed in Table 4.1.

The properties and functions of these proteins are considered in reviews (Ong *et al.*, 1994; Saari, 1994; Newcomer, 1995; Li and Norris, 1996). Cellular retinol-binding protein I (CRBP I) and II and cellular retinoic acid-binding protein I (CRABP I) and II are fairly small proteins (15.0–15.7 kDa) that vector the all-*trans* isomers. In the intestinal cell, CRBP I and II selectively direct retinol for esterification by lecithin:retinol acyltransferase (LRAT) over acyl-CoA:retinol acyltransferase (ARAT). CRABP I is found in many tissues, but CRABP II expression is localized to skin of the adult animal. Both CRBP I and II are influenced by retinoid nutritional status (Kato, 1985). In the small intestine of the retinoid-deficient rat, the mRNA for CRBP I is reduced, whereas that for CRBP II is increased. CRALBP (36 kDa) and IRBP (135 kDa) function as transporters of the 11-*cis* isomers within the retinal cells. The ERABP (18.5 kDa) vectors all-*trans* and 9-*cis*-retinoic acid within testicular cells. The CRBPs influence retinoid signalling pathways by modulating intracellular retinoid metabolism and by influencing ligand occupancy of the nuclear receptors.

Upon entry of retinoic acids (all-*trans* and 9-*cis*) and other retinoids into the nucleus, they are tightly bound to one or more of the three (α, β, γ) retinoic acid receptors (RARs) or retinoid X receptors (RXRs). These receptors, like others for hormones binding in the nucleus, possess six protein domains (designated A–F in progressing from the N- to the C-terminus) with specific functions. The E region binds the ligand. For both RAR and RXR, the hormone response element (HRE) in DNA is the consensus sequence AGGTCA. Many genes contain response elements for the retinoid receptors. Notable effects include: stimulation of both certain cytosolic and nuclear binding proteins for retinoids by retinoic acid via retinoic acid response elements; stimulation of *Hox a-1* (*Hox 1.6*) and *Hox b-1* (*Hox 2.9*) initiating genes of embryonic development; and enhancement of class I alcohol dehydrogenase type 3 inducing the conversion of retinol to more retinoic acid.

Vitamin D (Calciferols)

Processing of vitamin D, which is a prohormone, requires specific and successive events within different tissues (Collins and Norman, 2001). For D_3 (cholecalciferol), 7-dehydrocholesterol located primarily within the Malpighian layer of dermal cells is converted in a natural process that involves light-induced fission of the B ring at the 9,10 bond to form provitamin D_3. In the formation of D_2 (ergocalciferol), plant-derived ergosterol, which has a 22,23-Δ-24-methyl side chain, is altered photochemically in an artificial process that is similar to what transpires when 7-dehydrocholesterol is converted to D_3. The artificial D_2 is used for the enrichment of milk and dairy products. Both natural and artificial forms of vitamin D are hydroxylated and converted to hormonally active

Table 4.1. Cellular retinoid-binding proteins.

Name (abbrevation)	Major ligand
Cellular retinol-binding protein, type I (CRBP I)	All-*trans*-retinol
Cellular retinol-binding protein, type II (CRBP II)	All-*trans*-retinol
Cellular retinoic acid-binding protein, type I (CRABP I)	All-*trans*-retinoic acid
Cellular retinoic acid-binding protein, type II (CRABP II)	All-*trans*-retinoic acid
Epididymal retinoic acid-binding protein (ERABP)	All-*trans* and 9-*cis*-retinoic acid
Cellular retinaldehyde-binding protein (CRALBP)	11-*cis*-retinol and retinal
Interphotoreceptor retinol-binding protein (IRBP)	11-*cis*-retinal and all-*trans*-retinol

derivatives. Vitamin D is transported, bound to a specific protein in plasma, primarily to the liver, where 25-hydroxylation is catalysed by a P450-like hydroxylase in microsomes and mitochondria (Saarem et al., 1984). From the liver, the 25-hydroxy-D is carried as the plasma complex with the D-binding protein to the kidneys, where 1α-hydroxylation occurs in the mitochondria of proximal tubular cells. The 25-hydroxy-D 1α-hydroxylase is a mixed-function oxidase that uses molecular oxygen (Henry and Norman, 1974). The enzyme is comprised of three proteins (renal ferredoxin, its reductase and a cytochrome P450) that are integral components of the mitochondrial membrane. The control of the renal 1α-hydroxylase is the most important point of regulation in the vitamin D endocrine system (Henry, 1992). Major factors are the product 1α,25-dihydroxy-D, parathyroid hormone (PTH), and the serum concentrations of Ca^{2+} and phosphate (Henry et al., 1992). Besides the natural, hormonally active dihydroxy-D, there are nearly 40 known metabolites of D_3. Most are inactive and excreted in the faeces. However, the 24R,25-dihydroxy-D_3, which is also produced in the kidneys, is probably required along with 1α,25-dihydroxy-D_3 for some of the biological responses to vitamin D, for example in the mineralization of bone and egg shells (Collins and Norman, 2001). As for storage of vitamin D_3, there is some variation among species, but in the human, adipose tissue serves predominantly as a storage site for D_3, whereas muscle retains significant 25-hydroxy-D_3 (Mawer et al., 1972). The 1α,25-dihydroxy-D_3 is catabolized via a number of pathways that lead to its rapid removal from the organism (Kumar, 1986). Upon protein-vectored transport of 1α,25-dihydroxy-D through plasma to target tissues, the hormone interacts with specific, high-affinity, intracellular receptors, first in the cytosol and then in the nucleus. Over two dozen target tissues and cells have been reported to have high-affinity receptors for 1α,25-dihydroxy-D_3 (Collins and Norman, 2001). The nuclear receptor for this hormone (VDR), first discovered in the intestines of vitamin D-deficient chicks, was found to be a 50 kDa DNA-binding protein that belongs to a superfamily of homologous nuclear receptors. As in the case of the nuclear receptors for vitamin A, the E domain toward the C-terminus of VDR binds the dihydroxy-D. As a group II receptor, VDR (like RAR and RXR) can form heterodimers with other receptors. This enhances the diversity of physiological effects.

The nuclear receptor–hormone complex is activated and binds to an HRE on the DNA to modulate expression of hormone-sensitive genes. Modulation of gene transcription results in induction or repression of specific mRNAs, which leads to changes in protein expression ultimately reflected in biological responses. More than 50 genes are known to be regulated by 1α,25-dihydroxy-D_3 (Hannah and Norman, 1994). Some of those that were registered by changes in the mRNA level as well as in intestinal or renal tissue are noted in Table 4.2.

Among those proteins that are increased by D action is calbindin. The genomic induction of this

Table 4.2. Some genes in intestine and kidney regulated by 1α,25-dihydroxy-D_3.

Gene	Tissue	Regulation
α-Tubulin	Intestine	Down
Aldolase subunit B	Kidney	Up
Alkaline phosphatase	Intestine	Up
ATP synthase	Intestine, kidney	Up, down
Calbindin 9K and 28K	Intestine, kidney	Up
Cytochrome oxidase subunits I, II and III	Intestine, kidney	Up, down
Cytochrome b	Kidney	Down
Ferridoxin	Kidney	Down
1-Hydroxy-D 24-hydroxylase	Kidney	Up
Metallothionein	Kidney	Up
NADH dehydrogenase subunit I subunits III and IV	Kidney, intestine	Down, up
Plasma membrane Ca^{2+} pump	Intestine	Up
VDR	Intestine	Up

calcium-binding protein is one of the major effects of 1α,25-dihydroxy-D$_3$. In addition, there are non-genomic actions of the dihydroxy-D that precede its slower hormonal response. The rapid transport of Ca^{2+} mediated by dihydroxy-D is termed 'transcaltachia' (Nemere and Norman, 1987). In the intestine, this process appears to involve internalization of Ca^{2+} in endocytic vesicles at the brush border membrane, which then fuse with lysosomes and travel along microtubules to the basal lateral membrane for exocytosis. Other non-genomic actions of dihydroxy-D may include phosphoinositide breakdown. It is now clear that vitamin D, largely through the action of its 1α,25-dihydroxy metabolite, has a range of actions, with a central role in the intestinal absorption of Ca^{2+} and the mineralization of bone.

Vitamin E

The eight members of the vitamin E group, which includes four tocopherols (α, β, γ, δ) and four tocotrienols (α, β, γ, δ), are biosynthesized only by plants and are found at especially high levels in edible vegetable oils (Sheppard et al., 1993). Biologically, the most active form is RRR- (formerly designated d-) α-tocopherol, and most studies have dealt with the properties of this form of E. Absorption of E from the lumen of the small intestine is a passive diffusion, non-carrier-mediated process with relatively low efficiency (Traber and Sies, 1996; Chow, 2001). Within the enterocyte, vitamin E is incorporated into chylomicrons and secreted into the intracellular spaces and lymphatic system. Upon conversion of chylomicrons to remnant particles, E is distributed to circulating lipoproteins and ultimately to tissues. Liver parenchymal cells take up the remnants after chylomicrons are partially delipidated by lipoprotein lipase and there is acquisition of apoE. Liver is responsible for the control and packaging of tocopherol with very low-density lipoproteins (VLDLs). Cytosolic α-tocopherol transfer proteins, first identified in rat liver (Catignani and Bieri, 1977) and more recently characterized from the cDNA sequence in human liver (Arita et al., 1995), exhibit selectivity for the RRR-α-tocopherol that is transferred to VLDLs. The purified rat liver protein with a molecular mass of 30–36 kDa has two isoforms (Sato et al., 1991) and facilitates transfer of α-tocopherol between membranes. The human liver protein with similar action has a molecular mass of 36.6 kDa (Kuhlenkamp et al., 1993). A smaller (14.2 kDa) α-tocopherol-binding protein from liver and heart may be involved in intracellular transport and metabolism of α-tocopherol (Gordon et al., 1995).

From the liver, tocopherol-bearing VLDLs are secreted into plasma for delivery to peripheral tissues. Lipoproteins that are associated with E in plasma exchange tocopherol to cells at least partly via receptors, though specific mechanisms are not fully understood (Traber and Sies, 1996; Chow, 2001). Most tissues have the capacity to accumulate α-tocopherol, though none functions as a storage organ. Much of the vitamin E in the body is localized in the adipose tissue where tocopherol is mainly in the bulk lipid droplet from which turnover is slow. Turnover is also slow from muscle, testes, brain and the spinal cord. Adrenal glands have the highest concentration of α-tocopherol, although lungs and spleen also contain relatively high concentration. α-Tocopherol is taken up and located primarily in parenchymal cells (Bjorneboe et al., 1991), whereupon some transfer to non-parenchymal cells can occur. Parenchymal cells can store surplus tocopherol and are depleted less readily than non-parenchymal cells. Within a cell, light mitochondria have the highest concentration of α-tocopherol, whereas the concentration is low in cytosol. The majority of tocopherol is located in membranes, including those of erythrocytes from which turnover is relatively rapid. Approximately three-quarters of mitochondrial α-tocopherol is found in the outer membrane, and one-quarter is associated with the inner membrane.

Upon exerting its antioxidant action, tocopherol is converted to the tocopheryl chromanoxy radical (Chow, 2001). This latter can be reverted to tocopherol by such physiological reductants as ascorbate and glutathione. Some of the radical is oxidized further to α-tocopherol quinone, which can be converted to the hydroquinone by an NADPH-dependent reductase found in the mitochondria and microsomes of hepatocytes (Hayashi et al., 1992). Also an NADPH-cytochrome P450 reductase can catalyse reduction of the quinone to the hydroquinone. Organelle-localized, side chain oxidation of the hydroquinone leads to α-tocopheronic acid that is conjugated and excreted in the urine.

Vitamin K

The natural vitamin K group is comprised of phylloquinone (K_1) of plants and menaquinone-7 (K_2) and other bacterial and animal menaquinones in which the number of isoprenoid units of the side chain varies. The synthetic compound that has the fundamental 2-methyl-1,4-naphthoquinone structure common to all the K group is menadione (K_3); most commercial forms of K are water-soluble derivatives of this (Suttie, 2001). Absorption of natural K in the intestine generally is similar to that for other lipid and lipid-soluble components of the diet that are incorporated into mixed micelles. In the enterocytes, phylloquinone is packaged in chylomicrons, which exit to the lymphatic system. The tissue distribution of phylloquinone and menadione is significantly different with time. Menadione spreads over the whole body faster than phylloquinone, but the amount retained in tissue is low. Phylloquinone is found in the liver of those species ingesting plant material. Menaquinones containing 6–13 isoprenyl units in the alkyl chain occur in the liver of most species. Human liver reflects a K content of approximately 10% phylloquinone with a broad mixture of menaquinones (Matschiner, 1971). The menaquinones are absent in neonatal liver (Shearer et al., 1988), but the level rises with age of the infant (Kayata et al., 1989). In addition to liver, vitamin K concentrates in the adrenal glands, lungs, bone marrow, kidneys and lymph nodes.

Intracellularly, phylloquinone concentrates in the Golgi fraction and the microsomal fraction representative of the smooth endoplasmic reticulum (Nyquist et al., 1971). Menaquinones (specifically MK-9) preferentially localize in mitochondria (Reedstrom and Suttie, 1995). There is preliminary evidence that a K-binding cytosolic protein may facilitate intraorganelle movement of the vitamin (Kight et al., 1995). As a deficiency develops, vitamin K is lost more rapidly from the cytosol than the membrane fractions (Knauer et al., 1976).

Metabolic alterations and utilization of vitamin K occur in organelle membranes, especially in the endoplasmic reticulum. Animal tissue cannot synthesize the naphthoquinone ring, but menadione, when incubated with liver homogenates, undergoes side chain addition at position 3 to form menaquinone-4 (Martius, 1961). This activity resides in the microsomes (Dialameh et al., 1970). In addition to such alkylation of menadione to form a menaquinone, the 3-phytyl chain of phylloquinone can be dealkylated and then realkylated to form menaquinone-4 (Thijssen and Drittij-Reijnders, 1994). Among liver metabolites of K, recognition of its 2,3-epoxide (Matschiner et al., 1970) was an important step in understanding what is now recognized as the K cycle, which involves the K-dependent carboxylase (Suttie, 2001). The scheme of connected events localized to the microsomal fraction are shown in Fig. 4.1, which underscores the function of vitamin K in the formation of γ-carboxyglutamyl (Gla) residues from glutamyl (Glu) residues in certain proteins.

Reduction of the usual quinone forms of vitamin K is by microsomal K quinone reductases that are pyridine nucleotide dependent, and by a dithiol-dependent reductase that is sensitive to the anticoagulant, warfarin. This latter reductase appears to be the same as the activity that is the K

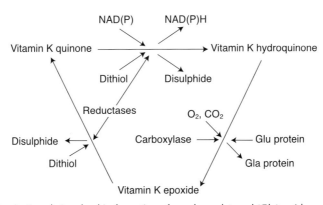

Fig. 4.1. The vitamin K cycle involved in formation of γ-carboxyglutamyl (Gla) residues.

epoxide reductase (Gardill and Suttie, 1990). It has been suggested that the thioredoxin/thioredoxin reductase system is the physiologically relevant dithiol/disulphide system involved (van Haarlem et al., 1987; Silverman and Nandi, 1988). The carboxylase step, first shown to form Gla residues in prothrombin synthesis (Esmon et al., 1975), is in the lumenal side of the rough endoplasmic reticulum (Carlisle and Suttie, 1980) and utilizes CO_2 and O_2. The human enzyme has been purified (Wu et al., 1991a) with cloning and expression of its cDNA (Wu et al., 1991b). The chemical mechanism by which the K-dependent carboxylase proceeds has been propounded by Dowd et al. (1995). This involves addition of O_2 to the K hydroquinone anion to form a dioxetane that generates an alkoxide. This latter is postulated to be the strong base necessary to abstract a hydrogen from the γ-methylene group of a Glu residue. The vitamin K-dependent carboxylations are found not only in liver where blood clotting factors such as prothrombin are formed but also in skeletal tissues where mineralizations result as a consequence of the Ca^{2+}-binding propensity of the Gla residues in proteins such as osteocalcin. Other Gla-containing proteins have been reported for other tissues, e.g. kidney (Griep and Friedman, 1980) and spermatozoa (Soute et al., 1985).

Thiamine (B_1)

As with most water-soluble vitamins, thiamine enters epithelial cells by two mechanisms, i.e. active or carrier-mediated uptake and passive diffusion (Bowman et al., 1989). Below 2 µM, thiamine is absorbed into enterocytes, mainly in jejunal and ileal portions of the small intestine, by a carrier-mediated process that is not dependent on ATPase (Tanpaichitr, 2001). It is unclear at present what the nature of the carrier is; however, there is a report of a thiamine-binding protein within liver and chicken egg white and yolk (Muniyappa et al., 1978). The entry of thiamine into most mammalian cells is a tandem process that is also linked with pyrophosphorylation (Rindi and Laforenza, 1997), a process that requires cytosolic thiamine pyrophosphokinase with ATP and Mg^{2+}. A high percentage of the thiamine in epithelial cells is (pyro)phosphorylated, whereas the thiamine arriving on the serosal side of the mucosa is largely free. Exit of thiamine from the serosal side is dependent on Na^+ and the normal function of ATPase at the serosal pole of the cell. Upon vectoring of thiamine through the blood, much as thiamine pyrophosphate (TPP) within erythrocytes, about half of the total thiamine is distributed in the skeletal muscles, with most of the rest in heart, liver, kidneys, and in brain and spinal cord, which have about twice the content of peripheral nerves. Leukocytes have a tenfold higher thiamine concentration than erythrocytes. Of the total body thiamine (~30 mg), about 80% is TPP, 10% the triphosphate (TTP) and the remainder the monophosphate (TMP) and thiamine. The three tissue enzymes known to participate in the formation of the phosphate esters are thiamine pyrophosphokinase responsible for TPP, TPP-ATP phosphoryltransferase, which catalyses the formation of TTP, and thiamine pyrophosphatase, which hydrolyses TPP to form TMP (Tanphaichitr, 2001).

Within mammalian cells, the operating coenzymic form of thiamine is TPP, which functions as the prosthetic group in two types of enzymic systems. One is cytosolic transketolase, important in the pentose phosphate pathway where reversible transketolations are catalysed between D-xylulose 5-phosphate and D-ribose 5-phosphate to form D-sedoheptulose 7-phosphate and D-glyceraldehyde 3-phosphate, or between D-xylulose 5-phosphate and D-erythrose 4-phosphate to form D-fructose 6-phosphate and D-glyceraldehyde 3-phosphate. The second type of coenzymic role of TPP is in the α-keto acid decarboxylase subunits of mitochondrial multienzymic dehydrogenase complexes that convert pyruvate to acetyl-CoA, α-ketoglutarate to succinyl-CoA, and branched chain amino acids to their several metabolites. In the dynamics of such multienzymic complexes, the α-keto acid substrate is decarboxylated and transferred via an α-hydroxyalkyl-TPP to the dihydrolipoyl moiety of a transacylase core. An acyl-CoA product is then released, and the oxidized lipoyl group is reduced by an FAD-dependent dehydrogenase that is linked to NAD^+ in the mitochondrion.

The numerous urinary catabolites of thiamine reflect the considerable cellular degradative events, which include thiaminase-catalysed cleavage of pyrimidine and thiazole portions of the vitamin, alcohol dehydrogenase-catalysed oxidation of the β-hydroxyethyl group on the thiazole, etc.

(McCormick, 1988). Loss of the vitamin and its acid metabolites is relatively rapid. There is a high turnover rate of about 2 weeks, and little storage for any period of time in any tissue.

Riboflavin (B$_2$)

Riboflavin, released from most of the dietary flavoproteins by dissociation followed by alkaline phosphatase and pyrophosphatase activities within the small intestinal lumen, enters the enterocytes where it is trapped metabolically by phosphorylation catalysed by flavokinase (McCormick, 1999; Rivlin and Pinto, 2001). Much of the riboflavin 5′-phosphate (FMN) so formed is converted further to FAD, and both flavocoenzymes function in metabolic processes. Upon exit at the serosal side of the lumen, riboflavin is complexed to both high- and low-affinity proteins (Whitehouse et al., 1991) for transport to the various cells of the body. Within most cells, the cytosolic flavokinase and FAD synthetase sequentially form the flavocoenzymes FMN and FAD, respectively. These ATP-utilizing enzymes, first purified from liver, are optimal with Zn^{2+} in the case of kinase (Merrill and McCormick, 1980) and with Mg^{2+} in the case of synthetase (Oka and McCormick, 1987). There are phosphatases that act on FMN and an FAD pyrophosphatase, but these hydrolytic enzymes are membrane separated and involved in turnover and release of the vitamin from the cell (McCormick, 1975). Flavocoenzyme biosynthesis is influenced by thyroid hormone (Rivlin, 1970; Lee and McCormick, 1985).

A small but important fraction of FAD is covalently bound to apoproteins within organelle membranes (Yagi et al., 1976; Addison and McCormick, 1978). Humans and other mammals have succinate dehydrogenase in the inner mitochondrial membrane, and sarcosine and dimethylglycine dehydrogenases in the mitochondrial matrix. In these enzymes, FAD is attached through its 8α-position to an imidazole N of a histidyl residue. Monoamine oxidase in the outer mitochondrial membrane contains 8α-(S-cysteinyl) FAD. In mammals capable of biosynthesizing vitamin C, L-gulonolactone oxidase, which has an 8α-(N^1-histidyl)FAD, is found in the microsomes, particularly in liver and kidneys. There are numerous non-covalent flavocoenzyme-dependent enzymes, some in the cytosol and some in the organelles. Examples include the FMN-dependent pyridoxine (pyridoxamine) 5′-phosphate oxidase of cytosol, the FAD-dependent fatty acyl-CoA dehydrogenase of the β-oxidative system of the mitochondrial matrix, and the microsomal NADPH-cytochrome P450 reductase that contains both FMN and FAD. The predominant flavocoenzyme within cells is FAD, which may constitute about 90% of the total flavin. Only approximately 5% of this becomes covalently bound to pre-formed apoenzymes as 8α-linked FAD.

The numerous urinary flavins reflect the extensive catabolism of side chain and ring methyl functions as well as dermal photochemical and intestinal microbial cleavage of the D-ribityl chain (McCormick, 1999). Among those catabolites that arise as a result of cellular oxidative actions are 7- and 8-hydroxymethyl (7α- and 8α-hydroxy) riboflavins, 10-formyl and 10-β-hydroxyethyl flavins, lumiflavin and lumichrome (Chastain and McCormick, 1991).

Niacin

Both forms of niacin (nicotinic acid and nicotinamide) can be absorbed from the stomach, but more rapid and extensive absorption is from the small intestine (Kirkland and Rawling, 2001). For both vitamers, the uptake is facilitated as well as involving passive diffusion. In the enterocyte (and erythrocytes and other cells), nicotinic acid and nicotinamide are converted to NAD, the former through the Preiss–Handler pathway (Preiss and Handler, 1958) and the latter through the Dietrich pathway involving a nicotinamide pyrophosphorylase (Dietrich et al., 1966). The steps are shown in Fig. 4.2 and entail phosphoribosyltransferases for nicotinic acid and nicotinamide, the nuclear adenylyltransferase for both nicotinic acid mononucleotide (NaMN) and nicotinamide mononucleotide (NMN), cytosolic NAD synthase, and the nicotinamide deamidase, which functions only at high concentrations of its substrate. The formation of NADP requires a cytosolic NAD kinase. Some nicotinic acid moves into the blood, as does nicotinamide, which is the principal circulating form released by NAD glycohydrolase activities important in controlling

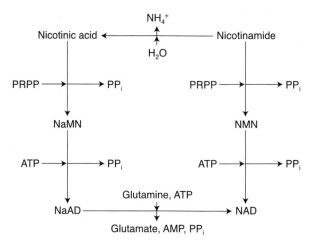

Fig. 4.2. The biosynthetic steps in converting niacin to NAD.

intracellular NAD levels. The liver is the central processing organ for niacin. The liver and kidneys can convert tryptophan to quinolinic acid, which reacts with phosphoribosyl pyrophosphate (PRPP) to form NaMN as catalysed by quinolinate phosphoribosyltransferase in these tissues (Moat and Foster, 1982).

There are two major roles played by NAD. Together with NADP, the pyridine nucleotides are redox coenzymes that participate as hydride ion trafficking agents in numerous oxidation–reduction reactions. Most NAD within the cell is oxidized, whereas most NADP is maintained in the reduced state as a result of the pentose phosphate pathway. Diverse enzymes utilizing either NAD or NADP are localized within cytosol and organelles. The other major role of NAD is as a substrate for enzyme systems, especially those that are membrane associated, that release nicotinamide and ADP-ribose as the mono-, poly- or cyclic product. The ADP-ribosylations affect diverse systems that include amino acid residues on acceptor proteins. Transferase proteins are found in nuclear, cytosolic and plasma membrane fractions. An NAD: arginine ADP-ribosyltransferase is anchored via glycosylphosphatidylinositol to the outer surface of the plasma membrane (Zolkiewska et al., 1994). Poly(ADP-ribose) is synthesized by a polymerase that affects DNA repair. The cyclic ADP-ribose affects calcium ion transport. Also a deamidated metabolite of NADP (NAADP) is a potent Ca^{2+}-releasing agent distinct from the cyclic ADP-ribose (Lee, 2000).

Liver has some capacity for NAD storage (Kirkland and Rawling, 2001). It is also the organ that forms most of the major methylated and hydroxylated catabolites excreted in urine. In humans, nicotinamide is methylated primarily to produce N^1-methylnicotinamide though some is oxidized to yield N^1-methyl-2-pyridone-5-carboxamide. Nicotinamide is conjugated mainly to glycine to form nicotinuric acid.

Vitamin B_6

The vitamin B_6 group is comprised of the three vitamers, i.e. pyridoxine (PN), pyridoxamine (PM) and pyridoxal (PL). In addition, there are three 5′-phosphates of these, i.e. PNP, PMP and PLP. The predominant coenzyme is PLP, which is also the major form in most natural foods. The 5′-phosphates are hydrolysed by alkaline phosphatase in the small intestinal lumen (Henderson, 1985). As with most other water-soluble vitamins, at physiological concentrations the absorption is facilitated with metabolic trapping. Upon entry into enterocytes and other cells, all three vitamers are phosphorylated (McCormick, 2001). Liver is the primary organ responsible for much of the metabolism of vitamin B_6 and its subsequent distribution, much of it as albumin-bound PLP. Cytosolic interconversions within the B_6 group require phosphorylation of the vitamers catalysed by pyridoxal phosphokinase, optimal with Zn^{2+}·ATP in eukaryotes (McCormick et al., 1961), and pyridoxine (pyridoxamine) 5′-phosphate oxidase (Kazarinoff and McCormick, 1975), which, because of its FMN dependency, is sensitive to riboflavin status (McCormick, 1989). These activities have been measured in human liver (Merrill et al., 1984) where the PLP formed is trapped within the cells not only by the anionic charge on the phosphate moiety but also by reaction of its aldehyde function with internal proteins, resulting in reversible Schiff bases (Li et al., 1974). Release of PLP from cells is mainly after hydrolysis catalysed by phosphatase bound to the plasma membrane (Merrill and Henderson, 1990). Distribution of B_6 and its metabolites is widespread in tissues (McCormick, 2001), but much is in skeletal muscle as PLP (Coburn et al., 1991), a considerable fraction associated with phosphorylase (Black et al., 1978). It is noteworthy that brain, rich in pyridoxal kinase (McCormick and Snell, 1959), requires PLP in the metabolism of glutamate and formation of amines active in the nervous system.

The enzymes that require PLP are found in the cytosol and organelles. Formation of coenzyme is in the former as is its attachment to apoenzymes, some of which remain cytosolic while some are transferred to specific organelles. Intracellular targeting of PLP-dependent enzymes commonly involves a sequence motif that 'directs' the ultimate location of the holoenzyme. With the pre-mitochondrial form of aspartate aminotransferase, there is an N-terminal sequence that interacts with a heat shock protein 70 (Hsp70) chaperone to guide the enzyme into the mitochondria (Artigues et al., 2000), whereas the cytosolic form of the enzyme lacks this motif. With the human ornithine decarboxylase that goes to the plasma membrane, a phosphorylation-regulated, p47phox-related, membrane-targeting motif surrounds Ser167 (Heiskala et al., 2000). As far as concerns non-enzymatic functions of PLP, reports that it may be involved in steroid hormone regulation and gene expression have been reviewed (Leklem, 2001).

The catabolism of B_6 in the human primarily is the oxidation of pyridoxal to 4-pyridoxic acid catalysed by aldehyde oxidase. The 4-pyridoxate is an end-product that is excreted in urine.

Pantothenic Acid

Pantothenate uptake by cells is reported to be dependent on Na^+ co-transport (Fenstermacher and Rose, 1986). Phosphorylation of pantothenate is catalysed by pantothenate kinase, the primary regulatory site in CoA biosynthesis (Robishaw et al., 1982). The main route for biosynthesis of the coenzymically functional 4′-phosphopantheine in humans and other mammals involves condensation of the pantothenate 4′-phosphate with cysteine in an ATP-dependent reaction to form 4′-phosphopantothenylcysteine, which is decarboxylated by a PLP enzyme to yield 4′-phosphopantheine (Brown, 1959). One use of this product is to be linked to the fatty acyl carrier protein and fatty acid synthase of mitochondria; another use is to be converted to CoA by addition of the AMP portion of ATP followed by phosphorylation of the ribosyl 3′-hydroxy group. All the enzymes required for CoA synthesis have been isolated from the cytosol, yet mitochondria must have a similar system because 95% of CoA is found there, but CoA itself cannot cross mitochondrial membranes (Plesofsky, 2001).

Biochemical functions of pantothenate are the synthesis of fatty acids, a process located in the mitochondrial outer membrane, and the multiple uses of acyl-CoAs, which occur in cytosol, the reticuloendothelial system, mitochondria and the nucleus. There are numerous metabolic events that use acyl-CoA substrates that are altered to products still bearing the CoA as a thioester, e.g. in β-oxidation of fatty acyl-CoAs in the mitochondria. Among acyl group transfers from CoA esters are the N-terminal modifications of proteins with acetyl, myristyl or palmityl residues. These types of modifications affect the location and activity of such proteins, including those that have a role in signal transduction. It has been estimated that over half of soluble eukaryotic proteins are acetylated (Driessen et al., 1985). Acetylation of histones within the nucleus markedly affects their actions. For the myristoylation of proteins, catalysed by the myristoyltransferase (Towler et al., 1987), an N-terminal glycyl residue, originally in the second position, is absolutely required, and there is a preference for a small neutral subterminal residue and a seryl residue in the sixth position. Myristoylation occurs co-translationally, is irreversible and leads to weak membrane associations or interactions with other proteins. Palmitoylation is less specific, occurs post-translationally with formation of a reversible ester bond with cysteinyl or seryl residues and targets the protein for strong association with membranes. Many membrane receptors, e.g. for rhodopsin, are palmitoylated.

CoA is degraded by hydrolysis to pantothenate in multiple steps, with the final, unique one being the conversion of panetheine to cysteamine and pantothenate. The latter is excreted in the urine.

Biotin

A biotin transporter is present in the intestinal brush border membrane (Mock, 2001). Most studies point to specificity for valerate chain and ureido portions of the vitamin as well as a 1:1 coupling with Na^+ in a electroneutral process. Exit of biotin from the enterocyte across the basolateral membrane is also carrier mediated but is independent of Na^+ and is electrogenic (Said et al., 1988). Although transport of biotin is most active in the proximal small intestine of rats, absorption from the proximal colon is significant (Bowman and Rosenberg, 1987). This is probably true for humans as well (Sorrell et al., 1971). Uptake of biotin in hepatocytes (Bowers-Komro and McCormick, 1985) and peripheral tissues appears to exhibit characteristics similar to those of the small intestine.

Inside cells, the vitamin is catabolized significantly as well as utilized for carboxylases and perhaps in the specific modifications of other proteins. Biotin is incorporated into all four of the mammalian biotin-dependent carboxylases by the action of holocarboxylase synthetase, which is present in both cytosol and mitochondria. The synthetase catalyses ATP-dependent formation of biotinyl 5′-adenylate, which is then condensed with the ε-amino function of a specifically recognized lysyl residue in the apocarboxylase to form the amide-linked biocytinyl prosthetic group of the holocarboxylase. Three of the biotin-dependent carboxylases are mitochondrial. These are the ones active on pyruvate to form oxalacetate, methylcrotonyl-CoA to form methylglutaconyl-CoA, and propionyl-CoA to form D-methylmalonyl-CoA. Acetyl-CoA carboxylase is found in both mitochondria and the cytosol. There are two forms of the mitochondrial enzyme that are less active than the cytosolic form, which catalyses formation of malonyl-CoA in the step committed to fatty acid biosynthesis. In all cases, these carboxylases require HCO_3^- with $Mg^{2+}\cdot ATP$ to form the transitory carbonyl phosphate that carboxylates the ureido N^1 of biotin attached to the enzyme, which then carboxylates the activated substrate.

A large fraction of biotin undergoes catabolism that involves β-oxidation of the valerate side chain and S-oxidations resulting largely in sulphoxides (McCormick and Wright, 1970; McCormick, 1976). Catabolites such as bisnorbiotin and its sulphoxides and sulphone, bisnorbiotin methyl ketone and tetranorbiotin sulphoxide have been identified in human urine (Zempleni et al., 1996). Biotin catabolites also represent a considerable fraction of biotin-related compounds in human milk (Stratton et al., 1996).

Folic Acid

Folic acid is the chemical parent of natural folacins that are reduced, polyglutamylated and substituted with one-carbon additions to the N^5 and N^{10} positions. It is only after oxidation and deconjugation that the free vitamin is absorbed, mainly in the jejunum via a saturable, carrier-mediated process (Brody and Shane, 2001). The intestinal transporter, encoded by the reduced folate carrier gene, is a transmembrane protein that is expressed in most tissues (Moscow et al., 1995; Said et al., 1996), but its specificity for various folates differs among tissues and their apical and basolateral membranes. A folate-binding protein (sometimes called the folate receptor) is also in the intestine and has been found to mediate endocytosis of folate in cells of kidneys and other tissue (Said et al., 1996). The binding protein and the membrane transporter are located on opposing membranes in some polarized cells (Chancy et al., 2000). Most folate is metabolized to 5-methyl-tetrahydrofolate (5-methyl-THF) during passage through the intestinal mucosa. Folate taken up in liver via the reduced folate carrier is metabolized to the reduced polyglutamates and retained. The cytosolic NADPH-utilizing dihydrofolate reductase reduces folate through the dihydro level to its tetrahydro form, and $Mg^{2+}\cdot ATP$-utilizing folylpolyglutamate synthetase converts the latter to the polyglutamates. The polyglutamates are more effective substrates than the monoglutamates for most folate-dependent enzymes. The major kinetic advantages are achieved by elongation to the triglutamate chain (Shane, 1989). Longer chains are required for the enzymes involved in the methione resynthesis cycle. Hexa- and heptaglutamates as well as longer chain derivatives are found in human cells (Foo et al., 1982). There are cytosolic and mitochondrial isozymes of folylpolyglutamate synthetase. Both forms are essential and supply the coenzymically optimal forms of reduced folyl oligo-γ-glutamates that are used for the enzymes listed in Table 4.3.

The major pathways for synthesis of methionine, thymidylate and purines are in the cytosol, but mitochondrial folate metabolism plays an important role in glycine metabolism and in providing one-carbon units for cytosolic one-carbon metabolism (Brody and Shane, 2001). Many of the enzymes participating in folate metabolism are multifunctional or part of multiprotein complexes; this allows channelling of polyglutamate intermediates between active sites without release of intermediate products from the complex (Paquin et al., 1985).

The major urinary excretory product of folate is N-acetyl-p-aminobenzoylglutamate (Murphy et al., 1976) with smaller amounts of p-aminobenzoylglutamate. Metabolic conditions that cause an accumulation of 5- or 10-formyl-THF-polyglutamates result in increased folate catabolism, and heavy chain ferritin has been isolated as an activity that cleaves folates to pterin derivatives (P.J. Stover, 2001, personal communication).

Table 4.3. Compartmentalization of enzymes involved in folate-dependent metabolism and interconversions.

Enzyme
Cytosol
Dihydrofolate reductase
Serine hydroxymethyltransferase
5,10-Methylene-THF dehydrogenase, cyclohydrolase, 10-formyl-THF synthetase[1]
Glycinamide ribonucleotide transformylase
Aminoimidazole carboxamide ribonucleotide transformylase
Thymidylate synthase
5,10-Methylene-THF reductase
Methionine synthetase
THF formiminotransferase, forminoTHF cyclodeaminase[2]
Glutamate transformylase
5-Formyl-THF isomerase
10-Formyl-THF dehydrogenase
Folyl-polyglutamate synthetase
γ-Glutamyl hydrolase
Mitochondria
Formyl-THF synthetase
10-Formyl-THF dehydrogenase
Methenyl-THF dehydrogenase
Methenyl-THF synthetase
Serine hydroxymethyltransferase
Glycine cleavage enzyme complex
Dimethylglycine dehydrogenase
Sarcosine dehydrogenase
Methionyl-tRNA transformylase

[1] A trifunctional enzyme.
[2] A bifunctional enzyme.

Vitamin B_{12} (Cobalamins)

Vitamin B_{12} is the natural hydroxocobalamin or the commercial cyanocobalamin in which the coordinated cyanide is replaced during cellular handling. The vitamin as a complex with intrinsic factor rapidly attaches to surface receptors, e.g. cubilin (Kozyraki et al., 1998), and then enters small intestinal cells in a slower, energy-requiring process (Hines et al., 1968). Upon entry, cobalamin is bound to a macromolecule immunologically similar to intrinsic factor (Rothenberg et al., 1972). After portal delivery, cobalamin bound to R-type binders is processed by hepatocytes before re-entering the blood plasma or being excreted in the bile (Burger et al., 1975). Within mammalian tissues, there are only two known coenzyme forms of B_{12}. One, 5′-deoxyadenosylcobalamin, is formed by a synthase that includes a flavoprotein that reduces B_{12a} through B_{12r} to B_{12s}, which then reacts with ATP to form the coenzyme and release triphosphate. The 5′-deoxyadenosylcobalamin functions with mitochondrial L-methylmalonyl-CoA mutase to convert its substrate to succinyl-CoA. The other coenzyme-like function of B_{12} is as methylcobalamin that participates in the cytosolic methionine synthase, which is a methyltetrahydrofolate-homocysteine methyltransferase. This important relationship of vitamin B_{12} to folic acid manifests clinically in the 'methyl trap hypothesis' when there is inadequate B_{12}

(Beck, 2001). A block in methionine synthase causes accumulation of folate at the 5-methyl-THF level. As this latter cannot release its methyl group without turnover of the methionine synthase reaction, folate becomes 'trapped' and there is a reduction in levels of other folate-dependent reactions.

Vitamin C (L-Ascorbic acid)

The absorption of L-ascorbate into cells of the small intestine is again representative of other water-soluble vitamins (Bowman *et al.*, 1989; McCormick and Zhang, 1992). There is rapid portal transport and a widespread distribution to tissues where it serves a general oxidant role as well as being specific as a redox cofactor in Fe^{2+}- and Cu^{2+}-containing oxygenases. Associations of L-ascorbate with enzymes are critical, and their functions are listed in Table 4.4.

The metabolism of L-ascorbate includes conversion to the 2-*O*-sulphate and methylation as well as hydrolysis and oxidations that lead to acid fragments appearing as catabolites in the urine.

Coenzyme Compartmentalization

Because most of the vitamins are converted to more polar, often anionic coenzymes that are the functional forms, and such charged species are not readily permeable to membranes of organelles, there are special means by which coenzymes gain access to the cellularly included bodies. As mentioned in the foregoing descriptions of the vitamins, a binding protein (sometimes apoenzyme) can associate with the cytosolically activated vitamin (coenzyme or hormone) and be transported to an organelle membrane where a receptor or chaperone assists entry. Examples are activated forms of vitamins A and D that gain access to the genomic material within the nucleus, or coenzymic B_6 that enters mitochondria as a PLP-dependent holotransaminase. In other cases, e.g. NAD(H), NADP(H), FAD(H_2) and acyl-CoAs, entry through the inner mitochondrial membrane requires devices such as 'shuttle' mechanisms. The operation of these requires that the appropriate enzymes are located on the correct side of the membrane and that appropriate transporters or translocases are present on/in the membrane to shuttle the various intermediates. Examples of these are the malate–aspartate shuttle in which redox turnover of NAD/NADH outside the mitochondrion is coupled with that inside by location of malate dehydrogenase and aspartate aminotransferase on both sides of the membrane and by allowed flux of malate and glutamate through the membrane transporters. A similar situation exists for the α-glycerol phosphate shuttle where the FAD-dehydrogenase is within the membrane. The major source of acetyl-CoA is from the pyruvate dehydrogenase complex operating within the mitochondria, yet a bypass mechanism is needed to move the equivalent of acetyl-CoA to the cytosol. This is provided by citrate, which is permeable, and upon exiting the mitochondria can form acetyl-CoA from the ATP- and CoA-utilizing citrate cleavage enzyme. Finally, the migration of the acyl moieties of long chain fatty acids into mitochondria where they can be β-oxidized depends on carnitine acyltransferases and the transferance of the acyl portion from a CoA ester to one with the vectoring carnitine.

Table 4.4. Enzyme activities associated with ascorbic acid function.

Enzyme	Function
Dioxygenases with Fe^{2+}	
Prolyl-3-hydroxylase	Extracellular matrix maturation
Prolyl-4-hydroxylase	Extracellular matrix maturation
Lysyl hydroxylase	Extracellular matrix maturation
6-*N*-Trimethyl-L-lysine hydroxylase	Carnitine biosynthesis
γ-Butyrobetaine hydroxylase	Carnitine biosynthesis
Tyrosine-4-hydroxyphenylpyruvate hydroxylase	Tyrosine metabolism
Monooxygenases with Cu^{2+}	
Dopamine-β-monooxygenase or hydroxylase	Norepinephrine biosynthesis
Peptidylglycine-α-amidating monooxygenase	Activation of hormones and hormone-releasing factors

References

Addison, R. and McCormick, D.B. (1978) Biogenesis of flavoprotein and cytochrome components in hepatic mitochondria from riboflavin-deficient rats. *Biochemical and Biophysical Research Communications* 81, 133–138.

Arita, M., Sato, Y., Miyata, A., Tanabe, T., Takahashi, E., Kayden, H.J., Arai, H. and Inoue, K. (1995) Human alpha-tocopherol transfer protein: cDNA cloning, expression and chromosomal localization. *Biochemical Journal* 306, 437–443.

Artigues, A., Bengoechea-Alonso, M.T., Crawford, D.L., Iriarte, A. and Martinez-Carrion, M. (2000) Biological implications of the different Hsp70 binding properties of mitochondrial and cytosolic aspartate aminotransferase. In: Iriarte, A., Kagan, H. and Martinez-Carrion, M. (eds) *Biochemistry and Molecular Biology of Vitamin B_6 and PQQ-dependent Proteins*. Birkhauser Verlag, Basel, pp. 111–116.

Beck, W.S. (2001) Cobalamin (vitamin B_{12}). In: Rucker, R.B., Suttie, J.W., McCormick, D.B. and Machlin, L.J. (eds) *Handbook of Vitamins*, 3rd edn. Marcel Dekker, New York, pp. 463–512.

Bjorneboe, A., Nenseter, M.S., Hagen, B.F., Bjorneboe, G.-E.A., Prydz, K. and Drevon, C.A. (1991) Effect of dietary deficiency and supplementation with all-rac-α-tocopherol in hepatic content in rat. *Journal of Nutrition* 121, 1208–1213.

Black, A.L., Guirard, B.M. and Snell, E.E. (1978) The behavior of muscle phosphorylase as a reservoir for vitamin B_6 in the rat. *Journal of Nutrition* 108, 670–677.

Bowers-Komro, D.M. and McCormick, D.B. (1985) Biotin uptake by isolated rat liver hepatocytes. *Annals of the New York Academy of Sciences* 447, 350–358.

Bowman, B.B. and Rosenberg, I. (1987) Biotin absorption by distal rat intestine. *Journal of Nutrition* 117, 2121–2126.

Bowman, B.B., McCormick, D.B. and Rosenberg, I.H. (1989) Epithelial transport of water-soluble vitamins. *Annual Review of Nutrition* 9, 187–199.

Brody, T. and Shane, B. (2001) Folic acid. In: Rucker, R.B., Suttie, J.W., McCormick, D.B. and Machlin, L.J. (eds) *Handbook of Vitamins*, 3rd edn. Marcel Dekker, New York, pp. 427–462.

Brown, G.E. (1959) The metabolism of pantothenic acid. *Journal of Biological Chemistry* 234, 370–378.

Burger, R.L., Schneider, C.S., Mehlman, C.S. and Allen, R.H. (1975) Human plasma R-type vitamin B_{12}-binding protein in the plasma transport of vitamin B_{12}. *Journal of Biological Chemistry* 250, 7707–7713.

Carlisle, T.L. and Suttie, J.W. (1980) Vitamin K dependent carboxylase: subcellular location of the carboxylase and enzymes involved in vitamin K metabolism in rat liver. *Biochemistry* 19, 1161–1167.

Catignani, G.L. and Bieri, J.G. (1977) Rat liver α-tocopherol binding protein. *Biochimica et Biophysica Acta* 497, 349–357.

Chancy, C.D., Kekuda, R. Huang, W., Prasad, P.D., Kuhnel, J., Sirotnak, F.M., Roon, P., Ganapathy, V. and Smith, S.B. (2000) Expression and differential polarization of the reduced-folate transporter I and the folate receptor alpha in mammalian retinal pigment epithelium. *Journal of Biological Chemistry* 275, 20676–20684.

Chastain, J.L. and McCormick, D.B. (1991) Flavin metabolites. In: Muller, F. (ed.) *Chemistry and Biochemistry of Flavins*. CRC Press, Boca Raton, Florida, pp. 195–200.

Chow, C.K. (2001) Vitamin E. In: Rucker, R.B., Suttie, J.W., McCormick, D.B. and Machlin, L.J. (eds) *Handbook of Vitamins*, 3rd edn. Marcel Dekker, New York, pp. 165–197.

Coburn, S.P., Ziegler, P.J., Costill, D.L., Mahuren, J.D., Fink, W.J., Schaltenbrand, W.E., Pauly, T.A., Pearson, D.R., Conn, P.J. and Guilarte, T.R. (1991) Response of vitamin B_6 content of muscle to changes in vitamin B_6 intake in men. *American Journal of Clinical Nutrition* 53, 1436–1442.

Collins, E.D. and Norman, A.W. (2001) Vitamin D. In: Rucker, R.B., Suttie, J.W., McCormick, D.B. and Machlin, L.J. (eds) *Handbook of Vitamins*, 3rd edn. Marcel Dekker, New York, pp. 51–113.

Dialameh, G.H., Yekundi, K.G. and Olson, R.E. (1970) Enzymatic akylation of menaquinone-O to menaquinones by microsomes from chick liver. *Biochimica et Biophysica Acta* 223, 332–338.

Dietrich, L.S., Fuller, L., Yero, I.L. and Martinez, L. (1966) Nicotinamide mononucleotide pyrophosphorylase activity in animal tissues. *Journal of Biological Chemistry* 241, 188–191.

Dowd, P., Ham, S.W., Naganathan, S. and Hershline, R. (1995) The mechanism of action of vitamin K. *Annual Review of Nutrition* 15, 419–440.

Driessen, H.P.C., de Jong, W.W., Tesser, G.I. and Bloemendal, H. (1985) The mechanism of N-terminal acetylation of proteins. *CRC Critical Reviews in Biochemistry* 18, 281–325.

Esmon, C.T., Sadowski, J.A. and Suttie, J.W. (1975) A new carboxylation reaction. The vitamin K-dependent incorporation of $H^{14}CO_3^-$ into prothrombin. *Journal of Biological Chemistry* 250, 4744–4748.

Fenstermacher, D.K. and Rose, R.C. (1986) Absorption of pantothenic in rat and chick intestine. *American Journal of Physiology* 250, G155.

Foo, S.K., McSloy, R.M., Rousseau, C. and Shane, B. (1982) Folate derivatives in human cells: studies on normal and 5,10-methylenetetrahydrofolate reductase-deficient fibroblasts. *Journal of Nutrition* 112, 1600–1608.

Gardill, S.L. and Suttie, J.W. (1990) Vitamin K epoxide and quinone reductase: evidence for reduction by a common enzyme. *Biochemical Pharmacology* 40, 1055–1061.

Gordon, M.J., Campbell, F.M., Duthie, G.G. and Dutta-Roy, A.K. (1995) Characterization of a novel alpha-tocopherol-binding protein from bovine heart cytosol. *Archives of Biochemistry and Biophysics* 318, 140–146.

Griep, A.E. and Friedman, P.A. (1980) Purification of a protein containing γ-carboxyglutamic acid from bovine kidney. In: Suttie, J.W. (ed.) *Vitamin K Metabolism and Vitamin K-dependent Proteins.* University Park Press, Baltimore, Maryland, pp. 307–310.

Hannah, S.S. and Norman, A.W. (1994) $1\alpha,25$-Dihydroxyvitamin D_3 regulated expression of the eukaryotic genome. *Nutrition Reviews* 52, 376–382.

Hayashi, T., Kanetoshi, A., Nakamura, M., Tamura, M. and Shirahama, H. (1992) Reduction of alpha-tocopherolquinone to alpha-tocopherolhydroquinone in rat hepatocytes. *Biochemical Pharmacology* 44, 489–493.

Heiskala, M., Zhang, J., Hayashi, S.-I., Holtta, E. and Anderson, L.C. (2000) Activation and transformation of cells induce translocation of ornithine decarboxylase (ODC) to the surface membrane. In: Iriarte, A., Kagan, H. and Martinez-Carrion, M. (eds) *Biochemistry and Molecular Biology of Vitamin B_6 and PQQ-dependent Proteins.* Birkhauser, Basel, pp. 227–232.

Henderson, L.M. (1985) Intestinal absorption of B_6 vitamers. In: Reynolds, R.D. and Leklem, J.E. (eds) *Vitamin B_6: Its Role in Health and Disease.* Alan R. Liss, New York, pp. 22–23.

Henry, H.L. (1992) Vitamin D hydroxylases. *Journal of Cellular Biochemistry* 49, 4–9.

Henry, H.L. and Norman, A.W. (1974) Studies on calciferol metabolism. IX. Renal 25-hydroxyvitamin D_3-1-hydroxylase. Involvement of cytochrome P-450 and other properties. *Journal of Biological Chemistry* 249, 7529–7535.

Henry, H.L., Dutta, C., Cunningham, N., Blanchard, R., Penny, R., Tang, G., Marchetto, G. and Chou, S.-Y. (1992) The cellular and molecular recognition of $1,25(OH)_2D_3$ production. *Journal of Steroid Biochemistry and Molecular Biology* 41, 401–407.

Hines, J.D., Rosenberg, A. and Harris, J.W. (1968) Intrinsic factor-mediated radio-B_{12} uptake in sequential incubator studies using everted sacs of guinea pig small intestine: evidence that IF is not absorbed into the intestinal cell. *Proceedings of the Society for Experimental Biology and Medicine* 129, 653–658.

Kato, M., Blaner, W.S., Mertz, J.R., Das, K., Kato, K. and Goodman, D.S. (1985) Influence of retinoid nutritional status on cellular retinol- and cellular retinoic acid-binding protein concentrations in various rat tissues. *Journal of Biological Chemistry* 260, 4832–4838.

Kayata, S., Kindberg, Greer, F.R. and Suttie, J.W. (1989) Vitamin K_1 and K_1 in infant human liver. *Journal of Pediatrics, Gastroenterology and Nutrition* 8, 304–307.

Kazarinoff, M.N. and McCormick, D.B. (1975) Rabbit liver pyridoxamine (pyridoxine) 5'-phosphate oxidase: purification and properties. *Journal of Biological Chemistry* 250, 3436–3442.

Kight, C.E., Reedstrom, C.K. and Suttie, J.W. (1995) Identification, isolation, and partial purification of a cytosolic binding protein for vitamin K from rat liver. *FASEB Journal* 9, A725.

Kirkland, J.B. and Rawling, J.M. (2000) Niacin. In: Rucker, R.B., Suttie, J.W., McCormick, D.B. and Machlin, L.J. (eds) *Handbook of Vitamins,* 3rd edn. Marcel Dekker, New York, pp. 213–254.

Knauer, T.E., Siegfried, C.M. and Matschiner, J.T. (1976) Vitamin K requirement and the concentration of vitamin K in rat liver. *Journal of Nutrition* 106, 1747–1756.

Kozyraki, R., Kristiansen, M., Silahtaroglu, A., Hansen, C., Jacobsen, C., Tommerup, N., Verroust, P.J. and Moestrup, S.K. (1998) The human intrinsic factor-vitamin B_{12} receptor, cubilin: molecular chacterization and chromosomal mapping of the gene to 10p within the autosomal recessive megaloblastic anemia (MGA1) region. *Blood* 91, 3593–3600.

Kuhlenkamp, J., Ronk, M., Yusin, M., Stolz, A. and Kaplowitz, N. (1993) Identification and purification of a human liver cytosolic tocopherol binding protein. *Protein Expression and Purification* 4, 382–389.

Kumar, R. (1986) The metabolism and mechanism of action of 1,25-dihydroxyvitamin D_3. *Kidney International* 30, 793–803.

Lee, H.C. (2000) NAADP: an emerging calcium signaling molecule. *Journal of Membrane Biology* 173, 1–8.

Lee, S.-S. and McCormick, D.B. (1985) Thyroid hormone regulation of flavocoenzyme biosynthesis. *Archives of Biochemistry and Biophysics* 237, 197–201.

Leklem, J.E. (2001) Vitamin B_6. In: Rucker, R.B., Suttie, J.W., McCormick, D.B. and Machlin, L.J. (eds) *Handbook of Vitamins,* 3rd edn. Marcel Dekker, New York, pp. 339–396.

Li, E. and Norris, A.W. (1996) Structure/function of cytoplasmic vitamin A-binding proteins. *Annual Review of Nutrition* 16, 205–234.

Li, T.-K., Lumeng, L. and Veitch, R.L. (1974) Regulation of pyridoxal 5'-phosphate metabolism in liver. *Biochemical and Biophysical Research Communications* 61, 627–634.

Martius, C. (1961) The metabolic relationships between the different K vitamins and the synthesis of the ubiquinones. *American Journal of Clinical Nutrition* 9, 97–103.

Matschiner, J.T. (1971) Isolation and identification of vitamin K from animal tissue. In: *Symposium*

Proceedings on the Biochemistry, Assay, and Nutritional Value of Vitamin K and Related Compounds. Association of Vitamin Chemists, Chicago, pp. 21–37.

Matschiner, J.T., Bell, R.G., Amelotti, J.M. and Knauer, T.E. (1970) Isolation and chacterization of a new metabolite of phylloquinone in the rat. *Biochimica et Biophysica Acta* 201, 309–315.

Mawer, E.B., Backhouse, J., Holman, C.A., Lumb, G.A. and Stanbury, S.W. (1972) The distribution and storage of vitamin D and its metabolites in human tissues. *Clinical Science* 43, 413–431.

McCormick, D.B. (1975) Metabolism of riboflavin. In: Rivlin, R.S. (ed.) *Riboflavin*. Plenum Press, New York, pp. 153–198.

McCormick, D.B. (1976) Biotin. In: Hegsted, D.M. (ed.) *Present Knowledge in Nutrition*. The Nutrition Foundation, New York, pp. 217–225.

McCormick, D.B. (1988) Thiamin. In: Shils, M.E. and Young, V.R. (eds) *Modern Nutrition in Health and Disease*. Lea and Febiger, Philadelphia, pp. 355–361.

McCormick, D.B. (1989) Two interconnected B vitamins: riboflavin and pyridoxine. *Physiology Reviews* 69, 1170–1198.

McCormick, D.B. (1999) Riboflavin. In: Shils, M.E., Olson, J.A., Shike, M. and Ross, A.C. (eds) *Modern Nutrition in Health and Disease*. Lea and Febiger, Malvern, Philadelphia, pp. 391–399.

McCormick, D.B. (2001) Vitamin B_6. In: Bowman, B.A. and Russell, R.M. (eds) *Present Knowledge in Nutrition*. ILSI-Nutrition Foundation, Washington, DC, pp. 207–213.

McCormick, D.B. and Snell, E.E. (1959) Pyridoxal kinase of human brain and its inhibition by hydrazine derivatives. *Proceedings of the National Academy of Sciences USA* 45, 1371–1379.

McCormick, D.B. and Wright, L.D. (1970) The metabolism of biotin and its analogues. *Comprehensive Biochemistry* 21, 81–110.

McCormick, D.B. and Zhang, Z. (1992) Cellular assimilation of water-soluble vitamins in the mammal: riboflavin, B_6, biotin, and C. *Proceedings of the Society for Experimental Biology and Medicine* 202, 265–270.

McCormick, D.B., Gregory, M.E. and Snell, E.E. (1961) Pyridoxal phosphokinases I. Assay, distribution, purification, and properties. *Journal of Biological Chemistry* 236, 2076–2084.

Merrill, A.H. and Henderson, J.M. (1990) Vitamin B_6 metabolism by human liver. *Annals of the New York Academy of Sciences* 585, 110–117.

Merrill, A.H. Jr and McCormick, D.B. (1980) Affinity chromatographic purification and properties of flavokinase (ATP: riboflavin 5′phosphotransferase) from rat liver. *Journal of Biological Chemistry* 255, 1335–1338.

Merrill, A.H., Henderson, J.M., Wang, E., McDonald, B.W. and Millikin, W.J. (1984) Metabolism of vitamin B_6 by human liver. *Journal of Nutrition* 114, 1664–1674.

Moat, A.G. and Foster, J.W. (1982) Biosynthesis and salvage pathways of pyridine nucleotides. In: Dolphin, D., Powanda, M. and Poulson, R. (eds) *Pyridine Nucleotide Coenzymes: Chemical, Biochemical and Medical Aspects*. John Wiley & Sons, New York, Part B, pp. 1–24.

Mock, D.M. (2001) Biotin. In: Rucker, R.B., Suttie, J.W., McCormick, D.B. and Machlin, L.J. (eds) *Handbook of Vitamins*, 3rd edn. Marcel Dekker, New York, pp. 427–462.

Moscow, J.A., Gong, M., He, R., Sgagias, M.K., Dixon, K.H., Anzick, S.L., Meltzer, P.S. and Cowan, K.H. (1995) Isolation of a gene encoding a human reduced folate carrier (RFC1) and analysis of its expression in transport-deficient methotrexate-resistant human breast cancer cells. *Cancer Research* 55, 3790–3794.

Muniyappa, K., Murphy, U.S. and Adiga, P.R. (1978) Estrogen induction of thiamin carrier protein in chicken liver. *Journal of Steroid Biochemistry* 9, 888.

Murphy, M., Keating, M., Boyle, P., Weir, D.G. and Scott, J.M. (1976) The elucidation of the mechanism of folate catabolism in the rat. *Biochemical and Biophysical Research Communications* 71, 1017–1024.

Nemere, I. and Norman, A.W. (1987) Studies on the mode of action of calciferol. LII. Rapid action of 1,25-dihydroxyvitamin D_3 on calcium transport in perfused chick duodenum: effect of inhibitors. *Journal of Bone and Mineral Research* 2, 99–107.

Newcomer, M.E. (1995) Retinoid-binding proteins: structural determinants important for function. *FASEB Journal* 9, 229–239.

Nyquist, S.E., Matschiner, J.T. and James Morre, D.J. (1971) Distribution of vitamin K among rat liver cell fractions. *Biochimica et Biophysica Acta* 244, 645–649.

Oka, M. and McCormick, D.B. (1987) Complete purification and general characterization of FAD synthetase from rat liver. *Journal of Biological Chemistry* 262, 7418–7422.

Olson, J.A. (2001) Vitamin A. In: Rucker, R.B., Suttie, J.W., McCormick, D.B. and Mzchlin, L.J. (eds) *Handbook of Vitamins*, 3rd edn. Marcel Dekker, New York, pp. 1–50.

Ong, D.E., Newcomer, M.E. and Chytil, F. (1994) Cellular retinoid-binding proteins. In: Sporn, M.B., Roberts, A.B. and Goodman, D.S. (eds) *The Retinoids: Biology, Chemistry, and Medicine*, 2nd edn. Raven Press, New York, pp. 283–317.

Paquin, J., Baugh, C.M. and MacKenzie, R.E. (1985) Channeling between the active sites of formiminotransferase-cyclodeaminase. Binding and kinetic studies. *Journal of Biological Chemistry* 260, 14925–14931.

Plesofsky, N.S. (2001) Pantothenic acid. In: Rucker, R.B., Suttie, J.W., McCormick, D.B. and Machlin, L.J.

(eds) *Handbook of Vitamins*, 3rd edn. Marcel Dekker, New York, pp. 317–337.

Preiss, J. and Handler, P. (1958) Biosynthesis of diphosphopyridine nucleotide I. Identification of intermediates. *Journal of Biological Chemistry* 233, 488–500.

Reedstrom, C.K. and Suttie, J.W. (1995) Comparative distribution, metabolism, and utilization of phylloquinone and menaquinone-9 in rat liver. *Proceedings of the Society for Experimental Biology and Medicine* 209, 403–409.

Rindi, G. and Laforenza, U. (1997) *In vitro* systems for studying thiamin transport in mammals. *Methods in Enzymology* 279, pp. 118–131.

Rivlin, R. (1970) Medical progress: riboflavin metabolism. *New England Journal of Medicine* 283, 463.

Rivlin, R. and Pinto, J.T. (2001) Riboflavin (Vitamin B_2). In: Rucker, R.B., Suttie, J.W., McCormick, D.B. and Machlin, L.J. (eds) *Handbook of Vitamins*, 3rd edn. Marcel Dekker, New York, pp. 255–273.

Robishaw, J.D., Berkich, D. and Neely, J.R. (1982) Rate-limiting step and control of coenzyme A synthesis in cardiac muscle. *Journal of Biological Chemistry* 257, 10967–10972.

Rothenberg, S.P., Weisberg, H. and Ficarra, A. (1972) Evidence for the absorption of immunoreactive intrinsic factor into the intestinal epithelial cell during vitamin B_{12} absorption. *Journal of Laboratory and Clinical Medicine* 79, 587–597.

Saarem, K., Bergseth, S., Oftebro, H. and Pedersen, J.I. (1984) Subcellular localization of vitamin D_3 25-hydroxylase in human liver. *Journal of Biological Chemistry* 259, 10936–10940.

Saari, J.C. (1994) Retinoids in photosensitive systems. In: Sporn, M.B., Roberts, A.B. and Goodman, D.E. (eds) *The Retinoids: Biology, Chemistry, and Medicine*, 2nd edn. Raven Press, New York, pp. 351–385.

Said, H.M., Redha, R. and Nylander, W. (1988) Biotin transport in basal lateral membrane vesicles of human intestine. *Gastroenterology* 94, 1157–1163.

Said, H.M., Nguyen, T.T., Dyer, D.L., Cowan, K.H. and Rubin, S.A. (1996) Intestinal folate transport: identifiction of a cDNA involved in folate transport and functional expression and distribution of its mRNA. *Biochimica et Biophysica Acta* 1281, 164–172.

Sato, Y., Hagiwara, K., Arai, H. and Inoue, K. (1991) Purification and characterization of the α-tocopherol transfer protein from rat liver. *FEBS Letters* 288, 41–45.

Shane, B. (1989) Folylpolyglutamate synthesis and role in the regulation of one-carbon metabolism. In: Aurbach, G.D. and McCormick, D.B. (eds) *Vitamins and Hormones*. Academic Press, Vol. 45, pp. 263–335.

Shearer, M.J., McCarthy, P.T., Crampton, O.E. and Mattock, M.B. (1988) The assessment of human vitamin K status from tissue measurements. In: Suttie, J.W. (ed.) *Current Advances in Vitamin K Research*. Elsevier, New York, pp. 437–452.

Sheppard, A.J., Pennington, J.A.T. and Weihrauch, J.L. (1993) Analysis and distribution of vitamin E in vegetable oils and foods. In: Packer, L. (ed.) *Vitamin E in Health and Disease*. Marcel Dekker, New York, pp. 9–31.

Silverman, R.B. and Nandi, D.L. (1988) Reduced thioredoxin: a possible physiological cofactor for vitamin K epoxide reductase. Further support for an active site disulfide. *Biochemical and Biophysical Research Communications* 155, 1248–1254.

Sorrell, M.F., Frank, O., Thomson, A.D., Aquino, H. and Baker, H. (1971) Absorption of vitamins from the large intestine *in vivo*. *Nutrition Reports International* 3, 143–148.

Soute, B.A.M., Muller-Ester, W., de Boer-van den Berg, M.A.G., Ulrich, M. and Vermeer, C. (1985) Discovery of a γ-carboxyglutamic acid-containing protein in human spermatozoa. *FEBS Letters* 190, 137–141.

Stratton, S., Mock, N. and Mock, D. (1996) Biotin and biotin metabolites in human milk: the metabolites are not negligible. *Journal of Investigative Medicine*

Suttie, J.W. (2001) Vitamin K. In: Rucker, R.B., Suttie, J.W., McCormick, D.B. and Machlin, L.J. (eds) *Handbook of Vitamins*, 3rd edn. Marcel Dekker, New York, pp. 115–164.

Tanphaichitr, V. (2001) Thiamine. In: Rucker, R.B., Suttie, J.W., McCormick, D.B. and Machlin, L.J. (eds) *Handbook of Vitamins*, 3rd edn. Marcel Dekker, New York, pp. 275–316.

Thijssen, H.H.W. and Drittij-Reijnders, M.J. (1994) Vitamin K distribution in rat tissues: dietary phylloquinone is a source of tissue menaquinone-4. *British Journal of Nutrition* 72, 415–425.

Towler, D.A., Adams. S.P., Eubanks, S.R., Towery, D.S., Jackson-Machelski, E., Glaser, L. and Gordon, J.I. (1987) Purification and characterization of yeast myristoyl CoA: protein *N*-myristoyltransferase. *Proceedings of the National Academy of Sciences USA* 84, 2708–2712.

Traber, M.G. and Sies, H. (1996) Vitamin E in humans: demand and delivery. *Annual Review of Nutrition* 16, 321–347.

Van Haarlem, L.J.M., Soute, B.A.M. and Vermeer, C. (1987) Vitamin K-dependent carboxylase. Possible role for thioredoxin in the reduction of vitamin K metabolites in liver. *FEBS Letters* 222, 353.

Whitehouse, W.S.A., Merrill, A.H. Jr and McCormick, D.B. (1991) Riboflavin-binding protein. In: Muller, F. (ed.) *Chemistry and Biochemistry of Flavins*. CRC Press, Boca Raton, Florida, pp. 287–292.

Wu, S.-M., Morris, D.P. and Stafford, D. (1991a) Identification and purification to near homogeneity of the vitamin K-dependent carboxylase. *Proceedings of the National Academy of Sciences USA* 88, 2236–2240.

Wu, S.-M., Cheung, W.-F., Frazier, D. and Stafford, D. (1991b) Cloning and expression of the cDNA for human γ-glutamyl carboxylase. *Science* 254, 1634–1636.

Yagi, K., Nakaga, Y., Suzuki, O. and Ohishi, N. (1976) Incorporation of riboflavin into covalently-bound flavins in rat liver. *Journal of Biochemistry* 79, 841–843.

Zempleni, J., McCormick, D.B. and Mock, D.M. (1996) The identification of biotin sulfone, bisnorbiotin methyl ketone, and tetranorbiotin-*l*-sulfoxide in human urine. *American Journal of Clinical Nutrition* 65, 508–511.

Zolkiewska, A., Okazaki, I.J. and Moss, J. (1994) Vertebrate mono-ADP-ribosyltransferases. *Molecular and Cellular Biochemistry* 138, 107–112.

5 Nutrient Homeostasis in Proliferating Cells

Janos Zempleni
Department of Nutritional Science and Dietetics, University of Nebraska-Lincoln, Lincoln, Nebraska, USA

The Mammalian Cell Cycle

The cell cycle is an ordered set of processes by which one cell grows and divides into two daughter cells: cell proliferation (Murray and Hunt, 1993). Two fundamentally different cell cycles exist in mammals: (i) the mitotic cell cycle, in which two diploid daughter cells are generated; and (ii) the meiotic cell cycle, in which four haploid daughter cells are produced (generation of sperm and egg). This chapter will focus exclusively on the mitotic cell cycle.

The absence of nutrients or growth factors causes cells to enter a specialized resting state called G_0, during which they do not divide (Fig. 5.1) (Murray and Hunt, 1993). If supply with nutrients is sufficient, cells are recruited into the cell cycle by stimulation with compounds such as hormones, antigens and mitogens. The mitotic cell cycle can be thought of as occurring in two major sections: interphase (which includes phases G_1, S and G_2) and mitosis. The cycle ends with the division of the cell.

1. *Interphase.* The interphase is composed of three phases (Murray and Hunt, 1993): during the G_1 phase, processes such as synthesis of membranes, mitochondria and most cellular proteins occur. During S phase, DNA is synthesized for duplication of chromosomes. Upon completion of S phase, cells enter G_2 phase in which growth processes prepare the cell for mitosis ('M').
2. *Mitosis.* Mitosis is subdivided into four phases (Murray and Hunt, 1993): during the first phase (prophase), the chromosomes condense (Fig. 5.2). In prometaphase, the membrane of the nucleus breaks down, and the chromosomes attach to microtubules, which are connected to centrosomes. In metaphase, the microtubule-bound chromosomes line up midway between the two centrosomes. Dissolution of the linkage between sister chromatids marks the beginning of anaphase; sister chromatids separate and move away from each other to opposite poles of the spindle. Finally, cytokinesis (the physical process of cell division) begins.

Cell proliferation is associated with a dramatic increase in the expression of certain genes. For example, mitogen-induced proliferation of human lymphocytes causes increased expression of genes such as c-*myc*, 4F1, JE-3, KC-1 (Kaczmarek *et al.*, 1985), HsRAD51 (Flygare *et al.*, 1996) and P120 (Wilson and Freeman, 1996). Likewise, cell proliferation is associated with increased rates of

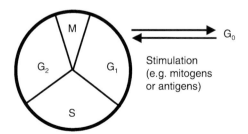

Fig. 5.1. The cell cycle. Pie sections approximately parallel the percentage duration of the phases of the cell cycle. G_0 = resting state; G_1 = gap 1 phase; S = DNA synthesis; G_2 = gap 2 phase; M = mitosis.

©CAB *International* 2003. *Molecular Nutrition*
(eds J. Zempleni and H. Daniel)

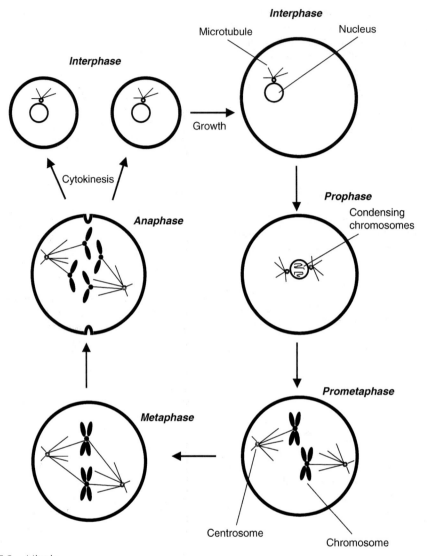

Fig. 5.2. Mitosis.

metabolic flux through many metabolic pathways. For example, cell proliferation requires cell growth (e.g. synthesis of cell membranes) and doubling of DNA (Murray and Hunt, 1993). Nutrients are essential coenzymes (e.g. B vitamins), energy-providing substrates (e.g. glucose) and precursors (e.g. essential fatty acids) in many of the metabolic pathways that are essential for cell proliferation. In this chapter, we will provide evidence that cell proliferation is associated with increased expression of genes involved in nutrient uptake and metabolism. Furthermore, we will provide evidence that cell proliferation may generate an increased demand for nutrients. If not met, this increased demand may reduce cellular proliferation rates and impair cellular function.

Biotin as a Model for Studies of Nutrient Homeostasis in Proliferating Cells

The vitamin biotin will be used as a model to illustrate the effects of cell proliferation on cellular nutrient uptake and utilization. We chose biotin as a model because both biotin transport and biotin utilization in mammalian cells have been characterized sufficiently well to provide a basis for the analysis of effects of cell proliferation on biotin homeostasis. In this section, the cellular transport and utilization of biotin in quiescent, non-proliferating cells are reviewed to provide this basis for the reader.

Cellular biotin transport

Pioneering studies of biotin transport in rat liver cells suggested that biotin uptake is a transporter-mediated process that is energy dependent, sodium dependent and Na-K-ATPase dependent (Bowers-Komro and McCormick, 1985). Similar characteristics of biotin uptake were described for human-derived HepG2 cells (Said *et al.*, 1994) and the placental brush border membrane (Karl and Fisher, 1992; Schenker *et al.*, 1993; Hu *et al.*, 1994). Recently, a protein responsible for the transport of biotin, pantothenic acid and lipoic acid across rat placenta (Prasad *et al.*, 1998) and human placenta (Wang *et al.*, 1999) has been cloned and functionally expressed. This transporter has been named the sodium-dependent multivitamin transporter (SMVT). mRNA encoding SMVT subsequently has been detected in various tissues such as intestine, kidney, liver, brain, lung, heart, pancreas and skeletal muscle (Prasad *et al.*, 1998; Wang *et al.*, 1999).

Notwithstanding the important role of SMVT in cellular uptake of biotin, other transporters in human tissues might also account for (some) biotin uptake. For example, human mononuclear cells accumulate biotin by a transporter that appears to be specific for biotin; pantothenic acid and lipoic acid do not compete with biotin for binding to this transporter (Zempleni and Mock, 1998, 1999b). The biotin transporter in human mononuclear cells has not yet been cloned and sequenced. Thus, the degree of homology between SMVT and the biotin transporter in mononuclear cells is unknown. Preliminary evidence has been provided that monocarboxylate transporters might account for some biotin transport in human mononuclear cells (see below). The same transporter that mediates biotin uptake into mononuclear cells also accounts for biotin efflux from these cells, as evidenced by countertransport experiments (Zempleni and Mock, 1999a).

Biotin-dependent carboxylases

In mammals, biotin serves as a covalently bound coenzyme for the following four carboxylases: acetyl-CoA carboxylase (EC 6.4.1.2), pyruvate carboxylase (EC 6.4.1.1), propionyl-CoA carboxylase (EC 6.4.1.3) and 3-methylcrotonyl-CoA carboxylase (EC 6.4.1.4) (Wood and Barden, 1977; Knowles, 1989). The covalent attachment of biotin to the ε-amino group of lysine in each of the apocarboxylases is catalysed by holocarboxylase synthetase (EC 6.3.4.10; Fig. 5.3) (Dakshinamurti and Chauhan, 1994). The four mammalian carboxylases catalyse essential steps in important metabolic pathways such as fatty acid synthesis, amino acid metabolism, odd chain fatty acids and gluconeogenesis (Zempleni, 2001).

Proteolytic degradation of holocarboxylases leads to the formation of biotinyl peptides. These peptides are degraded by biotinidase (EC 3.5.1.12) to release biocytin (biotinyl-ε-lysine). Finally, biotinidase releases biotin from biocytin for recycling into new holocarboxylases (Wolf *et al.*, 1985).

Biotinylation of histones

Histones are the primary proteins that mediate the folding of DNA into chromatin (Wolffe, 1998). The binding of DNA to histones is of an electrostatic nature; binding is mediated by the association of negatively charged phosphate groups of DNA with positively charged ε-amino groups (lysine residues) and guanidino groups (arginine residues) of histones.

In vivo, histones are modified post-translationally by acetylation (Ausio and Holde, 1986; Hebbes *et al.*, 1988; Lee *et al.*, 1993), methylation (Wolffe, 1998), phosphorylation (Wolffe, 1998), ubiquitination (Wolffe, 1998), and poly(ADP-ribosylation) (Chambon *et al.*, 1966; Boulikas, 1988; Boulikas *et al.*, 1990). These groups are attached covalently to amino acid residues of histones such as ε-amino groups in lysine residues and guanidino groups in arginine residues. Some of these modifications (e.g. acetylation and poly (ADP-ribosylation)) take away one positive charge from histones, leading to weakened association between DNA and histones. Evidence has been provided that some of these histone modifications correlate with increased transcription of DNA (e.g. acetylation of histones) and DNA repair mechanisms (poly(ADP-ribosylation) of histones) (Wolffe, 1998).

Recently, Hymes et al. (1995) have proposed a reaction mechanism by which the enzyme biotinidase mediates covalent binding of biotin to histones, leading to a decrease in the number of positively charged amino groups of histones. Biotinylation of histones suggests that the vitamin biotin might play some role in transcription, replication or repair of DNA in analogy to the effects of acetylation, methylation, phosphorylation, ubiquitination and ADP-ribosylation of histones. The enzymatic mechanism leading to biotinylation of histones and the potential physiological roles of this process are discussed in depth in Chapter 17.

Peripheral Blood Mononuclear Cells (PBMCs) as a Cell Model

This chapter provides a review of recent studies that have directly addressed the interaction between cell proliferation and biotin metabolism. Most of these studies used PBMCs as a cell model. PBMCs represent a heterogeneous population of immune cells (B cells, T cells and various granulocytes) that arise from pluripotent haematopoietic stem cells in the bone marrow (Janeway et al., 1999).

Why are PBMCs a good model to study the effects of cell proliferation on biotin homeostasis? PBMCs are easily obtainable from human peripheral blood; results from studies using these human cells directly apply to human nutrition. Freshly isolated PBMCs typically are quiescent; some PBMCs (B and T cells) reliably will proliferate rapidly after antigenic or mitogenic stimulation. Biotin homeostasis in quiescent human PBMCs (uptake, metabolism and efflux) has been well characterized (Velazquez et al., 1990, 1995; Zempleni and Mock, 1998, 1999a, b), providing a solid basis for studies of proliferating PBMCs.

Effects of cell proliferation on biotin uptake

Rates of biotin uptake

Mitogen-stimulated, proliferating PBMCs accumulate biotin at a rate five times faster than unstimulated controls (Zempleni and Mock,

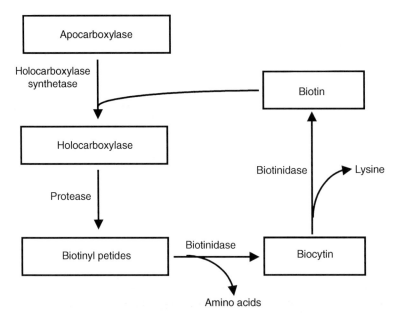

Fig. 5.3. Utilization and recycling of biotin.

1999c). For example, biotin uptake into PBMCs increased in a dose-dependent fashion from 481 to 722% of the control value in quiescent cells if proliferation was induced by incubation with pokeweed mitogen (lectin from *Phytolacca americana*) for 3 days (Fig. 5.4). A similar pattern of biotin uptake was observed when proliferation was induced by either concanavalin A or phytohaemagglutinin.

Next, PBMCs were incubated with mitogen for up to 4 days, and biotin transport rates were determined at timed intervals. Stimulation of biotin uptake paralleled the rate of cell proliferation as judged by uptake of the proliferation marker thymidine into PBMCs (Zempleni and Mock, 1999c). Uptake of both biotin and thymidine was maximal 48–72 h after addition of pokeweed mitogen to the culture medium (Zempleni and Mock, 1999c). These data provide evidence that increased biotin transport into mitogen-stimulated PBMCs is caused by increased biotin demand due to proliferation. Moreover, short-term exposure (<15 min) of PBMCs to mitogens did not affect transport rates of biotin. This suggests that mitogens do not increase biotin transport by chemically interacting with biotin or transporters on the cell surface, but that effects of long-term stimulation with mitogens (1–4 days) are mediated by induction of cell proliferation.

Biotin uptake at specific phases of the cell cycle

The studies described in the previous section were conducted using non-synchronized cultures of PBMCs, i.e. populations of cells from various phases of the cell cycle. Thus, these studies did not allow identification of a specific phase of the cell cycle during which biotin uptake is increased: G_1, S, G_2 or M phase. This gap in knowledge was addressed in another series of studies (J.S. Stanley *et al.*, 2002): proliferating PBMCs were arrested at designated phases of the cell cycle by incubation with one of the following chemicals: (i) wortmannin (100 nmol l^{-1}) to cause G_1 arrest; (ii) aphidicolin (118 μmol l^{-1}) to cause arrest S arrest; (iii) doxorubicin (1 μmol l^{-1}) to cause G_2 arrest; or (iv) colchicine (5 μmol l^{-1}) to cause M arrest; quiescent PBMCs (G_0 phase) served as controls. Biotin uptake rates into these cells were measured using [^{3}H]biotin.

Biotin uptake rates increased in the G_1 phase of the cell cycle (4.5 times the control value in G_0 phase) and remained increased during later phases of the cell cycle (S, G_2 and M phase; Fig. 5.5). Differences among G_1, S, G_2, and M phase were not significant. These findings suggest (i) that the number of biotin transporters on the cell surface increases early in the cell cycle (see below); and (ii) that the number of transporters is increased during the entire course of the cell cycle.

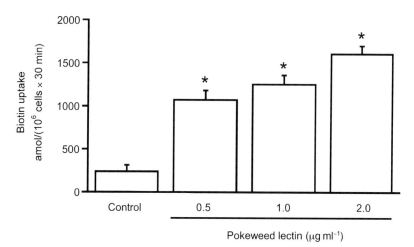

Fig. 5.4. Uptake of [^{3}H]biotin into proliferating human PBMCs and quiescent controls. Proliferation was induced by incubation with pokeweed lectin (0.5, 1.0 or 2.0 μg ml^{-1}) at 37°C for 3 days; controls were incubated without pokeweed lectin. Values are means ± 1 SD; $n = 6$. *$P < 0.01$ vs. control by ANOVA with Dunnett's procedure (Zempleni and Mock, 1999c).

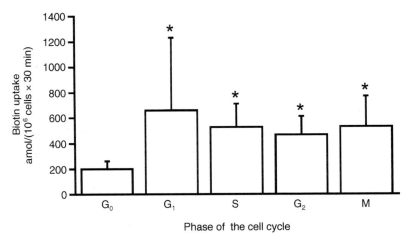

Fig. 5.5. Uptake of [^{3}H]biotin into cell cycle-arrested human PBMCs. Values are means ± SD; $n = 6$. *Significantly different from controls ($P < 0.01$ vs. control) (Stanley et al., 2002).

Transporter synthesis

Is the increased biotin uptake into proliferating PBMCs due to an increased number of biotin transporters on the cell surface or to an increased affinity of transporters for biotin? Kinetic data suggest that the increased uptake of biotin is caused by an increased number of transporters. The maximal transport rate (V_{max}) of biotin uptake provides an estimate for the number of biotin transporters on the PBMC surface. Theoretically, increased synthesis of biotin transporters in proliferating PBMCs is paralleled by increased V_{max} for biotin. V_{max} of biotin uptake in proliferating PBMCs was approximately four times greater compared with non-proliferating PBMCs: 9.1 ± 6.6 versus 2.3 ± 1.6 fmol/(10^6 cells × 30 min) (Zempleni and Mock, 1999c). Transporter affinity for biotin (as judged by the Michaelis–Menten constant) was not significantly different in proliferating and quiescent PBMCs. In summary, these findings suggest that the increased biotin uptake into proliferating PBMCs is mediated by increased synthesis of transporters rather than by increased affinity for substrate.

Analysis of SMVT gene expression as judged by mRNA levels is consistent with increased synthesis of SMVT in proliferating PBMCs. SMVT gene expression was determined by reverse transcriptase–polymerase chain reaction (RT–PCR) of total RNA isolated from quiescent and proliferating PBMCs; certain housekeeping genes were quantified for comparison (Zempleni et al., 2001). The concentration of RNA encoding SMVT in proliferating PBMCs was 10.6 ± 1.6 times the control value in quiescent cells (Fig. 5.6). The increased expression of SMVT in proliferating PBMCs exceeded the increased expression of the housekeeping genes encoding transferrin receptor (5.6 ± 6.4 times control), glyceraldehyde-3-phosphate dehydrogenase (1.0 ± 0.1 times control), histone H1.3 (1.8 ± 0.1 times control) and histone H4 (1.1 ± 0.1 times control); typical examples are depicted in Fig. 5.6.

Theoretically, the increased biotin uptake in proliferating PBMCs could also be caused (or supplemented) by increased synthesis of a transporter other than SMVT. However, the following lines of evidence suggest that proliferating and quiescent PBMCs express the same biotin transporter. (i) The transporter affinity for biotin (as judged by the Michaelis–Menten constant) is the same in proliferating and quiescent PBMCs (Zempleni and Mock, 1999c). (ii) The substrate specificity of the biotin transporter is the same in proliferating and quiescent PBMCs (Zempleni and Mock, 1999c). (iii) mRNA encoding SMVT has been identified in both proliferating and quiescent PBMCs (Zempleni et al., 2001).

Notwithstanding the important role of SMVT in biotin transport, preliminary evidence has been provided that other transporters also might account for some biotin uptake into PBMCs. For example, certain organic acids such as hexanoate, pyruvate and acetate compete with biotin for cellular uptake (J. Zempleni, unpublished observation).

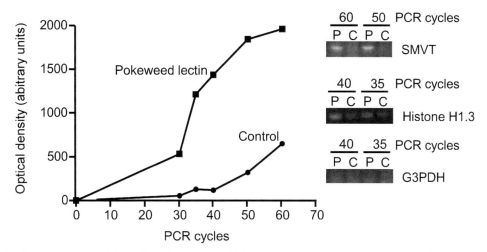

Fig. 5.6. Expression of the sodium-dependent multivitamin transporter (SMVT) gene in human PBMCs. Cells were incubated with 2.0 µg ml^{-1} pokeweed lectin (P) for 48 h; controls (C) were incubated without mitogen. Total RNA was extracted and gene sequences amplified by RT–PCR using primers that were specific for SMVT; histone H1.3 and glyceraldehyde-3-phosphate dehydrogenase (G3PDH) were quantitated by RT–PCR as controls (Zempleni et al., 2001).

These organic acids are substrates for monocarboxylate transporters in mammalian cells (Halestrap and Price, 1999). Moreover, two monocarboxylate transporters (MCT4 and MCT7) from human blood cells have been cloned and sequenced (Halestrap and Price, 1999). In summary, these preliminary findings are consistent with the hypothesis that monocarboxylate transporters might account for some biotin transport in PBMCs.

Effects of cell proliferation on expression of genes encoding biotin-dependent carboxylases and holocarboxylase synthetase

Carboxylase activities

PBMCs respond to mitogen-induced proliferation with increased activities of biotin-dependent carboxylases (Stanley et al., 2002). Three days after addition of pokeweed lectin to the culture medium, the activity of 3-methylcrotonyl-CoA carboxylase was 2.8 ± 0.8 times the activity in quiescent controls (Fig. 5.7). The increase of carboxylase activity was similar (3.1 ± 1.0 times control values) when proliferation was induced with concanavalin A. The peak activity of 3-methylcrotonyl-CoA carboxylase coincided with maximal rates of biotin uptake into proliferating PBMCs.

The activity of propionyl-CoA carboxylase in PBMCs also increased in response to mitogen-stimulated proliferation (Stanley et al., 2002). However, the increase of activity was less than the increase observed for 3-methylcrotonyl-CoA carboxylase. Three days after addition of pokeweed lectin to the culture medium, the activity of propionyl-CoA carboxylase was 1.5 ± 0.5 times the activity in quiescent controls (Fig. 5.8). Stimulation with a different mitogen had a similar effect: 3 days after addition of concanavalin A to the medium, the activity of propionyl-CoA carboxylase was 1.2 ± 0.2 times control values. The activities of two other carboxylases (acetyl-CoA carboxylase and pyruvate carboxylase) were too low to be detectable.

These findings are consistent with the hypothesis that proliferation of PBMCs generates an increased demand for biotin in order to increase carboxylase activities. The increased demand is met by increasing the cellular uptake of biotin.

Levels of mRNA encoding biotin-dependent carboxylases and holocarboxylase synthetase

Probably, cells increase carboxylase activities by increasing the biotinylation of apocarboxylases to produce holocarboxylases. Theoretically, the apocarboxylases could come from either one of the following two sources: (i) pre-formed apocarboxylases from the cellular carboxylase pool; or (ii) new apocarboxylases that have been synthesized specifically in response to cell proliferation. Most probably, the latter would be associated with an increase of mRNA levels encoding biotin-dependent carboxylases. The studies summarized here quantified levels of mRNA encoding

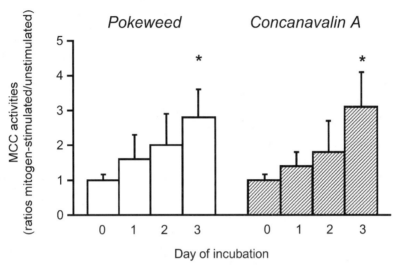

Fig. 5.7. Activity of 3-methylcrotonyl-CoA carboxylase (MCC) in mitogen-stimulated human PBMCs. PBMCs were incubated with pokeweed lectin (2.0 µg ml^{-1}) or concanavalin A (20 µg ml^{-1}) for the indicated times. At timed intervals, the activity of MCC was measured. Values are means ± 1 SD; $n = 6$. *Significantly different from controls ($P < 0.01$ vs. time zero) (Stanley et al., 2002).

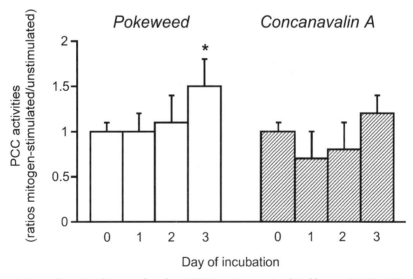

Fig. 5.8. Activity of propionyl-CoA carboxylase (PCC) in mitogen-stimulated human PBMCs. PBMCs were incubated with pokeweed lectin (2.0 µg ml^{-1}) or concanavalin A (20 µg ml^{-1}) for the indicated times. At timed intervals, the activity of PCC was measured. Values are means ± 1 SD; $n = 6$. *Significantly different from controls ($P < 0.05$ vs. time zero) (Stanley et al., 2002).

biotin-dependent carboxylases in order to determine whether PBMCs respond to cell proliferation with increased synthesis of new apocarboxylases.

Levels of mRNA encoding biotin-dependent carboxylases were greater in proliferating PBMCs than in quiescent controls (Stanley et al., 2002). For example, the level of mRNA encoding propionyl-CoA carboxylase in PBMCs that were stimulated with pokeweed lectin was approximately 4.2 times the mRNA levels in quiescent controls, as judged by quantitative PCR (Fig. 5.9). Likewise, mRNA encoding 3-methylcrotonyl-CoA carboxylase in stimulated PBMCs was 2.3 times that in quiescent controls (data not shown). It remains uncertain whether these increases were due to increased transcription, decreased mRNA degradation or both.

Holocarboxylase synthetase catalyzes the covalent binding of biotin to apocarboxylases to form holocarboxylases. Levels of mRNA encoding holocarboxylase synthetase in mitogen-stimulated, proliferating PBMCs were approximately 2.4 times the level in quiescent controls (Fig. 5.10). This finding is consistent with the hypothesis that PBMCs respond to proliferation with increased expression of the holocarboxylase synthetase gene in order to increase binding of biotin to carboxylases. Levels of mRNA encoding housekeeping genes (glyceraldehyde-3-phosphate dehydrogenase and histone H4) were the same in proliferating and quiescent PBMCs, suggesting that the effects observed for biotin-dependent carboxylases and holocarboxylase synthetase were not artefacts due to unequal sample loading on the gels.

In summary, the findings presented here provide evidence that PBMCs respond to proliferation with increased *de novo* synthesis of 3-methylcrotonyl-CoA carboxylase, propionyl-CoA carboxylase and holocarboxylase synthetase. Subsequently, holocarboxylase synthetase mediates biotinylation of the newly synthesized 3-methylcrotonyl-CoA carboxylase and propionyl-CoA carboxylase.

Effects of cell proliferation on biotinylation of histones

Wolf and co-workers proposed a reaction mechanism by which biotinidase catalyzes biotinylation of histones (Hymes et al., 1995). Subsequently, it was shown that biotinylation of histones occurs *in vivo* and that biotinylation of histones increases in response to cell proliferation (Stanley et al., 2001). Biotinylation of histones increases by approximately fourfold during G_1, S, G_2 and M

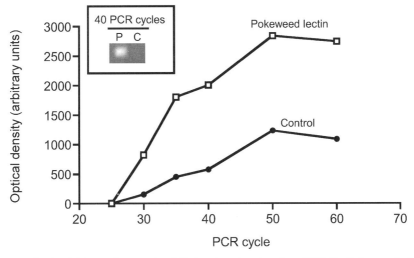

Fig. 5.9. Levels of mRNA encoding propionyl-CoA carboxylase in cultured PBMCs. Cells were isolated from healthy adults and cultured in mitogen-containing medium (2.0 μg ml^{-1} pokeweed lectin) for 3 days; controls were cultured without pokeweed lectin. Total RNA was isolated and mRNA encoding propionyl-CoA carboxylase was quantitated by RT–PCR. P = pokeweed lectin; C = control (Stanley et al., 2002).

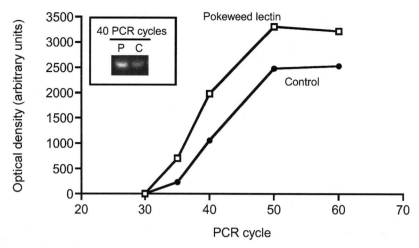

Fig. 5.10. Levels of mRNA encoding holocarboxylase synthetase in cultured PBMCs. Cells were isolated from healthy adults and cultured in mitogen-containing medium (2.0 µg ml^{-1} pokeweed lectin) for 3 days; controls were cultured without pokeweed lectin. Total RNA was isolated and mRNA encoding holocarboxylase synthetase was quantitated by RT–PCR. P = pokeweed lectin; C = control (Stanley et al., 2002).

phase of the cell cycle compared with G_0 phase. This finding is consistent with the hypothesis that increased biotinylation of histones in proliferating PBMCs generates an increased demand for biotin. Biotinylation of histones in quiescent and proliferating PBMCs is discussed extensively in Chapter 17.

Nutrient Status Affects Cell Proliferation

This chapter provides evidence that cell proliferation generates an increased demand for biotin due to increased biotinylation of carboxylases and histones in proliferating cells. Proliferating cells meet this increased demand by increasing cellular uptake of biotin. Likewise, cells respond to proliferation with increased uptake of other nutrients such as NAD (Williams et al., 1985), riboflavin (Zempleni and Mock, 2000), pantothenic acid (Zempleni et al., 2001), cobalamin (Hall, 1984) and various amino acids (Carter and Halvorson, 1973; van den Berg and Betel, 1973). These findings are consistent with an increased demand for these nutrients in proliferating cells. If not met, this increased demand may cause decreased cellular proliferation rates.

Cell culture experiments provide important information regarding cell cycle arrest in nutrient-deficient media. For example, HeLa cells arrest in G_0 phase if incubated in serine-free medium (Dakshinamurti et al., 1985). In mitogen-stimulated lymphocytes, 14 amino acids are essential to maintain protein synthesis and a normal rate of proliferation (Waithe et al., 1975). Likewise, biotin and linoleic acid are essential for cell proliferation (Dakshinamurti et al., 1985; Lernhardt, 1990); biotin stimulates the production of an unidentified growth factor in cells (Moskowitz and Cheng, 1985). Notwithstanding the essential role of nutrients in cell proliferation, supraphysiological concentrations of some nutrients in tissue culture medium may cause decreased rates of cell growth and proliferation (Tsai and Gardner, 1994). Taken together, cells that depend on rapid proliferation for normal function may be severely affected by nutrient deficiency or toxicity. Such cells include intestinal tissues, fetal cells, and immune cells after antigenic stimulation.

Acknowledgements

Supported by the National Institutes of Health (DK 60447) and USDA/CSREES project award 2001-35200-10187.

References

Ausio, J. and van Holde, K.E. (1986) Histone hyperacetylation: its effect on nucleosome conformation and stability. *Biochemistry* 25, 1421–1428.

Boulikas, T. (1988) At least 60 ADP-ribosylated variant histones are present in nuclei from dimethylsulfate-treated and untreated cells. *EMBO Journal* 7, 57–67.

Boulikas, T., Bastin, B., Boulikas, P. and Dupuis, G. (1990) Increase in histone poly(ADP-ribosylation) in mitogen-activated lymphoid cells. *Experimental Cell Research* 187, 77–84.

Bowers-Komro, D.M. and McCormick, D.B. (1985) Biotin uptake by isolated rat liver hepatocytes. In: Dakshinamurti, K. and Bhagavan, H.N. (eds) *Biotin*. New York Academy of Sciences, New York, pp. 350–358.

Carter, B.L.A. and Halvorson, H.O. (1973) Periodic changes in rate of amino acid uptake during yeast cell cycle. *Journal of Cell Biology* 58, 401–409.

Chambon, P., Weill, J.D., Doly, J., Strosser, M.T. and Mandel, P. (1966) On the formation of a novel adenylic compound by enzymatic extracts of liver nuclei. *Biochemical and Biophysical Research Communications* 25, 638–643.

Dakshinamurti, K. and Chauhan, J. (1994) Biotin-binding proteins. In: Dakshinamurti, K. (ed.) *Vitamin Receptors: Vitamins as Ligands in Cell Communication*. Cambridge University Press, Cambridge, pp. 200–249.

Dakshinamurti, K., Chalifour, L.E. and Bhullar, R.J. (1985) Requirement for biotin and the function of biotin in cells in culture. In: Dakshinamurti, K. and Bhagavan, H.N. (eds) *Biotin*. New York Academy of Sciences, New York, pp. 38–54.

Flygare, J., Benson, F. and Hellgren, D. (1996) Expression of the human RAD51 gene during the cell cycle in primary human peripheral blood lymphocytes. *Biochimica et Biophysica Acta* 1312, 231–236.

Halestrap, A.P. and Price, N.T. (1999) The proton-linked monocarboxylate transporter (MCT) family: structure, function and regulation. *Biochemical Journal* 343, 281–299.

Hall, C.E. (1984) The uptake of vitamin B12 by human lymphocytes and the relationships to the cell cycle. *Journal of Laboratory and Clinical Medicine* 103, 70–81.

Hebbes, T.R., Thorne, A.W. and Crane-Robinson, C. (1988) A direct link between core histone acetylation and transcriptionally active chromatin. *EMBO Journal* 7, 1395–1402.

Hu, Z.-Q., Henderson, G.I., Mock, D.M. and Schenker, S. (1994) Biotin uptake by basolateral membrane of human placenta: normal characteristics and role of ethanol. *Proceedings of the Society for Biology and Experimental Medicine* 206, 404–408.

Hymes, J., Fleischhauer, K. and Wolf, B. (1995) Biotinylation of histones by human serum biotinidase: assessment of biotinyl-transferase activity in sera from normal individuals and children with biotinidase deficiency. *Biochemical and Molecular Medicine* 56, 76–83.

Janeway, C.A., Travers, P., Walport, M. and Capra, J.D. (1999) *Immuno Biology*, Garland Publishing/Elsevier, London.

Kaczmarek, L., Calabretta, B. and Baserga, R. (1985) Expression of cell-cycle-dependent genes in phytohemagglutinin-stimulated human lymphocytes. *Proceedings of the National Academy of Sciences USA* 82, 5375–5379.

Karl, P. and Fisher, S.E. (1992) Biotin transport in microvillous membrane vesicles, cultured trophoblasts and the isolated perfused cotyledon of the human placenta. *American Journal of Physiology* 262, C302–C308.

Knowles, J.R. (1989) The mechanism of biotin-dependent enzymes. *Annual Review of Biochemistry* 58, 195–221.

Lee, D.Y., Hayes, J.J., Pruss, D. and Wolffe, A.P. (1993) A positive role for histone acetylation in transcription factor access to nucleosomal DNA. *Cell* 72, 73–84.

Lernhardt, W. (1990) Fatty acid requirement of B lymphocytes activated *in vitro*. *Biochemical and Biophysical Research Communications* 166, 879–885.

Moskowitz, M. and Cheng, D.K.S. (1985) Stimulation of growth factor production in cultured cells by biotin. In: Dakshinamurti, K. and Bhagavan, H.N. (eds) *Biotin*. New York Academy of Sciences, New York, pp. 212–221.

Murray, A. and Hunt, T. (1993) *The Cell Cycle*. Oxford University Press, New York.

Prasad, P.D., Wang, H., Kekuda, R., Fujita, T., Fei, Y.-J., Devoe, L.D., Leibach, F.H. and Ganapathy, V. (1998) Cloning and functional expression of a cDNA encoding a mammalian sodium-dependent vitamin transporter mediating the uptake of pantothenate, biotin, and lipoate. *Journal of Biological Chemistry* 273, 7501–7506.

Said, H.M., Ma, T.Y. and Kamanna, V.S. (1994) Uptake of biotin by human hepatoma cell line, Hep G(2): a carrier-mediated process similar to that of normal liver. *Journal of Cellular Physiology* 161, 483–489.

Schenker, S., Hu, Z., Johnson, R.F., Yang, Y., Frosto, T., Elliott, B.D., Henderson, G.I. and Mock, D.M. (1993) Human placental biotin transport: normal characteristics and effect of ethanol. *Alcohol Clinical and Experimental Research* 17, 566–575.

Stanley, J.S., Griffin, J.B. and Zempleni, J. (2001) Biotinylation of histones in human cells: effects of cell proliferation. *European Journal of Biochemistry* 268, 5424–5429.

Stanley, J.S., Mock, D.M., Griffin, J.B. and Zempleni, J. (2002) Biotin uptake into human peripheral blood mononuclear cells increases early in the cell cycle, increasing carboxylase activities. *Journal of Nutrition* 132, 1854–1859.

Tsai, F.C.H. and Gardner, D.K. (1994) Nicotinamide, a component of complex culture media, inhibits mouse embryo development *in vitro* and reduces subsequent developmental potential after transfer. *Fertility and Sterility* 61, 376–382.

van den Berg, K.J. and Betel, I. (1973) Selective early activation of a sodium dependent amino acid transport system in stimulated rat lymphocytes. *FEBS Letters* 29, 149–152.

Velazquez, A., Zamudio, S., Baez, A., Murguia-Corral, R., Rangel-Peniche, B. and Carrasco, A. (1990) Indicators of biotin status: a study of patients on prolonged total parenteral nutrition. *European Journal of Clinical Nutrition* 44, 11–16.

Velazquez, A., Teran, M., Baez, A., Gutierrez, J. and Rodriguez, R. (1995) Biotin supplementation affects lymphocyte carboxylases and plasma biotin in severe protein-energy malnutrition. *American Journal of Clinical Nutrition* 61, 385–391.

Waithe, W.I., Dauphinais, C., Hathaway, P. and Hirschhorn, K. (1975) Protein synthesis in stimulated lymphocytes. II. Amino acid requirements. *Cellular Immunology* 17, 323–334.

Wang, H., Huang, W., Fei, Y.-J., Xia, H., Fang-Yeng, T.L., Leibach, F.H., Devoe, L.D., Ganapathy, V. and Prasad, P.D. (1999) Human placental Na^+-dependent multivitamin transporter. *Journal of Biological Chemistry* 274, 14875–14883.

Williams, G.T., Lau, K.M.K., Coote, J.M. and Johnstone, A.P. (1985) NAD metabolism and mitogen stimulation of human lymphocytes. *Experimental Cell Research* 160, 419–426.

Wilson, A. and Freeman, J.W. (1996) Regulation of P120 mRNA levels during lymphocyte stimulation: evidence that the P120 gene shares properties with early and late genes. *Journal of Cellular Biochemistry* 60, 458–468.

Wolf, B., Heard, G.S., McVoy, J.R.S. and Grier, R.E. (1985) Biotinidase deficiency. *Annals of the New York Academy of Sciences* 447, 252–262.

Wolffe, A. (1998) *Chromatin*. Academic Press, San Diego, California.

Wood, H.G. and Barden, R.E. (1977) Biotin enzymes. *Annual Review of Biochemistry* 46, 385–413.

Zempleni, J. (2001) Biotin. In: Bowman, B.A. and Russell, R.M. (eds) *Present Knowledge in Nutrition*, 8th edn. International Life Sciences Institute, Washington, DC, pp. 241–252.

Zempleni, J. and Mock, D.M. (1998) Uptake and metabolism of biotin by human peripheral blood mononuclear cells. *American Journal of Physiology* 275, C382–C388.

Zempleni, J. and Mock, D.M. (1999a) The efflux of biotin from human peripheral blood mononuclear cells. *Journal of Nutritional Biochemistry* 10, 105–109.

Zempleni, J. and Mock, D.M. (1999b) Human peripheral blood mononuclear cells: inhibition of biotin transport by reversible competition with pantothenic acid is quantitatively minor. *Journal of Nutritional Biochemistry* 10, 427–432.

Zempleni, J. and Mock, D.M. (1999c) Mitogen-induced proliferation increases biotin uptake into human peripheral blood mononuclear cells. *American Journal of Physiology* 276, C1079–C1084.

Zempleni, J. and Mock, D.M. (2000) Proliferation of peripheral blood mononuclear cells increases riboflavin influx. *Proceedings of the Society for Experimental Biology and Medicine* 225, 72–79.

Zempleni, J., Stanley, J.S. and Mock, D.M. (2001) Proliferation of peripheral blood mononuclear cells causes increased expression of the sodium-dependent multivitamin transporter gene and increased uptake of pantothenic acid. *Journal of Nutritional Biochemistry* 12, 465–473.

6 Nutrients and Apoptosis

John C. Mathers
Human Nutrition Research Centre, School of Clinical Medical Sciences, University of Newcastle, Newcastle upon Tyne, UK

Life and Cell Death

The commonest multicellular organism on this planet is a nematode *Caenorhabditis elegans*, which flourishes in compost, lives in the human gut and spreads diseases including river blindness and elephantiasis. Because it is small (1 mm long), easy to grow and has a short life span (~2 weeks), the roundworm *C. elegans* is used widely in studies of developmental biology and of ageing. A total of 1090 cells develop initially, but the mature hermaphrodite animal contains only 959 somatic cells with the other 131 having been removed by apoptosis. Most of these cells die early in embryogenesis (250–450 min after fertilization) and many (105/131) are neurones (Xue *et al.*, 2002). Apoptosis is a form of programmed cell death identified readily by histological and biochemical features including cell membrane blebbing, cell shrinkage, condensation of chromatin and fragmentation of DNA. Unlike necrosis, there is no spillage of cell contents into the intercellular space and apoptosis does not evoke an immune response. At the end of the killing process, the cell remnants, described as apoptotic bodies, are engulfed rapidly by neighbouring cells. Aminophospholipids such as phosphatidylserine or phosphatidylethanolamine, which are located normally on the inner leaflet of the plasma membrane, are externalized on the cell surface during apoptosis. These aminophospholipids on apoptotic cells are bound by the glycoprotein milk fat globule epidermal growth factor (EGF)-8 and this signals engulfment by phagocytes (Hanayama *et al.*, 2002).

Apoptosis has been recognized as a form of cell death for about 30 years since the pioneering publication by Kerr *et al.* (1972), but its fundamental importance in normal growth and development and in the aetiology and treatment of many diseases has become apparent rather more recently. A second form of programmed cell death, which is distinct from apoptosis on morphological and biochemical criteria and which does not respond to apoptosis inhibitors, has been described and named paraptosis (Sperandio *et al.*, 2000). Paraptosis may have a role in neural development and degeneration.

Roles of Apoptosis in Health and Disease

The development of a human baby from a fertilized egg involves the four key processes of cell division, differentiation, morphogenesis and apoptosis. Cell death by apoptosis is an integral part of human embryonic development, just as it is in the development of *C. elegans*, and this inbuilt death programme responds to a complex array of external and internal signals. Obvious manifestations of apoptosis in the embryo include the sculpting of organs and the removal of webs between the digits on hands and feet. The nervous system develops through the overproduction of cells followed by apoptosis of those failing to establish functional synaptic connections (Renehan *et al.*, 2001). Apoptosis is an important mechanism in development of the endocrine pancreas and specifically in remodelling of the β-cell mass during neonatal life (Scaglia *et al.*, 1997).

Involution of the thymus has long been recognized as a sensitive marker of malnutrition (Prentice, 1999) and impaired immune function. Thymocytes from young rats made severely malnourished by doubling the number of young suckling each dam had seven times higher rates of spontaneous apoptosis compared with those from well-fed controls (Ortiz *et al.*, 2001). This greater rate of apoptosis may be responsible for the atrophy of the thymus in malnutrition (Ortiz *et al.*, 2001), with serious consequences for both short- and long-term health (Moore *et al.*, 1997). T-lymphocyte homeostasis occurs as a balance between apoptosis, mediated by local concentrations of interleukin-2 (IL-2), and proliferation in response to diverse antigens (Lenardo *et al.*, 1999) including those from food. Nur77 is one of a pair of orphan nuclear receptors (the other is Nurr1), which heterodimerizes with the nuclear receptor retinoid X receptor (RXR) to activate gene transcription after binding retinoic acid (the transcriptionally active derivative of vitamin A) (McCaffrey *et al.*, 2001). Nur77 appears to be important in apoptosis in the immune system and in negative selection of T cells (McCaffrey *et al.*, 2001).

With very few exceptions, such as the nervous system, cell division continues in all tissues throughout life and, as a result, apoptosis is an essential process in adulthood since maintenance of normal cell numbers and the integrity of tissues and organs depends on a fine balance between the numbers of new cells born and those dying. Indeed, it has been proposed that the cell's proliferative and apoptotic pathways are coupled (Evan and Littlewood, 1998) although the biochemical machinery for each is quite distinct. Apoptosis also continues to be important for tissue remodelling during wound healing, regression of the uterus after parturition and post-lactational changes in mammary tissue.

Apoptosis in Response to DNA Damage

DNA is exposed continuously to damage by agents including ionizing radiation (which causes single or double strand breaks), reactive oxygen species (ROS) (due to a deficiency of antioxidants resulting in attacks at purine and pyrimidine rings) and alkylating agents (such as heterocyclic amines from foods that form DNA adducts), and acquires errors during replication. For a unicellular organism, repair of DNA is the only option if the cell is to survive with normal function. In contrast, metazoans can 'choose' to kill off damaged cells by apoptosis if the cost or risks associated with repair are too great. Apoptosis may be the prudent course of action for damaged stem cells or other cells with substantial proliferative potential but, for post-mitotic cells, or those with a limited life span (gut epithelium or skin), repair may be a 'safe' option (Evan and Littlewood, 1998). The evolutionary origins of apoptosis remain obscure but it has been argued that single-celled organisms could have evolved a cell death programme 'as a contingent strategy to prevent infection of related individuals' (Vaux, 2002).

As a form of defence, cells with DNA damage arrest at the G_1–S cell cycle checkpoint when damage is sensed by the protein product of the *ATM* gene. Inherited mutations in this gene are responsible for the rare recessive disease ataxia telangiectasia, which is characterized by cerebellar ataxia, dilation of the blood vessels of the eye, immunodeficiency and growth retardation, and by a strong predisposition to cancer (Strachan and Read, 1999). By processes that are poorly understood, the ATM protein signals DNA damage to *TP53*, resulting in increased concentrations of the p53 tumour suppressor protein, which is often described as the 'guardian of the genome'. In its tetrameric form, p53 is a transcription factor normally present in low concentrations through interactions with the MDM-2 protein that signals its degradation (Evan and Littlewood, 1998). When p53 concentrations are raised, progression through the cell cycle is halted, providing time for DNA repair or for initiation of apoptosis. Little is known about why a given cell may respond to DNA damage by growth arrest and repair whilst another responds by inducing apoptosis, but it has been hypothesized that apoptosis would be favoured in cells with a reduced capacity for repair (Liu and Kulesz-Martin, 2001). The most primitive form of p53 identified is that in *Drosophila* where p53 mediates apoptosis but not growth arrest, which suggests that apoptosis signalling was the earliest role for p53 (Liu and Kulesz-Martin, 2001). p53-independent apoptosis appears to play a role in prevention of teratogenesis by facilitating DNA repair or causing fetal death (Norimura *et al.*, 1996). *TP53* is the gene most frequently mutated or lost in tumours, which indicates its central role in protection of the genome. More than 50% of human tumours contain a mutated *TP53*, with most of those mutations (>90%) being missense mutations within the evolutionarily conserved DNA-binding domain (Hollstein *et al.*, 1994).

The binding of p53 to DNA occurs by both sequence-specific and non-sequence-specific mechanisms, and is highly dependent upon the reduction state of the molecule since the binding involves a zinc finger and a further seven cysteine residues (Liu and Kulesz-Martin, 2001). Evidence is accumulating that signals from both oxidative stress pathways and DNA strand breaks are integrated by p53 (Liu and Kulesz-Martin, 2001), enhancing its central role in protecting the genome from diet-induced and other forms of damage. It is possible that the familial breast cancer genes *BRCA1* and *BRCA2* are also involved in the damage checkpoint at G_1–S (Strachan and Read, 1999). Evidence is accumulating (Hickman and Samson, 1999) that chemicals that alkylate DNA resulting in the formation of O^6-alkylguanine adducts initiate apoptosis using signals from the MutSα branch of the DNA mismatch repair (MMR) pathway and may be independent of the p53 status of the cell. Fishel (2001) has proposed that these MMR proteins function as specific 'direct sensors' linking pathways to DNA repair or to apoptosis.

The Apoptotic Process

Although its importance as a counterbalance to mitosis has been recognized for some time and the broad features of the apoptotic process were known for nearly 20 years (Raff, 1992), it is only over the last decade that apoptosis has attracted extensive research of an intensity previously devoted to cell proliferation. It is now apparent that most, if not all, human cells express the proteins necessary to undergo suicide by apoptosis and that this may be the default pathway in the absence of survival signals from surrounding cells (Raff, 1992; Thompson, 1995). Apoptosis can occur very rapidly, with the entire process from initiation to phagocytosis being complete within an hour (McCarthy, 2002). It is also a stochastic process, i.e. not all cells in a given tissue die at the same time so histological assessment of the magnitude/rate of apoptosis can be difficult (McCarthy, 2002). The apoptotic process can be considered from the perspective of the molecules involved in four distinct phases: (i) survival signals; (ii) death signals; (iii) regulatory genes and proteins; and (iv) effector molecules. An excellent review of the molecular biology of apoptosis is given by Jacobson and McCarthy (2002) and the processes are summarized in Fig. 6.1.

Survival signals

Raff (1992) pioneered the idea of social controls on cell survival, citing, for example, the importance of neurotrophic factors secreted by target cells for the survival of vertebrate neurons. Epithelial cells in the ventral prostate die by apoptosis in the absence of testosterone, as do cells of the adrenal cortex in the absence of adrenocorticotrophic hormone. T lymphocytes require IL-2, and endothelial cells require fibroblast growth factor (among other growth factors) to prevent apoptosis. As a means of ensuring that only the appropriate cells survive in a given body location, this inverse relationship between survival signals and apoptosis has universal utility (Fig. 6.2). If a cell loses contact with its appropriate neighbouring cells, that cell is likely to die (Ashkenazi and Dixit, 1998). Cancer cells can establish metastases only if they acquire by mutation the capacity to survive without external survival signals (Raff, 1992).

Death signals

Stimuli of various kinds can trigger the process of apoptosis, but these come together via two major signalling pathways of apoptosis, i.e. the death receptor pathway and the death receptor-independent or mitochondrial pathway (Gupta, 2001).

Death receptor pathway

Mammals have evolved a mechanism that enables the organism to direct the death by apoptosis of unwanted cells (Ashkenazi and Dixit, 1998). The instructions for this altruistic suicide are provided by ligands such as tumour necrosis

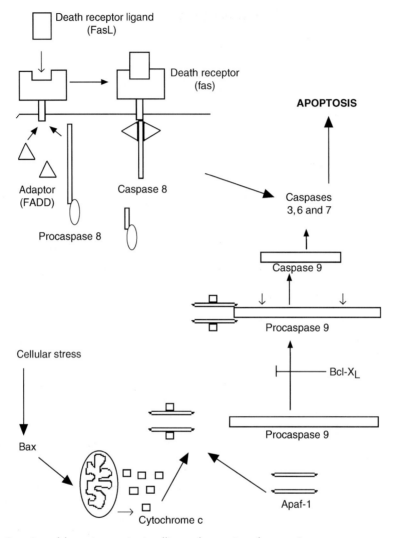

Fig. 6.1. Overview of the main steps in signalling and execution of apoptosis.

factor (TNF) and Fas ligand (FasL or CD95L). Receptors for these death signals, found on the cell's plasma membrane, are encoded by genes that belong to the TNF receptor gene superfamily, and the best characterized of these are TNF receptor 1 (TNFR1) and Fas (also called CD95 and Apo1) (Ashkenazi and Dixit, 1998). These receptors have both extracellular and cytoplasmic domains, with the latter acting as death domains (DDs). The sequence of events following binding of FasL or TNF to its respective receptor is subtly different (Ashkenazi and Dixit, 1998; Gupta, 2001), but the overall process is similar. FasL is an homotrimeric molecule, which, on binding to its receptor (three molecules of Fas), causes the clustering of the receptor's DD. This results in the binding of an adaptor protein called FADD (Fas-associated death domain), which in turn recruits procaspase-8 (the zymogen form of caspase-8; discussed below) (Ashkenazi and Dixit, 1998). The combination of the DD from Fas, FADD and procaspase-8 forms the death-inducing signalling complex (DISC), resulting in the activation of caspase-8 by the autocatalytic cleavage of procaspase-8 and so committing the cell to apoptosis (Gupta, 2001)

The Fas–FasL pathway of apoptosis is important in cellular immunity ensuring: (i) killing of

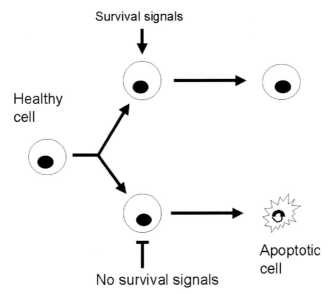

Fig. 6.2. Withdrawal of survival signals may initiate apoptosis.

target cells by cytotoxic T cells and natural killer (NK) cells; (ii) selection of T-cell repertoire and removal of self-recognizing T cells; (iii) deletion of activated lymphocytes when an immune response has been completed; and (iv) deletion of inflammatory/immune cells in 'immune privileged' tissues such as the testis and eye (Gupta, 2001). Mutations in either the *Fas* or the *FasL* gene are associated with autoimmune disorders (Lenardo *et al.*, 1999).

TNF is secreted by activated macrophages and T cells in response to infection and, via binding to TNFR1, activates the transcription factors NF-κB and AP-1, resulting in the transcription of pro-inflammatory and immunomodulatory genes (Ashkenazi and Dixit, 1998; Gupta, 2001). However, if protein synthesis is inhibited, TNF triggers apoptosis. With the trimeric TNF bound to three molecules of TNFR1, the DDS of TNFR1 associate and recruit a group of signalling proteins including TRADD, FADD, TRAF-2 and RIP. The latter two proteins appear to be required for activation of NF-κB, whereas FADD is required for signalling of apoptosis via recruitment and activation of effector and regulatory caspases (Ashkenazi and Dixit, 1998; Gupta, 2001).

Mitochondrial pathway

Exposure to stress molecules (ROS and reactive nitrogen species), chemotherapeutic drugs and UV irradiation causes mitochondrial damage and stimulates apoptosis by a death receptor-independent pathway (Gupta, 2001). In this case, apoptosis is signalled by the release from the mitochondrion of cytochrome c (also known as Apaf-2), which binds to apoptotic protease-activating factor-1 (Apaf-1) (Zou *et al.*, 1997) and to Apaf-3 (also known as caspase-9) to form an 'apoptosome', resulting in the catalytic activation of caspase-3 and leading to apoptosis (Li *et al.*, 1997). Expression of UCP2 (a member of the small family of mitochondrial uncoupling proteins) is up-regulated early in apoptosis in apoptosis-sensitive but not apoptosis-insensitive B-cell mouse lymphoma cells (Voehringer *et al.*, 2000), but it remains to be established whether UCPs are involved directly or indirectly in the induction of apoptosis (Collins *et al.*, 2001).

Regulatory genes and proteins

The Bcl-2 family of proteins appears to play a key role in deciding whether a cell lives or dies by apoptosis (Tsujimoto, 2002). They include both pro- and anti-apoptotic members, and in many cases have up to four conserved Bcl-2 homology (BH) domains designated BH1–4 (Reed, 1998; Gross *et al.*, 1999). In healthy cells, the anti-apoptotic members are found as integral membrane proteins in the mitochondria, nucleus and endoplasmic reticulum, whereas the pro-apoptotic members localize to various sites within the cytosol and cytoskeleton (Gross *et al.*, 1999). The mechanisms by which Bcl-2 family proteins regulate cell death are not fully understood, but it appears that the relative abundance of pro- and anti-apoptotic members and post-translational modifications such as phosphorylation or cleavage determine whether the proteins are activated. Following a cell death signal or cellular damage, pro-apoptotic proteins may undergo conformational changes that result in the 'exposure' of the BH3 domain, and translocate to and dock on the outer membrane of the mitochondria (Gross *et al.*, 1999). Here they may form dimers or higher-order oligomers with a similar molecule or an anti-apoptotic member of the family to induce changes in the permeabilization of the membrane and the release of cytochrome c (Gupta, 2001). Gross *et al.* (1999) suggested that the anti-apoptotic Bcl-2 family proteins may be 'guarding the mitochondrial gate' and that the pro-apoptotic proteins 'gain access' following a death signal. These pro-apoptotic molecules may be positioned at strategic locations throughout the cell to act as sentinels for cellular damage. In this way, the common apoptotic pathway could be activated in response to a diverse range of insults (Gross *et al.*, 1999).

Effector molecules

In *C. elegans*, the induction of apoptosis requires two genes, *ced-3* and *ced-4*, which normally are counteracted by a *bcl-2*-like gene, *ced-9* (White, 1996). The mammalian homologue of *ced-3* is interleukin-1β-converting enzyme (ICE or caspase-1), which processes IL-1β to its mature form as part of the inflammatory response (Thornberry *et al.*, 1992; Yuan *et al.*, 1993). During apoptosis, cells are killed by the cleavage of specific cellular proteins by caspases, a set of proteases that share similarities in amino acid sequence, structure and substrate specificity (Thornberry and Lazebnik, 1998). These proteases have a cysteine residue in their active sites and cleave proteins after an aspartic acid residue (Cohen, 1997). Caspases are expressed as zymogens, which are activated by proteolysis to yield a large (~20 kDa) and a small (~10 kDa) subunit that associate as a heterodimer, with two of these heterodimers then combining to form the catalytically active tetramer (Thornberry and Lazebnik, 1998). At least 14 mammalian caspases have been identified, and their classification on the basis of their optimal substrate cleavage sites has been described (Roy and Cardone, 2002).

Caspases kill cells by: (i) inactivating inhibitors of apoptosis, e.g. Bcl-2 proteins; (ii) destroying cell structures, e.g. lamins (which form the nuclear lamina); and (iii) deregulating proteins, e.g. DNA-PK_{CS} (involved in DNA repair), by separating the regulatory and catalytic domains (Thornberry and Lazebnik, 1998). Probably the best characterized substrate for caspase cleavage is the DNA repair enzyme poly(ADP-ribose) polymerase (PARP), but others include the cell cycle regulatory protein Rb, the sterol regulatory element-binding proteins (SREBP-1 and -2) (Cohen, 1997) and APC (a key component of the Wnt signalling pathway) (Browne *et al.*, 1994).

There is an activation cascade starting with conversion of procaspase-8 to caspase-8 (death receptor pathway) or procaspase-9 to caspase-9 (mitochondrial pathway) and resulting in the activation of the effector caspases caspase-3 and -7 (Cohen, 1997). Given that there is constitutive expression of caspases in (probably) all cells, including the life-long neurones, so that all have the potential to undergo apoptosis (the default pathway), the activation of caspases is closely regulated (Thornberry and Lazebnik, 1998). It appears as if this regulation is targeted at the early stages of apoptosis, with a major role being played by Bcl-2 and its family members (White, 1996; Gross *et al.*, 1999).

Modulation of Apoptosis in Disease Prevention and Treatment

Tissue homeostasis requires equality in the numbers of new cells produced by mitosis and those that die by apoptosis or are lost by other routes. It is increasingly apparent that many diseases involve an excess or a suppression of apoptosis (Table 6.1). For example, exposure to gluten in genetically susceptible individuals led to apoptosis of enterocytes at rates that exceeded the rate of cell proliferation within crypts of the

Table 6.1. Diseases associated with an imbalance between cell proliferation and apoptosis.

Diseases associated with suppressed apoptosis	Diseases associated with enhanced apoptosis
Cancer	AIDS
Autoimmune diseases	Neurodegenerative diseases
Viral infections	Myelodysplastic syndromes
	Ischaemic injury
	Toxin-induced liver disease

Adapted from Thompson (1995).

small bowel epithelium, itself increased fourfold, resulting in flattening of the villi (Moss *et al.*, 1996) and a range of symptoms characteristic of active coeliac disease, due to a failure to hydrolyse food macromolecules and absorb the constituent monomers. Consumption of gluten-free diets returned small intestinal apoptotic scores to normal (Moss *et al.*, 1996). Apoptosis may be de-regulated in other intestinal disorders characterized by 'flat' lesions and induced, for example, by chemotherapy or by deficiencies of folate or vitamin B12 (Lewin and Weinstein, 1996). An abnormally high rate of intestinal mucosal cell apoptosis may contribute to the pathophysiology of ulcerative colitis (Strater *et al.*, 1997). The development of AIDS (acquired immunodeficiency syndrome) is strongly correlated with the depletion of $CD4^+$ T cells, which undergo apoptosis when the soluble viral product gp120 binds to the CD4 receptor (Thompson, 1995). Increased cell death by apoptosis may be involved in several other diseases including Alzheimer's disease, osteoarthritis, osteoporosis (Thompson, 1995) and the age-related retinal degeneration known as retinitis pigmentosa (Davidson and Steller, 1998). Cancers are an excellent example of the abnormal growth of tissues because of a failure of apoptosis, and are discussed in detail below.

The recognition that deregulated apoptosis plays a fundamental role in many diseases and the elucidation of key steps in the pathways to this form of programmed cell death have stimulated extensive research to identify agents that will manipulate apoptosis and provide therapeutic benefit. Obvious targets include (i) the death signals; (ii) the *bcl-2* family (White, 1996); and (iii) the caspases (Thornberry and Lazebnik, 1998). One might prevent apoptosis by selectively blocking death signals by, for example, substances that bind to Fas or to TNFR1. Some tumours express decoy receptors that prevent tumour cell killing by FasL-bearing NK cells (Pitti *et al.*, 1998). Overexpression of Bcl-2 protects cells against apoptosis (White, 1996), so manipulation of expression of various members of this family offers a route to both up- and down-regulation of apoptosis. Angiotensin-converting enzyme inhibitors and HIV (human immodeficiency virus) protease inhibitors are two classes of drugs that work by inhibiting proteases (Thornberry and Lazebnik, 1998). However, there will be major problems to be overcome with respect to tissue and cellular specificity before effective drugs can be developed that activate or inhibit apoptosis selectively in disease tissue.

Cancer

At its most fundamental, cancer is a genetic disease due to the acquisition of mutations in, loss of or epigenetic silencing of both copies of about six specific genes. These changes are all forms of DNA damage that give the cells a selective growth advantage. Whilst most models of tumorigenesis assume that the tumour grows by increased cell division, Tomlinson and Bodmer (1995) proposed mathematical models that include changes in differentiation and in apoptosis, and showed that these could have profound effects on the growth trajectory of the tumour and may help explain the development of benign tumors, pre-malignant growths and variation in the growth rate (including regression) of tumours. This suggests that tumours grow through faster proliferation, by more cell divisions before reaching 'proliferative senescence' (possibly signalled by telomere shortening), and/or by suppression of apoptosis. As the tumour mass gets bigger, tumour cells often experience hypoxia, which can activate p53 to promote apoptosis (Graeber *et al.*, 1996), and p53 is required for apoptosis induced by telomere malfunction (Karlseder et al., 1999). Lowe and Lin (2000) have speculated that a cell acquiring a p53 mutation might not have a selective advantage until the developing tumour encounters hypoxic conditions or achieves sufficient telomere erosion, and that this might explain why p53 mutations are usually late events in tumour development.

Colorectal Cancer

Because of its architecture, relative accessibility and rapid cell kinetics, the intestinal epithelium has been a favourite tissue for studies of cell proliferation, cell migration, differentiation and, more recently, death. Colorectal cancer (CRC) is a disease of the colonic epithelium, which develops because of damage to genes encoding the regulatory systems for these fundamental cellular processes. Indeed, much of our understanding of the genetic events in tumorigenesis has arisen from studies of CRC. The development of CRC is believed to be a multistep process first observable as microscopic aberrant crypt foci and leading though benign adenomatous polyps (adenomas) to carcinomas and, in time, metastasis. These histological stages are caused by successive genetic changes (due to mutations or gene silencing by methylation of promoters) in at least five to six genes (Fearon and Vogelstein, 1990). According to Kinzler and Vogelstein's (1996) gatekeeper gene hypothesis, one gene expressed in a population of proliferating cells has overall responsibility for ensuring that the numbers of cells in that tissue remains constant. For the intestinal epithelium, the gatekeeper gene is *APC*, which encodes a large multifunctional protein central to regulation of the Wnt signalling pathway (Peifer and Polakis, 2000). By definition, mutations in the gatekeeper gene (*APC* in this case) would be expected to lead to an imbalance between cell birth and cell death in favour of cell proliferation. This should give the cell containing this mutation a selective advantage, and there is evidence for inhibition of apoptosis during the development of CRC (Bedi et al., 1995) due, probably, to loss of *APC*, which can be detected in almost all CRC tumours from the earliest stages (Fearon and Vogelstein, 1990). This idea is supported by studies in which transfection and expression of the full-length *APC* in HT29 cells (which lack a functional APC protein) reduced cell growth because of the induction of cell death by apoptosis (Morin et al., 1996). It may be significant that, early in apoptosis, the APC protein is cleaved by caspase-3 after Asp777 to yield a characteristic and stable 90 kDa fragment (Browne et al., 1994, 1998; Webb et al., 1999). The *TP53* gene is mutated very frequently in colorectal carcinomas but not in adenomas (Fearon and Vogelstein, 1990), resulting in loss of another important component of apoptosis regulation late in tumour progression.

Apoptosis as a Target in Chemoprevention of Colorectal Cancer

Epidemiological studies suggest that variation in diet is responsible for much of the variation in CRC incidence worldwide (Doll and Peto, 1981). However, given the considerable inter- and intra-individual variation in dietary habits and the chemical complexity of foods, classical epidemiology is a blunt instrument for dissecting out which components of foods and in which amounts are protective and which may be potentially harmful (Mathers and Burn, 1999). Progress in chemoprevention of CRC is being made through *in vitro* and animal studies, and by intervention studies in human volunteers (Burn et al., 1998; Mathers, 2000). Given that killing of cells with damaged DNA may be a prudent means of preventing intestinal cells from acquiring mutations in *APC* and other key genes and that apoptosis is suppressed during CRC tumorigenesis, the identification of nutrients or other food constituents capable of enhancing apoptosis may lead to the development of effective chemopreventative agents.

Long chain fatty acids

The amount, but more importantly the type, of fat in the diet influences the risk of CRC (Department of Health, 1998; Whelan and McEntee, 2001). Results from studies in rodent models indicate that higher fat diets increase tumorigenesis whilst, at a given fat intake, replacing fatty acids of the *n*-6 series (especially arachidonic acid) with those of the *n*-3 series (especially eicosapentaenoic acid; EPA) reduces intestinal tumorigenesis (Petrik et al.,

2000; Whelan and McEntee, 2001). Feeding fish oil (rich in n-3 polyunsaturated fatty acids (PUFAs) including EPA and docosahexaenoic acid (DHA)) suppresses dimethylhydrazine (DMH)-induced colonic tumours in rats (Latham et al., 1999). Curiously, feeding DHA to Apc$^{\Delta 716}$ mice resulted in fewer intestinal polyps in females but not in males (Oshima et al., 1995). Several PUFAs are cytotoxic to epithelial cells, and apoptosis may be an important element of this cytotoxicity (Johnson, 2001). EPA induces apoptosis in CRC cell lines, and the protection against CRC afforded by the fatty acids in fish oils may result from greater diversion of crypt cells with DNA damage to apoptosis (Latham et al., 1999). Induction of apoptosis by these n-3 PUFAs may be signalled by lipid peroxidation products especially in tumour cells that overexpress P450 (Stoll, 2002). The hypothesis that apoptosis is due to increased oxidative stress is supported by the observation that the effect is amplified by depletion of cellular glutathione and blocked by antioxidants (Clarke et al., 1999; Latham et al., 2001) but other mechanisms are possible (Johnson, 2001). Among these mechanisms, promising evidence is emerging about the central role of the peroxisome proliferator-activated receptor (PPAR)γ, a member of the superfamily of nuclear hormone receptors, which is expressed at high levels in colorectal adenomas (Sarraf et al., 1998). Long chain PUFAs are among the ligands that activate PPARγ (Kliewer et al., 1997), inducing terminal differentiation and apoptosis. PPARγ activation down-regulates cyclooxygenase (COX)-2 and induces apoptosis in the human colon cancer cell line HT29 (Yang and Frucht, 2001).

Conjugated linoleic acids (CLAs) are a collection of geometric and positional isomers of C18 fatty acids that occur naturally in the milk and tissues of ruminant animals due to partial hydrogenation of PUFAs by rumen bacteria. In the last few years, CLAs have emerged as potential anticancer agents (Parodi, 1997), but their mode of action is poorly understood. Park et al. (2001) hypothesized that CLAs may be a substrate for COX and may share the same chain elongation and desaturation systems as linoleic, linolenic and arachidonic acids. Exposure to CLAs could mean competition between the fatty acids and result in different patterns of eicosanoid end-products. This group demonstrated that including 1% CLA in the diet of DMH-treated rats reduced tumorigenesis in the colon and that this was associated with lower concentrations of both prostaglandin E_2 and thromboxane B_2 in the colonic mucosa (Park et al., 2001). The apoptotic index in flat mucosa in the terminal colon was 2.5 times greater in rats fed CLA than in controls, but Park et al. (2001) were unable to determine the signalling mechanism by which CLA treatment increased apoptosis. Higher rates of apoptosis in macroscopically normal colonic mucosa might be a mechanism for removal of cells damaged by the DMH treatment and, therefore, explain the tumour-suppressing action of CLAs. Alternatively, this putative damage could be a direct result of exposure to CLAs so that the increased apoptosis is a normal homeostatic process unrelated to tumour inhibition. CLA fed at 3 g 100 g^{-1} diet was ineffective in altering intestinal tumour multiplicity in Apc$^{Min/+}$ mice (Petrick et al., 2000) but it should be noted that the majority of tumours in this mouse model occur in the small bowel, and effects of CLAs on tumorigenesis may be site specific.

Fat-soluble vitamins

The epidemiological evidence for protection against CRC (and other cancers) by diets rich in vegetables and fruits is persuasive (Department of Health, 1998), but the components of such diets that confer protection remain elusive. Vegetables and fruits are rich in a wide variety of antioxidants, of which the most extensively studied is β-carotene. However, disappointingly, high doses of β-carotene given to middle-aged male smokers or those exposed previously to asbestos had no protective effect against CRC and increased the incidence of lung cancer (Alpha-Tocopherol, Beta Carotene Cancer Prevention Study Group, 1994). It is probable that, at the doses given, β-carotene acted as a pro- not an antioxidant in lung tissues experiencing a high oxidant stress. Recent evidence suggests that β-carotene suppresses the growth of CRC cells in a dose-dependent manner by inducing cell cycle arrest at G_2–M phase and promoting apoptosis (Palozzi et al., 2002). Cell cycle arrest was associated with down-regulation of expression of cyclin A (a key regulator of G_2–M progression), whilst suppression of the anti-apoptotic proteins Bcl-2 and of Bcl-xL (but no change in the pro-apoptotic Bax) accompanied the increase in apoptosis (Palozzi et al., 2002). For reasons that are not apparent, different CRC cell lines incorporated β-carotene into their cell membranes to differing extents, and the greater the β-carotene incorporation, the greater the effectiveness in suppressing cell growth and increasing apoptosis (Palozzi et al., 2002).

Apoptosis in colorectal adenocarcinoma cells exposed to high doses of β-carotene may be induced by formation of ROS (Palozza et al., 2001) as was discussed above for long chain PUFAs. This induction of apoptosis by β-carotene appeared to be by a p53-independent mechanism since several of the cell lines tested by Palozza et al. (2002) contained mutant forms of p53. Could it be that oxidative damage in these cells is being sensed by an MMR-dependent mechanism and resulting in death signals via the mitochondrial pathway? Of particular practical interest is the observation that the concentrations of β-carotene used could be achieved in human plasma (Palozza et al., 2002). The vitamin A derivative 9-cis-retinoic acid is a ligand for RXR, which heterodimerizes with PPARs, binds to specific response elements and induces gene expression. Simultaneous exposure of HT29 colon cancer cells to a specific PPARγ ligand and to 9-cis-retinoic acid potentiated the apoptotic effects of the PPARγ ligand although, on its own, 9-cis-retinoic acid had little effect (Yang and Frucht, 2001).

Vitamin E is also known to suppress the growth of, and induce apoptosis in, CRC cells (Chinery et al., 1997). As with the action of β-carotene, these effects result from a p53-independent mechanism and appear to be brought about by induction of cell cycle arrest through the activation of $p21^{WAF1/CIP1}$ and c/EBPβ (Chinery et al., 1997). This effect of vitamin E is somewhat paradoxical since ROS are involved in the induction of apoptosis possibly via activation of prostate apoptosis response-4 (Par-4) (Meydani et al., 2001) and pretreatment of cell cultures with vitamin E prevents Par-4 induction (Chan et al., 1999).

Plant secondary metabolites

Isothiocyanates

Cruciferous vegetables such as broccoli and cauliflower contain relatively large concentrations of secondary metabolites known as glucosinolates as well as the enzyme myrosinase (thioglucoside hydrolase; EC 3.2.3.1). Physical damage to the plant tissue, e.g. by cutting or chewing, results in the exposure of the glucosinolates to myrosinase and hydrolysis of the glucosinolates to release isothiocyanates and other products. These isothiocyanates are believed to be important mediators of the tumour-suppressing activity of brassicas (Hecht, 1999) and result in increased apoptosis *in vivo* (Smith et al., 1998). One of the most potent isothiocyanates in terms of anti-tumour effects is sulphoraphane derived from broccoli. Sulphoraphane inhibits the growth of HT29 tumour cells by causing G_2–M arrest (an effect that was correlated with increased expression of cyclins A and B1) (Gamet-Payrastre et al., 2000) and by increasing cell death through apoptosis (Gamet-Payrastre et al., 2000). In the latter study, the anti-apoptotic Bcl-2 protein was not detected, but expression of the pro-apoptotic Bax was increased by sulphoraphane treatment. These cell cycle arrest and apoptosis induction effects of sulphoraphane appear to be general phenomena since Fimognari et al. (2002) reported similar observations with human T-cell leukaemia cells.

Curcumin

Curcumin (1,7-bis(4-hydroxy-3-methoxyphenyl) 1,6-heptadiene-3,5-dione), the major pigment in the spice tumeric, which is derived from *Curcuma longa*, inhibits the growth of the colon adenocarcinoma cell lines HT29 and HCT15 (Hanif et al., 1997) and suppresses tumorigenesis in the intestine of both the azoxymethane-induced F344 rat model (Kawamori et al., 1999) and in the Apc^{min} mouse (Mahmoud et al., 2000). These antineoplastic effects are associated with decreased activation of the transcription factor NF-κB (Singh and Aggarwal, 1995) and promotion of apoptosis both *in vitro* (Kuo et al., 1996) and *in vivo* (Samaha et al., 1997). Exposure of Apc^{min} mice to PhIP (2-amino 1-methyl-6-phenylimidazo(4,5-*b*) pyridine – the most abundant heterocyclic amine in cooked meat) doubled tumour multiplicity in the proximal small bowel and was accompanied by only half the rate of mucosal cell apoptosis that was seen in wild-type C57Bl/6 animals (Figs 6.3 and 6.4) (Collett et al., 2001). Curcumin treatment partially attenuated the resistance to PhIP-induced apoptosis observed in the Apc^{min} mice and inhibited PhIP-induced tumorigenesis in the

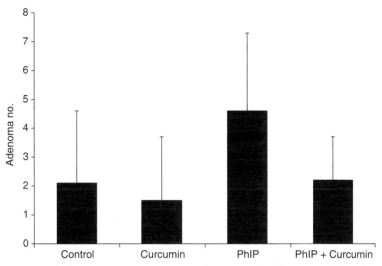

Fig. 6.3. Adenoma multiplicity (tumours per mouse) in the proximal small intestine of Apc^{min} mice treated with curcumin and/or PhIP for 10 weeks from weaning (Collett et al., 2001).

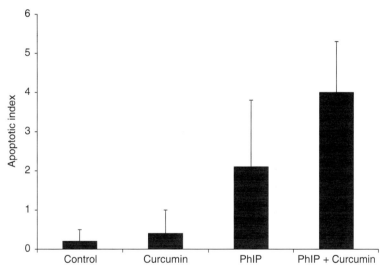

Fig. 6.4. Apoptotic index in the mucosa of the proximal small intestine of Apc^{min} mice treated with curcumin and/or PhIP for 10 weeks from weaning (Collett et al., 2001).

proximal small intestine (Figs 6.3 and 6.4) (Collett et al., 2001). Whilst the details of the anti-tumorigenic action of curcumin remain to be elucidated, it has been suggested that curcumin may ameliorate the apoptosis resistance that is associated with *Apc* mutation through alteration of the equilibrium of the transcription factors p53 and NF-κB (Collett et al., 2001) since it inhibits TNF-induced NF-κB activation (Plummer et al., 1999) and increases p53 expression (Jee et al., 1998).

Salicylates

There is strong epidemiological evidence that regular consumption of the non-steroidal anti-inflammatory drug (NSAID) aspirin (acetyl salicylate) is associated with reduced risk of CRC (Giovannucci et al., 1994). The effect is similar for both men and women and appears to be dose and time dependent. At first, it was assumed that this protection was mediated by COX-dependent events since aspirin is a well-known COX inhibitor and up-regulation of COX-2 occurs in tumours (DuBois and Smalley, 1996). Indeed, intervention with COX-2-specific inhibitors suppresses intestinal polyp formation in both mice (Oshima et al., 1996) and people with familial adenomatous polyposis (Steinback et al., 2000). However, this protection may not necessarily be dependent on effects on COX-2-related pathways. The aspirin metabolite salicylate inhibits the growth of CRC cell lines by enhancing apoptosis (Elder et al., 1996), and NSAIDs may suppress the growth of CRC tumour cell lines by a combination of inhibition of proliferation and induction of apoptosis independent of their activity towards either COX-1 or COX-2 (Richter et al., 2001). PPARδ is an APC-regulated target of NSAIDs (He et al., 1999), and genes induced via PPARδ may direct terminal differentiation and apoptosis. Salicylates are widespread in foods as a result of being important secondary plant metabolites central to defence systems, but are also acquired during food processing. Because of the absence of data on salicylate contents of foods in food composition databases, there is a paucity of information on salicylate intakes but likely intakes are only a few milligrams per day (Janssen et al., 1996). This is approximately an order of magnitude less than the amounts of aspirin associated with reduced CRC risk (Giovannuci et al., 1994) so that it is unlikely that food-derived salicylates offer much protection against CRC among typical Western consumers.

Butyrate

One of the most potent anti-neoplastic agents is the short chain carboxylic acid butyrate (D'Argenio, 1996) produced by fermentation of carbohydrates in the colon where concentrations can reach 25 mM (Cummings et al., 1987). At concentrations between 1 and 5 mM, butyrate treatment of CRC cell lines suppresses cell growth, results in accumulation of cells in G_1 and induces apoptosis by a p53-independent pathway (Hague et al., 1993, 1995; Heerdt et al., 1994). Exactly what this pathway is remains uncertain, but it is known that butyrate is an effective inhibitor of histone deacetylase (Wu et al., 2001) so that the conformational changes in the genome facilitated by acetylation may allow expression of pro-apoptotic genes that normally are silenced (Johnson, 2001). The use of cDNA microarrays is beginning to offer new insights into the pathways of differentiation and apoptosis in CRC cells treated with butyrate (Mariadason et al., 2000; Della Ragione et al., 2001). Adding wheat bran to the diet of rats increased caecal butyrate concentrations and was associated with increased histone acetylation in colonic epithelial cells (Boffa et al., 1992). This provides circumstantial evidence that the anti-neoplastic effect of butyrate in rats (D'Argenio, 1996) may be mediated by histone acetylation. Interestingly, treatment with a COX-2-selective agent enhanced the growth-inhibitory effect of butyrate in CRC tumour cells expressing COX-2 (Crew et al., 2000).

Whilst consumption of carbohydrates resistant to small bowel hydrolases (e.g. resistant starch) that are fermented to butyrate in the large bowel is the most obvious route to increasing exposure to butyrate, the quantitative relationship between consumption of such carbohydrates and colonic butyrate is not well understood (Mathers et al., 1997). Butyric acid and its common salt sodium butyrate have pungent odours, making them unsuited as direct chemopreventive agents, so there has been considerable interest in developing effective butyrate derivatives. Growth of HT29 cells is inhibited by both phenylbutyrate and tributyrin through a block in the G_1/S cell cycle traverse (associated with a decrease in expression of CDK2) and Rb protein hypophosphorylation (Clarke et al., 2001). These agents also induced apoptosis of HT29 cells mediated through activation of caspase-3 activity (Clarke et al., 2001). Like the parent compound, phenylbutyrate inhibits histone deacetyase and it is highly likely that tributyrin acts similarly since endogenous lipases will hydrolyse it to release butyrate.

Concluding Comments

It is now widely recognized that cell death by apoptosis is an essential component of normal growth and development and that derangement of apoptosis is important in the aetiology of many different diseases. Working out the molecular mechanisms of regulation and execution of apoptosis has been, and continues to be, a fertile area for basic cellular and molecular biological research. It is becoming clear that nutrition throughout the life cycle plays a fundamental role in health or its absence, and that many of these effects are mediated by modulation of apoptosis. There is good evidence, especially from the CRC area, that several food-derived substances induce apoptosis by cancer cells and so have considerable potential as chemoprevention agents. However, much of the research in this field is at the phenomenological stage, with only sketchy understanding of the mechanisms of action of these compounds. In addition, the doses used in *in vitro* and animal model studies are often many times higher than are likely to be consumed by humans as part of a conventional diet. The ready availability (albeit still at a relatively high cost) of post-genomic technologies for examining global gene expression, e.g. using microarrays and proteomics, offers nutrition researchers an exciting opportunity to begin to understand how nutrients and other food-derived bioactive compounds influence apoptosis and, ultimately, health.

Acknowledgements

My thanks to Dawn Marland for Fig. 6.1 and to Liz Williams for help with Figs 6.3 and 6.4. Studies of nutrition and apoptosis in my laboratory have been funded by the Food Standards Agency (Contract No. NO3002) and by the Biotechnology and Biological Sciences Research Council (Grant No. 13/DO9671).

References

Alpha-Tocopherol, Beta Carotene Cancer Prevention Study Group (1994) The effect of vitamin E and beta-carotene on lung cancer incidence and other cancers in male smokers. *New England Journal of Medicine* 330, 1029–1035.

Ashkenazi, A. and Dixit, V.M. (1998) Death receptors: signalling and modulation. *Science* 281, 1305–1308.

Bedi, A., Pasricha, P.J., Akhtar, A.J., Barber, J.B., Bedi, G.C., Giardello, F.M., Zehnbauer, B.A., Hamilton, S.R. and Jones, R.J. (1995) Inhibition of apoptosis during development of colorectal cancer. *Cancer Research* 55, 1811–1816.

Boffa, L.C., Lupton, J.R., Mariani, M.R., Ceppi, M., Newmark, H.L., Scalmati, A. and Lipkin, M. (1992) Modulation of epithelial cell proliferation, histone acetylation, and luminal short chain fatty acids by variation of dietary fiber (wheat bran) in rats. *Cancer Research* 52, 5906–5912.

Browne, S.J., Williams, A.C., Hague, A., Butt, A.J. and Paraskeva, C. (1994) Loss of APC protein expressed by human colonic epithelial cells and the appearance of a specific low-molecular-weight form is associated with apoptosis *in vitro*. *International Journal of Cancer* 59, 56–64.

Browne, S.J., Macfarlane, M., Cohen, G.M. and Paraskeva, C. (1998) The adenomatous polyposis coli protein and retinoblastoma protein are cleaved early in apoptosis and are potential substrates for caspases. *Cell Death and Differentiation* 5, 206–213.

Burn, J., Chapman, P.D., Bishop, D.T. and Mathers, J. (1998) Diet and cancer prevention: the Concerted Action Polyp Prevention (CAPP) Studies. *Proceedings of the Nutrition Society* 57, 183–186.

Chan, S.L., Tammariello, S.P., Estus, S. and Mattson, M.P. (1999) Prostate apoptosis response-4 mediates trophic factor withdrawal-induced apoptosis of hippocampal neurones: actions prior to mitochondrial dysfunction and caspase activation. *Journal of Neurochemistry* 73, 502–512.

Chinery, R., Brockman, J.A., Peeler, M.O., Shyr, Y., Beauchamp, R.D. and Coffey, R.J. (1997) Antioxidants enhance the cytotoxicity of chemotherapeutic agents in colorectal cancer: a p53-independent function of p21WAF1/CIP1 via c/EBP beta. *Nature Medicine* 3, 1233–1241.

Clarke, R.G., Lund, E.K., Latham, P., Pinder, A.C. and Johnson, I.T. (1999) Effect of eicosapentaenoic acid on the proliferation and incidence of apoptosis in the colorectal cancer cell line HT29. *Lipids* 34, 1287–1295.

Cohen, G.M. (1997) Caspases: the executioners of apoptosis. *Biochemical Journal* 326, 1–16.

Collett, G.P., Robson, C.N., Mathers, J.C. and Campbell, F.C. (2001) Curcumin modifies Apc^{min} apoptosis resistance and inhibits 2-amino 1-methyl-6-phenylimidazo[4,5-b] pyridine (PhIP) induced tumour formation in Apc^{min} mice. *Carcinogenesis* 22, 821–825.

Collins, S., Cao, W., Dixon, T.M., Daniel, K.W., Onuma, H. and Medvedev, A.V. (2001) Body weight regulation, uncoupling proteins, and energy metabolism. In: Moustaïd-Moussa, N. and Berdanier, C.D. (eds) *Nutrient–Gene Interactions in Health and Disease*. CRC Press, Boca Raton, Florida, pp. 262–281.

Crew, T.E., Elder, D.J.E. and Paraskeva, C. (2000) A cyclooxygenase-2 (COX-2) selective non-steroidal anti-inflammatory drug enhances the growth inhibitory effect of butyrate in colorectal carcinoma cells expressing COX-2 protein: regulation of COX-2 by butyrate. *Carcinogenesis* 21, 69–77.

Cummings, J.H., Pomare, E.W., Branch, W.J., Naylor, C.P.E. and Macfarlane, G.T. (1987) Short chain fatty acids in the human large intestine, portal, hepatic and venous blood. *Gut* 28, 1221–1227.

D'Argenio, G., Cosenza, V., Delle Cave, M., Iovino, P., Della Valle, N., Lombardi, G. and Mazacca, G. (1996) Butyrate enemas in experimental colitis and protection against large bowel cancer in a rat model. *Gastroenterology* 110, 1727–1734.

Davidson, F.F. and Steller, H. (1998) Blocking apoptosis prevents blindness in *Drosophila* retinal degeneration mutants. *Nature* 391, 587–591.

Della Ragione, F., Criniti, V., Della Pietra, V., Borriello, A., Oliva, A., Indaco, S., Yamamato, T. and Zappia, V. (2001) Genes modulated by histone acetylation as new effectors of butyrate activity. *FEBS Letters* 499, 199–204.

Department of Health (1998) *Nutritional Aspects of the Development of Cancer*. Report on Health and Social Subjects 48. The Stationery Office, London.

Doll, R. and Peto, R. (1981) The causes of cancer: quantitative estimates of avoidable risks of cancer in the United States today. *Journal of the National Cancer Institute* 66, 1191–1308.

DuBois, R.N. and Smalley, W.E. (1996) Cyclooxygenase, NSAIDs, and colorectal cancer. *Journal of Gastroenterology* 31, 898–906.

Elder, D.J., Hague, A., Hicks, D.J. and Paraskeva, C. (1996) Differential growth inhibition by the aspirin metabolite salicylate in human colorectal tumor cell lines: enhanced apoptosis and *in vitro*-transformed adenoma relative to adenoma cell lines. *Cancer Research* 56, 2273–2276.

Evan, G. and Littlewood, T. (1998) A matter of life and cell death. *Science* 281, 1317–1322.

Fearon, E.R. and Vogelstein, B. (1990) A genetic model for colorectal tumorigenesis. *Cell* 61, 759–767.

Fimognari, C., Nüsse, M., Cesari, R., Iori, R., Cantelli-Forti, G. and Hrelia, P. (2002) Growth inhibition, cell-cycle arrest and apoptosis in human T-cell leukemia by the isothiocyanate sulforaphane. *Carcinogenesis* 23, 581–586.

Fishel, R. (2001) The selection for mismatch repair defects in hereditary nonpolyposis colorectal cancer: revising the mutator hypothesis. *Cancer Research* 61, 7369–7374.

Gamet-Payrastre, L., Li, P., Lumeau, S., Cassar, G., Dupont, M.A., Chevolleau, S., Gasc, N., Tulliez, J. and Tercé, F. (2000) Sulforaphane, a naturally occurring isothiocyanate, induces cell cycle arrest and apoptosis in HT29 human colon cancer cells. *Cancer Research* 60, 1426–1433.

Giovannicci, E., Rimm, E.B., Stampfer, M.J., Colditz, G.A., Ascherio, A. and Willett, W.C. (1994) Aspirin use and risk for colorectal cancer and adenoma in male health professionals. *Annals of Internal Medicine* 121, 241–246.

Graeber, T.G., Osmanian, C., Jacks, T., Housman, D.E., Koch, C.J., Lowe, S.W. and Giaccia, A.J. (1996) Hypoxia-mediated selection of cells with diminished apoptotic potential in solid tumours. *Nature* 379, 88–91.

Gross, A., McDonnell, J.M. and Korsmeyer, S.J. (1999) BCL-2 family members and the mitochondria in apoptosis. *Genes and Development* 13, 1899–1911.

Gupta, S. (2001) Molecular steps of death receptor and mitochondrial pathways of apoptosis. *Life Sciences* 69, 2957–2964.

Hague, A., Manning, A.M., Hanlon, K.A., Huschtscha, L.I., Hart, D. and Paraskeva, C. (1993) Sodium butyrate induces apoptosis in human colonic tumour cell lines in a p53-independent pathway: implications for the possible role of dietary fibre in the prevention of large-bowel cancer. *International Journal of Cancer* 55, 498–505.

Hague, A., Elder, D.J., Hicks, D.J. and Paraskeva, C. (1995) Apoptosis in colorectal tumour cells: induction by the short chain fatty acids butyrate, propionate and acetate and by the bile salt deoxycholate. *International Journal of Cancer* 60, 400–406.

Hanayama, R., Tanaka, M., Miwa, K., Shinohara, A., Iwamatsu, A. and Nagata, S. (2002) Identification of a factor that links apoptotic cells to phagocytes. *Nature* 417, 182–187.

Hanif, R., Qaio, L., Shiff, S.J. and Rigas, B. (1997) Curcumin, a natural plant phenolic food additive, inhibits cell proliferation and induces cell cycle changes in colon adenocarcinoma cell lines by a prostaglandin-independent pathway. *Journal of Laboratory and Clinical Medicine* 130, 576–584.

He, T.-C., Chan, T.A., Vogelstein, B. and Kinzler, K.W. (1999) PPARδ is a APC-regulated target of non-steroidal anti-inflammatory drugs. *Cell* 99, 335–345.

Hecht, S.S. (1999) Chemoprevention of cancer by isothiocyanates, modifiers of carcinogen metabolism. *Journal of Nutrition* 129, 768S–774S.

Heerdt, B.G., Houston, M.A. and Augenlicht, L.H. (1994) Potentiation by specific short-chain fatty acids of differentiation and apoptosis in human colonic carcinoma cell lines. *Cancer Research* 54, 3288–3293.

Hickman, M.J. and Samson, L.D. (1999) Role of DNA mismatch repair and p53 in signalling induction of apoptosis by alkylating agents. *Proceedings of the National Academy of Sciences USA* 96, 10764–10769.

Hollstein, M., Rice, K., Greenblatt, M.S., Soussi, T., Fuchs, R., Sorlie, T., Horvig, E., Smith-Sorensen, B., Montesano, R. and Harris, C.C. (1994) Database of p53 gene somatic mutations in human tumours and cell lines. *Nucleic Acids Research* 22, 3551–3555.

Janssen. P.L., Hollman, P.C., Reichman, E., Venema, D.P., van Staveren, W.A. and Katan, M.B. (1996) Urinary salicylate excretion in subjects eating a variety of diets shows that amounts of bioavailable salicylates are low. *American Journal of Clinical Nutrition* 64, 743–747.

Jee, S.H., Shen, S.C., Tseng, C.R., Chiu, H.C. and Kuo, M.L. (1998) Curcumin induces a p53-dependent apoptosis in human basal cell carcinoma cells. *Journal of Investigative Dermatology* 111, 656–666.

Jacobson, M.D. and McCarthy, N. (eds) (2002) *Apoptosis*. Frontiers in Molecular Biology 40, Oxford University Press, Oxford.

Johnson, I.T. (2001) Mechanisms and anticarcinogenic effects of diet-related apoptosis in the intestinal mucosa. *Nutrition Research Reviews* 14, 229–256.

Karlseder, J., Broccoli, D., Dai, Y., Hardy, S. and de Lange, T. (1999) p53- and ATM-dependent apoptosis induced by telomeres lacking TRF2. *Science* 283, 1321–1325.

Kawamori, T., Lubert, R., Steele, V.E., Kelloff, G.K., Kaskey, R.B., Rao, C.V. and Reddy, B.S. (1999) Chemopreventive effect of curcumin, a naturally-occurring anti-inflammatory agent, during promotion/progression stages of colon cancer. *Cancer Research* 59, 597–601.

Kerr, J.F., Wyllie, A.H. and Currie, A.R. (1972) Apoptosis: a basic biological phenomenon with wide-ranging implications in tissue kinetics. *British Journal of Cancer* 26, 239–257.

Kinzler, K.W. and Vogelstein, B. (1996) Lessons from hereditary colorectal cancer. *Cell* 87, 159–170.

Kliewer, S.A., Sundseth, S.S., Jones, S.A., Bron, P.J., Wisely, G.B., Koble, C.S., Devchand, P., Wahli, W., Wilson, T.M., Lenhard, J.M. and Lehmann, J.M. (1997) Fatty acids and eicosanoids regulate gene expression through direct interactions with peroxisome proliferator-activated receptors α and γ. *Proceedings of the National Academy of Sciences USA* 94, 4318–4323.

Kuo, M.L., Huang, T.S. and Lin, J.K. (1996) Curcumin, an antioxidant and anti-tumour promoter, induces apoptosis in human leukemia cells. *Biochimica et Biophysica Acta* 1317, 95–100.

Latham, P., Lund, E.K. and Johnson, I.T. (1999) Dietary n-3 PUFA increases the apoptotic response to 1,2-dimethylhydrazine, reduces mitosis and suppresses the induction of carcinogenesis in the rat colon. *Carcinogenesis* 20, 645–650.

Latham, P., Lund, E.K., Brown, J.C. and Johnson, I.T. (2001) Effects of cellular redox balance on induction of apoptosis by eicosapentaenoic acid in HT29 colorectal adenocarcinoma cells and rat colon *in vivo*. *Gut* 49, 97–105.

Lenardo, M., Chan, K.M., Hornung, F., McFarland, H., Siegel, R., Wang, J. and Zheng, L. (1999) Mature lymphocyte apoptosis – immune regulation in a dynamic and unpredictable antigenic environment. *Annual Reviews in Immunology* 17, 221–253.

Lewin, D. and Weinstein, W.M. (1996) Cell death – where is thy sting? *Gut* 39, 883–884.

Li, P., Nijhawan, D., Budihardjo, I., Srinivasula, S.M., Ahmad, M., Alnemri, E.S. and Wang, X. (1997) Cytochrome c and dATP-dependent formation of Apaf-1/Caspase-9 complex initiates and apoptotic protease cascade. *Cell* 91, 479–489.

Liu, Y. and Kulesz-Martin, M. (2001) p53 protein at the hub of cellular DNA damage response pathways through sequence-specific and non-sequence-specific DNA binding. *Carcinogenesis* 22, 851–860.

Lowe, S.W. and Lin, A.W. (2000) Apoptosis in cancer. *Carcinogenesis* 21, 485–495.

Mahmoud, N.N., Carothers, A.M., Grunberger, D., Bilinski, R.T., Churchill, M.R., Martucci, C., Newmark, H. and Bertagnolli, M.M. (2000) Plant phenolics decrease intestinal tumours in a murine model of familial adenomatous polyposis. *Carcinogenesis* 21, 921–927.

Mariadason, J.M., Corner, G.A. and Augenlicht, L.H. (2000) Genetic reprogramming in pathways of colonic cell maturation induced by short chain fatty acids: comparison with trichostatin A, sulindac, and curcumin and implications for chemoprevention of colon cancer. *Cancer Research* 60, 4561–4572.

Mathers, J.C. (2000) Food and cancer prevention: human intervention studies. In: Johnson, I.T. and Fenwick G.R. (eds) *Dietary Anticarcinogens and Antimutagens: Chemical and Biological Aspects*. Royal Society of Chemistry, Cambridge, pp. 395–403.

Mathers, J.C. and Burn, J. (1999) Nutrition in cancer prevention. *Current Opinion in Oncology* 11, 402–407.

Mathers, J.C., Smith, H. and Carter, S. (1997) Dose–response effects of raw potato starch on small intestinal escape, large-bowel fermentation and gut transit time in the rat. *British Journal of Nutrition* 78, 1015–1029.

McCaffrey, P., Andreola, F., Giandomenico, V. and De Luca, L.M. (2001) Vitamin A and gene expression. In: Moustad-Moussa, N. and Berdanier, C.D. (eds) *Nutrient–Gene Interactions in Health and Disease*. CRC Press, Boca Raton, Florida, pp. 283–319.

McCarthy N.J. (2002) Why be interested in death? In: Jacobson, M.D and McCarthy, N.J. (eds) *Apoptosis*. Oxford University Press, Oxford, pp. 1–22.

Meydani, S.N., Claycombe, K.J. and Sacristán, C. (2001) Vitamin E and gene expression. In: Moustaïd-Moussa, N. and Berdanier, C.D. (eds) *Nutrient–Gene Interactions in Health and Disease*. CRC Press, Boca Raton, Florida, pp. 393–424.

Moore, S.E., Cole, T.J., Poskitt, E.M., Sonko, B.J., Whitehead, R.G., McGregor, I.A. and Prentice, A.M. (1997) Season of birth predicts mortality in rural Gambia. *Nature* 388, 434.

Morin, P.J., Vogelstein, B. and Kinzler, K.Z. (1996) Apoptosis and *APC* in colorectal tumorigenesis. *Proceedings of the National Academy of Sciences USA* 93, 7950–7954.

Moss, S.F., Attia, L., Scholes, J.V., Walters, J.R.F. and Holt, P.R. (1996) Increased small intestinal apoptosis in coeliac disease. *Gut* 39, 811–817.

Norimura, T., Nomoto, S., Katsuki, M., Gondo, Y. and Kondo, S. (1996) *p53*-independent apoptosis suppresses radiation-induced teratogenesis. *Nature Medicine* 2, 577–580.

Ortiz, R., Cortéz, L., González-Márquez, H., Gómez, J.L., González, C. and Cortéz, E. (2001) Flow cytometric analysis of spontaneous and dexamethazone-induced apoptosis in thymocytes from severely malnourished rats. *British Journal of Nutrition* 86, 545–548.

Oshima, M., Takahashi, M., Oshima, H., Tsutsumi, M., Yazawa, K., Sugimura, T., Nishimura, S., Wakabayashi, K. and Taketo, M.M. (1995) Effects of docosahexaenoic acid (DHA) on intestinal polyp development in Apc delta 716 knockout mice. *Carcinogenesis* 16, 2605–2607.

Oshima, M., Dinchuk, J.E., Kargman, S.L., Oshima, H., Hancock, B., Kwong, E., Trzaskos, J.M., Evans, J.F. and Taketo, M.M. (1996) Suppression of intestinal polyps in ApcΔ716 knockout mice by inhibition of cyclooxygenase-2 (COX-2). *Cell* 87, 803–809.

Palozza, P., Calviello, G., Serini, S., Maggiano, N., Lanza, P., Raneletti, F.O. and Bartoli, G.M. (2001) β-carotene at high concentrations induces apoptosis by enhancing oxy-radical production in human adenocarcinoma cells. *Free Radicals in Biology and Medicine* 30, 1000–1007.

Palozza, P., Serini, S., Maggiano, N., Angeline, M., Boninsegna, A., Di Nicuolo, F., Ranelletti, F.O. and Calviello, G. (2002) Induction of cell cycle arrest and apoptosis in human colon adenocarcinoma cell lines by β-carotene through down-regulation of cyclin A and Bcl-2 family proteins. *Carcinogenesis* 23, 11–18.

Park, H.S., Ryu, J.H., Ha, Y.L. and Park, J.H.Y. (2001) Dietary conjugated linoleic acid (CLA) induces apoptosis of colonic mucosa in 1,2-dimethlyhydrazine-treated rats: mechanism of the anticarcinogenic effect by CLA. *British Journal of Nutrition* 86, 549–555.

Parodi, P.W. (1997) Cows' milk fat components as potential anticarcinogenic agents. *Journal of Nutrition* 127, 1055–1060.

Peifer, M. and Polakis, P. (2000) Wnt signalling in oncogenesis and embryogenesis – a look outside the nucleus. *Science* 287, 1606–1609.

Petrik, M.B., McEntee, M.F., Johnson, B.T., Obukowicz, M.G. and Whelan, J. (2000) Highly-unsaturated (n-3) fatty acids, but not α-linolenic, conjugated linoleic and gamma-linolenic acids, reduce tumorigenesis in Apc$^{Min/-}$ mice. *Journal of Nutrition* 130, 2434–2443.

Plummer, S.M., Holloway, K.A., Manson, M.M., Munks, R.J., Kaptein, A., Farrow, S. and Howells, L. (1999) Inhibition of cyclooxygenase 2 expression in colon cells by the chemopreventive agent curcumin involves inhibition of NFκB activation via the NIK/IKK signalling complex. *Oncogene* 18, 6013–6020.

Pitti, R.M., Marsters, S.A., Lawrence, D.A., Roy, M., Kischkel, F.C., Dowd, P., Huang, A., Donahue, C.J., Sherwood, S.W., Baldwin, D.T., Godowski, P.J., Wood, W.I., Gurney, A.L., Hillan, K.J., Cohen, R.L., Goddard, A.D., Botstein, D. and Ashkenazi, A. (1998) Genomic amplification of a decoy receptor for Fas ligand in lung and colon cancer. *Nature* 396, 699–703.

Prentice, A.M. (1999) The thymus: a barometer of malnutrition. *British Journal of Nutrition* 81, 345–347.

Reed, J.C. (1998) Bcl-2 family proteins. *Oncogene* 17, 3225–3236.

Renehan, A.G., Booth, C. and Potten, C.S. (2001) What is apoptosis, and why is it important? *British Medical Journal* 322, 1536–1538.

Roy, N. and Cardone, M.H. (2002) The caspases: consequential cleavage. In: Jacobson, M.D and McCarthy, N.J. (eds) *Apoptosis*. Oxford University Press, Oxford, pp. 93–135.

Samaha, H.S., Kelloff, G.J., Stelle, V., Rao, C.V. and Reddy, B.S. (1997) Modulation of apoptosis by sulindac, curcumin, phenylethyl-3-methyl cafeate and 6-phenylhexyl isothiocyanate: apoptotic index as a biomarker in colon cancer chemoprevention and promotion. *Cancer Research* 59, 1301–1305.

Sarraf, P., Mueller, E., Jones, D., King, F.J., DeAngelo, D.J., Partridge, J.B., Holden, S.A., Chen, L.B., Singer, S., Fletcher, C. and Spiegelman, B.M. (1998) Differentiation and reversal of malignant changes in colon cancer through PPARγ. *Nature Medicine* 4, 1046–1052.

Scaglia, L., Cahill, C.J., Finegood, D.T. and Bonner-Weir, S. (1997) Apoptosis participates in the remodelling of the endocrine pancreas in the neonatal rat. *Endocrinology* 138, 1736–1741.

Singh, S. and Aggarwal, B.B. (1995) Activation of transcription factor NF-κB is suppressed by curcumin (diferuloylmethane). *Journal of Biological Chemistry* 270, 24995–25000.

Smith, T.K., Lund, E.K., Musk, S.S.R. and Johnson, I.T. (1998) Inhibition of DMH-induced aberrant crypt foci, and induction of apoptosis in rat colon, following oral administration of a naturally occurring glucosinolate. *Carcinogenesis* 19, 267–273.

Sperandio, S., de Belle, I. and Bredesen, D.E. (2000) An alternative, nonapoptotic form of programmed cell death. *Proceedings of the National Academy of Sciences USA* 97, 14376–14381.

Steinback, G., Lynch, P.M., Phillips, R.K., Wallace, M.H., Hawk, E., Gordon, G.B., Wakabayashi, N., Shen, Y., Fujimura, T., Su, L.K. and Levin, B. (2000) The effect of celecoxib, a cyclooxygenase-2 inhibitor, in familial adenomatous polyposis. *New England Journal of Nutrition* 342, 1946–1952.

Stoll, B.A. (2002) n-3 Fatty acids and lipid peroxidation in breast cancer inhibition. *British Journal of Nutrition* 87, 193–198.

Strachan. T. and Read, A.P. (1999) *Human Molecular Genetics 2*. BIOS Scientific Publishers, Oxford.

Sträter, J., Wellish, I., Riedl, S., Walczak, H., Koretz, K., Tandara, A., Krammer, P.H. and Möller, P. (1997) CD95 (APO-1/Fas)-mediated apoptosis in colon epithelial cells: a possible role in ulcerative colitis. *Gastroenterology* 113, 160–167.

Thompson, C.B. (1995) Apoptosis in the pathogenesis and treatment of disease. *Science* 267, 1456–1462.

Thornberry, N.A. and Lazebnik, Y. (1998) Caspases: enemies within. *Science* 281, 1312–1316.

Thornberry, N.A., Bull, H.G., Calaycay, J.R., Chapman, K.T., Howard, A.D., Kostura, M.J., Miller, D.K., Molineau, S.M., Weidner, J.R. and Auins, J. (1992) A novel heterodimeric cysteine protease is required for interleukin-1β processing in monocytes. *Nature* 356, 768–774.

Tomlinson, I.P.M. and Bodmer, W.F. (1995) Failure of programmed cell death and differentiation as causes of tumors: some simple mathematical models. *Proceedings of the National Academy of Sciences USA* 92, 11130–11134.

Tsujimoto, Y. (2002) Regulation of apoptosis by the Bcl-2 family of proteins. In: Jacobson, M.D and McCarthy, N.J. (eds) *Apoptosis*. Oxford University Press, Oxford, pp. 136–160.

Vaux, D.L. (2002) Viruses and apoptosis. In: Jacobson, M.D and McCarthy, N.J. (eds) *Apoptosis*. Oxford University Press, Oxford, pp. 262–277.

Voehringer, D.W., Hirschberg, D.L., Xiao, J., Lu, Q., Roederer, M., Lock C.B., Herzenberrg, L.A., Steinman, L. and Herzenberg, L.A. (2000) Gene microarray identification of redox and mitochondrial elements that control resistance or sensitivity to apoptosis. *Proceedings of the National Academy of Sciences USA* 97, 2680–2685.

Webb, S.J., Nicholson, D., Bubb, V.J. and Wyllie, A.H. (1999) Caspase-mediated cleavage of APC results in an amino-terminal fragment with an intact armadillo repeat domain. *FASEB Journal* 13, 339–346.

Whelan, J. and McEntee, M.F. (2001) Dietary fats and APC-driven intestinal tumorigenesis. In: Moustaïd-Moussa, N. and Berdanier, C.D. (eds) *Nutrient–Gene Interactions in Health and Disease*. CRC Press, Boca Raton, Florida, pp. 231–260.

White, E. (1996) Life, death, and the pursuit of apoptosis. *Genes and Development* 10, 1–15.

Wu, J.T., Archer, S.Y., Hinnebusch, B., Meng, S. and Hodin, R.A. (2001) Transient vs. prolonged histone hyperacetylation: effects on colon cancer cell growth, differentiation, and apoptosis. *American Journal of Physiology* 280, G482–G490.

Xue, D., Wu, C.-I. and Shah, M.S. (2002) Programmed cell death in *C. elegans*: the genetic framework. In: Jacobson, M.D and McCarthy, N.J. (eds) *Apoptosis*. Oxford University Press, Oxford, pp. 23–55.

Yang, W.-L. and Frucht, H. (2001) Activation of the PPAR pathway induces apoptosis and COX-2 inhibition in HT-29 human colon cancer cells. *Carcinogenesis* 22, 1379–1383.

Yuan, J., Shaham, S., Ledoux, S., Ellis, H.M. and Horvitz, H.R. (1993) The *C. elegans* death gene *ced-3* encodes a protein similar to mammalian interleukin-1β-converting enzyme. *Cell* 75, 641–652.

Zou, H., Henzel, W.J., Liu, X., Lutschg, A. and Wang, X. (1997) Apaf-1, a human protein homologous to *C. elegans* CED-4, participates in cytochrome c-dependent activation of caspase-3. *Cell* 90, 405–413.

7 Glucose Regulation of Gene Expression in Mammals

Fabienne Foufelle and Pascal Ferré
INSERM Unit 465, Paris 6 University, Centre de Recherches Biomédicales des Cordeliers, Paris, France

Introduction

Adaptation to the nutritional environment through the modulation of gene expression is a common requirement for all living organisms from bacteria to humans. In unicellular organisms, nutrients themselves are the signals that induce these adaptations. One of the first examples of gene regulation by nutrients was described for the lactose operon in *Escherichia coli*. In mammals, a potential effect of nutrients has been obscured by the importance devoted to hormones in the response to the nutritional environment. It is only recently that a more immediate role of nutrients has been appreciated.

In this review, we will focus on the regulation of gene expression by glucose, and more specifically on hepatic genes since the understanding of cellular and molecular mechanisms in hepatocytes recently has made important progress.

Our View of a Glucose-responsive Gene in a Nutritional Context

What are the criteria allowing classification of a gene as a specifically glucose-responsive gene in a nutritional context? First, it must be shown that, *in vivo*, the expression of the gene is modulated by glucose availability in the diet. Secondly, *in vitro* experiments must differentiate between a hormonal (insulin for instance) and a glucose effect on gene expression. Thirdly, the range of concentrations of glucose for it to have an effect on gene expression must be reasonable. Although it is acceptable to use extreme concentrations to detect a given phenomenon, the observation has to apply when more physiological variations of glucose concentrations are investigated (usually in the 4–15 mM range). One must be very cautious when comparing low (or zero) glucose with high glucose concentrations *in vitro*, since in this case gene regulation might be secondary to a stress phenomenon linked to energy depletion or glycosylation-defective processes, as seen for glucose-regulated proteins such as GRP78/Bip. The effects of glucose need to be distinguished from other kinds of cellular stress.

Physiological Basis for the Regulation of Hepatic Gene Expression by Glucose

In mammals, glucose is utilized continuously at high rates by organs such as the brain (120 g day^{-1} in humans), red blood cells and renal medulla. When a meal that contains carbohydrate is absorbed, it induces several metabolic events aimed at decreasing endogenous glucose production by the liver (glycogenolysis and gluconeogenesis) and increasing glucose uptake and storage in the form of glycogen in the liver and muscle. Absorbed glucose is delivered to the liver via the portal vein; once the hepatic glycogen

©CAB *International* 2003. *Molecular Nutrition*
(eds J. Zempleni and H. Daniel)

stores are repleted, hepatocytes convert glucose into lipids (lipogenesis), which are exported as very low-density lipoproteins (VLDLs) and ultimately stored as triglycerides in adipose tissue. Conversely, if glucose availability in the diet is reduced, glucose-utilizing pathways are inhibited and glucose-producing pathways are activated. The regulation of metabolic pathways involves the rapid modulation of the activity of specific proteins (enzymes, transporters) but also a (slower) change in the quantities of these proteins. This is achieved mainly by modulating their transcription rate.

Expression of Glycolytic/Lipogenic and Gluconeogenic Genes is Regulated by the Carbohydrate Availability in the Diet

The expression of several key glycolytic and lipogenic enzymes is induced by a high-carbohydrate diet in the liver: glucokinase (Iynedjian et al., 1987), 6-phosphofructo-1-kinase (Rongnoparut et al., 1991), 6-phosphofructo-2-kinase/fructose-2-6-bisphosphatase (Colosia et al., 1988), aldolase B (Weber et al., 1984) and L-pyruvate kinase (L-PK) (Vaulont et al., 1986) for glycolysis; ATP-citrate lyase (Elshourbagy et al., 1990), acetyl-CoA carboxylase (ACC) (Pape et al., 1988), fatty acid synthase (FAS) (Paulauskis and Sul, 1989; Katsurada et al., 1990), stearoyl-CoA desaturase (Ntambi, 1992) for lipogenesis; and glucose-6-phosphate dehydrogenase (Kletzien et al., 1985; Katsurada et al., 1989) and 6-phosphogluconate dehydrogenase (Miksicek and Towle, 1983) for the pentose-phosphate pathway. The induction of the S14 gene, which encodes a small acidic polypeptide that seems to be related to lipogenesis, should also be mentioned here. S14 is expressed in lipogenic tissues (white and brown adipose tissue, liver, lactating mammary gland) and shares a similar regulation with genes involved in the lipogenic pathway (Mariash et al., 1986; Kinlaw et al., 1987; Clarke et al., 1990). The specific function of the S14 protein remains unclear, although some studies have suggested that S14 might be involved in the machinery that regulates expression of lipogenic enzymes (Kinlaw et al., 1995; Zhu et al., 2001). For most of these genes involved in glucose carbon utilization, the induction of their mRNA synthesis by a carbohydrate-rich diet is substantial (from 4- to 25-fold), rapid (1–2 h after stimulation) and involves a transcriptional mechanism.

Conversely, a high-carbohydrate diet inhibits the expression of gluconeogenic enzymes such as phosphoenolpyruvate carboxykinase (PEPCK) (Granner et al., 1983) and glucose-6-phosphatase (Argaud et al., 1997). The inhibition is rapid, but also easily reversible once the carbohydrate availability decreases.

Absorption of carbohydrates causes increased concentrations of substrates such as glucose and lactate, and affects concentrations of the pancreatic hormones insulin and glucagon. Differentiating between the respective roles of hormones and substrates in the regulation of gene expression is not easy in *in vivo* experiments. Unravelling of the mechanisms by which hormones and substrates modulate gene transcription has benefited greatly from *in vitro* studies using either primary cultured hepatocytes or cell lines. From these studies, different kinds of gene regulation have emerged: some genes (such as glucokinase) can be induced by insulin in a glucose-independent fashion (Iynedjian et al., 1989); other genes require increased concentrations of both insulin and glucose in order to be induced, e.g. L-PK, FAS, ACC, S14 and stearoyl-CoA desaturase (Decaux et al., 1989; Waters and Ntambi, 1994; Prip-Buus et al., 1995; Koo et al., 2001; O'Callaghan et al., 2001). PEPCK expression can be down-regulated independently by insulin (Sasaki et al., 1984) and glucose (Scott et al., 1998; Cournarie et al., 1999). Finally, the expression of glucose-6-phosphatase is decreased by insulin but paradoxically increased by a high glucose concentration (Argaud et al., 1997; Van Schaftingen and Gerin, 2002). Here, we will focus on the mechanisms by which glucose alone modulates gene expression in the liver. The effects of insulin on gene expression have been reviewed elsewhere (Foufelle and Ferré, 2002).

Effects of Glucose on the Expression of Glycolytic and Lipogenic Enzymes

As stated above, some of the glycolytic and lipogenic genes in the liver require increased concentrations of both insulin and glucose to achieve maximal expression (Girard et al., 1997; Towle et al., 1997; Vaulont et al., 2000). This regulation involves the stimulation of their transcription rate.

In the absence of glucose, insulin is unable to induce their expression. In the absence of insulin, the effect of glucose is greatly reduced in adipose tissue and nearly absent in cultured hepatocytes. One potential reason why these genes require both a high insulin and a high glucose concentration could be that after a meal, the metabolic priority is to replenish glycogen stores, and it is only if glucose is particularly abundant that the glucose carbons are utilized in lipid synthesis.

The Glucose Effect: the Metabolic Hypothesis

Glucose enters hepatocytes by using a glucose transporter (GLUT2) that has a high Michaelis–Menten constant (K_m) for glucose. In contrast to the GLUT4 transporter that is found in muscle and adipose tissue, GLUT2 activity is independent of insulin. Once inside the cell, glucose is phosphorylated by glucokinase (a high K_m hexokinase) to produce glucose-6-phosphate. GLUT2 and glucokinase mediate a rate of phosphorylation that approximately parallels the plasma concentration of glucose. Glucose-6-phosphate can then enter the glycogen synthesis pathway, the glycolytic pathway or the pentose-phosphate pathway.

Evidence has been provided that glucose has to be metabolized in order to affect transcription (Foufelle et al., 1996). For instance, in the liver, the glucose effect requires the presence of glucokinase, the enzyme responsible for glucose phosphorylation into glucose-6-phosphate (Doiron et al., 1994; Prip-Buus et al., 1995). Glucokinase expression is strongly activated by insulin; thus, the insulin dependency of genes was first explained by the insulin-mediated induction of glucokinase, leading to metabolism of glucose. However, there is also evidence for a direct effect of insulin, mediated by the insulin-responsive transcription factor SREBP-1c (Foufelle and Ferré, 2002).

The questions which then arise concern: (i) the nature of the glucose metabolite that affects transcription (which metabolite?); (ii) the mechanism of transduction by which the glucose metabolite affects the transcriptional machinery; (iii) the glucose response element in gene promoters; and (iv) the nature of the transcription factor that is involved.

For a metabolite to qualify as a signal, its concentration must vary in parallel with the extracellular glucose concentration since the glucose effect on gene expression is concentration-dependent; also, changes in metabolite concentrations must precede changes in mRNA concentrations. In previous studies, we have proposed that glucose-6-phosphate might be the metabolite that affects transcription of FAS (Foufelle et al., 1992; Prip-Buus et al., 1995) and genes that belong to the same class, i.e. L-PK and S14. This proposal is based on the observations: (i) that in adipose tissue and a β-cell line (INS1), the effect of glucose can be mimicked by 2-deoxyglucose, a glucose analogue that cannot be metabolized beyond phosphorylation to 2-deoxyglucose-6-phosphate, which accumulates in cells (Foufelle et al., 1992; Brun et al., 1993; Marie et al., 1993); (ii) that the intracellular concentration of glucose-6-phosphate parallels the expression of genes (Foufelle et al., 1992; Prip-Buus et al., 1995; Mourrieras et al., 1997); and (iii) that, in vivo, the time course of hexose-phosphate concentrations fits the time-related pattern of gene induction (Munnich et al., 1987). Consistent with this hypothesis, glucose-6-phosphate serves as a signal of glucose availability in hepatic glycogen metabolism since it is an activator of glycogen synthase activity (Bollen et al., 1998). Interestingly, fasted GLUT2-null mice have a paradoxical increase in L-PK expression and glycogen content that is paralleled by concentrations of glucose-6-phosphate, which remain at a high level (Burcelin et al., 2000). Another interesting situation is found when rats are injected with an inhibitor of the glucose-6-phosphate translocator activity, a component of the glucose-6-phosphatase system. This induces a large increase in hepatic glucose-6-phosphate, glycogen and triglyceride concentrations and activation of FAS and ACC gene expression (Bandsma et al., 2001).

Xylulose-5-phosphate, an intermediate of the non-oxidative pathway of glucose metabolism, may also play a role in signalling (Doiron et al., 1996), based on the following observations: (i) in some cells, 2-deoxyglucose-6-phosphate can be metabolized further to intermediates of the pentose-phosphate pathway; (ii) xylitol, a precursor of xylulose-5-phosphate, stimulates the transcription of a reporter gene driven by the L-PK promoter in a hepatocyte cell-line (AT3F) at a low concentration (0.5 mM) without detectable changes in glucose-6-phosphate concentration;

(iii) in primary cultured hepatocytes, xylitol (5–10 mM) is able to induce L-PK mRNA concentrations by sixfold (Doiron et al., 1996); and (iv) xylulose-5-phosphate activates phosphatase 2A-mediated dephosphorylation, and this phosphatase is also involved in the dephosphorylation of transcription factors (Nishimura and Uyeda, 1995). Glucose, xylitol and dihydroxyacetone enter the glycolytic/pentose-phosphate pathways at various levels; in contrast to glucose-6-phosphate, induction of FAS and S14 gene expression did not parallel concentrations of xylulose-5-phosphate in hepatocytes that were cultured at various concentrations of glucose, xylitol and dihydroxyacetone (Mourrieras et al., 1997). Interestingly, xylitol causes a dose-dependent increase in glucose-6-phosphate concentrations. This finding is not unexpected since metabolites of the pentose-phosphate pathway can be recycled in glycolysis. Thus, in these experiments, xylulose-5-phosphate does not qualify as a signal metabolite for FAS and S14 induction. A clear role for glucose metabolites can only be identified after the full transduction machinery has been characterized.

The Glucose Effect: the Non-metabolic Hypothesis

Membrane glucose sensors have been described in yeast (Johnston, 1999). These proteins, called SnF3 and Rgt2, have a structure that is similar to that of glucose transporters although they appear unable to transport glucose. They have a large intracytoplasmic loop, which seems to be involved in the glucose signalling pathway. It has been hypothesized that the glucose transporter GLUT2 could be a glucose sensor in hepatocytes. It has been shown that the large intracytoplasmic loop of GLUT2 may play a role in the glucose-dependent control of L-PK expression (Guillemain et al., 2000, 2002). Interaction of glucose with GLUT2 might generate a signal through this loop, leading to modifications of signal transducers. Recently, using a two-hybrid system with the GLUT2 loop, karyopherin α2 (a receptor involved in nuclear import) was identified as a potential binding partner (Guillemain et al., 2002). Overexpression of an inactive karyopherin α2 in hepatoma cells led to decreased glucose-induced L-PK gene expression. Thus this mechanism could act in concert with the signalling system involving glucose metabolism in the liver.

What is the Cellular Cascade from Glucose to the Transcription Machinery?

In order to explain the glucose responsiveness of tissues, the following hypotheses have been considered: (i) the presence of glucose might modify the nuclear amount of transcription factors (including modifications of cellular localization); and (ii) transcription factors undergo post-translational modifications in the presence of glucose, leading to a modification of interactions with the basic transcription machinery or with co-activator proteins. Post-translational modifications of a transcription factor might involve allosteric modification of the protein by binding of the glucose-related signal metabolite, modifications through phosphorylation/dephosphorylation mechanisms by modulating the activity of a nuclear protein kinase/phosphatase, or both. In this context, it has been shown that protein phosphatase inhibitors prevent the effect of glucose on FAS, ACC and S14 gene expression (Sudo and Mariash, 1994; Daniel et al., 1996; Foretz et al., 1998), suggesting that the glucose stimulatory effect could involve a dephosphorylation mechanism.

The AMP-activated protein kinase (AMPK) provides a potential candidate for a protein kinase involved in the regulation of glucose-activated genes. AMPK is a serine/threonine kinase acting as a metabolic 'master switch' by phosphorylating key enzymes involved in cholesterol and fatty acid metabolism. Indeed, AMPK phosphorylates and inactivates ACC and 3-hydroxy-3-methylglutaryl-CoA reductase, resulting in the inhibition of both lipogenesis and cholesterol synthesis (Carling et al., 1987). AMPK is activated by stresses that deplete ATP and increase AMP within the cell, such as hypoxia or muscle contraction. AMP activates AMPK by two mechanisms: (i) allosteric activation of AMPK; and (ii) stimulation of an AMPK kinase, leading to phosphorylation of AMPK (Hardie et al., 1998).

A significant clue regarding a possible role for AMPK in the regulation of gene transcription came from the finding that it is structurally and functionally related to the yeast protein kinase complex Snf1 (sucrose non-fermenting) (Woods et al., 1994). In yeast, the transcription of a number of genes is repressed by high concentrations of glucose (Gancedo, 1998). The kinase activity of Snf1 is essential for the derepression of these genes in yeast grown under conditions of glucose limitation. AMPK and Snf1 both form heterotrimeric complexes consisting of one catalytic subunit and two regulatory subunits (Hardie and Carling, 1997; Hardie et al., 1998). The amino acid sequences of the mammalian AMPK subunits are highly homologous to their counterparts in the Snf1 complex, and the two kinases show functional similarities (Woods et al., 1994).

These findings led several groups to speculate that AMPK may be involved in the regulation of gene transcription by glucose in mammals. Evidence that this may be the case came from studies in which AMPK in hepatocytes was activated by incubation with 5-amino-imidazolecarboxamide (AICA) riboside, a cell-permeable activator of AMPK, leading to inhibition of the glucose-induced expression of FAS, L-PK and S14 (Foretz et al., 1998; Leclerc et al., 1998). A similar inhibition of FAS, L-PK and S14 gene expression was obtained when AMPK activity was increased in cultured hepatocytes by the overexpression of a constitutively active form of AMPK (Woods et al., 2000). These results imply that AMPK is involved in the repression of genes induced by glucose in hepatocytes.

Since an increase in AMPK activity inhibits glucose-activated gene transcription, a decrease in AMPK activity could be part of a mechanism involved in the stimulation of gene transcription by glucose. However, inhibition of AMPK activity using a dominant-negative form of the kinase has no detectable effect on any of the glucose-induced genes (Woods et al., 2000). In addition, changing the glucose concentration in the medium of cultured hepatocytes from 5 to 25 mM has no inhibitory effects on AMPK activity (Foretz et al., 1998). Taken together, these results suggest that in liver, glucose does not exert its effects on gene expression by directly inhibiting AMPK.

In contrast, the activity of AMPK increases in β-cell lines in response to a decrease in the extracellular glucose concentration from 30 to 3 mM (Salt et al., 1998; da Silva Xavier et al., 2000). This is paralleled by an increased promoter activity of a transfected L-PK promoter. In MIN6 cells cultured at 3 mM glucose, inhibiting the AMPK activity in both cytoplasm and nucleus by injection of an α2 AMPK antibody mimics the effect of a high glucose concentration on the L-PK promoter activity (da Silva Xavier et al., 2000). Thus, it is feasible that in cells that are particularly sensitive to a low concentration of glucose (such as β-cells), inhibition of AMPK activity during periods of increasing glucose concentrations is part of the mechanism leading to increased gene transcription.

In conclusion, AMPK inhibits the expression of genes encoding lipogenic and glycolytic enzymes in the liver. However, the inhibitory effect of AMPK does not affect induction of gene expression by glucose. Since other genes that are not controlled by glucose are also inhibited by AMPK (Hubert et al., 2000; Lochhead et al., 2000; Zhou et al., 2000; Zheng et al., 2001; MacLean et al., 2002), this suggests that AMPK could have a general inhibitory effect on genes encoding enzymes involved in ATP-consuming pathways.

Cis-acting DNA Sequences Involved in Glucose Response

The first glucose response element (GlRE) to be identified was that of the L-PK gene (Thompson and Towle, 1991; Bergot et al., 1992). This element is located within 183 bp upstream of the transcription start site of the L-PK gene, a region that mediates specificity of tissue expression as well as nutrient and hormone response in transgenic animals (Cuif et al., 1992). Subsequent studies localized GlRE to a region between bp −183 and −96 of the transcription start site, as judged by transfection experiments in isolated hepatocytes (Thompson and Towle, 1991; Bergot et al., 1992; Diaz-Guerra et al., 1993). Another glucose response element named ChoRE for carbohydrate response element was then characterized in the S14 gene. The region between bp −1439 and −1423 of the transcription start site of the rat S14 gene was found to be essential for glucose responsiveness in isolated hepatocytes (Shih and Towle, 1992). In the mouse S14 promoter, ChoRE is located between bp −1450 and −1425 of the transcription start site (Koo and Towle, 2000). The complex that binds to the GlRE and ChoRE of the L-PK and S14 genes, respectively, cooperates with an adjacent auxiliary site to confer full glucose responsiveness (Bergot et al., 1992; Diaz-Guerra et al., 1993; Shih et al., 1995). For the

L-PK gene, the glucose responsiveness is conferred through the close cooperation of L3 (bp −126 to −144) and L4 (bp −145 to −168) elements (Bergot et al., 1992; Liu et al., 1993). A multimer of the L4 element is able to confer glucose responsiveness to a glucose-unresponsive heterologous promoter (Bergot et al., 1992); the full glucose response is conferred only when both L3 and L4 elements are present. For the S14 gene, the presence of an element located between bp −1467 and −1448 of the transcription start site is necessary for the full glucose responsiveness (Shih et al., 1995; Kaytor et al., 1997). Comparison of ChoRE and G1RE in S14 and L-PK genes, respectively, revealed the following similarities. The glucose response element consists of two E-box or E-box-like sequences of the form CANNTG, separated by 5 bp. The presence of these two E-box motifs is critical for the glucose response. More recently, a ChoRE/GlRE has also been identified in the rat FAS promoter between bp −7214 and −7190 of the transcription start site (Rufo et al., 2001). This region contains a palindromic sequence (CATGTG(n)$_5$GGCGTG) that is similar to the ChoRE in the L-PK and S14 genes. As shown previously for L-PK and S14, this ChoRE confers glucose responsiveness to a heterologous promoter.

For the ACC gene, a region sensitive to glucose was first identified in the PII promoter −340 to −249 bp upstream of the transcription start site. The PII promoter is expressed uniformly in all tissues and does not respond to dietary changes (Kim et al., 1996). The transcriptional activity of this promoter is induced threefold by glucose in the 30A5 adipocyte cell line (Daniel and Kim, 1996). Within this region, binding sites for SREBP-1 and Sp1 have been identified (Daniel and Kim, 1996; Lopez et al., 1996). It has been shown that glucose increases the binding of Sp1 through a mechanism involving dephosphorylation by protein phosphatase 1 (Daniel et al., 1996). In contrast to PII, the ACC PI promoter is expressed predominantly in adipose tissue and liver and its activity is elevated markedly in response to diets inducing lipogenesis (Lopez-Casillas et al., 1991; Kim et al., 1996). The PI transcriptional activity is induced fivefold by glucose in cultured hepatocytes. A ChoRE has been identified in this promoter between bp −126 and −102 of the transcription start site (O'Callaghan et al., 2001). This element is homologous to those described in the FAS, S14 and L-PK

promoters. A comparison of the sequences of these various ChoRE/GlREs is shown in Table 7.1. The ChoRE is described as two E-box or E-box-like sequences of the form CANNTG, separated by a defined distance. The first 4 bp of each E-box and the length of the spacing sequence are critical for the glucose response (Shih and Towle, 1994; Shih et al., 1995).

Characterization of *Trans*-acting Factors

The E-box described in the L-PK and S14 glucose response element is a binding site for the basic helix–loop–helix (bHLH) class of transcription factors. All these proteins have (i) a basic domain, a stretch of amino acids rich in lysine and arginine residues; and (ii) two amphipatic α-helices connected by a loop (Gregor et al., 1990). These structures are involved in the binding to DNA and in protein–protein interactions. Among the transcription factors belonging to this class of proteins, the transcription factor USF (upstream stimulatory factor) was presented as a potential candidate for a carbohydrate responsive factor since it was the main protein binding to ChoRE/GlRE. USF was first found as a protein binding to the adenovirus late promoter (Sawadago and Roeder, 1985). Two USF proteins, named USF1 and USF2, with identical DNA-binding properties have been characterized (Sawadogo, 1988; Sirito et al., 1994); they differ only in their N-terminal domain. USF forms dimers (USF1–USF2 in the liver) prior to DNA binding (Viollet et al., 1996). USF tissue distribution is ubiquitous and does not

Table 7.1. Comparison of the sequences of various ChoRE/GlREs.

Genes	Position	Glucose-responsive sequence
Rat L-PK	−166	**CACGGG**GCACT**CCCGTG**
Rat S14	−1439	**CACGTG**GTGGC**CCTGTG**
Mouse S14	−1442	**CACGCT**GGAGT**CAGCCC**
Rat FAS	−7210	**CATGTG**CCACA**GGCGTG**
Rat ACC PI	−122	**CATGTG**AAAAC**GTCGTG**

S14 = Spot14; L-PK = liver pyruvate kinase; ACC = acetyl-CoA carboxylase; FAS = fatty acid synthase. From Bergot et al. (1992); Liu et al. (1993); Shih et al. (1995); Koo and Towle (2000); Rufo et al. (2001).

seem to be altered by the nutritional or hormonal status. Binding of USF to the E-box of GlRE has been demonstrated for L-PK and S14 (Vaulont et al., 1989; Diaz-Guerra et al., 1993; Shih and Towle, 1994; Lefrancois-Martinez et al., 1995). A series of studies tested this hypothesis, including overexpression of wild-type and dominant-negative forms of USF in cultured cells (Kaytor et al., 1997, 2000), deletions of USF1 and USF2 genes in mice (Vallet et al., 1997, 1998), investigations of USF synthesis and binding activity in response to diets and investigations of USF binding activities in wild-type and mutated forms of GlRE/ChoRE in parallel with their glucose responsiveness. From the bulk of these studies, it was concluded that these factors alone cannot explain the glucose responsiveness through GlRE/ChoRE (Kaytor et al., 2000).

Other proteins were then shown to bind to GlRE/ChoRE, and some of these proteins were presented as new candidates for a glucose-responsive transcription factor although they were not characterized further (Hasegawa et al., 1999; Lou et al., 1999; Koo and Towle, 2000). It is only recently that one of the proteins that binds to GlRE/ChoRE has been identified and emerged as a potential glucose-responsive transcription factor. Uyeda's group has purified a transcription factor from rat liver, based on its capacity to bind to ChoRE in the L-PK promoter; the protein has been named ChREBP (carbohydrate response element-binding protein) (Yamashita et al., 2001). This factor fulfils a number of criteria for a *bona fide* ChoRE-binding protein. ChREBP is a bHLH leucine zipper protein with a bipartite nuclear localization signal, and its capacity to bind to mutated ChoRE parallels the capacity of these ChoRE to respond to glucose concentrations in hepatocyte cultures. When transfected into hepatocytes, ChREBP is able to stimulate the L-PK promoter containing ChoRE; this effect is increased dramatically in the presence of glucose. Finally, although the mRNA of this factor is found in tissues such as the cerebellum, intestine, kidney and liver, binding of the factor to ChoRE is found only in liver (Yamashita et al., 2001).

The effects of glucose on gene transcription may be linked to a translocation of ChREBP from the cytoplasm to the nucleus in response to glucose. ChREBP is located in the cytoplasm when the glucose concentration is low, and enters the nucleus at high glucose concentrations (Kawaguchi et al., 2001). This would be secondary to a dephosphorylation of a specific serine (Ser166) that is phosphorylated by protein kinase A (PKA) when concentrations of cAMP are high, e.g. during fasting. A second PKA phosphorylation site near the DNA-binding domain precludes ChREBP binding to DNA if phosphorylated; this site also becomes dephosphorylated in the presence of high glucose concentrations (Kawaguchi et al., 2001). Thus, the main effect of glucose would be to activate a phosphatase counteracting the effect of cAMP, inducing the translocation of ChREBP into the nucleus and increasing its DNA binding activity. This would be consistent with the antagonistic effect of glucagon and insulin/glucose on L-PK and lipogenic genes.

Effect of Glucose on Gluconeogenic Enzyme Gene Expression

Concurrent with the well-known inhibitory effect of insulin, glucose inhibits transcription of PEPCK in the liver (Kahn et al., 1987; Meyer et al., 1991). This effect is independent of the presence of insulin. As described above for lipogenic enzymes, the effect of glucose requires phosphorylation of glucose to produce glucose-6-phosphate. The inhibitory effect of glucose is observed in hepatocytes from adult rats, but not in hepatocytes from suckling rats, which do not express glucokinase (Cournarie et al., 1999). Moreover, the effect of glucose on PEPCK expression is absent in H4IIE hepatoma cells, which express hexokinase I but not glucokinase (Scott et al., 1998). The overexpression of glucokinase by adenovirus is sufficient to recover the glucose effect (Scott et al., 1998). At the present time, the localization of the glucose-sensitive region in the PEPCK promoter and the identity of transcription factors involved remain unknown.

Glucose-6-phosphatase is another key gluconeogenic enzyme. As stated previously, its expression is increased in response to glucose shortage in the diet and is repressed by a high-carbohydrate diet and by insulin. Repression by insulin seems to involve a transcription factor of the forkhead family (Nakae et al., 2001; Foufelle and Ferré, 2002). Surprisingly, it has been shown that high glucose concentrations in culture media induce glucose-6-phosphatase expression in both

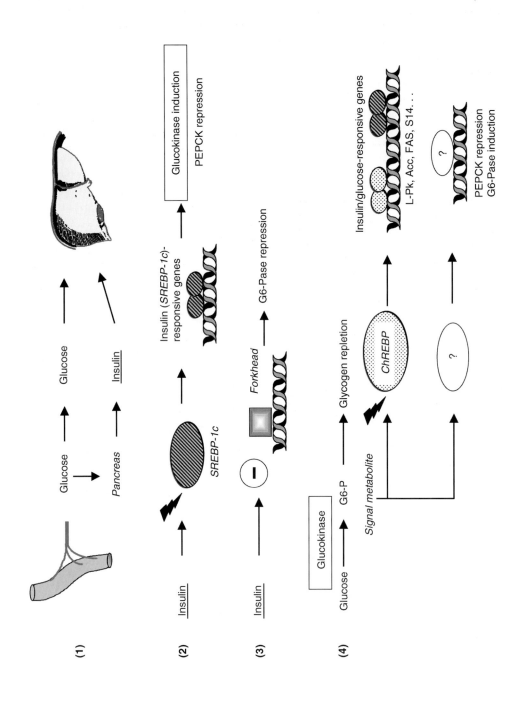

hepatoma cells and primary cultured hepatocytes; this effect is, at least in part, mediated by a transcriptional mechanism (Argaud et al., 1997). Again, a glucose metabolite seems to be involved in this stimulation since an increase in the concentration of the glycolytic activator fructose-2,6-bisphosphate as well as substrates (xylitol, fructose) entering the glycolytic pathway at different steps mimic the effect of glucose (Argaud et al., 1997; Massillon, 2001). In vivo effects of xylitol on the induction of the glucose-6-phosphatase gene have also been described (Massillon et al., 1998). This was paralleled by a doubling of the concentration of xylulose-5-phosphate but no change in glucose-6-phosphate concentration.

For all the genes sensitive to both insulin and glucose, both factors do act in synergy either as activators or as inhibitors. This is obviously not the case for the glucose-6-phosphatase gene. The physiological relevance of this peculiarity presently is not understood. This could indicate that glucose-6-phosphatase has an as yet unknown function in relation to glucose availability.

General Conclusion

Glucose regulation of gene transcription in the liver is an important component of the mechanisms that control glucose homeostasis (Fig. 7.1). It would be interesting to analyse whether these signalling cascades do exist in other tissues that are crucial for glucose and lipid metabolism such as muscle, adipose tissue and β-cells. In addition, it is important to determine whether some of the components of the glucose signalling pathways are involved in diseases such as type 2 diabetes or lipid disorders. Finally, it remains to be determined whether the mechanisms involved in the induction and repression of genes by glucose are identical.

Acknowledgements

We thank Dr Jean Girard for his major contribution during the course of these studies. Work in our laboratory was supported by INSERM and the European Community (CT Fair 97-3011).

References

Argaud, D., Kirby, T.L., Newgard, C.B. and Lange, A.J. (1997) Stimulation of glucose 6 phosphatase gene expression by glucose and fructose-2,6-biphosphatase. *Journal of Biological Chemistry* 272, 12854–12861.

Bandsma, R.H., Wiegman, C.H., Herling, A.W., Burger, H.J., ter Harmsel, A., Meijer, A.J., Romijn, J.A., Reijngoud, D.J. and Kuipers, F. (2001) Acute inhibition of glucose-6-phosphate translocator activity leads to increased *de novo* lipogenesis and development of hepatic steatosis without affecting VLDL production in rats. *Diabetes* 50, 2591–2597.

Bergot, M.O., Diaz-Guerra, M.J.H., Puizenat, N., Raymondjean, M. and Kahn, A. (1992) *Cis*-regulation of the L-type pyruvate kinase gene promoter by glucose, insulin and cyclic AMP. *Nucleic Acids Research* 20, 1871–1878.

Bollen, M., Keppens, S. and Stalmans, W. (1998) Specific features of glycogen metabolism in the liver. *Biochemical Journal* 336, 19–31.

Brun, T., Roche, E., Kim, K.H. and Prentki, M. (1993) Glucose regulates acetyl-CoA carboxylase gene

---

Fig. 7.1. (opposite)   Regulation of hepatic gene expression by insulin and glucose: (1) The plasma concentration of glucose increases in response to a meal rich in carbohydrates, leading to secretion of insulin by pancreatic β-cells; both glucose and insulin are transported to the liver. (2) Insulin stimulates SREBP-1c, leading to activation of glucokinase transcription and inhibition of PEPCK transcription (Foufelle and Ferré, 2002). (3) Insulin also inhibits a transcription factor of the forkhead family and, thus, represses expression of the glucose-6-phosphatase gene. (4) Increased synthesis of glucokinase leads to phosphorylation of glucose to glucose-6-phosphate. A signal metabolite downstream of the glucokinase step (glucose-6-phosphate or xylulose-5-phosphate) activates a transcription factor (ChREBP), possibly by a dephosphorylation-mediated mechanism. ChREBP then translocates to the nucleus where, together with SREBP-1c, ChREBP activates genes of the glycolytic/lipogenic pathway such as L-PK, ACC, FAS and S14. An as yet unidentified glucose signalling system represses PEPCK expression (in conjunction with insulin) and activates glucose-6-phosphatase expression (thus opposing the effects of insulin). This mechanism contributes to the inhibition of glucose production, the replenishment of glycogen stores, and the synthesis of lipids in response to a glucose overload.

expression in a pancreatic beta-cell line (INS-1). *Journal of Biological Chemistry* 268, 18905–18911.

Burcelin, R., del Carmen Munoz, M., Guillam, M.T. and Thorens, B. (2000) Liver hyperplasia and paradoxical regulation of glycogen metabolism and glucose-sensitive gene expression in GLUT2-null hepatocytes. Further evidence for the existence of a membrane-based glucose release pathway. *Journal of Biological Chemistry* 275, 10930–10936.

Carling, D., Zammit, V.A. and Hardie, D.G. (1987) A common bicyclic protein kinase cascade inactivates the regulatory enzymes of fatty acid and cholesterol biosynthesis. *FEBS Letters* 223, 217–222.

Clarke, S.D., Armstrong, M.K. and Jump, D.B. (1990) Nutritional control of rat liver fatty acid synthase and S14 mRNA abundance. *Journal of Nutrition* 120, 218–224.

Colosia, A.D., Marker, A.J., Lange, A.J., el-Maghrabi, M.R., Granner, D.K., Tauler, A., Pilkis, J. and Pilkis, S.J. (1988) Induction of rat liver 6-phosphofructo-2-kinase/fructose-2,6-bisphosphatase mRNA by refeeding and insulin. *Journal of Biological Chemistry* 263, 18669–18677.

Cournarie, F., Azzout-Marniche, D., Foretz, M., Guichard, C., Ferré, P. and Foufelle, F. (1999) The inhibitory effect of glucose on phosphoenolpyruvate carboxykinase gene expression in cultured hepatocytes is transcriptional and requires glucose metabolism. *FEBS Letters* 460, 527–532.

Cuif, M.H., Cognet, M., Boquet, D., Tremp, G., Kahn, A. and Vaulont, S. (1992) Elements responsible for hormonal control and tissue specificity of L-type pyruvate kinase gene expression in transgenic mice. *Molecular and Cellular Biology* 12, 4852–4861.

Daniel, S. and Kim, H.K. (1996) Sp1 mediates glucose activation of the acetyl-CoA carboxylase promoter. *Journal of Biological Chemistry* 271, 1385–1392.

Daniel, S., Zhang, S., DePaoli-Roach, A.A. and Kim, K.H. (1996) Dephosphorylation of sp1 by protein phosphatase 1 is involved in the glucose-mediated activation of the acetyl-CoA carboxylase gene. *Journal of Biological Chemistry* 271, 14692–14697.

da Silva Xavier, G., Leclerc, I., Salt, I.P., Doiron, B., Hardie, D.G., Kahn, A. and Rutter, G.A. (2000) Role of AMP-activated protein kinase in the regulation by glucose of islet beta cell gene expression. *Proceedings of the National Academy of Sciences USA* 97, 4023–4028.

Decaux, J.F., Antoine, B. and Kahn, A. (1989) Regulation of the expression of the L-type pyruvate kinase gene in adult rat hepatocytes in primary culture. *Journal of Biological Chemistry* 264, 11584–11590.

Diaz-Guerra, M.J.M., Bergot, M.O., Martinez, A., Cuif, M.H., Kahn, A. and Raymondjean, M. (1993) Functional characterisation of the L-type pyruvate kinase gene glucose response complex. *Molecular and Cellular Biology* 13, 7725–7733.

Doiron, B., Cuif, M.H., Kahn, A. and Diaz-Guerra, M.J. (1994) Respective roles of glucose, fructose, and insulin in the regulation of the liver-specific pyruvate kinase gene promoter. *Journal of Biological Chemistry* 269, 10213–10216.

Doiron, B., Cuif, M.H., Chen, R. and Kahn, A. (1996) Transcriptional glucose signaling through the glucose response element is mediated by the pentose phosphate pathway. *Journal of Biological Chemistry* 271, 5321–5324.

Elshourbagy, N.A., Near, J.C., Kmetz, P.J., Sathe, G.M., Southan, C., Strickler, J.E., Gross, M., Young, J.F., Wells, T.N. and Groot, P.H. (1990) Rat ATP citrate-lyase. Molecular cloning and sequence analysis of a full-length cDNA and mRNA abundance as a function of diet, organ, and age. *Journal of Biological Chemistry* 265, 1430–1435.

Foretz, M., Carling, D., Guichard, G., Ferré, P. and Foufelle, F. (1998) AMP-activated protein kinase inhibits the glucose-activated expression of fatty acid synthase gene in rat hepatocytes. *Journal of Biological Chemistry* 273, 14767–14771.

Foufelle, F. and Ferré, P. (2002) New perspectives in the regulation of hepatic glycolytic and lipogenic genes by insulin and glucose: a role for the transcription factor SREBP-1c. *Biochemical Journal* 366, 377–391.

Foufelle, F., Gouhot, B., Pégorier, J.P., Perdereau, D., Girard, J. and Ferré, P. (1992) Glucose stimulation of lipogenic enzyme gene expression in cultured white adipose tissue. *Journal of Biological Chemistry* 267, 20543–20546.

Foufelle, F., Girard, J. and Ferré, P. (1996) Regulation of lipogenic enzyme expression by glucose in liver and adipose tissue: a review of the potential cellular and molecular mechanisms. *Advances in Enzyme Regulation* 36, 199–226.

Gancedo, J.M. (1998) Yeast carbon catabolite repression. *Microbiology and Molecular Biology Reviews* 62, 334–361.

Girard, J., Ferré, P. and Foufelle, F. (1997) Mechanisms by which carbohydrates regulate expression of genes for glycolytic and lipogenic enzymes. *Annual Review of Nutrition* 17, 325–352.

Granner, D.K., Andreone, T.L., Sasaki, K. and Beale, E.G. (1983) Inhibition of transcription of PEPCK gene by insulin. *Nature* 305, 549–551.

Gregor, P.D., Sawadogo, M. and Roeder, R.G. (1990) The adenovirus major late transcription factor USF is a member of the helix–loop–helix group of regulatory proteins and binds to DNA as a dimer. *Genes and Development* 4, 1730–1740.

Guillemain, G., Loizeau, M., Pincon-Raymond, M., Girard, J. and Leturque, A. (2000) The large intracytoplasmic loop of the glucose transporter GLUT2 is involved in glucose signaling in hepatic cells. *Journal of Cell Science* 113, 841–847.

Guillemain, G., Munoz-Alonso, M.J., Cassany, A., Loizeau, M., Faussat, A.M., Burnol, A.F. and Leturque, A. (2002) Karyopherin alpha2: a control step of glucose-sensitive gene expression in hepatic cells. *Biochemical Journal* 364, 201–209.

Hardie, D.G. and Carling, D. (1997) The AMP-activated protein kinase – fuel gauge of the mammalian cell? *European Journal of Biochemistry* 246, 259–273.

Hardie, D.G., Carling, D. and Carlson, M. (1998) The AMP-activated/SNF1 protein kinase subfamily: metabolic sensors of the eukaryotic cell? *Annual Review of Biochemistry* 67, 821–855.

Hasegawa, J., Osatomi, K., Wu, R.F. and Uyeda, K. (1999) A novel factor binding to the glucose response elements of liver pyruvate kinase and fatty acid synthase genes. *Journal of Biological Chemistry* 274, 1100–1107.

Hubert, A., Husson, A., Chedeville, A. and Lavoinne, A. (2000) AMP-activated protein kinase counteracted the inhibitory effect of glucose on the phosphoenolpyruvate carboxykinase gene expression in rat hepatocytes. *FEBS Letters* 481, 209–212.

Iynedjian, P.B., Ucla, C. and Mach, B. (1987) Molecular cloning of glucokinase cDNA. Developmental and dietary regulation of glucokinase mRNA in rat liver. *Journal of Biological Chemistry* 262, 6032–6038.

Iynedjian, P.B., Jotterand, D., Nouspikel, T., Asfari, M. and Pilot, P.R. (1989) Transcriptional regulation of glucokinase gene by insulin in cultured liver cells and its repression by the glucagon–cAMP system. *Journal of Biological Chemistry* 264, 21824–21829.

Johnston, M. (1999) Feasting, fasting and fermenting. Glucose sensing in yeast and other cells. *Trends in Genetics* 15, 29–33.

Kahn, C.R., Lauris, V., Koch, S., Crettaz, M. and Granner, D.K. (1987) Acute and chronic regulation of phosphoenolpyruvate carboxykinase mRNA by insulin and glucose. *Molecular Endocrinology* 3, 840–845.

Katsurada, A., Iritani, N., Fukuda, H., Matsumura, Y., Noguchi, T. and Tanaka, T. (1989) Effects of nutrients and insulin on transcriptional and post-transcriptional regulation of glucose-6-phosphate dehydrogenase synthesis in rat liver. *Biochimica et Biophysica Acta* 1006, 104–110.

Katsurada, A., Iritani, N., Fukuda, H., Matsumura, Y., Nishimoto, N., Noguchi, T. and Tanaka, T. (1990) Effects of nutrients and hormones on transcriptional and post-transcriptional regulation of fatty acid synthase in rat liver. *European Journal of Biochemistry* 190, 427–435.

Kawaguchi, T., Takenoshita, M., Kabashima, T. and Uyeda, K. (2001) Glucose and cAMP regulate the L-type pyruvate kinase gene by phosphorylation/dephosphorylation of the carbohydrate response element binding protein. *Proceedings of the National Academy of Sciences USA* 98, 13710–13715.

Kaytor, E.N., Shih, H.M. and Towle, H.C. (1997) Carbohydrate regulation of hepatic gene expression: evidence against a role for the upstream stimulatory factor. *Journal of Biological Chemistry* 272, 7525–7531.

Kaytor, E.N., Qian, J., Towle, H.C. and Olson, L.K. (2000) An indirect role for upstream stimulatory factor in glucose-mediated induction of pyruvate kinase and S14 gene expression. *Molecular and Cellular Biochemistry* 210, 13–21.

Kim, T.S., Leahy, P. and Freake, H.C. (1996) Promoter usage determines tissue specific responsiveness of the rat acetyl-CoA carboxylase gene. *Biochemical and Biophysical Research Communications* 225, 647–653.

Kinlaw, W.B., Perez-Castillo, A.M., Fish, L.H., Mariash, C.N., Schwartz, H.L. and Oppenheimer, J.H. (1987) Interaction of dietary carbohydrate and glucagon in regulation of rat hepatic messenger ribonucleic acid S14 expression: role of circadian factors and 3′,5′-cyclic adenosine monophosphate. *Molecular Endocrinology* 1, 609–613.

Kinlaw, W.B., Church, J.L., Harmon, J. and Mariash, C.N. (1995) Direct evidence for a role of the spot 14 protein in the regulation of lipid synthesis. *Journal of Biological Chemistry* 270, 16615–16618.

Kletzien, R.F., Prostko, C.R., Stumpo, D.J., McClung, J.K. and Dreher, K.L. (1985) Molecular cloning of DNA sequences complementary to rat liver glucose-6-phosphate dehydrogenase mRNA. Nutritional regulation of mRNA levels. *Journal of Biological Chemistry* 260, 5621–5624.

Koo, S.H. and Towle, H.C. (2000) Glucose regulation of mouse S(14) gene expression in hepatocytes. Involvement of a novel transcription factor complex. *Journal of Biological Chemistry* 275, 5200–5207.

Koo, S.H., Dutcher, A.K. and Towle, H.C. (2001) Glucose and insulin function through two distinct transcription factors to stimulate expression of lipogenic enzyme genes in liver. *Journal of Biological Chemistry* 276, 9437–9445.

Leclerc, I., Kahn, A. and Doiron, B. (1998) The 5′-AMP-activated protein kinase inhibits the transcriptional stimulation by glucose in liver cells, acting through the glucose response complex. *FEBS Letters* 431, 180–184.

Lefrancois-Martinez, A.M., Martinez, A., Antoine, B., Raymondjean, M. and Kahn, A. (1995) Upstream stimulatory factor proteins are major components of the glucose response complex of the L-type pyruvate kinase gene promoter. *Journal of Biological Chemistry* 270, 2640–2643.

Liu, Z., Thompson, K.S. and Towle, H.C. (1993) Carbohydrate regulation of the rat L-type pyruvate kinase gene requires two nuclear factors: LF-A1 and a member of the c-myc family. *Journal of Biological Chemistry* 268, 12787–12795.

Lochhead, P.A., Salt, I.P., Walker, K.S., Hardie, D.G. and Sutherland, C. (2000)

5-Aminoimidazole-4-carboxamide riboside mimics the effects of insulin on the expression of the 2 key gluconeogenic genes PEPCK and glucose-6-phosphatase. *Diabetes* 49, 896–903.

Lopez, J., Bennett, M.K., Sanchez, H.B., Rosenfeld, J.M. and Osborne, T.F. (1996) Sterol regulation of acetyl CoA carboxylase: a mechanism for coordinate control of cellular lipid. *Proceedings of the National Academy of Sciences USA* 93, 1049–1053.

Lopez-Casillas, F., Ponce-Castaneda, M.V. and Kim, K.H. (1991) In vivo regulation of the activity of the two promoters of the rat acetyl coenzyme-A carboxylase gene. *Endocrinology* 129, 1049–1058.

Lou, D.Q., Tannour, M., Selig, L., Thomas, D., Kahn, A. and Vasseur-Cognet, M. (1999) Chicken ovalbumin upstream promoter-transcription factor II, a new partner of the glucose response element of the L-type pyruvate kinase gene, acts as an inhibitor of the glucose response. *Journal of Biological Chemistry* 274, 28385–28394.

MacLean, P.S., Zheng, D., Jones, J.P., Olson, A.L. and Dohm, G.L. (2002) Exercise-induced transcription of the muscle glucose transporter (GLUT 4) gene. *Biochemical and Biophysical Research Communications* 292, 409–414.

Mariash, C.N., Seelig, S., Schwartz, H.L. and Oppenheimer, J.H. (1986) Rapid synergistic interaction between thyroid hormone and carbohydrate on mRNAS14 induction. *Journal of Biological Chemistry* 261, 9583–9586.

Marie, S., Diaz-Guerra, M.J., Miquerol, L., Kahn, A. and Iynedjian, P.B. (1993) The pyruvate kinase gene as a model for studies of glucose-dependent regulation of gene expression in the endocrine pancreatic beta-cell type. *Journal of Biological Chemistry* 268, 23881–23890.

Massillon, D. (2001) Regulation of the glucose-6-phosphatase gene by glucose occurs by transcriptional and post-transcriptional mechanisms. Differential effect of glucose and xylitol. *Journal of Biological Chemistry* 276, 4055–4062.

Massillon, D., Chen, W., Barzilai, N., Prus-Wertheimer, D., Hawkins, M., Liu, R., Taub, R. and Rossetti, L. (1998) Carbon flux via the pentose phosphate pathway regulates the hepatic expression of the glucose-6-phosphatase and phosphoenolpyruvate carboxykinase genes in conscious rats. *Journal of Biological Chemistry* 273, 228–234.

Meyer, S., Höppner, W. and Seitz, H.J. (1991) Transcriptional and post-transcriptional effects of glucose in liver phosphoenolpyruvate carboxykinase gene expression. *European Journal of Biochemistry* 202, 985–991.

Miksicek, R.J. and Towle, H.C. (1983) Use of a cloned cDNA sequence to measure changes in 6-phosphogluconate dehydrogenase mRNA levels caused by thyroid hormone and dietary carbohydrate. *Journal of Biological Chemistry* 258, 9575–9579.

Mourrieras, F., Foufelle, F., Foretz, M., Morin, J., Bouché, S. and Ferré, P. (1997) Induction of fatty acid synthase and S14 gene expression by glucose, xylitol and dihydroxyacetone in cultured rat hepatocytes is closely correlated with glucose 6-phosphate concentrations. *Biochemical Journal* 323, 345–349.

Munnich, A., Lyonnet, S., Chauvet, D., Van Schaftingen, E. and Kahn, A. (1987) Differential effects of glucose and fructose on liver L-type pyruvate kinase gene expression in vivo. *Journal of Biological Chemistry* 262, 17065–17071.

Nakae, J., Kitamura, T., Silver, D.L. and Accili, D. (2001) The forkhead transcription factor Foxo1 (Fkhr) confers insulin sensitivity onto glucose-6-phosphatase expression. *Journal of Clinical Investigation* 108, 1359–1367.

Nishimura, M. and Uyeda, K. (1995) Purification and characterization of a novel xylulose 5-phosphate-activated protein phosphatase catalysing dephosphorylation of fructose-6-phosphate 2-kinase: fructose-2,6-biphosphatase. *Journal of Biological Chemistry* 270, 26341–26346.

Ntambi, J.M. (1992) Dietary regulation of stearoyl-CoA desaturase 1 gene expression in mouse liver. *Journal of Biological Chemistry* 267, 10925–10930.

O'Callaghan, B.L., Koo, S.H., Wu, Y., Freake, H.C. and Towle, H.C. (2001) Glucose regulation of the acetyl-CoA carboxylase promoter PI in rat hepatocytes. *Journal of Biological Chemistry* 276, 16033–16039.

Pape, M.E., Lopez-Casillas, F. and Kim, K.H. (1988) Physiological regulation of acetyl-CoA carboxylase gene expression: effects of diet, diabetes, and lactation on acetyl-CoA carboxylase mRNA. *Archives of Biochemistry and Biophysics* 267, 104–109.

Paulauskis, J.D. and Sul, H.S. (1989) Hormonal regulation of mouse fatty acid synthase transcription in liver. *Journal of Biological Chemistry* 264, 574–577.

Prip-Buus, K., Perdereau, D., Foufelle, F., Maury, J., Ferré, P. and Girard, J. (1995) Induction of fatty acid synthase gene expression by glucose in primary culture of rat hepatocytes. *European Journal of Biochemistry* 230, 309–315.

Rongnoparut, P., Verdon, C.P., Gehnrich, S.C. and Sul, H.S. (1991) Isolation and characterization of the transcriptionally regulated mouse liver (B-type) phosphofructokinase gene and its promoter. *Journal of Biological Chemistry* 266, 8086–8091.

Rufo, C., Teran-Garcia, M., Nakamura, M.T., Koo, S.H., Towle, H.C. and Clarke, S.D. (2001) Involvement of a unique carbohydrate-responsive factor in the glucose regulation of rat liver fatty-acid synthase gene transcription. *Journal of Biological Chemistry* 276, 21969–21975.

Salt, I.P., Johnson, G., Ashcroft, S.J. and Hardie, D.G. (1998) AMP-activated protein kinase is activated by low glucose in cell lines derived from pancreatic beta cells, and may regulate insulin release. *Biochemical Journal* 335, 533–539.

Sasaki, K., Cripe, T.P., Koch, S.R., Andreone, T.L., Petersen, D.D., Beale, E.G. and Granner, D.K. (1984) Multihormonal regulation of PEPCK gene transcription. The dominant role of insulin. *Journal of Biological Chemistry* 259, 15242–15251.

Sawadogo, M. (1988) Multiple forms of the human gene-specific transcription factor USF. *Journal of Biological Chemistry* 263, 11994–12001.

Sawadago, M. and Roeder, R. (1985) Interaction of a gene-specific transcription factor with the adenovirus major late promoter upstream of the TATA box region. *Cell* 43, 165–175.

Scott, D.K., O'Doherty, R.M., Stafford, J.M., Newgard, C.B. and Granner, D.K. (1998) The repression of hormone-activated PEPCK gene expression by glucose is insulin-independent but requires glucose metabolism. *Journal of Biological Chemistry* 273, 24145–24151.

Shih, H.M. and Towle, H.C. (1992) Definition of the carbohydrate response element of the rat S14 gene. *Journal of Biological Chemistry* 267, 13222–13228.

Shih, H.M. and Towle, H.C. (1994) Definition of the carbohydrate response element of the rat S14 gene. Context of the CACGTG motif determines the specificity of carbohydrate regulation. *Journal of Biological Chemistry* 269, 9380–9387.

Shih, H.M., Liu, Z. and Towle, H.C. (1995) Two CACGTG motifs with proper spacing dictate the carbohydrate regulation of hepatic gene transcription. *Journal of Biological Chemistry* 270, 21991–21997.

Sirito, M., Walker, S., Lin Q., Kozlowski, M., Klein, W. and Sawadago, M. (1994) Ubiquitous expression of the 43- and 44-kDa forms of transcription factor USF in mammalian cells. *Nucleic Acids Research* 22, 427–433.

Sudo, Y. and Mariash, C.N. (1994) Two glucose-signaling pathways in S14 gene transcription in primary hepatocytes: a common role of protein phosphorylation. *Endocrinology* 134, 2532–2540.

Thompson, K.S. and Towle, H.C. (1991) Localization of the carbohydrate response element of the rat L-type pyruvate kinase gene. *Journal of Biological Chemistry* 266, 8679–8682.

Towle, H.C., Kaytor, E.N. and Shih, H.M. (1997) Regulation of the expression of lipogenic enzyme genes by carbohydrate. *Annual Review of Nutrition* 17, 405–433.

Vallet, V.S., Henrion, A.A., Bucchini, D., Casado, M., Raymondjean, M., Kahn, A. and Vaulont, S. (1997) Glucose-dependent liver gene expression in USF2 −/− mice. *Journal of Biological Chemistry* 272, 21944–21949.

Vallet, V.S., Casado, M., Henrion, A.A., Bucchini, D., Raymondjean, M., Kahn, A. and Vaulont, S. (1998) Differential roles of upstream stimulatory factors 1 and 2 in the transcriptional response of liver genes to glucose. *Journal of Biological Chemistry* 273, 20175–20179.

Van Schaftingen, E. and Gerin, I. (2002) The glucose-6-phosphatase system. *Biochemical Journal* 362, 513–532.

Vaulont, S., Munnich, A., Decaux, J.F. and Kahn, A. (1986) Transcriptional and post-transcriptional regulation of L-type pyruvate kinase gene expression in rat liver. *Journal of Biological Chemistry* 261, 7621–7625.

Vaulont, S., Puzenat, N., Levrat, F., Cognet, M., Kahn, A. and Raymonjean, M. (1989) Proteins binding to the liver-specific pyruvate kinase gene promoter. *Journal of Molecular Biology* 209, 205–219.

Vaulont, S., Vasseur-Cognet, M. and Kahn, A. (2000) Glucose regulation of gene transcription. *Journal of Biological Chemistry* 275, 31555–31558.

Viollet, B., Lefrancois-Martinez, A.M., Henrion, A., Kahn, A., Raymondjean, M. and Martinez, A. (1996) Immunochemical characterization and transacting properties of upstream stimulatory factor isoforms. *Journal of Biological Chemistry* 271, 1405–1415.

Waters, K.M. and Ntambi, J.M. (1994) Insulin and dietary fructose induce stearoyl-CoA desaturase 1 gene expression of diabetic mice. *Journal of Biological Chemistry* 269, 27773–27777.

Weber, A., Marie, J., Cottreau, D., Simon, M.P., Besmond, C., Dreyfus, J.C. and Kahn, A. (1984) Dietary control of aldolase B and L-type pyruvate kinase mRNAs in rat. Study of translational activity and hybridization with cloned cDNA probes. *Journal of Biological Chemistry* 259, 1798–1802.

Woods, A., Munday, M.R., Scott, J., Yang, X., Carlson, M. and Carling, D. (1994) Yeast SNF1 is functionally related to mammalian AMP-activated protein kinase and regulates acetyl CoA carboxylase *in vivo*. *Journal of Biological Chemistry* 269, 19509–19515.

Woods, A., Azzout-Marniche, D., Foretz, M., Stein, S.C., Lemarchand, P., Ferre, P., Foufelle, F. and Carling, D. (2000) Characterization of the role of AMP-activated protein kinase in the regulation of glucose-activated gene expression using constitutively active and dominant negative forms of the kinase. *Molecular and Cellular Biology* 20, 6704–6711.

Yamashita, H., Takenoshita, M., Sakurai, M., Bruick, R.K., Henzel, W.J., Shillinglaw, W., Arnot, D. and Uyeda, K. (2001) A glucose-responsive transcription factor that regulates carbohydrate metabolism in the liver. *Proceedings of the National Academy of Sciences USA* 98, 9116–9121.

Zheng, D., MacLean, P.S., Pohnert, S.C., Knight, J.B., Olson, A.L., Winder, W.W. and Dohm, G.L. (2001)

Regulation of muscle GLUT-4 transcription by AMP-activated protein kinase. *Journal of Applied Physiology* 91, 1073–1083.

Zhou, M., Lin, B.Z., Coughlin, S., Vallega, G. and Pilch, P.F. (2000) UCP-3 expression in skeletal muscle: effects of exercise, hypoxia, and AMP-activated protein kinase. *American Journal of Physiology, Endocrinology and Metabolism* 279, E622–E629.

Zhu, Q., Mariash, A., Margosian, M.R., Gopinath, S., Fareed, M.T., Anderson, G.W. and Mariash, C.N. (2001) Spot 14 gene deletion increases hepatic *de novo* lipogenesis. *Endocrinology* 142, 4363–4370.

8 Amino Acid-dependent Control of Transcription in Mammalian Cells

Michael S. Kilberg, Van Leung-Pineda, and Chin Chen
Department of Biochemistry and Molecular Biology, University of Florida College of Medicine, Gainesville, Florida, USA

Amino Acids as Signal Molecules

Protein/amino acid availability is an important factor in general nutrition, particularly during development (Morgane *et al.*, 1993), in the progression of a wide range of diseases including diabetes (Hoffer, 1993), Kwashiorkor (Roediger, 1995) and hepatic encephalopathy (Mizock, 1999) and, as discussed below, in certain cases of cancer chemotherapy. Despite these obvious relationships, amino acid control of fundamental cellular processes is a topic about which we know very little in mammalian cells. Nutritional deprivation of intact animals is a necessary but complicated model for mechanistic studies because of the confounding effects of nutrient-induced changes in hormones that may themselves alter the expression of the activity under investigation. Therefore, a number of laboratories have used amino acid deprivation of cultured cells as a useful experimental model and, in so doing, have identified specific mRNAs, proteins and activities that are induced following amino acid deprivation (for examples of mRNAs, see Table 8.1). Under these circumstances, the amino acids are not serving their better known roles as metabolic or protein synthesis precursors, but rather as signal molecules that reflect the nutritional status of the organism. One of the many consequences of this amino acid-dependent signalling is a change in transcription rate for specific genes. From a mechanistic viewpoint, the goal is to understand each of the steps in this signal transduction process, which we will refer to as the amino acid response (AAR) pathway. While the importance of each of the AAR pathway steps is recognized, this review will focus on the role of amino acids as regulators of gene transcription.

Amino acid regulation of gene expression in yeast

The yeast *Saccharomyces cerevisiae* employs a general control mechanism (GCN, general control

Table 8.1. Examples of proteins for which the mRNA content is increased following amino acid limitation of mammalian cells.

Amino acid transporters	
CAT1	Hyatt *et al.* (1997)
ATA2	Gazzola *et al.* (2001)
Asparagine synthetase	Gong *et al.* (1991)
C/EBPα	Marten *et al.* (1994)
C/EBPβ	Marten *et al.* (1994)
CHOP	Marten *et al.* (1994)
IGFBP-1	Straus *et al.* (1993); Jousse *et al.* (1998)
Ribosomal proteins	
L17	Laine *et al.* (1991)
S25	Laine *et al.* (1994)
L35	Hitomi *et al.* (1993)
S13	Hitomi *et al.* (1993)

non-repressible) in which >1000 genes are regulated by starvation of the cell for a single amino acid (Natarajan et al., 2001). Identification of mammalian protein mediators that are analogous to those in the yeast GCN pathway indicates that we can learn a great deal from this more extensively studied organism. Translation of the mRNA for GCN4, a member of the bZIP family of transcriptional activators, is increased by amino acid starvation (Hinnebusch, 1997) and, subsequently, GCN4 binds to the sequence 5'-ATGA^C_GTCAT-3' in the promoter region of genes under its control (Hope and Struhl, 1985; Sellers et al., 1990). In mammalian cells, dimers of bZIP proteins bind to DNA sites such as the AP-1 site, 5'-ATGACTCAT-3' and the ATF/CREB site, 5'-ATGACGTCAT-3' (Sellers et al., 1990). As described below, the yeast GCN4 recognition sequence is not the same as that responsible for amino acid control in mammalian cells, and no mammalian GCN4 homologue has been reported, but at least some of the upstream events for amino acid signalling appear to be similar. Translational regulation of GCN4p synthesis is dependent on a scanning mechanism of translation initiation to identify AUG codons in front of the GCN4 protein-coding sequence. Within the GCN4 mRNA 5' region are four AUG codons, each of which is followed by one or two sense codons before an in-frame termination codon is reached (Hinnebusch, 1997). In amino acid-rich conditions, a repeated series of scanning initiation and termination events occurs at these open reading frames (ORFs) that results in little or no initiation at the actual GCN4-coding sequence. In response to amino acid starvation, the efficiency of reinitiation is slowed such that many ribosomes fail to reinitiate at the final short ORF and, therefore, initiate at the GCN4 protein start codon instead.

This translational control of GCN4 synthesis is regulated by the protein kinase GCN2, which once activated by elevated uncharged tRNA, phosphorylates the 'α' subunit of the yeast translation initiation factor eIF-2 on Ser51. When eIF-2α is phosphorylated, it binds more tightly to the guanine exchange factor eIF-2B, which results in a reduced level of eIF-2-GTP and, consequently, causes the 40S subunit to scan beyond upstream ORF4 (uORF4) before eIF-2-GTP associates with the subunit (Hinnebusch, 1997). These changes result in corresponding changes in translation initiation for GCN4 and, therefore, represent an important component of the starvation response signalling pathway. As additional evidence for a mechanism involving detection of uncharged tRNA, in both yeast (Wek et al., 1995) and mammalian cells (Andrulis et al., 1979) that contain temperature-sensitive tRNA synthetases, activation of the amino acid deprivation pathways occurs despite normal intracellular amino acid pools. Likewise, treating Fao hepatoma cells with the tRNA^{His} synthetase inhibitor histidinol results in activation of the AAR pathway (Hutson and Kilberg, 1994).

A mammalian counterpart of yeast GCN2 has been cloned and characterized independently by two groups (Berlanga et al., 1999; Sood et al., 2000). Although the mammalian GCN2 has yet to be characterized as well as its yeast counterpart, it is structurally homologous, including the presence of a histidinyl-tRNA synthetase domain and an eIF-2α kinase domain. A number of observations suggest that the mammalian GCN2 functions as an eIF-2α kinase (Berlanga et al., 1999; Sood et al., 2000) and that it serves to mediate amino acid-dependent signalling. With regard to the latter point, expression of mammalian GCN2 in yeast substitutes for the endogenous GCN2 protein, documenting that the mammalian GCN2 can be activated by amino acid deprivation. Furthermore, translational control of the transcription factor ATF-4, for which synthesis is enhanced by amino acid deprivation of mammalian cells, is deficient in GCN2 knockout mice (Harding et al., 2000).

Mammalian Activities Altered by Amino Acid Availability

A wide range of enzymatic and transport activities, protein content, mRNA content and transcription of specific genes are regulated by amino acid availability both *in vivo* and *in vitro*. Given that the focus of this summary is transcriptional regulation by amino acid deprivation, Table 8.1 presents a list of proteins for which the mRNA content has been shown to be increased following amino acid deprivation in mammalian cells. In most instances, the increase in mRNA content has been shown to be associated with transcription but, for some, mRNA stabilization may also occur. In this partial list of examples, C/EBPα, C/EBPβ, ATA2 transporter and the ribosomal proteins L35 and

S13 have been included despite direct proof for transcriptional regulation because their regulation by amino acid availability is particularly noteworthy given their biological function. Marten et al. (1994) showed that C/EBPα and C/EBPβ mRNA was induced by amino acid deprivation of rat hepatoma cells. Increased expression of C/EBP family members is of particular interest given the amino acid-dependent transcriptional regulation of the C/EBP-homologous protein, CHOP, discussed more extensively below. Also of note is the participation of C/EBPβ as one of the transcription factors that mediates induction of the human asparagine synthetase gene in response to activation of both the AAR and the endoplasmic reticulum stress response (ERSR) nutrient-sensing pathways (Siu et al., 2001). Future characterization of amino acid-dependent changes in transcription factor expression will provide valuable insight into the mechanisms of gene expression following amino acid limitation and the specific changes that occur during the cellular response.

Although increased transcription of ribosomal protein genes, in the face of amino acid limitation, appears to be counterintuitive, the mRNA content of several ribosomal proteins has been shown to be increased (Laine et al., 1994), and for ribosomal proteins L17 and S25 this increase has been demonstrated to be transcriptional in nature (Laine et al., 1994). Interestingly, the increased mRNA content of these two proteins is retained within the nucleus during amino acid deprivation, and only released into the cytoplasm for translation following amino acid refeeding (Laine et al., 1994). Interestingly, Adilakshmi and Laine (2002) have demonstrated that p53 binds to the S25 mRNA in the nucleus and may, in part, be responsible for this nuclear retention. Further investigation into how this nuclear retention is regulated by amino acid availability should provide mechanistic insight into an interesting and novel cellular process regulated by amino acids.

The importance of amino acid-dependent regulation of insulin-like growth factor-binding protein-1 (IGFBP-1) has been reviewed by Bruhat et al. (1999). Regulation of this particular gene product is of interest because nutrient-dependent control of its expression is likely to have far-reaching effects on a number of tissues and organs. Most of the other regulated genes identified so far are likely to affect the metabolic and growth status of a particular cell or a localized cell population within a single organ. A comprehensive investigation of the influence of amino acid availability on hormone and cytokine expression has not been reported, but such a study would contribute to our understanding of the interorgan effects that protein nutrition has on cell growth and metabolism. For example, it has been demonstrated that histidine deprivation of murine pancreatic cells suppresses the synthesis of glucagon (Paul et al., 1998).

The substrate-dependent regulation of the sodium-dependent zwitterionic amino acid transporter system A has been investigated for three decades (Gazzola et al., 1972), and the subject has been reviewed periodically during this time (Kilberg et al., 1993; Palacín et al., 1998). The more recent identification of multiple genes encoding this activity has permitted investigators to document that the ATA2 gene (also known as SAT2) is responsible for the modulation of system A transport activity in response to amino acid availability (Gazzola et al., 2001). Description of nutrient control for the cationic amino acid transporter CAT1 has emerged during the last 5 years (Hyatt et al., 1997; Aulak et al., 1999; Fernandez et al., 2001). An interesting feature of the nutrient control of this particular amino acid transporter is that the regulation occurs at both the transcriptional and translational levels. The preferential use of an internal ribosome entry site for enhanced translation of the CAT1 mRNA during amino acid limitation is likely to represent a prototypical model for amino acid control of the biosynthesis of specific proteins (Fernandez et al., 2001).

Given the new screening technologies using gene chips or arrays, it is anticipated that many more mammalian genes will be identified for which transcription (and translation) is regulated by amino acid availability. To illustrate the potential ramifications of amino acid limitation on global cellular processes, it has been documented that the yeast transcription factor GCN4p modulates the expression of >1000 genes (Natarajan et al., 2001).

Models for Transcriptional Control

CHOP

Function

Nutrient deprivation certainly results in the need for a cellular response with regard to the metabolism of that specific nutrient, but it is also becoming clear that limiting any single nutrient can often cause a broader stress response to occur. This stress response triggers activation of multiple signal transduction pathways that modify the status of the cell through transcriptional and post-transcriptional mechanisms. A large number of studies have documented the activation of individual transcription factors in response to a wide variety of cellular stress stimuli, and included among these is C/EBP homology protein/growth arrest and DNA damage protein 153 (CHOP/ GADD153). Originally identified as a gene that was induced by DNA damage, it is now recognized that CHOP is activated by a number of stress stimuli, including the ERSR, initiated by glucose starvation (Yoshida et al., 2000), and the AAR pathway, activated by amino acid limitation (Jousse et al., 1999; Fafournoux et al., 2000). Furthermore, deleting the CHOP gene has documented a link between CHOP expression and apoptosis induced by cellular stress (Zinszner et al., 1998). As mentioned above, CHOP is a member of the C/EBP family of transcription factors (Ron and Habener, 1992), and the expression of two other members (C/EBPα and C/EBPβ) is also modulated by amino acid availability (Marten et al., 1994). Originally, it was believed that CHOP formed heterodimers with other C/EBP family members only to inhibit their action and thus serve as a negative regulator (Ron and Habener, 1992). However, more recent evidence has documented that CHOP–C/EBP heterodimers are capable of activating a wide spectrum of genes (Wang et al., 1998).

AAR and ERSR pathways

Although amino acid limitation increases CHOP expression by post-transcriptional mechanisms (Bruhat et al., 1997; Abcouwer et al., 1999), there is also clear evidence for transcriptional control of the CHOP gene by either amino acid or glucose starvation (Price and Calderwood, 1992; Jousse et al., 1999). Glucose starvation of eukaryotic cells results in aberrant accumulation of glycoproteins in the endoplasmic reticulum (ER) that causes the ERSR, also known in yeast as the unfolded protein response (UPR) (Kaufman, 1999; Pahl, 1999; Patil and Walter, 2001). The ERSR signal transduction pathway culminates in increased transcription of a number of genes, many of which are involved in protein processing and trafficking within the ER. The prototypical example would be the ER resident chaperone GRP78. It is important to note that amino acid deprivation (i.e. the AAR pathway) does not induce ERSR-activated genes such as GRP78 (Barbosa-Tessmann et al., 1999a), presumably because mere slowing of protein synthesis does not result in a significant accumulation of misfolded proteins within the ER. The target genes for the ERSR pathway in mammalian cells contain one or more copies of a highly conserved cis-element (ER stress element, ERSE) for which the consensus sequence is 5′-CCAAT-N_9-CCACG-3′ (Yoshida et al., 1998; Roy and Lee, 1999). The human CHOP promoter contains two ERSE sequences (CHOP-ERSE1 and CHOP-ERSE2), which are orientated in opposite directions (Yoshida et al., 2000). Mutational analysis documented that CHOP-ERSE2 is non-functional, but that CHOP-ERSE1 (nucleotides −93 to −75) mediates activation of the gene by ER stress, which is the basis for induction of the gene following glucose deprivation (Yoshida et al., 2000). Jousse et al. (1999) used deletion analysis of the human CHOP promoter to provide evidence that the cis-element necessary for activation of transcription by the ERSR pathway was different from that responsible for increased transcription following amino acid deprivation and subsequent activation of the AAR pathway. Their deletion analysis was consistent with the identification of the ERSE sequence by Yoshida et al. (2000), and, as discussed below, the amino acid response element (AARE) in the CHOP gene was later shown to reside at nucleotides −302 to −310 (Bruhat et al., 2000; Fafournoux et al., 2000). Therefore, activation of the CHOP gene by the AAR pathway is mediated by a different set of genomic elements from those responsible for activation by glucose limitation via the ERSR pathway. The use of two independent genomic elements to respond to either amino acid or glucose limitation contrasts with the transcriptional control mechanisms functional within the human asparagine synthetase promoter, as described below.

CHOP AARE

After identifying the AARE core sequence as 5′-TGATGCAAT-3′, (nucleotides −302 to −310) within the human CHOP promoter, Bruhat et al. (2000) investigated its similarities to C/EBP- and

ATF/CREB-binding sites through *in vitro* electrophoretic mobility shift assay (EMSA). Those studies established that C/EBPβ and ATF-2 were present in protein–DNA complexes formed using the CHOP AARE sequence as a probe, but that the abundance of those complexes did not increase when extracts from amino acid-deprived cells were tested (Bruhat *et al.*, 2000). To determine whether or not both of these transcription factors functioned to activate the CHOP gene *in vivo*, the authors analysed CHOP mRNA content in mouse embryonic fibroblasts (MEFs) that were deficient for either ATF-2 or C/EBPβ. Those authors determined that activation of the CHOP gene in response to activation of the ERSR pathway was functional in both knockout cell types, whereas activation of the CHOP gene following amino acid limitation occurred in the C/EBPβ knockout cells, but not those deficient in ATF-2. They went on to demonstrate that transfection of the ATF-2-deficient cells with an ATF-2-expressing plasmid could restore amino acid control and that expression of a dominant-negative form of ATF-2 could suppress the induction in wild-type MEFs (Bruhat *et al.*, 2000). The results of Bruhat *et al.* (2000) illustrate that EMSA-based *in vitro* binding of a specific transcription factor, in this case C/EBPβ, does not provide conclusive evidence that this factor is functional *in vivo*.

Jousse *et al.* (2000) have used both Northern analysis of endogenous asparagine synthetase (AS) and CHOP mRNA, as well as transient transfection of promoter–luciferase reporter constructs, to document that the two genes are differentially regulated in response to limiting cells for specific amino acids. While the relative activation of the AS promoter was greater than that for CHOP when HeLa cells were deprived of leucine, cysteine, asparagine or histidine, activation of the CHOP promoter was nearly twice as great when the cells were starved for methionine (Jousse *et al.*, 2000). The differential regulatory effect of individual amino acids led Jousse *et al.* (2000) to propose that multiple pathways exist to sense the cellular availability of individual amino acids and that these pathways result in differential activation of individual genes. Those authors also showed that incubation of HeLa cells at decreased leucine concentrations (30 µM), amounts that did not cause a significant degree of protein synthesis inhibition, did result in increased CHOP and AS mRNA expression to a level just slightly less than that present in cells completely deprived of leucine. Furthermore, cycloheximide-induced inhibition of protein synthesis, to rates equal to those observed following total amino acid limitation, did not result in activation of either gene. These results led Jousse *et al.* (2000) to conclude that inhibition of protein synthesis is not a prerequisite for activation of the AAR pathway. While those authors used HeLa cells, our laboratory has obtained similar results, by treatment of rat Fao hepatoma cells with low levels of cycloheximide (R.G. Hutson and M.S. Kilberg, unpublished results).

Interestingly, there may also be tissue-specific differences with regard to the degree of activation of a particular gene by depletion of individual amino acids. Bruhat *et al.* (2000) demonstrated that the relative magnitude of transcriptional activation from the CHOP AARE following depletion of leucine, methionine or lysine was different in HeLa, Caco-2 and HepG2 hepatoma cells. How these results relate to possible *in vivo* tissue-specific effects on the early steps in the AAR pathway remains to be established. Entingh *et al.* (2001) studied the induction of CHOP by amino acid deprivation of mouse fibroblasts. Their results indicated that induction of the CHOP gene in these cells required serum, and they went on to document that the serum component required was insulin-like growth factor-1 (IGF-1). Inhibitor studies also led them to conclude that the pathway leading to CHOP induction involved both phosphatidylinositol-3-kinase (PI3K) and the mammalian target of rapamycin (mTOR). A serum requirement for induction of genes following amino acid limitation may depend on the target gene and, perhaps, the cell type under investigation. For example, induction of the system A amino acid transporter following amino acid starvation is dependent on serum in human fibroblasts (Gazzola *et al.*, 1981; Kilberg *et al.*, 1985), but not in rat hepatoma cells (Kilberg *et al.*, 1985). Likewise, induction of AS by amino acid limitation does not require the presence of serum (M.S. Kilberg *et al.*, unpublished results). Jousse *et al.* (1998) investigated the expression of IGF-I, IGF-II and IGFBP-1 in response to amino acid limitation of hepatoma cells and primary hepatocytes in culture. Their data indicated that IGFBP-1 mRNA and protein content is increased significantly following amino acid limitation, whereas the expression of the insulin-like growth factors themselves is either unaffected or slightly reduced. Protein malnutrition has been documented to decrease the circulating level of IGF-I and increase IGFBP-1, consistent with the *in vitro* data of Jousse *et al.* (1998). Although these observations appear to be in conflict with the requirement for IGF-I in the activation of the CHOP gene, Entingh *et al.* (2001) suggested that even a decreased level of IGF-I is still sufficient for induction of the CHOP gene. Obviously, the exact role of IGF-I in modulating the AAR pathway and the apparent cell specificity of this involvement will require further investigation.

Asparagine synthetase

Enzymology

Human AS catalyzes the synthesis of asparagine and glutamate from aspartate, ATP and glutamine. Based on sequence analysis, the AS protein belongs to the class II glutamine amidotransferase (GAT) superfamily. Structurally, the N-terminal portion of the AS protein contains the glutamine-utilizing (GAT) domain fused to a C-terminal region that contains the active site of the enzyme, which catalyses the activation of aspartate.

The cDNA for AS has been cloned from a number of species, and the analysis reveals a high degree of homology between species (Richards and Schuster, 1998). Rat, hamster and human cells each express a predominant AS mRNA species of approximately 2.0 kb. Hamster cells also express a larger mRNA of approximately 2.5 kb and, in the rat, three AS mRNA species of 2.0, 2.5 and 4.0 kb are observed. All three of these mRNAs are induced coordinately by amino acid starvation (Hutson and Kilberg, 1994). Northern analysis with the 3′-untranslated region of the 2.5 kb rat cDNA revealed hybridization to the 2.5 and a 4.0 kb species only, suggesting either alternative polyadenylation or splicing as explanations for multiple mRNA species.

Regulation of AS expression by amino acids

While screening for genes that could complement the cell cycle block in *ts11* mutant cells, Gong *et al.* (1990) identified AS. The same laboratory (Gong *et al.*, 1991) determined that the level of AS mRNA increased in cells deprived of asparagine, leucine, isoleucine or glutamine, consistent with earlier work monitoring enzymatic activity by Andrulis *et al.* (1979) using tRNA synthetase mutants. These observations indicate that the control of AS mRNA expression is not specific for asparagine, but rather modulated by the availability of other amino acids as well. Hutson and Kilberg (1994) demonstrated an increase in AS mRNA content in response to total amino acid deprivation in rat Fao hepatoma cells in culture and in normal rat liver tissue. Likewise, depletion of a single essential amino acid, such as histidine, threonine or tryptophan, from the culture medium also caused an elevation of the steady-state AS mRNA to a high level, and depletion of phenylalanine, leucine and isoleucine was effective, but to a lesser extent. These results also suggest a broad spectrum of regulation for the AAR pathway in mammalian cells.

Interestingly, the increased AS mRNA expression in Fao hepatoma cells can be repressed by the addition of a single amino acid to an otherwise amino acid-free culture medium (Hutson and Kilberg, 1994). When compared with refeeding the cells an amino acid-complete medium, glutamine repressed the induction of AS mRNA to the greatest degree, but asparagine, histidine and leucine were also effective. Proline, serine and threonine moderately repressed the induction of AS mRNA, and the poorest repressors were aspartate, glycine and glutamate. Culture of Fao cells for 12 h in medium lacking only histidine actually resulted in an increased intracellular concentration of several non-essential amino acids (aspartate, serine, glycine, alanine and proline), but the glutamine and histidine levels decreased by 50 and 35%, respectively (Hutson *et al.*, 1996). The observation that deletion of a single amino acid from an otherwise complete medium can cause induction of AS expression indicates that the presence of the other 19 amino acids is not sufficient to maintain the gene in the repressed state. On the other hand, repression of the gene by addition of only a single amino acid to amino acid-free Krebs–Ringer bicarbonate medium demonstrates that not all amino acids must be present in order to maintain the gene in the 'basal' or repressed state (Hutson and Kilberg, 1994). This paradox needs to be resolved and is likely to involve the early signalling events in the AAR pathway.

Following amino acid deprivation, the elevated AS mRNA is translated into protein as documented by increased AS mRNA association with polysomes (Hutson *et al.*, 1996) and also by pulse–chase labelling of AS protein synthesis

(Hutson et al., 1997). In agreement, Arfin et al. (1977) reported an increase in AS enzymatic activity following 24 h of amino acid deprivation in Chinese hamster ovary (CHO) cells. AS content of tissues varies widely, but the protein content is highest in the rat pancreas, testes, brain and spleen (Hongo et al., 1992). Even prior to isolation of a cDNA for AS, Arfin et al. (1977) showed that after transfer of CHO cells to medium lacking asparagine, the aminoacylation of tRNA^{Asn} decreased and the level of AS activity increased. Similarly, when a mutant cell line containing a temperature-sensitive asparaginyl-tRNA synthetase was transferred to the non-permissive temperature, the content of asparaginyl-tRNA^{Asn} declined while AS activity increased (Andrulis et al., 1979). CHO cell mutants with temperature-sensitive leucyl-, methionyl- and lysyl-tRNA synthetases also exhibit increased AS activity when grown at the non-permissive temperature, even though the level of asparaginyl-tRNA^{Asn} remained unchanged (Andrulis et al., 1979). These results are consistent with the proposal that the GCN2 kinase senses amino acid deprivation by binding a broad spectrum of uncharged tRNAs (Berlanga et al., 1999; Sood et al., 2000). Furthermore, the treatment of Fao hepatoma cells with 5 mM of the amino alcohol histidinol, which prevents the formation of histidinyl-tRNA^{His} by inhibition of the corresponding tRNA synthetase (Hansen et al., 1972), increased AS mRNA to a level equal to or greater than that observed when cells were starved for all amino acids (Hutson and Kilberg, 1994). The fact that histidinol treatment induces AS mRNA content without a decrease in the cytoplasmic free histidine concentration demonstrates that the AAR signalling pathway is not triggered by the level of free amino acids. Collectively, not only do the results document that the level of tRNA charging is important for the sensing of amino acid starvation, but they also illustrate that the sensing mechanism in mammalian cells is similar to the general control response of yeast, in that starvation of any one of a number of amino acids is effective.

AS expression and cell cycle control

Basilico and colleagues identified a human clone that could complement mutant hamster BHK *ts11* cells, which are blocked specifically in progression through the G_1 phase of the cell cycle when grown at the non-permissive temperature (Greco et al., 1989; Gong and Basilico, 1990). By sequence homology, the human clone was identified as AS and confirmed by the ability of exogenously added asparagine to bypass the *ts11* mutation. When grown at the non-permissive temperature, the BHK *ts11* cells produce an inactive asparagine synthetase (Gong and Basilico, 1990), resulting in depletion of cellular asparagine and a corresponding increase in AS mRNA. The addition of asparagine to the medium led to a decrease in AS mRNA levels (Greco et al., 1989). Also showing a link between asparagine content and the cell cycle, AS mRNA was increased substantially by refeeding serum to serum-deprived Balb/c 3T3 cells (Greco et al., 1987). The AS mRNA was induced at approximately mid-G_1 phase in human, mouse and hamster cells, and the addition of asparagine to the culture medium prevented the G_1 induction in serum-stimulated cells. Hongo et al. (1989) showed that AS activity is induced during lymphocyte activation by phytohaemagglutinin, and the increase in activity coincides with the rate of DNA synthesis. Additional evidence that the AS gene may be regulated in a cell cycle-dependent manner comes from the observation by Colletta and Cirafici (1992) who showed that thyroid-stimulating hormone treatment of quiescent rat thyroid cells causes entry into the S phase, and a concurrent increase in AS mRNA content.

Relationship of AS expression to asparaginase therapy of leukaemia

Cancer cells have an increased need for nutrients to sustain rapid growth and cell division, and, therefore, often express higher levels of enzymes involved in the corresponding metabolic pathways. Clinical treatment of certain cancers has been devised to take advantage of this need in instances in which critical enzymes are not expressed at sufficient levels. For example, childhood acute lymphoblastic leukaemia (ALL) is treated using combination chemotherapy, with L-asparaginase (ASNase) always present as one of the drugs (Capizzi and Holcenberg, 1993; Chabner and Loo, 1996; Chakrabarti and Schuster, 1996; Muller and Boos, 1998). This therapy is highly successful in inducing remission of the disease, but relapse due to drug resistance remains an issue. ASNase catalyses the reverse reaction of AS, and the logic of its therapeutic effectiveness is that the cells are depleted of asparagine and those with sufficient AS to counteract this starvation can survive. However, ALL cells express AS at a particularly low level and, therefore, treatment with ASNase is extremely effective in blocking growth of this form of leukaemia. In addition, the bacterial ASNase enzymes used clinically exhibit a low level of glutaminase activity, so glutamine depletion may also play a role.

Exposure to the ASNase can result in selection for a population of ALL cells that are drug resistant (Hutson et al., 1997). Aslanian and Kilberg (2001) showed that many adaptive metabolic changes occur in the ASNase resistant cells that collectively result in two effects: (i) to increase the cellular content of the AS substrates aspartate and glutamine, presumably to enhance conversion to asparagine; and (ii) to shift the plasma membrane flux of asparagine by minimizing the efflux through bidirectional Na^+-independent transporters and increase the influx of asparagine by Na^+-driven transporters. In addition, Aslanian et al. (2001) demonstrated that by overexpressing AS in ASNase-sensitive parental ALL cells, resistance to ASNase is conferred without drug selection, proving that elevated AS alone is sufficient to generate the drug-resistant phenotype. Furthermore, the authors showed that even after removal of the drug from the culture medium for 6 weeks or more, AS expression remained elevated, indicating that a permanent change occurs in the rate of AS gene expression. Consistent with this observation, it was also determined that ASNase resistance was not reversible following drug removal (Aslanian et al., 2001).

Analysis of the AS promoter

Guerrini et al. (1993) demonstrated that the regulation of AS mRNA by amino acids occurs at the level of transcription. A chloramphenicol acetyltransferase (CAT) reporter gene was placed under the control of a 3.4 kb fragment of the human AS genomic DNA containing the promoter region and the first two exon and intron sequences. When cells were transfected with this reporter construct, an increase in the level and activity of CAT mRNA was detected in response to either asparagine or leucine starvation. In contrast, the level of CAT mRNA and activity under the control of the simian virus 40 (SV40) early promoter decreased under the same conditions. Deletion analysis showed that the sequence spanning from nucleotide −164 to +44 within the AS gene retained full inducibility by amino acid deprivation (Guerrini et al., 1993). Further scanning of the AS promoter by mutagenesis revealed that an AARE was present at nucleotides −70 to −64 (5′-CATGATG-3′) and was essential for amino acid regulation of the AS gene. EMSAs, using double-stranded oligonucleotides containing this AARE sequence documented the formation of specific protein–DNA complexes in vitro, but the AARE sequence did not correspond precisely to a consensus sequence for any known transcription factors. As described below, subsequent research has defined the importance of this AS genomic sequence better with regard to nutrient regulation.

Complex nutrient sensing by the AS gene

Barbosa-Tessmann et al. (1999a) demonstrated that the human AS gene is also induced by glucose starvation and that this induction is mediated via the ERSR (UPR) pathway (Barbosa-Tessmann et al., 1999b). An increase in AS mRNA content in human hepatoma HepG2 cells was detectable following glucose starvation for 8 h, and reached a maximum by 12 h. Elevated AS protein expression following glucose deprivation was also documented (Barbosa-Tessmann et al., 1999a). To confirm that the induction of AS transcription following glucose limitation was the result of the ERSR pathway, other recognized activators for the pathway such as the protein glycosylation inhibitor tunicamycin and the proline analogue, azetidine-2-carboxylate (Aze), were shown to be activators (Barbosa-Tessmann et al., 1999b). These results documented that the human AS is a target gene for the ERSR pathway and that the gene represents a link between amino acid metabolism and the ERSR pathway. The exact metabolic function of increased asparagine biosynthesis during ER stress has not been determined.

The ERSR pathway induces the AS gene by a transcriptional mechanism. Barbosa-Tessmann et al. (1999b) showed that expression of a reporter gene was significantly enhanced by glucose starvation when it was driven by the human AS promoter. Deletion analysis indicated that the cis-elements responsible for the ERSR control of the AS gene were located within nucleotides −111 to −34 of the AS promoter, but the known mammalian ERSE consensus sequence (5'-CCAAT-N$_9$-CCACG-3'), present in all other ERSR-inducible genes previously identified, was not present. Barbosa-Tessmann *et al.* (2000) went on to demonstrate that activation of AS gene transcription by starvation for either amino acids (AAR pathway) or glucose (ERSR pathway) is mediated through a common and unique set of genomic elements within the AS proximal promoter.

Nutrient-sensing response elements

As shown in Fig. 8.1, dimethyl sulphate *in vivo* footprinting documented that the human AS promoter region immediately upstream of the major transcription start site contains six separate protein-binding sites (Barbosa-Tessmann *et al.*, 2000). Of these, five have been implicated in nutrient control of the human AS gene: three GC boxes (GC-I, GC-II and GC-III) and two nutrient-sensing response elements (NSRE-1 and -2). All three GC boxes are required to maintain basal transcription and to obtain maximal activation of the AS gene by amino acid limitation (Leung-Pineda and Kilberg, 2002). However, when functionally analysed individually, there is not complete redundancy among the three GC sequences and there is a difference in the degree of importance with regard to transcription (GC-III > GC-II > GC-I). *In vitro*, two of the GC sequences formed protein–DNA complexes (GC-II and GC-III) with either Sp1 or Sp3, but the absolute amount of these complexes and the total pool of either Sp1 or Sp3 protein did not increase following amino acid limitation. *In vivo* expression of Sp1 and Sp3 in *Drosophila* SL2 cells, which lack Sp proteins, increased AS promoter activity, but functional differences between the factors were observed (Leung-Pineda and Kilberg, 2002). Sp1 expression increased basal transcription from the AS promoter, but did not cause a further increase when SL2 cells were amino acid deprived. In contrast, Sp3 expression enhanced both the basal and the starvation-induced AS-driven transcription.

Two of the protein-binding sites identified by *in vivo* footprinting showed changes in protein protection in response to amino acid deprivation (Barbosa-Tessmann *et al.*, 2000). These two sites,

```
                            GC-I              GC-II
-173  CAAAAGAGCT CCTCCTTGCG CCCTTCCGCC GCCCCACTTA GTCCTGCTCC GCCCCGGACA
      GTTTTCTCGA GGAGGAACGC GGGAAGGCGG CGGGGTGAAT CAGGACGAGG CGGGGCCTGT

           GC-III                                        NSRE-1
-113  CCCCGCGGCC CCGCCCCTGT GCGCGCTGGT TGGTCCTCGC AGGCATGATG AAACTTCCCG
      GGGGCGCCGG GGCGGGGACA CGCGCGACCA ACCAGGAGCG TCCGTACTAC TTTGAAGGGC

         NSRE-2
-53   CACGCGTTAC AGGAGCCAGG TCGGTATAAG CGCCAGCGGC CTCGCCGCCC GTCaagctgt
      GTGCGCAATG TCCTCGGTCC AGCCATATTC GCGGTCGCCG GAGCGGCGGG CAGttcgaca


+8    ccacatccct ggcctcagcc cgccacatca ccctgacctg ctta
      ggtgtaggga ccggagtcgg gcggtgtagt gggactggac gaat
```

Fig. 8.1. The proximal promoter sequence of the human asparagine synthetase gene. Six potential protein-binding sites were identified by *in vivo* footprinting (Barbosa-Tessmann *et al.*, 2000). Of these, the five sites that are listed contribute to nutrient control of transcription, GC boxes I–III, NSRE-1 and NSRE-2. As measured by either *in vivo* footprinting or *in vitro* EMSA, protein binding at all three GC boxes was the same in control and nutrient-deprived cells, whereas protein binding at NSRE-1 and NSRE-2 was enhanced by activation of either the AAR or the ERSR pathways. The shaded boxes show the boundaries of NSRE-1 and NSRE-2 based on the single nucleotide mutagenesis shown in Fig. 8.2.

originally labelled sites V and VI, have been renamed NSRE-1 and NSRE-2 (see Fig. 8.1). Single nucleotide mutagenesis throughout the entire region of these two sites has defined the boundaries of these two elements more clearly (Fig. 8.2). The NSRE-1 sequence (5′-TGATGAAAC-3′) located from nucleotide −68 to −60 within the AS proximal promoter overlaps the sequence first identified by Guerrini *et al.* (1993) as having AARE activity. Given that the NSRE-1 sequence is necessary for induction of the AS gene following activation of not only the AAR pathway, but also the ERSR pathway (Barbosa-Tessmann *et al.*, 2000), it is clear that this element is functionally more important than simply as an AARE. To reflect this broader nutrient-detecting capability, the term NSRE-1 was coined. When nuclear extracts from amino acid-starved HepG2 cells were tested by EMSA, increased amounts of protein–NSRE-1 complexes were detected (Barbosa-Tessmann *et al.*, 2000). Furthermore, a second element 5′-GTTACA-3′ (nucleotides −48 to −43), positioned 11 nucleotides downstream of the NSRE-1, was shown also to be absolutely required for induction of the AS gene by both amino acid and glucose starvation (Barbosa-Tessmann *et al.*, 2000). Once again, single nucleotide mutagenesis has defined the boundaries of this site (Fig. 8.2), but the sequence does not correspond identically to any known transcription factor consensus sequence. Collectively, the promoter analysis documents that at least three separate *cis*-regulatory elements are minimally required for an optimal transcriptional response of the AS gene either to amino acid or glucose deprivation, one or more of the GC boxes, NSRE-1 and NSRE-2. The term nutrient-sensing response unit (NSRU) has been coined to describe the collective action of these elements.

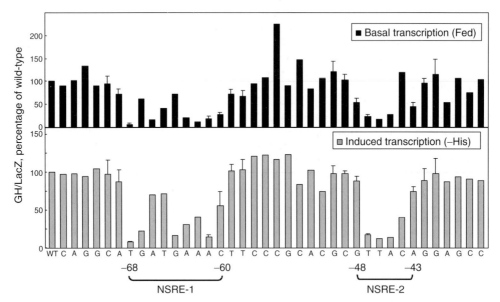

Fig. 8.2. Single nucleotide mutagenesis to define the boundaries of the NSRE-1 and NSRE-2 regulatory sites. Mutagenesis and transient expression were carried out as described by Barbosa-Tessmann *et al.* (2000). Nucleotides −173 to +51 of the human AS gene were linked to the human growth hormone gene (GH) as a reporter and transcription was determined by analysing the GH mRNA content relative to that for the co-transfection control, which was the *lacZ* gene driven by the cytomegalovirus promoter. The cells were incubated for 18 h in either complete minimal essential medium (MEM) for the data labelled 'Fed' or in MEM lacking histidine (labelled '−His'). For the bars without a standard deviation symbol, the data represent a single analysis and only one nucleotide substitution was tested, purine for purine, pyrimidine for pyrimidine. For the sites with standard deviation bars, the results are representative of three independent experiments, and the wild-type nucleotide was changed to at least two of the three other nucleotides. The latter sites were tested more extensively to define better the boundaries of the NSRE-1- and NSRE-2-binding sites.

As described above, the human CHOP gene is activated by the AAR and ERSR pathways through two completely independent sets of *cis*-acting elements within the CHOP promoter region, the ERSE-1 at −93 to −75 (Yoshida *et al.*, 2000) and the AARE at −302 to −310 (Bruhat *et al.*, 2000). As just described, the human AS gene is also transcriptionally activated in response to the AAR and ERSR pathways, but, in contrast to CHOP, this activation occurs through the same set of *cis*-acting elements that function together to make up the NSRU (Barbosa-Tessmann *et al.*, 2000). Interestingly, the NSRE-1 has a high degree of sequence identity with the AARE within the CHOP promoter, differing by only two nucleotides (Fig. 8.3). Although the AARE sequence in the CHOP promoter occurs on the opposite strand to that in the AS promoter, these sequences function as enhancer elements in that activation of the genes is observed regardless of orientation (Barbosa-Tessmann *et al.*, 2000). Beyond the two-nucleotide difference in sequence between the CHOP AARE and the AS NSRE-1, the major difference between the promoters that results in the mechanistic difference in response to the ERSR pathway appears to be the presence of the second *cis*-acting element, NSRE-2 (Barbosa-Tessmann *et al.*, 2000), which the CHOP promoter lacks. Given the similarity between the AS NSRE-1 sequence and the AARE in the CHOP promoter, one might speculate that elimination of the AS NSRE-2 sequence would block activation of the AS gene by the ERSR pathway, but permit retention of activation by the AAR pathway. However, mutagenesis of only the NSRE-2 sequence results in complete loss of responsiveness to both nutrient-regulated pathways (Barbosa-Tessmann *et al.*, 2000). For reasons unknown, the presence of the NSRE-1-like AARE sequence in the CHOP promoter is sufficient to permit transcriptional induction via the AAR pathway, whereas the closely related NSRE-1 sequence is not. Interestingly, insertion of the AS NSRE-2 sequence into the human CHOP promoter, 11 nucleotides downstream from the CHOP AARE, conveys responsiveness to the ERSR pathway (Bruhat *et al.*, 2002).

The CHOP AARE and the AS NSRE-1 also appear to bind different transcription factors. As described above, the CHOP binds both ATF-2 and C/EBPβ *in vitro* but, *in vivo*, only ATF-2 appears to be functionally associated with this sequence (Bruhat *et al.*, 2000). In contrast, the AS NSRE-1 sequence does not bind ATF-2 *in vitro*, but does bind C/EBPβ (Siu *et al.*, 2001). Furthermore, overexpression of C/EBPβ up-regulates both basal and induced transcription through the NSRE-1 sequence (Siu *et al.*, 2000). The significance of the differences in transcription factor binding at the CHOP AARE and the AS NSRE-1 is unknown, but it is suggested that there is heterogeneity in the upstream signalling steps in a single AAR pathway or, possibly, multiple AAR pathways. Future studies contrasting these two genes and their transcription control by amino acids will be interesting and informative.

Summary and Future Research Opportunities

The investigation of nutrient control in mammalian cells, especially by amino acids, is still in its infancy. The full complement of target genes that respond to protein/amino acid deprivation is not yet known, but should be delineated in the near future by DNA microchip and array technology. Of course, as is already clear, the list of genes

```
AS        nt -68      5´-TGATGAAAC-3´     nt -60
                      3´-ACTACTTTG-5´


CHOP      nt -310     5´-ATTGCATCA-3´     nt -302
                      3´-TAACGTAGT-5´
```

Fig. 8.3. Comparison of the sequences for the AS NSRE-1 and the CHOP AARE regulatory sites. The *cis*-elements that mediate the AAR in the human AS and CHOP promoters are shown. The double-stranded sequences and the corresponding nucleotide numbers relative to the transcription start sites are illustrated. Note that the two elements are orientated on opposite strands.

that are transcriptionally activated and the list of mRNA species that are translationally controlled will differ, possibly significantly. The mechanisms responsible for these transcriptional and translational events also remain to be elucidated more fully. Some of the steps are beginning to emerge and a few of the proteins required have been identified but, from the presentation above contrasting CHOP and AS, it is clear that many more proteins are involved and must still be identified. Finally, there are many questions unanswered regarding the initial sensors of amino acid deprivation and the subsequent signal transduction pathways that are activated. The investigation and characterization of these processes in normal homeostasis and in disease represents an interesting and important study in molecular nutrition.

Acknowledgements

This work was supported by grants to M.S.K. from the National Institutes of Health (DK-52064, DK-59315).

References

Abcouwer, S.F., Schwarz, C. and Meguid, R.A. (1999) Glutamine deprivation induces the expression of *GADD45* and *GADD153* primarily by mRNA stabilization. *Journal of Biological Chemistry* 274, 28645–28651.

Adilakshmi, T. and Laine, R.O. (2002) Ribosomal protein S25 mRNA partners with MTF-1 and La to provide a p53-mediated mechanism for survival or death. *Journal of Biological Chemistry* 277, 4147–4151.

Andrulis, I.L., Hatfield, G.W. and Arfin, S.M. (1979) Asparaginyl-tRNA aminoacylation levels and asparagine synthetase expression in cultured Chinese hamster ovary cells. *Journal of Biological Chemistry* 254, 10629–10633.

Arfin, S.M., Simpson, D.R., Chiang, C.S., Andrulis, I.L. and Hatfield, G.W. (1977) A role for asparaginyl-tRNA in the regulation of asparagine synthetase in a mammalian cell line. *Proceedings of the National Academy of Sciences USA* 74, 2367–2369.

Aslanian, A.M. and Kilberg, M.S. (2001) Multiple adaptive mechanisms affect asparagine synthetase substrate availability in asparaginase resistant MOLT-4 human leukemia cells. *Biochemical Journal* 358, 59–67.

Aslanian, A.M., Fletcher, B.S. and Kilberg, M.S. (2001) Asparagine synthetase expression alone is sufficient to induce L-asparaginase resistance in MOLT-4 human leukaemia cells. *Biochemical Journal* 357, 321–328.

Aulak, K.S., Mishra, R., Zhou, L., Hyatt, S.L., de Jonge, W., Lamers, W., Snider, M. and Hatzoglou, M. (1999) Post-transcriptional regulation of the arginine transporter Cat-1 by amino acid availability. *Journal of Biological Chemistry* 274, 30424–30432.

Barbosa-Tessmann, I.P., Pineda, V.L., Nick, H.S., Schuster, S.M. and Kilberg, M.S. (1999a) Transcriptional regulation of the human asparagine synthetase gene by carbohydrate availability. *Biochemical Journal* 339, 151–158.

Barbosa-Tessmann, I.P., Chen, C., Zhong, C., Schuster, S.M., Nick, H.S. and Kilberg, M.S. (1999b) Activation of the unfolded protein response pathway induces human asparagine synthetase gene expression. *Journal of Biological Chemistry* 274, 31139–31144.

Barbosa-Tessmann, I.P., Chen, C., Zhong, C., Siu, F., Schuster, S.M., Nick, H.S. and Kilberg, M.S. (2000) Activation of the human asparagine synthetase gene by the amino acid response and the endoplasmic reticulum stress response pathways occurs by common genomic elements. *Journal of Biological Chemistry* 275, 26976–26985.

Berlanga, J.J., Santoyo, J. and De Haro, C. (1999) Characterization of a mammalian homolog of the GCN2 eukaryotic initiation factor-2 alpha kinase. *European Journal of Biochemistry* 265, 754–762.

Bruhat, A., Jousse, C., Wang, X.-Z., Ron, D., Ferrara, M. and Fafournoux, P. (1997) Amino acid limitation induces expression of CHOP, a CCAAT/enhancer binding protein-related gene, at both transcriptional and post-transcriptional levels. *Journal of Biological Chemistry* 272, 17588–17593.

Bruhat, A., Jousse, C. and Fafournoux, P. (1999) Amino acid limitation regulates gene expression. *Proceedings of the Nutrition Society* 58, 625–632.

Bruhat, A., Jousse, C., Carraro, V., Reimold, A.M., Ferrara, M. and Fafournoux, P. (2000) Amino acids control mammalian gene transcription: activating transcription factor 2 is essential for the amino acid responsiveness of the CHOP promoter. *Molecular and Cellular Biology* 20, 7192–7204.

Bruhat, A., Averous, J., Carraro, V., Zhong, C., Reimold, A.M., Kilbery, M.A. and Fafournoux, P. (2002) Differences in the molecular mechanisms involved in the transcriptional activation of the CHOP and asparagine synthetase genes in response to amino acid deprivation or activation of the unfolded protein response. *Journal of Biological Chemistry* 277, 48107–48114.

Capizzi, R.L. and Holcenberg, J.S. (1993) Asparaginase. In: Holland, J.F., Frei, E. III, Bast, R.C. Jr,

Kufe, D.W., Morton, D.L. and Weichselbaum, R.R. (eds) *Cancer Medicine*, 3rd edn., R.R. Lea & Febiger, Philadelphia, pp. 796–805.

Chabner, B.A. and Loo, T.L. (1996) Enzyme therapy: L-asparaginase. In: Chabner, B.A. and Longo, D.L. (eds) *Cancer Chemotherapy and Biotherapy*, 2nd edn. Lippincott-Raven Publishers, Philadelphia, pp. 485–492.

Chakrabarti, R. and Schuster, S.M. (1996) L-Asparaginase: perspectives on the mechanisms of action and resistance. *International Journal of Pediatric Hematology/Oncology* 4, 597–611.

Colletta, G. and Cirafici, A.M. (1992) TSH is able to induce cell cycle-related gene expression in rat thyroid cell. *Biochemical and Biophysical Research Communications* 183, 265–272.

Entingh, A.J., Law, B.K. and Moses, H.L. (2001) Induction of the C/EBP homologous protein (CHOP) by amino acid deprivation requires insulin-like growth factor I, phosphatidylinositol 3-kinase, and mammalian target of rapamycin signaling. *Endocrinology* 142, 221–228.

Fafournoux, P., Bruhat, A. and Jousse, C. (2000) Amino acid regulation of gene expression. *Journal of Biochemistry* 351, 1–12.

Fernandez, J., Yaman, I., Mishra, R., Merrick, W.C., Snider, M.D., Lamers, W.H. and Hatzoglou, M. (2001) Internal ribosome entry site-mediated translation of a mammalian mRNA is regulated by amino acid availability. *Journal of Biological Chemistry* 276, 12285–12291.

Gazzola, G.C., Franchi-Gazzola, R., Saibene, V., Ronchi, P. and Guidotti, G.G. (1972) Regulation of amino acid transport in chick embryo heart cells I. Adaptive system of mediation for neutral amino acids. *Biochimica et Biophysica Acta* 266, 407–421.

Gazzola, G.C., Dall'Asta, V. and Guidotti, G.G. (1981) Adaptive regulation of amino acid transport in cultured human fibroblasts. *Journal of Biological Chemistry* 256, 3191–3198.

Gazzola, R.F., Sala, R., Bussolati, O., Visigalli, R., Dall'Asta, V., Ganapathy, V. and Gazzola, G.C. (2001) The adaptive regulation of amino acid transport system A is associated to changes in ATA2 expression. *FEBS Letters* 490, 11–14.

Gong, S.S. and Basilico, C. (1990) A mammalian temperature-sensitive mutation affecting G_1 progression results from a single amino acid substitution in asparagine synthetase. *Nucleic Acids Research* 18, 3509–3513.

Gong, S.S., Guerrini, L. and Basilico, C. (1991) Regulation of asparagine synthetase gene expression by amino acid starvation. *Molecular and Cellular Biology* 11, 6059–6066.

Greco, A., Ittmann, M. and Basilico, C. (1987) Molecular cloning of a gene that is necessary for G_1 progression in mammalian cells. *Proceedings of the National Academy of Sciences USA* 84, 1565–1569.

Greco, A., Gong, S.S., Ittmann, M. and Basilico, C. (1989) Organization and expression of the cell cycle gene, *ts 11*, that encodes asparagine synthetase. *Molecular and Cellular Biology* 9, 2350–2359.

Guerrini, L., Gong, S.S., Mangasarian, K. and Basilico, C. (1993) *Cis*- and *trans*-acting elements involved in amino acid regulation of asparagine synthetase gene expression. *Molecular and Cellular Biology* 13, 3202–3212.

Hansen, B.S., Vaughan, M.H. and Wang, L.-J. (1972) Reversible inhibition by histidinol of protein synthesis in human cells at the activation of histidine. *Journal of Biological Chemistry* 247, 3854–3857.

Harding, H.P., Novoa, I.I, Zhang, Y., Zeng, H., Wek, R., Schapira, M. and Ron, D. (2000) Regulated translation initiation controls stress-induced gene expression in mammalian cells. *Molecular Cell* 6, 1099–1108.

Hinnebusch, A.G. (1997) Translational regulation of yeast GCN4. *Journal of Biological Chemistry* 272, 21661–21664.

Hitomi, Y., Ito, A., Naito, Y. and Yoshida, A. (1993) Liver-specific induction of ribosomal protein gene expression by amino acid starvation in rats. *Bioscience, Biotechnology and Biochemistry* 57, 1216–1217.

Hoffer, L.J. (1993) Are dietary-protein requirements altered in diabetes-mellitus? *Canadian Journal of Physiology and Pharmacology* 71, 633–638.

Hongo, S., Takeda, M. and Sato, T. (1989) Induction of asparagine synthetase during lymphocyte activation by phytohemagglutinin. *Biochemistry International* 18, 661–666.

Hongo, S., Fujimori, M., Shioda, S., Nakai, Y., Takeda, M. and Sato, T. (1992) Immunochemical characterization of rat testicular asparagine synthetase. *Archives of Biochemistry and Biophysics* 295, 120–125.

Hope, I.A. and Struhl, K. (1985) GCN4 protein, synthesized *in vitro*, binds HIS3 regulatory sequences: implications for general control of amino acid biosynthetic genes in yeast. *Cell* 43, 177–188.

Hutson, R.G. and Kilberg, M.S. (1994) Cloning of rat asparagine synthetase and specificity of the amino acid-dependent control of its mRNA content. *Biochemical Journal* 303, 745–750.

Hutson, R.G., Warskulat, U. and Kilberg, M.S. (1996) An example of nutrient control of gene expression: amino acid-dependent regulation of asparagine synthetase. *Clinical Nutrition* 51, 327–331.

Hutson, R.G., Kitoh, T., Amador, D.A.M., Cosic, S., Schuster, S.M. and Kilberg, M.S. (1997) Amino acid control of asparagine synthetase: relation to asparaginase resistance in human leukemia cells. *American Journal of Physiology* 272, C1691–C1699.

Hyatt, S.L., Aulak, K.S., Malandro, M., Kilberg, M.S. and Hatzoglou, M. (1997) Adaptive regulation of the cationic amino acid transporter-1 (Cat-1) in Fao cells. *Journal of Biological Chemistry* 272, 19951–19957.

Jousse, C., Bruhat, A., Ferrara, M. and Fafournoux, P. (1998) Physiological concentration of amino acids regulates insulin-like-growth-factor-binding protein 1 expression. *Biochemical Journal* 334, 147–153.

Jousse, C., Bruhat, A., Harding, H.P., Ferrara, M., Ron, D. and Fafournoux, P. (1999) Amino acid limitation regulates CHOP expression through a specific pathway independent of the unfolded protein response. *FEBS Letters* 448, 211–216.

Jousse, C., Bruhat, A., Ferrara, M. and Fafournoux, P. (2000) Evidence for multiple signaling pathways in the regulation of gene expression by amino acids in human cell lines. *Journal of Nutrition* 130, 1555–1560.

Kaufman, R.J. (1999) Stress signaling from the lumen of the endoplasmic reticulum: coordination of gene transcriptional and translational controls. *Genes and Development* 13, 1211–1233.

Kilberg, M.S., Han, H.P., Barber, E.F. and Chiles, T.C. (1985) Adaptive regulation of neutral amino acid transport system A in rat H4 hepatoma cells. *American Journal of Physiology* 122, 290–298.

Kilberg, M.S., Stevens, B.R. and Novak, D. (1993) Recent advances in mammalian amino acid transport. *Annual Review of Nutrition* 13, 137–165.

Laine, R.O., Laipis, P.J., Shay, N.F. and Kilberg, M.S. (1991) Identification of an amino acid-regulated mRNA from rat liver as the mammalian equivalent of bacterial ribosomal protein L22. *Journal of Biological Chemistry* 266, 16969–16972.

Laine, R.O., Shay, N.F. and Kilberg, M.S. (1994) Nuclear retention of the induced mRNA following amino acid-dependent transcriptional regulation of mammalian ribosomal proteins L17 and S25. *Journal of Biological Chemistry* 269, 9693–9697.

Leung-Pineda, V. and Kilberg, M.S. (2002) Role of Sp1 and Sp3 in the nutrient-regulated expression of the human asparagine synthetase gene. *Journal of Biological Chemistry* 277, 16585–16591.

Marten, N.W., Burke, E.J., Hayden, J.M. and Straus, D.S. (1994) Effect of amino acid limitation on the expression of 19 genes in rat hepatoma cells. *FASEB Journal* 8, 538–544.

Mizock, B.A. (1999) Nutritional support in hepatic encephalopathy. *Nutrition* 15, 220–228.

Morgane, P.J., Austinlafrance, R., Bronzino, J., Tonkiss, J., Diazcintra, S., Cintra, L., Kemper, T. and Galler, J.R. (1993) Prenatal malnutrition and development of the brain. *Neuroscience and Behavioral Reviews* 17, 91–128.

Muller, H.J. and Boos, J. (1998) Use of L-asparaginase in childhood ALL. *CRC Critical Reviews in Oncology/Hematology* 28, 97–113.

Natarajan, K., Meyer, M.R., Jackson, B.M., Slade, D., Roberts, C., Hinnebusch, A.G. and Marton, M.J. (2001) Transcriptional profiling shows that Gcn4p is a master regulator of gene expression during amino acid starvation in yeast. *Molecular and Cell Biology* 21, 4347–4368.

Pahl, H.L. (1999) Signal transduction from the endoplasmic reticulum to the cell nucleus. *Physiological Reviews* 79, 683–701.

Palacin, M., Estevez, R., Bertran, J. and Zorzano, A. (1998) Molecular biology of mammalian plasma membrane amino acid transporters. *Physiological Reviews* 78, 969–1054.

Patil, C. and Walter, P. (2001) Intracellular signaling from the endoplasmic reticulum to the nucleus: the unfolded protein response in yeast and mammals. *Current Opinion in Cell Biology* 13, 349–355.

Paul, G.L., Waegner, A., Gaskins, H.R. and Shay, N.F. (1998) Histidine availability alters glucagon gene expression in murine alphaTC6 cells. *Journal of Nutrition* 128, 973–976.

Price, B.D. and Calderwood, S.K. (1992) Gadd45 and Gadd153 messenger RNA levels are increased during hypoxia and after exposure of cells to agents which elevate the levels of the glucose-regulated proteins. *Cancer Research* 52, 3814–3817.

Richards, N.G.J. and Schuster, S.M. (1998) Mechanistic issues in asparagine synthetase catalysis. *Advances in Enzymology* 72, 145–198.

Roediger, W.E.W. (1995) New views on the pathogenesis of Kwashiorkor – methionine and other amino-acids. *Journal of Pediatric Gastroenterology and Nutrition* 21, 130–136.

Ron, D. and Habener, J.F. (1992) CHOP, a novel developmentally regulated nuclear protein that dimerizes with transcription factors C/EBP and LAP and functions as a dominant-negative inhibitor of gene transcription. *Genes and Development* 6, 439–453.

Roy, B. and Lee, A.S. (1999) The mammalian endoplasmic reticulum stress response element consists of an evolutionarily conserved tripartite structure and interacts with a novel stress-inducible complex. *Nucleic Acids Research* 27, 1437–1443.

Sellers, J.W., Vincent, A.C. and Struhl, K. (1990) Mutations that define the optimal half-site for binding yeast GCN4 activator protein and identify an ATF/CREB-like repressor that recognizes similar DNA sites. *Molecular and Cellular Biology* 10, 5077–5086.

Siu, F.Y., Chen, C., Zhong, C. and Kilberg, M.S. (2001) CCAAT/enhancer-binding protein beta (C/EBPβ) is a mediator of the nutrient sensing response pathway that activates the human asparagine synthetase gene. *Journal of Biological Chemistry* 276, 48100–48107.

Sood, R., Porter, A.C., Olsen, D., Cavener, D.R. and Wek, R.C. (2000) A mammalian homologue of GCN2 protein kinase important for translational control by phosphorylation of eukaryotic initiation factor-2α. *Genetics* 154, 787–801.

Straus, D.S., Burke, E.J. and Marten, N.W. (1993) Induction of insulin-like growth factor binding protein-1 gene expression in liver of protein-restricted rats and in rat hepatoma cells limited for a single amino acid. *Endocrinology* 132, 1090–1100.

Wang, X.-Z., Kuroda, M., Sok, J., Batchvarova, N., Kimmel, R., Chung, P., Zinszner, H. and Ron, D. (1998) Identification of novel stress-induced genes downstream of *chop*. *EMBO Journal* 17, 3619–3630.

Wek, S.A., Zhu, S. and Wek, R.C. (1995) The histidyl-tRNA synthetase-related sequence in the eIF-2 alpha protein kinase GCN2 interacts with tRNA and is required for activation in response to starvation for different amino acids. *Molecular and Cellular Biology* 15, 4497–4506.

Yoshida, H., Haze, K., Yanagi, H., Yura, T. and Mori, K. (1998) Identification of the *cis*-acting endoplasmic reticulum stress response element responsible for transcriptional induction of mammalian glucose-regulated proteins. *Journal of Biological Chemistry* 273, 33741–33749.

Yoshida, H., Okada, T., Haze, K., Yanagi, H., Yura, T., Negishi, M. and Mori, K. (2000) ATF6 activated by proteolysis binds in the presence of NF-Y (CBF) directly to the *cis*-acting element responsible for the mammalian unfolded protein response. *Molecular and Cellular Biology* 20, 6755–6767.

Zinszner, H., Kuroda, M., Wang, X.Z., Batchvarova, N., Lightfoot, R.T., Remotti, H., Stevens, J.L. and Ron, D. (1998) CHOP is implicated in programmed cell death in response to impaired function of the endoplasmic reticulum. *Genes and Development* 12, 982–995.

9 Fatty Acids and Gene Expression

Ulrike Beisiegel, Joerg Heeren and Frank Schnieders
Institute of Molecular Cell Biology, University Hospital Hamburg-Eppendorf, Hamburg, Germany

Introduction

Fatty acids (FAs) are the most important molecules in the energy supply of all animals, and also serve as precursors for the synthesis of eicosanoids (prostaglandins, thromboxanes and leucotrienes). However, in recent years, it became evident that FAs are also involved in the transcriptional regulation of gene expression (for reviews, see Duplus *et al.*, 2000; Jump and Clarke, 1999; Jump, 2001; Hihi *et al.*, 2002). This chapter will focus on the role of FAs as ligands for transcription factors that control the expression of many genes involved in lipid and glucose metabolism, as well as cell differentiation, growth and inflammation.

Fatty acids and their active derivatives

The basic structure of FAs, the hydrophobic polycarbon chain, varies in chain length and the degree of saturation. Short and medium chain FAs, with ≤14 C-atoms, are mainly saturated and used predominantly for energy supply. Long chain FAs (LCFAs) can be saturated, as well as mono- or polyunsaturated, and can serve functions apart from energy supply. The most abundant monounsaturated LCFA is oleate (C18:1, ω-9)[1]. The two major classes of polyunsaturated LCFAs (PUFAs) are ω-3 and ω-6 FAs, where the ω-number indicates the position of the first double bond counted from the methyl group of the hydrocarbon chain. These FAs must be provided in the diet since animal cells cannot introduce double bonds beyond the C-9 position[2]. Linoleic acid (C18:2, ω-6) is an essential PUFA and precursor for arachidonic acid (C20:4, ω-6), while α-linolenic acid (C18:3, ω-3) is the precursor for eicosapentaenoic acid (C20:5, ω-3; EPA) and docosahexaenoic acid (C22:6, ω-3; DHA) (Goodridge, 1991). Arachidonic acid and EPA are both precursors for eicosanoids (leucotrienes, prostaglandins and thromboxanes), as well as FAs with longer chains (docosatetraenic acid and docosapentaenic acid). The numerous eicosanoids derived from the two different precursors are relevant factors in coagulation and inflammation, with independent functions for ω-3 and ω-6 eicosanoids (Serhan and Oliw, 2001).

---

[1] The numbering given in parentheses after trivial names of FAs is, for example, oleate (C18:1 ω-9): a fatty acid with 18 C-atoms and one double bond at bond 9 after the ω-atom. The ω-C atom is the first methyl group of the hydrocarbon chain in FAs.

[2] The omega numbering system starts from the methyl end, whereas the delta numbering system starts from the carboxyl (-COOH) group:

$$\begin{array}{cccc} & 3 & 2 & 1 \quad \text{delta numbering} \\ H_3C\text{-}(CH_2)_n\text{-}CH_2\text{-}CH_2\text{-}COOH & & & \\ \omega & \beta & \alpha & \quad \text{omega numbering} \end{array}$$

PUFAs are substrates for free radical-induced lipid peroxidation and can be oxidized easily at the C-atom of the methyl group adjacent to consecutive double bonds (Halliwell and Gutteridge, 1989). These oxidized FAs are not only catabolites but can also influence the gene expression (Nagy et al., 1998; Jump and Clarke, 1999; Delerive et al., 2000).

Conjugated FAs are derived from polyunsaturated C-18 FAs; most conjugated FAs are derivatives of linolenic acid and are therefore named 'conjugated linolenic acids' (CLAs). CLAs contain different patterns of conjugated double bonds, which determine their particular functions in metabolism. Phytanic acid has a branched structure and is derived from plant phytoles. This rare FA has become more relevant due to its effect on gene expression.

Metabolism of dietary lipids

After ingestion of dietary fat, intestinal cells synthesize triglyeride-rich chylomicrons, which enter the circulation via the ductus thoracicus. Chylomicrons are characterized by apolipoprotein B-48 as structural apoprotein and contain mainly the dietary LCFAs. Short and medium chain FAs are bound to albumin in the intestine and are transported directly into the blood. The hydrolysis of chylomicron triglycerides is facilitated by lipoprotein lipase (LPL), an endothelia-bound enzyme abundantly expressed in muscle and adipose tissue (Eckel, 1989). This lipolysis supplies LCFAs mainly to these two tissues, but also to all other tissues with LCFA consumption (Goldberg, 1996). Induced by the anabolic action of insulin, LCFAs are re-esterified and deposited as lipid droplets in adipose tissues. During fasting, catecholamine-stimulated activation of hormone-sensitive lipase in adipose tissue generates LCFAs, which are then transported to other tissues by albumin (Holm et al., 2000).

The predominant mechanism of cellular LCFA uptake under physiological conditions is a saturable process, probably mediated by as yet unidentified albumin receptors (Berk and Stump, 1999). LCFAs and their derivatives can, however, also been taken up via FA transporter (FAT) (Motojima et al., 1998). Intracellularly, LCFAs follow a number of different metabolic routes in various compartments, involving a number of different FA-binding proteins (FABPs) (Abumrad et al., 1998; McArthur et al., 1999; Norris and Spector, 2002). In the cytoplasm, FABP might release the FAs to allow an interaction with their soluble receptors (transcription factors), which then can effect gene expression in the nucleus. In microsomes, unsaturated LCFAs (MUFAs and PUFAs) are oxidized by cyclooxygenase, lipoxygenase or monooxygenase pathways to form eicosanoids and epoxy- or hydroxy-FAs (Sprecher, 2000), which in turn are capable of influencing gene expression by binding to different transcription factors (Jump and Clarke, 1999). In peroxisomes and mitochondria, activated PUFAs are the substrates for β-oxidation. In addition, activated FAs can act as precursors for the *de novo* synthesis of phospholipids and triglycerides, which subsequently are used for the formation of membranes and lipoproteins. All these pathways, as well as associated metabolic processes such as carbohydrate metabolism, are regulated directly by dietary lipids and are modulated by hormonal effects (e.g. insulin or T3) (Kaytor et al., 1997; Towle et al., 1997; Jump et al., 2001; Vaulont et al., 2000).

Transcription factors and gene expression

In the context of a healthy organism, the energy metabolism and 'health status' in a given cell need to be tightly regulated with regard to circumstances determined by the developmental state, the nutritional status, the microenvironment (pH, ion gradient) and the presence of growth factors or cytokines. To provide all necessary enzymes and proteins in any particular situation, the cell possesses a sophisticated system of transcription factors for the regulation of gene expression. More than 2000 known transcription factors are encoded in the human genome (Brivanlou and Darnell, 2002). Recently, Brivanlou and Dranell (2002) have proposed an alternative grouping of the eukaryotic transcription factors. This classification divides the positive-acting eukaryotic transcription factors into two groups: (i) constitutively active nuclear factors, which are present in the nuclei of all cells at all times and have transcriptional activating potential in *in vitro* assay systems; and (ii) regulatory transcription factors. The second group is divided into two major classes: the 'developmental' or 'cell type'-specific transcription factors (IIA) and the signal-dependent transcription factors (IIB). The signal-dependent transcription factors are divided further into three functional groups with different signal mechanisms: factors of the nuclear receptor superfamily; factors responding to internal signals; and factors that are activated by cell surface receptor–ligand interaction and signal cascades.

FA-activated transcription factors are members of the nuclear receptor superfamily, comprising endocrine receptors such as oestrogen receptor, glucocorticoid receptor (GR) and vitamin D receptor (VDR), as well as orphan receptors with unknown ligands and 'adopted' orphan receptors with recently identified ligands (Mangelsdorf and Evans, 1995; Schoonjans *et al.*, 1997). FAs probably activate genes after binding to members of the group of 'adopted' orphan receptors such as the retinoid X receptor (RXR) and the peroxisome proliferator-activated receptors (PPARs).

Fatty Acids as Ligands for the PPAR System

The identification of ligand-induced transcription factors causing proliferation of peroxisomes in 1990 was the starting point for a new area of research in regulation of gene expression (Issemann and Green, 1990). Then, evidence was provided that the PPAR system may be of importance for the regulation of lipid homeostasis. The PPARs, obviously recognized first by their effect on peroxisome proliferation, were found to have this effect mainly in rodents, while in human liver the regulatory effects on other genes were demonstrated later (Palmer *et al.*, 1998).

The first studies on the effects of PPAR agonists were performed with synthetic ligands, the hypolipidaemic fibrates, which are very potent activators of PPARα (Schoonjans *et al.*, 1996a), and the anti-diabetic thiazolidiones as high-affinity ligands for PPARγ (Henry, 1997; Berger and Moller, 2002; O'Moore-Sullivan and Prins, 2002). In 1992, Auwerx was able to show that FAs are physiologically relevant ligands for PPARs, and thus they regulate a large set of genes (Auwerx, 1992). All those genes revealed a PPAR-response element (PPRE) in their promoter region. There are at least three relevant types of this receptor: PPARα, PPARγ and PPARβ (or δ) (for reviews, see Michalik and Wahli, 1999; Berger and Moller, 2002). The tissue distribution and major effects of these receptors are shown in Fig. 9.1.

The major group of target genes for the PPARs are genes involved in lipid and glucose

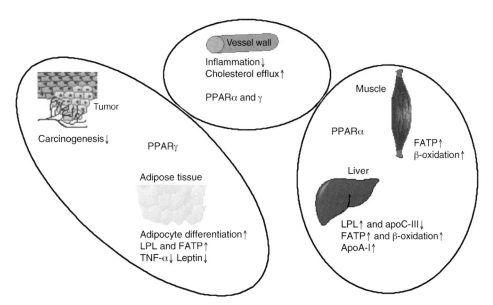

Fig. 9.1. PPAR effects in various tissues. PPARα exerts its main effects in the liver and muscle, whereas PPARγ influences gene expression mainly in adipose tissue and tumour cells. The endothelial cells and macrophages in the arterial wall are responsive to both PPARα and γ.

metabolism (Table 9.1). Thus, they regulate key functions of energy metabolism and establish an elaborate system to meet the various metabolic needs in an organism. By gaining more and more insight into the PPAR system, we are learning about the high level of complexity, which seems to be suited for this particular task (Berger and Moller, 2002). The complexity is established by the three distinct receptor isotypes α, β and γ, which bind many physiological ligands with distinct affinities, and by the formation of heterodimers with the RXR (Miyata *et al.*, 1994) to yield active transcription factors. Additionally, the heterodimers are dependent on cofactors for optimal activity; cofactors are recruited after activation by ligand binding (Xu *et al.*, 1999). Dependent on the nature of the receptor–ligand complex and on the availability of co-receptors and cofactors, the expression of the target genes will be regulated specifically (Fig. 9.2) (Berger and Moller, 2002).

In the last few years, FAs have received more and more attention as physiological ligands (Hihi *et al.*, 2002) for nuclear receptors, in particular for PPARs (Desvergne *et al.*, 1998). In Table 9.1, the physiological ligands and the major physiological functions of the three PPAR isoforms are listed.

Jump and Clarke (1999) summarized the current knowledge on regulation of gene expression by dietary fat and indicated the relative potency for PPAR activation of the various FAs compared with synthetic ligands. Among the physiological ligands, linoleic acid has the highest potency, followed by arachidonic acid and oleic acid. The saturated FAs have only low affinities for these receptors. Wolfrum *et al.* (2001) studied the specific effect of different FAs on the hepatic FABP and confirmed in his transactivation system the high activity of linoleic acid and γ-linolenic acid. Recent reviews further elucidate the distinct role of FAs and their derivatives in gene expression (Berger and Moller, 2002; Hihi *et al.*, 2002).

Many studies have been published on the particular role of ω-3 PUFAs in health and disease due to their special effects in lipid metabolism. These effects might also be regulated at the level of gene expression, where ω-3 PUFAs seem to be more potent than ω-6 PUFAs in functioning as PPARα ligands (Price *et al.*, 2000), resulting in increased lipid oxidation and thermogenesis (Baillie *et al.*, 1999). Thus ω-3 PUFAs direct FAs away from storage towards oxidation, which is essential for energy balance and insulin sensitivity (Mori *et al.*, 1999). CLAs have been demonstrated to be potent naturally occurring ligands and activators of PPARα (Moya-Camarena *et al.*, 1999). These data have to be seen in the light of the special implication of the CLAs as nutritional products of 'functional food' for human health, as summarized in a recent review (Whigham *et al.*, 2000). It has to be emphasized that most of the studies on the effect of CLA have been performed in rodents. When looking at the interspecies differences in the PPAR system between humans and rodents, the findings in rodents cannot be adapted directly to humans. There are data showing that different CLAs have different effects on PPARα (Moya-Camarena *et al.*, 1999), and studies with PPARα-null mice show that CLAs not only act via this transcription factor system but in addition can also use other regulatory pathways (Peters *et al.*, 2001).

Phytanic and pristanic acid are structurally closely related plant phytols. They recently were recognized as ligands of RXR, PPARα and PPARδ (Lampen *et al.*, 2001). Studies on their role in brown adipocyte differention revealed that phytanic acid is a differentiating agent, whereas pristanic acid is not (Schluter *et al.*, 2002).

The metabolites of arachidonic acid, known as eicosanoids, have been described to be ligands for the PPARs. This is true for prostaglandins such as PGA1 and 2, PGD1 and 2 and PGJ2, but cannot be confirmed for others (Yu *et al.*, 1995). The authors of this study describe an effect of the hydroxyeicostetraenoic acid (8S-HETE) on PPARα. The effect of PGJ2 on adipocyte differentiation via PPARγ was found in two groups independently in 1995 (Forman *et al.*, 1995; Kliewer *et al.*, 1995).

In the following sections, we focus on effects of FAs without further description of cofactors and co-activators.

Fatty Acid-regulated Gene Expression

Regulation of genes in lipid metabolism and atherosclerosis

The role of nuclear receptors in lipid metabolism has been reviewed recently by Chawla *et al.* (2001). The major target genes of PPARα play roles in conserving energy during fasting and feeding, whereas PPARγ is a key regulator of adipogenesis

Fatty Acids and Gene Expression 125

Table 9.1. Major physiological ligands and functions of lipid-activated transcription factors.

Nuclear receptor	Major lipid ligands	Mode of activation	Examples of responsive genes	Functions of responsive genes
PPARα	ω-3 PUFAs e.g. γ-linolenic acid ω-6 PUFAs e.g. arachidonic acid Eicosanoids e.g. 8S-HETE, LTB4	Direct interaction with lipids Heterodimerization with RXR	ApoA-I, apoA-II, apoC-III FABP CPTI Acyl-CoA oxidase Acyl-CoA dehydrogenase Cyp4A1/6 P450 family	Lipoprotein transport Intracellular FA transport FA entry into mitochondria Peroxisomal β-oxidation Mitochondrial β-oxidation Peroxisomal ω-oxidation
PPARγ	ω-3 PUFAs e.g. γ-linolenic acid ω-6 PUFAs e.g. arachidonic acid Eicosanoids e.g. prostaglandin J2	Direct interaction with lipids Heterodimerization with RXR	FABP, FATP, CD36 LPL Acyl-CoA synthetase UCP TNF-α Leptin	FA transport Lipoprotein hydrolysis Lipogenesis Thermogenesis Pro-inflammatory cytokine Regulator of food intake
PPARδ	ω-3 PUFAs e.g. γ-linolenic acid ω-6 PUFAs e.g. arachidonic acid Eiscosanoid e.g. prostanglandin J2	Direct interaction with lipids Heterodimerization with RXR	ABCA1 FABP Cyclooxygenase Complex pattern of genes during embryonic development	Reverse cholesterol transport FA transport Synthesis of prostaglandins Fetal development
HNF4α	Agonistic: SFA C14/C16 Antagonistic: SFA C18 ω-3/6 PUFAs	Direct interaction with activated lipids (CoA-FA) Homodimerization	ApoA-I, apoB, apoC-III MTP FABP Acyl-CoA dehydrogenase HNF1α/PXR Cyp3A4-6 P450 family	Lipoprotein transport Lipoprotein assembly FA transport Mitochondrial β-oxidation Transcription factors Peroxisomal hydroxylation
LXRs	Agonistic oxysterols Antagonistic ω-3/6 PUFAs	Direct interaction with oxysterols Competition by PUFAs Heterodimerization with RXR	Cyp7A CETP ABCA1 LXR affects SREBP1c influencing: FAs, stearyl-CoA desaturase ACL Malic enzyme Glucose-6P-dehydrogenase	Bile acid metabolism Cholesterol exchange Reverse cholesterol transport Transcription factor FA synthesis Lipogenic pathways

and plays an important role in cellular differentiation, insulin sensitization and atherosclerosis, as well as carcinogenesis. It needs to be emphasized here that most of the data obtained for these PPAR functions are derived from studies with synthetic PPAR agonists used in the pharmacological treatment of hyperlipidaemias and diabetes, i.e. fibrates and thiazolidinedions (Staels et al., 1997). Since PUFAs bind to PPARs at lower affinities, they are considered to exert the same effects on gene regulation, albeit at much lower levels, as revealed by the above synthetic ligands.

PPARα agonists stimulate the expression of the LPL gene in the liver (Schoonjans et al., 1996b), resulting in hydrolysis of triglycerides in chylomicrons and delivery of the nutritional FAs to the adipose tissue, as well as to muscle and other tissues. The chylomicron remnants will be taken up by hepatic remnant receptors to provide lipids for further distribution into other tissues via very low-density lipoproteins (VLDLs). ApoCIII is an important inhibitor of the remnant uptake. PPARα inhibits apoCIII expression (Staels et al., 1995) and thereby induces clearance of remnants. PPARα also affects apoAI gene expression (Vu-Dac et al., 1994) and increases the amount of liver-derived high-density lipoprotein (HDL) precursors, which mature under low-density lipoprotein (LDL)-mediated lipolysis of triglyceride-rich lipoproteins, leading to increased HDL_3. The effect of PPARα agonists on these three genes in hepatic lipoprotein metabolism thereby changes the lipoprotein profile in the direction of the desirable low triglyceride and high HDL concentration, to yield an antiatherosclerotic pattern (Fig. 9.1).

On a cellular level, PPARα activation up-regulates the transcription of FABP (Wolfrum et al., 2001) and the FA transporters ABC D2 and D3,

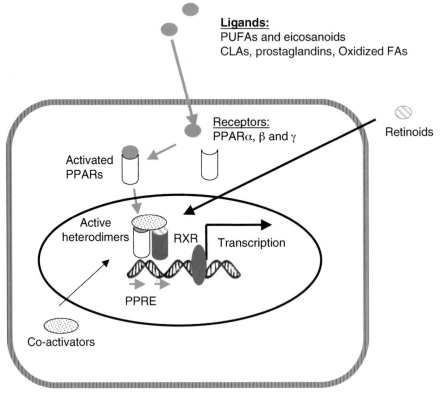

Fig. 9.2. Activation of PPAR by fatty acids. Fatty acids can be taken up into cells via fatty acid transport proteins. In the cytoplasm, they can activate the PPARs, while in parallel the retinoic acid receptor (RXR) is activated by retinoids or fatty acids. The two activated nuclear receptors form a heterodimer, which binds to the respective response elements on the promotor of the target genes. An additional activation can be achieved by the recruitment of co-activators, which bind to the heterodimer.

which promote the transport of the PUFAs into the peroxisomes (Fourcade et al., 2001). Enzymes involved in FA oxidation are stimulated by PPARα agonists and the cytochrome P450 enzyme CYP4A as the final catabolic step for the PPAR ligands to complete the metabolic cascade (Muerhoff et al., 1992) regulated via the PPAR system.

The effects of PPARγ on lipid metabolism (Walczak and Tontonoz, 2002) are mediated mainly by the function of genes in adipocytes, and will be described in the next section. However, PPARγ agonists have been found to be important also in atherosclerosis and the immune system.

The role of PPARs in regulation of genes involved in endothelial cell biology and inflammation is another important area in advancing our understanding on how dietary lipids might help to prevent atherosclerosis. The recent review by Plutzky (2001) provides insight into the PPARs expressed in the endothelial cells and their effect on vascular biology (Fig. 9.1). The inhibition of cytokine-induced gene expression plays an important role in the pathomechanisms of atherosclerosis, but a discussion of this field is beyond the limits of this review.

Most intriguing are the data on PPAR and inflammation (Clark, 2002), which might open up a whole new field not only in the dietary treatment of atherosclerosis but also for other degenerative diseases that are driven by inflammative processes, such as Alzheimer's disease (Landreth and Heneka, 2001) or arthritis (Fahmi et al., 2002).

Regulation of genes in adipose tissue

It has been shown by many authors that PPARγ is necessary and sufficient for adipogenesis (Brun et al., 1997; Spiegelman, 1997; Fajas et al., 1998; Rangwala and Lazar, 2000). In addition, it can influence the secretion of adipocyte-derived signalling molecules (Walczak and Tontonoz, 2002) (Figs 9.1 and 9.3).

The major genes regulated by PPARγ in white adipose tissue are summarized in a recent review by Berger and Moller (2002). The list includes genes that control cellular energy homeostasis and lipid metabolism such as LPL, FAT and the uncoupling proteins (UCPs). By stimulating the expression of acyl-CoA synthase in the adipocytes, PPARγ positively affects lipogenesis. PPARγ also affects genes encoding secretory 'hormones' such as leptin and tumour necrosis factor-α (TNFα), the latter being a potential mediator of insulin resistance (Walczak and Tontonoz, 2002). Another important group of genes influenced by PPARγ in the adipose tissue are those involved in glucose metabolism, such as the genes for insulin receptor substrate 2 (IRS-2) and pyruvate dehydrogenase kinase 4 (PDK4).

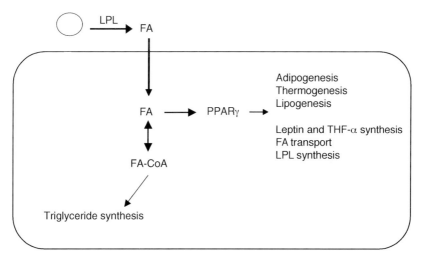

Fig. 9.3. Nuclear receptors and major metabolic pathways involved in FA-mediated adipose gene expression. The fatty acids released by LPL-mediated lipolysis are taken up into adipocytes and can either be activated and used for triglyceride synthesis or function as PPARγ activators. Several important genes involved in lipid metabolism in the adipocytes are regulated by PPARγ.

Rangwala and Lazar (2000) published an overview on the other transcription factors involved in adipogenesis, which, partly, are also lipid activated. In addition, Holst and Grimaldi (2002) discuss the role of PUFAs as agonists for PPARδ as a new transcription factor in adipocyte differentiation.

The spectrum of genes affected by PUFAs as PPARγ agonists, which cannot be discussed here in its full complexity, clearly indicates that stimulation of PPARγ by nutritional FAs as well as their derivatives can have important long-term effects on the development of hyperlipidaemia and diabetes, and thereby on prevention of atherosclerosis and coronary heart disease (Hsueh and Law, 2001).

Regulation of genes in the liver

Although the intracellular concentration of PUFAs and activated PUFAs (FA-CoA) is low (Knudsen et al., 1999; Elholm et al., 2000), it became evident that PUFAs and their metabolites exert some of their effects on hepatic gene expression by affecting the activity of nuclear transcription factors (Fig. 9.4). In addition to PPARα, these factors include the sterol element-binding protein type 1 (SREBP1), liver X receptors (LXRs) and hepatic nuclear factor 4α (HNF4α) (see Fig. 9.1) (for reviews, see Kersten et al., 2000; Osborne, 2000; Schultz et al., 2000; Hayhurst et al., 2001; Zannis et al., 2001; Berger and Moller, 2002). PPARα was the first transcription factor identified as a PUFA receptor that is highly expressed in parenchymal liver cells. The respective hepatic target genes are a relatively homogenous group participating in lipid catabolism, such as PUFA uptake through membranes (e.g. FAT; Motojima et al., 1998), intracellular PUFA transport (e.g. liver FABP; Issemann et al., 1992), PUFA oxidation in microsomes and mitochondria (e.g. acyl-CoA synthase (Schoonjans et al., 1995) and acyl-CoA oxidase (Dreyer et al., 1992)) as well as lipoprotein metabolism (e.g. LPL (Schoonjans et al., 1996b), apoAI (Vu-Dac et al., 1994) and apoCIII (Staels et al., 1995)) (Fig. 9.1).

In addition to their role in adaptation to fasting, oleic acid and PUFAs also suppress genes involved in de novo lipogenesis and carbohydrate metabolism. This is accounted for by LXRα- and LXRβ-mediated and FA-dependent suppression of hepatic gene expression (Repa et al., 2000). LXRs regulate transcription of SREBP1c (Peet et al.,

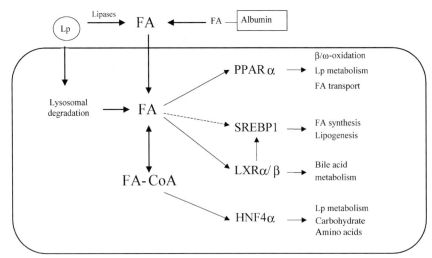

Fig. 9.4. Nuclear receptors and major metabolic pathways involved in FA-mediated hepatic gene expression. The regulation of lipid metabolism by fatty acids in the liver is not only regulated by PPARα, but HNF4α is also involved in the regulation of a variety of genes involved in the production and clearance of plasma lipoproteins. In addition, sterol binding to LXR controls cholesterol homeostasis and bile acid metabolism. LXR activity is inhibited by PUFAs, which thereby also reduce SREBP1c expression. Low concentrations of SREBP1c lead to reduced expression of key enzymes regulating hepatic lipogenic pathways.

1998; Repa and Mangelsdorf, 2000). PUFAs are able to antagonize the binding of oxysterols (the major physiological ligands of LXRs), resulting in a decrease of SREBP1c mRNA levels (Ou et al., 2001). SREBP1c is a membrane-bound transcription factor in the endoplasmic reticulum, which is released by two-step proteolysis, induced by low intracellular concentrations of cholesterol. The mature transcription factor enters the nucleus (Brown and Goldstein, 1997) and is a strong activator for the transcription of genes involved in anabolic FA and triglyceride pathways (e.g. FA synthase, stearyl-CoA desaturase, microsomal transfer protein) (Shimano et al., 1997; Sato et al., 1999). Thus, the competition of PUFAs with the LXR ligands results in a decreased level of active SREBP1c (Mater et al., 1999; Hannah et al., 2001; Ou et al., 2001), which can explain the suppressive effects of PUFAs on hepatic lipogenesis (Jump and Clarke, 1999).

In addition to PUFAs, the respective CoA-thioesters are able to bind the nuclear receptor HNF4α (Hertz et al., 1998). Activated HNF4α binds to DNA sequences as a homodimer (Jiang et al., 1995), triggered by ligand-induced conformational changes of the receptor (Petrescu et al., 2002). Responsive genes encode transcription factors (e.g. HNF1α), and proteins involved in FA, lipoprotein and lipid metabolism (e.g. apoAI, apoCIII and microsomal transfer protein) and others (Hayhurst et al., 2001; Zannis et al., 2001). Since expression of PPARα in human liver is rather low (Palmer et al., 1998), it is likely that HNF4α is important for some of the transcriptional effects mediated by PUFAs in humans (Jump and Clarke, 1999) (Fig. 9.4).

Regulation of genes in tumour cells

Lifestyle and nutrition affect a number of pathophysiological processes. As for the immune system, many reports underline nutritional influences on cancer development and growth. Recent studies by Helmlinger et al. (2000) support the paradigm that tumour metabolism is determined mainly by substrate availability but not by the metabolic demand of tumour cells. This and other findings have strong implications for the roles of dietary components in cancer treatment.

PPARs have been identified as the main mediators of the effects of FAs on cell proliferation. Interestingly, the different isoforms confer opposing effects. The published data indicate as a basic rule that PPARα and PPARγ have anti-proliferative effects, whereas PPARβ (also termed PPARδ) seems to promote certain neoplastic processes. As outlined below, there are exceptions to this rule.

FAs can influence the development of cancers in various tissues, having positive as well as negative effects on the patients' prognosis. In breast cancer, a growth-inhibiting effect of ω-3 FA has been observed (Fig. 9.1). In a study by Maillard et al. (2002), the relationship of the ω-3/ω-6 FA content in adipose breast tissue of patients with breast cancer was determined. The authors hypothesized that this content reflects the nutritional intake of these FAs and concluded that ω-3 FAs have a protective role against breast cancer growth. Ge et al. (2002) demonstrated anti-tumoural effects of ω-3 FAs by overexpressing a desaturase from the nematode *Caenorhabditis elegans* in MCF-7 cells; the endogenously synthesized ω-3 FAs promoted cancer cell death (Escrich et al., 2001). In summary, all these studies show that ω-3FAs have anti-proliferative effects on tumour growth. It should be noted that high-fat diets promote the growth of tumours derived from lipotrophic organs (e.g. breast and adipose tissue).

As agonists of PPARγ, FAs can influence the control of cell proliferation in development and growth of tumours (Rosen and Spiegelman, 2001). As a consequence of PPARγ activation, redifferentiation of cancer cells and induction of apoptosis were observed (Eibl et al., 2001; Takashima et al., 2001). PPARγ was also found to be up-regulated in human renal carcinoma cells, augmenting their susceptibility to PPAR-mediated effects of growth inhibition (Inoue et al., 2001). There were similar findings in liposarcomas (Tontonoz et al., 1997), and carcinomas of the breast (Mueller et al., 1998), colon and prostate (Mueller et al., 2000). A recent investigation supports a distinct pathway for PPARγ-mediated anti-proliferative effects. The PTEN tumour suppressor gene was found to be up-regulated by PPARγ and, in turn, this effect was abolished in PPARγ knockout mice (Patel et al., 2001).

Especially in the colon epithelia, PPARγ is found to be expressed at levels comparable with those in adipose tissue. Tumours of the colon also reveal elevated levels of this factor. These cells respond to PPAR agonist treatment by expression of differentiation markers, yielding a less malignant phenotype (Sarraf et al., 1998). Mutations in the PPARγ gene were found in colon carcinoma cells, all of which resulted in loss of function. Thus, PPARγ has tumour suppressor activity in the colon (Sarraf et al., 1999).

PPARs are expressed in prostate tissues. Segawa et al. (2002) used reverse transcription–polymerase chain reaction (RT–PCR) and immunohistological techniques to analyse normal, benign and malignant prostate tissues for the expression of PPARα, β and γ. Whereas PPARα and β were expressed in all tissues tested, compared with normal prostate tissue, PPARγ was induced in different neoplastic prostate tissues of benign as well as malignant origin. Shappell and colleagues showed that PPARγ activation by 15S-hydroxyeicosatetraenoicin (15S-HETE) in prostate cancer cells leads to the inhibition of tumour growth (Masamune et al., 2001). 15S-HETE is converted from dietary arachidonic acid by the 15-lipoxygenase, which was found to be down-regulated in the prostate cancer line. This finding matches the paradigm that tumours negatively regulate pathways of external proliferation control. It also shows that this down-regulation is reversible as long as PPARγ is present. In vascular smooth muscle cells, PPARα induced apoptosis by activating p38 protein kinase (Collett et al., 2000). In another study, Diep et al. (2001) found that PPARα is up-regulated in epithelia of undifferentiated prostatic adenocarcinomas. The function of this PPARα induction is unclear; however, PPARα-dependent regulatory effects such as up-regulation of the FAT point to a role in dietary energy support of tumours.

Taken together, dietary FAs can elicit a variety of effects on tumour cell proliferation. In many cases, PPARs trigger the effects with pro- as well as anti-proliferative results, depending on the tumour aetiology and metabolism.

Summary

After many epidemiological studies and population surveys suggested positive effects of PUFAs on human health, there are now numerous data published on the biochemical basis of these observations. Lipids, in particular PUFAs, have been shown to be important factors in regulating the expression of genes involved not only in lipid metabolism but also in the regulation of glucose levels, as well as cell differentiation, tumour growth and inflammation. Today we might see only the tip of the iceberg in terms of the importance of the composition of our diet for a healthy long life. We also have a long way to go to understand the possible danger of interactions between a particular genetic disposition and the effects of nutrients.

References

Abumrad, N., Harmon, C. and Ibrahimi, A. (1998) Membrane transport of long-chain fatty acids: evidence for a facilitated process. *Journal of Lipid Research* 39, 2309–2318.

Auwerx, J. (1992) Regulation of gene expression by fatty acids and fibric acid derivatives: an integrative role for peroxisome proliferator activated receptors. The Belgian Endocrine Society Lecture 1992. *Hormone Research* 38, 269–277.

Baillie, R.A., Takada, R., Nakamura, M. and Clarke, S.D. (1999) Coordinate induction of peroxisomal acyl-CoA oxidase and UCP-3 by dietary fish oil: a mechanism for decreased body fat deposition. *Prostaglandins, Leukotrienes and Essential Fatty Acids* 60, 351–356.

Berger, J. and Moller, D.E. (2002) The mechanisms of action of PPARs. *Annual Review of Medicine* 53, 409–435.

Berk, P.D. and Stump, D.D. (1999) Mechanisms of cellular uptake of long chain free fatty acids. *Molecular and Cellular Biochemistry* 192, 17–31.

Brivanlou, A.H. and Darnell, J.E. Jr (2002) Signal transduction and the control of gene expression. *Science* 295, 813–818.

Brown, M.S. and Goldstein, J.L. (1997) The SREBP pathway: regulation of cholesterol metabolism by proteolysis of a membrane-bound transcription factor. *Cell* 89, 331–340.

Brun, R.P., Kim, J.B., Hu, E. and Spiegelman, B.M. (1997) Peroxisome proliferator-activated receptor γ and the control of adipogenesis. *Current Opinion in Lipidology* 8, 212–218.

Chawla, A., Repa, J.J., Evans, R.M. and Mangelsdorf, D.J. (2001) Nuclear receptors and lipid physiology: opening the X-files. *Science* 294, 1866–1870.

Clark, R.B. (2002) The role of PPARs in inflammation and immunity. *Journal of Leukocyte Biology* 71, 388–400.

Collett, G.P., Betts, A.M., Johnson, M.I., Pulimood, A.B., Cook, S., Neal, D.E. and Robson, C.N. (2000) Peroxisome proliferator-activated receptor α is an androgen-responsive gene in human prostate and is highly expressed in prostatic adenocarcinoma. *Clinical Cancer Research* 6, 3241–3248.

Delerive, P., Furman, C., Teissier, E., Fruchart, J., Duriez, P. and Staels, B. (2000) Oxidized phospholipids activate PPARα in a phospholipase A2-dependent manner. *FEBS Letters* 471, 34–38.

Desvergne, B., Ijpenberg, A., Devchand, P.R. and Wahli, W. (1998) The peroxisome proliferator-activated receptors at the cross-road of diet and hormonal signalling. *Journal of Steroid Biochemistry and Molecular Biology* 65, 65–74.

Diep, Q.N., Touyz, R.M. and Schiffrin, E.L. (2000) Docosahexaenoic acid, a peroxisome proliferator-activated receptor-α ligand, induces apoptosis in vascular smooth muscle cells by stimulation of p38 mitogen-activated protein kinase. *Hypertension* 36, 851–855.

Dreyer, C., Krey, G., Keller, H., Givel, F., Helftenbein, G. and Wahli, W. (1992) Control of the peroxisomal β-oxidation pathway by a novel family of nuclear hormone receptors. *Cell* 68, 879–887.

Duplus, E., Glorian, M. and Forest, C. (2000) Fatty acid regulation of gene transcription. *Journal of Biological Chemistry* 275, 30749–30752.

Eckel, R.H. (1989) Lipoprotein lipase. A multifunctional enzyme relevant to common metabolic diseases. *New England Journal of Medicine* 320, 1060–1068.

Eibl, G., Wente, M.N., Reber, H.A. and Hines, O.J. (2001) Peroxisome proliferator-activated receptor gamma induces pancreatic cancer cell apoptosis. *Biochemical and Biophysical Research Communications* 287, 522–529.

Elholm, M., Garras, A., Neve, S., Tornehave, D., Lund, T.B., Skorve, J., Flatmark, T., Kristiansen, K. and Berge, R.K. (2000) Long-chain acyl-CoA esters and acyl-CoA binding protein are present in the nucleus of rat liver cells. *Journal of Lipid Research* 41, 538–545.

Escrich, E., Solanas, M., Soler, M., Ruiz de Villa, M.C., Sanchez, J.A. and Segura, R. (2001) Dietary polyunsaturated n-6 lipids effects on the growth and fatty acid composition of rat mammary tumors. *Journal of Nutritional Biochemistry* 12, 536–549.

Fahmi, H., Pelletier, J.P. and Martel-Pelletier, J. (2002) PPARγ ligands as modulators of inflammatory and catabolic responses in arthritis. An overview. *Journal of Rheumatology* 29, 3–14.

Fajas, L., Fruchart, J.C. and Auwerx, J. (1998) Transcriptional control of adipogenesis. *Current Opinion in Cell Biology* 10, 165–173.

Forman, B.M., Tontonoz, P., Chen, J., Brun, R.P., Spiegelman, B.M. and Evans, R.M. (1995) 15-Deoxy-delta 12, 14-prostaglandin J2 is a ligand for the adipocyte determination factor PPAR gamma. *Cell* 83, 803–812.

Fourcade, S., Savary, S., Albet, S., Gauthe, D., Gondcaille, C., Pineau, T., Bellenger, J., Bentejac, M., Holzinger, A., Berger, J. and Bugaut, M. (2001) Fibrate induction of the adrenoleukodystrophy-related gene (ABCD2): promoter analysis and role of the peroxisome proliferator-activated receptor PPARα. *European Journal of Biochemistry* 268, 3490–3500.

Ge, Y., Chen, Z., Kang, Z.B., Cluette-Brown, J., Laposata, M. and Kang, J.X. (2002) Effects of adenoviral gene transfer of *C. elegans* n-3 fatty acid desaturase on the lipid profile and growth of human breast cancer cells. *Anticancer Research* 22, 537–543.

Goldberg, I.J. (1996) Lipoprotein lipase and lipolysis: central roles in lipoprotein metabolism and atherogenesis. *Journal of Lipid Research* 37, 693–707.

Goodridge, A.G. (1991) Fatty acid desaturation and chain elongation in eucaryotes. In: Vance, D.E. and Vance, J. (eds) *Biochemistry of Lipids, Lipoproteins and Membranes*. Elsevier Science, Amsterdam, pp. 141–168.

Halliwell, B. and Gutteridge, J.M.C. (1989) Lipid peroxidation: a radical chain reaction. In: *Free Radicals in Biology and Medicine*. Oxford University Press, Oxford, pp. 188–276.

Hannah, V.C., Ou, J., Luong, A., Goldstein, J.L. and Brown, M.S. (2001) Unsaturated fatty acids down-regulate srebp isoforms 1a and 1c by two mechanisms in HEK-293 cells. *Journal of Biological Chemistry* 276, 4365–4372.

Hayhurst, G.P., Lee, Y.H., Lambert, G., Ward, J.M. and Gonzalez, F.J. (2001) Hepatocyte nuclear factor 4α (nuclear receptor 2A1) is essential for maintenance of hepatic gene expression and lipid homeostasis. *Molecular and Cellular Biology* 21, 1393–1403.

Helmlinger, G., Sckell, A., Dellian, M., Forbes, N.S. and Jain, R.K. (2002) Acid production in glycolysis-impaired tumors provides new insights into tumor metabolism. *Clinical Cancer Research* 8, 1284–1291.

Henry, R.R. (1997) Thiazolidinediones. *Endocrinology and Metabolism Clinics of North America* 26, 553–573.

Hertz, R., Magenheim, J., Berman, I. and Bar-Tana, J. (1998) Fatty acyl-CoA thioesters are ligands of hepatic nuclear factor-4α. *Nature* 392, 512–516.

Hihi, A.K., Michalik, L. and Wahli, W. (2002) PPARs: transcriptional effectors of fatty acids and their derivatives. *Cellular and Molecular Life Sciences* 59, 790–798.

Holm, C., Osterlund, T., Laurell, H. and Contreras, J.A. (2000) Molecular mechanisms regulating hormone-sensitive lipase and lipolysis. *Annual Review of Nutrition* 20, 365–393.

Holst, D. and Grimaldi, P.A. (2002) New factors in the regulation of adipose differentiation and metabolism. *Current Opinion in Lipidology* 13, 241–245.

Hsueh, W.A. and Law, R.E. (2001) PPARγ and atherosclerosis: effects on cell growth and movement. *Arteriosclerosis, Thrombosis and Vascular Biology* 21, 1891–1895.

Inoue, K., Kawahito, Y., Tsubouchi, Y., Kohno, M., Yoshimura, R., Yoshikawa, T. and Sano, H. (2001) Expression of peroxisome proliferator-activated receptor gamma in renal cell carcinoma and growth inhibition by its agonists. *Biochemical and Biophysical Research Communications* 287, 727–732.

Issemann, I. and Green, S. (1990) Activation of a member of the steroid hormone receptor superfamily by peroxisome proliferators. *Nature* 347, 645–650.

Issemann, I., Prince, R., Tugwood, J. and Green, S. (1992) A role for fatty acids and liver fatty acid binding protein in peroxisome proliferation? *Biochemical Society Transactions* 20, 824–827.

Jiang, G., Nepomuceno, L., Hopkins, K. and Sladek, F.M. (1995) Exclusive homodimerization of the orphan receptor hepatocyte nuclear factor 4 defines a new subclass of nuclear receptors. *Molecular and Cellular Biology* 15, 5131–5143.

Jump, D.B. (2002) The biochemistry of N3-polyunsaturated fatty acids. *Journal of Biological Chemistry* 277, 8755–8758.

Jump, D.B. and Clarke, S.D. (1999) Regulation of gene expression by dietary fat. *Annual Review of Nutrition* 19, 63–90.

Jump, D.B., Thelen, A.P. and Mater, M.K. (2001) Functional interaction between sterol regulatory element-binding protein-1c, nuclear factor Y, and 3,5,3′-triiodothyronine nuclear receptors. *Journal of Biological Chemistry* 276, 34419–34427.

Kaytor, E.N., Shih, H. and Towle, H.C. (1997) Carbohydrate regulation of hepatic gene expression. Evidence against a role for the upstream stimulatory factor. *Journal of Biological Chemistry* 272, 7525–7531.

Kersten, S., Desvergne, B. and Wahli, W. (2000) Roles of PPARs in health and disease. *Nature* 405, 421–424.

Kliewer, S.A., Lenhard, J.M., Willson, T.M., Patel, I., Morris, D.C. and Lehmann, J.M. (1995) A prostaglandin J2 metabolite binds peroxisome proliferator-activated receptor gamma and promotes adipocyte differentiation. *Cell* 83, 813–819.

Knudsen, J., Jensen, M.V., Hansen, J.K., Faergeman, N.J., Neergaard, T.B. and Gaigg, B. (1999) Role of acylCoA binding protein in acylCoA transport, metabolism and cell signaling. *Molecular and Cellular Biochemistry* 192, 95–103.

Lampen, A., Meyer, S. and Nau, H. (2001) Phytanic acid and docosahexaenoic acid increase the metabolism of all-*trans*-retinoic acid and CYP26 gene expression in intestinal cells. *Biochimica et Biophysica Acta* 1521, 97–106.

Landreth, G.E. and Heneka, M.T. (2001) Anti-inflammatory actions of peroxisome proliferator-activated receptor gamma agonists in Alzheimer's disease. *Neurobiology of Aging* 22, 937–944.

Maillard, V., Bougnoux, P., Ferrari, P., Jourdan, M.L., Pinault, M., Lavillonniere, F., Body, G., Le Floch, O. and Chajes, V. (2002) N-3 and N-6 fatty acids in breast adipose tissue and relative risk of breast cancer in a case–control study in Tours, France. *International Journal of Cancer* 98, 78–83.

Mangelsdorf, D.J. and Evans, R.M. (1995) The RXR heterodimers and orphan receptors. *Cell* 83, 841–850.

Masamune, A., Kikuta, K., Satoh, M., Sakai, Y., Satoh, A. and Shimosegawa, T. (2001) Ligands of peroxisome proliferator-activated receptor-γ block activation of pancreatic stellate cells. *Journal of Biological Chemistry* 277, 141–147.

Mater, M.K., Thelen, A.P., Pan, D.A. and Jump, D.B. (1999) Sterol response element-binding protein 1c (SREBP1c) is involved in the polyunsaturated fatty acid suppression of hepatic S14 gene transcription. *Journal of Biological Chemistry* 274, 32725–32732.

McArthur, M.J., Atshaves, B.P., Frolov, A., Foxworth, W.D., Kier, A.B. and Schroeder, F. (1999) Cellular uptake and intracellular trafficking of long chain fatty acids. *Journal of Lipid Research* 40, 1371–1383.

Michalik, L. and Wahli, W. (1999) Peroxisome proliferator-activated receptors: three isotypes for a multitude of functions. *Current Opinion in Biotechnology* 10, 564–570.

Miyata, K.S., McCaw, S.E., Marcus, S.L., Rachubinski, R.A. and Capone, J.P. (1994) The peroxisome proliferator-activated receptor interacts with the retinoid X receptor *in vivo*. *Gene* 148, 327–330.

Mori, T.A., Bao, D.Q., Burke, V., Puddey, I.B., Watts, G.F. and Beilin, L.J. (1999) Dietary fish as a major component of a weight-loss diet: effect on serum lipids, glucose, and insulin metabolism in overweight hypertensive subjects. *American Journal of Clinical Nutrition* 70, 817–825.

Motojima, K., Passilly, P., Peters, J.M., Gonzalez, F.J. and Latruffe, N. (1998) Expression of putative fatty acid transporter genes are regulated by peroxisome proliferator-activated receptor alpha and gamma activators in a tissue and inducer-specific manner. *Journal of Biological Chemistry* 273, 16710–16714.

Moya-Camarena, S.Y., Vanden Heuvel, J.P., Blanchard, S.G., Leesnitzer, L.A. and Belury, M.A. (1999) Conjugated linoleic acid is a potent naturally occurring ligand and activator of PPARα. *Journal of Lipid Research* 40, 1426–1433.

Mueller, E., Sarraf, P., Tontonoz, P., Evans, R.M., Martin, K.J., Zhang, M., Fletcher, C., Singer, S. and Spiegelman, B.M. (1998) Terminal differentiation of human breast cancer through PPAR gamma. *Molecular Cell* 1, 465–470.

Mueller, E., Smith, M., Sarraf, P., Kroll, T., Aiyer, A., Kaufman, D.S., Oh, W., Demetri, G., Figg, W.D., Zhou, X.P., Eng, C., Spiegelman, B.M. and Kantoff, P.W. (2000) Effects of ligand activation of peroxisome proliferator-activated receptor gamma in human prostate cancer. *Proceedings of the National Academy of Sciences USA* 97, 10990–10995.

Muerhoff, A.S., Griffin, K.J. and Johnson, E.F. (1992) The peroxisome proliferator-activated receptor mediates the induction of CYP4A6, a cytochrome P450 fatty acid omega-hydroxylase, by clofibric acid. *Journal of Biological Chemistry* 267, 19051–19053.

Nagy, L., Tontonoz, P., Alvarez, J.G., Chen, H. and Evans, R.M. (1998) Oxidized LDL regulates macrophage gene expression through ligand activation of PPARγ. *Cell* 93, 229–240.

Norris, A.W. and Spector, A.A. (2002) Very long chain *n*-3 and *n*-6 polyunsaturated fatty acids bind strongly to liver fatty acid-binding protein. *Journal of Lipid Research* 43, 646–653.

O'Moore-Sullivan, T.M. and Prins, J.B. (2002) Thiazolidinediones and type 2 diabetes: new drugs for an old disease. *Medical Journal of Australia* 176, 381–386.

Osborne, T.F. (2000) Sterol regulatory element-binding proteins (SREBPs): key regulators of nutritional homeostasis and insulin action. *Journal of Biological Chemistry* 275, 32379–32382.

Ou, J., Tu, H., Shan, B., Luk, A., DeBose-Boyd, R.A., Bashmakov, Y., Goldstein, J.L. and Brown, M.S. (2001) Unsaturated fatty acids inhibit transcription of the sterol regulatory element-binding protein-1c (SREBP-1c) gene by antagonizing ligand-dependent activation of the LXR. *Proceedings of the National Academy of Sciences USA* 98, 6027–6032.

Palmer, C.N., Hsu, M.H., Griffin, K.J., Raucy, J.L. and Johnson, E.F. (1998) Peroxisome proliferator activated receptor-α expression in human liver. *Molecular Pharmacology* 53, 14–22.

Patel, L., Pass, I., Coxon, P., Downes, C.P., Smith, S.A. and Macphee, C.H. (2001) Tumor suppressor and anti-inflammatory actions of PPARγ agonists are mediated via upregulation of PTEN. *Current Biology* 11, 764–768.

Peet, D.J., Turley, S.D., Ma, W., Janowski, B.A., Lobaccaro, J.M., Hammer, R.E. and Mangelsdorf, D.J. (1998) Cholesterol and bile acid metabolism are impaired in mice lacking the nuclear oxysterol receptor LXR alpha. *Cell* 93, 693–704.

Peters, J.M., Park, Y., Gonzalez, F.J. and Pariza, M.W. (2001) Influence of conjugated linoleic acid on body composition and target gene expression in peroxisome proliferator-activated receptor alpha-null mice. *Biochimica et Biophysica Acta* 1533, 233–242.

Petrescu, A.D., Hertz, R., Bar-Tana, J., Schroeder, F. and Kier, A.B. (2002) Ligand specificity and conformational dependence of the hepatic nuclear factor-4α (HNF-4α). *Journal of Biological Chemistry* 277, 23988–23999.

Plutzky, J. (2001) Peroxisome proliferator-activated receptors in endothelial cell biology. *Current Opinion in Lipidology* 12, 511–518.

Price, P.T., Nelson, C.M. and Clarke, S.D. (2000) Omega-3 polyunsaturated fatty acid regulation of gene expression. *Current Opinion in Lipidology* 11, 3–7.

Rangwala, S.M. and Lazar, M.A. (2000) Transcriptional control of adipogenesis. *Annual Review of Nutrition* 20, 535–559.

Repa, J.J. and Mangelsdorf, D.J. (2000) The role of orphan nuclear receptors in the regulation of cholesterol homeostasis. *Annual Review in Cell and Developmental Biology* 16, 459–481.

Repa, J.J., Liang, G., Ou, J., Bashmakov, Y., Lobaccaro, J.M., Shimomura, I., Shan, B., Brown, M.S., Goldstein, J.L. and Mangelsdorf, D.J. (2000) Regulation of mouse sterol regulatory element-binding protein-1c gene (SREBP-1c) by oxysterol receptors, LXRα and LXRβ. *Genes and Development* 14, 2819–2830.

Rosen, E.D. and Spiegelman, B.M. (2001) PPARγ: a nuclear regulator of metabolism, differentiation, and cell growth. *Journal of Biological Chemistry* 276, 37731–37734.

Sarraf, P., Mueller, E., Jones, D., King, F.J., DeAngelo, D.J., Partridge, J.B., Holden, S.A., Chen, L.B., Singer, S., Fletcher, C. and Spiegelman, B.M. (1998) Differentiation and reversal of malignant changes in colon cancer through PPARγ. *Nature Medicine* 4, 1046–1052.

Sarraf, P., Mueller, E., Smith, W.M., Wright, H.M., Kum, J.B., Aaltonen, L.A., de la Chapelle, A., Spiegelman, B.M. and Eng, C. (1999) Loss-of-function mutations in PPAR gamma associated with human colon cancer. *Molecular Cell* 3, 799–804.

Sato, R., Miyamoto, W., Inoue, J., Terada, T., Imanaka, T. and Maeda, M. (1999) Sterol regulatory element-binding protein negatively regulates microsomal triglyceride transfer protein gene transcription. *Journal of Biological Chemistry* 274, 24714–24720.

Schluter, A., Giralt, M., Iglesias, R. and Villarroya, F. (2002) Phytanic acid, but not pristanic acid, mediates the positive effects of phytol derivatives on brown adipocyte differentiation. *FEBS Letters* 517, 83–86.

Schoonjans, K., Watanabe, M., Suzuki, H., Mahfoudi, A., Krey, G., Wahli, W., Grimaldi, P., Staels, B., Yamamoto, T. and Auwerx, J. (1995) Induction of the acyl-coenzyme A synthetase gene by fibrates and fatty acids is mediated by a peroxisome proliferator response element in the C promoter. *Journal of Biological Chemistry* 270, 19269–19276.

Schoonjans, K., Staels, B. and Auwerx, J. (1996a) Role of the peroxisome proliferator-activated receptor (PPAR) in mediating the effects of fibrates and fatty acids on gene expression. *Journal of Lipid Research* 37, 907–925.

Schoonjans, K., Peinado-Onsurbe, J., Lefebvre, A.M., Heyman, R.A., Briggs, M., Deeb, S., Staels, B. and

Auwerx, J. (1996b) PPARα and PPARγ activators direct a distinct tissue-specific transcriptional response via a PPRE in the lipoprotein lipase gene. *EMBO Journal* 15, 5336–5348.

Schoonjans, K., Martin, G., Staels, B. and Auwerx, J. (1997) Peroxisome proliferator-activated receptors, orphans with ligands and functions. *Current Opinion in Lipidology* 8, 159–166.

Schultz, J.R., Tu, H., Luk, A., Repa, J.J., Medina, J.C., Li, L., Schwendner, S., Wang, S., Thoolen, M., Mangelsdorf, D.J., Lustig, K.D. and Shan, B. (2000) Role of LXRs in control of lipogenesis. *Genes and Development* 14, 2831–2838.

Segawa, Y., Yoshimura, R., Hase, T., Nakatani, T., Wada, S., Kawahito, Y., Kishimoto, T. and Sano, H. (2002) Expression of peroxisome proliferator-activated receptor (PPAR) in human prostate cancer. *Prostate* 51, 108–116.

Serhan, C.N. and Oliw, E. (2001) Unorthodox routes to prostanoid formation: new twists in cyclooxygenase-initiated pathways. *Journal of Clinical Investigation* 107, 1481–1489.

Shimano, H., Horton, J.D., Shimomura, I., Hammer, R.E., Brown, M.S. and Goldstein, J.L. (1997) Isoform 1c of sterol regulatory element binding protein is less active than isoform 1a in livers of transgenic mice and in cultured cells. *Journal of Clinical Investigation* 99, 846–854.

Spiegelman, B.M. (1997) Peroxisome proliferator-activated receptor gamma: a key regulator of adipogenesis and systemic insulin sensitivity. *European Journal of Medical Research* 2, 457–464.

Sprecher, H. (2000) Metabolism of highly unsaturated *n*-3 and *n*-6 fatty acids. *Biochimica et Biophysica Acta* 1486, 219–231.

Staels, B., Vu-Dac, N., Kosykh, V.A., Saladin, R., Fruchart, J.C., Dallongeville, J. and Auwerx, J. (1995) Fibrates downregulate apolipoprotein C-III expression independent of induction of peroxisomal acyl coenzyme A oxidase. A potential mechanism for the hypolipidemic action of fibrates. *Journal of Clinical Investigations* 95, 705–712.

Staels, B., Schoonjans, K., Fruchart, J.C. and Auwerx, J. (1997) The effects of fibrates and thiazolidinediones on plasma triglyceride metabolism are mediated by distinct peroxisome proliferator activated receptors (PPARs). *Biochimie* 79, 95–99.

Takashima, T., Fujiwara, Y., Higuchi, K., Arakawa, T., Yano, Y., Hasuma, T. and Otani, S. (2001) PPAR-γ ligands inhibit growth of human esophageal adenocarcinoma cells through induction of apoptosis, cell cycle arrest and reduction of ornithine decarboxylase activity. *International Journal of Oncology* 19, 465–471.

Tontonoz, P., Singer, S., Forman, B.M., Sarraf, P., Fletcher, J.A., Fletcher, C.D., Brun, R.P., Mueller, E., Altiok, S., Oppenheim, H., Evans, R.M. and Spiegelman, B.M. (1997) Terminal differentiation of human liposarcoma cells induced by ligands for peroxisome proliferator-activated receptor gamma and the retinoid X receptor. *Proceedings of the National Academy of Sciences USA* 94, 237–241.

Towle, H.C., Kaytor, E.N. and Shih, H.M. (1997) Regulation of the expression of lipogenic enzyme genes by carbohydrate. *Annual Review of Nutrition* 17, 405–433.

Vaulont, S., Vasseur-Cognet, M. and Kahn, A. (2000) Glucose regulation of gene transcription. *Journal of Biological Chemistry* 275, 31555–31558.

Vu-Dac, N., Schoonjans, K., Laine, B., Fruchart, J.C., Auwerx, J. and Staels, B. (1994) Negative regulation of the human apolipoprotein A-I promoter by fibrates can be attenuated by the interaction of the peroxisome proliferator-activated receptor with its response element. *Journal of Biological Chemistry* 269, 31012–31018.

Walczak, R. and Tontonoz, P. (2002) PPARadigms and PPARadoxes: expanding roles for PPARγ in the control of lipid metabolism. *Journal of Lipid Research* 43, 177–186.

Whigham, L.D., Cook, M.E. and Atkinson, R.L. (2000) Conjugated linoleic acid: implications for human health. *Pharmacology Research* 42, 503–510.

Wolfrum, C., Borrmann, C.M., Borchers, T. and Spener, F. (2001) Fatty acids and hypolipidemic drugs regulate peroxisome proliferator-activated receptors α- and γ-mediated gene expression via liver fatty acid binding protein: a signaling path to the nucleus. *Proceedings of the National Academy of Sciences USA* 98, 2323–2328.

Xu, L., Glass, C.K. and Rosenfeld, M.G. (1999) Coactivator and corepressor complexes in nuclear receptor function. *Current Opinion in Genetics and Development* 9, 140–147.

Yu, K., Bayona, W., Kallen, C.B., Harding, H.P., Ravera, C.P., McMahon, G., Brown, M. and Lazar, M.A. (1995) Differential activation of peroxisome proliferator-activated receptors by eicosanoids. *Journal of Biological Chemistry* 270, 23975–23983.

Zannis, V.I., Kan, H.Y., Kritis, A., Zanni, E. and Kardassis, D. (2001) Transcriptional regulation of the human apolipoprotein genes. *Frontiers in Biosciences* 6, D456–D504.

10 Role of RARs and RXRs in Mediating the Molecular Mechanism of Action of Vitamin A

Dianne Robert Soprano[1,2] and Kenneth J. Soprano[2,3]
[1]*Department of Biochemistry;* [2]*Fels Institute for Cancer Research and Molecular Biology; and* [3]*Department of Microbiology and Immunology, Temple University School of Medicine, Philadelphia, Pennsylvania, USA*

Introduction

Vitamin A was first demonstrated to be an essential nutrient in 1925 by Wolbach and Howe. Since that time, vitamin A (retinol) (Fig. 10.1) and its biologically active derivatives (collectively referred to as retinoids) have been shown to be essential for many diverse biological functions including growth, vision, reproduction, embryonic development, differentiation of epithelial tissues and immune responses (for a review, see Sporn *et al.*, 1994). With the exception of night vision, all of these functions are believed to be mediated by retinoic acid (RA) (Fig. 10.1).

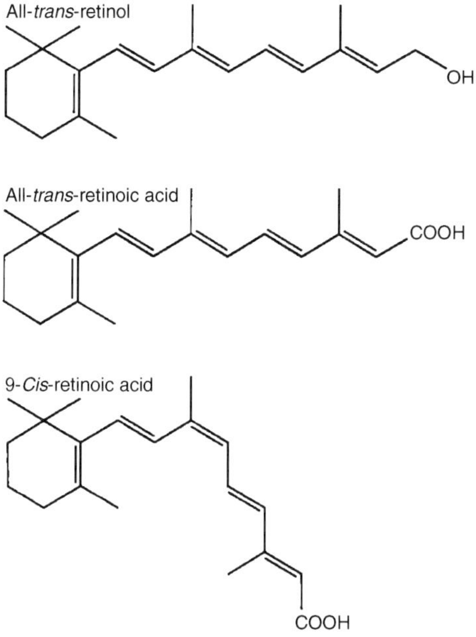

As is shown in Fig. 10.2, the bulk of retinol is transported in plasma bound to retinol-binding protein (RBP) to vitamin A-requiring cells in target tissues. Upon uptake by these cells, retinol can be stored as retinyl esters or can be oxidized to RA. Within the cytoplasm of these cells, retinol and RA can bind to their respective cellular binding proteins termed cellular retinol-binding protein (CRBP) and cellular retinoic acid-binding protein (CRABP). RA moves to the nucleus where it exerts most, if not all, of its biological effects at the level of control of gene expression. In order to regulate gene expression, RA must bind to nuclear proteins (ligand-dependent transcription factors) called retinoic acid receptors (RARs) and retinoid X receptors (RXRs). Since RARs and RXRs are critical to the understanding of the molecular mechanism of action of RA, this chapter will focus on RARs and RXRs and their role in mediating the pleiotropic effects of RA. Recent excellent review articles are available concerning RBP and retinol transport (Gottesman *et al.*, 2001), vitamin A metabolism (Napoli, 1999a,b; Duester, 2000) and cellular retinoid-binding proteins (Noy, 2000).

Fig. 10.1. Structure of retinol and retinoic acid.

Identification of RARs and RXRs

The identification of the first RAR was reported independently by two groups studying the superfamily of steroid hormone receptors in 1987 (Giguere *et al.*, 1987; Petkovich *et al.*, 1987). Shortly after the discovery of this RAR, two additional RARs were identified and named RARβ (Benbrook *et al.*, 1988; Brand *et al.*, 1988) and RARγ (Krust *et al.*, 1989; Zelent *et al.*, 1989), while the first RAR identified was termed RARα. Each of these three RAR subtypes are encoded by distinct genes located on different chromosomes in the genome.

A few years after the discovery of the first RAR, Mangelsdorf *et al.* (1990) characterized another orphan receptor, which led to the identification of a novel RA-responsive receptor, referred to as the retinoid X receptor (RXR). A short time later, two other highly homologous RXR subtypes encoded by distinct genetic loci were discovered (Mangelsdorf *et al.*, 1992); thus, the RXR family also includes three members, RXRα, RXRβ and RXRγ. Surprisingly, comparison of the amino acid sequences of the RAR family and RXR family of receptors demonstrates only weak homology over the entire length of the proteins, with the highest degree of similarity found in the DNA-binding domain (~60%), while the RA-binding domain displays only approximately 25% amino acid identity. Interestingly, there is an RXR homologue in *Drosophila* called ultraspiracle, which dimerizes with the ecdysone receptor, but there are no RAR homologues identified in *Drosophila* (Mangelsdorf *et al.*, 1992). From these findings, it has been suggested that RARs and RXRs arose separately during evolution and that receptor dimerization (see below) has been well conserved evolutionarily.

Anatomy of RARs and RXRs

Domain structure

The superfamily of nuclear receptors, including RARs and RXRs, share a common structure consisting of five or six functionally distinct domains,

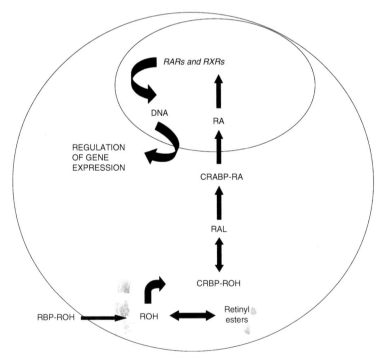

Fig. 10.2. Overview of the vitamin A signal transduction pathway. Abbreviations: CRABP, cellular retinoic acid-binding protein; CRBP, cellular retinol-binding protein; RA, retinoic acid; RAL, retinal; RAR, retinoic acid receptor; RBP, retinol-binding protein; ROH, retinol; RXR, retinoid X receptor.

domains A–F (Chambon, 1996) (Fig. 10.3). The N-terminal A/B domain is of variable length and is the least homologous between the receptors. The primary function of the A/B domain is ligand-independent transactivation, termed activation function-1 (AF-1) (Nagpal et al., 1993). This AF-1 has an autonomous function that can synergize efficiently with the AF-2 function of the E domain (see below) (Nagpal et al., 1993; Folkers et al., 1995). Domain C is the most conserved of the nuclear receptors (95% between three RAR subtypes and 92–95% between three RXR subtypes) and is responsible for DNA binding. Within the 66 amino acid central region of domain C are a pair of zinc-binding motifs termed zinc fingers, which are responsible for making contacts within the major groove of DNA and for discrimination of the DNA response element half-site spacing. Domain D is known as the hinge region. It has been implicated in defining a receptor's response element half-site spacing requirements (Stunnenberg, 1993) as well as the binding site for co-repressors (Chen and Evans, 1995; Kurokawa et al., 1995). Domain E or the ligand-binding domain (LBD) is a functionally complex region composed of 220 amino acid residues containing the ligand-binding region, receptor dimerization motifs (Nagpal et al., 1993) and ligand-dependent transactivation activity termed activation function-2 (AF-2) (Nagpal et al., 1993). The amino acid composition of domain E is also highly conserved between the three RAR subtypes (82–85%) and the three RXR subtypes (86–90%). Domain F is located at the C-terminal end of RARs; however, its function has not yet been determined. In addition, RXR does not contain an F domain.

Ligand specificity of RARs and RXRs

RXRs specifically bind 9-cis-RA ($K_d = 1–10$ nM), while RARs bind both all-trans-RA and 9-cis-RA with equal affinity ($K_d = 1–5$ nM) (Heyman et al., 1992; Levin et al., 1992). Interestingly, 9-cis-RA has high affinity for both RARs and RXRs; however 9-cis-RA has been difficult to detect in large quantities in vivo. This has raised the possibility that an as yet unknown RXR-specific ligand may exist (deUrquiza et al., 2000).

With the discovery of six different retinoid receptor subtypes, each of which binds RA with relatively equal affinity, extensive effort has been

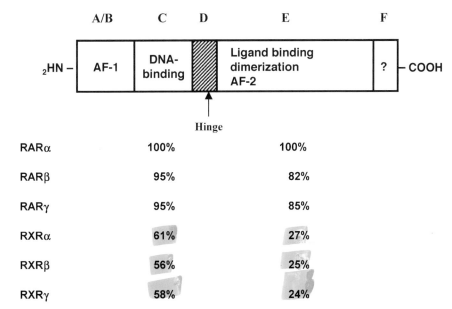

Fig. 10.3. Schematic representation of functional domains of the retinoic acid receptors and retinoid X receptors. The percentage amino acid identity for the DNA-binding and ligand-binding domains is indicated in relation to RARα.

put into the design and synthesis of selective retinoids that can target an individual RAR or RXR subtype or differentiate between the RAR and RXR subtypes. To date, a number of RAR subtype-selective retinoid agonists and antagonists along with several RXR-selective retinoid agonists and antagonists have been developed (see Table 10.1) (for reviews, see Chandraratna, 1998; Johnson and Chandraratna, 1999; Nagpal and Chandraratna, 2000). These selective retinoids have been very useful in the study of the structure and function of RARs and RXRs. More importantly, these retinoid prototypes along with additional more receptor-selective retinoid agonists, inverse agonists and antagonists to be developed have the potential of being of great clinical benefit for the treatment of a variety of disorders in the areas of dermatology, oncology, type 2 diabetes, atherosclerosis and obesity. These retinoids are likely to have a better therapeutic index and fewer toxicity problems.

Isoforms of RARs and RXRs

In addition to the three subtypes of RAR and RXR, several isoforms of each subtype have been described. The different isoforms of each RAR and RXR subtype vary only in the amino acid sequence of the N-terminal A domain and/or the nucleotide sequence of the 5'-untranslated region of its mRNA, while the amino acid sequence of domains B–F is identical. These isoforms arise due to alternative promoter usage, differential splicing and the use of an internal CUG codon for the initiation of translation. Four major isoforms of RARβ (termed β1, β2, β3 and β4) and two major isoforms of both RARα (termed α1 and α2) and RARγ (termed γ1 and 2) have been well characterized (Giguere et al., 1990; Kastner et al., 1990; Leroy et al., 1991; Zelent et al., 1991; Nagpal et al., 1992). In addition, five additional minor isoforms of both RARα (α3–α7) and RARγ (γ3–γ7) have been described. Finally, two isoforms for each of

Table 10.1. Nuclear retinoid receptors and representative receptor subtype-selective ligands.

Receptor	Isoforms	Ligands		
RARα	α1, α2	All-trans-RA	9-Cis-RA	AGN193312
		Am80	Am580/CD336	TD550
		AGN190521	CD2366	
		AGN193109 (antagonist)		
RARβ	β1, β2, β3, β4	All-trans-RA	9-Cis-RA	AGN193174
		CD2314	Ch55	AGN193078
		AGN193273	CD270	CD271
		CD2665	SR3985	TTNN
		AGN190521	CD2366	
		AGN193109 (antagonist)		
RARγ	γ1, γ2	All-trans-RA	9-Cis-RA	4-HPR
		AHPN/CD437	CD666	CD2325
		SR11254/MM11254	SR11363	SR11364
		MM11389	AGN193078	AGN193273
		CD270	CD271	CD2665
		SR3985	TTNN	AGN190521
		CD2366	MM11253 (antagonist)	
		AGN193109 (antagonist)		
RXRα	α1, α2	9-Cis-RA	LG1069	SR11203
		SR11217	SR11234	SR11236
		SR11246		
RXRβ	β1, β2	9-Cis-RA	LG1069	SR11203
		SR11217	SR11234	SR11236
		SR11246		
RXRγ	γ1, γ2	9-Cis-RA	LG1069	SR11203
		SR11217	SR11234	SR11236
		SR11246		

the three RXR subtypes ($\alpha 1$, $\alpha 2$, $\beta 1$, $\beta 2$, $\gamma 1$ and $\gamma 2$) have been reported (Liu and Linney, 1993; Brocard et al., 1994; Nagata et al., 1994).

Structure of the Ligand-binding Domain of RARs and RXRs

Within the last several years, high-resolution X-ray crystal structures of the LBD of apo-RXRα (Bourguet et al., 1995), holo-RARγ LBD (Renaud et al., 1995), a heterodimer between the LBDs of RARα bound to a selective antagonist and the constitutively active RXRαF318A mutant (Bourguet et al., 2000), and holo-RXRα LBD bound with 9-cis-retinoic acid (Egea et al., 2000) have been reported. Elucidation of the structure of the LBD of RARs and RXRs will lead to the design of novel subtype-selective retinoid agonists and antagonists with potential pharmacological application.

Examination of these crystal structures along with those from several other members of the steroid/thyroid hormone nuclear receptor families have demonstrated that each of these receptors shares a novel protein fold termed an antiparallel α-helical sandwich composed of 12 α-helices (numbered H1–H12) and a β-turn residing between helices H5 and H6 (Wurtz et al., 1996). RA within the ligand-binding pocket comes into contact with a large number of amino acid residues located on α-helices H1, H3, H5, H11 and H12, loops 6–7 and 11–12, and β-sheet s1. Comparison of the X-ray structures of apo-RXRα LBD and holo-RARγ LBD points to the fact that major structural changes occur upon binding of the ligand, leading to what has been termed the 'open' apo-form and 'closed' holo-form. This has been extended further recently with the ability to compare directly the X-ray structures of apo-RXRα LBD and holo-RXRα LBD bound to 9-cis-RA. Binding of 9-cis-RA to the LBD initially causes a rearrangement of helix H3, which then induces the repositioning of helix H12 by expelling helix H11. The most striking conformational change is seen in helix H12; in the apo-form, H12 protrudes from the protein core and is exposed to the solvent, while in the holo-form H12 rotates and folds back toward the ligand-binding pocket. This induces the compaction of the LBD and the formation of a new surface, which can bind co-activators involved in mediating transcriptional activity (see below).

Furthermore, comparison of agonist- versus antagonist-bound LBD structures has demonstrated the existence of two distinct conformations for helix H12, suggesting that helix H12 can adopt a holo-agonist-bound and a holo-antagonist-bound conformation. This conformational difference in helix H12 is at least partially responsible for mediating the different transcriptional activities of agonist-bound and antagonist-bound receptors. Thus, large conformational changes in helix H12 and the associated structural changes in helices H3, H6 and H11 that occur upon ligand binding to RARs and RXRs are critical in determining the transcriptional activity of these receptors by generating surfaces that allow subsequent binding of co-activators and removal of co-repressors.

In addition, numerous site-directed mutagenesis studies have examined the role of several conserved amino acid residues within RARα, RARβ and RARγ that appear from the crystal structure of holo-RARγ to be important for interaction between the ligand-binding pocket of RARγ and the carboxyl group of RA (Tairis et al., 1994, 1995; Scafonas et al., 1997; Wolfgang et al., 1997; Zhang et al., 1998, 2000). In all three RAR isotypes, mutations of the homologous arginine residue (αArg276, βArg269 and γArg278) to either an alanine or a glutamine demonstrates that this amino acid residue plays a critical role in the binding of RA by interacting with the carboxylate group of RA. This is consistent with the LBD holo-RARγ crystal structure, in which γArg278 has been demonstrated to form a salt bridge with the carboxylate O-22 of RA (Renaud et al., 1995). In RARβ, this arginine residue (Arg269) appears to function in conjunction with Lys220 and Ser278 in the binding of RA, while in both RARα and RARγ the homologous arginine residues (Arg276 and Arg278, respectively) function relatively independently in coordinating with the carboxyl group of the bound RA molecule. This suggests that the orientation of this amino acid residue in the ligand-binding site and the electronic environment associated with this arginine in each RAR subtype are different.

Both the analysis of the X-ray crystallographic structures of RAR LBDs and site-directed mutagenesis studies have identified three non-conserved amino acid residues located within the ligand-binding pocket of RARs that are important for RAR subtype ligand specificity (Ostrowski et al., 1995, 1998; Gehin et al., 1999). Mutation of Ala225 in RARβ to a serine can convert RARβ to an

RARα ligand-selective receptor (Ostrowski et al., 1998; Gehin et al., 1999). In addition, the mutation of Ile263 in RARβ to a methionine (Ostrowski et al., 1998) or the mutation of both Ile263 and Val388 in RARβ to a methionine and an alanine, respectively (Gehin et al., 1999), can convert RARβ to an RARγ ligand-selective receptor. Therefore, these three amino acid residues play a critical role in distinguishing RAR subtype-selective ligands but not RA. However, simultaneous mutation of both the carboxyl group coordinating amino acid of RARβ (Arg269) and either the RARγ ligand specificity amino acids (Ile263 and Val388) or the RARα ligand specificity amino acid (Ala225) of RARβ has only a small effect on the RA binding and RA-dependent transactivation activity of this receptor (Zhang et al., 2003). This suggests that the region of the ligand-binding pocket responsible for interaction with the carboxyl group of retinoids and the regions involved in subtype ligand specificity function independently.

Mechanism of Transcriptional Regulation by RARs and RXRs

Retinoic acid response elements

RAR- and RXR-dependent transcriptional transactivation of target genes requires the interaction of the receptors with a specific DNA sequence located in the promoter region of target genes. These regions are called retinoic acid response elements (RAREs) and retinoid X receptor response elements (RXREs). Analysis of the promoter of a large number of RA-responsive genes has resulted in the definition of the features of RAREs and RXREs. Figure 10.4 shows a model of the RAR–RXR heterodimer bound to a RARE, and an RXR homodimer bound to an RXRE.

RAREs and RXREs consist of a half-site with the consensus sequence AGGTCA; however, there is considerable pliancy for nucleotide substitution. This half-site sequence can form a direct repeat, a palindrome or can be found without any consensus structure; however, they are arranged most commonly as a direct repeat (Umesono et al., 1988, 1991; Umesono and Evans, 1989; Naar et al., 1991). The RAR–RXR heterodimer (see below) binds tightly to the half-site sequences separated by two or five nucleotides (termed DR-2 and DR-5, respectively). The DR-5 RARE, like that found in the promotor of the RARβ gene (de The et al., 1990; Sucov et al., 1990), is the one observed most often in the promoter of target genes. On the other hand, RXREs typically are direct repeats of the same consensus sequence AGGTCA with one nucleotide spacing the half-site sequence (DR-1).

Homodimerization and heterodimerization

RARs form heterodimers with the promiscuous RXRs and bind to the DR-5 or DR-2 RARE in the promoter of target genes (Fig. 10.4). Heterodimerization of RXR and RAR is required for tight binding and transactivation of gene expression through a RARE. RAR and RXR dimerize weakly in solution but, once they bind DNA, this interaction is strengthened. The heptad repeat motif, which is located in the LBD, is required for RAR–RXR and RXR–RXR dimerization (Forman et al., 1992). Also the zinc finger D box located in the DNA-binding domain (domain C) is

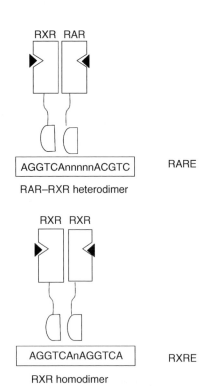

Fig. 10.4. Model of a RAR–RXR heterodimer and RXR homodimer.

required for dimerization and recognition of the RARE by the dimer (Umesono and Evans, 1988). The RAR–RXR heterodimer binds to the RARE DNA sequence with a specific orientation and polarity. RXR occupies the 5′ upstream half-site, while RAR occupies the 3′ downstream half-site (Predki *et al.*, 1994; Perlmann *et al.*, 1993; Kurokawa *et al.*, 1993).

Besides dimerizing with RARs, RXRs can homodimerize and bind to the DR-1 RXRE or can form heterodimers with other nuclear receptors including the vitamin D receptor, thyroid hormone receptor, peroxisome proliferator-activated receptor (PPAR) and several other orphan receptors.

Co-repressors and co-activators

As is the case with other nuclear receptors, binding of ligand to the receptor causes the release of co-repressor molecules that inhibit transcription and the recruitment of co-activators that facilitate the assembly of the transcriptionally active complex (Tong *et al.*, 1995) (Fig. 10.5). Two co-repressors that bind to unliganded RAR have been identified: N-CoR (nuclear receptor co-repressor) and SMRT (silencing mediator for RAR and TR) (Chen and Evans, 1995; Horlein *et al.*, 1995; Kurokawa *et al.*, 1995). Both of these co-repressors associate with the D domain of the receptor only in the absence of ligand. These co-repressors mediate their negative transcriptional effects by recruiting histone deacetylase (Alland *et al.*, 1997; Heinzel *et al.*, 1997). Histone deacetylase induces a change in the chromatin structure, causing DNA to be inaccessible to the transcriptional machinery. Upon ligand binding and a conformation change in the structure of the LBD (previously discussed), the co-repressors are released and co-activators associate with the RAR–RXR heterodimer. Several co-activators

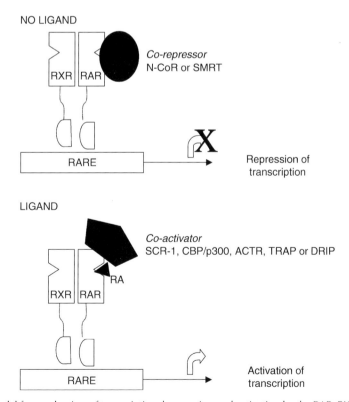

Fig. 10.5. Model for mechanism of transcriptional repression and activation by the RAR–RXR heterodimer. In the absence of RA, RAR–RXR heterodimers interact with the co-repressor complex that represses gene transcription. Upon RA binding, the co-repressor complex dissociates and is replaced by a co-activator complex, causing activation of transcription.

have been identified that interact with RARs and RXRs in a ligand-dependent manner. Among these co-activators are SRC-1, CBP/p300, ACTR, TRAP and DRIP (Bannister and Kouzarides, 1996; Chen et al., 1997; Spencer et al., 1997). These co-activators interact directly with the AF-2 region of the liganded receptor. Some of the coactivators possess an intrinsic histone acetyltransferse activity, causing greater accessibility of the basal transcription machinery to DNA.

Functional Role of the Different RAR and RXR Subtypes

One of the most intriguing unanswered questions is why six distinct retinoid receptor subtypes and several isoforms of each subtype exist. There are several lines of evidence that suggest that individual RAR and RXR subtypes and their respective isoforms are responsible for at least some specific, non-overlapping functions and thus play unique roles in mediating the large spectrum of biological processes affected by RA. These include the unique pattern of expression of individual RAR and RXR subtypes in the developing embryo and the adult, and the effect of inactivation of specific RAR–RXR subtypes or isoforms in transgenic mice and F9 teratocarcinoma cells.

Expression pattern of RARs and RXRs

Northern blot analysis of RNA isolated from various adult tissues has demonstrated that RARα mRNA is expressed ubiquitously, with the highest levels observed in specific regions of the brain including the hippocampus and cerebellum (Giguere et al., 1987; de The et al., 1989; Krezel et al., 1999). RARα1 mRNA is more abundant in the brain, skin, muscle, heart and kidney, while RARα2 mRNA is present at slightly higher levels in the lung, liver and intestines (Leroy et al., 1991). The mRNA distribution pattern of RARβ displays variability. High levels of RARβ expression are found in the kidney, prostrate, spinal cord, cerebral cortex, pituitary gland and adrenal gland; average levels are found in the liver, spleen, uterus, ovary and testis; and low levels are detected in breast and eye (Benbrook et al., 1988; de The et al., 1989). RARβ1 mRNA and RARβ3 mRNA, which are transcribed using the upstream P1 promoter, are expressed abundantly in the brain, skin and lung, while RARβ2 mRNA and RARβ4 mRNA, which are generated from a second downstream P2 promoter, are observed in the kidney, liver, lung, skin, heart, muscle, intestine and brain (Zelent et al., 1991; Nagpal et al., 1992). RARγ transcripts are the most restricted in their expression, with both RARγ1 and RARγ2 found predominantly in the skin and lung (Noji et al., 1989; Zelent et al., 1989; Kastner et al., 1990). The three RXR mRNAs are found to be present widely in adult tissues, with at least one RXR subytpe present in every tissue examined (Mangelsdorf et al., 1990, 1992). This is consistent with the role of RXR as a promiscuous dimerization partner for many nuclear receptor superfamily members.

Examination of the pattern of expression of the various RAR and RXR subtypes by *in situ* hybridization in mouse embryos of various developmental stages has also demonstrated that each subtype displays its own unique pattern of expression that may or may not overlap with other subtypes. RARα is expressed ubiquitiously during murine embryogenesis, whereas RARβ and RARγ are more restricted (Dolle et al., 1990a, b; Ruberte et al., 1990, 1991, 1993; Mendelsohn et al., 1994c). Similarly, RXRβ is expressed in a general fashion while RXRα and RXRγ are more restricted in their expression (Mangelsdorf et al., 1992; Dolle et al., 1994). Furthermore, the isoforms of each receptor subtype also display their own unique pattern of expression during embryogenesis.

Retinoid receptor knockout mutant mice

A second line of experimentation has involved the functional inactivation of specific RAR–RXR subtypes or their isoforms in mice by disrupting the specific gene of interest utilizing gene targeting technology. Since RA is essential for embryogenesis, and RARs and RXRs are believed to mediate the actions of RA, it seems reasonable that mice with null mutations in one of these receptors would show abnormalities similar to those resulting from lack of vitamin A. Some specific aberrations have been found to be associated with the lack of an individual nuclear receptor subtype in 'knockout' mice. However, overall, there appears to be a degree of forced functional redundancy between the individual RARs and RXRs. For example, homozygous RARα mutant mice display early postnatal lethality and testis degeneration (Lufkin et al., 1993); homozygous RARγ mutant mice display several abnormalities that previously have been associated with vitamin A deficiency including growth deficiency, early lethality and male sterility (Lohnes et al., 1993); homozygous RXRα mutant mice have been found to die between GD 13.5 and 16.5 (Sucov et al., 1994); and homozygous RXRβ mutant mice were normal except for male sterility (Kastner et al., 1996). The embryonic lethality observed in the RXRα-deficient mice is due to hypoplastic development of the ventricular chambers of the heart, which results in a very thin ventricular wall and defects in ventricular septation, linking RXRα function with important events associated with cardiac morphogenesis.

Unlike the single RAR mutant mice, the simultaneous inactivation of two RAR genes or both an RAR and an RXR gene results in mice with a variety of abnormalities, many of which resemble those previously reported in vitamin A-deficient fetuses including defects in the eyes, limbs, heart, head and nervous system (Wilson et al., 1953). Kastner et al. (1994) have reported a functional convergence when heterozygous RXRα mutant mice were crossed with mice carrying the null allele in the RARα or RARγ gene. This functional convergence suggests that the RXRα–RAR heterodimer mediates retinoid signalling in vivo. More severe abnormalities and death either in utero or shortly after birth have been reported in mice that carry a double homozygous RAR mutation resulting in the lack of expression of two different RAR subtypes or a homozygous mutation in both one RAR subtype and one RXR subtype (Lohnes et al., 1994; Mendelsohn et al., 1994b; Luo et al., 1996; Chiang et al., 1998; Dupe et al., 1999). Interestingly, analysis of hindbrain patterning in RARα/RARγ mutant mice and RARα/RARβ mutant mice suggests that RARα and/or RARγ plays an important role in mediating the RA-dependent events involved in specifying the prospective rhombomere 5/rhombomere 6 territory, whereas RARβ is important later in setting the caudal boundary of this territory (Wendling et al., 2001). Finally, mutant mice that lack both a functional RXRβ and RXRγ gene and have only one functional RXRα gene (RXRα+/−; RXRβ−/−; RXRγ−/−) are viable and display only growth deficiency and male sterility due to the lack of RXRβ, suggesting that only one copy of RXRα is sufficient to perform most of the functions of RXR (Krezel et al., 1996).

Less information is available concerning the individual isoforms of each RAR and RXR subtype. Mice in which both RARα1 genes, both RARβ2 genes or both RARγ2 genes were made functionally inactive by gene targeting techniques have been found to be viable and appear normal, with no apparent phenotypic abnormalities (Li et al., 1993; Lohnes et al., 1993; Mendelsohn et al., 1994a). On the other hand, retinal dysplasia and degeneration has been found in RARβ2 and RARγ2 compound mutant mice (Grondona et al., 1996). Recently, Mascrez et al. (2001) examined the role of the AF-1 region of RXRα1 by studying the effect of a mouse mutation that specifically deletes most of the RXRα1 A/B region, which includes the ligand-independent activation function AF-1. These studies demonstrate that the RXRα1 AF-1 region is functionally important (however, less than the ligand-dependent activation function AF-2) in transducing the retinoid signal through RAR–RXR during embryonic development.

Recently, the gene knockout technology has progressed to allow the efficient introduction of somatic mutations into a given gene in a site-specific and spatiotemporally controlled manner using the Cre/lox system (Metzger and Chambon, 2001). This allows the functional inactivation of a gene of interest within a specific cell type and at a particular time in the life span of a mouse. Initial studies using this approach have functionally inactivated RXRα in hepatocytes (Imai et al., 2001b; Wan et al., 2000a,b), adipocytes (Imai et al., 2001a) and the skin epidermis (Li et al., 2000, 2001) of mice at various times after birth. In all cases, the mice are viable; however, important information has already been obtained concerning the function of RXRα in specific tissues. In hepatoctyes, all pathways mediated by RXRα heterodimerization appear to be compromised. Furthermore, RXRα appears to play an important role in the mechanism(s) associated with the life span of hepatocytes in both adult and regenerating hepatocytes. In skin, RXRα is dispensable for prenatal epidermal development; however, it plays an important role in postnatal skin maturation particularly in hair follicle growth. Finally, RXRα appears to play a central role in adipogenesis, most probably as a heterodimeric partner with PPARγ. This approach shows great promise in the elucidation of specific functions of each of the RAR and RXR subtypes and their respective isoforms in specific tissues at different times during embryogenesis and in the adult mouse.

Retinoid receptor inactivation in F9 teratocarcinoma cells

An alternative approach has been functionally to inactivate specific RAR subtypes in F9 teratocarcinoma cells utilizing gene targeting technology, and to examine the pattern of expression of RA-dependent genes in these cells. F9 teratocarcinoma cells are a well-studied model system that is highly responsive to RA treatment differentiating to primitive endoderm, parietal endoderm, or visceral endoderm, depending on the culture conditions (Hogan et al., 1983). Furthermore, a large battery of genes that are known to be RA inducible have been well characterized in these cells. Examination of F9 cells in which either both RARγ genes or both RARα genes have been disrupted has demonstrated that each of these RARs exhibits specificity with respect to the regulation of RA-dependent differentiation-specific gene expression (Boylan et al., 1993, 1995; Taneja et al., 1995; Chiba et al., 1997). The absence of RARα has been demonstrated to be associated with a reduction in the RA-induced expression of both CRABP-II and Hoxb-1, while the absence of RARγ is associated with the loss of the RA-induced expression of the Hoxa-1, Hoxa-3, laminin B1, collagenase IV, GATA-4, CRBPI, Stra4, Stra6, Stra8, Cdx-1, GAP43 and BMP-2 genes. However, the expression of several RARγ-specific genes could be restored in RARγ-null cells by re-expressing RARγ, overexpressing RARα or, to a lesser extent, by overexpressing RARβ (Taneja et al., 1995). Therefore, a certain amount of functional redundancy can also be observed in these 'knockout' F9 cells; however, the physiological significance is unclear. Recently, Faria et al. (1999) have disrupted both allelles of RARβ2 in F9 cells and demonstrated that RARβ2 plays an important role in mediating late responses to RA. Several genes including CRABP II and laminin B1 have been shown to require RARβ2 expression for their extended and maximal expression during the process of RA-induced differentiation.

Summary

Taken together there is a considerable amount of evidence to support the concept that RA-dependent transcriptional regulation of at least some genes is mediated by specific RAR subtypes (functional specificity). However, functional redundancy has been observed in both RAR 'knockout' mice and F9 cells. Several reports (Chiba et al., 1997; Plassat et al., 2000; Rochette-Egly et al., 2000) have provided evidence that gene knockout in F9 cells generates arte-factual conditions within cells that unmask functional redundancies in a promoter-dependent context that actually do not normally occur. Thus, the functional redundancies that are observed in the gene knockout studies cannot be taken as evidence for a lack of functional specificity of each RAR subtype under wild-type physiological conditions. Additional studies utilizing different experimental approaches in which the levels of the endogenous retinoid receptor subtypes within the cell are not altered are necessary in order to resolve this problem.

References

Alland, L., Muhle, R., Hou, H., Potes, J., Chin, L., Schreiber-Agus, N. and DePinho, R.A. (1997) Role for N-CoR and histone deacetylase in Sin-3 mediated transcriptional repression. *Nature* 387, 49–55.

Bannister, A.J. and Kouzarides, T. (1996) The CBP co-activator is a histone acetyltransferase. *Nature* 384, 641–643.

Benbrook, D., Lernhardt, E. and Pfahl, M. (1988) A new retinoic acid receptor identified from a hepatocellular carcinoma. *Nature* 333, 669–672.

Bourguet, W., Ruff, M., Chambon, P., Gronemeyer, H. and Moras, D. (1995) Crystal structure of the ligand-binding domain of the human nuclear receptor RXR-α. *Nature* 375, 377–382.

Bourguet, W., Vivat, V., Wurtz, J.-M., Chambon, P., Gronemeyer, H. and Moras, D. (2000) Crystal structure of a heterodimeric complex of RAR and RXR ligand-binding domains. *Molecular Cell* 5, 289–298.

Boylan, J.F., Lohnes, D., Taneja, R., Chambon, P. and Gudas, L.J. (1993) Loss of retinoic receptor γ function in F9 cells by gene disruption results in aberrant *Hoxa-1* expression and differentiation upon retinoic acid treatment. *Proceedings of the National Academy of Sciences USA* 90, 9601–9605.

Boylan J.F., Lufkin, T., Achkar, C.C., Taneja, R., Chambon, P. and Gudas, L.J. (1995) Targeted disruption of retinoic acid receptor α (RARα) and RARγ results in receptor-specific alterations in retinoic acid-mediated differentiation and retinoic acid metabolism. *Molecular and Cellular Biology* 15, 843–851.

Brand, N., Petkovich, M., Krust, A., Chambon, P., de The, H., Marchio, A., Tiollais, P. and Dejean, A. (1988) Identification of a second human retinoic acid receptor. *Nature* 332, 850–853.

Brocard J., Kastner, P and Chambon, P. (1994) Two novel RXR alpha isoforms from mouse testis. *Biochemical and Biophysical Research Communications* 229, 211–218.

Chambon, P. (1996) A decade of molecular biology of retinoic acid receptors. *FASEB Journal* 10, 940–954.

Chandraratna, R.A. (1998) Future trends: a new generation of retinoids. *Journal of the American Academy of Dermatology* 39, S149–S152.

Chen, J.D. and Evans, R.M. (1995) A transcriptional co-repressor that interacts with nuclear hormone receptors. *Nature* 377, 454–457.

Chen, H., Lin, R.J., Schiltz, R.L., Chakravarti, D., Nash, A., Nagy, L., Privalsky, M.L., Nakatani, Y. and Evans, R.M. (1997) Nuclear receptor coactivator ACTR is a novel histone acetyltransferase and forms a multimeric activation complex with P/CAF and CBP/p300. *Cell* 90, 569–580.

Chiang, M.-Y., Misner, D., Kempermann, G., Schikorski, T., Giguere, V., Sucov, H.M., Gage, F.H., Stevens, C.F and Evans, R.M. (1998) An essential role for retinoid receptors RARβ and RXRγ in long-term potentiation and depression. *Neuron* 21, 1353–1361.

Chiba, H., Clifford, J., Metzger, D. and Chambon, P. (1997) Distinct retinoid X receptor–retinoic acid receptor heterodimers are differentially involved in the control of expression of retinoic target genes in F9 embryonal carcinoma cells. *Molecular and Cellular Biology* 17, 3013–3020.

de The, H., Marchio, A., Tiollais, P. and Dejean, A. (1989) Differential expression and ligand regulation of retinoic acid receptor α and β genes. *EMBO Journal* 8, 429–433.

de The, H., Vivanco-Ruiz, M.D., Tiollais, P., Stunnenberg, H. and Dejean, A. (1990) Identification of a retinoic acid responsive element in the retinoic acid receptor β gene. *Nature* 343, 177–180.

deUrquiza, A.M., Liu, S., Sjoberg, M., Zetterstrom, R.H., Griffiths, W., Sjovall, J. and Perlmann, T. (2000) Docosahexaenoic acid, a ligand for the retinoid X receptor in mouse brain. *Science* 290, 2140–2144.

Dolle, P., Ruberte, E., Kastner, P., Petkovich, M., Stoner, C.M., Gudas, L. and Chambon, P. (1990a) Differential expression of genes encoding α, β, and γ retinoic acid receptors and CRABP on the developing limbs of the mouse. *Nature* 342, 702–705.

Dolle, P., Ruberte, E., Leroy, P., Morriss-Kay, G. and Chambon, P. (1990b) Retinoic acid receptors and cellular retinoid binding proteins I. A systematic study of their differential pattern of transcription during mouse organogenesis. *Development* 110, 1133–1151.

Dolle, P., Fraulob, V., Kastner, P. and Chambon, P. (1994) Developmental expression of the murine retinoid X receptor (RXR) genes. *Mechanics of Development* 45, 91–104.

Duester, G. (2000) Families of retinoid dehydrogenases regulating vitamin A function. Production of visual pigment and retinoic acid. *FEBS Letters* 267, 4315–4324.

Dupe V., Ghyselinck, N.B., Wendling, O., Chambon, P. and Mark, M. (1999) Key roles of retinoic acid receptors alpha and beta in the patterning of the caudal hindbrain, pharyngeal arches and otocytst in the mouse. *Development* 126, 5051–5119.

Egea, P., Mitschler, A., Rochel, N., Ruff, M., Chambon, P. and Moras, D. (2000) Crystal structure of the human RXRα ligand-binding domain bound to its natural ligand: 9-*cis* retinoic acid. *EMBO Journal* 19, 2592–2601.

Faria, T.N., Mendelsohn, C., Chambon, P. and Gudas, L.J. (1999) The targeted disruption of both alleles of RARβ2 in F9 cells results in the loss of retinoic acid-associated growth arrest. *Journal of Biological Chemistry* 274, 26783–26788.

Folkers, G.E., van Heerde, E.C. and van der Saag, P.T. (1995) Activation function 1 of retinoic acid receptor beta 2 is an acidic activator resembling VP16. *Journal of Biological Chemistry* 270, 23552–23559.

Forman, B.M., Casanova, J., Raaka, B.M., Ghysdael, J. and Samuels, H.H. (1992) Half-site spacing and orientation determines whether thyroid hormone and retinoic acid receptors and relative factor bind to DNA response elements as monomers, homodimers or heterodimers. *Molecular Endocrinology* 6, 429–442.

Gehin, M., Vivat, V., Wurtz, J.M., Losson, R., Chambon, P., Moras, D. and Gronemeyer, H. (1999) Structural basis for engineering of retinoic acid receptor isotype-selective agonists and antagonists. *Chemistry and Biology* 6, 519–529.

Giguere, V., Ong, E.S., Segui, P. and Evans, R.M. (1987) Identification of a receptor for the morphogen retinoic acid. *Nature* 330, 624–629.

Giguere, V., Shago, M., Zirngibl, R., Tate, P., Rossant, J. and Varmuza, S. (1990) Identification of a novel isoform of the retinoic acid receptor expressed in the mouse embryo. *Molecular and Cellular Biology* 10, 2335–2340.

Gottesman, M.E., Quadro, L. and Blaner, W.S. (2001) Studies of vitamin A metabolism in mouse model systems. *BioEssays* 23, 409–419.

Grondona, J.M., Kastner, P., Gansmuller, A., Decimo, D. and Chambon, P. (1996) Retinal dysplasia and degeneration in RARβ2/RARγ2 compound mutant mice. *Development* 122, 2173–2188.

Heinzel, T., Lavinski, R.M., Mullen, T.M., Soderstrom, M., Laherty, C.D., Torchia, J., Yang, W.M., Brard, G., Ngo, S.D., Davie, J.R., Seto, E., Eisenman, R.N., Rose, D.W., Glass, C.K. and Rosenfeld, M.G. (1997) A complex containing N-CoR, mSin3 and histone deacetylase mediates transcriptional repression. *Nature* 387, 43–48.

Heyman, R.A., Mangelsdorf, D.J., Dyck, J.A., Stein, R.B., Eichele, G., Evans, R.M. and Thaller, C. (1992) 9-*Cis* retinoic acid is a high affinity ligand for the retinoid X receptor. *Cell* 68, 397–406.

Hogan, B.L.M., Barlow, D. and Tilly, R. (1983) F9 teratocarcinoma cells as a model for the differentiation of parietal and visceral endoderm in the mouse embryo. *Cancer Surveys* 2, 115–140.

Horlein, A.J., Naar, A.M., Heinzel, T., Torchia, J., Gloss, B., Kurokawa, R., Ryan, A., Kamei, Y., Soderstrom, M. and Glass, C.K. (1995) Ligand-independent repression by thyroid hormone receptor mediated by a nuclear receptor co-repressor. *Nature* 377, 397–404.

Imai, T., Jiang, M., Chambon, P. and Metzger, D. (2001a) Impaired adipogenesis and lipolysis in the mouse upon selective ablation of the retinoid X receptorα mediated by a tomoxifen-inducible chimeric Cre recombinase (Cre-ER^{T2}) in adipocytes. *Proceedings of the National Academy of Sciences USA* 98, 224–228.

Imai, T., Jiang, M., Kastner, P., Chambon, P and Metzger, D. (2001b) Selective ablation of retinoid X receptor alpha in hepatocytes impairs their lifespan and regenerative capacity. *Proceedings of the National Academy of Sciences USA* 98, 4581–4586.

Johnson, A. and Chandraratna, R.A. (1999) Novel retinoids with receptor selectivity and functional selectivity. *British Journal of Dermatology* 140, 12–17.

Kastner, P., Krust, A., Mendelsohn, C., Garnier, J.M., Zelent, A., Leroy, P., Staub, A. and Chambon, P. (1990) Murine isoforms of retinoic acid receptor gamma with specific patterns of expression. *Proceedings of the National Academy of Sciences USA* 87, 270–274.

Kastner, P., Grondona, J.M., Mark, M., Gansmuller, A., Le Meur, M., Decimo, D., Vonesch, J.-L., Dolle, P. and Chambon, P. (1994) Genetic analysis of RXR-α development function: convergence of RXR and RAR signaling pathways in heart and eye morphogenesis. *Cell* 78, 987–1003.

Kastner, P., Mark, M., Leid, M., Gansmuller, A., Chin, W., Grondona, J.M., Decimo, D., Krezel, W., Dierich, A and Chambon, P. (1996) Abnormal spermatogenesis in RXR beta mutant mice. *Genes and Development* 10, 80–92.

Krezel, W., Dupe, V., Mark, M., Dierich, A., Kastner, P. and Chambon, P. (1996) RXR gamma null mice are apparently normal and compound RXR alpha +/−/RXR beta −/−/RXR gamma −/− mutant mice are viable. *Proceedings of the National Academy of Sciences USA* 93, 9010–9014.

Krezel, W., Kastner, P. and Chambon, P. (1999) Differential expression of retinoid receptors in the adult mouse central nervous system. *Neuroscience* 89, 1291–1300.

Krust, A., Kastner, P., Petkovich, M., Zelent, A. and Chambon, P. (1989) A third human retinoic acid receptor, hRAR-γ. *Proceedings of the National Academy of Sciences USA* 86, 5310–5314.

Kurokawa, R., Yu, V.C., Naar, A., Kyakumoto, S., Han, Z., Silverman, S., Rosenfeld, M.G. and Glass, C.K. (1993) Differential orientations of the DNA-binding domain and carboxy-terminal dimerization regulate binding site selection by nuclear receptor heterodimers. *Genes and Development* 7, 1423–1435.

Kurokawa, R., Soderstrom, M., Horlein, A., Halachmi, S., Brown, M., Rosenfeld, M.G. and Glass, C.K. (1995) Polarity-specific activities of retinoic acid receptors by a co-repressor. *Nature* 377, 451–454.

Levin, A.A., Sturzenbecher, L.J., Kazmer, S., Bosakowski, T., Huselton, C., Allenby, G., Speck, J., Kratzeisen, C., Rosenberger, M., Lovey, A. and Grippo, J.F. (1992) 9-*Cis* retinoic acid stereoisomer binds and activates the nuclear receptor RXRα. *Nature* 355, 359–361.

Leroy, P., Krust, A., Zelent, A., Mendelsohn, C., Garnier, J.-M., Kastner, P., Dierich, A. and Chambon, P. (1991) Multiple isoforms of the mouse retinoic acid receptor α are generated by alternative splicing and differential induction by retinoic acid. *EMBO Journal* 10, 59–69.

Li, E., Sucov, H.M., Lee, K.-F., Evans, R.M. and Jaenisch, R. (1993) Normal development and growth of mice carrying a targeted disruption of the α1 retinoic acid receptor gene. *Proceedings of the National Academy of Sciences USA* 90, 1590–1594.

Li, M., Indra, A.K., Warot, X., Brocard, J., Messaddeq, N., Kato, S., Metzger, D. and Chambon, P. (2000) Skin abnormalities generated by temporally controlled RXRα mutations in mouse epidermis. *Nature* 407, 633–636.

Li, M., Chiba, H., Warot, X., Messaddeq, N., Gerard, C., Chambon, P. and Metzger, D. (2001) RXR-alpha ablation in skin keratinocytes results in alopecia and epidermal alteration. *Development* 128, 675–688.

Liu, Q. and Linney, E. (1993) The mouse retinoid-X receptor-gamma gene: genomic organization and evidence for functional isoforms. *Molecular Endocrinology* 7, 651–658.

Lohnes, D., Kastner, P., Dierich, A., Mark, M., LeMeur, M. and Chambon, P. (1993) Function of retinoic acid receptor γ in the mouse. *Cell* 73, 643–658.

Lohnes, D., Mark, M., Mendelsohn, C., Dolle, P., Dierich, A., Gorry, P., Gansmuller, A. and Chambon, P. (1994) Function of the retinoic acid receptors (RARs) during development. (I). Craniofacial and skeletal abnormalities in RAR double mutants. *Development* 120, 2723–2748.

Lufkin, T., Lohnes, D., Mark, M., Dierich, A., Gorry, P., Gaub, M.P., LeMeur, M. and Chambon, P. (1993) High postnatal lethality and testis degeneration in retinoic acid receptor alpha mutant mice. *Proceedings of the National Academy of Sciences USA* 90, 7225–7229.

Luo, J., Sucov, H.M., Bader, J.A., Evans, R.M. and Giguere, V. (1996) Compound mutants for retinoic acid receptor (RAR) beta and RAR alpha1 reveal developmental functions for multiple RAR beta isoforms. *Mechanics of Development* 55, 33–44.

Mangelsdorf, D.J., Ong, E.S., Dyck, J.A. and Evans, R.M. (1990) Nuclear receptor that identified a novel retinoic acid response pathway. *Nature* 345, 224–229.

Mangelsdorf, D.J., Borgmeyer, U., Heyman, R.A., Zhou, J.Y., Ong, E.S., Oro, A.E., Kakizuka, A. and Evans, R.M. (1992) Characterization of three RXR genes that mediate the action of 9-*cis* retinoic acid. *Genes and Development* 6, 329–344.

Mascrez, B., Mark, M., Krezel, W., Dupé, V., LeMeur, M., Ghyselinck, N.B. and Chambon, P.A. (2001) Differential contributions of AF-1 and AF-2 activities to the developmental functions of RXRα. *Development* 128, 2049–2062.

Mendelsohn, C., Mark, M., Dolle, P., Dierich, A., Gaub, M.P., Krust, A., Lampron, C. and Chambon, P. (1994a) Retinoic acid receptor beta2 (RAR beta 2) null mutant mice appear normal. *Developmental Biology* 166, 246–258.

Mendelsohn, C., Lohnes, D., Decimo, D., Lufkin, T., LeMeur, M., Chambon, P. and Mark, M. (1994b) Function of the retinoic acid receptors (RARs) during development (II). Multiple abnormalites at various stages of organogenesis in RAR double mutants. *Development* 120, 2749–2771.

Mendelsohn, C., Larkin, S., Mark, M., Lemeeur, M., Clifford, J., Zelent, A. and Chambon, P. (1994c) RARβ isoforms: distinct transcriptional control by retinoic acid and specific spatial patterns of promoter activity during mouse embryonic development. *Mechanics of Development* 45, 227–241.

Metzger, D. and Chambon, P. (2001) Site- and tissue-specific gene targeting in the mouse. *Methods* 24, 71–80.

Naar, A.M., Boutin, J.M., Lipkin, S.M., Yu, V.C., Holloway, J.M., Glass, C.K. and Rosenfeld, M.G. (1991) The orientation and spacing of core DNA-binding motifs dictate selective transcriptional responses to three nuclear receptors. *Cell* 65, 1267–1279.

Nagata, T., Kanno, Y., Ozato, K. and Taketo, M. (1994) The mouse RXRβ gene encoding RXR beta: genomic organization and two mRNA isoforms generated by alternative splicing of transcripts initiated by CpG island promoters. *Gene* 142, 183–189.

Nagpal, S. and Chandraratna, R.A. (2000) Recent developments in receptor-selective retinoids. *Current Pharmaceutical Design* 6, 919–931.

Nagpal, S., Zelent, A. and Chambon, P. (1992) RAR-β4, a retinoic acid receptor isoform is generated from RAR-β2 by alternative splicing and the usage of a CUG initiator codon. *Proceedings of the National Academy of Sciences USA* 89, 2718–2722.

Nagpal, S., Friant, S., Nakshatri, H. and Chambon, P. (1993) RARs and RXRs: evidence for two autonomous transactivation functions (AF-1 and AF-2) and heterodimerization *in vivo*. *EMBO Journal* 12, 2349–2360.

Napoli, J.L. (1999a) Retinoic acid: its biosynthesis and metabolism. *Progress in Nucleic Acid Research and Molecular Biology* 63, 139–188.

Napoli, J.L. (1999b) Interactions of retinoid binding proteins and enzymes in retinoid metabolism. *Biochimica et Biophysica Acta* 1440, 139–162.

Noji, S., Yamaii, T., Koyama, E., Nohno, T., Fujimoto, W., Arata, J. and Taniguichi, S. (1989) Expression of retinoic acid receptor genes in keratinizing front skin. *FEBS Letters* 259, 86–90.

Noy, N. (2000) Retinoid-binding proteins: mediators of retinoid action. *Biochemical Journal* 348, 481–495.

Ostrowski, J., Hammer, L., Roalsvig, T., Pokornowski, K. and Reczek, P.R. (1995) The N-terminal portion of the E domain of retinoic acid receptor alpha and beta is essential for the recognition of retinoic acid and various analogs. *Proceedings of the National Academy of Sciences USA* 92, 1812–1816.

Ostrowski, J., Roalsvig, T., Hammer, L., Starret, J., Yu, K.-L. and Reczek, P. (1998) Serine 232 and methionine 272 define the ligand binding pocket in retinoic acid receptor subtypes. *Journal of Biological Chemistry* 273, 3490–3495.

Perlmann, T., Rangarajan, P.N., Umesono, K. and Evans, R.M. (1993) Determinants for selective RAR and TR recognition of direct repeat HREs. *Genes and Development* 7, 1411–1422.

Petkovich, M., Brand, N.J., Krust, A. and Chambon, P. (1987) A human retinoic acid receptor which belongs to the family of nuclear receptors. *Nature* 330, 444–450.

Plassat, J.-L., Penna, L., Chambon, P. and Rochette-Egly, C. (2000) The conserved amphipatic α-helical core motif of RARγ and RARα activating domains is indispensable for RA-induced differentiation of F9 cells. *Journal of Cell Science* 113, 2887–2895.

Predki, P.F., Zamble, D., Sarkar, B. and Giguere, V. (1994) Ordered binding of retinoic acid and retinoid X receptors to asymmetric response elements involves determinants adjacent to the DNA-binding domain. *Molecular Endocrinology* 8, 31–39.

Renaud, J.P., Rochel, N., Ruff, M., Vivat, V., Chambon, P., Gronemeyer, H. and Moras, D. (1995) Crystal structure of the RAR-γ ligand-binding domain bound to all-*trans*-retinoic acid. *Nature* 378, 681–689.

Rochette-Egly, C., Plassat, J.-L., Taneja, R. and Chambon, P. (2000) The AF-1 and AF-2 activating domains of retinoic acid receptor-α (RARα) and their phosphorylation are differentially involved in parietal endodermal differentiation of F9 cells and retinoid-induced expression of target genes. *Molecular Endocrinology* 14, 1398–1410.

Ruberte, E., Dolle, P., Krust, A., Zelent, A., Morriss-Kay, G. and Chambon, P. (1990) Specific spatial and temporal distribution of retinoic acid receptor gamma transcripts during mouse embryogenesis. *Development* 108, 213–222.

Ruberte, E., Dolle, P., Chambon, P. and Morriss-Kay, G. (1991) Retinoic acid receptors and cellular retinoic acid binding proteins. II. Their differential pattern of transcription during early morphogenesis in mouse embryos. *Development* 111, 45–60.

Ruberte, E., Friederich, V., Chambon, P. and Morriss-Kay, G. (1993) Retinoic acid receptors and cellular retinoid binding proteins. III. Their differential transcript distribution during mouse nervous system development. *Development* 118, 267–282.

Scafonas, A., Wolfgang, C.L., Gabriel, J.L., Soprano, K.J. and Soprano, D.R. (1997) Differential role of homologous positively charged residues for ligand binding in retinoic acid receptor α compared with retinoic acid receptor β. *Journal of Biological Chemistry* 272, 11244–11249.

Spencer, T.E., Jenster, G., Burcin, M.M., Allis, C.D., Zhou, J., Mizzen, C.A., McKenna, N.J., Onate, S.A., Tsai, S.Y., Tsai, M.J. and O'Malley, B.W. (1997) Steroid receptor coactivator-1 is a histone acetyltransferase. *Nature* 389, 194–198.

Sporn, M.B., Roberts, A.B. and Goodman, D.S. (1994) *The Retinoids: Biology, Chemistry and Medicine*. Raven Press, New York.

Stunnenberg, H.G. (1993) Mechanism of transactivation by retinoic acid receptors. *BioEssays* 15, 309–315.

Sucov, H.M., Murakami, K.K. and Evans, R.M. (1990) Characterization of an autoregulated response element in the mouse retinoic acid receptor type b gene. *Proceedings of the National Academy of Sciences USA* 87, 5392–5396.

Sucov, H.M., Dyson, E., Gumeringer, C.L., Price, J., Chien, K.R. and Evans, R.M. (1994) RXR-α mutant mice establish a genetic basis for vitamin A signaling in heart morphogenesis. *Genes and Development* 8, 1007–1018.

Tairis, N., Gabriel, J.G., Gyda, M., Soprano, K.J. and Soprano, D.R. (1994) Arg269 and Lys220 of retinoic acid receptor-β are important for the binding of retinoic acid. *Journal of Biological Chemistry* 269, 19516–19522.

Tairis, N., Gabriel, J.L., Soprano, K.J. and Soprano, D.R. (1995) Alteration in the retinoid specificity of retinoic acid receptor-β by site-directed mutagenesis of Arg269 and Lys220. *Journal of Biological Chemistry* 270, 18380–18387.

Taneja, R., Bouillet, P., Boylan, J.F., Gaub, M.-P., Roy, B., Gudas, L.J. and Chambon, P. (1995) Reexpression of retinoic acid receptor (RAR) γ or overexpression of RARα or RARβ in RARγ-null F9 cells reveals a partial functional redundancy between the three RAR types. *Proceedings of the National Academy of Sciences USA* 92, 7854–7858.

Tong, G.X., Tanen, M.R. and Bagchi, M.K. (1995) Ligand modulates the interaction of thyroid hormone receptor beta with the basal transcriptional machinery. *Journal of Biological Chemistry* 270, 10601–10611.

Umesono, K. and Evans, R.M. (1989) Determinants of target gene specificity for steroid/thyroid hormone receptors. *Cell* 57, 1139–1146.

Umesono, K., Giguere, V., Glass, C.K., Rosenfeld, M.G. and Evans, R.M. (1988) Retinoic acid and thyroid hormone induce gene expression through a common responsive element. *Nature* 336, 262–265.

Umesono, K., Morakami, K.K., Thompson, C.C. and Evans, R.M. (1991) Direct repeats as selective response elements for the thyroid hormone, retinoic acid and vitamin D3 receptors. *Cell* 65, 1255–1266.

Wan, Y.J., Cai, Y., Lungo, W., Fu, P., Locker, J., French, S. and Sucov, H.M. (2000a) Peroxisome proliferator-activated receptor α-mediated pathways are altered in hepatocyte-specific retinoid X receptor α-deficient mice. *Journal of Biological Chemistry* 275, 28285–28290.

Wan, Y.J., An, D., Cai, Y., Repa, J.J., Hung-Po Chen, T., Flores, M., Postic, C., Magnuson, M.A., Chen, J., Chien, K.R., French, S., Mangelsdorf, D.T. and Sucov, H.M. (2000b) Hepatocyte-specific mutation establishes retinoid X receptor alpha as a heterodimeric integrator of multiple physiological processes in the liver. *Molecular and Cellular Biology* 20, 4436–4444.

Wendling, O., Ghyselinck, N.B., Chambon, P. and Mark, M. (2001) Roles of retinoic acid receptors in early embryonic morphogenesis and hindbrain patterning. *Development* 128, 2031–2038.

Wilson, J.G., Roth, C.B. and Warkany, J. (1953) An analysis of the syndrome of malformations induced by maternal vitamin A deficiency. Effects of restoration of vitamin A at various times during gestation. *American Journal of Anatomy* 92, 189–217.

Wolbach, S.B. and Howe, P.R. (1925) Tissue changes following deprivation of fat soluble A vitamin. *Journal of Experimental Medicine* 6, 753–757.

Wolfgang, C.L., Zhang, Z.-P., Gabriel, J.L., Pieringer, R.A., Soprano, K.J. and Soprano, D.R. (1997) Identification of sulfhydryl modified cysteine residues in the ligand binding pocket of retinoic acid receptor β. *Journal of Biological Chemistry* 272, 746–753.

Wurtz, J.-M., Bourquet, W., Renaud, J.-P., Vivat, V., Chambon, P., Moras, D. and Gronemeyer, H. (1996) A canonical structure for the ligand-binding domain of nuclear receptors. *Nature Structural Biology* 3, 87–94.

Zelent, A., Krust, A., Petkovich, M., Kastner, P. and Chambon, P. (1989) Cloning of a murine alpha and beta retinoic acid receptor and a novel receptor gamma predominantly expressed in the skin. *Nature* 339, 714–717.

Zelent, A., Mendelsohn, C., Kastner, P., Krust, A., Garnier, J.-M., Ruffenach, F., Leroy, P. and Chambon, P. (1991) Differentially expressed isoforms of the mouse retinoic acid receptor β are generated by usage of two promoters and alternative splicing. *EMBO Journal* 10, 71–81.

Zhang, Z.-P., Gambone, C.J., Gabriel, J.L., Wolfgang, C.L., Soprano, K.J. and Soprano, D.R. (1998) Arg^{278}, but not Lys^{229} or Lys^{236}, plays an important role in the binding of retinoic acid by retinoic acid receptor γ. *Journal of Biological Chemistry* 273, 4016–4021.

Zhang, Z.-P., Shukri, M., Gambone, C.J., Gabriel, J.L., Soprano, K.J. and Soprano, D.R. (2000) Role of Ser^{289} in RARγ and its homologous amino acid residue in RARα and RARβ in the binding of retinoic acid. *Archives of Biochemistry and Biophysics* 380, 339–346.

Zhang, Z.-P., Hutcheson, J.M., Poynton, H.C., Gabriel, J.L., Soprano, K.J. and Soprano, D.K. (2003) Arginine and retinoic acid receptor B which coordinates with the carboxyl group of retinoic acid functions independent of the amino acid residues responsible for retinoic acid receptor subtype ligand specificity. *Archives of Biochemistry and Biophysics* 409, 375–384

11 Regulation of Gene Expression by Biotin, Vitamin B_6 and Vitamin C

Krishnamurti Dakshinamurti
Department of Biochemistry and Medical Genetics, Faculty of Medicine, University of Manitoba and Division of Neurovirology and Neurodegenerative Disorders, St Boniface Hospital Research Centre, Winnipeg, Canada

Introduction

Only a small percentage of the genome in eukaryotes is transcriptionally active at any particular time. Response elements, mostly upstream from the structural genes affected, regulate the rates of transcription of individual genes. The regulation of gene expression during cellular differentiation and also through the influence of hormones, cytokines and growth factors is well established. Most of the specialized cells in multicellular organisms are capable of altering the pattern of their gene expression in response to extracellular signals. Different cell types respond in different ways to the same cues. Gene expression can be regulated at any of the steps in the pathway from the DNA to RNA to protein. The transcription of individual genes is switched on or off by gene regulatory proteins. Some genes are transcribed at constant levels and are turned on or off by post-transcriptional regulatory processes. These processes require recognition of specific sequences or structures, which is accomplished by either a regulatory protein or a regulatory RNA molecule.

Among the extracellular cues regulating gene expression are those that affect the hormonal, neuroregulatory or immune response systems in higher organisms. To this should be added the nutritional status of the organism as this impinges on various regulatory systems. In addition to such indirect routes, certain dietary constituents influence gene expression by interacting directly with regulatory elements in the genome, leading to changes in the rate of transcription of a given gene. Adaptive mechanisms to meet the challenge of a variable supply of nutrients have evolved. Thus, changes in nutrient supplies can either induce or repress transcription of specific genes or gene sets. In many cases, the regulation (up or down) of transcription is mediated by alteration in the availability of transcription factors or by changes in the factor's ability to bind to specific regulatory sequence motifs. The physiological effects of vitamins follow their entry into the cell, metabolic transformation and association with specific apoenzymes, resulting in the accentuation of a metabolic pathway. In this respect, vitamins are metabolic initiators. In some instances, the variation in the amount of a particular enzyme (apoenzyme) is the result of transcriptional or translational events in response to coenzyme (cofactor) supply. Such a response would require a mechanism able to sense variations in cofactor levels and to transduce this into a response of the level of gene expression. Where the vitamin function is other than as a coenzyme or a prosthetic group of an enzyme, its effect is through regulation of cellular protein synthesis (Dakshinamurti, 1994, 1997). The focus of this review is on the role of water-soluble vitamins such as biotin, vitamin B_6 and ascorbic acid on modulation of gene expression. The nutrient demands of proliferating cells for these vitamins will also be discussed.

Biotin

The best known and understood role of biotin is as the prosthetic group of the four biotin-containing carboxylases in higher organisms. The role of biotin in acetyl-CoA carboxylase, propionyl-CoA carboxylase, 3-methylcrotonyl-CoA carboxylase and pyruvate carboxylase explains its obligatory involvement in the metabolism of carbohydrates, lipids and deaminated residues of some amino acids. Many years ago, we hypothesized that biotin has a role in cellular function other than as just the prosthetic group of the four biotin-containing carboxylases (Dakshinamurti et al., 1985; Dakshinamurti and Chauhan, 1989). This was based on the requirement for biotin of cells in culture, which we had established. This concept has received steady support over the years from our work and from that of other investigators (Zempleni and Mock, 2001). It is possible that the synthesis and secretion of a single growth/differentiation factor or its receptor is initiated by biotin and that the other effects of biotin might flow from this. This is not likely in view of the diverse effects of biotin deficiency in animals. In addition to a role for biotin in cell proliferation and differentiation, its roles in embryological development, testicular development, immune defence mechanisms and the regulation of specific proteins have been identified (Dakshinamurti and Chauhan, 1994). Biotin regulates the synthesis of key glycolytic and gluconeogenic kinases. It induces the synthesis of biotin-binding proteins of egg yolk (White and Whitehead, 1987). Mitogen-induced proliferation increases biotin uptake into peripheral blood mononuclear cells (PBMCs) through an increase in the number of specific biotin transporters on the PBMC membrane (Zempleni and Mock, 1999). Biotin has been reported to have a role in the release of insulin by the pancreas (Sone et al., 1999), in the synthesis of some biotin carboxylases and biotin holocarboxylase synthetase (Rodriguez-Melendez et al., 2001), and in the expression of the asialoglycoprotein receptor in HepG2 cells (Collins et al., 1988). Various mechanisms such as a role for biotin in initiation of transcription, regulation of gene expression through its action on the cGMP-mediated guanylate kinase and replication of DNA through biotinylation of histones have been suggested.

Nuclear biotin-binding protein

Intracellular fractionation of biotin in various tissues of the rat and the chicken indicated that a significant amount of biotin was associated with the nuclear fraction (Dakshinamurti and Mistry, 1963a,b). The biotin content of biotin-deficient rat liver is about one-tenth that of normal rat liver, and a significant 20% of this is present in the nuclear fraction. Biotin is present in cell nuclei prepared from various tissues of cultured cells even though such nuclei do not possess any carbon dioxide-fixing capability, indicating that biotin in nuclei does not function as the prosthetic group of carboxylases. During biotin deficiency, biotin in the nuclear fraction seems to be conserved, whereas it is lost preferentially from other cellular organelles (Boeckx and Dakshinamurti, 1974, 1975). Biotin in nuclei has been shown to be bound non-covalently to a protein. A biotin-binding protein from rat liver nucleus has been isolated (Dakshinamurti et al., 1985). This protein binds reversibly to biotin in vitro, with a maximal binding of 3.54 pmol per µg of protein and with a dissociation constant for biotin of 2.2×10^{-7} M. Polyacrylamide gel electrophoresis in the presence of sodium dodecyl sulphate indicates an apparent subunit molecular mass of 60 kDa for this protein. A similar binding protein for pyridoxal phosphate has been reported by Meisler and Thannasi (1990) in rat hepatoma-derived HTC cells.

There have been other reports of the presence of high concentrations of biotin in optically clear nuclei (OCN). Solid cell nests resembling mulberries (morules) have been reported in colorectal adenomas and carcinomas, endometrial tissue and in pulmonary endodermal tumours resembling fetal lung (Nakatani et al., 1994; Yang et al., 1995). The nuclei in morules in several organs are referred to as OCN and have been found to contain significant amounts of biotin (Nakatani et al., 1994; Okamoto et al., 1995; Yang et al., 1995). OCN with significant accumulations of biotin unrelated to the development of a morule appear in endometrium during pregnancy and in ovarian endometroid carcinoma (Yokoyama et al., 1993; Sickel and

Sant'Agnese, 1994). The types of biotiylated polypeptide or biotin-binding proteins that accumulate, the mechanisms responsible for the accumulation and the clinicopathological significance of OCN remain to be determined (Sasaki et al., 1999).

Requirement for biotin by cells in culture

A requirement for biotin by cells in culture would be expected in view of the obligatory involvement of biotin in the metabolism of carbohydrates, lipids and deaminated residues of certain amino acids. However, various earlier studies indicated that cells in culture do not require biotin. Keranan (1972) reported that HeLa cells grown in biotin-deficient medium contained more biotin that those in a biotin-supplemented medium, due perhaps to the ability of transformed cells to synthesize biotin. Using biotin-depleted fetal bovine serum (FBS) and Eagle's minimum essential medium, we demonstrated a requirement for biotin by HeLa cells, human fibroblasts and Rous sarcoma virus-transformed (RST) baby hamster kidney (BHK) cells. This was based on the viability, biotin content and activities of biotin-dependent and -independent enzymes (Dakshinamurti and Chalifour, 1981; Chalifour and Dakshinamurti, 1982a,b). We also demonstrated that there was a significant decrease in the incorporation of leucine into protein of the homogenate or cytosol of biotin-deficient HeLa cells compared with cells grown in a biotin-supplemented medium. When biotin was added to the biotin-deficient medium, there was a twofold increase in the incorporation of leucine into proteins. Experiments using puromycin and cordycepin indicated that the appearance of new RNAs in the cytoplasm occurs when biotin-deficient cells are supplemented with biotin. Mammalian cultured cells, when they do not receive specific signals, which include specific growth factors, come to a halt in a quiescent, non-growing variant of the G_1 state, referred to as G_0. The synthesis of components of the cell cycle control system is switched off.

Normal cells in G_1 arrest due to serine starvation start incorporating [^{3}H]thymidine into DNA as soon as serine is restored to the medium. Biotin-deficient HeLa cells under similar conditions do not incorporate [^{3}H]thymidine into DNA even when serine is restored to the medium.

However, within 4 h of supplementation of biotin to the biotin-deficient medium, the incorporation of [^{3}H]thymidine into DNA reaches a maximum. By this time, there is stimulation of protein synthesis. The two phenomena are related and the growth-promoting effect of biotin might be achieved through the synthesis of certain proteins (Bhullar and Dakshinamurti, 1985). Moscowitz and Cheng (1985) claimed that RST BHK cells grown in a medium containing biotin produced a non-dialysable factor that stimulated cell multiplication.

A requirement for high-density lipoprotein (HDL) has been shown for the growth of Madin-Darby canine kidney cells grown in Dulbecco's modified Eagle's medium (DMEM) supplemented with transferrin when the cells were exposed to a mixture (1:1) of DMEM and F-12 medium (Gospodarowicz and Cohen, 1985). The components of the F-12 medium responsible for support of growth in the absence of HDL include biotin, which is absent from DMEM, and choline, which is present in insufficient concentrations in DMEM. It is significant that the HDL fraction of plasma has a considerable amount of biotin associated with it. This biotin is non-dialysable and is non-covalently attached to protein.

Role of biotin in cell differentiation

Various transformed cell lines produce transforming growth factors that enable them to grow in a serum-free medium supplemented with transferrin and insulin. The L1 subline derived from the 3T3 mouse fibroblast cell line has the capacity to differentiate in the resting state into a cell type having the characteristics of adipocytes. When they reach confluence and start to differentiate, they greatly increase their rate of triglyceride synthesis. The increase in lipogenic rate parallels a coordinated increase in the activities of key enzymes of the fatty acid biosynthetic pathway. This increase correlates with a marked rise in nuclear run-off transcription rates for these mRNAs during differentiation (Bernlohr et al., 1985). The process of differentiation can be accelerated in a number of ways, such as by increasing the amount of serum in the culture medium or by adding insulin or biotin. When serum in the cultures was dialysed extensively, the morphological and enzyme changes characteristic of adipocyte differentiation were induced without fat deposition (Kuri-Horcuch *et al.*, 1978). It was presumed that dialysis removed biotin from serum, and the role of biotin was linked to the increase in acetyl-CoA carboxylase. However, it took 24–48 h after the addition of biotin to the medium before deposition of triglyceride. The delay suggests that some factor required for the induction of the whole set of lipogenic enzymes (which, with the exception of acetyl-CoA carboxylase, are not biotin enzymes) might be formed under the influence of biotin. To express differentiation, the attainment of confluency is accompanied by the presence of an adipogenic factor in the culture medium (Kuri-Harcuch and Marsch-Morino, 1983). This is present in most animal sera but not in the serum of the domestic cat. It is significant that cat serum contains less than one-tenth of the biotin content of bovine serum.

Biotin and embryological development

Congenital malformations have been reported in embryos of domestic fowl maintained on a biotin-deficient diet (Watanabe, 1990). Maternal biotin deficiency was strongly teratogenic in mice even when dams did not exhibit any typical signs of biotin deficiency. At mid-gestation, biotin-deficient embryos weighed less than normal embryos and had external malformations such as micrognathia and micromelia. There was a marked reduction in the size of the palatal process on day 15.5 of gestation, which may be due to altered proliferation of mesenchyme. The same may be said for the proliferation of mesenchyme in the limb bud.

We investigated the development of the palatal process in culture (Watanabe *et al.*, 1995). After 72 h of organ culture, >90% of the explants from normal mouse embryos (biotin replete) were at stage 6 of development. The corresponding figure for explants from biotin-deficient embryos cultured in a biotin-deficient medium was 6.5%. If the deficient explants were cultured in a medium containing biotin (10^{-8} M), the percentage at stage 6 of development rose to >30. Administration of biotin (20 mg) to biotin-deficient dams 24 h prior to removal of the embryos resulted in 33% of the explants at stage 6 of development when cultured in a biotin-deficient medium, and >50% at stage 6 if cultured in a medium containing biotin (10^{-7} M).

In view of the role of biotin carboxylases in the major metabolic pathways, it was possible that the teratogenicity of biotin deficiency might be due to the accumulation of intermediary or secondary metabolites. However, there was no detrimental effect of any of the organic acid intermediates or secondary metabolites on palatal closure of the explants when these compounds were added to the organ culture medium at a concentration of 10^{-4} M. These results highlight the continuous requirement for biotin during the proliferation of the mesenchyme, perhaps for the synthesis of growth factors during organogenesis.

Biotin requirement for specific functions

Evidence of delayed spermatogenesis and decreased numbers of spermatozoa due to biotin deficiency in the rat has been reported in early literature. In mammals, spermatogenesis is dependent primarily upon testosterone, which is produced by Leydig cells and acts on the Sertoli and peritubular cells of the seminiferous tubules to drive spermatogenesis. We have shown (Paulose *et al.*, 1989) that testicular and serum levels of testosterone are decreased in the biotin-deficient rat. Biotin deficiency was accompanied by a significant degree of sloughing of the seminiferous tubule epithelium in these rats. Treatment of biotin-deficient rats with gonadotrophins or biotin increases serum levels of testosterone. However, even when testosterone levels are maintained at high levels in biotin-deficient rats by testosterone implants, the increase in serum testosterone does not result in normal spermatogenesis. The administration of biotin alone, or biotin in addition to testosterone, to biotin-deficient rats, leads to normal spermatogenesis, thus suggesting that biotin might be involved in the formation of local testicular factor(s) that are required in addition to testosterone and follicle-stimulating hormone for the normal interaction between Leydig, Sertoli and peritubular cells. The identity of these factors is not known, although we have shown that biotin is required for the synthesis of testicular proteins.

Brain-derived neurotrophic factor (BDNF) is a member of a family of cell signalling molecules (neurotrophins) that have an important role in neuronal development and plasticity. BDNF has been implicated in developmental and adult plasticity within telencephalic brain regions that control learned vocal behaviour in songbirds. BDNF expression seems to correlate with specific stages of songbird vocal learning. Johnson *et al.* (2000) have identified high levels of biotin in specific telencephalic nuclei (RA and HVC) among juvenile males. It is possible that there may be a specific up-regulation of biotinylated proteins within the HVA and RA nuclei in juvenile males. The developmental expression of this is correlated with vocal learning. High levels of biotin in the hippocampus, a brain region important for learning and memory, has also been reported (Wang and Pevsner, 1999), emphasizing the important role that biotin-regulated mechanisms might play in neuronal plasticity.

Biotin and the synthesis of biotin-related proteins

Biotin-binding proteins of egg yolk

Biotin in egg yolk is bound to two separate proteins, BBP I and BBP II, which are distinct from avidin. The concentrations of these proteins are directly related to dietary biotin content (White and Whitehead, 1987). At low concentrations of dietary biotin, BBP I is the major transporter, while at higher dietary biotin concentrations, BBP II predominates. Both these proteins are induced by sex hormones. Control by dietary biotin seems to override hormonal controls.

Biotin carboxylases

Rodriguez-Melandez *et al.* (1999) studied the effect of biotin on the genetic expression of rat liver mitochondrial carboxylases. In rats made biotin deficient, hepatic propionyl-CoA carboxylase (PPC) activity was <20% that in control rats. When biotin was added to primary cultures of hepatocytes from deficient rats, it took at least 24 h after the addition of biotin for PCC to reach control activity and biotinylation levels, whereas pyruvate carboxylase (PC) became active and fully biotinylated in the first hour after biotin addition to the culture. The enzyme mass of PC was minimally affected by biotin deficiency, whereas the masses of α-subunits of PCC and 3-methylcrotonyl CoA carboxylase (MCC) were substantially decreased in the deficient condition.

There was no significant change in the mRNA levels for these two proteins in deficient livers, suggesting that biotin regulates the expression of PCC and MCC at the post-transcriptional level. The amount of PC apocarboxylase was not affected in biotin deficiency.

Biotin holocarboxylase synthetase

Biotin, the prosthetic group of the biotin carboxylases, is covalently bound to the apocarboxylases by the holocarboxylase synthetase (HCS). HCS mRNA was significantly reduced in the biotin-deficient rat and increased following biotin administration to biotin-deficient rats, reaching control levels 24 h after biotin administration (Rodriguez-Melendez *et al.*, 2001). The results indicate the possibility of regulation by biotin at the transcriptional level. However, the long delay in the recovery of HCS mRNA levels might suggest an indirect action requiring metabolic processing affecting protein synthesis, endocrine or other signalling pathways. These reports provide evidence for biotin as a modulator of genetic expression of proteins involved in its transport or in its action as a prosthetic group.

Biotin and the immune system

Defects in the activity of biotin holocarboxylase synthetase or biotinidase, an enzyme that cleaves biotin off biotin proteins, result in multiple carboxylase deficiency syndrome in infants. This is an inherited disease of metabolism and is associated with clinical manifestations that include immune dysfunctions that respond to biotin administration (Roth *et al.*, 1982). Biotin-deficient rats had a weak antibody response to diphtheria toxoid. Deficient rats immunized with sheep erythrocytes had a decrease in the number of antibody-forming cells in the spleens. There was also a decrease in the size and cellularity of the thymus. In a study of the effects of biotin deficiency on indicators of immune function in mice Baez-Saldana *et al.* (1998) found significant changes in both the absolute number of spleen cells and in the proportion of spleen cells carrying different phenotypic markers. The percentage of cells (B cells) expressing surface immunoglobulin (sIg^+) decreased significantly. The mitogen-induced proliferation of spleen cells from deficient mice was lower compared with cells from control mice. Depressed T-cell proliferative response to mitogenic stimulus and a decrease in the population of B cells in the spleen affect the ability of the immune system to respond adequately to an antigenic challenge.

Lymphocytes are part of the immune system. They respond to antigenic stimulation with proliferation. Zempleni and Mock (1999, 2000, 2001) have studied the role of biotin in proliferating human lymphocytes. Biotin uptake into proliferating lymphocytes increased from three- to sevenfold compared with uptake in non-proliferating cells (see Chapter 5).

Biotinylation of histones

Proteolytic hydrolysis of biotin proteins yields biotin peptides, the smallest of them being biocytin, ε-N-biotinyl-L-lysine. Biocytin and synthetic biotin peptides are hydrolysed by biotinidase. The enzyme from human serum has been purified to homogeneity (Chauhan and Dakshinamurti, 1986). Although a role for biotinidase as a hydrolase for recycling biotin from small biotin peptides was generally accepted, our studies indicated that this may not be its major function because the optimum pH for biotinidase hydrolysis of biocytin is 4.5–5, whereas the pH of serum is 7.4. The concentration of biocytin in serum is in the nanomolar range, whereas the affinity constant of biotinidase for biocytin is in the micromolar range. We proposed (Chauhan and Dakshinamurti, 1988) that biotinidase functions as a biotin-carrier protein. Later work (Hymes et al., 1995) indicated that biotinidase itself is biotinylated by biocytin. They further showed that histone is a specific acceptor of biotin transferred from biotinylated biotinidase. This raised the possibility of histones being an endogenous substrate of biotinyl transfer (Hymes and Wolf, 1999). In further work, Stanley et al. (2001) have shown that biotinylation of histones occurs in vivo in human cells and that there is an increased biotinylation in response to cell proliferation (see Chapter 17). Histones in vivo are modified by actylation, methylation, phosphorylation, ubiquitination and poly(ADP-ribosylation) (Wolffe, 1998). Biotinylation of histones, by analogy with other modifications of histones, could lead to increased transcription of DNA (Sommerville et al., 1993; Pham and Sauer, 2000).

Regulation of the synthesis of specific proteins by biotin

Receptors for asialoglycoprotein and insulin

The asialoglycoprotein receptor (ASGR) is characteristic of fully differentiated hepatocytes. The human hepatoblastoma line HepG2 expresses maximal receptor activity only in confluent cultures. HepG2 cells grown to confluency in minimal essential medium (MEM) made 10% with respect to FBS demonstrated ASGR with ligand characteristics and molecular mass comparable with those of the receptor purified from human liver. However, dialysis or ultrafiltration, which removes the low molecular mass fraction of FBS, dramatically reduced expression of ASGR, under conditions where protein synthesis and total cellular protein content were comparable with those of control cells. The addition of biotin or biocytin to the medium supplemented with dialysed FBS during logarithmic growth supported normal expression of ASGR by confluent HepG2 cells (Collins et al., 1988). Biotin was most effective at a concentration of 10^{-8} M. Isolation of mRNA from HepG2 cells revealed no difference in ASGR transcripts when the cells were grown in MEM supplemented with 20% FBS or in MEM supplemented with 10% dialysed FBS, suggesting that a biotin-dependent post-translational event is responsible for the ultimate expression of ASGR by HepG2 cells.

De la Vega and Stockert (2000) have reported that the binding of ^{125}I-labelled insulin was decreased by >75% in human hepatoblastoma cell lines HepG2 or HuH-7 grown to confluence in MEM supplemented with dialysed FBS. The loss of cell surface insulin binding was due to a decrease in insulin receptor (IR). Similarly to the result for ASGR, repletion of dialysed FBS-supplemented medium with biotin fully restored expression of IR. Again, as for ASGR, there was no difference in the abundance of IR mRNA whether the cells were grown in MEM supplemented with FBS, dialysed FBS or dialysed FBS plus biotin. The results suggest that a biotin-mediated post-transcriptional event is responsible for the decrease in expression of IR.

The secondary messenger 8-bromo-cGMP mimicked the effect of biotin in a non-additive fashion, indicating that the effect of biotin was mediated through changes in cGMP levels via activation of guanylate cyclase (Vesely, 1982; Singh and Dakshinamurti, 1988). This is confirmed by other findings that the addition to the medium of atrial natriuretic factor or sodium nitroprusside, activators of particulate and soluble guanylate cyclase, respectively, resulted in restoration of ASGR synthesis (Stockert and Ren, 1997). The common downstream element in the biotin-mediated signal transduction pathway regulating the expressions of both ASGR and IR was identified as a cGMP-dependent protein kinase (cGk). De la Vega and Stockert (2000) postulate that phosphorylation of the coatomer protein α-COP in response to cGMP induction of cGk prevents the high-affinity binding of the coatomer complex to the 5′-untranslated region of the ASGR mRNA, allowing ribosomal scanning to the site of translation initiation. The molecular level at which cGk regulates IR expression is not yet known.

Induction of glucokinase

Dakshinamurti and Cheah-Tan (1968a,b) demonstrated that liver glucokinase activity was altered in response to the biotin status of rats. Biotin also played a role in the precocious development of glucokinase in young rats (Dakshinamurti and Hong, 1969). The synthesis of glucokinase is under developmental, nutritional and hormonal control (Meglasson and Matschinsky, 1984). The *in vivo* regulation of glucokinase mRNA by biotin was studied in starved rats, which have low glucokinase activity (Chauhan and Dakshinamurti, 1991). The concentration of glucokinase mRNA of biotin-injected starved rats increased about fourfold in comparison with normal-fed rats, as early as 1 h after the injection of biotin. This induction of glucokinase mRNA at 1 h following biotin administration to starved rats was 19-fold more than that of starved rats not receiving the biotin injection. Following insulin treatment of diabetic rats, the amount and activity of glucokinase increased, reaching 165% of non-diabetic control levels in 4 h. Although the actions of insulin and biotin are parallel, the effect of biotin on glucokinase induction seems to be more rapid than that of insulin. The relative rates of transcription of the glucokinase gene at various times following administration of biotin to starved rats were followed in 'run-on' transcription experiments. The transcription of the actin gene, used as an internal control, was not influencd by biotin. Biotin administration increased glucokinase gene transcription by about 6.7-fold.

Repression of phosphoenolpyruvate carboxykinase (PEPCK)

In liver and kidney, PEPCK participates in gluconeogenesis and is generally considered to be rate-limiting. Enzyme synthesis is subject to dietary and multihormonal controls (Granner and Pilkis, 1990). In both fasted and diabetic rats, hepatic PEPCK activities are markedly increased. The administration of biotin to starved rats decreased the PEPCK activity by 2.6-fold (Chauhan and Dakshinamurti, 1991). The regulation by biotin of PEPCK in the liver of diabetic rats was investigated (Dakshinamurti and Li, 1994). After biotin administration, hepatic PEPCK mRNA decreased to about 15% of the non-biotin-injected control level at 3 h. In parallel studies on insulin administration to diabetic rats, hepatic PEPCK mRNA decreased to about 10% of non-biotin-injected control levels at 3 h. In 'run-on' transcription experiments, biotin suppressed the transcription rates of the hepatic PEPCK gene by 55% at 30 min. The inhibition of transcription by biotin is dominant over other stimulatory events. The elevated plasma glucagon characteristic of the fasting or diabetic condition induces PEPCK synthesis through enhanced transcription of the PEPCK gene. There are many similarities between biotin and insulin in their action on enzymes of glucose metabolism. Both induce the mRNA that encodes glucokinase, a key glycolytic enzyme, and repress the mRNA that encodes PEPCK, a key gluconeogenic enzyme.

Sone *et al.* (1999) studied the effects of biotin on insulin secretion in biotin-deficient and control rats using isolated pancreatic perfusion. The insulin response to 20 mM glucose in biotin-deficient rats was about 20% of that seen in control rats. The insulin response to 20 mM glucose plus 1 mM biotin in biotin-deficient and control rats increased to 165 and 185%, respectively, over that to 20 mM glucose infusion. Biotin has an effect that reinforces the glucose-induced insulin response.

Romero-Navarro et al. (1999) have demonstrated that biotin stimulates glucokinase activity in rat pancreatic islets in culture. Using a branched DNA assay, they have shown relative increases in glucokinase mRNA levels of about 40 and 80% at 12 and 24 h, respectively, in islets treated with 10^{-6} M biotin. Treatment with biotin (10^{-6} M) also increased insulin secretion. Islet glucokinase activity and mRNA were reduced by 50% in biotin-deficient rats. Insulin secretion in response to glucose was also impaired in islets isolated from the deficient rat, indicating that biotin affects pancreatic islet glucokinase activity and expression as well as insulin secretion in cultured islets.

Glucose is the major insulin secretagogue. Its stimulation of insulin release follows its metabolism. Matchinsky (1996) has suggested that glucokinase is the glucose sensor and metabolic signal generator in pancreatic β-cells. The influence of glucokinase over the secretion of insulin by pancreatic β-cells seems to be increased by biotin through a transcriptional effect on the synthesis of glucokinase.

Vitamin B_6

Vitamin B_6 is metabolized to the active form, pyridoxal-5-phosphate (PLP), which serves as a coenzyme for >60 enzymes. In addition, the requirement for PLP for the activity of glycogen phosphorylase makes it a versatile factor in the metabolism of carbohydrates, lipids and amino acids. Vitamin B_6, through PLP, has a role in various systems such as the endocrine, neuroregulatory, immune and cardiovascular systems. Functions for PLP other than as just the cofactor of PLP-dependent enzymes have been recognized. Of particular significance is its role in gene expression.

Pyridoxal phosphate in the cell nucleus

Meisler and Thanassi (1990) demonstrated the presence of PLP as well as PLP-binding protein in the nucleus of rat hepatoma-derived HTC cells. In rats fed a diet adequate in vitamin B_6, the fraction of total cellular PLP found in the nucleus was 21%. This increased to 39% in rats fed a vitamin B_6-deficient diet, indicating a conservation of the vitamin in the nuclear compartment. This is analogous to our observation (Dakshinamurti and Mistry, 1963a) regarding the distribution of biotin in the cell nucleus. The physiological significance of the presence of PLP in the cell nucleus was emphasized by Meisler and Thanassi (1990) in view of earlier reports (Litwack, 1988; Maksymowyck et al., 1990) that PLP modulated steroid hormone activity by altering the interaction of steroid receptor complex with DNA, chromatin and nuclei.

Vitamin B_6 and cell proliferation

PBMCs and murine spleenocytes have been used as model systems for studying the nutritional and hormonal requirements for growth of cells in culture. Mathews et al. (1994) have developed a serum-free and protein-free medium that will support the short-term growth of human peripheral blood lymphocytes in response to mitogen stimulation. A similar medium for growth of murine splenocytes has also been defined. Using [^{3}H]thymidine incorporation into cells in vitro as a reflection of cell growth, they found that elimination of vitamin B_6 from the diet of mice resulted in diminished growth response that correlated only with the elimination of vitamin B_6 from the culture medium.

Vitamin B_6-deficient rodents have been reported to show impaired primary and secondary antibody production and a reduced number of antibody-producing cells. A significant decrease in thoracic duct lymphocytes, in vitro lymphocyte proliferation and T-cell mediated cytotoxicity have also been reported (Grimble, 1997).

PLP and steroid hormone action

Cells grown in the presence of 5 mM pyridoxine have a decreased glucocorticoid-dependent induction of enzymes such as tyrosine aminotransferase (Disorbo and Litwack, 1981). Allgood et al. (1990) have shown that vitamin B_6 regulates transcriptional activation by human glucocorticoid receptor in HeLa cells. This modulatory role in transcription is not restricted to glucocorticoid receptor but extends to other members of the steroid hormone superfamily (Allgood and

Cidlowski, 1992). The intracellular concentration of PLP could have a profound influence on steroid-induced gene expression, with increased PLP levels resulting in a decreased transcriptional response to various steroid hormones, and vice versa (Tully et al., 1994).

PLP and gene expression of liver enzymes

Evidence has been provided by Natori and colleagues that PLP interacts directly with glucocorticoid receptor and modulates cytosolic aspartate aminotransferase (cAST) gene expression in rat liver (Oka et al., 1995). They have shown that the amount of cAST mRNA in the liver of vitamin B_6-deficient rats is sevenfold higher than that in control rats although the immunochemically determined cAST protein content was the same in deficient and control livers. This has been shown by them to be related to the accelerated degradation of cAST by preferential sequestration of the apoenzyme into lysosomes. Glucocorticoid receptor binds to the genes of enzymes induced at glucocorticoid-responsive elements and increases their transcription. The binding of glucocorticoid receptor to glucocorticoid receptor elements was studied. Inactivation of the DNA-binding activity of liver extract upon incubation *in vitro* with PLP indicates that the modulation of cAST gene expression is due to the direct interaction of glucocorticoid with PLP.

In similar experiments, Sato et al. (1996) have shown that cystathionase mRNA was increased several fold in the liver of vitamin B_6-deficient rats. There was a corresponding increase in the amount of cystathionase in the lysosomes of deficient rat livers suggesting that the apoenzyme is recognized specifically for sequestration into lysosomes. This explains why the concentration of cystathionase protein is unchanged in vitamin B_6-deficient rat livers although the expression of the cAST gene is increased several fold.

Vitamin B_6 and expression of the albumin gene

A general increase in gene expression that includes housekeeping genes such as β-actin and glyceraldehyde-3-phosphate is seen in livers of vitamin B_6-deficient rats (Oka et al., 1993). This has been ascribed to the activation of RNA polymerases I and II in the deficient liver. Among the vitamin B_6-independent proteins, the expression of the albumin gene is increased in vitamin B_6 deficiency. Albumin mRNA content was increased sevenfold over controls in deficient rat liver. This increase was far greater than the increase in RNA polymerase activity (Oka et al., 1995). It has been proposed that vitamin B_6 modulates the expression of the albumin gene through a mechanism involving inactivation of tissue-specific transcription factors by direct interaction with PLP (Natori et al., 2000).

Vitamin B_6 and cancer

Natori et al. (2000) have shown that the growth of hepatoma HepG2 cells was inhibited by the addition of pyridoxine or pyridoxal to the culture medium. There was concurrent inhibition of protein synthesis and secretion of albumin. They also report that the growth of MH-134 hepatoma cells, transplanted into C3H/He mice, was significantly reduced by the administration of large amounts of pyridoxine to mice. Other reports indicate that vitamin B_6 suppressed azoxymethane-induced colon cancer (Komatsu et al., 2001).

Vitamin C

The antioxidant activity of ascorbic acid (vitamin C) is the best recognized function of this vitamin. Vitamin C and vitamin E eliminate cytotoxic free radicals by redox cycling. The role of ascorbic acid in the synthesis of extracellular matrix (ECM) proteins is well recognized. Vitamin C functions in the hydroxylation of proline and lysine residues in collagen, elastin, C1q of complement and acetylcholinesterase. Ascorbic acid also functions in other cellular processes such as neuromodulation, hormone and neurotransmitter syntheses and the immune system. Apart from its functions as an antioxidant and as a cofactor in hydroxylations, recent work has focused on its direct role in gene expression.

The role of vitamin C in extracellular matrix production

Ascorbic acid was recognized initially as an antiscorbutic vitamin. Vitamin C deficiency in humans leads to scurvy, a disease characterized by deficient ECM production. Delayed wound healing is a feature of vitamin C deficiency. Most animals, with the exception of primates, guinea pigs and the bat, are able to sythesize vitamin C and hence do not require vitamin C in their diet. In scorbutic guinea pigs, the administration of ascorbic acid stimulated proline hydroxylation in collagen. In these animals, proteoglycan synthesis is also impaired in the cartilage.

Collagen is a structural protein found in tissues such as bone, cartilage, tendon and skin. The collagens are characterized by a triple-helical structure. For the three peptides to form a stable helix, a significant number of proline residues in the peptide chain have to be hydroxylated by the enzyme prolyl hydroxylase, which requires ascorbic acid. Elastin, another ECM protein, is found in elastic tissues and also contains hydroxylated proline residues. In addition to its role in the hydroxylation reaction, vitamin C stimulates the synthesis of collagen in a number of cell types. Elastin synthesis is also affected by vitamin C.

Requirements for vitamin C for cell proliferation and cell differentiation

As vitamin C is not an essential micronutrient for most species and as most culture media contain a serum supplement, the requirement for vitamin C for cells in culture was not considered until it was demonstrated that mouse plasmacytoma cells responded to ascorbate added to the culture medium. Such a requirement is seen for cells derived from vitamin C-dependent species (Park and Kimler, 1991). There have been reports of cytotoxicity due to ascorbic acid added to the culture medium. Cytotoxicity is ascribed to the ascorbate-mediated generation of peroxides and oxygen free radicals in unprotected media. Stimulation of proliferation by ascorbic acid is not uniform even in closely related cells. The human promyeloic tumour cell line HL-60 responds positively to vitamin C (Alcain et al., 1990), whereas bone marrow cells from a variety of acute myeloid leukaemic patients responded unevenly (Park et al., 1992). Mesenchymal cells, which form connective tissue, respond to ascorbic acid with proliferation, whereas embryonic fibroblasts showed growth arrest. The addition of ascorbic acid to cultured osteoblast-like cells induces differentiation and expression of specific genes associated with the osteoblast phenotype, such as the L/K/B isoenzyme of alkaline phosphatase and osteocalcin. This follows the stimulation of collagen secretion (Franceshi et al., 1994).

Vitamin C and chrondrogenesis

Cartilage is replaced by bone-forming cells during endochondral longitudinal bone growth at the epiphysis and fracture healing. During the developmental process, cartilage tissue undergoes orderly morphological changes involving proliferation, cellular hypertrophy, production of ECM and mineralization (Brigelius-Flohe and Flohe, 1996). This is accompanied by biochemical changes (Castagnola et al., 1988). Immature cells produce type I and type II collagen and fibronectin. Proliferating chondrocytes synthesize and secrete large amounts of type II and IX collagen as well as cartilage-specific proteoglycans. Hypertrophic, non-dividing cells produce type X collagen. Mineralizing tissue expresses alkaline phosphatase and type X collagen. Those changes in collagen types are accompanied by changes in their mRNA level. Proliferation of pre-hypertrophic cells and their differentiation to hypertrophic cells require vitamin C (Gerstenfeld and Landis, 1991). The increase in alkaline phosphatase activity followed the expression of collagen type X. Mineralization in chondrocytes was also ascorbate dependent. Although vitamin C is not necessary for general cell growth, it is essential for the production of the ECM.

Vitamin C and collagen gene expression

Vitamin C increases the amount of procollagen mRNA by stimulating the transcription of the procollagen gene and by stabilizing the mRNA. Ascorbic acid treatment of human skin fibroblasts increased procollagen mRNA twofold. In primary avian tendon (PAT) cells, there was a similar sixfold increase of procollagen mRNA. In PAT cells treated with α,α-dipyridyl, an inhibitor of hydroxylase, there was a significant decrease in the rate of procollagen synthesis without any inhibitory effect on procollagen mRNA. These results suggest that ascorbic acid might directly increase the transcription of the procollagen gene through interaction with an ascorbic acid-specific *cis*-regulatory element on the procollagen genes (Kurata *et al.*, 1993).

Vitamin C and non-collagen gene expression

Related to the role of ascorbic acid in normal ECM production is the report of the action of ascorbic acid on type IV collagenase (matrix metalloproteinase-2; MMP-2). A low intake of vitamin C during pregnancy is associated with a high risk of premature rupture of the membrane (PROM). In addition to its role in the expression of the collagen gene, Pfeffer *et al.* (1998) have reported a new effect of ascorbic acid as a modulator of MMP-2 gene expression at the level of transcription, and suggest that collagen degradation might be a significant factor in PROM.

Skin fibroblasts from patients with Ehlers–Danlos syndrome type VI (EDS VI), an inherited connective tissue disorder, have decreased activity of lysyl hydroxylase (LH). LH hydroxylates lysine residues in peptide linkage, a key step in post-translational modification of collagen. EDS patients with LH deficiency have extreme joint hypermobility, skin hyperextensibility, bruisability and delayed wound healing. Treatment of skin fibroblasts from EDS patients with ascorbic acid increased LH levels, which paralleled LH mRNA levels. Ascorbate increased total collagen production twofold (Yeowell *et al.*, 1995).

Ascorbic acid deficiency was studied in ODS rats (genotype *od/od*) with an inherited defect in ascorbic acid synthesis (Ikeda *et al.*, 1996, 1997, 1998). Ascorbic acid deficiency decreased the serum apolipoprotein A-1 concentration mediated by a decrease in its mRNA levels. Ascorbic acid deficiency also decreased the renal level of kidney fatty acid-binding protein and the serum level of α_{2M}-globulin. The hepatic level of α_{2M}-globulin mRNA of the ascorbic acid-deficient rat was significantly reduced. The kidney fatty acid-binding protein is a proteolytic fragment of α_{2M}-globulin, which is synthesized in the liver. In further work, the effects of ascorbic acid deficiency in ODS rats on the hepatic gene expression of the positive acute phase proteins, haptoglobin and α_1-acid glycoprotein, as well as negative acute phase proteins, apolipoprotein A-1 and albumin, were studied. Ascorbic acid deficiency significantly elevated the serum concentration of haptoglobin and lowered the concentrations of apolipoprotein A-1 and albumin. The increases in haptoglobin and α_1-acid glycoprotein corresponded to increases in their mRNA levels. Correspondingly, the decrease in apolipoprotein A-1 and albumin matched decreases in hepatic mRNA levels of these proteins in ascorbic acid-deficient ODS rats.

Ascorbic acid has a significant role in catecholamine synthesis in the sympathetic nervous system as a cofactor of dopamine-β-hydroxylase. Seitz *et al.* (1998) have reported that incubation of the neuroblastoma cell line SK-N-SH with ascorbic acid for 2 h results in enhanced synthesis of 3,4-dihydroxyphenylalanine (DOPA) and dopamine. After 5 days of incubation of the cells with high levels of ascorbic acid, there was a threefold increase of tyrosine hydroxylase gene expression, whereas dopamine β-hydroxylase gene expression was not altered. The results suggest that ascorbic acid might be a useful adjunct in the treatment of early Parkinson's disease.

Mori *et al.* (1992, 1997) reported that ascorbic acid deficiency in guinea pigs caused decreases in the amount of various forms of cytochrome P450 in liver microsomes. Ascorbic acid deficiency decreases the expression of cytochrome P450 mRNA, specifically for the 1A1 and 1A2 subtypes, indicating that transcription of P450 is regulated by ascorbic acid in guinea pigs.

Bowie and O'Neill (2000) have shown that vitamin C inhibits NF-κB activation by tumour necrosis factor (TNF) through activation of p38 mitogen-activated protein kinase (MAPK). NF-κB is a eukaryotic transcription factor that regulates the expression of a number of genes involved in immune and inflammatory responses. It is activated by pathogenic signals including the pro-inflammatory cytokines interleukin-1 (IL-1) and TNF. In the cell cytosol, NF-κB exists in a latent form consisting of a transcriptionally active dimer (p50–p65) bound to an inhibitor protein IκB. Upon stimulation by cytokine inducers, IκB is phosphorylated by IκB kinase (IKK) and is then ubiquitinated and degraded by 26S proteosomes.

The released NF-κB dimer then translocates to the nucleus, binds to κB elements in the promoters and activates target gene expression. Vitamin C inhibits cytokine stimulation of NF-κB. This is not due to an antioxidant effect of vitamin C as other antioxidants do not mimic or augment the action of vitamin C. The effect of vitamin C is mediated through activation of stress-activated protein kinase p38 MAPK. It is not known yet how vitamin C activates p38 MAPK. Activated p38 MAPK inhibits TNF activation of NF-κB as well as the phosphorylation of IκB and activation of IKK.

A role for vitamin C in augmenting the chemotherapeutic reponse of cervical carcinoma HeLa cells has been proposed by Reddy et al. (2001). Human papilloma virus (HPV) is involved in the pathogenesis of cervical cancer. The oncopotential of high-risk HPV types is due to E6 and E7 oncoproteins, which transform the infected cell. The E6 oncoprotein targets p53 protein, which is ubiquitinated and degraded. Cell cycle regulation is affected. These oncoproteins also activate telomerase, contributing to the immortalization of transformed cells. HPV-positive HeLa cells were incubated with millimolar concentrations of vitamin C to study its effect on the transcriptional regulation of HPV E6/7 oncogene expression. Vitamin C treatment resulted in a down-regulation of the viral oncoprotein E6, which paralleled a decrease of activator protein (AP-1) members c-jun and c-fos in a time- and dose-dependent manner. The down-regulation of E6 was associated with an up-regulation of the proapoptotic p53 and Bax proteins, and a down-regulation of apoptosis inhibitor Bcl-2.

Catani et al. (2001) have shown that vitamin C mediates cellular responses to counter ultraviolet (UV) radiation-mediated cell damage and cell death. Ascorbic acid rescues keratinocytes from UV-mediated cytotoxicity directly by scavenging free radical species. In addition, ascorbic acid induces transcriptional down-regulation of IL-1α mRNA. Oxidative stress due to UV raditation activates redox-sensitive transcription factors including NF-κB and members of the AP-1 complex such as c-jun and c-fos. Ascorbate inhibits the AP-1-dependent transactivation of specific promoters. Ascorbic acid is a negative regulator of AP-1 and AP-1-dependent transcription.

Aromatic amine carcinogens undergo N-oxidation catalysed by hepatic cytochrome P450 to form reactive metabolites. These metabolites bind to tissue or organ macromolecules such as DNA to form adducts. The formation of such DNA–carcinogen adducts is an important step in the initiation of arylamine-induced carcinogenesis in the target tissue. Hung and Lu (2001) have shown that vitamin C decreased arylamine–DNA adduct formation in C6 glioma cells.

References

Alcain, F.J., Buron M.I., Rodriguez-Aguilera, J.C., Villalba, J.M.Z. and Navas, P. (1990) Ascorbate free radical stimulates the growth of a human promyelocytic leukemia cell line. *Cancer Research* 50, 5887–5891.

Allgood, V.E. and Cidlowski, J.A. (1992) Vitamin B6 modulates transcriptional activation by multiple members of the steroid hormone receptor superfamily. *Journal of Biological Chemistry* 267, 3819–3824.

Allgood, V.E., Powell-Oliver, F.E. and Cidlowski, J.A. (1990) Vitamin B6 influences glucocorticoid receptor-dependent gene expression. *Journal of Biological Chemistry* 265, 12424–12433.

Baez-Saldana, A. Diaz, G., Espinoza, B. and Ortega, E. (1998) Biotin deficiency induces changes in subpopulations of spleen lymphocytes in mice. *American Journal of Clinical Nutrition* 67, 431–437.

Bernlohr, D.A., Bolanowski, M.S., Kelly, T.J. Jr and Lane, M.D. (1985) Evidence for an increase in transcription of specific mRNA during differentiation of 373-L1 preadiposites. *Journal of Biological Chemistry* 260, 5563–5567.

Bhullar, R.P. and Dakshinamurti, K. (1985) The effects of biotin on cellular functions of HeLa cells. *Journal of Cellular Physiology* 122, 425–430.

Boeckx, R.L. and Dakshinamurti, K. (1974) Biotin-mediated protein synthesis. *Biochemical Journal* 140, 549–556.

Boeckx, R.L. and Dakshinamurti, K. (1975) Effect of biotin on RNA synthesis. *Biochimica et Biophysica Acta* 383, 282–289.

Bowie, A.G. and O'Neill, L.A.J. (2000) Vitamin C inhibits NF-κB activation by TNF via the activation of p38 mitogen-activated protein kinase. *Journal of Immunology* 165, 7180–7188.

Brigelius-Flohe, R. and Flohe, L. (1996) Ascorbic acid, cell proliferation and cell differentiation in culture. *Subcellular Biochemistry* 25, 83–107.

Castagnola, P., Dozin, B., Moro, G. and Cancedda, R. (1988) Changes in the expression of collagen genes show two stages in chondrocyte differentiation in vitro. *Journal of Cell Biology* 106, 461–467.

Catani, M.V., Ross, A., Costanzo, A., Sabatini, S. and Levrero, M. (2001) Induction of gene expression via activator protein-1 in the ascorbate protection against 4v-induced damage. *Biochemical Journal* 356, 77–85.

Chalifour, L.E. and Dakshinamurti, K. (1982a) The requirement of human fibroblasts in culture. *Biochemical and Biophysical Research Communications* 104, 1047–1053.

Chalifour, L.E. and Dakshinamurti, K. (1982b) The characterization of the uptake of avidin–biotin complex by HeLa cells. *Biochimica et Biophysica Acta* 721, 64–69.

Chauhan, J. and Dakshinamurti, K. (1986) Purification and characterization of human serum biotinidase. *Journal of Biological Chemistry* 261, 4268–4275.

Chauhan, J. and Dakshinamurti, K. (1988) Role of human serum biotinidase as biotin-binding protein. *Biochemical Journal* 256, 365–270.

Chauhan, J. and Dakshinamurti, K. (1991) Transcriptional regulation of the glucokinase gene by biotin in starved rats. *Journal of Biological Chemistry* 266, 10035–10038.

Collins, J.C., Paietta, E., Green, R., Morell, A.G. and Stockert, R.J. (1988) Biotin-dependent expression of the asialoglycoprotein receptor in HepG2. *Journal of Biological Chemistry* 263, 11280–11283.

Dakshinamurti, K. (1997) Vitamin receptors. In: Myers, R.A. (ed.) *Encyclopedia of Molecular Biology and Molecular Medicine*, Vol. 6. VCH Verlagsgesellschaft mbH, Weinheim, pp. 235–244.

Dakshinamurti, K. and Chalifour, L.E. (1981) The biotin requirement of HeLa cells. *Journal of Cellular Physiology* 107, 427–438.

Dakshinamurti, K. and Chauhan, J. (1989) Biotin. *Vitamins and Hormones* 45, 337–384.

Dakshinamurti, K. and Chauhan, J. (1994) Biotin binding proteins. In: Dakshinamurti, K. (ed.) *Vitamin Receptors*. Cambridge University Press, Cambridge, pp. 200–249.

Dakshinamurti, K. and Cheah-Tan, C. (1968a) Liver glucokinase of the biotin-deficient rat. *Canadian Journal of Biochemistry* 46, 75–80.

Dakshinamurti, K. and Cheah-Tan, C. (1968b) Biotin-mediated synthesis of hepatic glucokinase in the rat. *Archives of Biochemistry and Biophysics* 127, 17–21.

Dakshinamurti, K. and Hong, H.C. (1969) Regulation of key glycolytic enzymes. *Enzymologia Biologica Clinica* 11, 422–428.

Dakshinamurti, K. and Li, W. (1994) Transcriptional regulation of liver phosphoenolpyruvate carboxykinase by biotin in diabetic rats. *Molecular and Cellular Biochemistry* 132, 127–132.

Dakshinamurti, K. and Mistry, S.P. (1963a) Tissue and intracellular distribution of biotin-$C^{14}OOH$ in rats and chickens. *Journal of Biological Chemistry* 238, 294–296.

Dakshinamurti, K. and Mistry, S.P. (1963b) Amino acid incorporation in biotin deficiency. *Journal of Biological Chemistry* 238, 297–301.

Dakshinamurti, K., Chalifour, L. and Bhullar, R.P. (1985) Requirement for biotin and the function of biotin in cells in culture. *Annals of the New York Academy of Sciences* 447, 38–55.

De La Vega, L.A. and Stockert, R.J. (2000) Regulation of the insulin and asialoglycoprotein receptors via cGMP-dependent protein kinase. *American Journal of Physiology* 279, C2037–C2042.

Disorbo, D.M. and Litwack, G. (1981) Changes in the intracellular levels of pyridoxal-5-phosphate affect the induction of tyrosine aminotransferase by glucocorticoids. *Biochemical and Biophysical Research Communications* 99, 1203–1208.

Franceschi, R.T., Iyer, B.S. and Cui, Y. (1994) Effects of ascorbic acid on collagen matrix formation and osteoblast differentiation in murine MC3T3-F1 cells. *Journal of Bone and Mineral Research* 9, 843–854.

Gerstenfeld, L.C. and Landis, W.J. (1991) Gene expression and extracellular matrix ultrastructure of a mineralizing chondrocyte cell culture system. *Journal of Cell Biology* 112, 501–513.

Gospodarowicz, D. and Cohen, D.C. (1985) Biotin and choline replace the growth requirement of Madin-Darby canine kidney cells for high-density lipoprotein. *Journal of Cellular Physiology* 124, 96–106.

Granner, D. and Pilkes, S. (1990) The genes of hepatic glucokinase metabolism. *Journal of Biological Chemistry* 265, 10173–10176.

Grimble, R.F. (1996) Interaction between nutrients, proinflammatory cytokines and inflammation. *Clinical Science (Lond)* 91, 121–130.

Hung, C.F. and Lu, K.H. (2001) Vitamin C inhibited DNA adduct formation and arylamine N-acetyltransferase activity and gene expression in rat glial tumor cells. *Neurochemical Research* 26, 1107–1112.

Hymes, J. and Wolf, B. (1999) Human biotinidase isn't just for recycling biotin. *Journal of Nutrition* 129, 485S–489S.

Hymes, J., Fleischhauer, K. and Wolf, B. (1995) Biotinylation of biotinidase following incubation with biocytin. *Clinica Chimica Acta* 233, 39–45.

Ikeda, S., Horio, F., Yoshida, A. and Kakinuma, A. (1996) Ascorbic acid deficiency reduces hepatic apolipoprotein A-1 mRNA in scurvy-prone ODS rats. *Journal of Nutrition* 126, 2505–2511.

Ikeda, S., Takasu, M., Satsuda, T. and Kakinyma A. (1997) Ascorbic acid deficiency decreases the renal level of kidney fatty acid-binding protein by lowering the α_{2U}-globulin gene expression in liver in scurvy-prone ODS rats. *Journal of Nutrition* 127, 2173–2178.

Ikeda, S., Horio, F. and Kakinuma, A. (1998) Ascorbic acid deficiency changes hepatic gene expression of acute phase proteins in scurvy-prone ODS rats. *Journal of Nutrition* 128, 832–838.

Johnson, R., Norstrom, E. and Soderstrom, K. (2000) Increased expression of endogenous biotin, but not

BDNF, in telencephalic song regions during zebra finch vocal learning. *Developmental Brain Research* 120, 113–123.

Keranan, A.J.A. (1972) The biotin synthesis of HeLa cells in vivo. *Cancer Research* 32, 119–124.

Komatsu, S.I., Watanabe, H., Oka, T., Tsuge, H., Nii, H. and Kato, N. (2001) Vitamin B_6-supplemented diets compared with a low vitamin B_6 diet suppresses azomethane-induced colon tumorigenesis in mice by reducing cell proliferation. *Journal of Nutrition* 131, 2204–2207.

Kurata, S., Senoo, H. and Hata, R. (1993) Transcriptional activation of type I collagen genes by ascorbic acid-2-phosphates in human skin fibroblasts and its failure in cells from a patient with alpha 2(I)-chain-defective Ehlers–Danlos syndrome. *Experimental Cell Research* 206, 63–71.

Kuri-Harcuch, W. and Marsch-Morino, M. (1983) DNA synthesis and cell division related to adipose differentiation of 3T3 cells. *Journal of Cellular Physiology* 114, 39–44.

Kuri-Harcuch, W., Wise, L.S. and Green, H. (1978) Interruption of adipose conversion of 3T3 cells by biotin deficiency: differentiation without triglyceride accumulation. *Cell* 14, 53–59.

Litwack, G. (1988) The glucocorticoid receptor at the protein level. *Cancer Research* 48, 2636–2640.

Matchinsky, F.M. (1996) A lesson in metabolic regulation inspired by the glucokinase glucose sensor paradigm. *Diabetes* 45, 223–242.

Mathews, K.S., Mrowczynski, E. and Mathews, R. (1994) Dietary deprivation of B-vitamins reflected in murine splenocyte proliferation *in vitro*. *Biochemical and Biophysical Research Communications* 198, 451–458.

Meglasson, M.D. and Matchinsky, F.M. (1984) New perspective on pancreatic islet glucokinase. *American Journal of Physiology* 246, E1–E13.

Meisler, N.T. and Thanassi, J.W. (1990) Pyridoxine-derived vitamin B_6 vitamers and pyridoxal-5-phosphate-binding protein in cytosolic and nuclear fractions of HTC cells. *Journal of Biological Chemistry* 265, 1193–1198.

Mori, T., Kitamura, R., Imaoka, S., Funae, Y., Kitada, M. and Kamataki, T. (1992) Examination for lipid peroxidation in liver microsomes of guinea pigs as a causal factor in the decrease in the content of cytochrome P450 due to ascorbic acid deficiency. *Research Communications in Chemical Pathology and Pharmacology* 75, 209–219.

Mori, T., Itoh, S., Ohgiya, S., Ishizaki, K. and Kamataki, T. (1997) Regulation of CYP1A and CYP3A mRNAs by ascorbic acid in guinea pigs. *Archives of Biochemistry and Biophysics* 348, 268–277.

Moskowitz, M. and Cheng, D.K.S. (1985) Stimulation of growth factor production in cultured cells by biotin. *Annals of the New York Academy of Sciences* 447, 212–221.

Nakatani, Y., Kitamura, H., Inayama, Y. and Ogawa, N. (1994) Pulmonary endodermal tumor resembling fetal lung. The optically clear nucleus is rich in biotin. *American Journal of Surgical Pathology* 18, 637–642.

Natori, Y., Oka, T. and Kuwahata, M. (2000) Modulation of gene expression by vitamin B_6. In: Iriavte, A., Kagan, H.M. and Martinez-Carrion, M. (eds) *Biochemistry and Molecular Biology of Vitamin B_6 and PQQ-dependent Proteins*. Birkhauser Verlag, Basel, pp. 301–306.

Oka, T., Komori, N., Kuwahata, M., Sasa, T., Suzuki, I., Okada, M. and Natori, Y. (1993) Vitamin B6 deficiency causes activation of RNA polymerase and general enhancement of gene expression in rat liver. *FEBS Letters* 330, 409–413.

Oka, T., Komori, N., Kuwahata, M., Hiroi, Y., Shimoda, T., Okada, M. and Natori, Y. (1995) Pyridoxal 5-phosphate modulates expression of cytosolic aspartate amino transferase gene by inactivation of glucocorticoid receptor. *Journal of Nutritional Science and Vitaminology* 41, 363–375.

Okamoto, Y., Kashima, K., Daa, T., Yokyama, S., Nakayama, I. and Noguchi, S. (1995) Morule with biotin-containing intranuclear inclusions in thyroid carcinoma. *Pathology International* 45, 573–579.

Ozyhar, A., Kiltz, H.H. and Pongs, O. (1990) Pyridoxal phosphate inhibits the DNA-binding activity of the ecdysteroid receptor. *European Journal of Biochemistry* 192, 167–174.

Park, C.H. and Kimler, B.F. (1991) Growth modulation of human leukemic, pre-leukemic and myeloma progenitor cells by L-ascorbic acid. *American Journal of Clinical Nutrition* 54, 1241S–1246S.

Park, C.H., Kimler, B.F., Bodensteiner, D., Lynch, S.R. and Hassanien, R.S. (1992) *In vitro* growth modulation by L-ascorbic acid of colony forming cells from bone marrow of patients with myelodysplastic syndromes. *Cancer Research* 52, 4458–4466.

Paulose, C.S., Thliveris, J., Viswanathan, M. and Dakshinamurti, K. (1989) Testicular function in biotin-deficient adult rats. *Hormone and Metabolic Research* 21, 661–665.

Pfeffer, F., Casanuera, E., Kamer, J., Guerra, A. and Perichart, O. (1998) Modulation of 72 kilodalton type IV collagenase (matrix metalloproteinase-2) by ascorbic acid in cultured human amnion-derived cells. *Biology of Reproduction* 59, 326–329.

Pham, A.D. and Sauer, F. (2000) Ubiquitin-activating/conjugating activity of $TAF_{II}250$, a mediator of activation of gene expression in *Drosophila*. *Science* 289, 2357–2360.

Reddy, V.G., Khanna, N. and Singh, N. (2001) Vitamin C augments chemotherapeutic response of cervical carcinoma HeLa cells by stabilizing p53. *Biochemical and Biophysical Research Communications* 282, 409–415.

Rodriguez-Melandez, R., Perez-Andrade, M.E., Diaz, A., Deolarte, A., Camacho-Arroyo, Ciceron, I.,

Ibarra, I. and Velazquez, A. (1999) Different effects of biotin deficiency and replenishment on rat liver pyruvate and propionyl-CoA carboxylases and on their mRNAs. *Molecular Genetics and Metabolism* 66, 16–23.

Rodriguez-Melendez, R., Cano, S., Mendez, S.T. and Velazquez, A. (2001) Biotin regulates the genetic expression of holocarboxylase synthetase and mitochondrial carboxylases in rats. *Journal of Nutrition* 131, 1909–1913.

Romero-Navarro, G., Cabrera-Valladares, G., German, M.S., Matchinsky, F.M., Velazquez, A., Wang, J. and Fernandez-Mejia, C. (1999) Biotin regulation of pancreatic glucokinase and insulin in primary cultured rat islets and in biotin-deficient rats. *Endocrinology* 140, 4595–4600.

Roth, K.S., Yang, W., Alan, L., Saunders, J., Gravel, R.A. and Dakshinamurti, K. (1982) Prenatal administration of biotin in biotin responsive multiple carboxylase deficiency. *Pediatric Research* 16, 126–129.

Sasaki, A., Yokoyama, S., Arita, R., Inomata, M., Kashima, K. and Nakayama, I. (1999) Morules with biotin-containing optically clear nuclei in colonic tubular adenoma. *American Journal of Surgical Pathology* 23, 336–341.

Sato, A., Nishioka, M., Awata, S., Nakayama K., Okada, M., Horiuchi, S., Okabe, N., Sassa, T., Oka, T. and Natori, Y. (1996) Vitamin B6 deficiency accelerates metabolic turnover of cystathionase in rat liver. *Archives of Biochemistry and Biophysics* 330, 409–413.

Seitz, G., Gerbhardt, S., Beck, J.F., Bohm, W., Lode, H.N., Niethammer, D. and Bruchelt, G. (1998) Ascorbic acid stimulates DOPA synthesis and tyrosine hydroxylase gene expression in the human neuroblastoma cell line SK-N-SH. *Neuroscience Letters* 244, 33–36.

Sickel, J.Z. and Sant'Agnese, P.A. (1994) Ananalous immunostaining of 'optically clear' nuclei in gestational endometrium: a potential pitfall in the diagnosis of pregnancy-related herpesvirus infection. *Archives of Pathology and Laboratory Medicine* 118, 831–833.

Singh, I.N. and Dakshinamurti, K. (1988) Stimulation of guanylate cyclase and RNA polymerase II activities in HeLa cells and fibroblasts by biotin. *Molecular and Cellular Biochemistry* 79, 47–55.

Sommerville, J., Baird, J. and Turner, B.M. (1993) Histone H4 acetylation and transcription in amphibian chromatin. *Journal of Cell Biology* 120, 277–290.

Sone, H., Ito, M., Suguyama, K., Ohneda, M., Maebashi, M. and Furukawa, Y. (1999) Biotin enhances glucose stimulated insulin secretion in the isolated perfused pancreas of the rat. *Journal of Nutritional Biochemistry* 10, 237–243.

Stanley, J.S., Griffin, J.B. and Zempleni, J. (2001) Biotinylation of histones in human cells. Effects of cell proliferation. *European Journal of Biochemistry* 268, 5424–5429.

Stockert, R.J. and Ren, Q. (1997) Cytoplasmic protein mRNA interaction mediate cGMP-modulated translational control of the asialoglycoprotein receptor. *Journal of Biological Chemistry* 272, 9161–9165.

Tully, D.B., Allgood, V.E. and Cidlowski, J.A. (1994) Modulation of steroid receptor-mediated gene expression by vitamin B_6. *FASEB Journal* 8, 343–349.

Vesely, D.L. (1982) Biotin enhances guanylate kinase. *Science* 216, 1329–1330.

Wang, H. and Pevsner, J. (1999) Detection of endogenous biotin in various tissues: novel functions in the hippocampus and implications for its use in avidin–biotin technology. *Cell and Tissue Research* 296, 511–516.

Watanabe, T. (1990) Micronutrients and congenital anomalies. *Congenital Anomalies* 30, 79–92.

Watanabe, T., Dakshinamurti, K. and Persaud, T.V.N. (1995) Effect of biotin on palatal development of mouse embryos in organ culture. *Journal of Nutrition* 125, 2114–2121.

White, H.B. III and Whitehead, C.C. (1987) The role of avidin and other biotin-binding proteins in the deposition and distribution of biotin in chicken eggs: discovery of a new biotin-binding protein. *Biochemical Journal* 241, 677–684.

Wolffe, A. (1998) *Chromatin*. Academic Press, San Diego, California.

Yang, P., Morizumi, H. and Sato, T. (1995) Pulmonary blastoma: an ultrastructural and immunohistochemical study with special reference to nuclear filament aggregation. *Ultrastructural Pathology* 19, 501–509.

Yeowell, H.N., Walker, L.C., Marshall, M.K., Murad, S. and Pinnell, S.R. (1995) The mRNA and activity of lysyl hydroxylase are up-regulated by the administration of ascorbate and hydralazine to human skin fibroblasts from a patient with Ehlers-Danlos syndrome type VI. *Archives of Biochemistry and Biophysics* 321, 510–516.

Yokayama, S., Kashima, K., Inoue, S., Daa, T., Nakayama, I. and Moriuchi, A. (1993) Biotin-containing intranuclear inclusions in endometrial glands during gestation and puerperium. *American Journal of Clinical Pathology* 99, 13–17.

Zempleni, J. and Mock, D.M. (1999) Mitogen-induced proliferation increases biotin uptake into human peripheral blood mononuclear cells. *American Journal of Physiology* 276, C1079–C1084.

Zempleni, J. and Mock, D.M. (2000) Utilization of biotin in proliferating human lymphocytes. *Journal of Nutrition* 130, 335S–337S.

Zempleni, J. and Mock, D.M. (2001) Biotin homeostasis during the cell cycle. *Nutrition Research Reviews* 14, 45–63.

12   Selenium and Vitamin E

Alexandra Fischer[1], Josef Pallauf[1], Jonathan Majewicz[2],
Anne Marie Minihane[2] and Gerald Rimbach[2]
[1]Institute of Animal Nutrition and Nutrition Physiology, Justus-Liebig-University,
Giessen, Germany; [2]School of Food Biosciences, Hugh Sinclair Human Nutrition Unit,
University of Reading, UK

Vitamin E

Vitamin E, the most important lipid-soluble antioxidant, was discovered at the University of California at Berkeley in 1922 by Evans and Bishop (Evans and Bishop, 1922) (Table 12.1). Since its discovery, mainly antioxidant aspects of tocopherols and tocotrienols have been studied. In 1991 Angelo Azzi's group first described non-antioxidant, cell signalling functions for vitamin E, demonstrating that vitamin E regulates protein kinase C (PKC) activity in smooth muscle cells (Boscoboinik *et al.*, 1991a,b). At the transcriptional level, vitamin E modulates the expression of the hepatic α-tocopherol transfer protein (TTP) (Fechner *et al.*, 1998) as well as the expression of liver collagen α1 (Chojkier *et al.*, 1998), collagenase (Ricciarelli *et al.*, 1999) and α-tropomyosin genes (Aratri *et al.*, 1999). Recently, a tocopherol-dependent transcription factor (tocopherol-associated protein, TAP) has been discovered (Stocker *et al.*, 1999; Yamauchi *et al.*, 2001). Advances in molecular biology and genomic techniques have led to the discovery of novel vitamin E-sensitive genes and signal transduction pathways. In the current overview, we focus mainly on those transcription factors, signalling molecules, genes and proteins that play an important role in atherogenesis.

Chemistry and antioxidant properties of vitamin E

Tocopherols and tocotrienols are synthesized by plants from homogenestic acid. All are derivatives of 6-chromanol with an aliphatic isoprenoid side chain. The four tocopherol homologues (α, β, γ and δ) have a fully saturated 16-carbon phytol side chain, whereas tocotrienols contain a similar isoprenoid chain that has three double bonds. Individual tocopherols are named according to the position of number of the methyl groups on the phenol ring, with the α-, β-, γ- and δ-vitamins containing three, two, two and one methyl groups, respectively (Fig. 12.1). These structural differences determine biological activity, with α-homologues being the most biologically active.

The majority of the functionality of vitamin E is through its role as an antioxidant, maintaining the structural integrity of virtually all cells in the body. Its antioxidant function is mediated through the reduction of free radicals, thus protecting the body against the deleterious effects of highly reactive oxygen (ROS) and nitrogen species (RNS) that have been implicated in the pathophysiology of ageing and a number of chronic diseases such as atherosclerosis, cancers and rheumatoid arthritis (Halliwell, 1996; Parthasarathy *et al.*, 1999; Malins *et al.*, 2001). The ROS, which include hydrogen peroxide (H_2O_2), the superoxide radical (O_2^-) and the highly reactive hydroxyl radical (OH·), are by-products of normal aerobic metabolism formed during the respiratory and phagocytic processes and during microsomal cytochrome P450 metabolism.

The polyunsaturated fatty acids (PUFA) of biological membranes are particularly susceptible to free radical attack due to their high degree of unsaturation. In brief, the process is initiated by a free radical such as OH·, which extracts hydrogen from PUFA resulting in a PUFA· radical. Following molecular rearrangement to form a conjugated diene, the molecule is susceptible to attack by molecular oxygen (O_2), resulting in a peroxyl radical (PUFAOO·). Peroxyl radicals are capable of extracting a hydrogen atom from adjacent PUFAs, thus propagating a chain reaction. Such autooxidation continues, severely affecting the functionality of the tissue, unless the free radicals are scavenged. Due to its abundance, lipid solubility and its efficiency with respect to radical

Table 12.1. Important findings in experimental vitamin E research.

The early history
1922	The existence of vitamin E was recognized by Evans and Bishop when it became clear that this fat-soluble factor (named factor X) prevented fetal death in animals fed a diet containing rancid lard
1924	Sure gave factor X the name vitamin E as the fifth serial alphabetical designation for vitamins
1925	Evans proposed the word *tocopherol* from the Greek 'tos' for childbirth, 'phero' meaning to bring forth and 'ol' for the alcohol portion of the molecule

The early years of research (description of structural and functional features)
1930	Characterization of vitamin E deficiency symptoms (testicular atrophy, fetal resorption, encephalomalacia, paralysis associated with dystrophic muscle) in various animals
1938	Fernholz elucidates the structure of vitamin E
1938	Synthesis of vitamin E by Karrer
1955	Revelation by Gordon *et al.* that mature infants had low levels of blood tocopherol and abnormal haemolysis of erythrocytes, incubated in the presence of hydrogen peroxide
1967	Study by Bunyan *et al.* on the antioxidant impact of vitamin E on polyunsaturated fatty acids
1968	Vitamin E was recognized formally as an essential nutrient for humans by inclusion in the recommended dietary allowances table of the United States Food and Nutrition Board.

Recent vitamin E research (non-antioxidant function and influence of vitamin E on gene expression)
1991	Evidence by Boscoboinik *et al.* that smooth muscle cell proliferation is inhibited by α-tocopherol through protein kinase C modulation
1998	Discovery by Fechner *et al.* that the expression of α-tocopherol transfer protein in the liver is induced by α- and β-tocopherol
1998	Modulation of liver collagen α1 gene transcription by α-tocopherol (Chojkier *et al.*)
1999	Evidence by Aratri *et al.* that increased transcription level of α-tropomyosin is caused by α-tocopherol
2001	Discovery of α-tocopherol as a transcriptional regulator of gene expression via association with a transcription factor TAP (Yamauchi *et al.*)

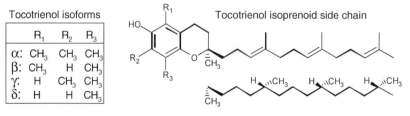

Fig. 12.1. Molecular structure of vitamin E stereoisomers.

quenching, vitamin E is considered to be the most important antioxidant in cell membranes (Ingold et al., 1987; Halliwell, 1996; Brigelius-Flohé and Traber, 1999).

The antioxidant property of vitamin E is exerted through the phenolic hydroxyl group, which readily donates its hydrogen to the peroxyl radical, resulting in the formation of a stable lipid species. In donating the hydrogen atom, vitamin E becomes a relatively unreactive free radical as the unpaired electron becomes delocalized into the aromatic ring. The efficiency of this protection depends upon two factors: first the mobility of the molecule in membranes, which is determined by the aliphatic tail; and secondly the number of methyl species on the chromanol ring, with each methyl group conferring additional antioxidant capacity. In addition, the proximity of the methyl species to the hydroxyl group is an important factor. Therefore α-homologues, which have the greatest number of methyl species, which also flank the hydroxyl group, are thought to be more effective than the other homologues.

α-Tocotrienol has been shown to be more effective in protecting against lipid peroxidation compared with α-tocopherol (Serbinova et al., 1991; Suzuki et al., 1993). A reason suggested for this is the nature of the aliphatic tail. The isoprenoid chain of α-tocotrienol has a stronger disordering effect on membranes compared with α-tocopherol. This leads to a greater mobility and more uniform distribution within the membrane. Nuclear magnetic resonance studies have also shown that the chromanol ring of α-tocotrienol is situated closer to the membrane surface. These factors contribute to a greater ability of tocotrienols to interact with radicals and allow for quicker recycling of the molecule to its active oxidized form (Serbinova et al., 1991; Suzuki and Packer, 1993).

Although vitamin E plays a unique role within membranes, it does not function in isolation. Protecting the cell from the deleterious effect of oxidative stress involves an array of other membrane and water-soluble antioxidants and antioxidant enzymes, which together form the 'antioxidant defence system' (Fig. 12.2). In this multifactorial system, the cytosolic metalloenzymes serve in the prevention of free radical formation. Superoxide dismutase serves to convert O_2^- to H_2O_2, whereas glutathione peroxidase and catalase further reduce H_2O_2, thus preventing the formation of the highly reactive OH·. The water-soluble antioxidants can act as cofactors for the antioxidant enzymes, can serve as independent antioxidants or can function in the recycling of vitamin E. Vitamin E exists in membranes at a concentration of one molecule per

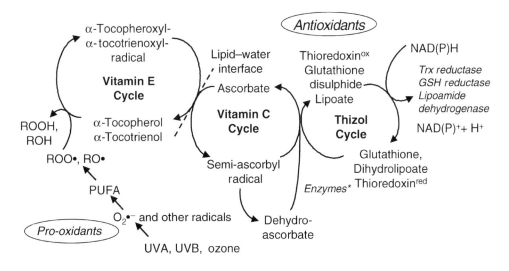

Fig. 12.2. The antioxidant network showing the interactions among vitamin E, vitamin C and thiol redox cycles.

2000–3000 phospholipids and therefore would become rapidly depleted unless it is regenerated to its active form. *In vitro* evidence suggests that ascorbate serves by regenerating membrane-bound vitamin E, converting the tocopheroxyl radical to its native form and resulting in the formation of an ascorbyl radical (Kagan and Tyurina, 1998; May *et al.*, 1998). However, *in vivo* evidence currently is lacking.

Absorption and transport

To date, the majority of information available on vitamin E absorption and transport is based on tocopherol (Cohn *et al.*, 1992; Kayden and Traber, 1993; Herrera and Barbas, 2001). In the small intestine, tocopherol esters hydrolysed to free vitamin E species are incorporated into mixed micelles due to the action of bile salts and pancreatic juices. Lack of these intestinal secretions, as occurs in individuals with conditions such as pancreatitis, cystic fibrosis or choleostatic liver disease, leads to vitamin E malabsorption, and resultant nutrient deficiency symptoms. The micelles enter the enterocyte via passive diffusion and the tocopherols are packaged into chylomicrons along with the phospholipid, cholesterol, triglyceride and apolipoprotein moieties. Upon entry into the circulation via the lymphatic system, the chylomicrons are hydrolysed sequentially due to the action of lipoprotein lipase attached to the capillary endothelium in the target tissue, such as muscle and adipose tissue, and a fraction of the tocopherol is released and taken up by the endothelium cells. Resultant chylomicron remnants are taken up by the liver by receptor-mediated endocytotic processes. In contrast to vitamins A and D, there does not appear to be a specific carrier protein for vitamin E in the circulation. Instead, vitamin E is incorporated into the lipoprotein particles in a non-specific manner. In hepatic cells, tocopherol from chylomicron remnants binds to cytosolic α-tocopherol transfer protein (α-TTP) (Catignani and Bieri, 1977; Hosomi *et al.*, 1997), which mediates its transfer to the site of very low-density lipoprotein (VLDL) synthesis (the rough endoplasmic reticulum and Golgi apparatus). This 32 kDa protein is expressed almost exclusively in the liver, and recent evidence from animal studies suggests that dietary α-tocopherol modulates hepatic α-TTP mRNA levels (Fechner *et al.*, 1998). Unlike tocopherol absorption, which is thought to be non-selective with respect to isomer, α-TTP displays stereoisomer specificity, with almost exclusive incorporation of α-tocopherol into the nascent VLDL particle. Relative affinities of tocopherol analogues for α-TTP, calculated by competition studies, are as follows: α-tocopherol 100%; β-tocopherol 38%; γ-tocopherol 9%; δ-tocopherol 2% (Hosomi *et al.*, 1997). The majority of non-α isomers are excreted via the bile (Traber and Kayden, 1989). α-TTP is now recognized to be the primary determinant of plasma tocopherol levels. Mutations of the α-TTP gene lead to reduced plasma and tissue α-tocopherol, which may ultimately lead to a severe condition known as ataxia with vitamin E deficiency (AVED), with associated neuronal and retinal damage (Traber *et al.*, 1990; Ben Hamida *et al.*, 1993). In a recent study, α-TTP knockout mice (Ttpa+/− and Ttpa−/−) were used as a model to examine the association between vitamin E deficiency and atherosclerosis (Terasawa *et al.*, 2000). Plasma and tissue α-tocopherol were reduced in a stepwise manner from control to Ttpa+/− to Ttpa−/−, with an absence of liver α-TTP in liver homogenates from Ttpa−/− and a 50% reduced protein level in the Ttpa+/− animals. This vitamin E deficiency was associated with increased lesions in the proximal aorta and increased rates of lipid peroxidation. These findings further demonstrate the role of this transfer protein on tocopherol metabolism and ultimate coronary heart disease (CHD) risk.

Approximately 50–70% of total secreted VLDL is hydrolysed to low-density lipoprotein (LDL) with associated transfer of tocopherols into the LDL fraction (Welty *et al.*, 2000). In the circulation, tocopherol exchanges rapidly between the lipoprotein particles although >90% is contained within the LDL and high-density lipoprotein (HDL) fractions (Behrens *et al.*, 1982). The 75 kDa plasma phospholipid transfer protein facilitates tocopherol exchange between HDL and LDL (Lagrost *et al.*, 1998).

The mechanisms of peripheral cellular uptake of vitamin E are poorly understood, although simultaneous uptake of tocopherol, via receptor-mediated lipoprotein endocytosis, or via fatty acid-binding proteins, may be involved. However, recent evidence suggests that specific membrane tocopherol-binding proteins (TBPpms) may also mediate tocopherol uptake (Dutta-Roy, 1999).

Information on intracellular tocopherol transport currently is lacking. Due to its strong hydrophobicity, transfer to cellular sites requires a specific transfer protein. However, it is still unclear how many other α-tocopherol-binding proteins exist and which mechanisms regulate tocopherol transfer within peripheral cells. Recently, a novel binding protein, TAP, has been identified (Stocker et al., 1999; Zimmer et al., 2000; Blatt et al., 2001; Yamauchi et al., 2001). This 46 kDa protein, which displays significant sequence homology to α-TTP, is expressed ubiquitously although the highest levels have been observed in the liver, brain and prostate (Zimmer et al., 2000). It is suggested that this protein plays a significant general role in intracellular tocopherol metabolism. Structural analysis of TAP suggested that it is a member of the widespread SEC14-like protein family, which plays a role in phospholipid exchange in the cell. Recent ligand competition studies suggest that TAP binds to α-tocopherol but not other tocopherol isomers (Blatt et al., 2001). Although research is at an early stage, it is likely that TAP will prove an important molecule with respect to cellular tocopherol events.

Protein kinase C and protein phosphatase 2A activity

Some of the cellular responses of α-tocopherol are regulated at the transcriptional and post-transcriptional level. For example, α-tocopherol inhibits smooth muscle cell (SMC) proliferation and PKC activity. The inhibition of SMC proliferation is specific for α-tocopherol, since trolox, phytol, β-tocopherol and α-tocopherol esters had no effect (Boscoboinik et al., 1991a,b). As α-tocopherol and β-tocopherol have very similar free radical-scavenging activities, the mechanism by which α-tocopherol acts on PKC is not related to its antioxidant properties. Subsequent studies have shown that PKC is inhibited in a number of other cell types, including monocytes (Devaraj et al., 1996), neutrophils (Kanno et al., 1995), fibroblasts (Hehenberger and Hansson, 1997) and mesangial cells (Tada et al., 1997). Most importantly, this inhibition of PKC by α-tocopherol occurs at concentrations close to those measured in human plasma (Azzi et al., 2001). Antiproliferative effects of vitamin E were not seen for HeLa cells, suggesting that there are different cell-specific pathways of cellular proliferation in which vitamin E can act (Fazzio et al., 1997). In addition, the inhibition of PKC was not related to a direct interaction of α-tocopherol with the enzyme nor with a diminution of its expression. However, PKC inhibition by α-tocopherol is linked to the activation of a protein phosphatase 2A, which in turn dephosphorylates PKC-α and thereby inhibits its activity (Clement et al., 1997; Ricciarelli et al., 1998). An inhibitory effect of α-tocopherol on PKC may be seen only at the cellular level and is not evident with recombinant PKC.

Cyclooxygenase

Cyclooxygenase has two isoforms, COX-1 and COX-2. COX-1 is constitutively expressed in most cells, whereas COX-2 is regulated by growth factors, tumour promoters, cytokines, glucocorticoids and lipopolysaccharide (LPS). Cyclooxygenases convert arachidonic acid (AA) into prostaglandin E_2 (PGE_2), the precursor of thromboxane and eicosanoid synthesis. High levels of COX-2 in epithelial cells are associated with the inhibition of apoptosis, and overexpression of COX-2 has been implicated in the pathogenesis of neoplastic diseases. An up-regulation of COX-2 transcription has been shown in most human colorectal cancers (Fosslien, 2001). Interestingly, changes in AA metabolism stimulate cell proliferation via activation of PKC, indicating that PKC might be one of the primary signalling pathways through which certain tumours are initiated or maintained. In recent years, a role for COX-2 in atherogenesis has been identified. Immunocytochemical studies using antibody to COX-2 showed that COX-2 was localized to macrophages in atherosclerotic lesions of patients with coronary artery disease (Baker et al., 1999). In monocytes derived from aged mice, it has been shown that a vitamin E-induced decrease in PGE_2 production is mediated via a decreased COX activity (Wu et al., 2001). However, vitamin E has no effect on COX mRNA and protein levels, indicating a post-translational regulation of the COX enzyme. Other non α-tocopherol homologues were also effective in inhibiting COX activity, but their degree of inhibition varied, indicating that the degree of inhibition was comparable with their antioxidant capacity, suggesting that alternative mechanisms may be involved.

It has been shown in LPS-stimulated RAW264.7 macrophages and interleukin (IL)-1β-treated A549 human epithelial cells that γ-tocopherols inhibited the production of PGE_2 due to a direct inhibition of COX-2 (Jiang et al., 2000). Furthermore, the major metabolite of dietary γ-tocopherol also exhibited an inhibitory effect in these cells. α-Tocopherol at 50 μM slightly reduced (25%) PGE_2 formation in macrophages, but had no effect in epithelial cells. Similar to the previously mentioned study, the inhibitory effect of γ-tocopherol and carboxyethyl hydroxychroman (γ-CEHC) stemmed from their inhibition of COX-2 activity, rather than from affecting protein expression or substrate availability, and appeared to be independent of their antioxidant activity.

Nuclear factor-κB

The transcription factors of the nuclear factor (NF)-κB/Rel family control the expression of various genes involved in inflammatory and proliferation responses. The typical NF-κB dimer is composed of the p50 and p65 subunits, and it is present in its inactive form in the cytosol bound to the inhibitory protein IκB. Following activation by various stimuli, including inflammatory or hyperproliferative cytokines, ROS and bacterial wall components, the phosphorylation and proteolytic removal of IκB from the complex occurs. Activated NF-κB then immediately enters the nucleus where it interacts with regulatory κB elements in the promoter and enhancer regions, thereby controlling the transcription of inducible genes (Baeuerle and Henkel, 1994; Baeuerle and Baltimore, 1996). Importantly, activated NF-κB has been identified *in situ* in human atherosclerotic plaques but not in cells of normal vessels devoid of atherosclerosis (Brand et al., 1996), as well as in an arterial injury model (Lindner and Collins, 1996). Furthermore, NF-κB is activated by an atherogenic diet (Liao et al., 1993) and by oxidized LDL (Brand et al., 1997) and advanced glycation end-products (Yan et al., 1994). Cumulatively, these observations suggest a key role for NF-κB in atherogenesis.

A spectrum of key genes known to be involved in atherosclerosis development has been shown to be regulated by NF-κB, including those coding for tumour necrosis factor-α (TNF-α), IL-1, macrophage or granulocyte colony-stimulating factor (M/G-CSF), monocyte chemoattractant protein-1 (MCP-1), c-myc and the adhesion molecules vascular cell adhesion molecule-1 (VCAM-1) and intracellular adhesion molecule-1 (ICAM-1) (Rimbach et al., 2000; Collins and Cybulski, 2001). In the early stages of an atherosclerotic lesion, different types of cells (macrophages, SMCs and endothelial cells) interact, causing a loss of homeostasis and a self-propagating system leading to dysfunction and lesion development in the artery wall. Figure 12.3 shows a sketch of the regulation of NF-κB activation. Some of the major genes involved in atherogenesis are also listed.

Several lines of evidence, including the inhibition by various antioxidants, suggest that NF-κB is subject to redox regulation. Because of its pivotal role in the inflammatory response, a significant effort has been focused on developing therapeutic agents that regulate NF-κB activity. In this scenario, vitamin E may play an important role, either by directly affecting key steps in the activation pathway of NF-κB, or by modulating the intracellular redox status, which is, in turn, one of the major determinants of NF-κB activation. Consistent experimental data are accumulating that suggest that the anti-inflammatory properties of vitamin E are due, in part, to its ability to downregulate NF-κB. Suzuki and Packer (1993) examined the effect of vitamin E derivatives on TNF-α-induced NF-κB activation. Incubation of human Jurkat cells with vitamin E acetate or α-tocopheryl succinate produced a concentration-dependent inhibition of NF-κB activation. Similarly, gel shift studies with the macrophage cell line THP-1 pre-treated with α-tocopheryl succinate and then activated with LPS showed an inhibition of NF-κB activity by 43% at 50 μM versus untreated controls (Nakamura et al., 1998). However, α-tocopherol had no effect on NF-κB activity. Cellular vitamin E uptake was analysed in this study by simultaneous determination of vitamin E and its derivatives using high-performance liquid chromatography. The vitamin E recovered from culture pellets revealed that both α-tocopherol and α-tocopheryl succinate were accumulated with equal efficiency and were not metabolized. These observations indicate that α-tocopheryl succinate possibly inhibits NF-κB activation and/or translocation to the nuclei in its unchanged form.

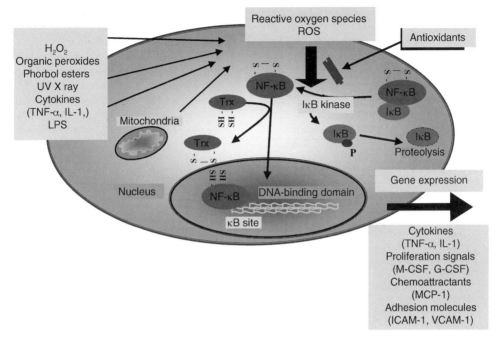

Fig. 12.3. Regulation of NF-κB activity.

α-Tropomyosin, cell adhesion proteins, chemokines and scavenger receptors

An involvement of tropomyosin in the progression of atherosclerosis has been suggested (Kocher et al., 1991). Early after balloon injury, SMCs of the media and those that have migrated into the intima contain decreased amounts of tropomyosin, and late after balloon injury tropomyosin returns towards normal values. In 1999, Aratri et al. discovered the induction of α-tropomyosin expression in rat vascular SMCs by α-tocopherol, using the differential display technique. No significant changes in mRNA levels were observed when β-tocopherol was used. The authors thus rate the observed overexpression of tropomyosin, induced by α-tocopherol, as a relevant finding in the understanding of the diminution of blood pressure by vitamin E at the molecular level, because this effect may be the consequence of a lesser contractility of SMCs containing more tropomyosin.

Activation of endothelial cells results in release of vascular cytokines such as IL-1 and TNF-α. These cytokines, in turn, induce the expression of cell surface adhesion molecules such as VCAM-1 and ICAM-1, which are centrally involved in the endothelial recruitment of neutrophils (Cybulski and Gimbrone, 1991). Focal expression of ICAM-1 and VACM-1 has been reported in arterial endothelium overlying early foam cell lesions in both dietary and genetic models of atherosclerosis in rabbits (Thiery et al., 1996). This expression, together with the activation of MCP-1 leads to infiltration of mononuclear cells into the wall and to an increase in the oxidation and scavenging of LDL, formation of lipid-laden foam cells, and development or progression of atherosclerotic plaques (Rubanyi, 1993).

As mentioned previously, transcription of ICAM-1, VCAM-1 and MCP-1 is dependent, at least in part, on the activation of NF-κB. Cell culture studies have shown that treatment of endothelial cells with oxidized LDL significantly increased expression of mRNA and proteins levels of ICAM-1 and VCAM-1 (Yoshida et al., 2000). However, pre-treatment with α-tocopherol reduced cell adhesion protein expression in a dose-dependent manner. Furthermore, adherence of polymorphonuclear leukocytes (PMNs) or mononuclear leukocytes (MNCs) to oxidized LDL-activated endothelial cells was much higher than in unstimulated endothelial cells. Supplementation of endothelial cells with α-tocopherol inhibited the adherence of PMNs or MNCs. Furthermore, IL-1β-induced production of MCP-1 was suppressed dose-dependently by enrichment of human endothelial cells with vitamin E (Zapolska-Downar et al., 2000). From this and other studies, it is suggested that the anti-atherogenic effect of α-tocopherol may be due in part to a down-regulation of cell adhesion proteins and chemokines. Despite evidence that vitamin E down-regulates cell adhesion proteins in vitro, in vivo evidence currently is lacking. Ricciarelli et al. (2000) have demonstrated recently that the CD36 scavenger receptor, which transports oxidized LDL into the cytoplasm, is expressed in human smooth muscle cells. Interestingly, α-tocopherol inhibited oxidized LDL uptake by a mechanism involving down-regulation of CD36 mRNA and protein expression. It is hypothesized that beneficial cardiovascular effects of α-tocopherol are mediated, at least in part, by lowering the uptake of LDL, which subsequently results in a reduction of foam cell formation.

Nitric oxide and platelet aggregation

Nitric oxide represents a pivotal molecule in the regulation of the vascular tone. Its production by the endothelial NO synthase (eNOS) suppresses expression of pro-inflammatory cytokines, adhesion molecules (De Caerina et al., 1995) and MCP-1 (Busse and Fleming, 1995). It also inhibits platelet adhesion to the endothelium (de Graaf et al., 1992), can reduce the permeability of the arterial wall (Cardonna-Sanclemente and Born, 1995; Forster and Weinberg, 1997), suppresses vascular SMC proliferation and migration (Garg and Hassid, 1989), and can act as an antioxidant (Patel et al., 2000). In experimental models, inhibition of NO production accelerates disease (Naruse et al., 1994) whilst increases in NO synthesis reduce it (Cooke et al., 1992). The major risk factors for atherosclerosis – age (Matz et al., 2000), hypercholesterolaemia (Stroes et al., 1995), diabetes (Williams et al., 1996), hypertension (Panza et al., 1995), smoking (Celermajer et al., 1993) and low birth weight (Leeson et al., 1997) – are all associated with impaired NO activity. Importantly, NO significantly inhibits NF-κB (Matthews et al., 1996). This may account for its influence on the transcription of genes for adhesion proteins,

MCP-1 and others. Key roles for both NO and NF-κB are not mutually exclusive.

Li et al. (2001) studied the effects of different isoforms of vitamin E on NO activity and platelet aggregation in human platelet-rich plasma. All three isoforms of vitamin E (α-, β- and δ-tocopherol) markedly decreased ADP-induced platelet aggregation and increased NO release in a dose-dependent manner. The isoforms did not affect cNOS protein expression, but increased cNOS phosphorylation. Furthermore it has been demonstrated that oral supplementation with α-tocopherol (400–1200 IU day^{-1}) resulted in an increase in platelet tocopherol concentration that correlated with marked inhibition of phorbol 12-myristate 13-acetate (PMA)-mediated platelet aggregation in humans (Freedman et al., 1996). Platelets derived from these subjects also demonstrated apparent complete inhibition of PKC. These findings may represent one potential mechanism of tocopherol in preventing the development of coronary artery disease.

Selenium

Biosynthesis of selenoproteins

Most known biological functions of selenium are exerted by selenoproteins, which have been identified in bacteria, archaebacteria and eukaryotes. For mammals, selenoproteins are essential for life, because a knockout in the tRNA^{Sec} gene, which is important for selenoprotein synthesis, is lethal early in embryogenesis (Bösl et al., 1997). Selenium is incorporated into selenoproteins as selenocysteine. The underlying mechanisms in eukaryotes are poorly understood and are under intensive investigation, but some insights can be drawn from the genetic and biochemical studies of selenoprotein synthesis in Escherichia coli, carried out by Böck and co-workers (Ehrenreich et al., 1992). Four genes (selA–selD) were identified as being involved in this process. The gene products have been identified as selenocysteine synthase (SELA), a selenocysteine-specific elongation factor (SELB), a selenocysteine-specific tRNA (tRNA^{Sec}, selC gene product) and selenophosphate synthetase (SELD).

A model of selenocysteine incorporation in eukaryotes is given in Fig. 12.4. tRNA^{Sec} is first

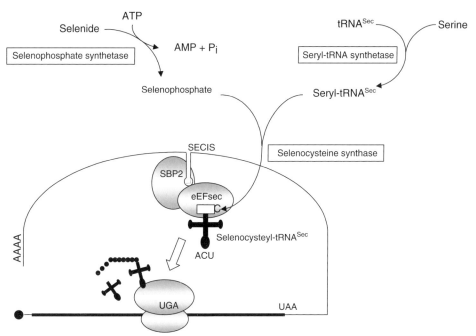

Fig. 12.4. Model of selenocysteine incorporation in eukaryotes modified according to Mansell and Berry (2001).

charged with serine by seryl-tRNA synthetase. Selenocysteine synthase then catalyses the conversion of seryl-tRNA^{Sec} to selenocysteyl-tRNA^{Sec} using selenophosphate as the active selenium donor (Low and Berry, 1996). Selenophosphate synthetase catalyses the synthesis of selenophosphate from selenid and ATP. The insertion of selenocysteine is encoded by the triplet UGA, which usually functions as a stop codon. Recoding of UGA as a selenocysteine codon requires a mRNA secondary structure called SECIS (selenocysteine insertion sequence), which is localized in eukaryotes in the 3′-non-translated region, and thus decoding from an appreciable distance is necessary (Low and Berry, 1996). For the incorporation, two further proteins are involved, the elongation factor eEFsec (SELB homologue) and SBP2 (Copeland and Driscoll, 1999). It has been postulated that SBP2 binds to SECIS, followed by binding of the tRNA^{Sec}–eEFsec complex, which subsequently can interact with the ribosomes and provoke the insertion of selenocysteine (Mansell and Berry, 2001). Presumably, UGA is the most recently evolved stop codon and it may have functioned previously solely as a sense codon for selenocysteine (Leinfelder *et al.*, 1988). The introduction of oxygen into the atmosphere may have counterselected against selenocysteine, since it is readily oxidized. Selenoprotein expression is now restricted to either anaerobic growth conditions or chemical environments well protected from oxygen. Most of the selenoenzymes identified to date catalyse oxido-reduction reactions in which the selenocysteine residue is in the active site. Substitution of the structurally related amino acid cysteine for selenocysteine produced functional enzymes, but significantly decreased the k_{cat}, indicating that the presence of selenocysteine confers specific biochemical properties. Selenoproteins can be labelled selectively by ^{75}Se in selenium-deficient animals and visualized autoradiographically after electrophoretic separation (Behne *et al.*, 1996). According to such experiments, the number of selenoproteins in mammals has been estimated to amount to 30–50 (Köhrle *et al.*, 2000). Fewer than 20 of such bands have been characterized by sequence analysis to date, and an enzymatic function has been assigned to >10 of them. Among the identified and relatively well-characterized selenoproteins are four glutathione peroxidases (GPxs), three thioredoxin reductases (TrxRs), three deiodinases (DIs), the selenophosphate synthetase-2, the selenoprotein P (SelP), the selenoprotein W in muscle and some others of unknown function (Table 12.2).

Regulation of selenoprotein expression

The biosynthesis of selenoproteins is dependent primarily on the bioavailability of selenium and, by consequence, on the formation of tRNA^{Sec}, which regulates the level of all selenoproteins. In selenium deficiency, UGA is interpreted as a stop codon, which results in premature chain termination at the UGA codon. The decrease in selenoprotein synthesis is accompanied further by a fall in the levels of the respective mRNAs. This is not due to a lower transcription rate but rather to a loss of stability, i.e. enhanced degradation of the mRNA (Bermano *et al.*, 1996).

For an optimum translation of a selenoprotein, apart from selenium supply, its mRNA appears to require an optimum UGA location, an optimum UGA context and the SECIS element to optimize selenocysteine incorporation. Surprisingly, individual selenoproteins vary in their specific responses to selenium deprivation, a phenomenon referred to as 'the hierarchy of selenoproteins' (Flohé *et al.*, 2000). Some of the selenoproteins, such as cGPx, disappear rapidly in

Table 12.2. Mammalian selenocysteine-containing proteins.

Selenoprotein	Common abbreviations	Expression	Function
Glutathione peroxidases ROOH + 2RSH → ROH + H$_2$O + RSSR			
Cytosolic GPx	cGPx, GPx1	Ubiquitous	Protection against oxidation
Phospholipid hydroperoxide GPx	PHGPx, GPx4	Gastrointestinal tract	Redox regulation, sperm maturation
Plasma GPx	pGPx, GPx3	Ubiquitous, especially kidney	Redox buffer? Regulation of prostanoid metabolism?
Gastrointestinal GPx	GI-GPx, GPx2	Ubiquitous	Local redox protection?
Iodothyronine deiodinases			
5′-Deiodinase, type I	5′DI-I	Thyroid gland, liver, kidney, central nervous system	T3 synthesis: T4 + 2e$^-$ + H$^+$ → T3 + I$^-$ Degradation of T3 and T4
5′-Deiodinase, type II	5′DI-II	Pituitary and thyroid glands, placenta, heart and skeletal muscles, central nervous system, brown fat	T3 synthesis: T4 + 2e$^-$ + H$^+$ → T3 + I$^-$
5′-Deiodinase, type III	5′DI-III	Placenta, central nervous system, skin	Degradation of T3 and T4
Thioredoxin reductase			
Thioredoxin reductase	TR, TrxR	Ubiquitous	Reduction of thioredoxin, DNA synthesis, thiol–disulphide balance
Thioredoxin reductase 2	TR2	Testis	Unknown
Mitochondrial thioredoxin reductase	TR3	Ubiquitous	Unknown
Selenophosphate synthetase-2	SPS2	Ubiquitous	Selenoprotein synthesis
15 kDa selenoprotein	Sep15	Ubiquitous	Unknown, protein folding?
Selenoprotein P	SelP	Plasma	Unknown, redox protection?
Selenoprotein W	SelW	Ubiquitous	Unknown, redox protection?
Selenoproteins X, N, R, T	SelX, SelN, SelR, SelT	?	Unknown

Modified according to Gladyshev and Hatfield (1999) and Flohé *et al.* (2000).

selenium deprivation (Knight and Sunde, 1987; Li et al., 1990; Müller and Pallauf, 2002) and are resynthesized slowly when resupplemented (Knight and Sunde, 1988). Others, however, such as PHGPx, decline gradually in the presence of a limited selenium supply and rebound immediately at resupplementation (Weitzel et al., 1990; Bermano et al., 1996). This unique and specific regulation of selenoprotein expression has been suggested to be regulated at least partially by the SECIS element, because selenoproteins with similar positions in the hierarchy have similar structures in the SECIS region. Differences in the selenoprotein mRNA SECIS element could alter relative affinities for the selenocysteine translation complex and thus confer differential translatability, especially when selenium is limiting.

Irrespective of the mechanisms involved, the hierarchy in the biosynthesis of selenoproteins is considered to reflect their biological importance. Sunde (1997) suggests the following ranking:

GI-GPx > TrxR, SelP, 5'DI-I > cGPx

Apart from selenium-dependent factors influencing the selenoprotein expression, selenium-independent factors also exist. For example, tissue-specific expression of 5'DI-I is found primarily in the thyroid, liver, kidney and pituitary, of 5'DI-II in the brown fat tissue of rodents, in the placenta, thyroid, pituitary and in the central nervous system, and of 5'DI-III in the placenta, skin and also in the fetal and neonatal liver, but not in the adult liver (Köhrle et al., 2000). The highest amount of PHGPx can be found in the testicle (Maiorino et al., 1998). GI-GPx is expressed only in the gastrointestinal tract (Wingler et al., 2000), whereas pGPx is secreted and formed preferentially at metabolic surfaces (proximal tubulus of the kidney, intestinal epithelium, skin, lung). Among the selenoproteins that are not yet defined functionally or structurally, selenoprotein W is found in the heart and skeletal muscle; other poorly characterized selenoproteins are found in testis, prostate and pancreas (Behne et al., 1997). The occasionally unusual expression pattern points to highly specialized roles, which have been identified for PHGPx (spermatogenesis), GI-GPx (barrier against hydroperoxide absorption) and DIs (thyroid hormone synthesis) (Köhrle et al., 2000). Yet, in none of these cases has the molecular basis for tissue-specific expression been elucidated so far. Furthermore, there is evidence that the selenoenzymes decrease in certain tissues faster under selenium-deprived conditions than in other tissues. In rats deprived of selenium for two generations, total GPx activity was <5% of normal in liver, kidney, lung, heart, adrenal gland, pancreas and muscle, <10% of normal in male thymus and testes, but decreased to only about 50% in the brain and to 25% in the ovary and female thymus (Thompson et al., 1995). From these results, it can be concluded that mechanisms must exist that guide selenium to a specific enzyme within a particular tissue, probably via tissue-specific regulation of mRNA stability. cGPx and PHGPx are retained preferentially under selenium-limiting conditions in the brain, the reproductive organs and endocrinological tissues. Brigelius-Flohé (1999) proposed the following ranking order of tissue-specific stabilities of cGPx (not to be confused with absolute activities):

brain >> thymus > thyroid > heart > liver, kidney, lung

For some selenoproteins, a hormone-dependent regulation could be observed. In monocytes, human TrxR is induced rapidly through 1,25-dihydroxy vitamin D3 (Schütze et al., 1999), and in fetal osteoblast-like cells through certain cytokines and growth factors (Schütze et al., 1998). The three DI isoenzymes are regulated by thyroid hormones, retinoids, sexual hormones, gluco- and corticosteroids, and a series of growth factors and cytokines. For PHGPx, a sexual hormone-dependent expression has been suggested. An induction of GPx genes by oxidative stress has often been postulated but has never been demonstrated convincingly *in vivo*. The induction of cGPx through an oxygen-responsive element has only been described *in vitro* (Cowan et al., 1993).

Why are selenoproteins important?

GPxs are found in all mammalian tissues in which oxidative processes occur. By reduction of hydroperoxides to the corresponding alcohols, these enzymes can prevent the production of reactive oxygen radicals and thus may contribute to the protection of the organism's macromolecules and biomembranes against oxidation. GPx1 catalyses the glutathione-dependent reduction of hydrogen peroxide and various organic hydroperoxides. Although GPx1 is considered, together with catalase and superoxide dismutase, a major antioxidant enzyme, its role as an essential antioxidant has been questioned. Dramatic losses in GPx1 activity and expression level occur in selenium-deficient animals without an apparent effect on cellular metabolism. In addition, no phenotypic changes were observed in GPx1 knockout mice, even under hyperoxide conditions (Ho et al., 1997). However, when exposed to avirulent Coxsackievirus strains, cGPx-deficient mice develop myocarditis. Surprisingly, the previously avirulent viral strain, after passage through cGPx (−/−) mice, proved also to be virulent in the control mice (Beck et al., 1998). It appears that an increased peroxide level as a result of deficiency in cGPx increases the rate of mutations, which could not be prevented by the role of any of the other selenoproteins or vitamin E. These findings suggest that cGPx may be important in rescuing cells from environmental stress, while its function under normal conditions may be compensated by other cellular components.

Although PHGPx, pGPx, GI-GPx and TrxR by way of reduced thioredoxin catalyse redox reactions, they are not competent in backing up cGPx in the detoxification of hydroperoxides. Therefore, their possible role might be in the modulation of redox-sensitive enzyme cascades.

GPxs together with reduced glutathione prevent in vitro arachidonate utilization by cyclo-oxygenase, 5-lipoxygenase and 15-lipoxygenase because these key enzymes of prostaglandin and leukotriene synthesis require a certain peroxide tone to become active (Flohé et al., 2000). Accordingly, H_2O_2 and alkylhydroperoxides can reactivate these enzymes. It has been shown that overexpression of PHGPx in RBL-2H3 cells significantly reduces leukotriene C4 and leukotriene B4 synthesis (Imai et al., 1998). Moreover, pGPx might act as a regulator of inflammatory responses by scavenging H_2O_2 and lipid hydroperoxides in the extracellular space. Selenoprotein P binds to endothelial cells and presumably protects them against pro-inflammatory hydroperoxides, including peroxynitrite (Sies et al., 1998).

The reduction of peroxide tonus can, furthermore, lead to a reduced activation of redox-sensitive transcription factors, such as NF-κB (Viita et al., 1999). The IL-1 and TNF-α-dependent activation of NF-κB can be inhibited by overexpression of cGPx (Kretz-Remy et al., 1996) and PHGPx in ECV 304 cells (Brigelius-Flohé et al., 1997). Thioredoxin (Trx), the main substrate for TrxR, seems, on the one hand, to inhibit NF-κB activation because overexpression of Trx in cells significantly reduces NF-κB activity. On the other hand, Trx seems to be required for the reduction of cysteine residues in NF-κB, which is essential for DNA binding of NF-κB (Hayashi et al., 1993).

The apoptotic process, another example of redox-regulated signalling, can be inhibited in vitro by stimulated GPx activity through selenium supplementation (Kayanoki et al., 1996) and overexpression of cGPx and PHGPx (Packham et al., 1996). These results are consistent with the perception that the upper part of the apoptotic signalling cascade may be inspired by hydroperoxides. Stressing cells with UV light results in 80% cell death by apoptosis, while selenite supplementation significantly improved cell survival. This anti-apoptotic function of selenium is believed to be mediated by inhibition of caspases-3, presumably by modulating cysteine residues (Park et al., 2000). Infection of T lymphocytes with human immunodeficiency virus (HIV) results in reduced GPx activity, which has been associated with higher susceptibility to lipid hydroperoxide- and H_2O_2-induced apoptosis (Sandstrom et al., 1994).

Selenium and its function in preventing apoptosis have been linked to an involvement in the cell cycle and thus cancer development. Early epidemiological studies suggested an inverse relationship between alimentary selenium intake and incidence of certain malignancies (Clark, 1985). In the early phase of tumour initiation, oxidative damage to DNA through ROS plays an important role. Cells that are adequately supplied with selenium and vitamin E seem less vulnerable to carcinogenic influences (Halliwell, 2000). Accordingly, UV-induced skin tumorigenesis correlates inversely with GPx activity in selenium-supplemented hairless mice (Pence et al., 1994). Incubation of cells with selenite or α-tocopherol 24 h before UV exposure or addition of the carcinogens benzoapyrene or tryptophanpyrolysate led in all cases to an inhibition of cell transformation, whereby the supplementation with both nutrients showed the greatest effect. Selenium incubation was accompanied by an increase in GPx activity and a decrease in peroxides (Borek et al., 1986). However, the protective effect of selenium is not explained readily in terms of prevention of oxidative DNA damage since it appears to be anti-tumorigenic at intakes that are substantially greater than those associated with maximal expression of the known selenoproteins. Current discussions therefore presume a direct anti-proliferative effect of pharmacological selenium concentrations by the enhancement of apoptosis. The excess selenium supply probably could lead to a depletion of intracellular reduced glutathione (Combs, 1999). Furthermore, certain selenium metabolites might inhibit cell proliferation. Selenodiglutathione can induce the tumour suppressor gene p53 (Lanfear et al., 1994), and Se-methylselenocysteine can arrest mammary cell growth in the S phase (Sinha and Medina, 1997).

Moreover, the incidence of CHD has been reported to be inversely related to selenium supply. Atherogenesis belongs to the pathological processes, probably connected to oxidative stress. Upon uptake of oxidized LDL (oxLDL), macrophages are transformed into foam cells, and SMCs undergo proliferation, processes that are considered to initiate atherogenesis. Accordingly, any antioxidant system that is able to prevent or antagonize the oxidation of LDL should inhibit the process, including the GPxs. Furthermore, lipoxygenases, which require a certain peroxide tone for activity, can generate hydroperoxy lipids within LDL particles, which then may propagate further lipid peroxidation. pGPx as an extracellular enzyme and PHGPx, which efficiently reduces hydroperoxides in oxidized LDL, would be possible candidates for the prevention of artherosclerosis. However, pGPx does not have the optimum specificity as it acts on H_2O_2 and to some degree on peroxidized complex lipids, but cannot reduce cholesterol ester hydroperoxides (Yamamoto and Takahashi, 1993). PHGPx does display the optimum specificity but is not present extracellularly, where LDL is oxidized (Köhrle et al., 2000). Therefore, some controversial findings still remain to be explained, and further research is required to elucidated fully the role of selenium in atherogenesis.

Consequences of selenium and/or vitamin E deficiency on gene expression

To examine the molecular events associated with selenium and vitamin E deficiency in rats, cDNA array technology has been applied to define the transcriptional response in rat liver after 7 weeks on a selenium- and/or vitamin E-deficient diet (Fischer et al., 2001). Atlas™ Rat cDNA Toxicology Array II from Clontech was used to monitor simultaneously the expression of 465 genes, whereby a change of twofold or more was considered as significant (Table 12.3).

Vitamin E deficiency alone did not significantly affect any of the genes monitored. Probably, other genes not present on the cDNA membrane could have been regulated differentially by vitamin E. Additionally, tissues other than liver might be more susceptible to vitamin E induced changes in differential gene expression.

Besides a 13.9-fold down-regulation of the cGPx gene, selenium deficiency alone was accompanied by an increase in the expression of UDP-glucuronosyltransferase 1 and bilirubin UDP-glucuronosyltransferase isoenzyme 2. These two enzymes are known to have an important function in the detoxification of xenobiotics in liver. Likewise, rat liver cytochrome P450 4B1, also involved in xenobiotic metabolism and inducible by glucocorticoids, was induced 2.3-fold. The mRNA levels of arachidonate 12-lipoxygenase (ALOX 12) were 2.4-fold higher in selenium-deficient animals as compared with the controls. It has been shown that ALOX12 and PHGPx are opposing enzymes balancing the intracellular concentration of hydroperoxy lipids (Schnurr et al., 1999) whereby an inhibition of PHGPx activity increases the enzymatic catalysis of ALOX 12 (Chen et al., 2000). Deficiency of selenium and vitamin E affected 5% of all genes monitored. Deficiency was characterized by a significant down-regulation of genes that inhibit programmed cell death, including defender against cell death 1 protein, inhibitor of apoptosis protein 1 and Bcl2-L1. Furthermore, the expression level of early growth response protein 1, known as a suppressor of growth and transformation and an inducer of apoptosis, was increased twofold. Accordingly, carbonic anhydrase III (CAIII), which recently was reported to play a role as an antioxidant preventing H_2O_2-inducible apoptosis (Raisanen et al., 1999), was down-regulated twofold. A stronger tendency towards negative cell cycle progression in livers of double-deficient rats was suggested further by the down-regulation of nucleophosmin and G_1/S-specific cyclin D1, which has been characterized as an important signal in anti-apoptotic mechanisms.

Furthermore, selenium plus vitamin E deficiency resulted in an induction of acute phase proteins (metallothionein, DT-diaphorase, α-1 acid

Table 12.3. Selection of selenium and vitamin E deficiency-related changes in gene expression in rat liver.

GenBank accession no.	Δ-Se-E (fold)		Gene	Function
Apoptosis/cell cycle				
Y13336	↓	2.0	Defender against cell death 1 protein (DAD1)	Protection against apoptosis
AF081503	↓	2.6	Inhibitor of apoptosis protein 1	Protection against apoptosis
U72350	↓	3.2	Bcl2-L1	Protection against apoptosis
M22413	↓	2.0	Carbonic anhydrase III (CAIII)	Antioxidant, protection against apoptosis
D90345	↓	2.2	T-complex protein 1 (CCT) α-subunit	Chaperone, folding of proteins
X82021	↓	2.2	HSC70-interacting protein (HIP)	Stabilization of the chaperone HSC70
J03969	↓	2.9	Nucleophosmin (NPM)	Stimulation of normal cell growth
D14014	↓	3.1	G_1/S-specific cyclin D1 (CCND1)	Initiation of cell cycle, oncogene
J04154	↑	2.1	Early growth response protein 1	Suppression of growth and induction of apoptosis
U77129	↑	2.0[1]	SPS1/Ste20 homologue KHS1	Transducer of signals in MAP kinase pathway
Antioxidant defence/stress response/inflammation				
X12367	↓	18.8	Cellular glutathione peroxidase I	Peroxide detoxification
J05181	↓	3.4	γ-Glutamylcysteine synthetase (γ-GCS)	Glutathione synthesis
U22424	↓	2.2	11-β-Hydroxysteroid dehydrogenase 2	Conversion of corticosterone into 11-dehydrocorticosterone
L49379	↓	2.3	Multispecific organic anion exporter (cMOAT)	Detoxification, export of leukotriene C4
J02608	↑	15.3[1]	DT-diaphorase	Xenobiotic metabolism
D00753	↑	2.1	SPI-3 serine protease inhibitor	Acute phase protein
J00696	↑	2.3	α-1 Acid glycoprotein	Acute phase protein
J00734	↑	2.3	Fibrinogen γ chain	Acute phase protein
S65838	↑	3.6	Metallothionein 1	Acute phase protein, antioxidant

[1]Gene signal at background level in one array.

glycoprotein) and SPI-3 serine proteinase inhibitor. Hence a further indication of pro-inflammatory response in rats fed diets deficient in selenium plus vitamin E is that they exhibited higher expression of the fibrinogen γ-chain, which has been shown to be up-regulated significantly in the rat liver during inflammation (Simpson-Haidaris *et al.*, 1995). The induction of pro-inflammatory genes was accompanied by a concerted depression of the anti-inflammatory enzyme 11-β-hydroxysteroid dehydrogenase 2, which converts the glucocorticoid corticosterone to its inactive 11-dehydro form in the rat, thereby controlling glucocorticoid access to receptors. These results show that selenium deficiency has both a negative impact on selenoprotein expression and, probably as a secondary effect, a modulating effect on inflammatory and cell cycle-dependent genes.

References

Aratri, E., Spycher, S.E., Breyer, I. and Azzi, A. (1999) Modulation of alpha-tropomyosin expression by alpha-tocopherol in rat vascular smooth muscle cells. *FEBS Letters* 447, 91–94.

Azzi, A., Breyer, I., Feher, M., Ricciarelli, R., Stocker, A., Zimmer, S. and Zingg, J. (2001) Nonantioxidant functions of alpha-tocopherol in smooth muscle cells. *Journal of Nutrition* 131, 378S–381S.

Baeuerle, P.A. and Baltimore, D. (1996) NF-kappa B: ten years after. *Cell* 87, 13–20.

Baeuerle, P.A. and Henkel, T. (1994) Function and activation of NF-kappa B in the immune system. *Annual Review of Immunology* 12, 141–179.

Baker, C.S., Hall, R.J., Evans, T.J., Pomerance, A., Maclouf, J., Creminon, C., Yacoub, M.H. and Polak, J.M. (1999) Cyclooxygenase-2 is widely expressed in atherosclerotic lesions affecting native and transplanted human coronary arteries and colocalizes with inducible nitric oxide synthase and nitrotyrosine particularly in macrophages. *Arteriosclerosis, Thrombosis, and Vascular Biology* 19, 646–655.

Beck, M.A., Esworthy, R.S., Ho, Y.S. and Chu, F.F. (1998) Glutathione peroxidase protects mice from viral-induced myocarditis. *FASEB Journal* 12, 1143–1149.

Behne, D., Kyriakopoeulos, A., Weiss-Nowak, C., Kalckloesch, M., Westphal, C. and Gessner, H. (1996) Newly found selenium-containing proteins in the tissues of the rat. *Biological Trace Element Research* 55, 99–110.

Behne, D., Kyriakopoulos, A., Kalcklosch, M., Weiss-Nowak, C., Pfeifer, H., Gessner, H. and Hammel, C. (1997) Two new selenoproteins found in the prostatic glandular epithelium and in the spermatid nuclei. *Biomedical and Environmental Science* 10, 340–345.

Behrens, W.A., Thompson, J.N. and Madere, R. (1982) Distribution of alpha-tocopherol in human plasma lipoproteins. *American Journal of Clinical Nutrition* 35, 691–696.

Ben Hamida, C., Doerflinger, N., Belal, S., Linder, C., Reutenauer, L., Dib, C., Gyapay, G., Vignal, A., Le Paslier, D. and Cohen, D. (1993) Localization of Friedreich ataxia phenotype with selective vitamin E deficiency to chromosome 8q by homozygosity mapping. *Nature Genetics* 5, 195–200.

Bermano, G., Arthur, J.R. and Hesketh, J.E. (1996) Selective control of cytosolic glutathione peroxidase and phospholipid hydroperoxide glutathione peroxidase mRNA stability by selenium supply. *FEBS Letters* 387, 157–160.

Blatt, D.H., Leonard, S.W. and Traber, M.G. (2001) Vitamin E kinetics and the function of tocopherol regulatory proteins. *Nutrition* 17, 799–805.

Borek, C., Ong, A., Mason, H., Donahue, L. and Biaglow, J.E. (1986) Selenium and vitamin E inhibit radiogenic and chemically induced transformation *in vitro* via different mechanisms. *Proceedings of the National Academy of Sciences USA* 83, 1490–1494.

Boscoboinik, D., Szewczyk, A. and Azzi, A. (1991a) Alpha-tocopherol (vitamin E) regulates vascular smooth muscle cell proliferation and protein kinase C activity. *Archives of Biochemistry and Biophysics* 286, 264–269.

Boscoboinik, D., Szewczyk, A., Hensey, C. and Azzi, A. (1991b) Inhibition of cell proliferation by alpha-tocopherol. Role of protein kinase C. *Journal of Biological Chemistry* 266, 6188–6194.

Bösl, M.R., Takaku, K., Oshima, M., Nishimura, S. and Taketo, M.M. (1997) Early embryonic lethality caused by targeted disruption of the mouse selenocysteine tRNA gene. *Proceedings of the National Academy of Sciences USA* 94, 5531–5534.

Brand, K., Page, S., Rogler, G., Bartsch, A., Brandl, R., Knuechel, R., Page, M., Kaltschmidt, C., Baeuerle, P.A. and Neumeier, D. (1996) Activated transcription factor nuclear factor-kappa B is present in the atherosclerotic lesion. *Journal of Clinical Investigation* 97, 1715–1722.

Brand, K., Page, S., Walli, A.K., Neumeier, D. and Baeuerle, P.A. (1997) Role of nuclear factor-kappa B in atherogenesis. *Experimental Physiology* 82, 297–304.

Brigelius-Flohé, R. (1999) Tissue-specific functions of individual glutathione peroxidases. *Free Radical Biology and Medicine* 27, 951–965.

Brigelius-Flohé, R. and Traber, M.G. (1999) Vitamin E: function and metabolism. *FASEB Journal* 13, 1145–1155.

Brigelius-Flohé, R., Friedrichs, B., Maurer, S., Schultz, M. and Streicher, R. (1997) Interleukin-1-induced nuclear factor kappa B activation is inhibited by overexpression of phospholipid hydroperoxide glutathione peroxidase in a human endothelial cell line. *Biochemical Journal* 328, 199–203.

Busse, R. and Fleming, I. (1995) Regulation and functional consequences of endothelial nitric oxide formation. *Annals of Medicine* 27, 331–340.

Cardona-Sanclemente, L.E. and Born, G.V. (1995) Effect of inhibition of nitric oxide synthesis on the uptake of LDL and fibrinogen by arterial walls and other organs of the rat. *British Journal of Pharmacology* 114, 1490–1494.

Catignani, G.L. and Bieri, J.G. (1977) Rat liver alpha-tocopherol binding protein. *Biochimica et Biophysica Acta* 497, 349–357.

Celermajer, D.S., Sorensen, K.E., Georgakopoulos, D., Bull, C., Thomas, O., Robinson, J. and Deanfield, J.E. (1993) Cigarette smoking is associated with dose-related and potentially reversible impairment of endothelium-dependent dilation in healthy young adults. *Circulation* 88, 2149–2155.

Chen, C.J., Huang, H.S., Lin, S.B. and Chang, W.C. (2000) Regulation of cyclooxygenase and 12-lipoxygenase catalysis by phospholipid hydroperoxide glutathione peroxidase in A431 cells. *Prostaglandins, Leukotrienes and Essential Fatty Acids* 62, 261–268.

Chojkier, M., Houglum, K., Lee, K.S. and Buck, M. (1998) Long- and short-term D-alpha-tocopherol supplementation inhibits liver collagen alpha1(I) gene expression. *American Journal of Physiology* 275, G1480–G1485.

Clark, L.C. (1985) The epidemiology of selenium and cancer. *Federation Proceedings* 44, 2584–2589.

Clement, S., Tasinato, A., Boscoboinik, D. and Azzi, A. (1997) The effect of alpha-tocopherol on the synthesis, phosphorylation and activity of protein kinase C in smooth muscle cells after phorbol 12-myristate

13-acetate down-regulation. *European Journal of Biochemistry* 246, 745–749.
Cohn, W., Gross, P., Grun, H., Loechleiter, F., Muller, D.P. and Zulauf, M. (1992) Tocopherol transport and absorption. *Proceedings of the Nutrition Society* 51, 179–188.
Collins, T. and Cybulski, M.I. (2001) NF-κB: pivotal mediator or innocent bystander in atherogenesis. *Journal of Clinical Investigation* 107, 255–264.
Combs, G.F. Jr (1999) Chemopreventive mechanisms of selenium. *Medizinische Klinik* 94 Supplement 3, 18–24.
Cooke, J.P., Singer, A.H., Tsao, P., Zera, P., Rowan, R.A. and Bilingam, M.E. (1992) Antiatherogenic effects of L-arginine in the hypercholesterolemic rabbit. *Journal of Clinical Investigation* 90, 1168–1172.
Copeland, P.R. and Driscoll, D.M. (1999) Purification, redox sensitivity, and RNA binding properties of SECIS-binding protein 2, a protein involved in selenoprotein biosynthesis. *Journal of Biological Chemistry* 274, 25447–25454.
Cowan, D.B., Weisel, R.D., Williams, W.G. and Mickle, D.A. (1993) Identification of oxygen responsive elements in the 5′-flanking region of the human glutathione peroxidase gene. *Journal of Biological Chemistry* 268, 26904–26910.
Cybulsky, M.I. and Gimbrone, M.A. (1991) Endothelial expression of a mononuclear leukocyte adhesion molecule during atherogenesis. *Science* 251, 788–791.
De Caerina, R., Libby, P., Peng, H.B., Thannickal, V.J., Rajavashisth, T.B., Gimbrone, M.A.J., Shin, W.S. and Liao, J.K. (1995) Nitric oxide decreases cytokine-induced endothelial activation: nitric oxide selectively reduces endothelial expression of adhesion molecules and proinflammatorry cytokines. *Journal of Clinical Investigation* 96, 60–68.
de Graaf, J.C., Banga, J.D., Moncado, S., Palmer, R.M., de Groot, P.G. and Sixma, J.J. (1992) Nitric oxide functions as an inhibitor of platelet adhesion under flow conditions. *Circulation* 85, 2284–2290.
Devaraj, S., Li, D. and Jialal, I. (1996) The effects of alpha tocopherol supplementation on monocyte function. Decreased lipid oxidation, interleukin 1 beta secretion, and monocyte adhesion to endothelium. *Journal of Clinical Investigation* 98, 756–763.
Dutta-Roy, A.K. (1999) Molecular mechanism of cellular uptake and intracellular translocation of alpha-tocopherol: role of tocopherol-binding proteins. *Food Chemistry and Toxicology* 37, 967–971.
Ehrenreich, A., Forchhammer, K., Tormay, P., Veprek, B. and Böck, A. (1992) Selenoprotein synthesis in *E. coli*. Purification and characterisation of the enzyme catalysing selenium activation. *European Journal of Biochemistry* 206, 767–773.
Evans, H.M. and Bishop, K.S. (1922) On the existence of a hitherto unrecognized dietary factor essential for reproduction. *Science* 56, 650–651.

Fazzio, A., Marilley, D. and Azzi, A. (1997) The effect of alpha-tocopherol and beta-tocopherol on proliferation, protein kinase C activity and gene expression in different cell lines. *Biochemistry and Molecular Biology International* 41, 93–101.
Fechner, H., Schlame, M., Guthmann, F., Stevens, P.A. and Rustow, B. (1998) Alpha- and delta-tocopherol induce expression of hepatic alpha-tocopherol-transfer-protein mRNA. *Biochemical Journal* 331, 577–581.
Fischer, A., Pallauf, J., Gohil, K., Weber, S.U., Packer, L. and Rimbach, G. (2001) Effect of selenium and vitamin E deficiency on differential gene expression in rat liver. *Biochemical and Biophysical Research Communications* 285, 470–475.
Flohé, L., Andreesen, J.R., Brigelius-Flohé, R., Maiorino, M. and Ursini, F. (2000) Selenium, the element of the moon, in life on earth. *IUBMB Life* 49, 411–420.
Forster, B.A. and Weinberg, P.D. (1997) Changes with age in the influence of endogenous nitric oxide on transport properties of the rabbit aortic wall near branches. *Arteriosclerosis, Thrombosis, and Vascular Biology* 17, 1361–1368.
Fosslien, E. (2001) Review: molecular pathology of cyclooxygenase-2 in cancer-induced angiogenesis. *Annals of Clinical Laboratory Science* 31, 325–348.
Freedman, J.E., Farhat, J.H., Loscalzo, J. and Keaney, J.F. Jr (1996) Alpha-tocopherol inhibits aggregation of human platelets by a protein kinase C-dependent mechanism. *Circulation* 94, 2434–2440.
Garg, U.C. and Hassid, A. (1989) Nitric oxide generation vasodilators and 8-bromo-cyclic guanosine monophosphate inhibit mitogenesis and proliferation of cultured rat vascular smooth muscle cells. *Journal of Clinical Investigation* 83, 1774–1777.
Gladyshev, V.N. and Hatfield, D.L. (1999) Selenocysteine-containing proteins in mammals. *Journal of Biomedical Sciences* 6, 151–160.
Halliwell, B. (1996) Antioxidants in human health and disease. *Annual Review of Nutrition* 16, 33–50.
Halliwell, B. (2000) Why and how should we measure oxidative DNA damage in nutritional studies? How far have we come? *American Journal of Clinical Nutrition* 72, 1082–1087.
Hayashi, T., Ueno, Y. and Okamoto, T. (1993) Oxidoreductive regulation of nuclear factor kappa B. Involvement of a cellular reducing catalyst thioredoxin. *Journal of Biological Chemistry* 268, 11380–11388.
Hehenberger, K. and Hansson, A. (1997) High glucose-induced growth factor resistance in human fibroblasts can be reversed by antioxidants and protein kinase C-inhibitors. *Cell Biochemistry and Function* 15, 197–201.
Herrera, E. and Barbas, C. (2001) Vitamin E: action, metabolism and perspectives. *Journal of Physiological Biochemistry* 57, 43–56.

Ho, Y.S., Magnenat, J.L., Bronson, R.T., Cao, J., Gargano, M., Sugawara, M. and Funk, C.D. (1997) Mice deficient in cellular glutathione peroxidase develop normally and show no increased sensitivity to hyperoxia. *Journal of Biological Chemistry* 272, 16644–16651.

Hosomi, A., Arita, M., Sato, Y., Kiyose, C., Ueda, T., Igarashi, O., Arai, H. and Inoue, K. (1997) Affinity for alpha-tocopherol transfer protein as a determinant of the biological activities of vitamin E analogs. *FEBS Letters* 409, 105–108.

Imai, H., Narashima, K., Arai, M., Sakamoto, H., Chiba, N. and Nakagawa, Y. (1998) Suppression of leukotriene formation in RBL-2H3 cells that overexpressed phospholipid hydroperoxide glutathione peroxidase. *Journal of Biological Chemistry* 273, 1990–1997.

Ingold, K.U., Webb, A.C., Witter, D., Burton, G.W., Metcalfe, T.A. and Muller, D.P. (1987) Vitamin E remains the major lipid-soluble, chain-breaking antioxidant in human plasma even in individuals suffering severe vitamin E deficiency. *Archives of Biochemistry and Biophysics* 259, 224–225.

Jiang, Q., Elson-Schwab, I., Courtemanche, C. and Ames, B.N. (2000) γ-Tocopherol and its major metabolite, in contrast to α-tocopherol, inhibit cyclooxygenase activity in macrophages and epithelial cells. *Proceedings of the National Academy of Sciences USA* 97, 11494–11499.

Kagan, V.E. and Tyurina, Y.Y. (1998) Recycling and redox cycling of phenolic antioxidants. *Annals of the New York Academy of Sciences* 854, 425–434.

Kanno, T., Utsumi, T., Kobuchi, H., Takehara, Y., Akiyama, J., Yoshioka, T., Horton, A.A. and Utsumi, K. (1995) Inhibition of stimulus-specific neutrophil superoxide generation by alpha-tocopherol. *Free Radical Research* 22, 431–440.

Kayanoki, Y., Fujii, J., Islam, K.N., Suzuki, K., Kawata, S., Matsuzawa, Y. and Taniguchi, N. (1996) The protective role of glutathione peroxidase in apoptosis induced by reactive oxygen species. *Journal of Biochemistry* 119, 817–822.

Kayden, H.J. and Traber, M.G. (1993) Absorption, lipoprotein transport, and regulation of plasma concentrations of vitamin E in humans. *Journal of Lipid Research* 34, 343–358.

Knight, S.A. and Sunde, R.A. (1987) The effect of progressive selenium deficiency on anti-glutathione peroxidase antibody reactive protein in rat liver. *Journal of Nutrition* 117, 732–738.

Knight, S.A. and Sunde, R.A. (1988) Effect of selenium repletion on glutathione peroxidase protein level in rat liver. *Journal of Nutrition* 118, 853–858.

Kocher, O., Gabbiani, F., Gabbiani, G., Reidy, M.A., Cokay, M.S., Peters, H. and Huttner, I. (1991) Phenotypic features of smooth muscle cells during the evolution of experimental carotid artery intimal thickening. Biochemical and morphologic studies. *Laboratory Investigation* 65, 459–470.

Köhrle, J., Brigelius-Flohé, R., Böck, A., Gartner, R., Meyer, O. and Flohé, L. (2000) Selenium in biology: facts and medical perspectives. *Biological Chemistry* 381, 849–864.

Kretz-Remy, C., Mehlen, P., Mirault, M.E. and Arrigo, A.P. (1996) Inhibition of IκB-α phosphorylation and degradation and subsequent NF-κB activation by glutathione peroxidase overexpression. *Journal of Cell Biology* 133, 1083–1093.

Lagrost, L., Desrumaux, C., Masson, D., Deckert, V. and Gambert, P. (1998) Structure and function of the plasma phospholipid transfer protein. *Current Opinion in Lipidology* 9, 203–209.

Lanfear, J., Fleming, J., Wu, L., Webster, G. and Harrison, P.R. (1994) The selenium metabolite selenodiglutathione induces p53 and apoptosis: relevance to the chemopreventive effects of selenium? *Carcinogenesis* 15, 1387–1392.

Leeson, C.P.M., Whincup, P.H., Cook, D.G., Donald, A.E., Papacosta, O., Lucas, A. and Deanfield, J.E. (1997) Flow mediated dilation in 9- to 11-year old children – the influence of intrauterine and childhood factors. *Circulation* 96, 2233–2238.

Leinfelder, W., Zehelein, E., Mandrand-Berthelot, M.A. and Böck, A. (1988) Gene for a novel tRNA species that accepts L-serine and cotranslationally inserts selenocysteine. *Nature* 331, 723–725.

Li, D., Saldeen, T., Romeo, F. and Mehta, J.L. (2001) Different isoforms of tocopherols enhance nitric oxide synthase phosphorylation and inhibit human platelet aggregation and lipid peroxidation: implications in therapy with vitamin E. *Journal of Cardiovascular and Pharmacological Therapy* 6, 155–161.

Li, N.Q., Reddy, P.S., Thyagaraju, K., Reddy, A.P., Hsu, B.L., Scholz, R.W., Tu, C.P. and Reddy, C.C. (1990) Elevation of rat liver mRNA for selenium-dependent glutathione peroxidase by selenium deficiency. *Journal of Biological Chemistry* 265, 108–113.

Liao, F., Andalibi, A., deBeer, F.C., Fogelman, A.M. and Lusis, A.J. (1993) Genetic control of inflammatory gene induction and NF-kappa B-like transcription factor activation in response to an atherogenic diet in mice. *Journal of Clinical Investigation* 91, 2572–5279.

Lindner, V. and Collins, T. (1996) Expression of NF-κB and IκB-α by aortic endothelium in an arterial injury model. *American Journal of Pathology* 148, 427–438.

Low, S.C. and Berry, M.J. (1996) Knowing when not to stop: selenocysteine incorporation in eukaryotes. *Trends in Biochemical Sciences* 21, 203–208.

Maiorino, M., Wissing, J.B., Brigelius-Flohé, R., Calabrese, F., Roveri, A., Steinert, P., Ursini, F. and Flohé, L. (1998) Testosterone mediates expression of the selenoprotein PHGPx by induction of spermatogenesis and not by direct transcriptional gene activation. *FASEB Journal* 12, 1359–1370.

Malins, D.C., Johnson, P.M., Wheeler, T.M., Barker, E.A., Polissar, N.L. and Vinson, M.A. (2001) Age-related radical-induced DNA damage is linked to prostate cancer. *Cancer Research* 61, 6025–6028.

Mansell, J.B. and Berry, M.J. (2001) Towards a mechanism for selenocysteine incorporation in eukaryotes. In: Hatfield, D.L. (ed.) *Selenium. Its Molecular Biology and Role in Human Health*. Kluwer Academic Publishers, Boston, pp. 69–80.

Matthews, J.R., Botting, C.H., Panico, M., Morris, H.R. and Hay, R.T. (1996) Inhibition of NF-κB DNA binding by nitric oxide. *Nucleic Acids Research* 24, 2236–2242.

Matz, R.L., Schott, C., Stoclet, J.C. and Andriantsitohaina, R. (2000) Age-related endothelial dysfunction with respect to nitric oxide, endothelium-derived hyperpolarizing factor and cyclooxygenase products. *Physiological Research* 49, 11–18.

May, J.M., Qu, Z.C. and Mendiratta, S. (1998) Protection and recycling of alpha-tocopherol in human erythrocytes by intracellular ascorbic acid. *Archives of Biochemistry and Biophysics* 349, 281–289.

Müller, A.S. and Pallauf, J. (2002) Downregulation of GPx1 mRNA and the loss of GPx1 activity causes cellular damage in the liver of selenium deficient rabbits. *Journal of Animal Physiology and Animal Nutrition* 86, 273–287.

Nakamura, T., Goto, M., Matsumoto, A. and Tanaka, I. (1998) Inhibition of NF-kappa B transcriptional activity by alpha-tocopheryl succinate. *Biofactors* 7, 21–30.

Naruse, K., Shimuzu, K., Maramatsu, M., Toky, Y., Miyazaki, Y., Okamura, K., Hashimoto, H. and Ito, T. (1994) Long-term inhibition on NO synthesis promoter atherosclerosis in the hypercholesterolemic rabbit thoracic aorta: PGH2 does not contribute to impaired endothelium-dependent relaxation. *Arteriosclerosis and Thrombosis* 14, 746–752.

Packham, G., Ashmun, R.A. and Cleveland, J.L. (1996) Cytokines suppress apoptosis independent of increases in reactive oxygen levels. *Journal of Immunology* 156, 2792–2800.

Panza, J.A., Garcia, C.E., Kilcoyne, C.M., Quyymi, A.A. and Cannon, R.O. (1995) Impaired endothelium-dependent vasodilatation in patients with essential hypertension: evidence that nitric oxide abnormality is not localized to a single signal transduction pathway. *Circulation* 91, 1732–1738.

Park, H.S., Huh, S.H., Kim, Y., Shim, J., Lee, S.H., Park, I.S., Jung, Y.K., Kim, I.Y. and Choi, E.J. (2000) Selenite negatively regulates caspase-3 through a redox mechanism. *Journal of Biological Chemistry* 275, 8487–8491.

Parthasarathy, S., Santanam, N., Ramachandran, S. and Meilhac, O. (1999) Oxidants and antioxidants in atherogenesis. An appraisal. *Journal of Lipid Research* 40, 2143–2157.

Patel, R.P., Levonen, A.L., Crawford, J.H. and Darley-Usma, V.M. (2000) Mechanisms of the pro- and antioxidant actions of nitric oxide in atherosclerosis. *Cardiovascular Research* 47, 465–474.

Pence, B.C., Delver, E. and Dunn, D.M. (1994) Effects of dietary selenium on UVB-induced skin carcinogenesis and epidermal antioxidant status. *Journal of Investigative Dermatology* 102, 759–761.

Raisanen, S.R., Lehenkari, P., Tasanen, M., Rahkila, P., Harkonen, P.L. and Vaananen, H.K. (1999) Carbonic anhydrase III protects cells from hydrogen peroxide-induced apoptosis. *FASEB Journal* 13, 513–522.

Ricciarelli, R., Tasinato, A., Clement, S., Ozer, N.K., Boscoboinik, D. and Azzi, A. (1998) Alpha-tocopherol specifically inactivates cellular protein kinase C alpha by changing its phosphorylation state. *Biochemical Journal* 334, 243–249.

Ricciarelli, R., Maroni, P., Ozer, N., Zingg, J.M. and Azzi, A. (1999) Age-dependent increase of collagenase expression can be reduced by alpha-tocopherol via protein kinase C inhibition. *Free Radical Biology and Medicine* 27, 729–737.

Ricciarelli, R., Zingg, J.M. and Azzi, A. (2000) Vitamin E reduces the uptake of oxidized LDL by inhibiting CD36 scavenger receptor expression in cultured aortic smooth muscle cells. *Circulation* 102, 82–87.

Rimbach, G., Valacchi, G., Canali, R. and Virgili, F. (2000) Macrophages stimulated with IFN-γ activate NF-κB and induce MCP-1 gene expression in primary human endothelial cells. *Molecular Cell Biology Research Communications* 3, 238–242.

Rubanyi, G.M. (1993) The role of endothelium in cardiovascular homeostasis and diseases. *Cardiovascular Pharmacology* 22, S1–S14.

Sandstrom, P.A., Tebbey, P.W., Van Cleave, S. and Buttke, T.M. (1994) Lipid hydroperoxides induce apoptosis in T cells displaying a HIV-associated glutathione peroxidase deficiency. *Journal of Biological Chemistry* 269, 798–801.

Schnurr, K., Borchert, A. and Kuhn, H. (1999) Inverse regulation of lipid-peroxidizing and hydroperoxyl lipid-reducing enzymes by interleukins 4 and 13. *FASEB Journal* 13, 143–154.

Schütze, N., Bachthaler, M., Lechner, A., Köhrle, J. and Jakob, F. (1998) Identification by differential display PCR of the selenoprotein thioredoxin reductase as a 1 alpha,25(OH)2-vitamin D3-responsive gene in human osteoblasts – regulation by selenite. *Biofactors* 7, 299–310.

Schütze, N., Fritsche, J., Ebert-Dumig, R., Schneider, D., Köhrle, J., Andreesen, R., Kreutz, M. and Jakob, F. (1999) The selenoprotein thioredoxin reductase is expressed in peripheral blood monocytes and THP1 human myeloid leukemia cells – regulation by 1,25-dihydroxyvitamin D3 and selenite. *Biofactors* 10, 329–338.

Serbinova, E., Kagan, V., Han, D. and Packer, L. (1991) Free radical recycling and intramembrane mobility in the antioxidant properties of alpha-tocopherol and alpha-tocotrienol. *Free Radical Biology and Medicine* 10, 263–275.

Sies, H., Klotz, L.O., Sharov, V.S., Assmann, A. and Briviba, K. (1998) Protection against peroxynitrite by selenoproteins. *Zeitschrift für Naturforschung* 53, 228–232.

Simpson-Haidaris, P.J., Wright, T.W., Earnest, B.J., Hui, Z., Neroni, L.A. and Courtney, M.A. (1995) Cloning and characterization of a lung-specific cDNA corresponding to the gamma chain of hepatic fibrinogen. *Gene* 167, 273–278.

Sinha, R. and Medina, D. (1997) Inhibition of cdk2 kinase activity by methylselenocysteine in synchronized mouse mammary epithelial tumor cells. *Carcinogenesis* 18, 1541–1547.

Stocker, A., Zimmer, S., Spycher, S.E. and Azzi, A. (1999) Identification of a novel cytosolic tocopherol-binding protein: structure, specificity, and tissue distribution. *IUBMB Life* 48, 49–55.

Stroes, E.S., Koomans, H.A., de Bruin, T.W. and Rabelink, T.J. (1995) Vascular function in the forearm of vasodilation in patients with non-insulin-dependent diabetes mellitus. *Journal of the American College of Cardiology* 27, 567–574.

Sunde, R.A. (1997) Selenium. In: O'Dell, B.L. and Sunde, R.A. (eds) *Handbook of Nutritionally Essential Mineral Elements*. Marcel Dekker, New York, pp. 493–556.

Suzuki, Y.J. and Packer, L. (1993) Inhibition of NF-kappa B activation by vitamin E derivatives. *Biochemical and Biophysical Research Communications* 193, 277–283.

Suzuki, Y.J., Tsuchiya, M., Wassall, S.R., Choo, Y.M., Govil, G., Kagan, V.E. and Packer, L. (1993) Structural and dynamic membrane properties of alpha-tocopherol and alpha-tocotrienol: implication to the molecular mechanism of their antioxidant potency. *Biochemistry* 32, 10692–10699.

Tada, H., Ishii, H. and Isogai, S. (1997) Protective effect of D-alpha-tocopherol on the function of human mesangial cells exposed to high glucose concentrations. *Metabolism* 46, 779–784.

Terasawa, Y., Ladha, Z., Leonard, S.W., Morrow, J.D., Newland, D., Sanan, D., Packer, L., Traber, M.G. and Farese, R.V. Jr (2000) Increased atherosclerosis in hyperlipidemic mice deficient in alpha-tocopherol transfer protein and vitamin E. *Proceedings of the National Academy of Sciences USA* 97, 13830–13834.

Thiery, J., Teupser, D., Walli, A.K., Ivandic, B., Nebendahl, K., Stein, O., Stein, Y. and Seidel, D. (1996) Study of causes underlying the low atherosclerotic response to dietary hypercholesterolemia in a selected strain of rabbits. *Atherosclerosis* 121, 63–73.

Thompson, K.M., Haibach, H. and Sunde, R.A. (1995) Growth and plasma triiodothyronine concentrations are modified by selenium deficiency and repletion in second-generation selenium-deficient rats. *Journal of Nutrition* 125, 864–873.

Traber, M.G. and Kayden, H.J. (1989) Preferential incorporation of alpha-tocopherol vs gamma-tocopherol in human lipoproteins. *American Journal of Clinical Nutrition* 49, 517–526.

Traber, M.G., Sokol, R.J., Burton, G.W., Ingold, K.U., Papas, A.M., Huffaker, J.E. and Kayden, H.J. (1990) Impaired ability of patients with familial isolated vitamin E deficiency to incorporate alpha-tocopherol into lipoproteins secreted by the liver. *Journal of Clinical Investigation* 85, 397–407.

Viita, H., Sen, C.K., Roy, S., Siljamaki, T., Nikkari, T. and Yla-Herttuala, S. (1999) High expression of human 15-lipoxygenase induces NF-κB-mediated expression of vascular cell adhesion molecule 1, intercellular adhesion molecule 1, and T-cell adhesion on human endothelial cells. *Antioxidants and Redox Signalling* 1, 83–96.

Weitzel, F., Ursini, F. and Wendel, A. (1990) Phospholipid hydroperoxide glutathione peroxidase in various mouse organs during selenium deficiency and repletion. *Biochimica et Biophysica Acta* 1036, 88–94.

Welty, F.K., Lichtenstein, A.H., Barrett, P.H., Jenner, J.L., Dolnikowski, G.G. and Schaefer, E.J. (2000) Effects of ApoE genotype on ApoB-48 and ApoB-100 kinetics with stable isotopes in humans. *Arteriosclerosis, Thrombosis and Vascular Biology* 20, 1807–1810.

Williams, S.B., Cusco, J.A., Roddy, M.A., Johnstone, M.T. and Creager, M.A. (1996) Impaired nitric oxide-mediated vasodilation in patients with non-insulin-dependent diabetes mellitus. *Journal of the American College of Cardiology* 27, 567–574.

Wingler, K., Müller, C., Schmehl, K., Florian, S. and Brigelius-Flohé, R. (2000) Gastrointestinal glutathione peroxidase prevents transport of lipid hydroperoxides in CaCo-2 cells. *Gastroenterology* 119, 420–430.

Wu, D., Hayek, M.G. and Meydani, S. (2001) Vitamin E and macrophage cyclooxygenase regulation in the aged. *Journal of Nutrition* 131, 382S–388S.

Yamamoto, Y. and Takahashi, K. (1993) Glutathione peroxidase isolated from plasma reduces phospholipid hydroperoxides. *Archives of Biochemistry and Biophysics* 305, 541–545.

Yamauchi, J., Iwamoto, T., Kida, S., Masushige, S., Yamada, K. and Esashi, T. (2001) Tocopherol-associated protein is a ligand-dependent transcriptional activator. *Biochemical and Biophysical Research Communications* 285, 295–299.

Yan, S.D., Schmidt, A.M., Anderson, G.M., Zhang, J., Brett, J., Zou, Y.S., Pinsky, D. and Stern, D. (1994) Enhanced cellular oxidant stress by the interaction of advanced glycation end products with their receptors/binding proteins. *Journal of Biological Chemistry* 269, 9889–9897.

Yoshida, N., Manabe, H., Terasawa, Y., Nishimura, H., Enjo, F., Nishino, H. and Yoshikawa, T. (2000) Inhibitory effects of vitamin E on endothelial-dependent adhesive interactions with leukocytes induced by oxidized low density lipoprotein. *Biofactors* 13, 279–288.

Zapolska-Downar, D., Zapolski-Downar, A., Markiewski, M., Ciechanowicz, A., Kaczmarczyk, M. and Naruszewicz, M. (2000) Selective inhibition by alpha-tocopherol of vascular cell adhesion molecule-1 expression in human vascular endothelial cells. *Biochemical and Biophysical Research Communications* 274, 609–615.

Zimmer, S., Stocker, A., Sarbolouki, M.N., Spycher, S.E., Sassoon, J. and Azzi, A. (2000) A novel human tocopherol-associated protein: cloning, *in vitro* expression, and characterization. *Journal of Biological Chemistry* 275, 25672–25680.

13 Sphingolipids: a New Strategy for Cancer Treatment and Prevention

Eva M. Schmelz
Karmanos Cancer Institute, Wayne State University School of Medicine, Detroit, Michigan, USA

Introduction

Sphingolipids are important structural components found in all eukaryotic and some prokaryotic cells. They are localized mainly in membranes, and thus influence membrane stability and fluidity, and modify the behaviour of proteins localized in the membranes. It is only in the last few decades that the role of sphingolipid metabolites as lipid second messengers has been discovered, and it has been shown that sphingolipids are involved in the regulation of a wide spectrum of processes that regulate cell growth and cell death. This review will summarize known effects of dietary sphingolipids on cell regulation, specifically in colon cancer cells, and propose signalling pathways that dietary sphingolipids utilize in the suppression of colon cancer.

Sphingolipid Structure and Metabolism

Sphingolipids are composed of a sphingoid base (in mammalian cells mostly sphingosine), an amide-bound fatty acid and a headgroup. As shown in Fig. 13.1, the *de novo* synthesis of sphingolipids begins with the synthesis of sphinganine from serine and palmitoyl-CoA, catalysed by serine-palmitoyl transferase, with ketosphinganine as an intermediate. Acylation of the amino group by ceramide synthase results in the formation of dihydroceramide. At this point, the 4,5-*trans* double bond can be introduced by a desaturase to form ceramide, or a headgroup is added for the synthesis of more complex sphingolipids. In this case, the double bond is introduced after the turnover of this complex sphingolipid (for more information, see the review by Huwiler *et al.*, 2000). Variations in these components make sphingolipids the structurally most diverse class of membrane lipids. The degradation of sphingolipids requires the phosphorylation of sphingosine (resulting from the hydrolysis of complex sphingolipids) to sphingosine-1-phosphate by sphingosine kinase, and the activity of sphingosine lyase to produce ethanolamine and hexadecenal. The intermediates of cellular sphingolipid metabolism have a significant impact on cell behaviour; this will be discussed in the following section.

Sphingolipids as Second Messengers

One of the first demonstrations that sphingolipid metabolites are involved in signal transduction came from a report that sphingosine inhibits protein kinase C (PKC) activity (Hannun *et al.*, 1986). Subsequently, the hydrolysis of sphingomyelin followed by ceramide accumulation was shown after tumour necrosis factor-α (TNF-α) or interferon-γ (IFN-γ) treatment of cells via activation of sphingomyelinases. Today the list of agents or events that activate this pathway includes growth factors, compounds inducing differentiation or

©CAB *International* 2003. *Molecular Nutrition*
(eds J. Zempleni and H. Daniel)

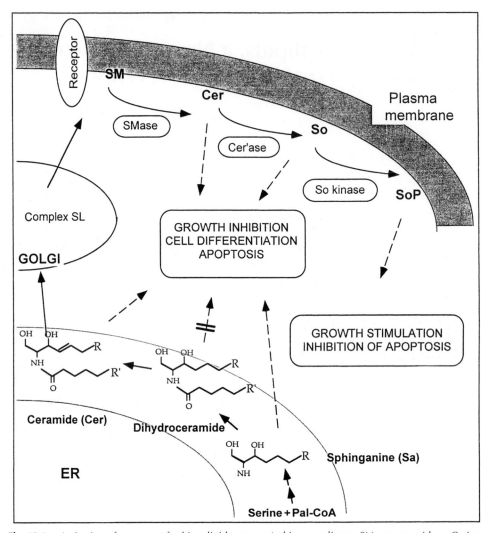

Fig. 13.1. Activation of enzymes of sphingolipid turnover (sphingomyelinase, SMase; ceramidase, Cer'ase; sphingosine kinase, So kinase) by extracellular agents or stress-activating receptors, or *de novo* synthesis can contribute to the accumulation of intracellular bioactive sphingolipid metabolites.

apoptosis (such as vitamin D_3, toxins, chemotherapeutic agents), and cellular stresses such as hypoxia and γ-irradiation. Some of the agents activate not only sphingomyelinases but also ceramidases that hydrolyse ceramide and release free sphingoid bases. Both metabolites generally are growth inhibitory and cytotoxic, and have been shown to induce apoptosis in numerous cell lines (Spiegel and Merrill, 1996; Mathias *et al.*, 1998). Furthermore, a combination of both seems to be more effective (Jarvis *et al.*, 1996). Growth factors such as platelet-derived growth factor (PDGF) also activate sphingosine kinase (Fig. 13.1) and, in contrast to ceramides and sphingoid bases, the effect of sphingosine-1-phosphate is mitotic; it can also block ceramide-induced apoptosis (Spiegel, 1999). The response of a cell to exogenous or endogenous sphingolipids is therefore dependent on the metabolite that is accumulating in the cell or the combination of metabolites that is generated by extracellular stimuli. Some agents increase intracellular ceramide concentrations not by activating the degradation of complex membrane sphingolipids but by activating *de novo* synthesis. This has been shown for daunorubicin (Bose et al., 1995), TNF-α (Xu et al., 1998), etoposide (Perry et al., 2000) and photodynamic therapy (Separovic et al., 1997). Importantly, the effects of the endogenous metabolites can be mimicked by cell-permeable exogenous sphingolipids. This has been widely used in cell culture, and *in vivo* using dietary sphingolipids for the treatment of chemically induced colon cancer.

Sphingolipids in Cancer

The roles of sphingolipids in the regulation of cell growth, differentiation and cell death that have been demonstrated in many *in vitro* studies suggested a possible impact of exogenous and/or endogenous sphingolipids on cells that lack tight regulation, such as cancer cells. Cancerous growth is a result of hyperproliferation, or a reduction of apoptosis (or a combination of both), and reduced differentiation. Sphingolipids are growth inhibitory, are cytotoxic to most cell lines and can induce apoptosis or differentiation. Furthermore, they have been shown to inhibit or circumvent multidrug resistance, inhibit cell motility and inhibit angiogenesis. These functions, together with the availability in food and safety of sphingolipids in the diet, make sphingolipids good candidates for cancer intervention and/or cancer prevention studies.

Sphingolipids in foods

Analyses specifically directed towards the identification of sphingolipid species in foods and their quantitation have yet to be performed. However, dairy and meat products, eggs and soybeans are especially rich in sphingolipids (Vesper et al., 1999). In animal products, the major sphingolipid is sphingomyelin, usually with a sphingosine backbone and amide-bound fatty acids that are mostly saturated and contain 16, or 22–24 carbons. In plants, the prominent sphingolipid class is cerebrosides, containing a variety of different sugar headgroups (glucose, galactose, mannose), sphingoid base backbones (different number and location of double bonds) and fatty acids (also saturated, but often α-hydroxylated).

Sphingolipids are minor components of food, and they do not contribute significantly to the energy content of foods. Although they are lipids, they are also found in 'fat-free' foods such as skimmed milk, since they are localized in the membranes of the milk fat globules that partially remain in the milk after the skimming process (Jensen, 1995). There are no known nutrient requirements for sphingolipids. None the less, their effect on experimental colon cancer in rodents suggests that they should be categorized as 'functional foods'. Evidence that dietary sphingolipids are safe to use as chemopreventive or protective agents comes from a study by Kobayashi et al. (1997), who showed that even high amounts of sphingolipids in the diet (1% by weight or 100 times more than average consumption) did not affect body weight, or blood lipid levels in mature rats or their offspring. This may be due to the limited digestion of complex sphingolipids (Nyberg et al., 1997). Moreover, feeding sphingolipids in the diet to rodents did not permanently increase blood levels of sphingolipid metabolites (Schmelz et al., 2001).

Epidemiological evidence that sphingolipid-rich foods alter cancer incidence

Numerous studies have been conducted to assess the effect of foods on the risk of cancer. The consumption of milk and milk products has been correlated with a decreased risk in colon cancer in some studies (Jarvinen et al., 2001), whereas other studies found no effect. However, it is very difficult to compare the results from these studies because of the very different designs and problematic influences of variables such as total fat or caloric intake. In studies that showed a decease in cancer risk, this usually was attributed to the calcium content (see Lipkin, 1999), although other components in milk, i.e. conjugated linoleic acid, butyrate, vitamin D and lactoferrin, have also been shown to inhibit colon and other cancers (see Holt, 1999; Parodi, 1999). Since there is no database available to correlate the sphingolipid content in foods to the incidence of (colon) cancer, the role of dietary sphingolipids has yet to be addressed in epidemiological studies.

Sphingolipids in skin cancer

The physicochemical properties of the sphingolipid metabolites and their considerable toxicity make the delivery of sphingolipids to *in vivo* targets problematic. This has limited the application of exogenous sphingolipids in cancer treatment/prevention to very few experimental models. Topical application has been used for measuring the effect of sphingosine, methylsphingosine and *N*-acetylsphingosine on skin cancer in Sencar mice (Enkvetchakul *et al.*, 1989, 1992; Birt *et al.*, 1998). Sphingosine did not inhibit the development of dimethylbenz-[*a*]-anthracene-induced papillomas (Enkvetchakul *et al.*, 1989, 1992). At high doses, the formation of papillomas was even enhanced (Enkvetchakul *et al.*, 1992). This was also seen in a follow-up study measuring the efficacy of sphingosine, *N*-methylsphingosine and *N*-acetylsphingosine; again, the sphingolipids did not change papilloma incidence (Birt *et al.*, 1998). However, both *N*-methylsphingosine and *N*-acetylsphingosine increased cancer-free survival. Furthermore, weekly application of sphingosine and *N*-acetylsphingosine for 10 weeks after treatment with phorbol esters suppressed tumour progression in these mice (Birt *et al.*, 1998).

Another sphingolipid derivative, safingol, the L-threo isomer of sphinganine, a potent inhibitor of PKC, has been developed to treat dermatoses and cancer. Topical application of safingol for 6 weeks caused increases in serum enzymes and liver damage that was more pronounced in female than in male rats although initial plasma levels were comparable. This indicates that topical safingol was taken up by the rats and induced a higher rate of apoptosis in the livers of female rats (characterized by the lack of inflammation and DNA laddering) because they showed insufficient clearance of safingol by cytochrome P450 isozymes (Carfagna *et al.*, 1996).

Dietary sphingomyelin suppresses early stages of chemically induced colon cancer in CF1 mice

Dietary complex sphingolipids are digested in all parts of the small intestine and the colon, and the bioactive metabolites ceramide and sphingosine are taken up by the cells (Schmelz *et al.*, 1994, 2001). Most of the metabolites are taken up by the small intestine, but about 10% reaches the colon (Schmelz *et al.*, 1994). Thus, by feeding complex sphingolipids, the bioactive metabolites are delivered directly to the colonic cells. It was therefore pertinent to conduct studies on the efficacy of dietary sphingolipids on colon carcinogenesis. In these studies, the effects of the sphingolipids on the appearance of early morphological markers of colon tumorigenesis, the aberrant crypt foci (ACF), were evaluated. It is thought that adenomas and adenocarcinomas will develop over time from these early lesions, making ACF a widely used cost- and time effective-end point for studies assessing the effects of dietary compounds on colon carcinogenesis.

CF1 mice were injected with 1,2-dimethylhydrazine (DMH) to induce colon tumours. After tumour initiation, the mice were fed a semi-purified, essentially sphingolipid-free AIN 76A diet alone (control group), or supplemented with 0.025–0.1% (by weight) complex sphingolipids. Sphingomyelin purified from skimmed milk powder reduced ACF formation by 70% (Fig. 13.2) and reduced tumour formation after long-term feeding (Dillehay *et al.*, 1994). In a larger follow-up study, the suppression of ACF by dietary sphingomyelin was confirmed, and a suppression of tumour progression from benign adenomas to malignant adenocarcinomas was noted (Schmelz *et al.*, 1996). This was indeed due to the sphingomyelin in the diet and not altered by a co-purified contaminant since a synthetic sphingomyelin showed a suppression of ACF comparable with the 'natural' compound from milk (Schmelz *et al.*, 1997).

Dietary sphingolipids reduce tumour formation in Min mice

Clinical trials to assess the effect of dietary sphingolipids on human colon cancer have yet to be performed. None the less, rodent models are available that closely resemble the human disease. C57/B6J$^{Min/+}$ mice (multiple intestinal neoplasia, Min mice) carry a mutation in the APC (adenomatous polyposis coli) gene that is found in almost all patients with familial adenomatous polyposis (FAP). These patients spontaneously develop thousands of polyps in the colon, some of which

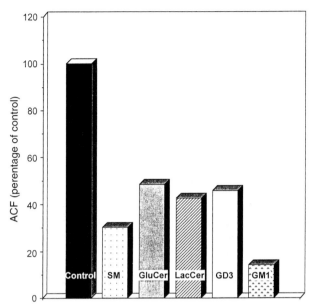

Fig. 13.2. Inhibtion of aberrant crypt foci (ACF) formation by dietary sphingolipids. CF1 mice were treated with a colon carcinogen and fed diets that are essentially sphingolipid free (control), or supplemented with sphingomyelin (SM), glucosylceramide (GluCer), lactosylceramide (LacCer) or the gangliosides G_{D3} or G_{M1}. After 4 weeks, the number of one of the earliest morphological changes in colon carcinogenesis, the ACF, was determined.

inevitably will progress into adenocarcinomas. The APC gene mutation is also found in 40–80% of sporadic colon cancer (Nagase and Nakamura, 1993; Sparks et al., 1998). Whereas the location of the tumours in Min mice is different from that of FAP patients, where they are located mostly in the colon, both have the same phenotype (fully autosomal dominantly inherited, development of multiple intestinal tumours, induced by *Apc* mutations) (Kinzler and Vogelstein, 1996). We supplemented the diet of Min mice with a mixture of complex sphingolipids as they appear in milk (70% sphingomyelin, 5% lactosylceramide, 7.5% glucosylceramide and 7.5% ganglioside G_{D3}) at 0.1% of the diet after weaning. This again was a late intervention study since tumors have already developed throughout the intestinal tract at this time. Feeding the sphingolipids for 65 days reduced tumour formation throughout the intestinal tract by 40% (Fig. 13.3). Adding ceramide to this mixture but maintaining 0.1% of the diet increased tumour suppression to 50%. This was seen in all regions of the intestinal tract (Schmelz et al., 2001).

Structure dependency of sphingolipid effects

As mentioned above, complex dietary sphingolipids are hydrolysed to ceramide and free sphingoid bases. The structure of the ceramide backbones has been shown to determine its biological effects. Ceramides that contain sphinganine (no 4,5-*trans* double bond, termed dihydroceramides) are biologically inactive compounds (Bielawska et al., 1993). Complex dietary sphingolipids containing the dihydroceramide backbone would therefore be inactive if the main metabolite responsible for ACF suppression was ceramide. However, feeding DMH-treated CF1 mice with sphingomyelin with a sphinganine backbone, dihydrosphingomyelin, reduced ACF formation even further (Schmelz et al., 1997). This suggests that the most important bioactive metabolites in ACF suppression may be the free sphingoid bases rather than ceramide, since there is no difference in the biological activity of sphingosine and sphinganine. Still, the introduction of a double bond by an intracellular desaturase, as is seen in the *de novo* synthesis of endogenous sphingolipids

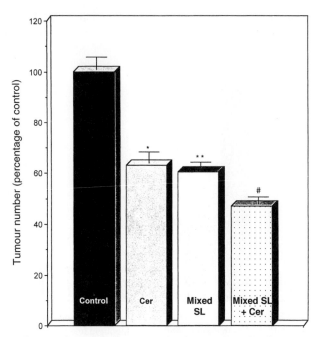

Fig. 13.3. Inhibition of tumour formation in Min mice by dietary sphingolipids. Min mice were fed a sphingolipid-free diet alone (control) or supplemented with 0.1% ceramide (Cer), complex sphingolipids (SL) as they appear in milk (mixed SL), or a combination of both (40% ceramide, 60% mixed SL). After 65 days, the number and localization of tumours were determined. (*$P < 0.05$; **$P < 0.01$; #$P < 0.001$).

after the uptake of dihydroceramide, cannot be ruled out at this point.

The complexity or structure of sphingolipid headgroups may affect digestion and thus the release of bioactive molecules necessary for ACF or tumour suppression. Therefore, we compared the effect of more complex sphingolipids on the appearance of ACF. Complex sphingolipids found in milk in appreciable amounts, glucosylceramide, lactosylceramide and ganglioside G_{D3}, also reduced ACF formation in CF1 mice (Schmelz et al., 2000). As shown in Fig. 13.2, ganglioside G_{M1} from brain was even more potent than sphingomyelin (Dillehay et al., 1994). This presumably is due to the (perhaps non-specific) removal of the headgroup and fatty acids in all of these complex sphingolipids (Schmelz et al., 2000), and the release of sphingoid bases from these compounds. Thus, it is feasible that most or all complex sphingolipids found in foods may be digestible, and could play a role in suppression of colon cancer. This is supported by our finding that glucosylceramide from soybeans (different from milk by the number and/or localization of the double bonds on the sphingosine moiety) also reduced tumour formation in the Min mice (E.M. Schmelz and A.H. Merrill, unpublished results).

Use of sphingolipid metabolites in chemotherapy

Dietary sphingolipids, to our knowledge, have not been used as chemotherapeutic agents in studies other than those mentioned above. However, intravenous application was used to determine the pharmacokinetics and toxicity of sphingolipid metabolites. Single doses of safingol (L-threo-sphinganine, see above) up to 5 mg kg^{-1} body weight did not cause adverse effects in either rats or dogs (Kedderis et al., 1995). Systemic exposure increased the plasma concentration of safingol more than proportionally, and caused haemolysis, renal degeneration and necrosis, and marked hepatotoxicity at higher dosages. Furthermore, safingol caused increased degeneration and necrosis of the intima and medial portions of the veins at the injection sites during infusion, and focal degeneration and necrosis of the seminiferous tubules in male animals. A combination of doxorubicin and safingol did not alter doxorubicin pharmacokinetics (Kedderis *et al.*, 1995; Schwartz *et al.*, 1997), suggesting a possible intravenous use of sphingolipids as chemotherapeutic agents if the initial haemolysis that probably caused the organ damage can be inhibited.

In another study, lipophilic ceramide derivatives dissolved in soybean oil were injected i.v. into mice at 1 mg kg^{-1} body weight prior to injection with highly metastatic Meth A-T tumour cells. This single injection did not change the incidence of lung metastasis; however, repeated injections (seven in total) reduced the number of pulmonary nodules significantly in a structure-dependent manner (Takenaga *et al.*, 1999). Due to solubility problems, the short chain ceramides could not be tested in this study.

Injection of sphingosine, dimethylsphingosine or trimethylsphingosine in mice that have been inoculated with MKN74 human gastric cancer cells showed a pronounced inhibition of tumour growth by the methylated sphingosine derivatives, that was sustained even after administration of the sphingolipids was stopped. However, the concentration required for a 50% reduction of tumour growth was approximately 8 μM (Endo *et al.*, 1991).

Possible Mechanisms of Cancer Suppression by Exogenous Sphingolipids

Defects in sphingolipid signalling in cancer cells have been reported in numerous *in vivo* and *in vitro* studies. A reduced sphingomyelinase activity was demonstrated in rat colonocytes treated with a colon carcinogen (Dudeja *et al.*, 1986); the same reduction was shown in human colon tumours (Hertervig *et al.*, 1997) and has been confirmed recently via oligonucleotide array analysis where a fourfold reduction in sphingomyelinase expression was found in colon tumours compared with normal tissue (Notterman *et al.*, 2001). Due to this low activity/expression, there is not sufficient generation of intracellular ceramide or sphingosine as the appropriate result of extracellular stimuli. Consequently, the cells do not respond with the modulation of cell behaviour. This has been shown for cell lines that do not undergo apoptosis after treatment with TNF-α (Wiegman *et al.*, 1994), camptothecin (X.Z. Wang *et al.*, 1999) or γ-irradiation (Wright

et al., 1996; Chmura *et al.*, 1997). Another mechanism to reduce the intracellular levels of bioactive sphingolipid metabolites and suppress the subsequent regulation of cell growth is the increased synthesis of sphingomyelin (via activation of sphingomyelin synthase; Luberto and Hannun, 1998) or glucosylceramide (via activation of glucosylceramide synthase; Lavie *et al.*, 1996). This has also been correlated to the development of multidrug resistance in cell lines (Lavie *et al.*, 1997) and human tumours of the breast, and in melanoma (Lucci *et al.*, 1998). All these aberrations of sphingolipid signalling pathways reduce the amount of bioactive metabolites in the cells, and abrogate the induction of apoptosis and differentiation. By feeding sphingolipids, we directly deliver the bioactive metabolites to the cells, and this may by-pass the defects in the cancer cell (Fig. 13.4).

Regulation of colonic proliferation

Treatment with a colon carcinogen such as DMH increases the rate of proliferation in rodents. Also, an initial 'wave' of apoptosis after the injections has been reported (Hirose *et al.*, 1996). In our studies with DMH-treated CF1 mice (see above), there was no increase in apoptosis after 4 weeks in the groups fed either the control diet or sphingolipid supplements (Schmelz *et al.*, 2000), suggesting that the induction of apoptosis may not play a role in ACF suppression by dietary sphingolipids. This does not exclude the possibility of an early apoptotic response to the sphingolipids. The rate of proliferation was increased in mice injected with the carcinogen, and the area of proliferation was elevated towards the lumen (where only differentiated non-proliferative cells are seen in normal colons). All complex sphingolipids that were used in our studies reduced proliferation approximately to the levels of control animals that had not been treated with the carcinogen. Furthermore, the proliferative area was more restricted towards the lower half of the colonic crypts (Schmelz *et al.*, 2000).

Modulation of β-catenin by exogenous sphingolipids

The APC gene product is located at a main 'intersection' of intracellular metabolism, regulating a

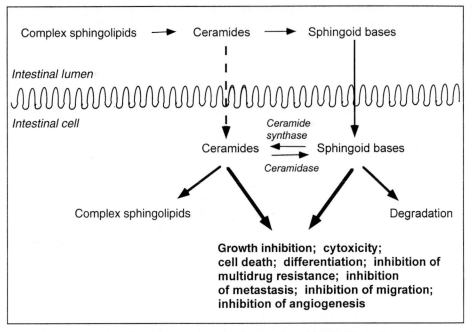

Fig. 13.4. Digestion and uptake of dietary sphingolipids. Complex sphingolipids are digested to ceramides, but more probably to free sphingoid bases throughout the intestinal tract, and are taken up by the intestinal cells. The effects of both these metabolites are mostly growth inhibitory and cytotoxic.

variety of processes. One protein that is regulated by APC and is especially important in colon cancer is β-catenin. β-Catenin is a cell adhesion protein that connects E-cadherin to the actin cytoskeleton via α-catenin (Pfeifer, 1995; Sacco et al., 1995), and to other transmembrane and peripheral cytoplasmic proteins (McNeill et al., 1990). This function is clearly distinguishable from its function as a signalling molecule in developmental systems (Fagotto et al., 1996; Orsulic and Pfeifer, 1996), and its role in cellular response to growth stimulation (i.e. Wnt/Wingless pathway, epidermal growth factor and hepatocyte growth factor receptor activation; Hoschuetzky et al., 1994; Shibamoto et al., 1994; Pfeifer, 1995). In colon cancer, a mutation of APC seems to be the critical event for the development of dysplastic ACF, and the progression towards adenoma and adenocarcinoma (Jen et al., 1994). As a result of a mutation in this pathway, stabilized β-catenin accumulates in the cytosol, and subsequently translocates to the nucleus where a complex with transcription factors of the Tcf/Lef family (T-cell factor/lymphocyte enhancer factor) is formed. This complex can modulate gene transcription processes and activate transcription (see review by Behrens, 1999).

As mentioned above, dietary sphingolipids in amounts that can be found in foods suppressed tumour formation in Min mice. Tumour formation in this model is preceded by the loss of the APC wild-type allele, and the concomitant accumulation of cytosolic β-catenin. Thus, in Min mice that were fed the control AIN 76A diet alone and developed a large number of intestinal tumours, we found a high expression of cytosolic β-catenin in intestinal sections as determined by fluorescence immunohistochemistry. Min mice that were fed sphingolipid supplements in their diet and that had developed only a small number of tumours displayed mostly membrane-associated β-catenin. This is the localization found in the genetic background mice, suggesting a normalization of β-catenin expression/localization by dietary sphingolipids (Schmelz et al., 2001). The same effect of exogenous sphingolipids could be seen *in vitro*, using human colon cancer cell lines that also carry an APC mutation, and stably overexpress cytosolic β-catenin. Both sphingosine and ceramides reduced the amount of cytosolic β-catenin after only 6 h of incubation with non-toxic concentrations; longer incubations were necessary for a visible reduction of nuclear β-catenin (Schmelz et al., 2001). It cannot be concluded from these data that this is the only pathway by which dietary sphingolipids down-regulate tumour formation in Min mice, since all proteins and lipids involved have multiple functions. None the less, given the importance of increased cytosolic β-catenin not only in early colon cancer but also in breast, skin, prostate and kidney cancer, the down-regulation of cytosolic β-catenin is clearly an important event, and may contribute directly to the anti-colon cancer effects of sphingolipids.

Sphingolipids and metastasis

Chemotactic cell migration plays an important role in embryogenesis, reproduction, wound healing, inflammation and angiogenesis. Cell migration is also a crucial process in metastasis. Sphingosine-1-phosphate has been shown to inhibit *in vitro* motility and invasiveness of tumour cells (F. Wang et al., 1999) and non-transformed cells (Kawa et al., 1997); this has also been demonstrated using ceramide (Metz et al., 1991). For this inhibition, only nanomolar concentrations of sphingosine-1-phosphate are necessary (F. Wang et al., 1999), but it is not known if dietary sphingolipids contribute to the circulating pool of sphingosine-1-phosphate.

The glycosphingolipid α-galactosylceramide has been correlated to the prevention of liver and lung metastasis (Toura et al., 1999; Nakagawa et al., 2001). Mice that were inoculated with melanoma or Lewis lung cancer cells were injected i.v. with dendritic cells prepared from mouse spleen that had been treated with galactosylceramide *in vitro*. The already established metastatic foci in liver and lung were almost completely eradicated by this treatment (Toura et al., 1999). This anti-tumour and anti-metastatic effect is caused by the induction of apoptosis in the micrometastasis mediated by galactosylceramide-activated NKT cells that produce IFN-γ, which in turn induces apoptosis by activating the Fas–Fas ligand signalling system (Nakagawa et al., 2001). It is not known, but may be an interesting concept, whether dietary galactosylceramide could activate gut-associated anti-tumour immunity.

Tumour gangliosides

A drastic change of cell surface gangliosides (neuraminic acid-containing glycosphingolipids) has been noted in cancer cells. This could be due to the incomplete synthesis and concomitant accumulation of precursor gangliosides (GM3, GM2, GD3 and GD2) as seen in melanoma and neuroblastoma. Also, new ganglioside species can be expressed, or structurally modified in their headgroup or ceramide backbone. This may cause conformational changes in the membranes, creating or masking antigenic epitopes (Ritter and Livingston, 1991), and alter ganglioside activity (Ladisch et al., 1994). Cells that proliferate rapidly, such as cancer cells, are shedding gangliosides as micelles, monomers and membrane vesicles into the cellular microenvironment as a prominent part of their ganglioside metabolism (Kong et al., 1998). Shed gangliosides have been demonstrated to interfere markedly with several steps of the anti-tumour host response, i.e. inhibition of IL-2-dependent lymphocyte proliferation, and suppression of $CD4^+$ on T lymphocytes (Valentino et al., 1990), and reduction of the number of macrophages and $CD8^+$ T lymphocytes (Chu and Sharom, 1993). This creates a tumour microenvironment that favours tumour formation and progression. Furthermore, gangliosides enhance binding of tumour cells to endothelial cells either directly or via modulation of adhesion receptors (Taki et al., 1997). Binding of gangliosides to blood cells leads to aggregation and formation of large clumps that may cause embolisms in the lung, thereby promoting metastasis (Ito et al., 2001). On the other hand, these shed gangliosides are targets for anti-cancer therapy via anti-ganglioside monoclonal antibodies, or ganglioside vaccines. It is unclear, however, if dietary gangliosides can also increase the gangliosides in blood, thereby possibly exerting the same host and tumour responses. Although we and others have not seen an increase of complex sphingolipids in blood, lymph or tissue after oral administration of sphingolipids (Nilsson, 1968, 1969; Schmelz et al., 1994, 2001; Kobayashi et al., 1997), in the absence of intestinal endoceramidases, smaller gangliosides could be absorbed and directly exposed to the immune system associated with the gut (Ravindra et al., 2000), causing an immunization that has been seen early on in life via breast milk G_{D3} (Takamizawa et al., 1986).

Furthermore, the anti-ganglioside antibody titre was found to be lower in older (>50 years) people. This may be of physiological significance because the antibodies could neutralize circulating tumour gangliosides from the serum (Ravindra *et al.*, 2000). A nutritional intervention that could raise the level of anti-tumour gangliosides may therefore be of interest.

Future directions of sphingolipids in cancer prevention and treatment

Our rodent studies show clearly a beneficial effect of dietary sphingolipids on colon carcinogenesis. Very important in this aspect is that this can be achieved in amounts that appear not to be toxic, and can be found in foods commonly associated with the Western diet. However, much more research is necessary to determine the effects of dietary sphingolipids on human colon cancer, and possibly on cancer of other organs. None the less, these early *in vivo* studies are very promising and, together with the available data from numerous *in vitro* studies, several signalling pathways and downstream targets of exogenous sphingolipids have been identified (Fig. 13.5). This still growing list of intracellular sphingolipid targets will direct future research towards the mechanism of how sphingolipids affect carcinogenesis, and may help to establish dietary sphingolipids as mechanism-driven chemotherapeutic and, perhaps, chemopreventive agents.

References

Behrens, J. (1999) Cadherins and catenins: role in signal transduction and tumor progression. *Cancer and Metastasis Reviews* 18, 15–30.

Bielawska, A., Crane, H., Liotta, D.C., Obeid, L.M. and Hannun, Y.A. (1993) Selectivity of ceramide-mediated biology. Lack of activity of erythro-dihydro-ceramide. *Journal of Biological Chemistry* 268, 26226–26232.

Birt, D.F., Merrill, A.H. Jr, Barnett, T., Enkvetchakul, B., Pour, P.M., Liotta, D.C., Geisler, V., Menaldino,

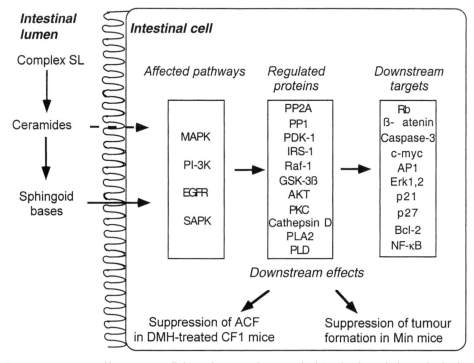

Fig. 13.5. Summary of known intracellular pathways and targets of sphingolipid metabolites. This by far incomplete list demonstrates the multiple possible pathways sphingolipids can modulate to regulate cell growth and cell death that may be involved in the down-regulation of ACF and tumour formation in rodents.

D.S. and Schwartzbauer, J. (1998) Inhibition of skin papillomas by sphingosine, *N*-methyl sphingosine, and *N*-acetyl sphingosine. *Nutrition and Cancer* 31, 119–126.

Bose, R., Verheil, M., Haimovitz-Friedman, A., Scotto, K., Fuks, Z. and Kolesnick, R. (1995) Ceramide synthase mediates daunorubicin-induced apoptosis: an alternative mechanism for generating death signals. *Cell* 82, 405–414.

Carfagna, M.A., Young, K.M. and Susick, R.L. (1996) Sex differences in rat hepatic cytolethality of the protein kinase C inhibitor safingol: role of biotransformation. *Toxicology and Applied Pharmacology* 137, 173–181.

Chmura, S.J., Nodzenski, E., Beckett, M.A., Kufe, D.W., Quintans, J. and Weichselbaum, R.R. (1997) Loss of ceramide production confers resistance to radiation-induced apoptosis. *Cancer Research* 57, 1270–1275.

Chu, J.W. and Sharom, F.J. (1993) Gangliosides inhibit T-lymphocyte proliferation by preventing the interaction of interleukin-2 with its cell surface receptors. *Immunology* 79, 10–17.

Dillehay, D.L., Webb, S.K., Schmelz, E.M. and Merrill, A.H. Jr. (1994) Dietary sphingomyelin inhibits 1,2-dimethylhydrazine-induced colon cancer in CF1 mice. *Journal of Nutrition* 124, 615–620.

Dudeja, P.K., Dahiya, R. and Brasitus, T.A. (1986) The role of sphingomyelin synthase and sphingomyelinase in 1,2-dimethylhydrazine-induced lipid alterations of rat colonic membranes. *Biochimica et Biophysica Acta* 863, 309–312.

Endo, K., Igarashi, Y., Nisar, M., Zhou, Q. and Hakomori, S.-I. (1991) Cell membrane signaling as target in cancer therapy: inhibitory effect of N,N-dimethyl and N,N,N-trimethyl sphingosine derivatives on *in vitro* and *in vivo* growth of human tumor cells in nude mice. *Cancer Research* 51, 1613–1618.

Enkvetchakul, B., Merrill, A.H. Jr and Birt, D.F. (1989) Inhibition of the induction of ornithine decarboxylase activity by 12-*O*-tetradecanoylphorbol-13-acetate in the mouse skin by sphingosine sulfate. *Carcinogenesis* 10, 379–381.

Enkvetchakul, B., Barnett, T., Liotta, D.C., Geisler, V., Menaldino, D.S., Merrill, A.H. Jr. and Birt, D.F. (1992) Influences of sphingosine on two-stage skin tumorigenesis in SENCAR mice. *Cancer Letters* 62, 35–42.

Fagotto, F., Funayama, N., Gluck, U. and Gumbiner, B.M. (1996) Binding to cadherins antagonizes the signaling activity of beta-catenin during axis formation in *Xenopus*. *Journal of Cell Biology* 132, 1105–1114

Hannun, Y.A., Loomis, C.R., Merrill, A.H. Jr and Bell, R.M. (1986) Sphingosine inhibition of protein kinase C activity and of phorbol dibutyrate binding *in vitro* and in human platelets. *Journal of Biological Chemistry* 261, 12604–12609.

Hertervig, E., Nilsson, Å., Nyberg, L. and Duan, R.D. (1997) Alkaline sphingomyelinase activity is decreased in human colorectal carcinoma. *Cancer* 79, 448–453.

Hirose, Y., Yoshimi, N., Makita, H., Hara, A., Tanaka, T. and Mori, H. (1996) Early apoptosis and cell proliferation in azoxymethane-initiated rat colonic epithelium. *Japanese Journal of Cancer Research* 87, 575–582.

Holt, P.R. (1999) Dairy foods and prevention of colon cancer: human studies. *Journal of the American College of Nutrition* 18 (Supplement 5), 379S–391S.

Hoschuetzky, H., Aberle, H. and Kemler, R. (1994) β-Catenin mediates the interaction of the cadherin–catenin complex with epidermal growth factor receptor. *Journal of Cell Biology* 127, 1375–1380.

Huwiler, A., Kolter, T., Pfeilschifter, J. and Sandhoff, K. (2000) Physiology and pathophysiology of sphingolipid metabolism and signaling. *Biochimica et Biophysica Acta* 1485, 63–99.

Ito, A., Handa, K., Withers, D.A., Satoh, M. and Hakomori, S. (2001) Binding specificity of siglec7 to disialogangliosides of renal cell carcinoma: possible role of disialogangliosides in tumor progression. *FEBS Letters* 504, 82–86.

Jarvinen, R., Knekt, P., Hakulinen, T. and Aromaa, A. (2001) Prospective study on milk products, calcium and cancers of the colon and rectum. *European Journal of Clinical Nutrition* 55, 1000–1007.

Jarvis, W.D., Fornari, F.A., Taylor, R.S., Martin, H.A., Kramer, L.B., Erukulla, R.K., Bittman, R. and Grant, S. (1996) Induction of apoptosis and potentiation of ceramide-mediated cytotoxicity by sphingoid bases in human myeloid leukemia cells. *Journal of Biological Chemistry* 271, 8275–8284.

Jen, J., Powell, S.M., Papdopoulos, N., Smith, K.J., Hamilton, S.R., Vogelstein, B. and Kinzler, K.W. (1994) Molecular determinants of dysplasia in colorectal lesions. *Cancer Research* 54, 5523–5526.

Jensen, R.G. (ed.) (1995) *Handbook of Milk Composition*. Academic Press, New York.

Kawa, S., Kimura, S., Hakomori, S.-I. and Igarashi, Y. (1997) Inhibition of chemotactic and trans-endothelial migration of human neutrophils by sphingosine-1-phosphate. *FEBS Letters* 420, 196–200.

Kedderis, L.B., Bozigan, H.P., Kleeman, J.M., Hall, R.L., Palmer, T.E., Harrison, S.D. Jr and Susick, R.L. Jr (1995) Toxicity of the protein kinase C inhibitor safingol administered alone and in combination with chemotherapeutic agents. *Fundamental and Applied Toxicology* 25, 201–217.

Kinzler, K.W. and Vogelstein, B. (1996) Lessons from hereditary colon cancer. *Cell* 87, 159–170.

Kobayashi, T., Shimizugawa, T., Osakabe, T., Watanabe, S. and Okuyama, H. (1997) A long-term feeding of sphingolipids affected the level of plasma cholesterol and hepatic triacylglycerol but not tissue phospholipids and sphingolipids. *Nutrition Research* 17, 111–114.

Kong, Y., Li, R. and Ladisch, S. (1998) Natural forms of shed tumor gangliosides. *Biochimica et Biophysica Acta* 1394, 43–56.

Ladisch, S., Li, R. and Olson, E. (1994) Ceramide structure predicts tumor ganglioside immunosuppressive activity. *Proceedings of the National Academy of Sciences USA* 91, 1974–1978.

Lavie, Y., Cao, H., Bursten, S.L., Giuliano, A.E. and Cabot, M.C. (1996) Accumulation of glucosylceramides in multi-drug resistant cancer cells. *Journal of Biological Chemistry* 27, 19530–19536.

Lavie, Y., Cao, H.T., Volner, A., Lucci, A., Han, T.Y., Geffen, V., Giuliano, A.E. and Cabot, M.C. (1997) Agents that reverse multidrug resistance, tamoxifen, verapamil, and cyclosporin A, block glycosphingolipid metabolism by inhibiting ceramide glycosylation in human cancer cells. *Journal of Biological Chemistry* 272, 1682–1687.

Lipkin, M. (1999) Preclinical and early human studies of calcium and colon cancer prevention. *Annals of the New York Academy of Sciences* 889, 120–127.

Luberto, C. and Hannun, Y.A. (1998) Sphingomyelin synthase, a potential regulator of intracellular levels of ceramide and diacylglycerol during SV40 transformation. Does sphingomyelin synthase account for the putative phosphatidylcholine-specific phospholipase C? *Journal of Biological Chemistry* 273, 14550–14559.

Lucci, A.E., Cho, W.I., Han, T.Y., Guiliani, A.E., Morton, D.L. and Cabot, M.C. (1998) Glucosylceramide: a marker for multiple-drug resistant cancers. *Anticancer Research* 18, 475–480.

Mathias, S., Peña, L.A. and Kolesnick, R.N. (1998) Signal transduction of stress via ceramide. *Biochemical Journal* 335, 465–480.

McNeill, H., Ozawa, M., Kemler, R. and Nelson, W.J. (1990) Novel function of the cell adhesion molecule uvomorulin as an inducer of cell surface polarity. *Cell* 62, 309–316.

Metz, R.J., Vellody, K., Patel, S., Bergstrom, R., Meisinger, J., Jackson, J., Wright, M.A. and Young, M.R. (1991) Vitamin D3 and ceramide reduce the invasion of tumor cells through extracellular matrix components by elevating protein phosphatase-2A. *Invasion and Metastasis* 16, 280–290.

Nagase, H. and Nakamura, Y. (1993) Mutations of the *APC* (adenomatous polyposis coli) gene. *Human Mutations* 2, 425–434.

Nakagawa, R., Nagafune, I., Tazunoki, Y., Ehara, H., Tomura, H., Iijima, R., Motoki, K., Kamishohara, M. and Seki, S. (2001) Mechanisms of the antimetastatic effect in the liver and of the hepatocyte injury induced by α-galactosylceramide in mice. *Journal of Immunology* 166, 6578–6584.

Nilsson, Å. (1968) Metabolism of sphingomyelin in the intestinal tract of the rat. *Biochimica et Biophysica Acta* 76, 575–584.

Nilsson, Å. (1969) Metabolism of cerebrosides in the intestinal tract of the rat. *Biochimica et Biophysica Acta* 187, 113–121.

Notterman, D.R., Alon, U., Sierk, A.J. and Levine, A.J. (2001) Transcriptional gene expression profiles of colorectal adenoma, adenocarcinoma, and normal tissue examined by oligonucleotide arrays. *Cancer Research* 61, 3124–3130.

Nyberg, L., Nilsson, Å., Lundgren, P. and Duan, R.-D. (1997) Localization and capacity of sphingomyelin digestion in the rat intestinal tract. *Journal of Nutritional Biochemistry* 8, 112–118.

Orsulic, S. and Pfeifer, M. (1996) An *in vitro* structure–function study of armadillo, the β-catenin homologue, reveals both separate and overlapping regions of the protein required for cell adhesion and for wingless signaling. *Journal of Cell Biology* 134, 1283–1300.

Parodi, P.W. (1999) Conjugated linoleic acid and other anticarcinogenic agents of bovine milk fat. *Journal of Dairy Sciences* 82, 1339–1349.

Perry, D.K., Carton, J., Shah, A.K., Meredith, F., Uhlinger, D.J. and Hannun, Y.A. (2000) Serine palmitoyl transferase regulates *de novo* ceramide generation during etoposide-induced apoptosis. *Journal of Biological Chemistry* 275, 9078–9084.

Pfeifer, M. (1995) Cell adhesion and signal transduction: the armadillo connection. *Trends in Cell Biology* 5, 224–229.

Ravindra, M.H., Gonzales, A.M., Nishimoto, K., Tam, W.-Y., Soh, D. and Morton, D.L. (2000) Immunology of gangliosides. *Indian Journal of Experimental Biology* 38, 301–312.

Ritter, G. and Livingston, P.O. (1991) Ganglioside antigens expressed by human cancer cells. *Seminars in Cancer Biology* 2, 401–409.

Sacco, P.A., McGranahan, M.J., Wheelock, M.J. and Johnson, K.R. (1995) Identification of plakoglobin domains required for association with N-cadherin and α-catenin. *Journal of Biological Chemistry* 270, 20201–20206.

Schmelz, E.M., Crall, K.L., LaRocque, R., Dillehay, D.L. and Merrill, A.H. Jr (1994) Uptake and metabolism of sphingolipids in isolated intestinal loops of mice. *Journal of Nutrition* 124, 702–712.

Schmelz, E.M., Dillehay, D.L., Webb, S.K., Reiter, A., Adams, J. and Merrill, A.H. Jr (1996) Sphingomyelin consumption suppresses aberrant colonic crypt foci and increases the proportion of adenomas versus adenocarcinomas in CF1 mice treated with 1,2-dimethylhydrazine: implications for dietary sphingolipids and colon carcinogenesis. *Cancer Research* 56, 4936–4941.

Schmelz, E.M., Bushnev, A.B., Dillehay, D.L., Liotta, D.C. and Merrill, A.H. Jr (1997) Suppression of aberrant colonic crypt foci by synthetic sphingomyelins with saturated or unsaturated sphingoid base backbones. *Nutrition and Cancer* 28, 81–85.

Schmelz, E.M., Sullards, M.C., Dillehay, D.L. and Merrill, A.H. Jr (2000) Inhibition of colonic cell proliferation and aberrant crypt foci formation by dairy glycosphingolipids in 1,2-dimethylhydrazine-treated CF1 mice. *Journal of Nutrition* 130, 522–527.

Schmelz, E.M.. Roberts, P.C., Kustin, E.M., Lemonnier, L.A., Sullards, M.C., Dillehay, D.L. and Merrill, A.H. Jr (2001) Modulation of β-catenin localization and intestinal tumorigenesis *in vitro* and *in vivo* by sphingolipids. *Cancer Research* 61, 6723–6729.

Schwartz, G.K., Ward, D., Saltz, L., Casper, E.S., Spiess, T., Mullen, E., Woodworth, J., Venuti, R., Zervos, P., Storniolo, A.M. and Kelsen, D.P. (1997) A pilot clinical/pharmacological study of the protein kinase C-specific inhibitor safingol alone and in combination with doxorubicin. *Clinical Cancer Research* 3, 537–543.

Separovic, D., He, J. and Oleinick, N.L. (1997) Ceramide generation in response to photodynamic treatment of L5178Y mouse lymphoma cells. *Cancer Research* 57, 1717–1721.

Shibamoto, S., Hayakawa, M., Takeuchi, K., Hori, T., Oku, N., Miyazawa, K., Kitamura, N., Takeichi, M. and Ito, F. (1994) Tyrosine phosphorylation of β-catenin and plakoglobin enhanced by hepatocyte growth factor and epidermal growth factor in human carcinoma cells. *Cellular Adhesion and Communication* 1, 295–305.

Sparks, A.B., Morin, P.J. and Kinzler, K.W. (1998) Mutational analysis of the APC/β-catenin/TCF pathway in colorectal cancer. *Cancer Research* 58, 1130–1134.

Spiegel, S. (1999) Sphingosine-1-phosphate: a prototype of a new class of second messengers. *Journal of Leukocyte Biology* 65, 341–344.

Spiegel, S. and Merrill, A.H. Jr (1996) Sphingolipid metabolism and growth regulation: a state-of-the-art review. *FASEB Journal* 10, 1388–1397.

Takamizawa, K., Iwamori, M., Mutai, M. and Nagai, Y. (1986) Selective changes in gangliosides of human milk during lactation: a molecular indicator for the period of lactation. *Biochimica et Biophysica Acta* 879, 73–77.

Takenaga, M., Igarashi, R., Matsumoto, K., Takeuchi, J., Mizushima, N., Nakayama, T., Morizawa, Y. and Mizushima, Y. (1999) Lipid microsphere preparation of a lipophilic ceramide derivative suppresses colony formation in a murine experimental metastasis model. *Journal of Drug Targeting* 7, 187–195.

Taki, T., Ishikawa, D., Ogura, M., Kakajima, M. and Handa, S. (1997) Ganglioside GD1a functions in the adhesion of metastatic tumor cells to endothelial cells of the target tissue. *Cancer Research* 57, 1882–1888.

Toura, I., Kawano, T., Akutsu, Y., Nkayama, T., Ochiai, T. and Taniguchi, M. (1999) Cutting edge: inhibition of experimental tumor metastasis by dendritic cells pulsed with α-galactosylceramide. *Journal of Immunology* 163, 2387–2391.

Valentino, L., Moss, T., Olson, E., Wang, H.J., Elashoff, R. and Ladisch, S. (1990) Shed tumor gangliosides and progression of human neuroblastoma. *Blood* 75, 1564–1567.

Vesper, H., Schmelz, E.M., Nikolova-Karakashian, M., Dillehay, D.L., Lynch, D.V. and Merrill, A.H. Jr (1999) Sphingolipids in food and the emerging importance of sphingolipids to nutrition. *Journal of Nutrition* 129, 1239–1250.

Wiegmann, K., Schütze, S., Machleid, T., Witte, D. and Krönke, M. (1994) Functional dichotomy of neutral and acidic sphingomyelinases in tumor necrosis factor signaling. *Cell* 78, 1005–1015.

Wang, F., Van Brocklyn, J.R., Edsall, L., Nava, V.E. and Spiegel, S. (1999) Sphingosine-1-phosphate inhibits motility of human breast cancer cells independently of cell surface receptors. *Cancer Research* 59, 6185–6191.

Wang, X.Z., Beebe, J.R., Pwiti, L., Bielawska, A. and Smyth, M.J. (1999) Aberrant sphingolipid signaling is involved in the resistance of prostate cancer cell lines to chemotherapy. *Cancer Research* 59, 5842–5848.

Wright, S.C., Zheng, H. and Zhong, J. (1996) Tumor cell resistance to apoptosis due to a defect in the activation of sphingomyelinase and the 24 kDa apoptotic protease (AP24). *FASEB Journal* 10, 325–332.

Xu, J., Yeh, C.H., Chen, S., He, L., Sensi, S.L., Canzoniero, L.M., Choi, D.W. and Hsu, C.Y. (1998) Involvement of *de novo* ceramide biosynthesis in tumor necrosis factor-alpha/cycloheximide-induced cerebral endothelial cell death. *Journal of Biological Chemistry* 1273, 16521–16526.

14 The Health Effects of Dietary Isoflavones

Thomas M. Badger, Martin J.J. Ronis and Nianbai Fang
Arkansas Children's Nutrition Center, Little Rock, Arkansas, USA

Introduction

This chapter will provide an introduction to the field of isoflavones and health. It will focus on providing the reader with an understanding level sufficient to evaluate critically the vast and rapidly growing literature on the actions of isoflavones, especially as they relate to nutrition and health status. The molecular structures of isoflavones in plants and in the body and the methods most widely used to identify and quantitate isoflavones will be described. Soy protein isolate will be used as a model soyfood to demonstrate the human urine and plasma pharmacokinetics of the two most abundant isoflavones in soy, and animal models will be used to demonstrate the molecular forms and concentrations in target tissues. Human exposure to isoflavones throughout the life cycle will be discussed, and some of the potential beneficial and adverse health effects will be explored, as well as the possible molecular mechanisms underlying these effects. In addition, the controversy related to safety of soy infant formulas will be addressed.

The soybean is an excellent source of high-quality dietary protein, and it has several constituents used in food processing and cooking. Soy is processed into tofu, miso, flour, oils, textured fibres, protein concentrates and isolates, and is extracted for several phytochemicals such as Bowman–Birk inhibitor of trypsin/chymotrypsin, saponins and isoflavonoids. The isoflavonoids are a subcategory of flavonoids, which are a large group of naturally occurring plant constituents distinguished by the basic structure of two aromatic rings joined by a three-carbon link. Figure 14.1 illustrates this structure, plus that of the isoflavanones (which have an oxygen bridge to position 9 of the A ring) and their C ring reduction metabolites, the isoflavanols and isoflavans. Isoflavones in soybeans are characterized by the hydroxylation of carbon 7 of the A ring and carbon 4′ on the B ring. The three most abundant soy isoflavones are represented by the aglycones daidzein, genistein and glycitein, which are present mainly in the glycoside forms, genistin, diadzin and glycitin, as well as several acetylated and malonylated derivatives (Fig. 14.2). It should be noted that there are other well known naturally occurring classes of isoflavonoids, such as the coumestans, rotenoids, pterocarpans and their isoflavonoid oligomers. We will focus only on the major soybean isoflavones in this chapter.

The isoflavone glycosides are sequestered within the protein matrix of the soybean, and the aglycones (genistein, daidzein and glycitein) are liberated in the intestine and colon, primarily by the actions of β-glycosidase present in enterocytes and bacterial enzymes, respectively. A fourth and very potent aglycone with potentially important health effects, equol, is derived from daidzein by bacterial enzymes. Since only about 10 and 50% of the total ingested genistein and daidzein, respectively, are accounted for in the urine (Shelnutt *et al.*, 2000), the apparent bioavailability of isoflavones in soy protein isolate (SPI[+]) is very low. The majority of absorbed aglycones are thought to be conjugated by glucuronidation and sulphation during first-pass metabolism. An unknown percentage of aglycones is absorbed and pass directly to the liver by the portal circulation where they are also conjugated,

©CAB *International* 2003. *Molecular Nutrition*
(eds J. Zempleni and H. Daniel)

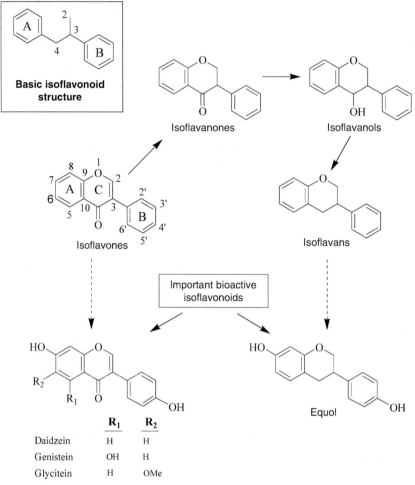

Fig. 14.1. The basic molecular structure of isoflavones is presented in the insert. The three major soy isoflavones (daidzein, genistein and glycitein) are formed with modification to carbons at the 5 and 6 positions on the A ring, and equol is the major metabolite of daidzein. The relationship of the isoflavones, isoflavanols and isoflavans involves carbon 4 of the link between rings A and B.

leaving a small percentage (<5%) of aglycone escaping to the general circulation. Furthermore, there are several reduced and/or oxidized isoflavone metabolites that are also conjugated. Thus, the soy isoflavones circulate and/or are excreted in several molecular forms, including glucuronide and sulphate conjugates, freely circulating aglycones and protein-bound aglycones (Setchell, 1988, 1995; Coward et al., 1993; Barnes et al., 1996a). The major urinary isoflavone metabolites of female rats fed diets made with SPI$^+$ are presented in Figs 14.3 and 14.4 (Fang et al., 2002).

There are two major conjugation sites on genistein or daidzein (positions 7 and 4′, see Fig. 14.2) and each can be sulphated and/or glucuronidated. As a consequence, there are monoglucuronides, monosulphates, diglucuronides, disulphates, and mixed conjugates with one site glucuronidated and one site sulphated. The majority of absorbed isoflavones are excreted as conjugates to the urine, but a small percentage undergo enterohepatic recycling (Setchell and Adlercruetz, 1988; Setchell, 1995). Isoflavones are similar to the endogenous gonadal and adrenal steroids with respect to conjugation and excretion into the urine. Because there is such a high percentage of intracellular conjugated isoflavones, it is assumed that these conjugates can be transported into and

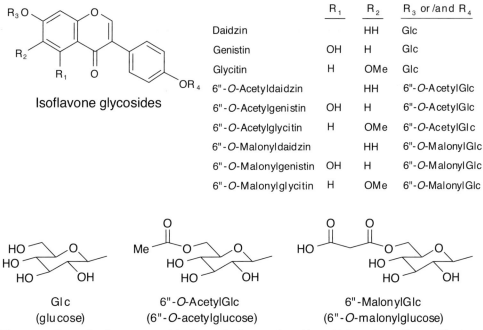

Fig. 14.2. The molecular structure of the known isoflavone glycosides, which are formed by modifications to carbons 5, 6 and 7.

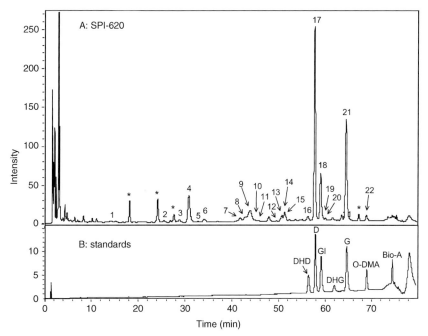

Fig. 14.3. The MS-MS profile of urinary isoflavone metabolites in rats fed an isoflavone-rich diet is shown in the top panel. SPI-620 is a soy protein isolate-containing diet, and the commercially available standards used to identify major metabolites are shown in the bottom panel. The identification of each metabolite is listed in Fig. 14.4, and details of the procedures appear in Fang et al. (2002). *Indicates the peaks present in both urine profile A and control profile (not shown here).

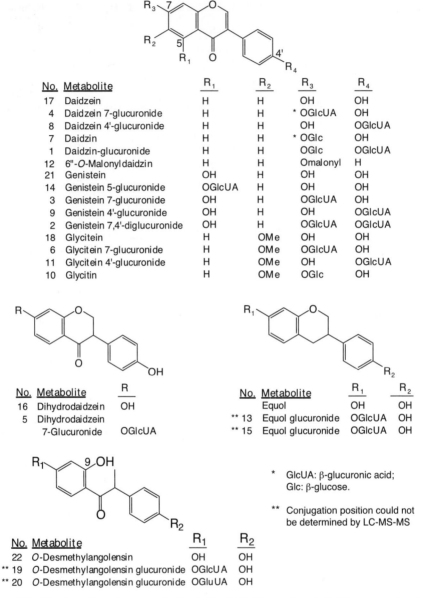

Fig. 14.4. Identification of the urinary metabolites in Fig. 14.3. The numbers in bold correspond to the peak numbers in Fig. 14.3, and details of the procedures appear in Fang et al. (2002).

out of cells in a similar fashion to conjugated endogenous steroids. This transport would provide a mechanism by which the monoglucuronides or monosulphates are to be conjugated at the second site in the liver in preparation for renal excretion. It also would explain how most target tissues that are not rich in conjugating enzymes, such as the mammary gland, secondary sex organs, kidney and brain, have a substantial percentage of total isoflavones in the conjugated form (Chang et al., 2002). The intracellular presence of these complex molecular isoflavone species raises important questions. (i) Which are the most biologically active molecules? (ii) Do these conjugates serve as a target tissue pool of bioactive isoflavone aglycone ligands? (iii) Were these conjugates actually transported into the cells, or were they formed intracellularly from aglycones?

Figure 14.5 and Table 14.1 illustrate the mean serum isoflavone concentrations (genistein plus daidzein equivalents) of 12 healthy adult men and women following ingestion of a beverage containing SPI⁺ that provided a dose of 1.0 mg kg^{-1} genistein (aglycone) equivalents and 0.6 mg kg^{-1} daidzein (aglycone) equivalents. These data demonstrate that consumption of a single commercially available soy beverage (similar to a soy meal) can result in rather high adult blood concentrations of isoflavones that have a characteristic time course. Isoflavones are absorbed quickly after consumption, with total plasma or urine genistein and daidzein peaking between 3 and 5.4 h post-ingestion and returning towards basal levels with an apparent half-life of 3–8 h (Cimino et al., 1999; Shelnutt et al., 2000, 2002).

The identification of isoflavonoids has centred largely on the application of spectroscopic methods including ultraviolet (UV) spectroscopy (Mabry et al., 1970), nuclear magnetic resonance (NMR) spectroscopy (Markham and Geiger, 1986) and mass spectrometry (MS) (Harbone et al., 1975). The UV spectra of isoflavones are characterized by two major absorption maxima, one of which occurs in the range 245–270 nm (band II) and the other in the range 300–340 nm (band I) (Harbone et al., 1975). Several methods have been reported to measure isoflavones and their metabolites in body fluids and foods, and these include: high-pressure liquid chromatography (HPLC) (Franke et al., 1995; Supko and Phillips, 1995; Kulling et al., 2001), gas chromatography/mass spectrometry (GC/MS) (Adlercruetz et al., 1993b), high-performance liquid chromatography/mass spectrometry (LC/MS) (Cimino et al., 1999), LC/MS/MS (Coward et al., 1996; Fang et al., 2002), radioimmunoassay (RIA) (Lapick et al., 1998), and HPLC with colometric array detection (Gamache and Acworth, 1998). Since most isoflavones in

Table 14.1. Plasma and urine isoflavone pharmacokinetics.

Isoflavone	Sample	T_{max} (h)[1]	$t_{½}$ (h)[2]
Genistein	Plasma	5.2 ± 0.4	8.2 ± 0.7
glucuronide	Urine	4.3 ± 0.5	6.0 ± 0.4
Sulphate	Plasma	4.5 ± 0.3	5.7 ± 0.4
	Urine	5.4 ± 1.1	4.5 ± 0.7
Aglycone	Urine	3.1 ± 0.8	ND[3]
Daidzein	Plasma	5.0 ± 0.5	3.3 ± 0.3
glucuronide	Urine	5.2 ± 0.4	3.8 ± 0.4
Sulphate	Plasma	4.5 ± 0.3	3.1 ± 0.4
	Urine	5.4 ± 1.1	3.9 ± 0.5
Aglycone	Urine	4.9 ± 1.9	ND[3]

[1]T_{max} is the time to peak rate for urine and the time to maximal concentration for plasma.
[2]$t_{½}$ is the apparent half-life.
[3]ND = not determined because concentrations were too low.

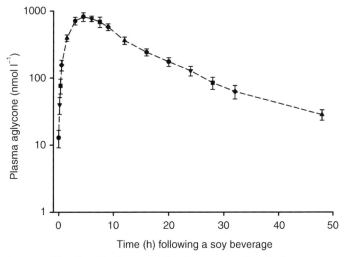

Fig. 14.5. The plasma profile of total isoflavones following consumption of a beverage containing soy protein isolate. The values are the mean (± SEM) of the total aglycones released upon deconjugation with sulphatases and glucuronidases. Details appear in Shelnutt et al. (2002).

biological samples are conjugated, enzymatic digestion into aglycones and subsequent detection with LC/MS, GC/MS or coulometric electrode array technology (CEAT) have been the methods of choice (Holder et al., 1999; Shelnutt et al., 2002). As more conjugated standards become available, enzyme digestion will become unnecessary.

The purified isoflavones have been studied extensively *in vitro* and found to affect several important intracellular systems (Polkowski and Mazyrek, 2000), some of which are listed in Table 14.2. Most of the *in vitro* effects occur at relatively high micromolar concentrations that are not thought to occur *in vivo* and most of the *in vitro* and many *in vivo* studies have been conducted with the purified isoflavones (Chang et al., 2000). These *in vitro* effects are due to the parent compound, since most of the cells studied do not have phase I or phase II enzymes. Therefore, very little is known about the biological activities of isoflavone metabolites. Consequently, the true biologically active molecular species for the *in vivo* actions of either dietary isoflavones, purified genistein or purified daidzein have not been confirmed. Similarly, metabolites of equol are less studied. Thus, the questions remaining to be answered are: what are the biologically active molecular species of isoflavones; what concentrations are required for their actions; and what are the intracellular concentrations of these species after various levels of isoflavone-containing meals?

Isoflavones are an excellent example of a category of phytochemicals that are not required for normal growth, normal development or maintenance of normal adult homeostasis, but are biologically active in mammals and potentially can affect health. In that regard, isoflavones are not nutrients, but rather are potentially important modulators of cellular functions that affect health. They have been demonstrated to have both beneficial and adverse health effects in animal and human studies (Gilani and Anderson, 2002). Most studies on isoflavones have produced results that suggest a beneficial influence on human health, including heart disease, cancer, kidney disease, osteoporosis and relief of menopausal symptoms. The latter two effects have led to studies on soy and isoflavones as possible alternatives to hormone replacement therapy. There are, however, several studies that point to potential adverse health effects of isoflavones, and much more research is required before the health effects of isoflavones under various physiological and environmental conditions are understood. The proceedings of the three most recent International Symposia on the Role of Soy on Preventing and Treating Chronic Disease (Messina, 1998, 2000; Messina et al., 2002) are excellent sources for the reader to obtain a historical view of research into the health effects of isoflavones.

Human Exposure to Isoflavones

Marketing data and hospital discharge records suggest that approximately 25% of the nearly 4 million newborns in the USA each year consume soy infant formula (American Academy of Pediatrics Committee on Nutrition, 1998). Commercially available soy formula marketed in the US is made with SPI[+], which has several phytochemicals associated with it, including isoflavones. Reasons for feeding infants soy formula range from medical indications (such as post-diarrhoea lactose intolerance, galactosaemia and primary lactase deficiency) to the desire to maintain a vegetarian lifestyle (Hill and Stuart, 1929; Setchell et al., 1997, 1998). Although the American Academy of Pediatrics recommends breastfeeding over formula feeding, it also recommends isolated soy protein-based formulas as a safe and effective alternative for providing appropriate nutrition for normal

Table 14.2. *In vitro* and *in vivo* biological effects of isoflavones.

Inhibition of ATP-utilizing enzymes, tyrosine-specific protein kinases, topoisomerase II
Inhibition of enzymes involved in phosphoinositol turnover
Induces apoptosis and differentiation in cancer cells
Inhibits cell proliferation
Inhibits angiogenesis and metastasis
Exerts antioxidant effects
Modulates cell cycling
Alters oestrogen metabolism
Modulates signalling pathways
 EGF, TGF-α, TGF-β
Differentially activates ERα and ERβ
Alters gonadal steroid metabolism
Increases immune responses and enhances host resistance

growth and development of term infants whose nutritional needs are not being met from maternal breast milk or cow's milk-based formulas.

The dietary habits of Americans, especially with respect to soyfoods, differ substantially from those of Asians. The exposure to soy isoflavones may actually begin at conception in countries where soy intake is a daily event. Data from the cord blood and amniotic fluid suggest that the Asian fetus may be exposed to isoflavone concentrations as high as 0.8 µmol l^{-1}, levels close to those reported in the maternal circulation (0.7 µmol l^{-1}) during pregnancy (Adlercreutz et al., 1999). However, immediately after birth, the isoflavone concentrations of the Asian newborn decline to near zero, because Asian infants are more likely to consume either milk formula or breast milk for the first few months of life. Since both cow's milk and breast milk from soy-consuming women have very low concentrations of isoflavones (5–50 nmol l^{-1}), the isoflavone exposure of Asian newborns through breast milk or cow milk formula is very low (Franke and Caster, 1996; Setchell et al., 1997, 1998). The circulating and tissue isoflavone concentrations remain low until soyfoods (or other isoflavone-containing foods) are introduced and increase to the maternal levels if the same dietary habits are acquired by the child. Thereafter, those who continue to consume the typical Asian diet containing substantial levels of soyfoods maintain an isoflavone body burden that is high relative to people in non-soy-consuming nations.

In the USA, where most pregnant women do not consume significant isoflavone-containing foods, fetal exposure is low, and 75% of American infants will not be exposed to soy formula, so that their plasma isoflavone levels remain low. This same very low soy consumption holds true throughout development and into adult life, since other than the soy protein hidden in processed foods, most Americans do not consume appreciable levels of soyfoods, and their plasma isoflavone concentrations remain generally low over their entire life.

However, 25% of American infants are fed soy formula (American Academy of Pediatrics Committee on Nutrition, 1998) that contains approximately 45 µg ml^{-1} (166 µmol l^{-1}) total isoflavones or >3000 times the concentration of human breast milk (Setchell et al., 1997, 1998). Infants fed soy formula have plasma isoflavone concentrations in the 1–10 µmol l^{-1} range (Setchell et al., 1997), sometimes starting as early as birth and continuing until weaning. This is the same general concentration range of the adults fed a soy beverage for the pharmacokinetics studies in Fig. 14.5 and Table 14.1 (Shelnutt et al., 2002). The total isoflavone intake of those subjects was 1.64 mg kg^{-1} body weight, and the mean plasma total isoflavone concentration was 1.2 µmol l^{-1}. Thus, infants fed soy formula containing the same SPI$^+$ used in the adult pharmacokinetic studies described in Fig. 14.5 consume 6–9 mg kg^{-1} total isoflavones, about five times more than the adult subjects, and had plasma total isoflavone concentrations that are about five times greater (Setchell et al., 1997). This compares with an average reported intake of total isoflavones from the traditional Japanese diet of 0.7 mg kg^{-1} day^{-1} (Messina, 1995) that results in a serum level of approximately 0.9 µmol l^{-1} total isoflavones (Patisaul et al., 2001). Thus, infants fed soy infant formula may be the segment of the human population most exposed to soyfood isoflavones. Upon weaning, American children who are fed soy formula are not likely to consume appreciable levels of soyfoods, and their plasma concentrations drop to less than 20 nmol l^{-1} and probably remain low for the rest of their lives. Therefore, typical Asians and Americans differ substantially in the isoflavone exposure profile and daily dose throughout their lifetime, and the health implications of these exposure profiles have not been fully explored.

Oestrogen-mediated Isoflavone Effects

Although the isoflavones have effects that are both oestrogen receptor-mediated and non-oestrogen receptor-mediated, the former has captured the most attention of both scientists and the lay public. Soy isoflavones, often referred to as phyto-oestrogens, bind to oestrogen receptors (ERs) and function as oestrogen agonists, antagonists or selective oestrogen receptor modulators (SERMs), depending on the tissue, cell type, isoflavone concentration and other conditions such as hormonal status, age, etc. (Kupier et al., 1997; Barkhem et al., 1998; An et al., 2001; Setchell, 2001).

The steroid estrogen hormones, such as 17β-oestradiol (E_2), affect growth and development by influencing differentiation and function of many target tissues, including those of the male and

Table 14.3. Functional differences between oestrogens and isoflavones.

Recruitment of transcription factors (co-regulators)
Activation or repression of gene transcription
Isoflavones preferentially trigger repression more than activation pathways
Isoflavones are less potent than oestrogen
ER–ligand conformation

female reproductive tract, brain, bone and cardiovascular system. Oestrogens are highly lipophilic and diffuse easily across cellular and intracellular membranes and bind intranuclear proteins, the ERs. The structural conformation of the ER changes when an oestrogen ligand binds, and this conformational change permits the receptor–ligand complex to interact with chromatin and modulate transcription of target genes. Although ERs have been studied for decades, it was not until 1987 that the ERα was cloned (Koike et al., 1987), and it took more than another decade to discover and clone the ERβ (Kupier et al., 1996). The ERα and ERβ proteins share a high degree of homology, but their target cell distribution and density differ. Although most ligands bind with very similar affinities for both receptors, the relative affinities for some synthetic oestrogens and naturally occurring steroid ligands differ greatly (Koike et al., 1987; Kupier et al., 1996). Human diets contain non-steroidal, weakly oestrogenic phytochemicals, including the flavonoids, coumestans and lignans, which compete with E_2 for binding to both ERα and ERβ proteins (Markiewicz, 1993; Miksicek, 1993; Korach, 1994; Kupier et al., 1998).

The high isoflavone concentrations of blood (and presumably tissue) of infants fed soy formula are of interest because humans do not normally have high concentrations of oestrogens until after puberty. This raises the question of short- and long-term health effects of such 'hormonal' actions on development of infants. Genistein has been reported to be 3000–400,000 times less potent than E_2 (Farmakalidis and Murphy, 1985; Kupier et al., 1997; An et al., 2001;), the wide variation most likely being due to methodology. A value of 1:1000 is used frequently. However, potency has been studied mainly in in vitro systems or in animals, and the true potency in target cells of human infants has never been reported. Thus, given the uncertainty about the potency of isoflavones in infants and the high plasma isoflavone concentrations during postnatal life when oestrogenic factors are usually very low, a scientific dialogue is occurring about the possible adverse effects of isoflavones in soy formula (Irvine et al., 1995; Robertson, 1995; Sheehan, 1997). Since this topic has been discussed at some length previously (Badger et al., 2002a,b), it will only be summarized here. The two adverse effects most often mentioned are cancer and reproductive effects, and these will be discussed briefly here.

Cancer

Epidemiological and experimental animal data suggest that the consumption of isoflavones, whether in soyfoods or as supplement in the diet, is associated with reduced risk of cancer incidence (Adlercreutz and Mazur, 1997; Barnes, 1997). For example, countries that consume high levels of soy have a lower incidence of breast and colon cancer than countries where soy intake is low (Nomura et al., 1978; Hirayama, 1985; Lee et al., 1991). The incidence of chemically induced mammary and colon cancers is lower in animals injected with isoflavones (Lamartiniere et al., 1995), fed isoflavones (Fritz et al., 1998; Lamartiniere et al., 2001), or fed soy-containing diets (Hakkak et al., 2000, 2001). On the other hand, promotion of the human breast cancer MCF-7 cell line occurs in athymic mice fed diets with either isoflavones or soy (Hsieh et al., 1998; Santell et al., 2000; Alfred et al., 2001), results that are purported to apply to women with either previously treated breast cancer or occult breast tumours, or to postmenopausal women. An interesting paradoxical feature of these studies is the strong implication that isoflavones are responsible for both the reduction of cancer risk, perhaps by reduced tumour initiation and promotion in the first set of studies (Nomura et al., 1978; Hirayama, 1985; Lee et al., 1991; Lamartiniere et al., 1995, 2001; Adlercreutz and Mazur, 1997; Fritz et al., 1998; Hakkak et al., 2000, 2001), and the increased risk of promotion of cancers in the latter studies (Hsieh et al., 1998; Santell et al., 2000; Alfred et al., 2001). How can a single class of dietary factors have both cancer-promoting and cancer-preventing features?

Consider the following information: (i) isoflavones bind to and activate the ERs, thus the appellation phyto-oestrogens; (ii) diets containing high phyto-oestrogen levels are associated with reduced cancer risks in human studies and reduced experimentally induced cancers in animals but, paradoxically, promotion of oestrogen-positive human breast cancer cells; and (iii) oestrogens used in hormone replacement therapy regimens may increase the risk of certain cancers (Collaborative Group on Hormonal Factors in Breast Cancer, 1997). The isoflavones are interesting because they share certain biological properties with oestrogens, but they also have bioactions that are distinctly non-oestrogenic. If the mechanisms of action of both oestrogens and isoflavones involve the ERs, what is the basis for their unique properties? The answer is complex, but may lie in the specificity of these ligands for the ERs and the dose-dependent ability and differential regulation of gene activation or repression by ERα and ERβ.

Evidence from interesting studies by An *et al.* (2001) demonstrates that E_2 effectively triggers both ERα- and ERβ-mediated transcriptional activation or repression pathways, whereas isoflavones are effective at triggering these pathways primarily through ERβ. Thus, the transcriptional actions elicited by isoflavones differ from those of oestrogens by virtue of unique gene activation/repression activities. This can explain the paradoxical effects discussed above, whereby even though oestrogen therapy appears to be associated with increased risks of certain cancers, dietary phyto-oestrogens have been associated with reduced cancer risks.

There are several ways that isoflavones and oestrogens differ with respect to ER activation/repression.

1. There is a differential potency of isoflavones for ERα and ERβ. For example, genistein is a weak agonist for triggering ERα-mediated gene activation or repression, but a potent agonist for ERβ-mediated gene transcriptional activity. In fact, in reporter gene studies, activation or repression of gene transcription was 1000-fold greater with ERβ than ERα (An *et al.*, 2001). The reasons or underlying mechanisms are complex and multifaceted, but may be due in part to the nearly 30-fold greater affinity of ERβ for genistein compared with ERα.

2. It is now thought that co-regulator proteins are required for both transcriptional activation and repression by ERs (An *et al.*, 1999; McKenna *et al.*, 1999a,b; Glass and Rosenfeld, 2000; Katzenellbogen *et al.*, 2000; Klinge, 2000). Thus, the ability of a ligand to 'recruit' these co-regulator proteins will greatly affect transcriptional activity. Co-regulator proteins bind to the E_2–ER complexes and differ from isoflavone–ER complexes, and this is associated with differential gene transcriptional activation or repression. For example, under certain experimental conditions, genistein appears to recruit co-regulators only to the ERβ, and triggers ERβ-mediated pathways and would result in eliciting the clinical effects associated with ERβ, and not with ERα. This would mean that the beneficial effects of ERβ-mediated pathways (such as reduced cancer risk) would be realized, while the often negative effects of the ERα pathways (such as increased cancer risk) would not.

3. The ligand dose is a very important factor in regulatory effects. At target cell concentrations that can be achieved easily through diet (0.5–1 µM), genistein is only very weakly capable of recruiting GRIP1 to ERα, whereas this transcription factor is easily recruited to the ERβ (Barnes *et al.*, 1996b). These co-regulators have several effects, including reducing dissociation of agonists from the ER–co-regulator complex, thus increasing the transcriptional effects. As a consequence, this stabilizing action may be an important factor in isoflavone actions.

4. Isoflavones are 10- to 300-fold more effective at triggering transcriptional repression of genes than they are with transcriptional gene activation (An *et al.*, 2001). Thus, the beneficial actions of isoflavones are probably due to 'turning off' pathways leading to adverse effects rather than to activating other pathways. This could provide the conceptual framework for lower incidence of menopausal symptoms and certain diseases such as osteoporosis, cardiovascular disease and cancers of the breast and endometrium in Asian countries where there is a high intake of soy isoflavones (Ingram *et al.*, 1997; Adlercreutz and Baillieres, 1998; Davis *et al.*, 1999).

5. The structural confirmation of ERβ is different when genistein is bound to it than when E_2 is bound (Pike *et al.*, 1999). The isoflavone–ERβ confirmation when isoflavones bind models that of SERMs. It is this confirmation that is thought to be responsible for the selective and specific attraction of co-activator and co-repressor proteins mentioned in point 2 above (McKenna *et al.*, 1999a,b; Klinge, 2000). Isoflavones may act to elicit clinical effects that are clearly distinct from those of endogenous oestrogens by selectively recruiting co-regulatory proteins to ERβ, and this triggers specific transcriptional pathways (An *et al.*, 2001). Thus, there is a physical basis to help explain the biological differences between actions of oestrogens and isoflavones on ER-dependent gene activation and repression, and this helps to explain the seemingly paradoxical effects of: (i) increased breast cancer risk associated with oestrogens and the decreased risk associated with high intake of soy isoflavones; and (ii) decreased menopausal symptoms without oestrogenic effects on the endometrium in post-menopausal women (Albertazzi *et al.*, 1998; Duncan, *et al.*, 1999; Upmalis *et al.*, 2000) or in macaques (Foth and Cline, 1998).

6. The endogenous gonadal and adrenal steroids have been studied for years, and the actions of the 'parent' compounds and their metabolites are well established. Similarly, the primary interactions between these endogenous hormones are well known. However, the biological actions of the parent soy isoflavones or their metabolites on specific target tissues have not been studied carefully. This may be an important nutritional consideration, because foods that contain isoflavones are complex in their composition. For example, there are three major isoflavones in soy, geinistin, daidzin and glycitin, and several metabolites. These isoflavones do not act in the same way in terms of ER binding nor other biological actions. The combined effects of a mixture of isoflavones could be quite different from the effects of a single isoflavone or gonadal steroid. Furthermore, there are other phytochemicals in soy and in other foods of a complex diet that may alter the response to isoflavones. Thus, a major future challenge is to determine the tissue-specific biological effects of these isoflavones, their metabolites (including conjugates) and other phytochemicals, and how they interact with each other to affect health.

7. Recent reports suggest that cytosolic ERα–ligand and ERβ–ligand complexes activate kinase cascades (Manolagas and Kousteni, 2001). These are non-genomic, oestrogen-like signalling actions of various ligands that bind cytosolic ER. The isoflavones clearly are potentially important ligands for these pathways and, although not as yet well studied, would be expected to differ from oestrogens in ways similar to other ligands (including androgens) and, therefore, represent another mechanism by which oestrogens and isoflavones can produce different and sometime diverse actions.

Reproduction

Because of the isoflavone–gonadal oestrogen link, great interest has developed in the effects of isoflavones on the reproductive system. The single most influential reported finding of isoflavones on reproduction is the impaired reproductive performance of Australian seasonal breeding sheep that ate large amounts of clover rich in potent isoflavones (Bennetts *et al.*, 1946). This was an important observation, because it demonstrated that isoflavones can have profound effects on reproductive outcomes under special circumstances. While this was and still is a significant factor in the sheep industry (Adams, 1990), it has little bearing on reproductive performance of people consuming soyfoods, as discussed previously (Badger *et al.*, 2002). These toxicological effects occur at consumption levels that would be unlikely ever to occur in humans through diet, but with the increased marketing of isoflavone supplements it is possible for toxic levels of purified isoflavones to be achieved and for eventual reproductive toxicity in humans.

Consumption of a soy protein has been reported to alter several hormonal aspects of the menstrual cycle, the most striking of which was lengthening of the follicular phase with no changes in the length of the luteal phase (Cassidy *et al.*, 1994). Similar studies in both pre- and post-menopausal women demonstrated that soy protein consumption reduced urinary oestrogen excretion and decreased putative genotoxic oestrogen metabolites (Xu *et al.*, 1998, 2000). Thus, the effects of isoflavones contained in soy protein diets consumed by humans are consistent with minor endocrinological effects that would not be considered to reduce reproductive capacity, but which would be cancer protective by: (i) lengthening the menstrual cycle periods of lowest oestrogen exposure; and (ii) decreasing oestrogen synthesis and altering oestrogen metabolism away from potential genotoxic metabolites. These data are consistent with much data on the association between menstrual cycle lengthening, oestrogen exposure and differences in the dynamics of the menstrual cycle of Western women who do not consume appreciable levels of soy and who have higher breast cancer rates compared with Asian women who consume significant amounts of soyfoods daily (Treolar *et al.*, 1970; Olsson *et al.*, 1983; World Health Organization, 1983; Henderson *et al.*, 1985;

Key et al., 1990; Lee et al., 1991; Munster et al., 1992).

There are no reports of reproductive toxicity of soyfoods as marketed in the USA, especially regarding reproductive competency. Similarly, there are no reports of soyfood-associated adverse health effects of Japanese newborns, suggesting that perinatal exposure to the high isoflavone concentrations achieved *in utero* (Adlercreutz et al., 1999) or by soy infant formula (Setchell et al., 1997) are not likely to result in adverse health effects, especially not the type of abnormalities ascribed to diethylstilbestrol or fetal alcohol syndrome. Women who consume soyfoods that result in high circulating concentrations of isoflavones are capable of conceiving, taking the pregnancy to term, delivering normal infants, having normal lactation and otherwise caring for their infants. Because these women consumed soyfoods before pregnancy and continued eating soy during pregnancy and lactation, soyfood isoflavones do not appear to have adverse effects on early human development or later reproductive performance.

Infant growth and development

Millions of American infants have been fed soy formula over the past three decades. Several studies have demonstrated that soy formula supports normal growth and development in term infants (Graham et al., 1970; Kohler et al., 1984; Businco et al., 1992; Churella et al., 1994), with body weight gains and body length of infants being virtually the same whether the infants were fed soy formula or cow's milk-based formula, or breastfed (Lasekan et al., 1999). Similar results were reported for young adults who were fed soy formula as infants (Strom et al., 2001). Thus, the available data suggest that soy formula is safe and effective in promoting normal growth and development of term infants.

Animal growth and development

One issue related to soy infant formula is the long-term health consequences of early consumption of these formulas. The effects of feeding the same SPI$^+$ used in infant formulas to several generations of rats was studied with the idea of establishing a situation similar to Asians who have high levels of soy intake throughout their lives (Badger et al., 2001). In those studies, AIN-93G diets (made with SPI$^+$) were fed throughout life, and both male and female rats had the same breeding efficiency as rats fed commercial diets or AIN-93G diets made with casein. The numbers of offspring, gender ratios, birth weights, birth lengths, health and general appearance of soy-fed rats were the same as casein-fed rats. Indices of oestrogenicity, such as weights of secondary sex organs, plasma oestrogen concentrations and mammary gland development, were found to be normal. The only major effect was vaginal opening being 1 day earlier in soy-fed rats; the practical consequence of this finding is unclear, because earlier puberty has not been recognized as a health or reproductive issue in Asia. However, this latter point has not been well studied.

Non-oestrogen-mediated Isoflavone Effects

The so-called phase I enzymes (cytochrome P450 enzymes, CYPs) are the major catalysts in biotransformation of xenobiotics (exogenous compounds), such as dietary factors, drugs, pesticides and carcinogens. They are also important in metabolizing endogenous compounds (endobiotics), such as gonadal and adrenal steroids, lipids and fat-soluble vitamins. Thus, regulation or modulation of these enzymes can be important to health, because each of the substances mentioned as substrates for these enzymes are important to physiological, endocrinological, metabolic or pharmacological outcomes. These enzymes often are regulated by their substrates, thereby amplifying substrate action on cellular function. There are numerous examples of substrate induction or inhibition of phase I enzymes by dietary factors, medications and other exogenous and potentially toxic chemicals (Guengerich, 1995). Regulation of phase I metabolism is a potential major mechanism for producing either beneficial or adverse health outcomes through diet.

Phase II enzymes have two main actions, each of which promotes their excretion from the body: (i) they promote conjugation of xenobiotics and endobiotics; or (ii) they destroy the reactive nature of the compound (Talalay et al., 1988; Prestera et al.,

1993). These enzymes include the glutathione transferases (GSTs), UDP-glucuronosyltransferases (UDP-GTs), sulphotransferases (STs), epoxide hydrolase and quinone reductase (QR). A number of *in vitro* and *in vivo* rodent studies have suggested that treatment with purified flavinoids and coumarins found in soy may induce the phase II enzymes QR, GSTs and UDP-GT (Appelt and Reicks, 1997; Wang *et al.*, 1998).

Isoflavones and/or soy diets high in isoflavones have been reported to modulate phase I metabolism significantly by altering expression of enzymes in CYP gene families 1, 2 and 3 (Backlund *et al.*, 1997; Heisby *et al.*, 1998; Ronis *et al.*, 1999, 2001). An excellent example of how isoflavone-rich diets can exert protective effects on lowering the risk of breast cancer incidence was reported by Rowlands *et al.* (2001) in which they studied the phase I enzymes involved in activation of the procarcinogen DMBA (7,12-dimethylbenz-[*a*] anthracene). One possible mechanism by which soyfoods could protect against toxic compounds such as procarcinogens would be to prevent activation by inhibiting enzymes necessary for the conversion from a procarcinogen to a carcinogen. This in turn would be expected to reduce the numbers of adducts and subsequent mutations that lead to mammary cancer. Rats fed SPI^+ diets have reduced levels of hepatic CYP1A1 and mammary gland CYP1B1 and CYP1A1 at the time of DMBA treatment (Ronis *et al.*, 2001). The expression of CYP1A1 and CYP1B1 is regulated by the DMBA-activated aryl hydrocarbon receptor (AhR). The AhR binds to a ligand and interacts with the AhR nuclear translocator (ARNT) to form a nuclear AhR–ANRT heterodimeric complex that acts as a ligand-activated transcription factor, which in turn binds to the xenobiotic response elements in the regulatory region of CYP1 genes (Evans, 1988). At the time of DMBA treatment, there were lower AhR and ARNT protein levels in the cytosol and nucleus, respectively. The downstream consequence of reduced DMBA activation would be lower target tissue carcinogen concentrations and fewer DNA adducts.

For example, the numbers of DMBA–DNA adducts in the ovary and adrenal glands, tissues where CYP1B1 levels are reported to be extremely high relative to other organs (Bhattacharyya *et al.*, 1995; Brake *et al.*, 1999; McFadyen *et al.*, 2001; Muskhelishvili *et al.*, 2001), were reduced ($P < 0.05$) in SPI^+-fed rats. These data complement those of Upadhyaya and el-Bayoumy (1998), who reported reduced DMBA–DNA adducts in the mammary gland of rats fed SPI^+-containing diets. Taken together, the existing data suggest that SPI^+ is working to reduce the incidence of DMBA-induced mammary gland cancer by down-regulating CYP1A1 and CYP1B1, reducing DMBA–DNA adducts and cancer incidence (Rowlands, *et al.*, 2001).

Thus, it appears that consumption of diets containing high levels of isoflavone, such as diets rich in soy, will decrease phase I and increase phase II metabolism, which would be expected to increase clearance and reduce carcinogenicity of ingested carcinogens. Maximizing beneficial health effects from diet, such as prevention of cancer caused by exposure to procarcinogens or carcinogens, would be expected to occur with suppression of phase I enzymes to inhibit activation of toxic xenobiotics and induction of phase II enzymes to enhance detoxication.

Acknowledgements

Much of the work presented in this chapter was supported by the USDA ARS Arkansas Children's Nutrition Center.

References

Adams, N.R. (1990) Permanent infertility in ewes exposed to plant oestrogens. *Australian Veterinary Journal* 67, 197–201.

Adlercreutz, H. and Baillieres, J.H. (1998) Epidemiology of phytoestrogens. *Clinical Endocrinology and Metabolism* 12, 605–623.

Adlercreutz, H. and Mazur, W. (1997) Phyto-oestrogens and Western diseases. *Annals of Medicine* 29, 95–120.

Adlercreutz, H., van der Widt, J. and Lampe, J. (1993) Quantitative determination of ligands and isoflavones in plasma of omivorous and vegetarian women by isotope dilution gas-chromatography–mass spectrometry. *Scandinavian Journal of Clinical Laboratory Investigation* 52, 97–102.

Adlercreutz, H., Yamad, T., Wahala, K. and Watanabe, S. (1999) Maternal and neonatal phytoestrogens in Japanese women during birth. *American Journal of Obstetrics Gynecology* 180, 737–743.

Albertazzi, P., Pansini, F., Bonaccorsi, G., Zanotti, L., Forini, E. and De Aloysio, D. (1998) The effect of dietary soy supplementation on hot flushes. *Obstetrics and Gynecology* 91, 6–11.

Allred, C.D., Allred, K.F., Young, H.J., Suzanne, M.V. and Helferich, W.G. (2001) Soy diets containing varying amounts of genistein stimulate growth of estrogen-dependent (MCF-7) tumors in a dose-dependent manner. *Cancer Research* 61, 5045–5050.

American Academy of Pediatrics Committee on Nutrition (1998) Soy protein-based formulas: recommendations for use in infant feeding. *Pediatrics* 101, 148–153.

An, J., Ribeiro, R.C., Webb, P., Gustafsson, J.A., Kushner, P.J., Baxter, J.D. and Leitman, D.C. (1999) Estradiol repression of tumor necrosis factor-alpha transcription requires estrogen receptor activation function-2 and is enhanced by coactivators. *Proceedings of the National Academy of Sciences USA* 96, 15161–15166.

An, J., Tzagarakis-Foster, C., Scharschmidt, T.C., Lomri, N. and Leitman, D.C. (2001) Estrogen receptor β-selective transcriptional activity and recruitment of coregulators by phytoestrogens. *Journal of Biological Chemistry* 276, 17808–17814.

Appelt, L.C. and Reicks, M.M. (1997) Soy feeding induces phase II enzymes in rat tissues. *Nutrition and Cancer* 28, 270–275.

Backlund, M., Johansson, I., Mkrtchian, S. and Ingelman-Sundberg, M. (1997) Signal transduction-mediated activation of the aryl hydrocarbon receptor in rat hepatoma H4IIE cells. *Journal of Biological Chemistry* 272, 31755–31763.

Badger, T.M., Ronis, M.J.J. and Hakkak, R. (2001) Developmental effects and health aspects of soy protein isolate, casein, and whey in male and female rats. *International Journal of Toxicology* 20, 165–174.

Badger, T.M., Ronis, M.J.J., Hakkak, R. and Korourian, S. (2002a) The health consequences of early soy consumption. *Journal of Nutrition* 132, 559S–565S.

Badger, T.M., Ronis, M.J.J., Hakkak, R. and Korourian, S. (2002b) The health consequences of soy infant formula, soy protein and isoflavones. In: Gilani, C.S. and Anderson, J.J.B. (eds) *Phytoestrogens and Health*. AOCS Press, Champaign, Illinois, pp. 586–605.

Barkhem, T., Carlson, B., Nilsson, Y., Enmark, E., Gustafsson, J. and Nilsson, S. (1998) Differential response of estrogen receptor α and estrogen receptor β to partial estrogen agonist/antagonists. *Molecular Pharmacology* 54, 105–112.

Barnes, S. (1997) The chemopreventive properties of soy isoflavonoids in animal models of breast cancer. *Breast Cancer Research and Treatment* 46, 169–179.

Barnes, S., Kirk, M. and Coward, L. (1996a) Isoflavones and their conjugates in soy foods – extraction conditions and analysis by HPLC mass spectrometry. *Journal of Agriculture and Food Chemistry* 442, 2466–2474.

Barnes, S., Sfakianos, J., Coward, L. and Kirk, M. (1996b) Soy isoflavonoids and cancer prevention. Underlying biochemical and pharmacological issues. *Advances in Experimental Medicine and Biology* 401, 87–100.

Bennetts, H.W., Underwood, E.J. and Shier, F.L. (1946) A specific breeding problem of sheep on subterranean clover pastures in Western Australia. *Australian Veterinary Journal* 22, 2–12.

Bhattacharyya, K.K., Brake, P.B., Eltom, S.E., Otto, S.A. and Jefcoate, C.R. (1995) Identification of a rat adrenal cytochrome P450 active in polycyclic hydrocarbon metabolism as rat CYP1B1. *Journal of Biological Chemistry* 270, 11595–11602.

Brake, P.B., Arai, M., As-Sanie, S., Jecoate, C.R. and Widmaier, E.P. (1999) Developmental expression and regulation of adrenocortical cytochrome P4501B1 in the rat. *Endocrinology* 140, 1672–1680.

Businco, I., Bruno, G. and Giampieto, P.G. (1992) Allergenicity and nutritional adequacy of soy protein formulas. *Journal of Pediatrics* 121, 821–828.

Cassidy, A., Bingham, S. and Setchell, K.D.R. (1994) Biological effects of a diet of soy protein rich in isoflavones on the menstrual cycle of premenopausal women. *American Journal of Clinical Nutrition* 60, 333–340.

Chang, H.C., Churchwell, M.I., Delclos, K.B., Newbold, R.R. and Doerge, D.R. (2000) Mass spectrometric determination of genistein tissue distribution in diet-exposed Sprague Dawley rats. *Journal of Nutrition* 130, 1963–1970.

Chang, H.C., Fletcher, T., Ferguson, M., Hale, K., Fang, N., Ronis, M., Prior, R. and Badger, T.M. (2002) Serum and tissue profiles of isoflavone aglycones and conjugates in rats fed diets containing soy protein isolate (SPI). *FASEB Journal* 16, A1008.

Churella, H.R., Borschel, M.W., Thomas, M.R., Breem, M. and Jacobs, J. (1994) Growth and protein status of term infants fed soy protein formulas differing in protein content. *Journal of the American College of Nutrition* 13, 262–267.

Cimino, C.O., Shelnutt, S.R., Ronis, M.J. and Badger, T.M. (1999) An LC-MS method to determine concentrations of isoflavones and their sulfate and glucuronide conjugates in urine. *Clinica Chimica Acta* 287, 69–82.

Collaborative Group on Hormonal Factors in Breast Cancer (1997) Breast cancer and hormone replacement therapy: collaborative reanalysis of data from 51 epidemiological studies of 52,705 women with breast cancer and 108,411 women without breast cancer. *Lancet* 350, 1047–1059.

Coward, L., Barnes, N.C., Setchell, K.D.R. and Barnes, S. (1993) Genistein and daidzein and their β-glycosides conjugates: anti-tumor isoflavones in soybean foods from American and Asian diets. *Journal of Agriculture and Food Chemistry* 41, 1961–1967.

Coward, L., Kirk, M., Albin, N. and Barnes, S. (1996) Analysis of plasma isoflavones by reversed-phase HPLC-multiple reaction ion monitoring-mass spectrometry. *Clinica Chimica Acta* 247, 121–142.

Davis, S.R., Dalais, F.S., Simpson, E.R. and Murkies, A.L. (1999) Phytoestrogens in health and disease. *Recent Progress in Hormone Research* 54, 185–210.

Duncan, A.M., Underhill, K.E., Xu, X., Lavalleur, J., Phipps, W.R. and Kurzer, M.S. (1999) Modest hormonal effects of soy isoflavones in post menopausal women. *Journal of Clinical Endocrinology and Metabolism* 84, 3479–3484.

Evans, R.M. (1988) The steroid and thyroid hormone super-family. *Science* 240, 889–895.

Fang, N., Yu, S. and Badger, T.M. (2002) Characterization of isoflavones and their conjugates in female rat urine using LC/MS/MS. *Journal of Agriculture and Food Chemistry* 50, 2700–2707.

Farmakalidis, E. and Murphy, P.A. (1985) Isolation of 6″-O-acetylgenistein from toasted defatted soyflakes. *Journal of Agriculture and Food Chemistry* 33, 385–389.

Foth, D. and Cline, J.M. (1998) Effects of mammalian and plant estrogens on mammary glands and uteri of macques. *American Journal of Clinical Nutrition* 6S, 1413–1417.

Franke, A.A. and Custer, L.J. (1996) Daidzein and genistein concentrations in human milk after soy consumption. *Clinical Chemistry* 42, 955–964.

Franke, A.A., Custer, L.J., Cerna, C.M. and Narala, K. (1995) Rapid HPLC analysis of dietary phytoestrogens from legumes and from human urine. *Proceedings of the Society for Experimental Biology and Medicine* 208, 18–26.

Fritz, W., Coward, L., Wang, J. and Lamartiniere, C.A. (1998) Genistein: perinatal mammary cancer prevention, bioavailability and toxicity testing in the rat. *Carcinogenesis* 19, 2151–2158.

Gamache, P.H. and Acworth, I.N. (1998) Analysis of phytoestrogens and polyphenols in plasma, tissue, and urine using HPLC with coulometric array detection. *Proceedings of the Society for Experimental Biology and Medicine* 217, 274–280.

Gilani, G. and Anderson, J.J.B. (eds) (2002) *Phytoestrogens and Health*. AOCS Press, Champaign, Illinois, 660 pp.

Glass, C.K. and Rosenfeld, M.G. (2000) The coregulator exchange in transcriptional functions of nuclear receptors. *Genes and Development* 14, 121–141.

Graham, G.G., Placko, R.P. and Morals, E. (1970) Dietary protein quality in infants and children. VI. Isolated soy protein milk. *American Journal of Disabled Children* 120, 419–423.

Guengerich, F.G. (1995) Influence of nutrients and other dietary materials on cytochrome P-450 enzymes. *Journal of Nutrition* 61S, 651S–658S.

Hakkak, R., Korourian, S., Shelnutt, S.R., Lensing, S., Ronis, M.J.J. and Badger, T.M. (2000) Diets containing whey proteins or soy protein isolate protect against 7,12-dimethylbenz(a)anthracene-induced mammary tumors in female rats. *Cancer Epidemiology Biomarkers and Prevention* 9, 113–117.

Hakkak, R., Korourian, S., Ronis, M.J.J., Johnson, J. and Badger, T.M. (2001) Soy protein isolate consumption protects against azoxymethane-induced colon tumors in male rats. *Cancer Letters* 166, 27–32.

Hakkak, R., Korourian, S., Fletcher, T., Ferguson, M., Hale, K., Holder, D., Parker, J., Ronis, M., Treadaway, P. and Badger, T. (2002) The effects of soy protein containing negligible isoflavones and casein on DMBA-induced mammary tumors in rats. *American Association of Cancer Research* 43, 823.

Harborne, J.B., Mabry, T.J. and Mabry, H. (1975) *The Flavonoids*. Academic Press, New York.

Heisby, N.A., Chipman, J.K., Gescher, A. and Kerr, D. (1998) Inhibition of mouse and human CYP1A- and CYP2E1-dependent substrate metabolism by the isoflavonoids genistein and equol. *Food Chemistry and Toxicology* 36, 375–382.

Henderson, B.E., Ross, R.K., Judd, H.L., Frailo, M.D. and Pike, M.C. (1985) Do regulatory ovulatory cycles increase breast cancer risk? *Cancer* 56, 1206–1208.

Hill, L.W. and Stuart, H.C. (1929) A soy bean food preparation for feeding infants with milk idiosyncrasy. *Journal of the American Medical Association* 93, 985–987.

Hirayama, T. (1985) A large scale cohort study on cancer risks by diet, with special reference to the risk reducing effects of green–yellow vegetable consumption. *Princess Takamatsu Symposium* 160, 41–53.

Holder, C.L., Churchwell, M.I. and Doerge, D.R. (1999) Quantitation of soy isoflavones, genestein and daidzein, and conjugates in rat blood using LC/ES-MS. *Journal of Agriculture and Food Chemistry* 47, 3764–3770.

Hsieh, C.Y., Santell, R.C., Haslam, S.Z. and Helferich, W.G. (1998) Estrogenic effects of genistein on the growth of estrogen receptor-positive human breast cancer (MCF-7) cells *in vitro* and *in vivo*. *Cancer Research* 58, 3833–3838.

Ingram, D., Sanders, K., Kolybaba, M. and Lopez, D. (1997) Case–control study of phyto-oestrogens and breast cancer. *Lancet* 350, 990–994.

Irvine, C., Fitzpatrick, M., Robertson, I. and Woodhams, D. (1995) The potential adverse effects of soybean phytoestrogens in infant feeding. *New Zealand Medical Journal* 108, 208–209.

Katzenellebogen, B.S., Montano, M.M., Ediger, T.R., Sun, J., Wlkena, K., Lazzennec, G., Martini, P.G., McInerney, E.M., Delage-Mourroux, R., Weis, K. and Katzenellebogen, J.A. (2000) Estrogen receptors: selective ligands, partners, and distinctive pharmacology. *Recent Progress in Hormone Research* 55, 163–193.

Key, T.J.A., Chen, D.Y., Wang, D.Y., Pike, M.C. and Boreham, J. (1990) Sex hormones in rural China and Britain. *British Journal of Cancer* 37, 467–480.

Klinge, C.M. (2000) Estrogen receptor interaction with co-activators and co-repressors. *Steroids* 65, 227–251.

Kohler, L., Meeuwisse, G. and Mortensson, W. (1984) Food intake and growth of infants between six and twenty-six weeks of age on breast milk, cow's milk formula, or soy formula. *Acta Paediatrica Scandinavica* 73, 40–48.

Koike, S., Sakai, M. and Muramatsu, M. (1987) Molecular cloning and characterization of rat estrogen receptor cDNA. *Nucleic Acids Research* 15, 2499–2513.

Kulling, S.E., Honig, D.M. and Metzler, M. (2001) Oxidative metabolism of the soy isoflavone daidzein and genistein in humans *in vitro* and *in vivo*. *Journal of Agriculture and Food Chemistry* 49, 3024–3033.

Kupier, G.G., Enmark, E., Pelton-Huikko, M.-H., Nilsson, S. and Gustafsson, J.-A. (1996) Cloning of a novel estrogen receptor expressed in rat prostate and ovary. *Proceedings of the National Academy of Sciences USA* 93, 5925–5930.

Kupier, G.G., Carlsson, B., Grandien, K., Enmark, E., Haggblad, J., Nilsson, S. and Gustafsson, J.-A. (1997) Comparison of the ligand binding specificity and transcript tissue distribution of estrogen receptor α and β. *Endocrinology* 138, 863–870.

Kupier, G.G., Lemmen, J.G., Carlsson, B., Corton, J.C., Safe, S.H., Van Der Saag, P.T., Van Der Burg, B. and Gustafsson, J.-A. (1998) Interaction of estrogenic chemicals and phytoestrogens with estrogen receptor β. *Endocrinology* 139, 4252–4263.

Lamartiniere, C.A., Moore, J., Holland, M. and Barnes, S. (1995) Genistein and chemoprotection of breast cancer. *Proceedings of the Society for Experimental Biology and Medicine* 208, 120–123.

Lamartiniere, C.A., Cotroneo, M.S., Frizt, W.A., Wang, J., Mentor-Marcel, R. and Elgavish, A. (2001) Genistein chemoprotection: timing and mechanisms of action in murine mammary and prostate. *Journal of Nutrition* 132, 552S–585S.

Lapcik, O., Hampl, R., Hill, M., Wahala, K., Maharik, N.A. and Adlercreutz, H. (1998) Radioimmunoassay of free genistein in human serum. *Journal of Steroid Biochemistry and Molecular Biology* 64, 261–268.

Lasekan, J.B., Ostrom, K.M., Jacobs, J.R., Blatter, M.M., Ndife, L.I., Gooch, W.M. and Cho, S. (1999) Growth of newborn, term infants fed soy formulas for 1 year. *Clinical Pediatrics* 38, 563–571.

Lee, H.P., Gourley, L., Diffy, S.W., Esteve, J., Lee, J. and Day, N.E. (1991) Dietary effects on breast cancer risk in Singapore. *Lancet* 337, 1197–1200.

Mabry, T.J., Markham, K.R. and Thomas, M.B. (1970) *The Systematic Identification of Flavonoids*. Springer, New York.

Manolagas, S.C. and Kousteni, S. (2001) Perspective: nonreproductive sites of action of reproductive hormones. *Endocrinology* 142, 2200–2204.

Markham, K.R. and Geiger, H. (1986) ^{1}H nuclear magnetic resonance spectroscopy of flavonoids and their glycosides in hexadeuterodimethylsulfoxide. In: Harborne, J.B. (ed.) *Flavonoids*. Chapman & Hall, London, pp. 321–335.

Markiewicz, L., Garvey, J., Adlercreutz, H. and Gurpide, E. (1993) *In vitro* bioassays of non steroidal phytoestrogens. *Journal of Steroid Biochemistry and Molecular Biology* 45, 399–405.

McFadyen, M.C., Crickshank, M.E., Miller, I.D., McLeod, H.L., Melvin, W.T., Haites, N., Parin, D. and Murray, G.I. (2001) Cytochrome P450 CYP1B1 over-expression in primary and metastatic ovarian cancer. *British Journal of Cancer* 85, 242–246.

McKenna, N., Xu, J., Nawaz, Z., Tsai, S., Tsai, M.-J. and O'Malley, B. (1999a) Nuclear receptor coactivators: multiple functions. *Journal of Steroid Biochemistry and Molecular Biology* 69, 3–12.

McKenna, N.J., Lanz, R.B. and O'Malley, B.W. (1999b) Nuclear receptor coregulators: cellular and molecular biology. *Endocrine Review* 20, 321–344.

Messina, M. (1995) Isoflavone intakes by Japanese were overestimated. *American Journal of Clinical Nutrition* 62, 645.

Messina, M. (1998) The role of soy in preventing and treating chronic disease. *American Journal of Clinical Nutrition* 68, 1329S–1544S.

Messina, M. (2000) The role of soy in preventing and treating chronic disease. *American Journal of Clinical Nutrition* 70, 1329S–1541S.

Messina, M., Gardner, C. and Barnes, S. (2002) Gaining insight into the health effects of soy but a long way still to go: commentary on the Fourth International Symposium on the Role of Soy in Preventing and Treating Chronic Disease. *Journal of Nutrition* 132, 547S–551S.

Miksicek, R.J. (1993) Commonly occurring plant flavonoids have estrogenic activity. *Molecular Pharmacology* 44, 37–43.

Munster, K., Schmidt, L. and Helm, P. (1992) Length and variation in the menstrual cycle – a cross sectional study from Danish county. *British Journal of Obstetrics and Gynaecology* 99, 422–429.

Muskhelishvili, L., Thompson, P.A., Kusewitt, D.F., Wang, C. and Kadlubar, F.F. (2001) *In situ* hybridization and immunohistochemical analysis of cytochrome analysis of cytochrome P4501B1 expression in human normal tissues. *Journal of Histochemistry and Cytochemistry* 49, 229–236.

Nomura, A., Henderson, B.E. and Lee, J. (1978) Breast cancer and diet among the Japanese in Hawaii. *American Journal of Clinical Nutrition* 31, 2020–2025.

Olsson, H., Landin-Olsson, M. and Gullberg, B. (1983) Retrospective assessment of menstrual cycle length in patients with breast cancer, in patients with benign breast disease and in women without breast disease. *Journal of the National Cancer Institute* 70, 17–20.

Patisaul, H.B., Dindo, M., Whitten, P.L. and Young, L.J. (2001) Soy isoflavone supplements antagonize reproductive behavior and estrogen receptor alpha and beta-dependent gene expression in the brain. *Endocrinology* 142, 2946–2952.

Pike, A.C., Brzozowski, A.M., Hubbard, R.E., Bonn, T., Thorsell, A.G., Engstrom, O., Ljunggren, J., Gustafsson, J.A. and Carlqvist, M. (1999) Structure of the ligand-binding domain of oestrogen receptor β in the presence of a partial agonist and a full agonist. *EMBO Journal* 18, 4608–4618.

Polkowski, K. and Mazyrek, A.P. (2000) Biological properties of genistein. A review of *in vitro* and *in vivo* data. *Acta Polica Pharmica* 57, 135–155.

Prestera, Y., Holtzclaw, Y., Zhang, Y. and Talalay, P. (1993) Chemical and molecular regulation of enzymes that detoxify carcinogens. *Proceedings of the National Academy of Sciences USA* 92, 8965–8969.

Robertson, I.G.C. (1995) Phytoestrogens: toxicity and regulatory recommendations. *Proceedings of the Nutrition Society of New Zealand* 20, 35–42.

Ronis, M.J., Rowlands, J.C., Hakkak, R. and Badger, T.M. (1999) Altered expression and glucocorticoid-inducibility of hepatic CYP3A and CYP2B enzymes in male rats fed diets containing soy protein isolate. *Journal of Nutrition* 129, 1958–1965.

Ronis, M.J., Rowlands, J.C., Hakkak, R. and Badger, T.M. (2001) Inducibility of hepatic CYP1A enzymes by 3-methylcholanthrene and isosafrole differs in male rats fed diets containing casein, soy protein isolate or whey from conception to adulthood. *Journal of Nutrition* 131, 1180–1188.

Rowlands, J.C., He, J., Hakkak, R., Ronis, M.J.J. and Badger, T.M. (2001) Soy and whey proteins downregulate DMBA-induced liver and mammary gland CYP1 expression in female rats. *Journal of Nutrition* 131, 3281–3287.

Santell, R.C., Kieu, N. and Helferich, W.G. (2000) Genistein inhibits growth of estrogen-independent human breast cancer cells in culture but not in athymic mice. *Journal of Nutrition* 30, 1665–1669.

Setchell, K.D.R. (1995) Non-steroidal estrogens of dietary origin: possible roles in health and disease, metabolism and physiological effects. *Proceedings of the Nutrition Society of New Zealand* 20, 1–21.

Setchell, K.D.R. (2001) Soy isoflavones-benefits and risks from nature's selective estrogen receptor modulators (SERMs). *Journal of the American College of Nutrition* 20, 354S–362S.

Setchell, K.D.R. and Aldercreutz, H. (1988) Mammalian ligands and phytochemicals: recent studies on their formation, metabolism and biological role in health and disease. In: Rowland I.A. (ed.) *The Role of Gut Microflora in Toxicity and Cancer*. Academic Press, New York, pp. 315–345.

Setchell, K.D.R., Zimmer-Nechemias, L., Cai, J. and Heubi, J.E. (1997) Exposure of infants to phyto-oestrogens from soy-based infant formula. *Lancet* 350, 23–27.

Setchell, K.D.R., Zimmer-Nechemias, L., Cai, J. and Heubi, J.E. (1998) Isoflavone content of infant formulas and the metabolic fate of these phytoestrogens in early life. *American Journal of Clinical Nutrition* 68, 1453S–1461S.

Sheehan, D.M. (1997) Isoflavone content of breast milk and soy formula: benefits and risks. *Clinical Chemistry* 43, 850–852.

Shelnutt, S.R., Cimino, C.O., Wiggins, P.A. and Badger, T.M. (2000) Urinary pharmacokinetics of the glucuronide and sulfate conjugates of genistein and daidzein. *Cancer Epidemology Biomarkers and Prevention* 9, 413–419.

Shelnutt, S.R., Cimino, C.O., Wiggins, P.A., Ronis, M.J.J. and Badger, T.M. (2002) Pharmacokinetics of the glucuronide and sulfate conjugates of genistein and daidzein following a soy beverage in men and women. *American Journal of Clinical Nutrition* 76, 588–594.

Strom, B.L., Schinnar, R., Ziegler, E.E., Barnhart, K., Sammel, M., Macones, G., Stallings, V., Hanson, S.A. and Nelson, S.E. (2001) Follow-up study of a cohort fed soybased formula during infancy. *Journal of the American Medical Association* 286, 807–814.

Supko, J.G. and Phillips, L.R. (1995) High performance liquid chromatography assay for genistein in biological fluids. *Journal of Chromatography, Biomedical Applications* 666, 157–167.

Talalay, P., DeLong, M.J. and Prochaska, H.J. (1988) Identification of a common chemical signal regulating the induction of enzymes that protect against chemical carcinogenesis. *Proceedings of the National Academy of Sciences USA* 85, 8261–8265.

Treolar, A.E., Boynton, R.E., Behn, B.G. and Brown, B.W. (1970) Variation of the human menstrual cycle through reproductive life. *International Fertility* 12, 77–126.

Upahyaya, P. and el-Bayoumy, K. (1998) Effect of dietary soy protein isolate, genistein and 1,4-phenylenebis (methylene)selenocyanate on DNA binding of 7,12-dimethylbenz[a]anthracene in mammary glands of CD rats. *Oncology Reports* 5, 1541–1545.

Upmalis, D.H., Lobo, R., Bradley, L., Warren, M., Conne, F.L. and Lamia, C.A. (2000) Vasomotor symptom relief by soy isoflavone extract tablets in postmenopausal women: a multicenter, double-blind, randomized, placebo-controlled study. *Menopause* 7, 236–242.

Wang, W., Liu, L.Q., Higuchi, C.M., and Chen, H.W. (1998) Induction of NADPH-quinone reductase by dietary phytoestrogens in colonic colo205 cells. *Biochemical Pharmacology* 56, 189–195.

World Health Organization (1983) A prospective multicenter trial of the ovulation method of nature family planning. III Characteristics of the menstrual cycle and of the fertile phase. *Fertility and Sterility* 40, 773–778.

Xu, X., Duncan, A.M., Merz, B.E. and Kurzer, M.S. (1998) Effects of soy isoflavones on estrogen and phytoestrogen metabolism in premenopausal women. *Cancer Epidemiology Biomarkers and Prevention* 7, 1101–1108.

Xu, X., Duncan, A.M., Wangen, K.E. and Kurzer, M.S. (2000) Soy consumption alters endogenous estrogen metabolism in postmenopausal women. *Cancer Epidemiology Biomarkers and Prevention* 9, 781–786.

15 Mechanisms of Ubiquitination and Proteasome-dependent Proteolysis in Skeletal Muscle

Didier Attaix, Lydie Combaret, Anthony J. Kee[1] and Daniel Taillandier

Human Nutrition Research Center of Clermont-Ferrand and INRA, Nutrition and Protein Metabolism Unit, Ceyrat, France

Introduction

Proteins in skeletal muscle, as in all mammalian tissues, undergo a continuous process of synthesis and degradation, which regulates both the overall muscle protein mass and the levels of specific proteins (Waterlow *et al.*, 1978). Increased proteolysis contributes to the muscle wasting seen in several pathological conditions (e.g. cancer, sepsis, diabetes, burn injury, trauma, etc.) and during fasting. Enhanced proteolysis in skeletal muscle, the major protein reservoir in the body, is a key metabolic adaptation providing the organism with free amino acids for energy production via gluconeogenesis and direct oxidation. Net mobilization of muscle protein in pathological situations also provides free amino acids for acute phase protein synthesis in the liver, and for protein synthesis in vital organs (e.g. brain and heart).

The lysosomal, Ca^{2+}-activated and ubiquitin–proteasome-dependent pathways are the most important processes responsible for the breakdown of the bulk of skeletal muscle protein (Attaix and Taillandier, 1998). However, there are also many other proteases, e.g. caspases (see Belizario *et al.*, 2001) or matrix metalloproteinases (see Balcerzak *et al.*, 2001), that are active in muscle. Nevertheless,

it is now clear that both lysosomal and Ca^{2+}-activated proteases (e.g. cathepsins and calpains, respectively) do not play a major role in skeletal muscle proteolysis. These enzymes contribute <15–20% of total protein breakdown in muscles from both control and cachectic animals, and are not directly responsible for the breakdown of myofibrillar proteins (Lowell *et al.*, 1986; Tiao *et al.*, 1994). Furthermore, a systematic activation of either cathepsins or calpains is not observed in muscle wasting conditions (for detailed information, see Attaix and Taillandier, 1998). In contrast, the activation of the ubiquitin–proteasome pathway is mainly responsible for the muscle wasting that occurs in various animal models of cachexia (Attaix and Taillandier, 1998; Attaix *et al.*, 1998; Jagoe and Goldberg, 2001; Hasselgren *et al.*, 2002). This pathway is also activated in human patients when muscle wasting is rapid and pronounced (Mansoor *et al.*, 1996; Tiao *et al.*, 1997), but not in chronic situations such as muscular dystrophy (Combaret *et al.*, 1996) or Cushing syndrome (Rallière *et al.*, 1997). In this chapter, we first critically review the mechanisms of ubiquitination and proteasome-dependent proteolysis, and then discuss the regulation of this proteolytic machinery in muscle, with a particular emphasis on its nutritional and hormonal control.

---

[1] Present address: Muscle Development Unit, Children's Medical Research Institute, Locked Bag 23, Wentworthville, NSW 2145, Australia.

©CAB *International* 2003. *Molecular Nutrition*
(eds J. Zempleni and H. Daniel)

Mechanisms of Ubiquitin–Proteasome-dependent Proteolysis

The ubiquitin–proteasome system is the major non-lysosomal process responsible for the breakdown of most short- and long-lived proteins in mammalian cells (Rock *et al.*, 1994). This complex proteolytic machinery is under the control of several hundreds of genes, including genes of the ubiquitination–deubiquitination system (~200–300), genes (~50) encoding different proteasome subunits, and endogenous proteasome activators (see below) and inhibitors. The pathway plays housekeeping functions in basal protein turnover and the elimination of miscoded, misfolded or mislocalized abnormal proteins. In addition, the system also controls several major biological functions such as class I antigen presentation, and the control of the cell cycle and transcription, signal transduction, protein sorting, etc. (for detailed information, see Glickman and Ciechanover, 2002).

Very schematically there are two main steps in the pathway: (i) covalent attachment of a polyubiquitin chain to the protein substrate, under the control of ubiquitination and deubiquitination enzymes; and (ii) specific recognition of the polyubiquitin chain and degradation of the targeted protein by the 26S proteasome (Fig. 15.1).

Ubiquitination

Ubiquitin is a 76-amino-acid polypeptide extremely conserved in eukaryotic cells. Ubiquitination is defined as the covalent attachment of ubiquitin to a protein substrate. Ubiquitination is a widespread post-translational modification akin to phosphorylation, and as such has major and diverse proteolytic and non-proteolytic functions. In order to be degraded by the proteasome system, the substrate must be tagged by a polyubiquitin degradation signal that consists of at least four ubiquitin moieties (Thrower *et al.*, 2000). Mono-, di- or

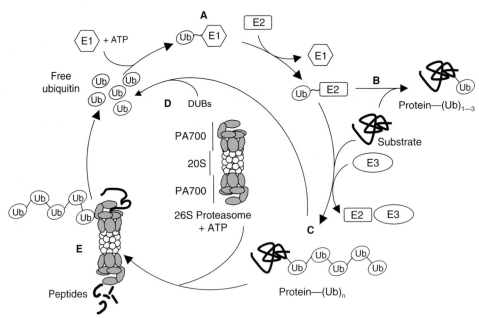

Fig. 15.1. Schematic representation of the ubiquitin–proteasome pathway. (A) Ubiquitin (Ub) is first activated by the ubiquitin-activating enzyme (E1) and transferred to a ubiquitin-conjugating enzyme (E2). (B) E2 mono-, di- or triubiquitinate substrates (Protein–(Ub)$_{1-3}$) that are not targeted for breakdown. (C) In contrast, when E2 forms a polyubiquitinated protein (Protein–(Ub)$_n$), with or without an E3, (D) the substrate can either be deubiquitinated by the deubiquitinating enzymes (DUBs) or (E) is recognized, unfolded, injected into the 26S proteasome and degraded into peptides. Note that ATP hydrolysis is required in both steps A and E.

triubiquitinated proteins are not usually degraded but are targeted for other fates. For example, monoubiquitination is a signal of endocytosis of a number of receptors (Shih et al., 2000).

Polyubiquitination is a complex multiple step process (Pickart, 2001). In brief, ubiquitin initially is activated by a single ubiquitin-activating enzyme, E1 (Fig. 15.1). E1 then transfers the activated ubiquitin to one of the ubiquitin-conjugating enzymes, E2s. E2s transfer the first ubiquitin molecule to protein substrates through the formation of an isopeptide bond between the activated C-terminal glycine residue of ubiquitin and the ε-amino group of a lysine residue in the substrate. This results in the formation of a monoubiquitinated protein. Some E2s can catalyse the continued addition of ubiquitin molecules to form polyubiquitinated conjugates. This is usually achieved by transfer of additional activated ubiquitin moieties to Lys48 of the preceding conjugated ubiquitin molecule. In most cases, however, the formation of polyubiquitin chains requires the presence of a ubiquitin–protein ligase (E3), which also has the important role of recognizing protein substrates.

Ubiquitin-conjugating enzyme, E1

The sequence of E1 is highly conserved across many species. The human gene for E1 possesses an alternative start codon, resulting in the expression of a 110 kDa nuclear form and a 117 kDa cytoplasmic form (Haas and Siepmann, 1997). The reaction catalysed by E1 starts with the binding of ATP-Mg^{2+}, and then ubiquitin to the enzyme, leading to the formation of a ubiquitin adenylate intermediate that serves as the donor of ubiquitin to the critical cysteine residue in the E1 active site. When fully loaded, E1 carries two molecules of activated ubiquitin (as a thiol ester and as an adenylate, respectively), so that the thiol-linked ubiquitin is transferred to one of the E2s. This very efficient reaction results in the production of activated ubiquitin for the entire downstream ubiquitin conjugation pathways.

Ubiquitin-conjugating enzymes, E2s

The ubiquitin-conjugating enzymes, E2s, represent a superfamily of related proteins, with a molecular weight range of usually 14–35 kDa. There are four classes of E2s, which all possess a central catalytic domain of approximately 150 amino acids with a critical cysteine residue that also forms a thiol ester intermediate with ubiquitin, and variable N- and/or C-terminal extension(s). In mammals, there are at least 20–30 E2s (Scheffner et al., 1998). Despite their structural similarities, E2s are responsible for distinct biological functions so that only a limited number of E2s (e.g. three out of 11 E2s in yeast) play a role in the formation of the polyubiquitin degradation signal. Some E2s – that possess undefined substrate recognition sites – form a polyubiquitin degradation signal in the presence of E1 and ATP. However, the formation of polyubiquitin chains generally is achieved in the presence of an E3 enzyme (Pickart, 2001). Most E2s interact with several E3s (and vice versa), which in turn recognize their specific protein substrates. Furthermore, a given protein substrate can be ubiquitinated by different combinations of E2s and E3s. This results in a wide range of alternative ubiquitination pathways (Glickman and Ciechanover, 2002).

Ubiquitin–protein ligases, E3s

E3s play a pivotal role in polyubiquitination, as they are responsible for the selective recognition of protein substrates. All known E3s are HECT (homologous to E6-AP C-terminus) domain E3s, RING (really interesting new genes) finger E3s (Pickart, 2001), or U-box-containing E3s.

The first major group of E3s are monomeric enzymes of the HECT domain family. This large domain (~350 amino acids), located in the C-terminal region of the enzyme, mediates E2 binding and ubiquitination of the target protein via thiol ester linkage formation with ubiquitin. The N-terminal region of HECT E3s binds to specific substrate(s). The human genome encodes at least 20 HECT domain proteins. However, mammalian genome sequencing projects have identified numerous potential uncharacterized HECT E3s (Pickart, 2001).

Eight cysteine and histidine residues that coordinate two Zn^{2+} ions define the RING finger structure. There are several hundred cDNAs encoding RING finger proteins in the GenBank database, and many unrelated RING finger proteins with unknown functions behave *in vitro* as E3s (Attaix et al., 2001).

The most simple RING finger E3s are the monomeric 'N-end rule' enzymes E3α and E3β. E3α binds to proteins bearing basic or bulky hydrophobic N-terminal amino acid residues, while E3β binds to small uncharged N-terminal amino acids (Varshavsky, 1996). A more complex route by which eukaryotic cells recognize various specific substrates is by conscripting numerous substrate-specific adaptor proteins that recruit protein substrates to core ubiquitination complexes (Attaix et al., 2001; Pickart, 2001). These high molecular weight multisubunit complexes form the most abundant E3 families. There are at least three different types of RING finger E3 complexes: (i) cyclosomes or APC (anaphase-promoting complexes); (ii) SCF (Skp1–Cdc53–F-box protein family) E3 ligases; and (iii) VCB-like (von Hippel–Lindau tumour suppressor-elonginC/B) ligases. These complexes contain adaptor subunits, e.g. F-box proteins (by analogy with the cyclin F sequence) in the SCF family. F-box proteins recognize different substrates through specific protein–protein interaction domains such as leucine-rich repeats or WD40 domains. F-box proteins are themselves rapidly degraded in a ubiquitin–proteasome-dependent fashion. This allows SCF complexes to be switched and to recognize numerous alternative substrates (Fig. 15.2). Finally, the U-box is a domain of ~70 amino acids with a predicted three-dimensional structure similar to that of the RING finger. A few U-box E3s have been recently characterized.

Deubiquitination

Eukaryotic cells also contain DUBs (deubiquitinating enzymes), which are encoded by the UCH (ubiquitin carboxyl-terminal hydrolases) and the UBP (ubiquitin-specific processing proteases) gene families. Like the E3 ligases, the DUB enzymes form a huge family with at least 90 member proteins based on the data from the genome sequencing projects (Chung and Baek, 1999). UCHs are relatively small proteins (<40 kDa) and only few isoforms have been characterized in higher

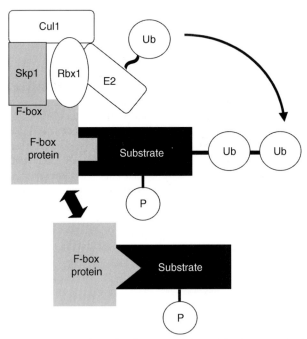

Fig. 15.2. Schematic representation of a multisubunit RING finger SCF E3 (see text). The three subunits in the catalytic core are boxed in white. Rbx1 (Hrt1/Roc1) is the RING finger protein, and an E2 of the Ubc3 family is part of the catalytic core. Cul, cullin1. Adaptor proteins are boxed in grey. Skp1 binds to an F-box protein through the F-box motif. The exchange of F-box proteins is responsible for the recognition of multiple specific substrates. Substrates bind to their respective F-box proteins by protein–protein interactions (see text). Note that phosphorylated (P) substrates are ubiquitinated by this E3 family.

eukaryotes. UCHs mainly hydrolyse small amides and esters at the C-terminus of ubiquitin. In contrast, UBPs are a structurally much larger (50–250 kDa) and diverse group of DUBs. Deubiquitination can be compared with dephosphorylation. Thus, UBPs are involved in several biological processes, including the control of growth, differentiation and genome integrity (Chung and Baek, 1999). The putative roles of DUBs in proteasome-dependent proteolysis are: (i) to maintain free ubiquitin levels by processing polyubiquitin degradation signals and ubiquitin precursors into free monomers; (ii) to deubiquitinate substrates erroneously tagged for degradation (proofreading); and (iii) to keep 26S proteasomes free of bound polyubiquitin chains.

Molecular basis of ubiquitination and of substrate recognition by the 26S proteasome

Some signals in proteins that mark them for polyubiquitination and proteasome-dependent proteolysis have been identified. These signals are usually determined by short regions in the primary sequence of the targeted protein (Pickart, 2001). First, the nature of the N-terminal amino acid of a protein can determine its rate of ubiquitination and subsequent degradation (Varshavsky, 1996). This 'N-end rule' pathway is up-regulated in the increased breakdown of soluble muscle proteins in muscle wasting (Solomon *et al.*, 1998), and controls several biological functions (e.g. peptide import into cells; see Turner *et al.*, 2000).

Secondly, efficient ubiquitination of proteins by the SCF E3 ligases requires phosphorylation (Pickart, 2001). The major recognition site of G_1 cyclins is present within a 100–200 amino acid C-terminal region rich in Pro-Glu-Ser-Thr (PEST) motifs. PEST motifs, which are characteristic of rapidly degraded proteins, are minimum consensus phosphorylation sites for several protein kinases (Wilkinson, 2000). The ability of an E3 to recognize a phosphorylated signal may be due to the presence of phosphoamino acid-binding motifs, such as WW or WD40 domains (Pickart, 2001). Unfortunately, there is no clear pattern of phosphorylation that targets substrates for ubiquitination. Phosphorylation has been reported at single or multiple sites. In addition, phosphorylation of some proteins (e.g. c-Fos, c-Jun) actually prevents their ubiquitination and degradation (Attaix *et al.*, 2001), and recognition of substrates that depends on dephosphorylation has also been reported (Pickart, 2001). A third, structural motif that appears to target proteins for ubiquitination *in vivo* and *in vitro* is the so-called destruction box. This very degenerate nine amino acid motif (as only argine and leucine are invariable residues in position 1 and 4, respectively) is a crucial signal for the ubiquitination and breakdown of mitotic cyclins and other cell cycle regulators. The destruction box itself is not an ubiquitination site, but is a transferable degradation signal. Reporter proteins containing such motifs are degraded rapidly in a cell cycle-dependent manner. A specific conformation of the destruction box may be required for efficient E3 recognition (Pickart, 2001).

Finally, the 26S proteasome also degrades a growing number of non-ubiquitinated protein substrates (reviewed in Attaix *et al.*, 2001; Benaroudj *et al.*, 2001). For example, some skeletal muscle-specific proteins (e.g. troponin C) are degraded by the 26S proteasome via a ubiquitin-independent mechanism (Benaroudj *et al.*, 2001). In addition misfolded, oxidized and mutant abnormal proteins are very good substrates of the 26S proteasome, which can recognize non-ubiquitinated misfolded proteins (Strickland *et al.*, 2000). Hydrophobicity plays a major role in polyubiquitin chain recognition by the 26S proteasome (Thrower *et al.*, 2000). Hydrophobic stretches of amino acids in the primary sequence of some non-ubiquitinated substrates of the 26S proteasome (e.g. calmodulin and troponin C) may substitute for ubiquitin and be sufficient for recognition by the 26S proteasome (Benaroudj *et al.*, 2001). Alternatively, other adaptor proteins play a role in substrate recognition. For example, ornithine decarboxylase binds to the 26S proteasome via antizyme, and is degraded in a ubiquitin-independent fashion (see Glickman and Ciechanover, 2002).

Proteolysis

The second major step in the ubiquitin–proteasome pathway is the degradation of polyubiquitinated proteins by the 26S proteasome complex (Fig. 15.1). The 26S proteasome is formed by the association of the 20S proteasome with two 19S regulatory complexes (Fig. 15.3) (reviewed in Voges et al., 1999; Attaix et al., 2001; Glickman and Ciechanover, 2002).

The 20S proteasome

The 20S proteasome is the core of the proteolytic machinery. This barrel-shaped complex is organized as a stack of four rings, each ring having seven subunits (Fig. 15.3A). The non-catalytic α-subunits form the two outer rings, and the catalytic β-subunits the two inner rings. The β-subunit active sites are located inside the cylinder and thus the proteasome is a self-compartmentalizing protease (Voges et al., 1999), as substrates must enter the catalytic chamber delimited by the β-rings to be degraded into peptides. Substrates access the active sites by traversing a narrow opening in the α-ring that is blocked in the unliganded free 20S proteasome by N-terminal sequences of α-subunits. The gating of this channel is due to the binding of proteasome activators (see Kohler et al., 2001, and Fig. 15.3C). In eukaryotes, the 20S proteasome contains at least two chymotrypsin-, two trypsin- and two caspase-like active sites (and see below). These activities are allosterically regulated, and it has been suggested that there is an ordered, cyclical mechanism for protein degradation

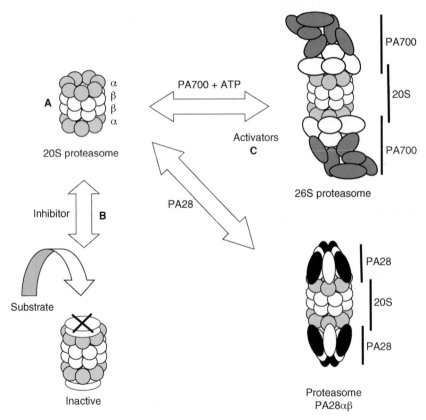

Fig. 15.3. Schematic representation of the 20S proteasome, 26S proteasome and proteasome–PA28αβ. (A) α and β denote the α and β subunits of the 20S proteasome, respectively. (B) Proteasome inhibitors compete with proteasome activators (PA700 and PA28) for the binding with 20S proteasomes; when a proteasome inhibitor binds to the 20S proteasome, substrates cannot enter the catalytic chamber. (C) In contrast, binding of proteasome activators PA700 or PA28αβ to the 20S proteasome gates the narrow 20S proteasome channel delimited by the α-rings. The resulting 26S proteasome exhibits increased proteolytic and peptidase activities, while the resulting proteasome–PA28αβ exhibits only stimulated peptidase activities. The lid and the base of PA700 are shown in dark grey and white, respectively. PA28α and PA28β subunits of PA28 are shown in black and white, respectively.

(Kisselev et al., 1999). The confinement of multiple active sites within a nano-compartment has another advantage. Proteasomes hydrolyse most peptide bonds and generate peptides that are typically 3–22 amino acids long and do not retain biological properties, except antigenic peptides.

The 26S proteasome

The 19S complexes, also called PA700 (proteasome activator 700), bind to the two α-rings of the 20S proteasome in the presence of ATP to form the 26S proteasome (Fig. 15.3C). PA700 contains at least 18 different subunits assembled within two subcomplexes called the base and the lid. The base contains six ATPase and three non-ATPase subunits in human proteasomes. The ATPase subunits are members of the AAA-ATPase family (ATPases associated with a variety of cellular activities) and have very specific roles despite their sequence similarity. They provide energy for assembly of the 26S proteasome, gating of the 20S proteasome channel, unfolding and injection of protein substrates into the catalytic chamber of the proteasome and for peptide release (Voges et al., 1999; Kohler et al., 2001; Navon and Goldberg, 2001). The binding of the 20S proteasome to the base alone supports ATP-dependent peptide hydrolysis. In contrast, both the base and the lid are required for ubiquitin-dependent proteolysis.

The non-ATPase subunit S5a stabilizes the base and the lid subcomplexes and binds avidly to the polyubiquitin degradation signal (Deveraux et al., 1994). However, the ATPase S6' subunit within the base also recognizes polyubiquitin chains, and this interaction depends on ATP hydrolysis (Lam et al., 2002).

Other forms of proteasomes

Immunoproteasomes are 20S or 26S proteasomes in which interferon-γ induces the replacement of three different catalytic β-subunits by three alternative subunits. The resulting particles generate class I antigenic peptides very efficiently (Table 15.1).

Apart from the PA700, several other proteasome activators that bind to the 20S core particle have been isolated. One of these, the PA28 (proteasome activator 28), exists as a hexameric or heptameric complex of two distinct subunits, PA28α and β, that binds to either α-ring of the 20S proteasome in an ATP-independent fashion (Fig. 15.3C). Binding of this complex to the 20S proteasome stimulates peptidase activities, but not proteolytic activity (Table 15.1). The PA28–proteasome complex plays a role in class I antigen presentation, possibly in connection with 26S immunoproteasomes (Rechsteiner et al., 2000).

PA28γ, which is closely related to subunits PA28α and β, can form homopolymers that bind

Table 15.1. Roles, substrates and ATP dependence of proteasomes.

	Role	Substrates	ATP[1]
20S Proteasome	Protein breakdown[2] and/or peptidasic activity	Non-ubiquitinated proteins and/or peptides	–
26S Proteasome	Protein breakdown	Ubiquitinated proteins and non-ubiquitinated proteins (e.g. ODC)[3]	+
20S and 26S immunoproteasomes	Class I antigen presentation	Non-ubiquitinated (20S) or ubiquitinated (26S) proteins and peptides	– (20S) + (26S)
Proteasome PA28αβ	Class I antigen presentation	Peptides	–
Proteasome PA28γ	Growth	Peptides	–
Hybrid proteasome PA28αβ–20S–PA700	Class I antigen presentation and protein breakdown (e.g. ODC)	Non-ubiquitinated (or ubiquitinated?) proteins and peptides	+
PA200–20S–PA200	DNA repair	Peptides	–
Hybrid proteasome PA200–20S–PA700	?	?	?

[1]ATP-dependent substrate hydrolysis.
[2]20S proteasomes are believed to degrade oxidized proteins.
[3]ODC, ornithine decarboxylase.

to the 20S proteasome particles. The resulting proteasome–PA28γ complex is involved in growth control (Murata et al., 1999).

Recently, a fourth proteasome activator (PA200) has been characterized (Ustrell et al., 2002). This nuclear activator is involved in DNA repair.

Finally, there are also hybrid proteasomes. The PA28–20S proteasome–PA700 is an example of such a hybrid proteasome formed by the binding of one PA28 and one PA700 to each α-ring of a 20S proteasome (Hendil et al., 1998). These hybrid complexes are induced by interferon-γ and play a role both in antigen presentation and in the breakdown of some proteins (Tanahashi et al., 2000). The existence of a hybrid PA200–20S proteasome–PA700 is also strongly suspected (Ustrell et al., 2002).

Thus, various proteasome populations coexist in mammalian cells. The different forms of proteasomes result from the binding of various activators to the 20S catalytic core, and have specific proteolytic and non-proteolytic functions (Table 15.1). In addition, it should be pointed out that 19S proteasome subcomplexes, which are *a priori* part of the proteolytic machinery of the 26S proteasome, also perform non-proteolytic functions. For example, 19S proteins play a role in nucleotide excision repair and transcription elongation, and can be recruited to an active promoter (Gonzalez et al., 2002).

Regulation of Ubiquitin–Proteasome-dependent Proteolysis in Skeletal Muscle

In this section, we review the limited data available concerning the hormonal and nutritional control of the ubiquitin–proteasome pathway in muscle (Combaret et al., 2001). Other conditions that have been shown to regulate the ubiquitin–proteasome pathway will also be discussed.

Regulation of ubiquitin conjugation

Ubiquitin

Increased skeletal muscle mRNA levels for ubiquitin have been reported in several catabolic states in rodents (reviewed in Attaix et al., 1998; Jagoe and Goldberg, 2001) and humans (Mansoor et al., 1996; Tiao et al., 1997). This adaptation appears to be a general response of muscle as it is observed in both type I and type II skeletal muscle fibres from starved rats (Medina et al., 1995) and other catabolic models. It should be pointed out that elevated mRNA levels for ubiquitin do not systematically reflect increased muscle proteolysis. For example, aged rats treated with dexamethasone exhibit increased ubiquitin expression, without any change in muscle proteolytic rates (Dardevet et al., 1995). A key mediator of increased ubiquitin expression in wasting conditions appears to be elevated glucocorticoid serum levels. There is increased production of corticosterone in several catabolic states in rodents (Tiao et al., 1996; Marinovic et al., 2002), including starvation (Wing and Goldberg, 1993). Cortisol is also elevated in head trauma weight-losing patients, who exhibit increased muscle expression of ubiquitin and other proteolytic genes (Mansoor et al., 1996). Elevated ubiquitin mRNA levels are induced by glucocorticoids and reflect increased transcription in muscles from acidotic (Bailey et al., 1996) and diabetic (Price et al., 1996) rats. In vitro studies in glucocorticoid-treated L6 muscle cells demonstrated that transcription of ubiquitin is regulated by Sp1 and MEK1 (Marinovic et al., 2002). Ubiquitin expression is also, but presumably indirectly, up-regulated by several cytokines including tumour necrosis factor-α (TNF-α) (see Combaret et al., 2002).

Food intake following a period of starvation leads to decreased ubiquitin mRNA levels in rat muscles (Medina et al., 1995). Insulin clearly has a role in this process as there was a rapid (within 3–4 h) and marked down-regulation of ubiquitin mRNA levels in type II skeletal muscle fibres during hyperinsulinaemic clamps in the presence (Larbaud et al., 1996) and the absence (Larbaud et al., 2001) of hyperaminoacidaemia. These observations suggest furthermore that amino acids poorly regulate ubiquitin expression *in vivo*. Indeed, in related experiments, we have shown that infusing amino acids into starved rats did not rapidly restore muscle ubiquitin expression to the control fed level (Kee et al., 2003). Finally, the role of insulin-like growth factor-I (IGF-I) in the *in vivo* regulation of ubiquitin expression is still unclear. A down-regulatory effect (Fang et al., 2000) or no effect (Kee et al., 2002) has been reported.

Ubiquitin-activating enzyme, E1

E1 has low expression in skeletal muscle and its mRNA level is not regulated in catabolic states (Lecker et al., 1999). This is not surprising because: (i) E1 is an extremely active enzyme capable of charging excess amounts of E2s with ubiquitin (K_m values for E2s of ~100 pM); and (ii) E1 is a common element in all pathways of ubiquitin conjugation (Rajapurohitam et al., 2002). Thus, any E1 impairment would affect the whole downstream ubiquitination cascade.

Ubiquitin-conjugating enzymes, E2s

Various E2s have been found in skeletal muscle. These include the 14, 17, and 20 kDa E2 species (Attaix and Taillandier, 1998), E2-F1 (Gonen et al., 1996), Ub-E2G (Chrysis and Underwood, 1999), UBC4 (Rajapurohitam et al., 2002), and a huge 230 kDa protein (E2-230k) that is expressed predominantly in skeletal muscle and heart (Yokota et al., 2001). Immunodepletion of either the 14 kDa E2 of the UBC2 family or UBC4 inhibited ubiquitination rates by >50% in soluble muscle extracts (Rajapurohitam et al., 2002). The expression of the 1.2 kb transcript of the 14 kDa E2 is up-regulated in most, but not all, muscle wasting conditions (reviewed in Attaix and Taillandier, 1998), including starvation (Wing and Banville, 1994). Taillandier et al. (1996) have shown that the increased mRNA levels for the 14 kDa E2 entered active translation in the unweighted soleus muscle. However, in several studies, increased expression of the 14 kDa E2 correlated neither with muscle proteolytic rates (Temparis et al., 1994; Fang et al., 2000; Combaret et al., 2002) nor with altered muscle protein content of the enzyme (Hobler et al., 1999a; Lecker et al., 1999). The expression of the 14 kDa E2 is up-regulated by glucocorticoids in adult rats (Dardevet et al., 1995). In vitro, insulin and IGF-I decrease the stability of the 14 kDa E2 mRNAs (Wing and Bedard, 1996). Such down-regulation, however, does not prevail in insulin-infused animals (Larbaud et al., 1996, 2001). Similarly, the expression of the 14 kDa E2 was not altered by IGF-I treatment in rat muscle (Kee et al., 2002). To our knowledge, only Chrysis and Underwood (1999) have reported changes in mRNA levels for other E2s in rat muscle. These authors showed that dexamethasone increased the transcripts of 14 kDa E2 and Ub-E2G, and the 2E isoform of the 17 kDa E2 mRNA. They also report that injections of IGF-I in dexamethasone-treated animals resulted in the suppression of elevated transcripts for both the 14 kDa E2 and Ub-E2G, whereas the 2E isoform of the 17 kDa E2 was only affected modestly.

Ubiquitin–protein ligases, E3s

A single report has described a very large E3 (E3L), which was involved with E2-F1 in the *in vitro* breakdown of ubiquitinated actin, troponin T and MyoD (Gonen et al., 1996). In catabolic muscles, several groups (e.g. Lecker et al., 1999) have reported increases in mRNA levels for E3α, the ubiquitous N-end rule RING finger ligase that functions with the 14 kDa E2. However, such changes were not associated with altered protein levels of E3α. Recent studies have demonstrated the existence of multiple muscle-specific RING finger E3s. These include MuRF-1, -2 and -3 (muscle RING finger proteins; Centner et al., 2001), SMRZ (striated muscle RING zinc finger; Dai and Liew, 2001), ANAPC11 (anaphase-promoting complex; Chan et al., 2001), atrogin-1/MAFbx (atrophy gene; Gomes et al., 2001) also called MAFbx (muscle atrophy F-box; Bodine et al., 2001). Atrogin-1/MAFbx and MuRF-1 are differentially over-expressed in atrophying muscles, including fasting and diabetes for atrogin-1/MAFbx. These findings are very important since E3α, the only E3 previously reported to play a role in muscle wasting, presumably has little significant physiological role. First, E3α is involved in the ubiquitination of soluble muscle proteins, not of myofibrillar proteins (Lecker et al., 1999). Secondly, mice lacking the E3α gene are viable and fertile, and only exhibited smaller skeletal muscles than control animals (Kwon et al., 2001). In striking contrast, mice deficient in either Atrogin-1/MAFbx or MuRF1 were resistant to atrophy (Bodine et al., 2001).

Ubiquitin conjugates and rates of ubiquitination/deubiquitination

An increased amount of muscle ubiquitin conjugates has been observed in some wasting conditions (see Attaix et al., 1998). This suggests that the breakdown of ubiquitin conjugates by the 26S proteasome was rate limiting. However, several studies have given conflicting results. For example, no accumulation of ubiquitin conjugates was observed in the muscles from septic animals (Tiao et al., 1994), although ubiquitin–proteasome-dependent proteolysis was elevated (Tiao et al., 1994; Voisin et al., 1996). The reasons for such discrepancies are unclear. However, levels of ubiquitin conjugates also depend on ubiquitination rates, deubiquitinating activities and on their rates of breakdown. Tilignac et al. (2002) reported unchanged amounts of ubiquitin conjugates in both soluble and myofibrillar fractions from mice where proteasome activities were depressed below basal control levels, suggesting that the rate of ubiquitination was also impaired. Ubiquitinated proteins accumulated preferentially in the myofibrillar fraction (Wing et al., 1995; Combaret et al., 2002), although others report conflicting findings (Tiao et al., 1994). These discrepancies must be clarified, and the contractile proteins that are ubiquitinated in skeletal muscle remain to be clearly identified.

In vitro rates of ubiquitination of exogenous ^{125}I-labelled ubiquitin and of a model substrate of the N-end rule pathway (e.g. [^{125}I]α-lactalbumin) increased in soluble muscle extracts from rats in various catabolic conditions (Solomon et al., 1998; Lecker et al., 1999). In muscle extracts of animals where overall protein breakdown falls because of deficiency of thyroid hormones and rises after treatment with these hormones, there were parallel changes in ubiquitination rates, suggesting hormonal regulation (Solomon et al., 1998).

There are many deubiquitinating enzymes in skeletal muscle with unknown functions (Woo et al., 1995). Differential expression of UBP45 and UBP69 is involved in the regulation of muscle cell differentiation (Park et al., 2002). Deubiquitinating enzymes are very active in soluble muscle extracts (Rajapurohitam et al., 2002). Whether deubiquitination plays a role in muscle wasting remains to be elucidated. However, mRNA levels for the deubiquitinating enzyme USP1 (ubiquitin-specific protease) increased in the atrophying muscles from fasted rats and decreased upon refeeding (L. Combaret and S.S. Wing, unpublished data).

Regulation of the proteasome

The proteolytic activities of the 26S proteasome are regulated by a variety of subtle mechanisms that include the synthesis and processing of proteasome subunits, the assembly of 20S and 26S proteasomes, post-translational modifications (e.g. phosphorylation of some 20S and PA700 subunits), the binding of proteasome activators and inhibitors, etc. A detailed review of each of these different levels of regulation is beyond the scope of this chapter, and the reader is referred to some excellent and extensive reviews for detailed information (e.g. Voges et al., 1999; Glickman and Ciechanover, 2002).

Rates of proteolysis

Goldberg and colleagues provided the first evidence for a role for the ubiquitin–proteasome pathway in the muscle wasting seen during starvation and denervation atrophy (reviewed in Attaix and Taillandier, 1998). In these conditions, most of the increased proteolysis was found to be ATP-dependent. ATP depletion (but not inhibition of the calpains and/or the cathepsins) suppresses the elevated rates of proteolysis in atrophying incubated muscles from starved and denervated rats, and also in other rodent catabolic models (Temparis et al., 1994; Price et al., 1996; Taillandier et al., 1996). ATP is required for the activation of ubiquitin by E1, and proteolysis by the 26S proteasome (see above).

Strong further support for a major role for the proteasome in the breakdown of muscle proteins is the demonstration that only proteasome inhibitors such as lactacystin or MG132 suppress the enhanced rates of overall proteolysis in wasting conditions (Bailey et al., 1996; Price et al., 1996; Hobler et al., 1998; Combaret et al., 2002). More importantly, only proteasome inhibitors suppressed the elevated 3-methylhistidine release by incubated atrophying muscles (Hobler et al., 1998). 3-Methylhistidine is formed by post-translational modifications of actin and pale myosins, and its rate of appearance in incubation media reflects the breakdown of these myofibrillar proteins (Attaix and Taillandier, 1998). Lobster muscle 20S proteasome has been studied extensively and contains five peptidase activities (Mykles, 1993). In vitro studies have shown that the branched chain amino acid-preferring, the peptidylglutamyl peptide hydrolase and trypsin-like activities were involved in the degradation of contractile proteins. Compared with other tissues, both peptidase activities and proteolytic activity are low in rat skeletal muscles (Farout et al., 2000). Chymotrypsin-like peptidase activity of muscle proteasomes increased in some wasting conditions (Hobler et al., 1999b; Ordway et al., 1999; Kee et al., 2002). In contrast, this peptidase activity was unchanged in diabetes (Liu et al., 2000). Indeed, the 20S proteasome population comprises at least six distinct subtypes in skeletal muscle, including constitutive proteasomes, immunoproteasomes and their intermediate forms (Dahlmann et al., 2001). Therefore, the properties of a 20S proteasome population isolated from a given tissue represent the average properties of the whole set of proteasome subtypes. This may explain discrepancies between rates of overall muscle proteolysis and some specific proteasome activities. However, and in contrast, both chymotrypsin- and trypsin-like activities were depressed when proteasome-dependent proteolysis, measured in vitro, was impaired by chemotherapy (Tilignac et al., 2002). Increased ATP- and ubiquitin-dependent breakdown of exogenous protein substrates (e.g. [methyl-^{14}C]casein) in muscle extracts correlated with increased chymotrypsin-like activity of the proteasome in the exercised and electrically chronically stimulated muscles from rats (Kee et al., 2002) and rabbits (Ordway et al., 1999), respectively.

Finally, a single group has reported that insulin may have a direct inhibitory effect on proteasome activities through the insulin-degrading enzyme (Duckworth et al., 1998) and/or changes in the proteasome conformation (Hamel et al., 1998).

Expression and protein content of 20S proteasome subunits

Many groups have shown that enhanced ATP- or proteasome-dependent rates of muscle proteolysis correlate with elevated mRNA levels for several subunits of the 20S proteasome, including in starved and starved/refed animals (reviewed in Attaix et al., 1998; Jagoe and Goldberg, 2001). The increased protein abundance of the 27 kDa 20S proteasome subunit correlated with enhanced expression of other subunits in cancer cachexia (Baracos et al., 1995). Conversely, when proteasome-dependent proteolysis was depressed by chemotherapy, mRNA levels for 20S proteasome subunits correlated with reduced amounts of two subunits (Tilignac et al., 2002). Taillandier et al. (1996) have shown that the overexpressed RC9 subunit entered active translation in atrophying muscles. An increase in transcribed RC3 proteasome subunit mRNA was observed in acidosis (Bailey et al., 1996). Glucocorticoids (Price et al., 1994; Dardevet et al., 1995) and tumour necrosis factor-α (TNF-α; Combaret et al., 2002) up-regulate mRNA levels for 20S proteasome subunits. Glucocorticoids induce proteasome C3 subunit transcription in L6 muscle cells by opposing the suppression of its transcription by nuclear factor (NF)-κB (Du et al., 2000), while the glucocorticoid-dependent increased transcription of ubiquitin involves Sp1 and MEK1 (Marinovic et al., 2002). Thus, the increased coordinate transcription of several genes in the ubiquitin–proteasome system results from the activation of alternative signalling pathways. In contrast to glucocorticoids, insulin (Larbaud et al., 1996, 2001) and IGF-I (Kee et al., 2002) do not regulate the in vivo expression of either α or β 20S muscle proteasome subunits. Finally, the muscle content of proteasomes fell in animals fed a protein-deficient diet and with hypothyroidism (Tawa and Goldberg, 1993).

Expression and protein content of PA700 subunits

Some, but not all, mRNA levels for subunits of PA700 are also up-regulated in muscle wasting (Attaix et al., 1997), including starvation (Kee et al., 2003). The expression of ATPase and non-ATPase subunits of PA700 is regulated independently of the expression of 20S proteasome subunits, and clearly depends on a given catabolic state (Attaix et al., 1997; Combaret et al., 2002). Furthermore, the mRNA levels and protein contents of the individual regulatory subunits are regulated independently, and do not correlate systematically with rates of proteolysis (Combaret et al., 2002; Tilignac et al., 2002). These data are similar to the observations of Dawson et al. (1995) in abdominal intersegmental muscles of Manduca sexta during programmed cell death.

Glucocorticoids (L. Combaret, D. Taillandier and D. Attaix, unpublished data), insulin (Attaix et al., 1997) and TNF-α (Combaret et al., 2002) have all been shown to regulate mRNA levels for several ATPase and non-ATPase subunits of PA700. However, the specific changes in subunit expression and/or protein content that are crucial for altered proteolytic activities have not yet been identified.

Finally, two recent reports indicated that eicosapentaenoic acid attenuated the increased chymotrypsin-like activity of the proteasome and the enhanced expression of both 20S proteasome α-subunits and p42 (an ATPase subunit of PA700) in muscles from fasted (Whitehouse and Tisdale, 2001) and tumour-bearing (Whitehouse et al., 2001) mice. The effect was not shown in starved mice treated with the related (n-3) fatty acid docosahexaenoic acid or with linoleic acid. However, 2,3,5-trimethyl-6-(3-pyridylmethyl)1,4-benzoquinone (CV-6504), an inhibitor of 5-, 12- and 15-lipoxygenases, also attenuated the catabolic response. These results suggest that protein catabolism in starvation and cancer cachexia is mediated through a common pathway, which is inhibited by eicosapentaenoic acid and is likely to involve a lipoxygenase metabolite as a signal transducer.

Functional relationships with other proteolytic pathways

The 26S proteasome degrades proteins only into peptides. Except when presented on major histocompatibility complex (MHC) class I molecules, these peptides must undergo further hydrolysis into free amino acids (Attaix et al., 2001). Recent studies showed that the extralysosomal peptidase tripeptidyl-peptidase II (TPP II) degrades peptides generated by the proteasome (Fig. 15.4, and see Hasselgren et al., 2002). TPP II expression, protein content and activity increased in septic muscles. In addition, the glucocorticoid receptor antagonist RU 38486 blunted these adaptations, indicating that glucocorticoids participate in the upregulation of TPP II (Wray et al., 2002).

Conversely, other proteases may act upstream of the proteasome (Fig. 15.4). Specific interactions between the myofibrillar proteins appear to protect them from ubiquitin-dependent degradation, and the rate-limiting step in their degradation is probably their dissociation from the myofibril (Solomon and Goldberg, 1996). Calpains play key roles in the disassembly of sarcomeric proteins and in Z-band disintegration, resulting in the release of myofilaments (Williams et al., 1999). These data suggest that calpains are acting upstream of the proteasome (Hasselgren et al., 2002). In addition, Deval et al. (2001) recently have shown that cathepsin L is differentially up-regulated in various wasting conditions, and that changes in its expression closely follow adaptations in ubiquitin-proteasome-dependent proteolysis. It remains to be demonstrated whether there is a functional connection between calpains, cathepsin L and the ubiquitin pathway. However, the expression of several proteolytic genes (including cathepsin L and several components of the ubiquitin-proteasome system) was down-regulated in mice knocked out for the muscle-specific calpain p94 (Combaret et al., 2003). In any case, it seems very important to elucidate proteolytic mechanisms both upstream and downstream of the proteasome that result in the complete degradation of muscle proteins.

Concluding Remarks

How the ubiquitin-proteasome pathway degrades muscle proteins, and more particularly contractile proteins, remains largely unknown. Neither the signals that target myofibrillar proteins for breakdown, nor the precise substrates of the pathway have been identified. Detailed studies that aim to explore the role of nutrients, hormones and their signalling pathways are also clearly needed. The complexity of the ubiquitin pathway clearly will impede the identification of the precise mechanisms that are important in the control of muscle proteolysis. However, microarray analyses of transcription profiles and proteomic surveys of control and atrophying muscles should provide extremely valuable information. Finally, the precise elucidation of the different steps of the breakdown of myofibrillar proteins upstream and downstream of the proteasome has major importance in devising new strategies to prevent muscle wasting.

Acknowledgements

We thank Dr Daniel Béchet for critical review of the manuscript. Studies in the laboratory of the authors are supported by the Association pour la Recherche sur le Cancer, the Conseil Régional d'Auvergne, the Institut National de la Recherche Agronomique, the Institut National de la Santé et de la Recherche Médicale, the French Ministère de la Recherche and Nestlé. A.J. Kee was supported by a post-doctoral fellowship from INRA.

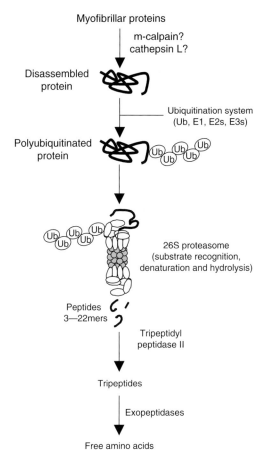

Fig. 15.4. Schematic representation of the breakdown of myofibrillar proteins in skeletal muscle. Contractile proteins are first disassembled (possibly by m-calpain and/or cathepsin L, see text). These proteins (and/or fragments) are then polyubiquitinated and degraded by the 26S proteasome, which generates peptides that are 3–22 amino acids long. Peptides greater than tripeptides are hydrolysed by tripeptidyl peptidase II, and the resulting tripeptides finally hydrolysed into free amino acids by exopeptidases.

References

Attaix, D. and Taillandier, D. (1998) The critical role of the ubiquitin–proteasome pathway in muscle wasting in comparison to lysosomal and Ca^{2+}-dependent systems. In: Bittar, E.E. and Rivett, A.J. (eds) *Intracellular Protein Degradation*. JAI Press, Greenwich, Connecticut, pp. 235–266.

Attaix, D., Taillandier, D., Combaret, L., Rallière, C., Larbaud, D., Aurousseau, E. and Tanaka, K. (1997) Expression of subunits of the 19S complex and of the PA28 activator in rat skeletal muscle. *Molecular Biology Reports* 24, 95–98.

Attaix, D., Aurousseau, E., Combaret, L., Kee, A., Larbaud, D., Rallière, C., Souweine, B., Taillandier, D. and Tilignac, T. (1998) Ubiquitin–proteasome-dependent proteolysis in skeletal muscle. *Reproduction Nutrition Development* 38, 153–165.

Attaix, D., Combaret, L., Pouch, M.-N. and Taillandier, D. (2001) Regulation of proteolysis. *Current Opinion in Clinical Nutrition and Metabolic Care* 4, 45–49.

Bailey, J.L., Wang, W., England, B.K., Price, S.R., Ding, X. and Mitch, W.E. (1996) The acidosis of chronic renal failure activates muscle proteolysis in rats by augmenting transcription of genes encoding proteins of the ATP-dependent ubiquitin–proteasome pathway. *Journal of Clinical Investigation* 97, 1447–1453.

Balcerzak, D., Querengesser, L., Dixon, W.T. and Baracos, V.E. (2001) Coordinate expression of matrix-degrading proteinases and their activators and inhibitors in bovine skeletal muscle. *Journal of Animal Science* 79, 94–107.

Baracos, V.E., DeVivo, C., Hoyle, D.H.R. and Goldberg, A.L. (1995) Activation of the ATP–ubiquitin–proteasome pathway in skeletal muscle of cachectic rats bearing a hepatoma. *American Journal of Physiology* 268, E996–E1006.

Belizario, J.E., Lorite, M.J. and Tisdale, M.J. (2001) Cleavage of caspases-1, -3, -6, -8 and -9 substrates by proteases in skeletal muscles from mice undergoing cancer cachexia. *British Journal of Cancer* 84, 1135–1140.

Benaroudj, N., Tarcsa, E., Cascio, P. and Goldberg, A.L. (2001) The unfolding of substrates and ubiquitin-independent protein degradation by proteasomes. *Biochimie* 83, 311–318.

Bodine, S.C., Latres, E., Baumhueter, S., Lai, V.K., Nunez, L., Clarke, B.A., Poueymirou, W.T., Panaro, F.J., Na, E., Dharmarajan, K., Pan, Z.Q., Valenzuela, D.M., DeChiara, T.M., Stitt, T.N., Yancopoulos, G.D. and Glass, D.J. (2001) Identification of ubiquitin ligases required for skeletal muscle atrophy. *Science* 294, 1704–1708.

Centner, T., Yano, J., Kimura, E., McElhinny, A.S., Pelin, K., Witt, C.C., Bang, M.L., Trombitas, K., Granzier, H., Gregorio, C.C., Sorimachi, H. and Labeit, S. (2001) Identification of muscle specific ring finger proteins as potential regulators of the titin kinase domain. *Journal of Molecular Biology* 306, 717–726.

Chan, A.H., Lee, S.M., Chim, S.S., Kok, L.D., Waye, M.M., Lee, C.Y., Fung, K.P. and Tsui, S.K. (2001) Molecular cloning and characterization of a RING-H2 finger protein, ANAPC11, the human homolog of yeast Apc11p. *Journal of Cellular Biochemistry* 83, 249–258.

Chrysis, D. and Underwood, L.E. (1999) Regulation of components of the ubiquitin system by insulin-like growth factor I and growth hormone in skeletal muscle of rats made catabolic with dexamethasone. *Endocrinology* 140, 5635–5641.

Chung, C.H. and Baek, S.H. (1999) Deubiquitinating enzymes: their diversity and emerging roles. *Biochemical and Biophysical Research Communications* 266, 633–640.

Combaret, L., Taillandier, D., Voisin, L., Samuels, S.E., Boespflug-Tanguy, O. and Attaix, D. (1996) No alteration in gene expression of components of the ubiquitin–proteasome proteolytic pathway in dystrophin-deficient muscles. *FEBS Letters* 393, 292–296.

Combaret, L., Taillandier, D. and Attaix, D. (2001) Nutritional and hormonal control of protein breakdown. *American Journal of Kidney Diseases* 37, Supplement 2, S108–S111.

Combaret, L., Béchet, D., Claustre, A., Taillandier, D., Richard, I. and Attaix, D. (2003) Down-regulation of genes in the lysosomal and ubiquitin-proteasome proteolytic pathway in calpain-3-deficient muscle. *International Journal of Biochemistry and Cell Biology* (in press).

Combaret, L., Tilignac, T., Claustre, A., Voisin, L., Taillandier, D., Obled, C., Tanaka, K. and Attaix, D. (2002) Torbafylline (HWA 448) inhibits enhanced skeletal muscle ubiquitin–proteasome-dependent proteolysis in cancer and septic rats. *Biochemical Journal* 361, 185–192.

Dahlmann, B., Ruppert, T., Kloetzel, P.M. and Kuehn, L. (2001) Subtypes of 20S proteasomes from skeletal muscle. *Biochimie* 83, 295–299.

Dai, K.S. and Liew, C.C. (2001) A novel human striated muscle RING zinc finger protein, SMRZ, interacts with SMT3b via its RING domain. *Journal of Biological Chemistry* 276, 23992–23999.

Dardevet, D., Sornet, C., Taillandier, D., Savary, I., Attaix, D. and Grizard, J. (1995) Sensitivity and protein turnover response to glucorticoids are different in skeletal muscle from adult and old rats. Lack of regulation of the ubiquitin–proteasome proteolytic pathway in aging. *Journal of Clinical Investigation* 96, 2113–2119.

Dawson, S.P., Arnold, J.E., Mayer, N.J., Reynolds, S.E., Billett, M.A., Gordon, C., Colleaux, L., Kloetzel, P.M., Tanaka, K. and Mayer, R.J. (1995) Developmental changes of the 26S proteasome in abdominal intersegmental muscles of *Manduca sexta* during programmed cell death. *Journal of Biological Chemistry* 270, 1850–1858.

Deval, C., Mordier, S., Obled, C., Béchet, D., Combaret, L., Attaix, D. and Ferrara, M. (2001) Identification of cathepsin L as a differentially-expressed message associated with skeletal muscle wasting. *Biochemical Journal* 360, 143–150.

Deveraux, Q., Ustrell, V., Pickart, C. and Rechsteiner, M. (1994) A 26S protease subunit that binds ubiquitin conjugates. *Journal of Biological Chemistry* 269, 7059–7061.

Du, J., Mitch, W.E., Wang, X. and Price, S.R. (2000) Glucocorticoids induce proteasome C3 subunit expression in L6 muscle cells by opposing the suppression of its transcription by NF-κB. *Journal of Biological Chemistry* 275, 19661–19666.

Duckworth, W.C., Bennett, R.G. and Hamel, F.G. (1998) Insulin acts intracellularly on proteasomes through insulin-degrading enzyme. *Biochemical and Biophysical Research Communications* 244, 390–394.

Fang, C.H., Li, B.G., Sun, X. and Hasselgren, P.O. (2000) Insulin-like growth factor I reduces ubiquitin and ubiquitin-conjugating enzyme gene expression but does not inhibit muscle proteolysis in septic rats. *Endocrinology* 141, 2743–2751.

Farout, L., Lamare, M.C., Cardozo, C., Harrisson, M., Briand, Y. and Briand, M. (2000) Distribution of proteasomes and of the five proteolytic activities in rat tissues. *Archives of Biochemistry and Biophysics* 374, 207–212.

Glickman, M.H. and Ciechanover, A. (2002) The ubiquitin–proteasome proteolytic pathway: destruction for the sake of construction. *Physiological Reviews* 82, 373–428.

Gomes, M.D., Lecker, S.H., Jagoe, R.T., Navon, A. and Goldberg, A.L. (2001) Atrogin-1, a muscle-specific F-box protein highly expressed during muscle atrophy. *Proceedings of the National Academy of Sciences USA* 98, 14440–14445.

Gonen, H., Stancovski, I., Shkedy, D., Hadari, T., Bercovich, B., Bengal, E., Mesilati, S., Abu-Atoum, O., Schwartz, A.L. and Ciechanover, A. (1996) Isolation, characterization, and partial purification of a novel ubiquitin-protein ligase, E$_3$. *Journal of Biological Chemistry* 271, 302–310.

Gonzalez, F., Delahodde, A., Kodadek, T. and Johnston, S.A. (2002) Recruitment of a 19S proteasome subcomplex to an activated promoter. *Science* 296, 548–550.

Haas, A.L. and Siepmann, T.J. (1997) Pathways of ubiquitin conjugation. *FASEB Journal* 11, 1257–1268.

Hamel, F.G., Bennett, R.G. and Duckworth, W.C. (1998) Regulation of multicatalytic enzyme activity by insulin and the insulin-degrading enzyme. *Endocrinology* 139, 4061–4066.

Hasselgren, P.O., Wray, C. and Mammen, J. (2002) Molecular regulation of muscle cachexia: it may be more than the proteasome. *Biochemical and Biophysical Research Communications* 290, 1–10.

Hendil, K.B., Khan, S. and Tanaka, K. (1998) Simultaneous binding of PA28 and PA700 activators to 20S proteasomes. *Biochemical Journal* 332, 749–754.

Hobler, S.C., Tiao, G., Fischer, J.E., Monaco, J. and Hasselgren, P.O. (1998) Sepsis-induced increase in muscle proteolysis is blocked by specific proteasome inhibitors. *American Journal of Physiology* 274, R30–R37.

Hobler, S.C., Wang, J.J., Williams, A.B., Melandri, F., Sun, X., Fischer, J.E. and Hasselgren, P.O. (1999a) Sepsis is associated with increased ubiquitin conjugating enzyme E2 14k mRNA in skeletal muscle. *American Journal of Physiology* 276, R468–R473.

Hobler, S.C., Williams, A., Fischer, D., Wang, J.J., Sun, X., Fischer, J.E., Monaco, J.J. and Hasselgren, P.O. (1999b) Activity and expression of the 20S proteasome are increased in skeletal muscle during sepsis. *American Journal of Physiology* 277, R434–R440.

Jagoe, R.T. and Goldberg, A.L. (2001) What do we really know about the ubiquitin–proteasome pathway in muscle atrophy? *Current Opinion in Clinical Nutrition and Metabolic Care* 4, 183–190.

Kee, A.J., Taylor, A.J., Carlsson, A.R., Sevette, A., Smith, R.C. and Thompson, M.W. (2002) IGF-I has no effect on postexercise suppression of the ubiquitin–proteasome system in rat skeletal muscle. *Journal of Applied Physiology* 92, 2277–2284.

Kee, A.J., Combaret, L., Tilignac, T., Souweine, B., Aurousseau, E., Dalle, M., Taillandier, D. and Attaix, D. (2003) Ubiquitin-proteasome-dependent muscle proteolysis responds slowly to insulin release and refeeding n starved rats. *Journal of Physiology* 546, 765–776.

Kisselev, A.F., Akopian, T.N., Castillo, V. and Goldberg, A.L. (1999) Proteasome active sites allosterically regulate each other, suggesting a cyclical bite–chew mechanism for protein breakdown. *Molecular Cell* 4, 395–402.

Kohler, A., Cascio, P., Leggett, D.S., Woo, K.M., Goldberg, A.L. and Finley, D. (2001) The axial channel of the proteasome core particle is gated by the Rpt2 ATPase and controls both substrate entry and product release. *Molecular Cell* 7, 1143–1152.

Kwon, Y.T., Xia, Z., Davydov, I.V., Lecker, S.H. and Varshavsky, A. (2001) Construction and analysis of mouse strains lacking the ubiquitin ligase UBR1 (E3α) of the N-end rule pathway. *Molecular and Cellular Biology* 21, 8007–8021.

Lam, Y.A., Lawson, T.G., Velayutham, M., Zweier, J.L. and Pickart, C.M. (2002) A proteasomal ATPase subunit recognizes the polyubiquitin degradation signal. *Nature* 416, 763–767.

Larbaud, D., Debras, E., Taillandier, D., Samuels, S.E., Temparis, S., Champredon, C., Grizard, J. and Attaix, D. (1996) Euglycemic hyperinsulinemia and hyperaminoacidemia decrease skeletal muscle ubiquitin mRNA in goats. *American Journal of Physiology* 271, E505–E512.

Larbaud, D., Balage, M., Taillandier, D., Combaret, L., Grizard, J. and Attaix, D. (2001) Differential regulation of the lysosomal, Ca^{2+}-dependent and ubiquitin–proteasome-dependent proteolytic pathways in fast-twitch and slow-twitch rat muscle following hyperinsulinemia. *Clinical Science* 101, 551–558.

Lecker, S.H., Solomon, V., Price, S.R., Kwon, Y.T., Mitch, W.E. and Goldberg, A.L. (1999) Ubiquitin conjugation by the N-end rule pathway and mRNAs for its components increase in muscles of diabetic rats. *Journal of Clinical Investigation* 104, 1411–1420.

Liu, Z., Miers, W.R., Wei, L. and Barrett, E.J. (2000) The ubiquitin–proteasome proteolytic pathway in heart vs. skeletal muscle: effects of acute diabetes. *Biochemical and Biophysical Research Communications* 276, 1255–1260.

Lowell, B.B., Ruderman, N.B. and Goodman, M.N. (1986) Evidence that lysosomes are not involved in the degradation of myofibrillar proteins in rat skeletal muscle. *Biochemical Journal* 234, 237–240.

Mansoor, O., Beaufrère, B., Boirie, Y., Rallière, C., Taillandier, D., Aurousseau, E., Schoeffler, P., Arnal, M. and Attaix, D. (1996) Increased mRNA levels for components of the lysosomal, Ca^{2+}-activated and ATP-ubiquitin-dependent proteolytic pathways in skeletal muscle from head trauma patients. *Proceedings of the National Academy of Sciences USA* 93, 2714–2718.

Marinovic, A.C., Zheng, B., Mitch, W.E. and Price, S.R. (2002) Ubiquitin (UbC) expression in muscle cells is increased by glucocorticoids through a mechanism involving Sp1 and MEK1. *Journal of Biological Chemistry* 277, 16673–16681.

Medina, R., Wing, S.S. and Goldberg, A.L. (1995) Increase in levels of polyubiquitin and proteasome mRNA in skeletal muscle during starvation and denervation atrophy. *Biochemical Journal* 307, 631–637.

Murata, S., Kawahara, H., Tohma, S., Yamamoto, K., Kasahara, M., Nabeshima, Y., Tanaka, K. and Chiba, T. (1999) Growth retardation in mice lacking the proteasome activator PA28γ. *Journal of Biological Chemistry* 274, 38211–38215.

Mykles, D.L. (1993) Lobster muscle proteasome and the degradation of myofibrillar proteins. *Enzyme Protein* 47, 220–231.

Navon, A. and Goldberg, A.L. (2001) Proteins are unfolded on the surface of the ATPase ring before transport into the proteasome. *Molecular Cell* 8, 1339–1349.

Ordway, G.A., Neufer, P.D., Chin, E.R. and DeMartino, G.N. (2000) Chronic contractile activity upregulates the proteasome system in rabbit skeletal muscle. *Journal of Applied Physiology* 88, 1134–1141.

Park, K.C., Kim, J.H., Choi, E.J., Min, S.W., Rhee, S., Baek, S.H., Chung, S.S., Bang, O., Park, D., Chiba, T., Tanaka, K. and Chung, C.H. (2002) Antagonistic regulation of myogenesis by two deubiquitinating enzymes, UBP45 and UBP69. *Proceedings of the National Academy of Sciences USA* 99, 9733–9738.

Pickart, C.M. (2001) Mechanisms underlying ubiquitination. *Annual Review of Biochemistry* 70, 503–533.

Price, S.R., England, B.K., Bailey, J.L., Van Vreede, K. and Mitch, W.E. (1994) Acidosis and glucocorticoids concomitantly increase ubiquitin and proteasome subunit mRNA levels in rat muscles. *American Journal of Physiology* 267, C955–C960.

Price, S.R., Bailey, J.L., Wang, X., Jurkovitz, C., England, B.K., Ding, X., Phillips, L.S. and Mitch, W.E. (1996) Muscle wasting in insulinopenic rats results from activation of the ATP-dependent, ubiquitin–proteasome proteolytic pathway by a mechanism including gene transcription. *Journal of Clinical Investigation* 98, 1703–1708.

Rajapurohitam, V., Bedard, N. and Wing, S.S. (2002) Control of ubiquitination of proteins in rat tissues by ubiquitin conjugating enzymes and isopeptidases. *American Journal of Physiology* 282, E739–E745.

Rallière, C., Tauveron, I., Taillandier, D., Guy, L., Boiteux, J.-P., Giraud, B., Attaix, D. and Thiéblot, P. (1997) Glucocorticoids do not regulate the expression of proteolytic genes in skeletal muscle from Cushing's syndrome patients. *Journal of Clinical Endocrinology and Metabolism* 82, 3161–3164.

Rechsteiner, M., Realini, C. and Ustrell, V. (2000) The proteasome activator 11S REG (PA28) and class I antigen presentation. *Biochemical Journal* 345, 1–15.

Rock, K.L., Gramm, C., Rothstein, L., Clark, K., Stein, R., Dick, L., Hwang, D. and Goldberg, A.L. (1994) Inhibitors of the proteasome block the degradation of most cell proteins and the generation of peptides presented on MHC class I molecules. *Cell* 78, 761–771.

Scheffner, M., Smith, S. and Jentsch, S. (1998) The ubiquitin-conjugation system. In: Peters, J.-M., Harris, J.R. and Finley, D. (eds) *Ubiquitin and the Biology of the Cell*. Plenum Press, New York, pp. 65–98.

Shih, S.C., Sloper-Mould, K.E. and Hicke, L. (2000) Monoubiquitin carries a novel internalization signal that is appended to activated receptors. *EMBO Journal* 19, 187–198.

Solomon, V. and Goldberg, A.L. (1996) Importance of the ATP–ubiquitin–proteasome pathway in the degradation of soluble and myofibrillar proteins in rabbit muscle extracts. *Journal of Biological Chemistry* 271, 26690–26697.

Solomon, V., Baracos, V., Sarraf, P. and Goldberg, A.L. (1998) Rates of ubiquitin conjugation increase when muscles atrophy, largely through activation of the N-end rule pathway. *Proceedings of the National Academy of Sciences USA* 95, 12602–12607.

Strickland, E., Hakala, K., Thomas, P.J. and DeMartino, G.N. (2000) Recognition of misfolding proteins by PA700, the regulatory subcomplex of the 26S proteasome. *Journal of Biological Chemistry* 275, 5565–5572.

Taillandier, D., Aurousseau, E., Meynial-Denis, D., Béchet, D., Ferrara, M., Cottin, P., Ducastaing, A., Bigard, X., Guezennec, C.-Y., Schmid, H.-P. and Attaix, D. (1996) Coordinate activation of lysosomal, Ca^{2+}-activated and ATP-ubiquitin-dependent proteinases in the unweighted rat soleus muscle. *Biochemical Journal* 316, 65–72.

Tanahashi, N., Murakami, Y., Minami, Y., Shimbara, N., Hendil, K.B. and Tanaka, K. (2000) Hybrid proteasomes. Induction by interferon-gamma and contribution to ATP-dependent proteolysis. *Journal of Biological Chemistry* 275, 14336–14345.

Tawa, N.E. Jr and Goldberg, A.L. (1993) Protein and amino acid metabolism in muscle. In: Engel, A.G. and Franzini-Armstrong, C. (eds) *Myology*. McGraw-Hill, New York, pp. 683–707.

Temparis, S., Asensi, M., Taillandier, D., Aurousseau, E., Larbaud, D., Obled, A., Béchet, D., Ferrara, M., Estrela, J.M. and Attaix, D. (1994) Increased ATP-ubiquitin-dependent proteolysis in skeletal muscles of tumor-bearing rats. *Cancer Research* 54, 5568–5573.

Thrower, J.S., Hoffman, L., Rechsteiner, M. and Pickart, C.M. (2000) Recognition of the polyubiquitin proteolytic signal. *EMBO Journal* 19, 94–102.

Tiao, G., Fagan, J.M., Samuels, N., James, J.H., Hudson, K., Lieberman, M., Fischer, J.E. and Hasselgren, P.O. (1994) Sepsis stimulates nonlysosomal, energy-dependent proteolysis and increases ubiquitin mRNA levels in rat skeletal muscle. *Journal of Clinical Investigation* 94, 2255–2264.

Tiao, G., Fagan, J.M., Roegner, V., Lieberman, M., Wang, J.J., Fischer, J.E. and Hasselgren, P.O. (1996) Energy-ubiquitin-dependent muscle proteolysis during sepsis is regulated by glucocorticoids. *Journal of Clinical Investigation* 97, 339–348.

Tiao, G., Hobler, S., Wang, J.J., Meyer, T.A., Luchette, F.A., Fischer, J.E. and Hasselgren, P.O. (1997) Sepsis is associated with increased mRNAs of the ubiquitin–proteasome proteolytic pathway in human skeletal muscle. *Journal of Clinical Investigation* 99, 163–168.

Tilignac, T., Temparis, S., Combaret, L., Taillandier, D., Pouch, M.-N., Cervek, M., Cardenas, D.M., Le Bricon, T., Debiton, E., Samuels, S.E., Madelmont, J.C. and Attaix, D. (2002) Chemotherapy inhibits skeletal muscle ubiquitin–proteasome-dependent proteolysis. *Cancer Research* 62, 2771–2777.

Turner, G.C., Du, F. and Varshavsky, A. (2000) Peptides accelerate their uptake by activating a ubiquitin-dependent proteolytic pathway. *Nature* 405, 579–583.

Ustrell, V., Hoffman, L., Pratt, G. and Rechsteiner, M. (2002) PA200, a nuclear proteasome activator involved in DNA repair. *EMBO Journal* 21, 3516–3525.

Varshavsky, A. (1996) The N-end rule: functions, mysteries, uses. *Proceedings of the National Academy of Sciences USA* 93, 12142–12149.

Voges, D., Zwickl, P. and Baumeister, W. (1999) The 26S proteasome: a molecular machine designed for controlled proteolysis. *Annual Review of Biochemistry* 68, 1015–1068.

Voisin, L., Breuillé, D., Combaret, L., Pouyet, C., Taillandier, D., Aurousseau, E., Obled, C. and Attaix, D. (1996) Muscle wasting in a rat model of long lasting sepsis results from the activation of lysosomal, Ca^{2+}-activated and ubiquitin–proteasome proteolytic pathways. *Journal of Clinical Investigation* 97, 1610–1617.

Waterlow, J.C., Garlick, P.J. and Millward, D.J. (1978) *Protein Turnover in Mammalian Tissues and in the Whole Body*. Elsevier-North Holland, Amsterdam.

Whitehouse, A.S. and Tisdale, M.J. (2001) Down-regulation of ubiquitin-dependent proteolysis by eicosapentaenoic acid in acute starvation. *Biochemical and Biophysical Research Communications* 285, 598–602.

Whitehouse, A.S., Smith, H.J., Drake, J.L. and Tisdale, M.J. (2001) Mechanism of attenuation of skeletal muscle protein catabolism in cancer cachexia by eicosapentaenoic acid. *Cancer Research* 61, 3604–3609.

Wilkinson, K.D. (2000) Ubiquitination and deubiquitination: targeting of proteins for degradation by the proteasome. *Seminars in Cell and Developmental Biology* 11, 141–148.

Williams, A.B., Decourten-Myers, G.M., Fischer, J.E., Luo, G., Sun, X. and Hasselgren, P.O. (1999) Sepsis stimulates release of myofilaments in skeletal muscle by a calcium-dependent mechanism. *FASEB Journal* 13, 1435–1443.

Wing, S.S. and Banville, D. (1994) 14-kDa ubiquitin-conjugating enzyme: structure of the rat gene and regulation upon fasting and by insulin. *American Journal of Physiology* 267, E39–E48.

Wing, S.S. and Bedard, N. (1996) Insulin-like growth factor I stimulates degradation of a mRNA transcript encoding the 14 kDa ubiquitin-conjugating enzyme. *Biochemical Journal* 319, 455–461.

Wing, S.S. and Goldberg, A.L. (1993) Glucocorticoids activate the ATP-ubiquitin-dependent proteolytic system in skeletal muscle during fasting. *American Journal of Physiology* 264, E668–E676.

Wing, S.S., Haas, A.L. and Goldberg, A.L. (1995) Increase in ubiquitin–protein conjugates concomitant with the increase in proteolysis in rat skeletal muscle during starvation and atrophy denervation. *Biochemical Journal* 307, 639–645.

Woo, S.K., Lee, J.I., Park, I.K., Yoo, Y.J., Cho, C.M., Kang, M.S., Ha, D.B., Tanaka, K. and Chung, C.H. (1995) Multiple ubiquitin C-terminal hydrolases from chick skeletal muscle. *Journal of Biological Chemistry* 270, 18766–18773.

Wray, C.J., Tomkinson, B., Robb, B.W. and Hasselgren, P.O. (2002) Tripeptidyl-peptidase II expression and activity are increased in skeletal muscle during sepsis. *Biochemical and Biophysical Research Communications* 296, 41–47.

Yokota, T., Nagai, H., Harada, H., Mine, N., Terada, Y., Fujiwara, H., Yabe, A., Miyazaki, K. and Emi, M. (2001) Identification, tissue expression, and chromosomal position of a novel gene encoding human ubiquitin-conjugating enzyme E2-230k. *Gene* 267, 95–100.

16 Diet, DNA Methylation and Cancer

Judith K. Christman
Department of Biochemistry and Molecular Biology and Eppley Cancer Center, University of Nebraska Medical Center, Omaha, Nebraska, USA

Introduction

The first indication that limited availability of methyl donors and other factors necessary for single-carbon metabolism can influence development of cancer was obtained >50 years ago in a rat model and has been supported by many additional studies in rodent models. More recently, it has been demonstrated that a variety of other disorders including birth defects, immune deficiency, cardiovascular disease, diabetes and some forms of mental retardation may also arise from methyl donor insufficiency (reviewed in Maier and Olek, 2002; Issa, 2002a; Van den Veyver, 2002). In some, but not all, of these diseases, dietary methyl deficiency has been linked to alterations in DNA methylation.

There is a large body of evidence supporting the concept that methylation of cytosine residues in DNA plays an important role in regulating gene expression (reviewed in Singal and Ginder, 1999; Leonhardt and Cardoso, 2000). It is clear, from studies in many species, ranging from bacterial restriction/modification systems to regulation of gene expression in mammalian cells, that methylation of bases in DNA can enhance or block binding of proteins to DNA. In mammalian cells, the bulk of DNA methylation occurs on carbon 5 of cytosine in CpG dinucleotides. Depending on cell type or tissue, 3–4% of all C residues in vertebrate DNA may be present as 5-methylcytosine (5mC) and approximately 70% of C residues in CpG dinucleotides are methylated (Razin and Riggs, 1980; Ehrlich and Wang, 1981). Studies of mice with knockouts of DNA cytosine (C5) methyltransferases have proven that DNA methylation is critical during normal fetal development in mammals (Li *et al.*, 1992; Okano *et al.*, 1999), playing a role in X inactivation and imprinting (Goto and Monk, 1998; Mann *et al.*, 2000).

The accepted paradigm for cancer development was first proposed by Nowell (1976), i.e. that cancer is the sum of a series of genetic alterations that lead to dysregulated growth, i.e. that tumour progression is a product of selection for cells with a growth advantage (drug resistance, increased production and/or response to growth signals or loss of response to growth limitations imposed by tumour-suppressor genes). The specific DNA methylation patterns established during embryonic development are severely dysregulated during the process of tumorigenesis, leading to an overall reduction in the level of genomic DNA methylation, with selective loss of methylation in specific genes as well as repetitive DNA. Despite the overall loss in 5mC content of DNA during tumour progression, increased methylation also occurs. Hyper- and hypomethylation of DNA appear to occur independently of each other on a temporal basis and target different regions of the genome. Loss of methylation occurs in repeated sequences such as satellite DNA, retroposons, endogenous retroviral long terminal repeats and single-copy genes. Loss of methylation in these sequences has been linked to genomic instability and activation of genes involved in tumour progression (reviewed in Ehrlich, 2002). Hypermethylation occurs primarily in GC-rich regions (CpG islands) of the genome that do

©CAB *International* 2003. *Molecular Nutrition*
(eds J. Zempleni and H. Daniel)

not have the typical under-representation of CpG common to vertebrate genomes (Bird, 1986), and has been associated with silencing of a variety of tumour-suppressor genes. It has been cogently argued that methylation-mediated (epigenetic) gene silencing could act as the 'second hit' (Knudson, 1971) during development of hereditary cancers, as well as a means of inactivating the normal allele of a tumour suppressor in sporadic cancers.

If dysregulation in DNA methylation plays a causal role in cancer initiation and progression, one should be able to provide evidence for the following.

1. Methylation changes that occur in pre-neoplastic cells and/or are already present during the early stages of cancer.
2. Additional methylation changes or defects in regulation of methylation that occur during progression when the initial changes are not sufficient for optimal expression of the tumour phenotype.
3. A mechanistic link exists between the methylation changes and initiation/progression. The changes should either confer a susceptibility to additional genetic changes, such as mutation or chromosome loss, or should lead to loss or gain of expression of proteins involved in processes critical to initiation/progression such as growth regulation, apoptosis and metastatic potential.
4. When silencing of a specific gene or set of genes is critical to the tumour phenotype, then reactivation of epigenetically silenced gene(s) through use of drugs that inactivate the enzymes catalysing DNA methylation, such as decitabine (5-aza-2-deoxycytidine (ZdCyd)), or introduction and expression of appropriate transgenes will lead to restoration of normal growth controls, restore sensitivity to drugs that induce apoptosis and/or restore expression of a normal phenotype.
5. When activation of specific genes or processes through loss of methylation is critical to the tumour phenotype, then blocking expression of these genes will also restore a more normal phenotype. Although it is not yet possible experimentally to manipulate and recreate loss or gain of methylation patterns in specific gene regions in cells or tissues, silencing could possibly be achieved through use of siRNAs or antisense DNAs.

Of all the cancers studied, familial adenomatous polyposis coli (APC) comes closest to fulfilling these criteria. The total 5mC content of pre-neoplastic polyps is lower than that of normal colonic epithelium and there is extensive hypomethylation of specific genes in pre-malignant lesions. In the linear progression model developed by Vogelstein and his colleagues, the only genetic event preceding DNA hypomethylation is loss or mutation of the *Apc* gene (Fearon and Vogelstein, 1990). Although there does not appear to be a progressive increase in DNA hypomethylation with progression in this disease (Feinberg et al., 1988) and many of the genes that become hypomethylated do not exhibit increased expression or appear irrelevant to progression of the cancer (Feinberg and Vogelstein, 1983; Goelz et al., 1985), there is ample evidence for epigenetic silencing of genes during progression. All of the topics in this introduction have been reviewed extensively. A limited list of additional relevant reviews includes: Doerfler (1983); Riggs and Jones (1983); Christman (1984); Zingg and Jones (1997); Baylin et al. (1998); Tyeko (2000); Issa (2000b); Costello and Plass (2001); Jones and Takai (2001); Reik et al. (2001); Robertson (2001); Rountree et al. (2001); Esteller et al. (2002); Feinberg et al. (2002).

The aim of this chapter is to focus specifically on how the availability of dietary sources of methyl groups affects DNA methylation. The enzymes involved in methylation of DNA utilize the 'universal methyl donor', *S*-adenosylmethionine (AdoMet). AdoMet is synthesized either directly from methionine and adenosine or through utilization of methyl groups derived from single-carbon metabolism. Thus, it is logical to propose that decreased availability of AdoMet and/or alterations in the ratio of AdoMet to *S*-adenosylhomocysteine (AdoHcy) caused by inadequate intake of folate, methionine, choline and other dietary factors involved in providing methyl groups for AdoMet synthesis cause disease by inhibiting DNA methylation. The epidemiological evidence (or lack thereof) for this linkage in human cancers, particularly colorectal cancer, will be reviewed, as will our current knowledge of the properties of the enzymes involved in establishing and maintaining DNA methylation in mammalian cells. Insights into the link between methyl deficiency, altered DNA methylation and cancer of the liver and colon obtained from animal studies will be presented, followed by a discussion of current theories as to how interactions of DNA methyltransferases (DNMTs) with chromatin proteins modulate their activity and how dietary methyl deficiency may impinge on these processes.

Diet and Cellular Methylation Status

AdoMet is synthesized directly from dietary methionine and adenosine by methionine adenosyl transferase (MAT; EC 2.5.1.6; Fig. 16.1). In most tissues and fetal liver, this reaction is carried out by MAT II, which is a heterodimer composed of a catalytic (α2) and a regulatory (β) domain. The liver differs, having two forms, MAT I and MAT III, which consist of tetramers and dimers of α1, respectively (Kotb et al., 1997). The methionine required for formation of AdoMet can be synthesized by two additional inter-related pathways: methylation of homocysteine by folate-dependent homocysteine:methionine synthase (MTR, also methionine synthase, MS; EC 2.1.1.13) using the methyl group from 5-methyltetrahydrofolate (5MTHF) and methylation of homocysteine by betaine-homocysteine S-methyltransferase (BHMT; EC 2.1.1.5) using the methyl group from betaine, an oxidation product of choline. MTR is expressed and active in all mammalian tissues. Two forms of BHMT are expressed in liver and kidney. However, BHMT2 differs from BHMT in that it is also expressed at low levels in adult brain, heart, skeletal muscle and a variety of fetal tissues including heart, liver, lung, kidney and eye (Chadwick et al., 2000).

The major sources of methyl groups in the diet are choline ($\sim$30 mmol methyl day^{-1}) and methionine ($\sim$10 mmol methyl day^{-1}). An additional 5–10 mmol methyl day^{-1} are provided via single-carbon metabolism, with serine and glycine providing the single carbon units (Institute of Medicine and National Academy of Science, 2000). This process relies on availability of dietary folate to supply tetrahydrofolate (THF) and vitamin B_6 (pyridoxin) as a cofactor for glycine hydroxymethyltransferase (GHMT) which carries out transfer of the β-carbon of serine to 5-formimino-THF to yield 5,10-methylenetetrahydrofolate (5,10MTHF). In addition, vitamin B_2 (riboflavin) is required as a cofactor for methylenetetrahydrofolate reductase (MTHFR), vitamin B_{12} (cobalamin) as a methylated intermediate required for methyl transfer by methionine synthase, and Zn^{2+} as a cofactor for BHMT.

Single-carbon metabolism is also adversely affected by alcohol intake (Finkelstein et al., 1974). Chronic overconsumption of alcohol can cause maladsorption of folate, leading to systemic folate depletion. Alcohol also leads to decreased uptake and release of folate by the liver and increased urinary excretion of folate (Weir et al., 1985). Acetaldehyde, a metabolite of alcohol, may inactivate 5MTHF and/or inhibit MTR (Kenyon et al., 1998; Homann et al., 2000). Inhibition of MTR leads to trapping of folate as 5MTHF, blocking regeneration of 5,10MTHF with resultant inhibition of nucleotide synthesis. 5MTHF is a poor substrate for polyglutamation, a process required for retention of folate in cells leading to an overall cellular depletion of folate (Shane, 1995).

Although this review is focused on the effects of diet on DNA methylation, it should be kept in mind that there are three additional 'smoking guns' that link dietary methyl deficiency with cancer and other human diseases: (i) effects on nucleotide metabolism and DNA synthesis; (ii) effects on homocysteine accumulation; and (iii) effects of AdoMet depletion related to its role in modification of molecules other than DNA.

Purine metabolism and DNA synthesis

Availability of 5,10MTHF regulates these pathways. 5,10MTHF is either oxidized to formyl-THF, which is required for synthesis of purines, or is utilized directly for transfer of a methyl group to deoxyuridine monophosphate (dUMP) to provide thymidine monophosphate (TMP) for DNA synthesis. The latter reaction is a rate-limiting step in DNA synthesis in mammalian cells (Blount et al., 1997). Failure to methylate TMP can also lead to incorporation of dUMP into DNA (Pogribny et al., 1997). In severe cases, this may lead to a 'futile cycle' of removal of uracil from DNA by a repair glycosylase and its reintroduction by repeated cycles of DNA repair. The transient introduction of strand breaks during the repair process has been hypothesized to lead to irreparable damage and/or chromosome instability, and may also lead to hypomethylation of DNA (Ryan and Weir, 2001). Since AdoMet levels are reduced, methylation of DNA 'patches' synthesized during repair may be inefficient. A simultaneous increase in DNA strand breakage and hypomethylation of DNA has been noted in a variety of folate-deficient tissue culture and animal models (Choi, 1999; James and Yin, 1989; Kim et al., 1995, 1997) as well as in white blood cells of folate-deficient human volunteers (Jacob et al., 1998).

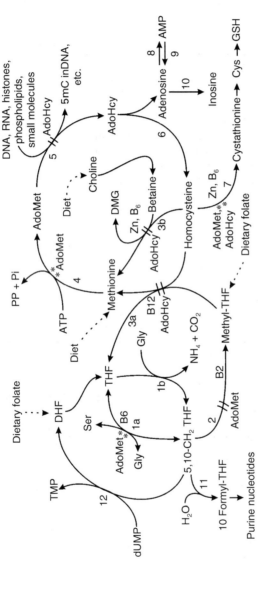

Fig. 16.1. Methyl metabolism. Dietary factors, enzymes and substrates involved in methyl metabolism. Enzymes in methyl metabolism identified by number: 1a, glycine hydroxymethyltransferase (GHMT); 1b, serine transhydroxymethylase; 2, methylenetetrahydrofolate reductase (MTHFR); 3a, 5-methyltetrahydrofolate-homocysteine S-methyltransferase (methionine synthase or MS); 3b, betaine-homocysteine S-methyltransferase (BHMT); 4, methionine adenosyltransferase (MAT); 5, various methyltransferases, including DNA methyltransferase (DNMT); 6, S-adenosylhomocysteine hydrolase (SAHH); 7, cystathionine-β-synthase (CBS); 8, adenosine kinase (AK); 9, 5-nucleotidase (5-NT); 10, adenosine deaminase (ADA); 11, methylenetetrahydrofolate cyclohydrolase; 12, thymidylate synthetase (TS). Abbreviations: DHF = dihydrofolate; Ser = serine; Gly = glycine; Cys = cysteine; THF = tetrahydrofolate; B6 = vitamin B_6; B12 = vitamin B_{12} or cobalamin; B2 = vitamin B_2 or riboflavin; 5,10-CH$_2$ THF = 5,10-methylenetetrahydrofolate; Methyl-THF = 5-methyltetrahydrofolate; Zn = zinc; DMG = dimethylglycine; AdoMet = S-adenosylmethionine; AdoHcy = S-adenosylhomocysteine; GSH = glutathione (reduced); ATP = adenosine triphosphate; Pi = inorganic phosphate; PP = pyrophosphate; TMP = thymidine monophosphate; dUMP = deoxyuridylate monophosphate; AMP = adenosine monophosphate; // = inhibits; ** = activates.

Homocysteine

Dietary deficiency of methyl donors and vitamins (choline, methionine, folate, vitamin B_{12} and vitamin B_6) can lead to accumulation of homocysteine. S-Adenosylhomocysteine hydrolase (SAHH) catalyses a reversible reaction that either breaks down AdoHcy to adenosine and homocysteine or condenses homocysteine and adenosine to synthesize AdoHcy. Even though equilibrium favours the synthetic reaction, under normal physiological conditions, three irreversible reactions prevent resynthesis of AdoHcy. Adenosine is removed by conversion to inosine by adenosine deaminase, and homocysteine is recycled to AdoMet by remethylation to methionine by MTR or removed by entry into the trans-sulphuration pathway. Homocysteine is condensed with serine to form cystathionine by the vitamin B_6-dependent enzyme, cystathionine β synthase (CBS). Cystathionine is cleaved to 2-oxoglutarate and cysteine by γ-cystathionase (GCT), which is also dependent on vitamin B_6. Due to the difference in K_m value on MTR (0.1 mM) and CBS (1 mM) for homocysteine, remethylation predominates at low homocysteine levels; at high levels removal by the trans-sulphuration pathway is activated. Since both reactions are irreversible, homocysteine levels are tightly regulated (Finkelstein, 1998, 2000). This is critical because AdoHcy is a competitive inhibitor of most of the great variety of methyltransferases that utilize AdoMet. However, with dietary deficiency of folate and vitamin B_{12} and possibly B_6, homocysteine levels rise. When intracellular levels of adenosine are not limiting, SAHH can increase levels of AdoHcy and inhibit the methylation of a wide variety of biologically important substrates including DNA. In addition, homocysteine itself has been found to affect cell growth and apoptosis, and to mediate cellular damage by homocysteinylation of proteins, generation of reactive oxygen species and a decrease in cellular glutathione (reviewed in Medina *et al.*, 2001). It has also been shown to alter signalling through its action as an antagonist of the *N*-methyl-D-asparate (NMDA)-type glutamate receptor (Rosenquist *et al.*, 1999), and to have cell-specific effects on the binding of the transcription factor AP-1 to its recognition site in DNA (Torres *et al.*, 1999). These activities have implicated the elevation of homocysteine levels in development of cardiovascular disease, Alzheimer's and other neurodegenerative diseases, and increased risk of developmental abnormalities, pregnancy disorders and cancer.

The complexity of dietary interactions in the function of single-carbon metabolism, the synthesis of AdoMet and the regulation of homocysteine levels make it extremely difficult to tease out the mechanism(s) by which dietary methyl donor deficiencies lead to human diseases. However, it is clear that AdoMet and AdoHcy play a central role in orchestrating the flow of methyl groups from the single-carbon cycle to methionine and regulating homocysteine metabolism (Finkelstein, 1990). AdoMet is an allosteric inhibitor of MTHFR, and this inhibition is reversed by AdoHcy. CBS is activated by both AdoMet and AdoHcy. BHMT is inhibited by AdoHcy which is competitive with homocysteine and non-competitive with betaine. When AdoMet levels are low, 5MTHF synthesis will occur at an optimal level and homocysteine levels will rise due to reduced activity of CBS favouring methionine synthesis and increased levels of AdoMet. At high AdoMet levels, 5MTHF synthesis is inhibited, methionine levels go down and excess homocysteine is removed by activated CBS. Since the expression of CBS is limited to liver, pancreas, kidney and brain, removal of excess homocysteine from most tissues depends on transport out of the cell and removal from the plasma by the kidney or other organs.

AdoMet as the 'Universal methyl donor'

It is important to recognize that AdoMet is the predominant methyl donor in biological reactions, while the various forms of folate are a distant second (reviewed in Fauman, 1999). For example, in mammals, 5MTHF is utilized as a methyl donor by only one enzyme (MTR), while AdoMet is utilized by about 100 enzymes. This reflects the fact that the charged sulphur atom in AdoMet greatly reduces the thermodynamic barrier for transfer of the methyl group to a variety of acceptors. It has been estimated that AdoMet is three orders of magnitude more reactive as methyl donor to polarizable nucleophiles (N, O and S) than methylated folates. It is important to note that AdoMet can also serve as a donor of either the adenosyl or the decarboxylated aminobutyryl side chains, which also display enhanced reactivity. In mammals, it has been estimated that up to 10% of AdoMet is decarboxylated to provide propylamine moieties for polyamine synthesis (Eloranta and Raina, 1977). Methylation of guanidoacetic acid to creatinine in the liver consumes >80% of the flow of methionine through AdoMet (Mudd and Poole, 1975; Im et al., 1979). Another major use of AdoMet is for synthesis of membrane phospholipids, i.e. phosphatidylcholine (PC) synthesis from phosphoethanolamine (PE) by PE methyltransferase (PEMT; EC2.1.1.17). In addition AdoMet is utilized to methylate DNA, tRNA, rRNA and mRNA, nuclear proteins including histones as well as a variety of proteins and small molecules involved in signal transduction and neurotransmission (Fauman, 1999). Thus, it is obvious that even when changes in DNA methylation are noted in a particular disease process related to dietary insufficiency of methyl donors, it is difficult to prove that DNA methylation changes are necessary and sufficient for causation of the disease. At a minimum, such proof requires the demonstration that the disease process can be replicated in an animal model by specific inhibition of DNA methyltransferases rather than inhibition of global methyl group metabolism.

Mammalian DNA Cytosine (C5) Methyltransferases: Mechanism, Structure and Function

The catalytic mechanism of DNA (C5) methyltransferases (MTases) involves formation of a covalent bond between a cysteine residue in the active site of the enzyme and carbon 6 of cytosine in DNA, increasing flow of electrons to carbon 5, with subsequent attack on the methyl group of AdoMet. Abstraction of a proton from carbon 5 then allows reformation of the 5–6 double bond and release of the enzyme by β-elimination (Fig. 16.2) (Santi et al., 1984). It is assumed that the mammalian DNA (C5) MTases accomplish this transfer in the same manner as the bacterial DNA (C5) MTase, M.HhaI, since they share a set of conserved catalytic domain motifs (Lauster et al., 1989; Posfai et al., 1989). These include the prolylcysteinyl active site dipeptide (region IV), a group of separate domains that come together to form the binding pocket for AdoMet (I, X) and a region (VI) containing the glutamyl residue that protonates nitrogen 3 of the target cytosine. The recognition domain that makes base-specific contacts in the major groove of DNA usually lies between motifs VIII and IX. A crystal structure of M.HhaI with the cysteine of the prolylcysteinyl dipeptide covalently linked to methylated 5-fluorocytosine in DNA demonstrated that this irreversibly linked target is flipped out of the DNA helix into the catalytic pocket of the methyltransferase. Carbon 5 of the target cytosine is aligned with a bound AdoMet molecule and carbon 6 with the thiolate residue of the active site cysteine in a manner that allows a concerted attack and methyl transfer (Santi et al., 1984).

Four separately encoded proteins containing the conserved motifs characteristic of DNA (C5) MTases have been identified (DNMT1, 2, 3a and 3b) in mammalian cells. DNMT2 does not have detectable MTase activity and homozygous *DnmtT2* knockout mice have no detectable developmental defects (Okano et al., 1998b; Yoder and Bestor, 1998). A fifth DNA MTase-like protein, DNMT3L, closely related to DNMT3a, has regions with similarity to conserved motifs I, IV and VI, but lacks the active site domains (Aapola et al., 2000). However, despite its lack of enzymatic activity, disruption of the DNMTL gene leads to failure of normal methylation at imprinted regions during development (Bourc'his et al., 2001). This may reflect a requirement for an interaction

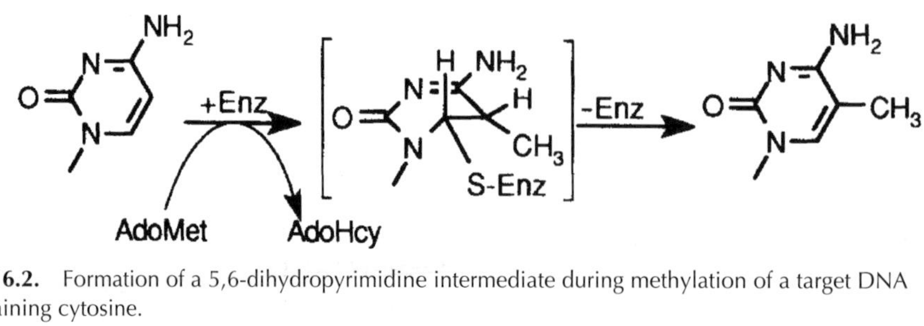

Fig. 16.2. Formation of a 5,6-dihydropyrimidine intermediate during methylation of a target DNA containing cytosine.

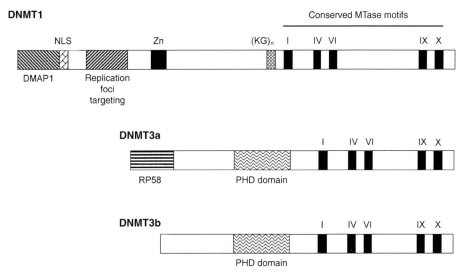

Fig. 16.3. Structure of the catalytically active mammalian DNA methyltransferases. All three enzymes have a regulatory and catalytic domain. The conserved motifs of the catalytic domain (Roman numerals) are discussed in the text. Other structural features of DNMT1 are the domains for nuclear localization, targeting to replication foci and zinc binding. Binding regions for proteins involved in linkage to the nuclear matrix (p23 (Zhang and Verdine, 1996)), annexin V or transcriptional repression (HDAC1 (Fuks et al., 2000), HDAC2-DMAP1 (Rountree et al., 2000), pRb/DMAP1 (Robertson et al., 2000)) and targeting to replication and repair sites (PCNA (Chuang et al., 1997)) are indicated. DNMT3a and 3b share a cysteine-rich PHD (plant homeodomain) region (Aasland et al., 1995) and regulatory regions binding to transcriptional repressors interacting with HDAC1 (Bachman et al., 2001; Fuks et al., 2001).

between DNMT3a and DNMT3L to stimulate *de novo* methylation during imprinting (Chedin et al., 2002). The properties of the known, enzymatically active, mammalian DNMTs (Fig. 16.3) are summarized below.

DNMT1

DNMT1, the first mammalian DNA MTase cloned (Bestor et al., 1988), is the most abundant DNA MTase in mammalian cells and the best studied. It is encoded by a single gene mapped to chromosome 19p13.2. It has three promoters for alternative first exons, each of which is spliced to the same downstream exon 2. The most distal promoter, 1o, which is expressed only in oocytes, gives rise to the shortest form of active DNMT1 (170 kDa), translated from an AUG in the fourth exon. This enzyme is localized primarily in the cytoplasm and only enters the nucleus during the eight-cell stage, suggesting that it is essential for establishing normal methylation patterns in imprinting. The second promoter, 1s, gives rise to an mRNA that is expressed in all somatic tissues (Mertineit et al., 1998). Although mRNAs containing somatic exon 1 have four potential AUG transcription start sites, initiation appears to be most efficient at the third AUG, yielding a 184 kDa protein (Pradhan et al., 1999). Additional promoters have been reported 5′ to exon 4, upstream of the third AUG (Bigey et al., 2000), as have splice variants of somatic DNMT1, including a larger form of the enzyme with an additional 18 amino acids encoded by a 48 nucleotide insert of Alu repeat sequence between exons 4 and 5 (Hsu et al., 1999; Bonfils et al., 2000). It is not clear how these variants affect the catalytic function of DNMT1 *in vivo*, although it has been shown that absence of the amino acids encoded upstream of the fourth AUG has no significant effect. The 1p promoter, which is closest to exon 2, is expressed only in pachytene sperm and differentiated myotubes (Mertinet et al., 1998).

Although the 500 amino acid C-terminal domain of DNMT1 has great structural similarity to bacterial type II DNA (C5) MTases, it differs from these enzymes in several aspects. First, although the catalytic efficiency of DNMT1 is influenced by the structure of DNA (below), it has little or no sequence specificity beyond recognition of CpG dinucleotides. Bacterial enzymes minimally require a four-base recognition sequence. Secondly, DNMT1 has a large N-terminal regulatory region that contains motifs involved in association with the cellular replication machinery (Fig. 16.3) that are absent in the bacterial enzymes. These motifs include a sequence that targets the protein to replication foci, a region that interacts with PCNA (proliferating cell nuclear antigen) and a region similar to the polybromo-1 protein that also interacts with proteins at replication foci (Leonhardt et al., 1992; Chuang et al., 1997; Liu et al., 1998; Rountree et al., 2000). In addition, there are several regions that act as transcriptional repressor domains through interactions with pRB, E2F, a newly identified repressor DMAP1, and histone deacetylases, HDAC1 and HDAC2 (Robertson 2001). In vitro, pRB interaction inhibited DNMT1 binding to DNA and reduced catalytic activity (Pradhan and Kim, 2002). The domain structure of the DNMT1 regulatory region is consistent with the concept that the primary function of DNMT1 is maintenance methylation. The interactions of the N-terminus of DNMT1 with specific nuclear proteins ensures that it is in the right place at the right time to carry out maintenance methylation and provide it with the potential for repressing transcription in these regions through direct interaction with proteins that bind HDACs (see below).

There is some disagreement as to whether DNMT1 is also a *de novo* methyltransferase. *In vitro* assays with purified DNMT1 or nuclear extracts enriched with DNMT1 reveal a high degree of preference (5- to >100-fold) for hemi-methylated DNA as a substrate. Nevertheless, intact purified full-length DNMT1 is as capable, on a molar basis, of methylating completely unmethylated substrates *in vitro* as the known *de novo* DNMTs (DNMT3a and DNMT3b) (Okano et al., 1998a). However, *in vivo*, DNMT1 did not detectably methylate a replicating episome that was a substrate for *de novo* DNMTs (Hsieh, 1999). This difference most likely reflects the difference in interaction between DNMT1 and unmethylated CpG sites in replicating DNA and/or non-replicating chromatin regions as compared with its interactions with the variety of naked DNA substrates utilized for *in vitro* studies (Bestor, 1992; Christman et al., 1995; Pradhan et al., 1999; Brank et al., 2002). In this regard, it is of interest that a recent report suggests that a complex of DNMT1 with methylated DNA-binding proteins (MBD2 and MBD3) specifically binds hemi-methylated DNA (Tatematsu et al., 2000), a property that could further constrain the enzyme's ability to *de novo* methylate fully unmethylated sites *in vivo*, but would enhance its ability to methylate newly replicated hemi-methylated DNA.

DNMT3a and DNMT3b

DNMT3a and 3b were discovered by TBLASTN searching using full-length bacterial type II (C5) MTase sequences as queries (Okano, et al., 1998a). The genes have been mapped to chromosomes 2p23 and 20q11.2, respectively. Similarly to DNMT1, there are multiple splice variants and alternative upstream exons for both DNMT3a and 3b mRNAs (Chen et al., 2002; Weissenberger et al., 2002). The largest forms of 3a and 3b mRNAs encode proteins 908 (3a) and 859 (3b1) amino acids long. Alternative splicing of mRNA leads to loss of conserved catalytic domains and introduction of novel sequences resulting from frameshifts occurs in two of the short forms of DNMT3b4 and 3b5, and there is evidence that DNMT3b4 may act as a dominant-negative inhibitor of DNA methylation (Saito et al., 2002). *In vitro*, DNMT3a, 3b1 and 3b2 show no preference for hemi-methylated versus fully unmethylated sites, and have catalytic efficiencies similar to that of DNMT1 with unmethylated sites. This means that they methylate hemi-methylated sites in DNA with at least one log lower efficiency than DNMT1 *in vitro* and are, thus, unlikely to play a significant role in maintaining global methylation patterns. DNMT3a and 3b also have the unusual property of methylating C residues that are not in CpG sites, with a preference for CpA sites, although at a lower rate than CpG sites (Gowher and Jertsch, 2001; Aoki et al., 2001). DNMT1 neither methylates nor maintains methylation at Cs that are not next to Gs (Wigler et al., 1981), so methylation at these sites would be lost unless maintained by enzymes other than DNMT1. Methylation at non-CpG sites may, however, play a role in spreading of methylation, as looped DNA structures that form intrastrand 5mCpA/CpG sites are particularly good substrates for both DNMT1 and DNMT3b and could lead to creation of hemi-methylated CpG sites at a distance from the 5mCpA site (Christman et al., 1995; Pogribny et al., 2000; A. Farrell and J.K. Christman, unpublished data). Cooperativity of DNMT1 with DNMT3a and 3b has been reported, and the enzymes have been shown to interact through their N-terminal domains (Fatemi et al., 2001; Liang et al., 2002; Rhee et al., 2002). Interestingly, in one colorectal tumour cell line, significant loss of genomic DNA methylation did not occur when either *Dnmt1* or *Dnmt3b* genes were removed by knockout. However, knockout of both genes led to significant loss of methylation in repeated DNA sequences, loss of imprinting and silencing of the p16^{INK4} gene (Rhee et al., 2002).

The N-termini of DNMT3a and 3b do not possess any domains homologous to the replication focus targeting domains of DNMT1 (Leondardt et al.,1992; Liu et al., 1998) nor do they co-localize with DNMT1 and PCNA at replication foci in the nuclei of replicating cells (Chuang et al., 1997). Rather, DNMT3a localizes in discrete foci that appear to be late replicating heterochromatin in pericentromeric DNA (Bachman et al., 2001). DNMT3b has a diffuse localization. The protein domains that direct this localization have not yet been identified, but there is evidence that an ATRX-like cysteine-rich domain in both DNMT3a and 3b can function as a transcriptional repressor through binding of HDACs (see below).

Passive vs. active loss of DNA methylation

Our understanding of the mechanisms by which methylation patterns are established and changed during development is still in its infancy. Candidate proteins for *de novo* methylation have been identified, and a theoretical basis is being built that will stimulate new research into how interactions between DNMTs and nuclear proteins lead to gene activation or silencing. However, even though it is known that massive loss of methylation of cytosine residues in DNA occurs early in both mammalian embryonic development and tumorigenesis, the mechanisms for removing 5mC from DNA remain to be resolved. Loss of 5mC could occur by a passive mechanism (Fig. 16.4) whenever DNMT1 either is not produced in sufficient quantity to maintain methylation or is inactivated by endogenous or exogenous inhibitors. As outlined in pathway (b), a minimum of two rounds of DNA replication without maintenance methylation is required to complete 'demethylation' of a given site in 50% of newly synthesized DNA molecules. Thus, extensive loss of methylation by a passive mechanism requires repeated replication and/or repair of DNA in the absence of maintenance methylation. Passive loss could be a factor in altering methylation patterns in rapidly growing tumours, if the rate of DNA synthesis exceeds the methyl transfer capacity of available DNMT1. However, in most tumours, the expression of DNMT1 is slightly elevated (~two- to threefold) when normalized against expression of other genes expressed during the S phase of the cell cycle (Lee et al., 1996; Eads et al., 1999). This makes it more likely that passive loss of methylation is the result of alterations in actual and/or relative levels of AdoHcy and AdoMet. Alternatively, interactions of DNMT1 with pRB or other proteins that prevent its binding to DNA or its localization to replication foci, or attenuate its catalytic activity could lead to passive loss of methylation in dividing cells (Pradhan and Kim, 2002).

Recent studies on the loss of DNA methylation early in embryonic development of the mouse suggest that loss of methylation from maternal chromosomes occurs gradually after multiple rounds of cell division in the developing zygote. However, the demonstration of massive, if not complete, loss of 5mC from paternal chromosomes during the first cell division of the fertilized oocyte (Mayer et al., 2000) strongly supports the concept that mechanisms exist for active removal of 5mC from DNA, as do studies indicating loss of DNA methylation in non-dividing cells (Jost et al., 2001; Pradhan and Kim, 2002). Enzymes have been well characterized that actively remove 5mC residues from either hemi-methylated or fully methylated DNA (Gallinari and Jiricny, 1996; Bird and Wolffe, 1999; Jost et al., 2001), and 5mC can also be removed from DNA after oxidative deamination to T followed by G:T mismatch repair (Hardeland et al., 2001). There is no evidence that these enzymes could account for targeted massive removal of 5mC from paternal DNA. Although evidence has been published suggesting that a methylated DNA-binding protein actively removes the methyl group from C5 of cytosine in DNA (Bhattacharya et al., 1999), the existence of this

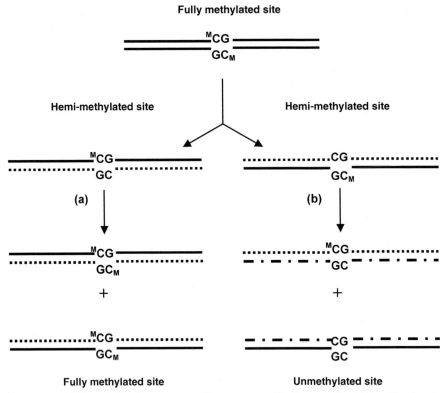

Fig. 16.4. Fate of fully methylated sites in DNA during DNA replication. (a) After DNA replication, hemi-methylated sites are remethylated by the maintenance methyltransferase, DNMT1, restoring pre-existing patterns of methylation. (b) Methylation is lost passively in one of the daughter DNA molecules because DNMT1 was not available to remethylate hemi-methylated sites created during the first round of DNA replication. This occurs in ZCyd- and ZdCyd-treated cells because DNMT1 is inactivated due to covalent linkage to ZCyt residues at CpG sites in DNA.

activity has been questioned (Magdinier and Wolffe, 2001).

Epidemiological Evidence for a Link Between Dietary Methyl Donor Deficiency and Cancer in Humans

There is ample evidence that diets rich in fruits and vegetables correlate with reduced risk of incidence of cancer of the breast, uterus, ovary, prostate, testis and kidney (American Institute for Cancer Research, 1997). Since these diets are also rich in sources of folic acid (legumes, cruciferous and some non-cruciferous vegetables, some fruits and whole grains), the proposed link between folic acid and cancer prevention is reasonable. However, on the basis of studies carried out prior to 1995, the AICR Panel on Food, Nutrition and the Prevention of Cancer came to the conclusion that the only type of cancer where a reasonable case has been made for a link between risk and dietary methyl donor deficiency is colorectal cancer.

Dietary folate and colon cancer

As has been reviewed extensively (Giovannucci and Willett, 1994; Giovannucci et al., 1995; Garay and Engstrom 1999; Tomeo et al., 1999; Potter, 1999; Fuchs et al., 2002), there is a consistent relationship between low dietary folate and increased occurrence of colorectal adenoma, an early stage in the progression to colon cancer. In the Health Professionals Follow-up and the Nurses' Health

Study, a comparison of high and low quantities of folate intake gave a relative risk of 0.66 for adenoma in women and 0.44 for men after adjustment for age, family history of colorectal cancer, total energy, saturated fat, dietary fibre and body mass index (Giovannucci et al., 1993). With some variation in relative odds ratios for men and women, four additional studies in Spain, France and the USA found a significant inverse association between folate intake and adenoma (Benito et al., 1993; Bird et al., 1995; Boutron-Ruault et al., 1996; Tseng et al., 1996).

Studies of colorectal cancer risk related to folate intake generally support this inverse association. Five case–control studies found lower odds ratios for development of colon cancer in women with high folate intake (Benito et al., 1991; Freudenheim et al., 1991; Ferraroni et al., 1994; Slattery et al., 1997; White et al., 1997), although two studies (Slattery et al., 1997; White et al., 1997) failed to detect this effect in men. A sixth study found no association of folate intake with risk for colon cancer (Boutron-Ruault et al., 1996). Out of five prospective studies, two found a significant inverse relationship between dietary folate and colon cancer risk in men, but not in women (Glynn et al., 1996; Su and Arab, 2001). A third found no appreciable relationship between total folate intake and colon cancer risk in men although there was some indication of reduced risk after long-term consumption of multivitamins (Giovannucci et al., 1995). A fourth study found a significant reduction in colon cancer risk with >400 µg day^{-1} folate or long-term consumption of multivitamins. One study found an inverse relationship between pre-diagnostic serum folate and colon cancer risk in women (Kato et al., 1999) while another found correlation between baseline plasma folate and colon cancer in men (Ma et al., 1997).

Interestingly, many studies have detected a greatly increased relative risk for colon cancer when a comparison was made between individuals with low folate intake and high alcohol consumption and those with high folate and low alcohol intake (Benito et al., 1991; Giovannucci et al., 1993, 1995; Ferraroni et al., 1994; Boutron-Ruault et al., 1996; White et al., 1997). These ranged from 2.67 to 5.07 in men with high alcohol consumption and from 1.2 to 4 in women. As outlined above, this is probably due in great part to alcohol's effect on folate uptake and its other effects on single-carbon metabolism. In addition, the levels of alcohol in the colonic mucosa of alcohol drinkers are high enough to allow substantial production of acetaldehyde after metabolism by colonic bacteria (Salaspuro, 1996), which could also be a major factor. In the rat, ethanol consumption has been shown to reduce folate levels by 48% in the colonic mucosa (Homann et al., 2000) and chronic alcohol consumption has been found to induce DNA hypomethylation (Choi et al., 1999).

Dietary folate and other cancers

While there have been a number of studies aimed at detecting a link between folic acid and prevention of cervical cancer, the AICR panel considered the results gathered to be 'null' (AICR, 1997). For example, the strongest statement from a large, multiethnic case control study that examined blood samples collected from 245 women with histologically confirmed invasive cervical cancer and 545 controls and evaluated appropriate risk factors by questionaire was that the risk of invasive cervical cancer was 'moderately, but non-significantly elevated in the lowest folate quartile compared to the highest' (Weinstein et al., 2001a).

One of the pitfalls in linking dietary methyl deficiency to cancer development is that evaluation of folate levels generally relies on the use of questionnaires dependent on diet recall. This may be quite imprecise, and there are limitations to quantifying dietary folate from databases due to multiple forms, instability and variations in bioavailability. Similarly, measurement of folate in plasma or red blood cells as a marker for dietary intake suffers from inherent limitations of available assays and provides no information as to the folate concentrations in the tissues of interest (Gunter, 1996). Recently, it has been proposed that serum homocysteine levels may provide a more sensitive measure of folate status. In this regard, it is of interest that evaluation of the relationship between elevated serum homocysteine levels and risk of invasive cervical cancer using data from 183 cases and 540 controls indicated statistically significantly elevated risk in each of the three highest quartiles (RR = 2.4–3.2, $P = 0.01$) (Weinstein et al., 2001b).

Ongoing studies of the relationship between dietary vitamin intake and development of squamous cell carcinoma (SSC) of the lung in smokers suggest that localized deficiency of folate and

B$_{12}$ in tumour tissue is accompanied by lower levels of cytosine methylation in SSC DNA measured as an increase in acceptance of methyl groups from *Sss*I (see Endnote; Piyathilake *et al.*, 2000). A comparison of serum and buccal tissue folate levels in smokers and non-smokers suggested that reduced buccal folate levels might be related to direct exposure to cigarette smoke (Piyathilake *et al.*, 1992). Finally, in a group of 14 subjects, methyl acceptance was found to be significantly higher in DNA isolated from buccal mucosa and malignant lung tissues in both primary non-small cell lung cancer and metastatic lung cancer than in leukocytes or non-malignant tissues ($P < 0.05$) (Piyathilake and Johanning, 2002).

Several recent case–control studies on folate and breast cancer found significantly decreased risk of pre- and post-menopausal breast cancer with higher folate intake (Graham *et al.*, 1991; Freudenheim *et al.*, 1996; Ronco *et al.*, 1999; Negri *et al.*, 2000). However, in two of these studies, the odds ratios were not significant after adjustment for vegetable uptake (Freudenheim *et al.*, 1996; Ronco *et al.*, 1999). In one study, the use of folic acid supplements was not associated with decreased risk of menopausal breast cancer, and in another no association was found with either dietary or supplemental folate for either pre- or post-menopausal breast cancer (Freudenheim *et al.*, 1996; Potischman *et al.*, 1999). Similarly, no association between total folate intake and overall risk of breast cancer was found in the Nurses' health study (3483 cases) or a Canadian cohort trial with 1469 cases (Zhang *et al.*, 1999; Rohan *et al.*, 2000). Interestingly, both of these trials, as well as a large case–control study (Negri *et al.*, 2000), found a significant effect of folate on risk of breast cancer in women consuming relatively high levels of alcohol. This is consistent with the known effects of alcohol on single-carbon metabolism summarized above.

There is also one study indicating an inverse relationship between risk for pancreatic cancer and serum folate levels in male smokers (Stolzenberg-Solomon *et al.*, 1999).

Influence of MTHFR function on risk of colon cancer

A polymorphism (C677T: cytosine at position 677 in the coding region is replaced with thymine) in the *mthfr* gene reduces MTHFR temperature stability, resulting in loss of MHTFR activity (Todesco *et al.*, 1999). TT homozygosity occurs in about 10% of the population. Individuals with a TT genotype have been found to have approximately 30% of the MTHFR activity of a CC genotype and moderate elevation in plasma homocysteine (Frosst *et al.*, 1995). In males, this genotype is associated with decreased risk of colorectal cancer compared with CC and CT genotypes but increased risk relative to the CC genotype on a high alcohol–low folate, low methionine diet (Chen *et al.*, 1996; Ma *et al.*, 1997; Slattery *et al.*, 1999). Several studies report that alcohol is a statistically significant factor in colorectal cancer risk in TT individuals (Chen *et al.*, 1996; Ulrich *et al.*, 1999; Levine *et al.*, 2000). The TT genotype has also been associated with a lower level of DNA methylation in peripheral leukocytes than the CC genotype (Stern *et al.*, 2000), and the level of leukocyte DNA methylation as determined by direct mass spectrophotometric measurement of 5mC content of DNA was directly related to the folate status and inversely related to homocysteine levels (Friso *et al.*, 2002). Finally, there is some evidence of elevated risk of cancers of the endometrium, oesophagus, stomach, breast, ovary and bone marrow in TT individuals (Esteller *et al.*, 1997; Gershoni-Baruch *et al.*, 2000; Shen *et al.*, 2001; Song *et al.*, 2001; Wiemels *et al.*, 2001; Sharp *et al.*, 2002). However, the TT genotype appears to have little effect on development of lung cancer and is protective in some leukaemias (Skibola *et al*, 1999; Franco *et al.*, 2001).

Taken together, many of these studies show a moderate increase in colon cancer risk in folate-deficient individuals that becomes more significant when accompanied by high alcohol intake or involves dietary methyl deficiency in individuals with constitutively low levels of MTHFR. Does this mean that dietary methyl insufficiency plays a causal role in cancer development? As noted above, most tumour tissues have undergone significant loss of methylation relative to the normal tissue of origin, and a decrease in DNA methylation is already detectable in pre-neoplastic lesions of the colon. In other tissues (breast, ovary, cervix, prostate and brain), progressive increases in the extent of hypomethylation have been correlated with increasing tumour grade (Kim *et al.*, 1994; Fowler *et al.*, 1998; Narayan *et al.*, 1998; Qu, 1999; Shen *et al.*, 1998; Santourlidis *et al.*, 1999). The first steps toward proof of a causal role of dietary methyl deficiency in colon cancer have been taken. Dietary folate deficiency has been shown to reduce folate levels, induce strand breaks, and to decrease DNA methylation in leukocytes (Jacob *et al.*, 1998). Folate levels in normal colonic tissue reflect dietary folate intake (Kim *et al.*, 2001) and there is some evidence that folate supplementation increases the level of DNA methylation in rectal mucosa (Cravo *et al.*, 1998). In familial APC, where methylation status can be followed at sequential stages in progression to malignancy, the global loss of total 5mC that occurs in pre-neoplastic tissues of the colon does not seem to increase with progression (Feinberg *et al.*, 1988). However, the focus of colon cancer studies has been on epigenetic silencing of tumour-suppressor genes with progression, which has been amply demonstrated (Issa, 2000a). Thus, the links between dietary methyl deficiency, inhibition of DNA methylation and activation of critical genes has yet to be proven. It is possible that the major role of an adequate supply of dietary methyl groups is to mitigate the effects of pre-existing alterations in regulation of DNA methylation in pre-neoplastic cells. Conversely, dietary methyl deficiency could be more important as a promoter of progression in cells that have already undergone genetic alterations rather than as an initiator of carcinogenesis.

Evidence from Animal Models that Dietary Deficiency of Methyl Groups Can Cause or Promote Cancer via Inhibition of DNA Methylation

Methyl deficiency as a promoter of carcinogen-induced liver cancer in rodents

Copeland and Salmon (1946) reported that long-term feeding of a choline-deficient diet increased the incidence of tumours in liver and other organs of the rat. As shown by later studies, groundnut meal used in formulating the diet was contaminated with aflatoxin, suggesting that choline deficiency promoted tumour growth rather than having a direct carcinogenic effect (Newberne *et al.*, 1964). The tumour-promoting effect of deficiency of choline and/or other lipotropes (methionine, folate and vitamin B_{12}) was demonstrated subsequently with a wide variety of carcinogens (Christman, 1995b). The proof that diet alone, in the absence of added carcinogens, could cause tumours came from the demonstration that feeding of an amino acid-defined (AAD) diet lacking methionine and choline but supplemented with folate, vitamin B_{12} and homocysteine led to a high incidence of liver tumours (Mikol *et al.*, 1983). Two additional aspects of these studies are of interest: (i) unlike low methionine/folate-deficient diets, which are used commonly for tumour studies, no tumours were obtained in the presence of folate/B_{12} deficiency alone; and (ii) the methionine and choline deficiency actually afforded 'slight protection against spontaneous tumor formation in extrahepatic tissues'. The first finding suggests that only severe methyl donor insufficiency is causal for tumour formation in this model, and the second that there are tissue-specific differences in susceptibility to tumour induction by methyl deficiency.

As previously summarized (Christman, 1995a,b), the link between dietary methyl deficiency and what is essentially liver-specific carcinogenesis in the rat may depend on the liver-specific response to the deficiency. Availability of phosphatidylcholine in the liver regulates secretion of very-low density lipoproteins. Since the majority of plasma lipoproteins are synthesized in the liver, and phosphatidylcholine accounts for approximately 75% of the phospholipid in rat lipoproteins, it is not surprising that depletion of liver choline leads to rapid and massive accumulation of triglycerides in the liver and not in other tissues (Yao and Vance, 1988). Although a consensus has yet to be reached as to the mechanism by which methyl deficiency leads to hepatocarcinogenesis (Christman, 1995a,b), several groups have shown that lipid accumulation occurs within days and is accompanied by a rapid increase (≤ 1 week) in DNA synthesis and cell division. At the same time, levels of AdoMet decrease and ratios of AdoHcy:AdoMet increase (Table 16.1). The combination of increased DNA replication with decreased AdoMet and a relative increase in the level of AdoHcy, its competitive inhibitor in methyl transfer reactions, means not only a decrease in overall availability of AdoMet for all cellular methyltransferases but also an increased need for AdoMet to maintain DNA methylation patterns in newly replicated hepatocyte DNA. As diagrammed in Fig. 16.4b, inhibition of methyl transfer at the same time as repeated rounds of DNA synthesis are occurring can easily lead to passive 'demethylation'.

Table 16.1. Effects of short-term feeding of an amino acid-defined lipotrope-deficient diet on rat liver.

Effect	Onset	References[1]
Increased rate of DNA synthesis	≤1 week	Wainfan et al. (1989)
		Christman et al. (1993b)
Increased numbers of mitotic cells	≤1 week	Christman et al. (1993b)
	<6 weeks	Hoover et al. (1984)
Increased activity of DNA methyltransferase	1–2 weeks	Christman et al. (1993a)
		Dizik et al. (1991)
Increased activity of tRNA methyltransferase	1–2 weeks	Christman et al. (1993a)
		Wainfan et al. (1984, 1988)
Increased levels of specific mRNAs	≤1 week	Dizik et al. (1991)
Decreased levels of AdoMet or decreased ratio of AdoMet:AdoHcy	≤1 week	Shivapurkar and Poirier (1983)
Decreased levels of specific mRNAs	≤1 week	Dizik et al. (1991)
Decreased methylation of DNA	1 week	Wainfan et al. (1989)
Decreased methylation of tRNA	5 days	Wainfan et al. (1986, 1988, 1989)

[1]Diets are defined in Shivapurkar and Poirier (1983) and Wainfan et al. (1989).

The studies of Poirier and colleagues demonstrated that DNA in liver tumours induced by methyl deficiency had reduced 5mC content (Wilson et al., 1984). However, studies on the extent of DNA methylation of carcinogen-induced tumours in animals fed methyl-sufficient diets also showed reduced 5mC content in DNA relative to that in normal liver even though the level of DNMTs was elevated (Lapeyre and Becker, 1979). Our studies demonstrated that loss of DNA methylation, as detected by the ability of the DNA to be methylated by DNMT1, occurred within a week of initiating total lipotrope deficiency and that loss of methylation at specific sites in several oncogenes (c-*myc*, c-*fos* and c-Ha-*ras*) occurred within 4 weeks and was associated with increased gene expression (Wainfan et al., 1989; Dizik et al., 1991). Loss of methylation also occurred in the coding region of the *p53* gene, but no increase in *p53* mRNA was detected (Christman et al., 1993a). tRNA methylation was inhibited while levels of tRNA methyltransferase were elevated. The same results were obtained with diets lacking choline and methionine but containing vitamin B_{12} and folate, while no effect of folate deprivation alone could be detected (Chen et al., 1993). When adequate levels of lipotropes were restored to the diet after 4–8 weeks, DNA synthesis and the number of mitotic cells decreased to control levels within a week. Methylation of newly synthesized DNA (hemi-methylated sites) returned to normal within 1–2 weeks. However, *de novo* methylation of CpG sites in the c-*myc* gene that had lost methylation on both strands, as determined by restriction endonuclease digestion, was not detected until at least 10 months after refeeding a methyl-sufficient diet (Table 16.2).

These results support several conclusions relevant to the link between cancer development, DNA methylation and dietary methyl deficiency. (i) Methyl deficiency resulting in lower AdoMet levels and AdoMet:AdoHcy ratios affects multiple methyltransfer processes in the liver (DNA methylation, tRNA methylation, phosphatidylcholine synthesis from PE and presumably methylation of histones and other critical substrates). (ii) All of the processes induced by short-term methyl deficiency were reversible within 1–2 weeks of restoring normal lipotrope levels in the diet, with the exception of DNA methylation that had been lost on both strands of a CpG site. This is consistent with inefficient *de novo* methylation in somatic cells. (iii) At least in the genes we examined, loss of DNA methylation was not sufficient to confer a stable increase in gene expression once normal (low) levels of DNA synthesis and cell division had been restored. (iv) Complete absence of folate in the diet with sufficient levels of other dietary methyl groups does not cause changes in liver DNA methylation, presumably because sufficient folate is provided by gut flora.

A detailed examination of the time course of changes in the levels of the *p53* gene in hepatocyte DNA methylation in rats fed a choline, methyl-deficient diet suggested a progressive loss of CpG methylation in exons 6–7. These sites were slowly remethylated and were methylated fully in tumour

Table 16.2. Reversibility of the effects of short-term feeding of an amino acid-defined lipotrope-deficient diet.

Effect	Time to reach control levels[1]
Increased rate of DNA synthesis	≤1 week
Increased numbers of mitotic cells	≤1 week
Increased lipid accumulation	8–10 weeks
Increased activity of DNA methyltransferase	1–2 weeks
Increased activity of tRNA methyltransferase	1–2 weeks
Increased (or decreased) levels of specific mRNAs	1–3 weeks
Decreased methylation of DNA and tRNA (ability to act as a substrate for *in vitro* methylation – see text)	1–2 weeks
Decreased methylation of CCGG sites in specific genes	>12 weeks–<10 months

[1]All parameters are compared with those in livers from age-matched animals fed a lipotrope-sufficient diet. Data are presented in Christman *et al.* (1993a,b), Chen *et al.* (1993).

DNA. Remethylation of sites that had lost methylation on both strands correlated with increased *de novo* methyltransferase activity in the liver and tumour. Interestingly, loss of p53 expression in the tumours (Pogribny *et al.*, 1997) was associated with *de novo* methylation of CpG sites in the non-CpG island promoter (Pogribny and James, 2002). However, there is little evidence that deficiency of multiple lipotropes is sufficient to produce the unique combination of increased DNA synthesis and decreased AdoMet levels in tissues other than liver. Our unpublished studies (M.L. Chen and J.K. Christman) indicated that combined deficiency of choline, methionine, folate and vitamin B_{12} actually suppresses DNA synthesis in rat lung, testes, kidney and colon, an effect similar to that observed in haematopoietic tissue in animals and humans with folate deficiency. Lipotrope deficiency also fails to cause significant alterations in AdoMet levels and/or increases in AdoHcy: AdoMet ratios in a variety of tissues including colon. The exceptions are lung and testes (Shivapurkar and Poirier, 1983). It should be noted that two non-dietary inhibitors of DNA methylation are also hepatocarcinogens (Cox and Irving, 1977; Carr *et al.*, 1984). L-Ethionine, which is non-genotoxic in the liver, inhibits DNA methylation and acts as an inhibitory analogue of AdoMet (Cox and Irving, 1977). 5-Azacytidine (ZCyd) is a cytidine analogue that can form inhibitory complexes with DNMT1 when it is incorporated into a hemi-methylated CpG site (Christman, 2002). It also has the ability to form potentially mutagenic breakdown products in DNA (Jackson-Grosby *et al.*, 1997).

Folate deficiency and inhibition of DNA methylation in liver and colon

Complete dietary folate depletion, caused by feeding of a defined folate-deficient diet along with antibiotic to repress folate synthesis by intestinal microflora, can lead to a decrease of AdoMet and increase of AdoHcy in the liver that is accompanied by loss of DNA methylation detectable within 4 weeks, but does not allow survival to a cancer end point (Wagner, 1982). Absence of dietary folate without antibiotic administration produces a more moderate folate deficiency, without the overall growth suppression noted in severe folate deficiency. Moderate folate deprivation also induces alterations in AdoMet:AdoHcy ratios in the liver, although no significant changes are observed in the colon. Even after 24 weeks moderate folate deficiency, loss of methylation, measured by the capacity of the DNA to be methylated by *SssI*, was not detected in the liver. In the colon, this degree of deficiency had no significant effect on levels of AdoMet or AdoHcy, nor was there any evidence of global or c-*myc*-specific loss of methylation (Kim *et al.*, 1995). Nevertheless, induction of DNA strand breaks and exon-specific changes in patterns of methylation of the *p53* gene were observed in both the liver and colon of these folate-depleted rats (Kim *et al.*, 1997). Moderate dietary folate deprivation for 20 weeks was not sufficient to induce pre-neoplastic changes in rat colon, but was able to promote development of dimethylhydrazine (DMH)-induced tumours (Cravo *et al.*, 1996). However, DMH treatment alone caused exon-specific loss of methylation in colonic *p53* genes. Decreasing or increasing dietary folate levels was variably effective in enhancing or preventing loss of methylation observed at these specific sites in colon DNA, but folate supplementation after DMH treatment was sufficient to provide some protection against development of macroscopic colonic neoplasia (Kim et al., 1996). These results are consistent with the epidemiological evidence that folate depletion is a risk factor in colon cancer and that elevated dietary folate can protect against colon carcinogenesis. They also suggest that the effect of folate depletion on tumour formation in colonic tissue is not due to a simple change in methylation capacity due to altered AdoMet and AdoHcy levels or ratios. The overall effect of moderate folate deficiency may be similar to that of severe methyl-deficiency in non-hepatic tissues of rodents fed a methyl deficient diet in the absence of antibiotics. If the diet does not stimulate cell division or actually reduces the rate of division of normal cells, moderate methyl deficiency would be predicted to favour inhibition of DNA methylation and enhancement of progression of cells that have already undergone genetic changes that enhance growth rates and increase resistance to apoptosis, predisposing to tumour development.

Intestinal cancer in the Apc^{min} mouse model, a different paradigm

In contrast to the studies above, which strongly support loss of DNA methylation as a primary player in cancer development, it has been demonstrated that spontaneous intestinal tumour formation in heterozygous Apc^{min} mice is inhibited when DNMT1 levels are reduced by about 50% after knockout of one $Dnmt1$ allele. This inhibition was enhanced by administration of ZdCyd. Since the frequency of loss of the intact Apc allele was the same in mice with normal and reduced levels of DNMT1, is was concluded that an event other than loss of heterozygosity (LOH) must be rate limiting (Laird et al., 1995).

This result supports the hypothesis that methylation of CpG islands leading to inactivation of tumour suppressor genes is a causal event in tumorigenesis and that inhibition of DNMT1 is sufficient to reduce or delay tumour progression. However, the effect of ZdCyd was constrained to a fairly narrow time window in postnatal development. This left open the possibility that the drug might be inducing loss of cells present at a specific stage in gut maturation that play a role in polyp formation and progression. However, later studies using Apc^{min} mice with genetically modulated expression of DNMT1 demonstrated that levels of DNMT1 lower than that found in heterozygous $Dnmt1$ knockouts completely suppressed polyp formation, providing evidence that the anti-cancer effect was due solely to the lower level of DNMT1 and/or reduced CpG island methylation (Eads et al., 2002).

The effects of dietary methyl donors were less obvious. One study of folate supplementation in weanling Apc^{min} mice suggested that increased dietary folate reduced ileal polyp formation. At 3 months, supplementation reduced aberrant crypt foci by 75–100% compared with a folate-deficient diet. However, at later time points, supplementation appeared to increase the number of ileal polyps (Song et al., 2000a,b). A second study found a positive linear correlation between AdoMet and AdoHcy in the pre-neoplastic intestine and tumour numbers in Apc^{min} mice fed a diet with adequate folate levels. A positive correlation between global DNA hypomethylation and tumour multiplicity was also found in these animals. However, the effect of folate/choline deficiency on these animals was not consistent, although increased tumour numbers were observed in mice with low levels of AdoMet and DNA hypomethylation (Sibani et al., 2002). Both studies suggest that the modulatory role of dietary methyl donors may be dependent on the transformation state of the cell.

DNMT Function: a More Complex View

It has long been recognized that DNMTs in mammals do not interact with naked DNA but with chromatin, and that histones and other non-histone proteins associated with chromatin might play an important role in temporal and spatial regulation of DNA methylation. This concept was supported by the finding that even though the bulk of nuclear DNMTs was associated with inactive chromatin, unmethylated CpG sites in inactive chromatin were not accessible to the DNMTs until chromatin proteins were removed (Creusot and Christman, 1981). The recent identification of non-histone nuclear proteins that interact with binding domains in the N-termini of DNMT1, 3a, 3b and 3L has greatly increased our understanding of how genes are silenced by DNA methylation and chromatin compaction. The picture that is emerging is that DNMTs function as components in large multiprotein complexes. The proteins within these complexes can determine where DNMTs localize in chromatin, help to enhance the silencing function of CpG methylation and influence the ability of DNMT1 to carry out maintenance methylation. There is also accumulating evidence that DNMTs may act to suppress gene expression independently of their methyltransferase function (Bachman *et al.*, 2001; Fuks *et al.*, 2001).

DNMTs, histone deacetylases (HDACs) and methylated DNA-binding proteins

Proteins with methyl-binding domains (MBDs) bind to methylated DNA and recruit or are part of complexes that contain HDACs. Histone deacetylation, in turn, is critical for chromatin compaction, and tightly packed chromatin makes DNA inaccessible to the transcription machinery. Inhibitors of histone deacetylation can enhance gene expression induced by ZdCyd and often can relieve DNMT1-mediated silencing (reviewed in Robertson, 2001; Burgers *et al.*, 2002). From the vantage point of interactions of diet and cancer, it is of interest that butyrate, which is produced by colonic flora, is an inhibitor of HDACs. In addition, at least in some cultured cells, butyrate induces differentiation (Sealy and Chalkley, 1978; Kruh, 1982) accompanied by decreased DNA methylation (Christman *et al.*, 1980).

DNMT1, 3a and 3b also bind directly to HDACs (Fig. 16.3) and can repress gene expression (Robertson, 2001). To close the circle, DNMT1 also interacts directly with MBD2 and 3 (Tatematsu *et al.*, 2000) in the MeCP1 repressor complex, which contains HDACs (Feng and Zhang, 2001). Since DNMT1 localized with HDAC2 at replication foci in late S phase, this or another complex could maintain silencing by ensuing deacetylation of histones and propagation of gene silencing in late replicating genes as nucleosomes are reassembled or repositioned on newly replicated DNA (Rountree *et al.*, 2000).

DNMT1 also binds to DMAP1 and HDAC2 to form a complex that co-localizes with DNMT1 at replication foci throughout S phase, and could perform a similar function (Rountree *et al.*, 2000).

DNMTs and chromatin-associated proteins

In addition to its effects on DNMT1 activity, Rb cooperates with DNMT1 to repress transcription from E2F-responsive promoters (Robertson *et al.*, 2000). Rb recruits DNMT1, HDACs and histone methyltransferases (HMTs) to these promoters, followed by binding of methylated lysine-binding proteins of the HP1 family.

DNMT3a binds RP58, a transcriptional repressor associated with heterochromatin (Fuks *et al.*, 2001). The N-terminal fragment of 3a also co-localizes with the heterochromatin-associated proteins HP12 and MCP2, which could be important in targeting DNA methylation in densely methylated pericentromeric chromatin or to specific DNA sequences destined for *de novo* methylation (Bachman *et al.*, 2001). However, as yet, there is no direct evidence that DNMT3a functions in these complexes since gene silencing can occur in the absence of catalytic activity.

The only complex for which methyltransferase targeting has been demonstrated contains PML–RAR, a fusion protein arising from a reciprocal translocation between the retinoic acid receptor α (RARα) on chromosome 17 and the promyelocytic leukaemia gene (PML) on chromosome 15. PML–RAR co-immunoprecipitates with DNMT1 and DNMT3a. It also co-localizes with them in cells that overexpress PML–RAR and has been shown to mediate silencing and *de novo* methylation of the RARβ2 promoter (Di Croce *et al.*, 2002). A transcription repressor, Daxx, which binds to DNMT1 and is known to co-localize with PML, has been suggested as a possible bridge for this interaction.

DNMTs and chromatin remodelling

If compaction of chromatin limits access of DNMTs to remaining unmethylated sites in DNA, then it would seem logical that chromatin-remodelling proteins could play a role in allowing DNMTs to maintain and/or initiate methylation at specific sites.

Mammalian cells contain a number of homologues of the yeast sucrose non-fermenter gene (SNF2) family. SNF2 family members disrupt histone–DNA interactions using energy from hydrolysis of ATP, allowing nucleosomes to slide along DNA to new positions (Wolffe and Pruss, 1996; Vignali et al., 2000). In humans, mutations in the plant homeodomain (PHD) of a structural homologue of SNF2, domain ATRX, leads to X-linked α-thalassaemia mental retardation (ATR-X) syndrome (Gibbons et al., 1997). Although ATRX has not yet been shown actually to be active in chromatin remodelling, it does lead to loss of methylation in ribosomal DNA repeats. In addition, normal ATRX has been localized to these repeat sequences. However, a Y chromosome-specific repeat was found to be hypermethylated (Gibbons et al., 2000).

The interactions of ATRX seem highly selective for a few regions of the genome, and it is not unlikely that it is interacting with different DNMT–HDAC complexes in these regions. The homology of its PHD domain to that of DNMT3a suggests that ATRX could act to block 3a–RP58–HDAC compaction of specific pericentromeric loci. Conversely, its role in rRNA repeat regions might be restricted to chromatin remodelling to allow maintenance methylation by a complex containing DNMT1, which does not have a PHD domain.

The second SNF2-like chromatin-remodelling protein shown to affect DNA methylation in mammals is LSH (lymphocyte-specific helicase). Although it was identified initially in lymphocytes (Jarvis et al., 1996), LSH is widely distributed in mammalian cells, with expression tightly correlated to proliferation (Geiman and Muegge, 2000; Raabe et al., 2001). LSH homozygous knockouts are born live but die within a few days. They have reduced birth weight, renal lesions and reduced numbers of lymphoid cells. Most strikingly, total genomic 5mC levels are reduced by 50–60% in all tissues. All classes of repetitive DNA sequences, which normally are highly methylated, are significantly affected. All single-copy genes examined including tissue-specific, housekeeping and imprinted genes lose methylation relative to those in normal tissue (Dennis et al., 2001). These results not only demonstrate the importance of chromatin structure in establishment and maintenance of methylation patterns but also suggest that, at least during fetal development, severe inhibition of DNA methylation does not preclude extensive division. Although it remains to be determined whether there are some regions in the DNA of $Lsh^{-/-}$ embryos where methylation is selectively retained, extensive loss of methylation also fails to preclude near normal completion of the complex series of events involved determining cell fate and establishing tissue-specific patterns of gene expression during embryogenesis.

Dietary Methyl Deficiency and Cancer: Implications of Regulatory Interaction of Chromatin and Chromatin-associated Proteins with DNMTs

The interactions between DNMTs, lysine methylated histones, locus-targeting proteins and transcriptional repression may actually provide an important clue as to why the results obtained by lowering DNA methylation levels in Apc^{min} heterozygotes and $Dnmt1$ knockout mice are so markedly different from those obtained by feeding a folate- or choline and folate-deficient diet. The knockout of $Dnmt1$ lowers the level of catalytically active DNMT1 but would not be predicted to affect the methylation status of other cellular components.

Our studies demonstrated that L-ethionine induced differentiation of human (HL-60) and murine cells (Friend erythroleukaemia, FLC) (Christman et al., 1977; Mendelsohn et al., 1980). L-Ethionine can be converted to S-adenosylethionine in some cell types including hepatocytes (Brada and Bulba, 1987) and is a potent inhibitor of AdoMet binding (Cox and Irving, 1977). Thus, its mode of action mimics the effects of dietary methyl deficiency in blocking a variety of methyltransferase reactions that utilize AdoMet. Growth in the presence of L-ethionine caused global hypomethylation of DNA in FLC, but also inhibited RNA and histone methylation (Christman et al., 1977). Interestingly lysine methylation in histones was inhibited to a much greater extent than arginine methylation (Copp, R., Ph.D. Thesis, New York University, 1981).

These results suggest that dietary methyl deficiency should have the potential to affect gene expression not only through its ability to interfere with DNA methylation but also by interfering with the assembly of repressor complexes that can function with an inactive DNMT1 but require methylated histones or other as yet unidentified methylated nuclear proteins for assembly.

In contrast, in $Dnmt1^{+/-}$ knockout mice, the level of functional DNMT1 is reduced, but methylation of other proteins and RNA, etc., is unaffected, and some formation of repressor complexes can occur.

A similar situation may account for the difference between $Lsh^{-/-}$ mice and $Dnmt^{-/-}$ mice. The $Lsh^{-/-}$ mice have extensive loss of DNA methylation but presumably maintain normal levels of wild-type DNMT1 protein as well as normal methylation of RNA, histones and other proteins, and can still carry out the changes in gene expression required for embryonic development. $Dnmt1^{-/-}$ mice do not express wild-type DNMT1 protein and lose the capacity to carry out gene silencing. They die at day 10.5 of embryogenesis (Li et al., 1992).

While there is still much to learn about the role of DNMT interactions with nuclear proteins and the establishment of methylation patterns and the regulation of gene expression, it is clear that the next few years should provide many opportunities for defining the links between dietary folate/methyl deficiency and the development of birth defects, cancer and other human diseases.

Acknowledgements

I wish to express my gratitude to Elsie Wainfan, friend and mentor, for an introduction to the importance of dietary methyl deficiency in carcinogenesis and her invaluable contributions to our joint studies on dietary methyl deficiency and nucleic acid methylation. I am indebted to James Finkelstein for his help in understanding the regulatory role played by AdoMet and AdoHcy in one-carbon metabolism and homeostasis of AdoHcy and homocysteine levels. I am also grateful to the members of my research group for their assistance in preparation of tables and figures (D. Klinkebiel and M. Boland) and in preparing the bibliography (E. Uzvolgyi, M. Boland, A. Farrell and D. Van Bemmel).

J.K.C. has been supported for studies on diet, DNA methylation and cancer by grants from the American Institute for Cancer Research (94A37 REN and 96A115-REV) and for studies on DNA methylation, DNA methyltransferase inhibitors and cancer by the Department of the Army (DAMD-17-98-1-8215 and DAMD-17-98-1-8215) and the NIH (R21 CA91315).

Endnote

When reading the literature on DNA methylation, it is important to carefully evaluate the methods used to determine that a gene has become 'hypermethylated' or has 'lost methylation' or that 'global' changes in DNA methylation have occurred and been reversed. The methods used to define these changes may allow examination of only a few residues and/or may bias detection toward loss or gain of methylation in CpG-rich regions and regions that contain cleavage sites for methylation-sensitive endonucleases. The 'gold standard' for evaluating DNA methylation in a specific region in DNA is bisulphite sequencing (Warnecke et al., 2002) which allows a detailed analysis of the methylation status of each cytosine over a region of approximately 300 bp at single molecule resolution. With careful microdissection or laser capture methods to ensure homogeneity of cell type, one can draw meaningful conclusions as to whether any particular region in the DNA of a pre-neoplastic or tumour cell has become hypo- or hypermethylated relative to normal tissue, and determine which of the observed changes commonly are associated with tumour initiation progression and the temporal relationship of these changes to loss or gain of gene function. Many other methods have been developed that allow rapid throughput of large numbers of samples. These methods are used commonly to determine whether specific DNA sequences are hypo- or hypermethylated, and may provide valuable diagnostic markers for cancer staging. However, most focus on a few sites in a 'likely' region such as a CpG island whose methylation status may or may not be related to gene activity or cell function.

'Global loss of methylation' means that the total level of 5mC in DNA is lower in one sample relative to a reference sample. Although the 5mC content of DNA may actually be measured by what are now fairly sensitive methods (mass spectrometry), the surrogate approach often is to compare the amount of [^{3}H]methyl from AdoMet by $SssI$, an enzyme that is specific for CpG sites and methylates both completely unmethylated and hemimethylated sites. Without further fractionation of DNA or chromatin, detected 'loss of methylation' by both methods gives no information as to the regions of the genome that are affected. Similarly, these methods cannot distinguish between loss of 5mC that occurs on both strands of a CpG site (complete 'demethylation') or loss of 5mC that reflects a failure to complete maintenance methylation leaving the 5mC on one strand to mark a site where remethylation can occur (Fig. 16.4). Maintenance methyltransferase (DNMT1) has been used to evaluate the extent to which incomplete remethylation has occurred (Christman et al., 1993b). Use of methods dependent on methylation-sensitive restriction endonucleases allows detection of loss of methylation on both strands or gain of methylation at a subset of previously unmethylated sites, but in most studies is limited to examination of a few sites (Christman, 1984). Newer methods for examining the methylation status of large numbers of CpG island sequences by microarray or two-dimensional gel analysis (Yan et al., 2001) have greatly improved our ability to examine global methylation changes but may underestimate the contribution of methylation changes in less CpG-rich regions (Rein et al., 1998, provide a good review of this area). These considerations are particularly relevant to studies of DNA methylation in cells that undergo a normal cycle of maturation from stem cell precursors, terminal differentiation and removal (leukocytes, epithelial cells, etc.) When such cells become hypomethylated as a result of dietary methyl deficiency and remethylated after supplementation, it is important to distinguish not only whether the 'hypomethylation' is due to failure to methylate newly replicated DNA or represents an actual change in the pattern of methylation, but also whether 'remethylation' is actually completion of maintenance methylation, restoration of the original pattern of methylation, replacement of the cells with 'hypomethylated' DNA by a new population of cells that have been differentiated under conditions of dietary methyl sufficiency, or selective outgrowth of a population of cells that have maintained critical methylation patterns

References

Aapola, U., Kawasaki, K., Scott, H.S., Ollila, J., Vihinen, M., Heino, M., Shintani, A., Minoshima, S., Krohn, K., Antonarakis, S.E., Shimizu, N., Kudoh, J. and Peterson, P. (2000) Isolation and initial characterization of a novel zinc finger gene, DNMT3L, on 21q22.3, related to the cytosine-5-methyltransferase 3 gene family. *Genomics* 65, 293–298.

Aasland, R., Gibson, T.J. and Stewart, A.F. (1995) The PHD finger: implications for chromatin-mediated transcriptional regulation. PG-56-9. *Trends in Biochemical Science* 20(2).

AICR (1997) *Food, Nutrition and the Prevention of Cancer: a Global Perspective.* American Institute for Cancer Research, Washington, DC.

Aoki, K., Meng, G., Suzuki, K., Takashi, T., Kameoka, Y., Nakahara, K., Ishida, R. and Kasai, M. (1998) RP58 associates with condensed chromatin and mediates a sequence-specific transcriptional repression. *Journal of Biological Chemistry* 273, 26698–26704.

Aoki, A., Svetake, I., Miyagawa, J., Fujio, T., Chijiwa, T., Sasaki, H. and Tajima, S. (2001) Enzymatic properties of denovotype mouse (cystosine-5) methyltransferases. *Nucleic Acids Research* 29, 2506–3512.

Bachman, K.E., Rountree, M.R. and Baylin, S.B. (2001) Dnmt3a and Dnmt3b are transcriptional repressors that exhibit unique localization properties to heterochromatin. *Journal of Biological Chemistry* 276, 32282–32287.

Baylin, S.B., Herman, J.G., Graff, J.R., Vertino, P.M. and Issa, J.P. (1998) Alterations in DNA methylation: a fundamental aspect of neoplasia. *Advances in Cancer Research* 72, 141–196.

Benito, E., Stiggelbout, A., Bosch, F.X., Obrador, A., Kaldor, J., Mulet, M. and Munoz, N. (1991) Nutritional factors in colorectal cancer risk: a case-control study in Majorca. *International Journal of Cancer* 49, 161–167.

Benito, E., Cabeza, E., Moreno, V., Obrador, A. and Bosch, F.X. (1993) Diet and colorectal adenomas: a case-control study in Majorca. *International Journal of Cancer* 55, 213–219.

Bestor, T., Laudano, A., Mattaliano, R. and Ingram, V. (1988) Cloning and sequencing of a cDNA encoding DNA methyltransferase of mouse cells. The carboxyl-terminal domain of the mammalian enzymes is related to bacterial restriction methyltransferases. *Journal of Molecular Biology* 203, 971–983.

Bestor, T.H. (1992) Activation of mammalian DNA methyltransferase by cleavage of a Zn binding regulatory domain. *EMBO Journal* 11, 2611–2617.

Bhattacharya, S.K., Ramchandani, S., Cervoni, N. and Szyf, M. (1999) A mammalian protein with specific demethylase activity for mCpG DNA. *Nature* 397, 579–583.

Bigey, P., Ramchandani, S., Theberge, J., Araujo, F.D. and Szyf, M. (2000) Transcriptional regulation of the human DNA methyltransferase (dnmt1) gene. *Gene* 242, 407–418.

Bird, A.P. (1986) CpG-rich islands and the function of DNA methylation. *Nature* 321, 209–213.

Bird, A.P. and Wolffe, A.P. (1999) Methylation-induced repression – belts, braces, and chromatin. *Cell* 99, 451–454.

Bird, C.L., Swendseid, M.E., Witte, J.S., Shikany, J.M., Hunt, I.F., Frankl, H.D., Lee, E.R., Longnecker, M.P. and Haile, R.W. (1995) Red cell and plasma folate, folate consumption, and the risk of colorectal adenomatous polyps. *Cancer Epidemiology Biomarkers Prev* 4, 709–714.

Blount, B.C., Mack, M.M., Wehr, C.M., MacGregor, J.T., Hiatt, R.A., Wang, G., Wickramasinghe, S.N., Everson, R.B. and Ames, B.N. (1997) Folate deficiency causes uracil misincorporation into human DNA and chromosome breakage: implications for cancer and neuronal damage. *Proceedings of the National Academy of Sciences USA* 94, 3290–3295.

Bonfils, C., Beaulieu, N., Chan, E., Cotton-Montpetit, J. and MacLeod, A.R. (2000) Characterization of the human DNA methyltransferase splice variant Dnmt1b. *Journal of Biological Chemistry* 275, 10754–10760.

Bourc'his, D., Xu, G.L., Lin, C.S., Bollman, B. and Bestor, T.H. (2001) Dnmt3L and the establishment of maternal genomic imprints. *Science* 294, 2536–2539.

Boutron-Ruault, M.C., Senesse, P., Faivre, J., Couillault, C. and Belghiti, C. (1996) Folate and alcohol intakes: related or independent roles in the adenoma carcinoma sequence? *Nutrition and Cancer* 26, 337–346.

Brada, Z. and Bulba, S. (1987) The synthesis of the adenosyl-moiety of S-adenosylethionine in liver of rats fed DL-ethionine. *Research Communications in Chemical Pathology and Pharmacology* 56, 133–136.

Brank, A.S., Van Bemmel, D.M. and Christman, J.K. (2002) Optimization of baculovirus-mediated expression and purification of hexahistidine-tagged murine DNA (cytosine-C5)-methyltransferase-1 in *Spodoptera frugiperda* 9 cells. *Protein Expression and Purification* 25, 31–40.

Burgers, W.A., Fuks, F. and Kouzarides, T. (2002) DNA methyltransferases get connected to chromatin. *Trends in Genetics* 18, 275–277.

Carr, B.I., Reilly, J.G., Smith, S.S., Winberg, C. and Riggs, A. (1984) The tumorigenicity of 5-azacytidine in the male Fischer rat. *Carcinogenesis* 5, 1583–1590.

Chadwick, L.H., McCandless, S.E., Silverman, G.L., Schwartz, S., Westaway, D. and Nadeau, J.H. (2000) Betaine-homocysteine methyltransferase-2: cDNA cloning, gene sequence, physical mapping, and expression of the human and mouse genes. *Genomics* 70, 66–73.

Chedin, F., Lieber, M.R. and Hsieh, C.L. (2002) The DNA methyltransferase-like protein DNMT3L stimulates de novo methylation by Dnmt3a. *Proceedings of the National Academy of Sciences USA* 99, 16916–16921.

Chen, J., Giovannucci, E., Kelsey, K., Rimm, E.B., Stampfer, M.J., Colditz, G.A., Spiegelman, D., Willett, W.C. and Hunter, D.J. (1996) A methylenetetrahydrofolate reductase polymorphism and the risk of colorectal cancer. *Cancer Research* 56, 4862–4864.

Chen, M.-L., Abileah, S., Wanfan, E. and Christman, J.K. (1993) Influence of folate and vitamin B12 on the effects of dietary lipotrope deficiency. *Proceedings of the American Association for Cancer Research* 34, 131.

Chen, T., Ueda, Y., Xie, S. and Li, E. (2002) A novel Dnmt3a isoform produced from an alternative promoter localizes to euchromatin and its expression correlates with active de novo methylation. *Journal of Biological Chemistry*.

Choi, S.-W., Stickel, F., Baik, H., Kim, Y., Seitz, H., amd Mason, J. (1999) Chronic alcohol consumption induces genomic but not p53-specific DNA hypomethylation in rat colon. *Journal of Nutrition* 129, 1945–1950.

Christman, J.K. (1984) DNA methylation in Friend erythroleukemia cells: the effects of chemically induced differentiation and of treatment with inhibitors of DNA methylation. *Current Topics in Microbiology and Immunology* 108, 49–78.

Christman, J.K. (1995a) Dietary effects on DNA methylation: Do they account for the hepatocarcinogenic properties of lipotrope deficient diets? *Advances in Experimental Medicine and Biology* 369, 141–154.

Christman, J.K. (1995b) Lipotrope deficiency and persistent changes in DNA methylation. *Advances in Experimental Medicine and Biology* 375, 97–106.

Christman, J.K. (2002) 5-Azacytidine and 5-aza-2'-deoxycytidine as inhibitors of DNA methylation: mechanistic studies and their implications for cancer therapy. *Oncogene* 21, 5483–5495.

Christman, J.K., Price, P., Pedrinan, L. and Acs, G. (1977) Correlation between hypomethylation of DNA and expression of globin genes in Friend erythroleukemia cells. *European Journal of Biochemistry* 81, 53–61.

Christman, J.K., Weich, N., Schoenbrun, B., Schneiderman, N. and Acs, G. (1980) Hypomethylation of DNA during differentiation of Friend erythroleukemia cells. *Journal of Cell Biology* 86, 366–370.

Christman, J.K., Sheikhnejad, G., Dizik, M., Abileah, S. and Wainfan, E. (1993a) Reversibility of changes in nucleic acid methylation and gene expression induced in rat liver by severe dietary methyl deficiency. *Carcinogenesis* 14, 551–557.

Christman, J.K., Cheng, M.-L., Sheikhnejad, G., Dizik, M., Alileah, S., and Wainfan, E. (1993b) Methyl deficiency, DNA methylation, and cancer: studies on the reversibility of the effects of lipotrope-deficient diet. *Journal of Nutrition and Biochemistry* 4, 672–680.

Christman, J.K., Sheikhnejad, G., Marasco, C.J. and Sufrin, J.R. (1995) 5-Methyl-2'-deoxycytidine in single-stranded DNA can act in cis to signal de novo DNA methylation. *Proceedings of the National Academy of Sciences USA* 92, 7347–7351.

Chuang, L.S., Ian, H.I., Koh, T.W., Ng, H.H., Xu, G. and Li, B.F. (1997) Human DNA-(cytosine-5) methyltransferase-PCNA complex as a target for p21WAF1. *Science* 277, 1996–2000.

Copeland, D.H. and Salmon, W.D. (1946) The occurrence of enoplasms in the liver, lungs, and other tissues of rats as a result of prolonged choline deficiency. *American Journal of Pathology* 22, 1059.

Costello, J.F. and Plass, C. (2001) Methylation matters. *Journal of Medical Genetics* 38, 285–303.

Costello, J.F., Smiraglia, D.J. and Plass, C. (2002) Restriction landmark genome scanning. *Methods* 27, 144–149.

Cox, R. and Irving, C.C. (1977) Inhibition of DNA methylation by S-adenosylethionine with the production of methyl-deficient DNA in regenerating rat liver. *Cancer Research* 37, 222–225.

Cravo, M.L., Mason, J.B., Dayal, Y., Hutchinson, M., Smith, D., Selhub, J. and Rosenberg, I.H. (1992) Folate deficiency enhances the development of colonic neoplasia in dimethylhydrazine-treated rats. *Cancer Research* 52, 5002–5006.

Cravo, M.L., Pinto, A.G., Chaves, P., Cruz, J.A., Lage, P., Nobre Leitao, C. and Costa Mira, F. (1998) Effect of folate supplementation on DNA methylation of rectal mucosa in patients with colonic adenomas: correlation with nutrient intake. *Clinical Nutrition* 17, 45–49.

Creusot, F. and Christman, J.K. (1981) Localization of DNA methyltransferase in the chromatin of Friend erythroleukemia cells. *Nucleic Acids Research* 9, 5359–5381.

Dennis, K., Fan, T., Geiman, T., Yan, Q. and Muegge, K. (2001) Lsh, a member of the SNF2 family, is required for genome-wide methylation. *Genes and Development* 15, 2940–2944.

Di Croce, L., Raker, V.A., Corsaro, M., Fazi, F., Fanelli, M., Faretta, M., Fuks, F., Lo Coco, F., Kouzarides, T., Nervi, C., Minucci, S. and Pelicci, P.G. (2002) Methyltransferase recruitment and DNA hypermethylation of target promoters by an oncogenic transcription factor. *Science* 295, 1079–1082.

Dizik, M., Christman, J.K. and Wainfan, E. (1991) Alterations in expression and methylation of specific genes in livers of rats fed a cancer promoting methyl-deficient diet. *Carcinogenesis* 12, 1307–1312.

Doerfler, W. (1983) DNA methylation and gene activity. *Annals of Biochemistry* 52, 93–124.

Eads, C.A., Danenberg, K.D., Kawakami, K., Saltz, L.B., Danenberg, P.V. and Laird, P.W. (1999) CpG island hypermethylation in human colorectal tumors is not associated with DNA methyltransferase overexpression. *Cancer Research* 59, 2302–2306.

Eads, C.A., Nickel, A.E. and Laird, P.W. (2002) Complete genetic suppression of polyp formation and reduction of CpG-island hypermethylation in Apc(Min/+) Dnmt1-hypomorphic mice. *Cancer Research* 62, 1296–1299.

Ehrlich, M. (2002) DNA methylation in cancer: Too much, but also too little. *Oncogene* 21, 5400–5413.

Ehrlich, M. and Wang, R.Y. (1981) 5-Methylcytosine in eukaryotic DNA. *Science* 212, 1350–1357.

Eloranta, T.O. and Raina, A.M. (1977) S-adenosylmethionine metabolism and its relation to polyamine synthesis in rat liver. Effect of nutritional state, adrenal function, some drugs and partial hepatectomy. *Biochemistry Journal* 168, 179–185.

Esteller, M., Garcia, A., Martinez-Palones, J.M., Xercavins, J. and Reventos, J. (1997) Germ line polymorphisms in cytochrome-P450 1A1 (C4887 CYP1A1) and methylenetetrahydrofolate reductase (MTHFR) genes and endometrial cancer susceptibility. *Carcinogenesis* 18, 2307–2311.

Esteller, M., Fraga, M.F., Paz, M.F., Campo, E., Colomer, D., Novo, F.J., Calasanz, M.J., Galm, O., Guo, M., Benitez, J. and Herman, J.G. (2002) Cancer epigenetics and methylation. *Science* 297, 1807–1808.

Fatemi, M., Hermann, A., Pradhan, S. and Jeltsch, A. (2001) The activity of the murine DNA methyltransferase Dnmt1 is controlled by interaction of the catalytic domain with the N-terminal part of the enzyme leading to an allosteric activation of the enzyme after binding to methylated DNA. *Journal of Molecular Biology* 309, 1189–1199.

Fauman, E.B. (1999) *S-Adenosylmethionine-Dependent Methyltransferases: Structures and Functions*. World Scientific Publishing.

Fearon, E.R. and Vogelstein, B. (1990) A genetic model for colorectal tumorigenesis. *Cell* 61, 759–767.

Feinberg, A.P. and Vogelstein, B. (1983) Hypomethylation of ras oncogenes in primary human cancers. *Biochemistry and Biophysics Research Communications* 111, 47–54.

Feinberg, A.P., Gehrke, C.W., Kuo, K.C. and Ehrlich, M. (1988) Reduced genomic 5-methylcytosine content in human colonic neoplasia. *Cancer Research* 48, 1159–1161.

Feinberg, A., Cui, H. and Ohlsson, R. (2002) DNA methylation and genomic imprinting: insights from cancer into epigenetic mechanisms. *Seminars in Cancer Biology* 12, 389–395.

Fenech, M. (2001) The role of folic acid and Vitamin B12 in genomic stability of human cells. *Mutation Research* 475, 57–67.

Feng, Q. and Zhang, Y. (2001) The MeCP1 complex represses transcription through preferential binding, remodeling, and deacetylating methylated nucleosomes. *Genes and Development* 15, 827–832.

Ferraroni, M., La Vecchia, C., D'Avanzo, B., Negri, E., Franceschi, S. and Decarli, A. (1994) Selected micronutrient intake and the risk of colorectal cancer. *British Journal of Cancer* 70, 1150–1155.

Finkelstein, J.D. (1990) Methionine metabolism in mammals. *Journal of Nutritional Biochemistry* 1, 228–237.

Finkelstein, J.D. (1998) The metabolism of homocysteine: pathways and regulation. *European Journal of Pediatrics* 157 (Supplement 2), S40–S44.

Finkelstein, J.D. (2000) Pathways and regulation of homocysteine metabolism in mammals. *Seminars in Thrombosis and Hemostasis* 26, 219–225.

Finkelstein, J.D., Cello, J.P. and Kyle, W.E. (1974) Ethanol-induced changes in methionine metabolism in rat liver. *Biochemistry and Biophysics Research Communications* 61, 525–531.

Fowler, B.M., Giuliano, A.R., Piyathilake, C., Nour, M. and Hatch, K. (1998) Hypomethylation in cervical tissue: is there a correlation with folate status?' *Cancer Epidemiology Biomarkers Prev* 7, 901–906.

Franco, R.F., Simoes, B.P., Tone, L.G., Gabellini, S.M., Zago, M.A. and Falcao, R.P. (2001) The methylenetetrahydrofolate reductase C677T gene polymorphism decreases the risk of childhood acute lymphocytic leukaemia. *British Journal of Haematology* 115, 616–618.

Freudenheim, J.L., Graham, S., Marshall, J.R., Haughey, B.P., Cholewinski, S. and Wilkinson, G. (1991) Folate intake and carcinogenesis of the colon and rectum. *International Journal of Epidemiology* 20, 368–374.

Freudenheim, J.L., Marshall, J.R., Vena, J.E., Laughlin, R., Brasure, J.R., Swanson, M.K., Nemoto, T. and Graham, S. (1996) Premenopausal breast cancer risk and intake of vegetables, fruits, and related nutrients. *Journal of National Cancer Institute* 88, 340–348.

Friso, S., Choi, S.W., Girelli, D., Mason, J.B., Dolnikowski, G.G., Bagley, P.J., Olivieri, O., Jacques, P.F., Rosenberg, I.H., Corrocher, R. and Selhub, J. (2002) A common mutation in the 5,10-methylenetetrahydrofolate reductase gene affects genomic DNA methylation through an interaction with folate status. *Proceedings of the National Academy of Sciences USA* 99, 5606–5611.

Frosst, P., Blom, H.J., Milos, R., Goyette, P., Sheppard, C.A., Matthews, R.G., Boers, G.J., den Heijer, M., Kluijtmans, L.A., van den Heuvel, L.P. and Rozen, R. (1995) A candidate genetic risk factor for vascular disease: a common mutation in methylenetetrahydrofolate reductase. *Nature Genetics* 10, 111–113.

Fuchs, C.S., Willett, W.C., Colditz, G.A., Hunter, D.J., Stampfer, M.J., Speizer, F.E. and Giovannucci, E.L. (2002) The influence of folate and multivitamin use on the familial risk of colon cancer in women. *Cancer Epidemiology Biomarkers Prev* 11, 227–234.

Fuks, F., Burgers, W.A., Brehm, A., Hughes-Davies, L. and Kouzarides, T. (2000) DNA methyltransferase Dnmt1 associates with histone deacetylase activity. *Nature Genetics* 24, 88–91.

Fuks, F., Burgers, W.A., Godin, N., Kasai, M. and Kouzarides, T. (2001) Dnmt3a binds deacetylases and is recruited by a sequence-specific repressor to silence transcription. *EMBO Journal* 20, 2536–2544.

Gallinari, P. and Jiricny, J. (1996) A new class of uracil-DNA glycosylases related to human thymine-DNA glycosylase. *Nature* 383, 735–738.

Garay, C.A. and Engstrom, P.F. (1999) Chemoprevention of colorectal cancer: dietary and pharmacologic approaches. *Oncology (Huntingt)* 13, 89–97.

Geiman, T.M. and Muegge, K. (2000) Lsh, an SNF2/helicase family member, is required for proliferation of mature T lymphocytes. *Proceedings of the National Academy of Sciences USA* 97, 4772–4777.

Gershoni-Baruch, R., Dagan, E., Israeli, D., Kasinetz, L., Kadouri, E. and Friedman, E. (2000) Association of the C677T polymorphism in the MTHFR gene with breast and/or ovarian cancer risk in Jewish women. *European Journal of Cancer* 36, 2313–2316.

Gibbons, R.J., Bachoo, S., Picketts, D.J., Aftimos, S., Asenbauer, B., Bergoffen, J., Berry, S.A., Dahl, N., Fryer, A., Keppler, K., Kurosawa, K., Levin, M.L., Masuno, M., Neri, G., Pierpont, M.E., Slaney, S.F. and Higgs, D.R. (1997) Mutations in transcriptional regulator ATRX establish the functional significance of a PHD-like domain. *Nature Genetics* 17, 146–148.

Gibbons, R.J., McDowell, T.L., Raman, S., O'Rourke, D.M., Garrick, D., Ayyub, H. and Higgs, D.R. (2000) Mutations in ATRX, encoding a SWI/SNF-like protein, cause diverse changes in the pattern of DNA methylation. *Nature Genetics* 24, 368–371.

Giovannucci, E. and Willett, W.C. (1994) Dietary factors and risk of colon cancer. *Annals of Medicine* 26, 443–452.

Giovannucci, E., Stampfer, M.J., Colditz, G.A., Rimm, E.B., Trichopoulos, D., Rosner, B.A., Speizer, F.E. and Willett, W.C. (1993) Folate, methionine, and alcohol intake and risk of colorectal adenoma. *Journal of the National Cancer Institute* 85, 875–884.

Giovannucci, E., Rimm, E.B., Ascherio, A., Stampfer, M.J., Colditz, G.A. and Willett, W.C. (1995) Alcohol, low-methionine–low-folate diets, and risk of colon cancer in men. *Journal of the National Cancer Institute* 87, 265–273.

Giovannucci, E., Stampfer, M.J., Colditz, G.A., Hunter, D.J., Fuchs, C., Rosner, B.A., Speizer, F.E. and Willett, W.C. (1998) Multivitamin use, folate, and colon cancer in women in the Nurses' Health Study. *Annals of Internal Medicine* 129, 517–524.

Glynn, S.A., Albanes, D., Pietinen, P., Brown, C.C., Rautalahti, M., Tangrea, J.A., Gunter, E.W., Barrett, M.J., Virtamo, J. and Taylor, P.R. (1996) Colorectal cancer and folate status: a nested case-control study among male smokers. *Cancer Epidemiological Biomarkers Prev* 5, 487–494.

Goelz, S.E., Vogelstein, B., Hamilton, S.R. and Feinberg, A.P. (1985) Hypomethylation of DNA from benign and malignant human colon neoplasms. *Science* 228, 187–190.

Goto, T. and Monk, M. (1998) Regulation of X-chromosome inactivation in development in mice and humans. *Microbiology and Molecular Biology Reviews* 62, 362–378.

Gowher, H. and Jeltsch, A. (2001) Enzymatic properties of recombinant Dnmt3a DNA methyltransferase from mouse: the enzyme modifies DNA in a non-processive manner and also methylates non-CpG sites. *Journal of Molecular Biology* 309, 1201–1208.

Graham, S., Hellmann, R., Marshall, J., Freudenheim, J., Vena, J., Swanson, M., Zielezny, M., Nemoto, T., Stubbe, N. and Raimondo, T. (1991) Nutritional epidemiology of postmenopausal breast cancer in western New York. *American Journal of Epidemiology* 134, 552–566.

Gunter, E., Bowman, B., Caudill, S., Twite, D., Adams, M. and Sampson, E. (1996) Results of an international round robin for serum and whole-blood folate. *Clinical Chemistry* 42, 1689–1694.

Hardeland, U., Bentele, M., Lettieri, T., Steinacher, R., Jiricny, J. and Schar, P. (2001) Thymine DNA glycosylase. *Progress in Nucleic Acid Research and Molecular Biology* 68, 235–253.

Homann, N., Tillonen, J. and Salaspuro, M. (2000) Microbially produced acetaldehyde from ethanol may increase the risk of colon cancer via folate deficiency. *International Journal of Cancer* 86, 169–173.

Hoover, K.L., Lynch, P.H. and Poirier, L.A. (1984) Profound postinitiation enhancement by short-term severe methionine, choline, vitamin B12, and folate deficiency of hepatocarcinogenesis in F344 rats given a single low-dose diethylnitrosamine injection. *Journal of the National Cancer Institute* 73, 1327–1336.

Hsieh, C.L. (1999) In vivo activity of murine de novo methyltransferases, Dnmt3a and Dnmt3b. *Molecular Cell Biology* 19, 8211–8218.

Hsu, D.W., Lin, M.J., Lee, T.L., Wen, S.C., Chen, X. and Shen, C.K. (1999) Two major forms of DNA (cytosine-5) methyltransferase in human somatic tissues. *Proceedings of the National Academy of Sciences USA* 96, 9751–9756.

Im, Y.S., Chiang, P.K. and Cantoni, G.L. (1979) Guanidoacetate methyltransferase. Purification and molecular properties. *Journal of Biological Chemistry* 254, 11047–11050.

Institute of Medicine and National Academy of Science, USA (2000) *Dietary Reference Intakes for Thiamin, Riboflavin, Niacin, Vitamin B6, Folate, Vitamin B12, Pantothenic Acid, Biotin and Choline*. National Academy Press, Washington, DC.

Issa, J.P. (2000a) CpG-island methylation in aging and cancer. *Current Topics in Microbiology and Immunology* 249, 101–118.

Issa, J.P. (2000b) The epigenetics of colorectal cancer. *Annals of the New York Academy of Science* 910, 140–53; discussion 153–155.

Jackson-Grusby, L., Laird, P.W., Magge, S.N., Moeller, B.J. and Jaenisch, R. (1997) Mutagenicity of 5-aza-2′-deoxycytidine is mediated by the mammalian DNA methyltransferase. *Proceedings of the National Academy of Sciences USA* 94, 4681–4685.

Jacob, R.A., Gretz, D.M., Taylor, P.C., James, S.J., Pogribny, I.P., Miller, B.J., Henning, S.M. and Swendseid, M.E. (1998) Moderate folate depletion increases plasma homocysteine and decreases lymphocyte DNA methylation in postmenopausal women. *Journal of Nutrition* 128, 1204–1212.

James, S.J. and Yin, L. (1989) Diet-induced DNA damage and altered nucleotide metabolism in lymphocytes from methyl-donor-deficient rats. *Carcinogenesis* 10, 1209–1214.

Jarvis, C.D., Geiman, T., Vila-Storm, M.P., Osipovich, O., Akella, U., Candeias, S., Nathan, I., Durum, S.K. and Muegge, K. (1996) A novel putative helicase produced in early murine lymphocytes. *Gene* 169, 203–207.

Jones, P.A. and Takai, D. (2001) The role of DNA methylation in mammalian epigenetics. *Science* 293, 1068–1070.

Jost, J.P., Oakeley, E.J., Zhu, B., Benjamin, D., Thiry, S., Siegmann, M. and Jost, Y.C. (2001) 5-Methylcytosine DNA glycosylase participates in the genome-wide loss of DNA methylation occurring during mouse myoblast differentiation. *Nucleic Acids Research* 29, 4452–4461.

Kato, I., Dnistrian, A.M., Schwartz, M., Toniolo, P., Koenig, K., Shore, R.E., Akhmedkhanov, A., Zeleniuch-Jacquotte, A. and Riboli, E. (1999) Serum folate, homocysteine and colorectal cancer risk in women: a nested case-control study. *British Journal of Cancer* 79, 1917–1922.

Kenyon, S.H., Nicolaou, A. and Gibbons, W.A. (1998) The effect of ethanol and its metabolites upon methionine synthase activity in vitro. *Alcohol* 15, 305–309.

Kim, Y.I., Giuliano, A., Hatch, K.D., Schneider, A., Nour, M.A., Dallal, G.E., Selhub, J. and Mason, J.B. (1994) Global DNA hypomethylation increases progressively in cervical dysplasia and carcinoma. *Cancer* 74, 893–899.

Kim, Y.I., Christman, J.K., Fleet, J.C., Cravo, M.L., Salomon, R.N., Smith, D., Ordovas, J., Selhub, J.

and Mason, J.B. (1995) Moderate folate deficiency does not cause global hypomethylation of hepatic and colonic DNA or c-myc-specific hypomethylation of colonic DNA in rats. *American Journal of Clinical Nutrition* 61, 1083–1090.

Kim, Y.I., Salomon, R.N., Graeme-Cook, F., Choi, S.W., Smith, D.E., Dallal, G.E. and Mason, J.B. (1996) Dietary folate protects against the development of macroscopic colonic neoplasia in a dose responsive manner in rats. *Gut* 39, 732–740.

Kim, Y.-I., Pogribny, I., Basnakian, A., Miller, J., Selhub, J., James, S., and Mason, J. (1997) Folate deficiency in rats induces DNA strand breaks and hypomethylation within the p53 tumor suppressor gene. *American Journal of Clinical Nutrition* 65, 46–52.

Kim, Y.I., Fawaz, K., Knox, T., Lee, Y.M., Norton, R., Libby, E. and Mason, J.B. (2001) Colonic mucosal concentrations of folate are accurately predicted by blood measurements of folate status among individuals ingesting physiologic quantities of folate. *Cancer Epidemiology Biomarkers Prev* 10, 715–719.

Knudson, A.G. Jr (1971) Mutation and cancer: statistical study of retinoblastoma. *Proceedings of the National Academy of Sciences USA* 68, 820–823.

Kotb, M., Mudd, S.H., Mato, J.M., Geller, A.M., Kredich, N.M., Chou, J.Y. and Cantoni, G.L. (1997) Consensus nomenclature for the mammalian methionine adenosyltransferase genes and gene products. *Trends in Genetics* 13, 51–52.

Kruh, J. (1982) Effects of sodium butyrate, a new pharmacological agent, on cells in culture. *Molecular Cell Biochemistry* 42, 65–82.

Laird, P.W., Jackson-Grusby, L., Fazeli, A., Dickinson, S.L., Jung, W.E., Li, E., Weinberg, R.A. and Jaenisch, R. (1995) Suppression of intestinal neoplasia by DNA hypomethylation. *Cell* 81, 197–205.

Lapeyre, J.N. and Becker, F.F. (1979) 5-Methylcytosine content of nuclear DNA during chemical hepatocarcinogenesis and in carcinomas which result. *Biochemical and Biophysical Research Communications* 87, 698–705.

Lauster, R., Trautner, T.A. and Noyer-Weidner, M. (1989) Cytosine-specific type II DNA methyltransferases. A conserved enzyme core with variable target-recognizing domains. *Journal of Molecular Biology* 206, 305–312.

Lee, P.J., Washer, L.L., Law, D.J., Boland, C.R., Horon, I.L. and Feinberg, A.P. (1996) Limited up-regulation of DNA methyltransferase in human colon cancer reflecting increased cell proliferation. *Proceedings of the National Academy of Sciences USA* 93, 10366–10370.

Leonhardt, H. and Cardoso, M.C. (2000) DNA methylation, nuclear structure, gene expression and cancer. *Journal of Cell Biochemistry* (Supplement 35), 78–83.

Leonhardt, H., Page, A.W., Weier, H.U. and Bestor, T.H. (1992) A targeting sequence directs DNA methyltransferase to sites of DNA replication in mammalian nuclei. *Cell* 71, 865–873.

Levine, A.J., Siegmund, K.D., Ervin, C.M., Diep, A., Lee, E.R., Frankl, H.D. and Haile, R.W. (2000) The methylenetetrahydrofolate reductase 677C→T polymorphism and distal colorectal adenoma risk. *Cancer Epidemiology Biomarkers Prev* 9, 657–663.

Li, E., Bestor, T.H. and Jaenisch, R. (1992) Targeted mutation of the DNA methyltransferase gene results in embryonic lethality. *Cell* 69, 915–926.

Liang, G., Chan, M.F., Tomigahara, Y., Tsai, Y.C., Gonzales, F.A., Li, E., Laird, P.W. and Jones, P.A. (2002) Cooperativity between DNA methyltransferases in the maintenance methylation of repetitive elements. *Molecular Cellular Biology* 22, 480–491.

Liu, Y., Oakeley, E.J., Sun, L. and Jost, J.P. (1998) Multiple domains are involved in the targeting of the mouse DNA methyltransferase to the DNA replication foci. *Nucleic Acids Research* 26, 1038–1045.

Ma, J., Stampfer, M.J., Giovannucci, E., Artigas, C., Hunter, D.J., Fuchs, C., Willett, W.C., Selhub, J., Hennekens, C.H. and Rozen, R. (1997) Methylenetetrahydrofolate reductase polymorphism, dietary interactions, and risk of colorectal cancer. *Cancer Research* 57, 1098–1102.

Magdinier, F. and Wolffe, A.P. (2001) Selective association of the methyl-CpG binding protein MBD2 with the silent p14/p16 locus in human neoplasia. *Proceedings of the National Academy of Sciences USA* 98, 4990–4995.

Maier, S. and Olek, A. (2002) Diabetes: a candidate disease for efficient DNA methylation profiling. *Journal of Nutrition* 132 (Supplement 8), 2440S–2443S.

Mann, J.R., Szabo, P.E., Reed, M.R. and Singer-Sam, J. (2000) Methylated DNA sequences in genomic imprinting. *Critical Reviews in Eukaryotic Gene Expression* 10, 241–257.

Mayer, W., Niveleau, A., Walter, J., Fundele, R. and Haaf, T. (2000) Demethylation of the zygotic paternal genome. *Nature* 403, 501–502.

Medina, M., Urdiales, J.L. and Amores-Sanchez, M.I. (2001) Roles of homocysteine in cell metabolism: old and new functions. *European Journal of Biochemistry* 268, 3871–3882.

Mendelsohn, N., Michl, J., Gilbert, H.S., Acs, G. and Christman, J.K. (1980) L-Ethionine as an inducer of differentiation in human promyelocytic leukemia cells (HL-60). *Cancer Research* 40, 3206–3210.

Mertineit, C., Yoder, J.A., Taketo, T., Laird, D.W., Trasler, J.M. and Bestor, T.H. (1998) Sex-specific exons control DNA methyltransferase in mammalian germ cells. *Development* 125, 889–897.

Mikol, Y.B., Hoover, K.L., Creasia, D. and Poirier, L.A. (1983) Hepatocarcinogenesis in rats fed methyl-deficient, amino acid-defined diets. *Carcinogenesis* 4, 1619–1629.

Mudd, S.H. and Poole, J.R. (1975) Labile methyl balances for normal humans on various dietary regimens. *Metabolism* 24, 721–735.

Narayan, A., Ji, W., Zhang, X.Y., Marrogi, A., Graff, J.R., Baylin, S.B. and Ehrlich, M. (1998) Hypomethylation of pericentromeric DNA in breast adenocarcinomas. *International Journal of Cancer* 77, 833–838.

Negri, E., La Vecchia, C. and Franceschi, S. (2000) Re: dietary folate consumption and breast cancer risk. *Journal of the National Cancer Institute* 92, 1270–1271.

Newberne, P.M., Carltob, W.W. and Wong, G.N. (1964) Hepatomas in rats and hepatorenal injury induced by peanut meal in *Aspergillus flavus* extract. *Path Vet* 1, 105–132.

Nowell, P. (1976) The clonal evolution of tumor cell populations. *Science* 194, 23–28.

Okano, M., Xie, S. and Li, E. (1998a) Cloning and characterization of a family of novel mammalian DNA (cytosine-5) methyltransferases. *Nature Genetics* 19, 219–220.

Okano, M., Xie, S. and Li, E. (1998b) Dnmt2 is not required for de novo and maintenance methylation of viral DNA in embryonic stem cells. *Nucleic Acids Research* 26, 2536–2540.

Okano, M., Bell, D.W., Haber, D.A. and Li, E. (1999) DNA methyltransferases Dnmt3a and Dnmt3b are essential for de novo methylation and mammalian development. *Cell* 99, 247–257.

Piyathilake, C.J. and Johanning, G.L. (2002) Cellular vitamins, DNA methylation and cancer risk. *Journal of Nutrition* 132 (Supplement 8), 2340S–2344S.

Piyathilake, C.J., Hine, R.J., Dasanayake, A.P., Richards, E.W., Freeberg, L.E., Vaughn, W.H. and Krumdieck, C.L. (1992) Effect of smoking on folate levels in buccal mucosal cells. *International Journal of Cancer* 52, 566–569.

Piyathilake, C.J., Johanning, G.L., Macaluso, M., Whiteside, M., Oelschlager, D.K., Heimburger, D.C. and Grizzle, W.E. (2000) Localized folate and vitamin B-12 deficiency in squamous cell lung cancer is associated with global DNA hypomethylation. *Nutrition and Cancer* 37, 99–107.

Pogribny, I P. and James, S.J. (2002) Reduction of p53 gene expression in human primary hepatocellular carcinoma is associated with promoter region methylation without coding region mutation. *Cancer Letters* 176, 169–174.

Pogribny, I.P., Miller, B.J. and James, S.J. (1997a) Alterations in hepatic p53 gene methylation patterns during tumor progression with folate/methyl deficiency in the rat. *Cancer Letters* 115, 31–38.

Pogribny, I.P., Muskhelishvili, L., Miller, B.J. and James, S.J. (1997b) Presence and consequence of uracil in preneoplastic DNA from folate/methyl-deficient rats. *Carcinogenesis* 18, 2071–2076.

Pogribny, I.P., Pogribna, M., Christman, J.K. and James, S.J. (2000) Single-site methylation within the p53 promoter region reduces gene expression in a reporter gene construct: possible in vivo relevance during tumorigenesis. *Cancer Research* 60, 588–594.

Posfai, J., Bhagwat, A.S., Posfai, G. and Roberts, R.J. (1989) Predictive motifs derived from cytosine methyltransferases. *Nucleic Acids Research* 17, 2421–2435.

Potischman, N., Swanson, C.A., Coates, R.J., Gammon, M.D., Brogan, D.R., Curtin, J. and Brinton, L.A. (1999) Intake of food groups and associated micronutrients in relation to risk of early-stage breast cancer. *International Journal of Cancer* 82, 315–321.

Potter, J.D. (1999) Colorectal cancer: molecules and populations. *Journal of the National Cancer Institute* 91, 916–932.

Pradhan, S., Bacolla, A., Wells, R.D. and Roberts, R.J. (1999) Recombinant human DNA (cytosine-5) methyltransferase. I. Expression, purification, and comparison of de novo and maintenance methylation. *Journal of Biological Chemistry* 274, 33002–33010.

Pradhan, S. and Kim, G.D. (2002) The retinoblastoma gene product interacts with maintenance human DNA (cytosine-5) methyltransferase and modulates its activity. *EMBO Journal* 21, 779–788.

Qu, G., Dubeau, L., Narayan, A., Yu, M.C., Ehrlich, M. (1999) Satellite DNA hypomethylation vs overall genomic hypomethylation in ovarian epithelial tumors of different malignant potential. *Mutation Research Fund Molecular Mech Mutagens* 423, 91–101.

Raabe, E.H., Abdurrahman, L., Behbehani, G. and Arceci, R.J. (2001) An SNF2 factor involved in mammalian development and cellular proliferation. *Developmental Dynamics* 221, 92–105.

Razin, A. and Riggs, A.D. (1980) DNA methylation and gene function. *Science* 210, 604–610.

Reik, W., Dean, W. and Walter, J. (2001) Epigenetic reprogramming in mammalian development. *Science* 293, 1089–1093.

Rein, T., DePamphilis, M.L. and Zorbas, H. (1998) Identifying 5-methylcytosine and related modifications in DNA genomes. *Nucleic Acids Research* 26, 2255–2264.

Rhee, I., Bachman, K.E., Park, B.H., Jair, K.W., Yen, R.W., Schuebel, K.E., Cui, H., Feinberg, A.P., Lengauer, C., Kinzler, K.W., Baylin, S.B. and Vogelstein, B. (2002) DNMT1 and DNMT3b cooperate to silence genes in human cancer cells. *Nature* 416, 552–556.

Riggs, A.D. and Jones, P.A. (1983) 5-Methylcytosine, gene regulation, and cancer. *Advances in Cancer Research* 40, 1–30.

Robertson, K.D. (2001) DNA methylation, methyltransferases, and cancer. *Oncogene* 20, 3139–3155.

Robertson, K.D. (2002) DNA methylation and chromatin – unraveling the tangled web. *Oncogene* 21, 5361–5379.

Robertson, K.D., Ait-Si-Ali, S., Yokochi, T., Wade, P.A., Jones, P.L. and Wolffe, A.P. (2000) DNMT1 forms a complex with Rb, E2F1 and HDAC1 and represses transcription from E2F-responsive promoters. *Nature Genetics* 25, 338–342.

Rohan, T.E., Jain, M.G., Howe, G.R. and Miller, A.B. (2000) Dietary folate consumption and breast cancer risk. *Journal of the National Cancer Institute* 92, 266–269.

Ronco, A., De Stefani, E., Boffetta, P., Deneo-Pellegrini, H., Mendilaharsu, M. and Leborgne, F. (1999) Vegetables, fruits, and related nutrients and risk of breast cancer: a case-control study in Uruguay. *Nutrition and Cancer* 35, 111–119.

Rosenquist, T.H., Schneider, A.M. and Monaghan, D.T. (1999) *N*-methyl-D-aspartate receptor agonists modulate homocysteine-induced developmental abnormalities. *FASEB Journal* 13, 1523–1531.

Rountree, M.R., Bachman, K.E. and Baylin, S.B. (2000) DNMT1 binds HDAC2 and a new co-repressor, DMAP1, to form a complex at replication foci. *Nature Genetics* 25, 269–277.

Rountree, M.R., Bachman, K.E., Herman, J.G. and Baylin, S.B. (2001) DNA methylation, chromatin inheritance, and cancer. *Oncogene* 20, 3156–3165.

Ryan, B.M. and Weir, D.G. (2001) Relevance of folate metabolism in the pathogenesis of colorectal cancer. *Journal of Laboratory and Clinical Medicine* 138, 164–176.

Saito, Y., Kanai, Y., Sakamoto, M., Saito, H., Ishii, H. and Hirohashi, S. (2002) Overexpression of a splice variant of DNA methyltransferase 3b, DNMT3b4, associated with DNA hypomethylation on pericentromeric satellite regions during human hepatocarcinogenesis. *Proceedings of the National Academy of Sciences USA* 99, 10060–10065.

Salaspuro, M. (1996) Bacteriocolonic pathway for ethanol oxidation: characteristics and implications. *Annals of Medicine* 28, 195–200.

Santi, D.V., Norment, A. and Garrett, C. E. (1984) Covalent bond formation between a DNA-cytosine methyltransferase and DNA containing 5-azacytosine. *Proceedings of the National Academy of Sciences USA* 81, 6993–6997.

Santourlidis, S., Florl, A., Ackermann, R., Wirtz, H.C. and Schulz, W.A. (1999) High frequency of alterations in DNA methylation in adenocarcinoma of the prostate. *Prostate* 39, 166–174.

Sealy, L. and Chalkley, R. (1978) The effect of sodium butyrate on histone modification. *Cell* 14, 115–121.

Shane, B. (1995) Folate chemistry and metabolism. In: *Folate in Health and Disease*. Marcel Dekker, New York, pp. 1–22.

Sharp, L., Little, J., Schofield, A.C., Pavlidou, E., Cotton, S.C., Miedzybrodzka, Z., Baird, J.O., Haites, N.E., Heys, S.D. and Grubb, D.A. (2002) Folate and breast cancer: the role of polymorphisms in methylenetetrahydrofolate reductase (MTHFR). *Cancer Letters* 181, 65–71.

Shen, L., Fang, J., Qiu, D., Zhang, T., Yang, J., Chen, S. and Xiao, S. (1998) Correlation between DNA methylation and pathological changes in human hepatocellular carcinoma. *Hepatogastroenterology* 45, 1753–1759.

Shen, H., Spitz, M.R., Wang, L.E., Hong, W.K. and Wei, Q. (2001) Polymorphisms of methylenetetrahydrofolate reductase and risk of lung cancer: a case-control study. *Cancer Epidemiology Biomarkers Prev* 10, 397–401.

Shivapurkar, N. and Poirier, L.A. (1983) Tissue levels of S-adenosylmethionine and S-adenosylhomocysteine in rats fed methyl-deficient, amino acid-defined diets for one to five weeks. *Carcinogenesis* 4, 1051–1057.

Sibani, S., Melnyk, S., Pogribny, I.P., Wang, W., Hiou-Tim, F., Deng, L., Trasler, J., James, S.J. and Rozen, R. (2002) Studies of methionine cycle intermediates (SAM, SAH), DNA methylation and the impact of folate deficiency on tumor numbers in Min mice. *Carcinogenesis* 23, 61–65.

Singal, R. and Ginder, G.D. (1999) DNA methylation. *Blood* 93, 4059–4070.

Skibola, C.F., Smith, M.T., Kane, E., Roman, E., Rollinson, S., Cartwright, R.A. and Morgan, G. (1999) Polymorphisms in the methylenetetrahydrofolate reductase gene are associated with susceptibility to acute leukemia in adults. *Proceedings of the National Academy of Sciences USA* 96, 12810–12815.

Slattery, M.L., Schaffer, D., Edwards, S.L., Ma, K.N. and Potter, J.D. (1997) Are dietary factors involved in DNA methylation associated with colon cancer?' *Nutrition and Cancer* 28, 52–62.

Slattery, M.L., Potter, J.D., Samowitz, W., Schaffer, D. and Leppert, M. (1999) Methylenetetrahydrofolate reductase, diet, and risk of colon cancer. *Cancer Epidemiology Biomarkers Prev* 8, 513–518.

Song, C., Xing, D., Tan, W., Wei, Q. and Lin, D. (2001) Methylenetetrahydrofolate reductase polymorphisms increase risk of esophageal squamous cell carcinoma in a Chinese population. *Cancer Research* 61, 3272–3275.

Song, J., Sohn, K.-J., Medline, A., Ash, C., Gallinger, S. and Kim, Y.-I. (2000b) Chemopreventive effects of dietary folate on intestinal polyps in Apc+/− Msh2−/− mice. *Cancer Research* 60, 3191–3199.

Song, J., Medline, A., Mason, J.B., Gallinger, S. and Kim, Y.I. (2000a) Effects of dietary folate on intestinal tumorigenesis in the apcMin mouse. *Cancer Research* 60, 5434–5440.

Stern, L.L., Mason, J.B., Selhub, J. and Choi, S.W. (2000) Genomic DNA hypomethylation, a characteristic of most cancers, is present in peripheral leukocytes of individuals who are homozygous for the C677T polymorphism in the methylenetetrahydrofolate reductase gene. *Cancer Epidemiology Biomarkers Prev* 9, 849–853.

Stolzenberg-Solomon, R.Z., Albanes, D., Nieto, F.J., Hartman, T.J., Tangrea, J.A., Rautalahti, M., Sehlub, J., Virtamo, J. and Taylor, P.R. (1999) Pancreatic cancer risk and nutrition-related methyl-group availability indicators in male smokers. *Journal of the National Cancer Institute* 91, 535–541.

Su, L.J. and Arab, L. (2001) Nutritional status of folate and colon cancer risk: evidence from NHANES I epidemiologic follow-up study. *Annals of Epidemiology* 11, 65–72.

Tatematsu, K.I., Yamazaki, T. and Ishikawa, F. (2000) MBD2-MBD3 complex binds to hemi-methylated DNA and forms a complex containing DNMT1 at the replication foci in late S phase. *Genes and Cells* 5, 677–688.

Todesco, L., Angst, C., Litynski, P., Loehrer, F., Fowler, B. and Haefeli, W.E. (1999) Methylenetetrahydrofolate reductase polymorphism, plasma homocysteine and age. *European Journal of Clinical Investigations* 29, 1003–1009.

Tomeo, C.A., Colditz, G.A., Willett, W.C., Giovannucci, E., Platz, E., Rockhill, B., Dart, H. and Hunter, D.J. (1999) *Harvard Report on Cancer Prevention*, Vol. 3, *Prevention of Colon Cancer in the United States*. *Cancer Causes Control* 10, 167–180.

Torres, L., Garcia-Trevijano, E.R., Rodriguez, J.A., Carretero, M.V., Bustos, M., Fernandez, E., Eguinoa, E., Mato, J.M. and Avila, M.A. (1999) Induction of TIMP-1 expression in rat hepatic stellate cells and hepatocytes: a new role for homocysteine in liver fibrosis. *Biochimica Biophysica Acta* 1455, 12–22.

Tseng, M., Murray, S.C., Kupper, L.L. and Sandler, R.S. (1996) Micronutrients and the risk of colorectal adenomas. *American Journal of Epidemiology* 144, 1005–1014.

Tycko, B. (2000) Epigenetic gene silencing in cancer. *Journal of Clinical Investigations* 105, 401–407.

Ulrich, C.M., Kampman, E., Bigler, J., Schwartz, S.M., Chen, C., Bostick, R., Fosdick, L., Beresford, S.A., Yasui, Y. and Potter, J.D. (1999) Colorectal adenomas and the C677T MTHFR polymorphism: evidence for gene–environment interaction? *Cancer Epidemiology Biomarkers Prev* 8, 659–668.

Van den Veyver, I.B. (2002) Genetic effects of methylation diets. *Annual Review of Nutrition* 22, 255–282.

Vignali, M., Hassan, A.H., Neely, K.E. and Workman, J.L. (2000) ATP-dependent chromatin-remodeling complexes. *Molecular Cell Biology* 20, 1899–1910.

Wagner, C. (1982) Cellular folate binding proteins; function and significance. *Annual Review of Nutrition* 2, 229–248.

Wainfan, E., Dizik, M. and Balis, M.E. (1984) Increased activity of rat liver N2-guanine tRNA methyltransferase II in response to liver damage. *Biochimica et Biophysica Acta* 799, 282–290.

Wainfan, E., Dizik, M., Hluboky, M. and Balis, M.E. (1986) Altered tRNA methylation in rats and mice fed lipotrope-deficient diets. *Carcinogenesis* 7, 473–476.

Wainfan, E., Kilkenny, M. and Dizik, M. (1988) Comparison of methyltransferase activities of pair-fed rats given adequate or methyl-deficient diets. *Carcinogenesis* 9, 861–863.

Wainfan, E., Dizik, M., Stender, M. and Christman, J.K. (1989) Rapid appearance of hypomethylated DNA in livers of rats fed cancer-promoting, methyl-deficient diets. *Cancer Research* 49, 4094–4097.

Warnecke, P.M., Stirzaker, C., Song, J., Grunau, C., Melki, J.R. and Clark, S.J. (2002) Identification and resolution of artifacts in bisulfite sequencing. *Methods* 27, 101–107.

Weinstein, S.J., Ziegler, R.G., Frongillo, E.A. Jr, Colman, N., Sauberlich, H.E., Brinton, L.A., Hamman, R.F., Levine, R.S., Mallin, K., Stolley, P.D. and Bisogni, C.A. (2001a) Low serum and red blood cell folate are moderately, but nonsignificantly associated with increased risk of invasive cervical cancer in U.S. women. *Journal of Nutrition* 131, 2040–2048.

Weinstein, S.J., Ziegler, R.G., Selhub, J., Fears, T.R., Strickler, H.D., Brinton, L.A., Hamman, R.F., R.S., L., Malin, K. and Stolley, P.D. (2001b) Elevated serum homocysteine levels and increased risk of invasive cervical cancer in U.S. women. *Cancer Causes Control* 12, 317–324.

Weir, D.G., McGing, P.G. and Scott, J.M. (1985) Folate metabolism, the enterohepatic circulation and alcohol. *Biochemical Pharmacology* 34, 1–7.

Weisenberger, D.J., Velicescu, M., Preciado-Lopez, M.A., Gonzales, F.A., Tsai, Y.C., Liang, G. and Jones, P.A. (2002) Identification and characterization of alternatively spliced variants of DNA methyltransferase 3a in mammalian cells. *Gene* 298, 91–99.

White, E., Shannon, J.S. and Patterson, R.E. (1997) Relationship between vitamin and calcium supplement use and colon cancer. *Cancer Epidemiology Biomarkers Prev* 6, 769–774.

Wiemels, J.L., Smith, R.N., Taylor, G.M., Eden, O.B., Alexander, F.E. and Greaves, M.F. (2001) Methylenetetrahydrofolate reductase (MTHFR) polymorphisms and risk of molecularly defined subtypes of childhood acute leukemia. *Proceedings of the National Academy of Sciences USA* 98, 4004–4009.

Wigler, M., Levy, D. and Perucho, M. (1981) The somatic replication of DNA methylation. *Cell* 24, 33–40.

Wilson, M.J., Shivapurkar, N. and Poirier, L.A. (1984) Hypomethylation of hepatic nuclear DNA in rats fed with a carcinogenic methyl-deficient diet. *Biochemistry Journal* 218, 987–990.

Wolffe, A.P. and Pruss, D. (1996) Hanging on to histones. Chromatin. *Current Biology* 6, 234–237.

Yan, P.S., Chen, C.M., Shi, H., Rahmatpanah, F., Wei, S.H., Caldwell, C.W. and Huang, T.H. (2001) Dissecting complex epigenetic alterations in breast cancer using CpG island microarrays. *Cancer Research* 61, 8375–8380.

Yao, Z.M. and Vance, D.E. (1988) The active synthesis of phosphatidylcholine is required for very low density lipoprotein secretion from rat hepatocytes. *Journal of Biological Chemistry* 263, 2998–3004.

Yoder, J.A. and Bestor, T.H. (1998) A candidate mammalian DNA methyltransferase related to pmt1p of fission yeast. *Human Molecular Genetics* 7, 279–284.

Zhang, S., Hunter, D.J., Hankinson, S.E., Giovannucci, E.L., Rosner, B.A., Colditz, G.A., Speizer, F.E. and Willett, W.C. (1999) A prospective study of folate intake and the risk of breast cancer. *Journal of the American Medical Association* 281, 1632–1637.

Zhang, X. and Verdine, G.L. (1996) Mammalian DNA cytosine-5 methyltransferase interacts with p23 protein. *FEBS Letters* 392, 179–183.

Zingg, J.M. and Jones, P.A. (1997) Genetic and epigenetic aspects of DNA methylation on genome expression, evolution, mutation and carcinogenesis. *Carcinogenesis* 18, 869–882.

17 Biotinylation of Histones in Human Cells

Janos Zempleni
Department of Nutritional Sciences and Dietetics, University of Nebraska at Lincoln, Lincoln, Nebraska, USA

Introduction

Post-translational modifications of histones

Histones are the primary proteins that mediate the folding of DNA into chromatin (Wolffe, 1998). Chromatin consists of a repetitive nucleoprotein complex, the nucleosome. Each nucleosome consists of DNA that is wrapped around an octamer of core histones (two molecules each of histones H2a, H2b, H3 and H4); nucleosome assembly is completed by incorporation of a linker histone such as histone H1. The binding of DNA to histones is of electrostatic nature; binding is mediated by the association of negatively charged phosphate groups of DNA with positively charged ε-amino groups and guanidino groups of histones.

In vivo, histones are modified post-translationally by acetylation (Ausio and Holde, 1986; Hebbes *et al.*, 1988; Lee *et al.*, 1993), methylation (Wolffe, 1998), phosphorylation (Wolffe, 1998), ubiquitination (Wolffe, 1998) and poly(ADP-ribosylation) (Chambon *et al.*, 1966; Boulikas, 1988; Boulikas *et al.*, 1990). These groups are attached covalently to amino acid residues of histones such as ε-amino groups in lysine residues and guanidino groups in arginine residues. Some of these modifications (e.g. acetylation and poly (ADP-ribosylation)) take away one positive charge from histones, leading to weakened association between DNA and histones.

Transcriptional activity of DNA

Evidence has been provided that the transcriptional activity of chromatin is regulated by covalent modification of histones. In particular, the role of acetylation of histones in transcription of DNA has been investigated extensively. These studies suggested that transcriptional activators have histone acetyltransferase activity (Brownell *et al.*, 1996) and that transcriptional repressors are histone deacetylases (Taunton *et al.*, 1996). Histone acetylation and transcriptional activity of chromatin are strongly correlated (Allfrey *et al.*, 1964; Gorovsky, 1973; Mathis *et al.*, 1978). Likewise, ubiquitination of histones correlates with the transcriptional activity of chromatin (Pham and Sauer, 2000). In contrast, methylation of lysine residues in core histones might contribute to transcriptional repression (Wolffe, 1998).

DNA repair

Evidence has been provided that poly(ADP-ribosylation) of histones is linked to DNA repair mechanisms (Durkacz *et al.*, 1980; Althaus, 1992). Induction of DNA damage in mammalian cells by chemical treatment or UV irradiation causes a dramatic increase of poly(ADP-ribosylation) of histones, i.e. covalent binding of ADP-ribosyl moieties (Juarez-Salinas et al., 1979). Likewise, the increased poly(ADP-ribosylation) after mitogen stimulation of lymphoid cells has been attributed to an increase in DNA strand breaks caused by stimulation with phytohaemagglutinin (Boulikas et al., 1990).

Poly(ADP-ribosylation) is catalysed by the enzyme poly(ADP-ribose) polymerase (EC 2.4.2.30) (Chambon et al., 1966; Nishizuka et al., 1971; Boulikas et al., 1990). Inhibitors of poly (ADP-ribose) polymerase stimulate unscheduled DNA synthesis in UV-irradiated human lymphocytes (Sims et al., 1982).

Among core histones, H2B is poly(ADP-ribosylated) to the greatest extent (Boulikas et al., 1990). However, H2A, H3 and H4, as well as linker histone H1, are also poly(ADP-ribosylated) (Boulikas et al., 1990). In addition, poly (ADP-ribose) polymerase catalyses auto-ADP-ribosylation in response to DNA damage (Althaus, 1992).

Poly(ADP-ribosylation) of histones catalyses a temporary dissociation from and reassociation of histones with DNA (Althaus, 1992). It has been proposed that this mechanism may guide enzymes to sites of DNA repair (Althaus, 1992). In addition, poly(ADP-ribosylation) might be involved in nucleosomal unfolding of chromatin in DNA excision repair (Althaus, 1992). It has been proposed that poly(ADP-ribosylation) of histone H1 might be involved in early apoptotic events (Yoon et al., 1996).

Nutritional status might be important for DNA repair. The rejoining of DNA strand breaks is prevented by nutritionally depleting cells of NAD, the substrate for poly(ADP-ribosylations) (Durkacz et al., 1980). Niacin and tryptophan are dietary precursors to NAD. Even a mild niacin deficiency decreases poly(ADP-ribose) concentrations in rat liver (Rawling et al., 1994).

Cell proliferation

Post-translational modifications of histones are essential to ensure normal cell proliferation. For example, replacement of all four acetylatable lysines in histone H4 of *Saccharomyces cerevisiae* with arginine or glutamine leads to extremely slow growth or a delay in G_2/M progression (Megee et al., 1995). The pattern of histone modification changes dramatically during the course of the cell cycle in eukaryotic cells (Wolffe, 1998), consistent with a role for these modifications in cell proliferation. For example, core histones are deacetylated during mitosis. Post-translational modifications of histones during the cell cycle might regulate transcription of genes that are involved in cell cycle progression (Wolffe, 1998).

Spermatogenesis

Spermatogenesis requires the packaging of DNA into an inert chromatin structure. During spermatogenesis, the histones are heavily acetylated, and transiently replaced by 'transition proteins' (Wolffe, 1998). Subsequently, transition proteins are replaced by highly basic protamines, leading to complete condensation of the nucleus. Binding of protamines to DNA is regulated by post-translational modification of protamines (e.g. by phosphorylation). Condensation makes the DNA in the nucleus of a spermatozoon much more resistant to enzymatic, physical and chemical degradation. Together, these findings suggest that post-translational modification of histones and other nucleic acid-binding proteins is important during spermatogenesis.

Enzymatic Biotinylation of Histones

Recently, evidence has been provided for another post-translational modification of histones: biotinylation of lysine residues. Hymes et al. (1995) have proposed a reaction mechanism by which the enzyme biotinidase (EC 3.5.1.12) mediates covalent binding of biotin to histones. The authors proposed that cleavage of biocytin (biotin-ε-lysine) by biotinidase leads to the formation of a biotinyl-thioester intermediate (cysteine-bound biotin) at or near the active site of biotinidase (Fig. 17.1) (Hymes et al., 1995; Hymes and Wolf, 1999). In a next step, the biotinyl moiety is transferred from the thioester to the ε-amino group of lysine (or other amino groups) in histones (Fig. 17.2). Biotinidase is ubiquitous in mammalian cells, and 26% of the cellular biotinidase activity is located in the nuclear fraction (Pispa, 1965).

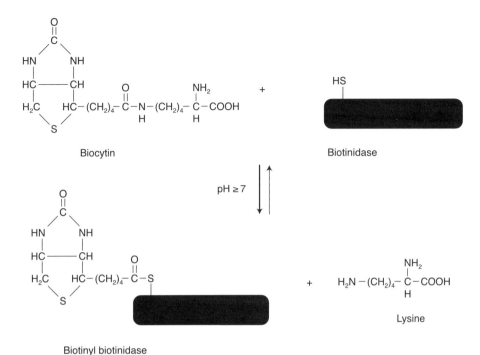

Fig. 17.1. Cleavage of biocytin (biotin-ε-lysine) by biotinidase leads to the formation of a biotinyl-thioester intermediate (cysteine-bound biotin; biotinyl biotinidase) at or near the active site of biotinidase (Hymes and Wolf, 1999).

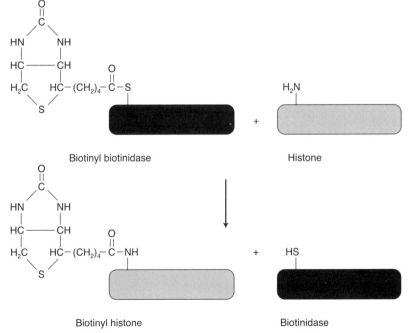

Fig. 17.2. The biotinyl moiety from the biotinyl biotinidase intermediate is transferred on to the ε-amino group of lysine (or other amino groups) in histones (Hymes and Wolf, 1999).

Biotinylated Histones in Human Cells

Identification of biotinylated histones

In our laboratory, we determined whether human cells contain biotinylated histones. Histones were isolated from nuclei of human peripheral blood mononuclear cells (PBMCs) by acid extraction (Stanley et al., 2001). This procedure yielded a pure preparation of the five major classes of histones (H1, H2a, H2b, H3 and H4), as judged by co-migration with commercially available histones on polyacrylamide gels and by the absence of quantitatively important non-histone bands after Coomassie blue staining (Fig. 17.3, lanes 1 and 2). The same extracts from PBMC nuclei were used to determine whether histones contain covalently bound biotin. Using Western blots of histones, biotin was probed with streptavidin-conjugated peroxidase; peroxidase activity was visualized by using chemiluminescence. Histones H1, H3 and H4 contained streptavidin-binding substances, suggesting that these proteins were biotinylated (Fig. 17.3, lane 3). Histones H2A and H2B electrophoresed as one single band. Thus, it remains uncertain whether streptavidin bound to histone H2A, H2B or both.

The following data suggest that streptavidin specifically bound to biotinylated histones rather than unspecifically to non-biotinylated proteins. The (non-biotinylated) proteins aprotinin, α-lactalbumin, β-lactoglobulin, trypsin inhibitor and trypsinogen have molecular weights similar to those of the histones; no visible bands were observed when these proteins were probed with streptavidin-conjugated peroxidase (Fig. 17.3, lane 4). The synthetic polypeptides poly-L-lysine and poly-L-arginine mimic lysine- and arginine-rich histones; no visible bands were observed when these synthetic polypeptides were probed with streptavidin-conjugated peroxidase (Fig. 17.3, lanes 5 and 6). Next, histone H1 was biotinylated chemically as described previously (Zempleni and Mock, 1999); this synthetic conjugate was used as a positive control (Stanley et al., 2001). The chemically biotinylated histone H1 produced a strong band if probed with streptavidin-conjugated peroxidase (Fig. 17.3, lane 7). Finally, extracts from PBMC nuclei were probed with a monoclonal antibody against biotin (Fig. 17.3, lane 8). The antibody bound to histones, providing additional evidence that histones contain biotin.

Effects of cell proliferation

Increased expression of various genes is essential to ensure normal cell proliferation. Theoretically, biotinylation of histones might lead to increased transcription of DNA, in analogy to transcriptional activation of DNA by acetylation and

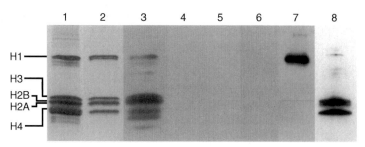

Fig. 17.3. Nuclei from human PBMCs contain biotinylated histones. Histones were extracted from PBMC nuclei and chromatographed using SDS–gel electrophoresis. Non-histone proteins and synthetic polypeptides (poly-L-arginine and poly-L-lysine) were purchased from Sigma and were chromatographed as described for histones. Samples were either stained with Gelcode blue (lanes 1 and 2), probed with streptavidin-conjugated peroxidase (lanes 3–7) or probed with a monoclonal antibody against biotin (lane 8). Lane 1 = histones from human PBMC nuclei; lane 2 = commercially available histones from calf thymus; lane 3 = histones from human PBMC nuclei; lane 4 = mixture of aprotinin, α-lactalbumin, β-lactoglobulin, trypsin inhibitor and trypsinogen; lane 5 = poly-L-arginine; lane 6 = poly-L-lysine; lane 7 = chemically biotinylated histone H1; lane 8 = histones from human PBMC nuclei. Equimolar amounts of histones, poly-L-lysine, poly-L-arginine and non-histone proteins were used, except for the chemically biotinylated histone H1, which was diluted approximately 20,000-fold (Stanley et al., 2001).

ubiquitination of histones (Lee et al., 1993; Sommerville et al., 1993; Pham and Sauer, 2000). Is biotinylation of histones linked to cell proliferation? In our laboratory, proliferation of PBMCs was induced by incubation with concanavalin A; controls were incubated without concanavalin A (Stanley et al., 2001). The culture medium contained [^{3}H]biotin at a physiological concentration (475 pmol l^{-1}) to track biotinylation of histones both by probing of Western blots with streptavidin-conjugated peroxidase and by liquid scintillation counting.

PBMCs responded to proliferation with increased biotinylation of histones. Analysis of PBMC histones by gel electrophoresis provided evidence that biotinylation increased in all classes of histones, i.e. H1, H2a, H2b, H3 and H4 (Fig. 17.4A). Biotinylation of histones was also detectable in quiescent controls, suggesting that histones maintain a certain level of biotinylation regardless of their state of proliferation. Liquid scintillation counting allowed a more accurate quantitation of ^{3}H-biotinylation of histones compared with Western blotting. PBMCs responded to proliferation by increasing ^{3}H-biotinylation of histones from 3.9 ± 1.3 amol biotin per 10^6 cells (quiescent state) to 20.1 ± 2.9 amol biotin per 10^6 cells (Fig. 17.4B). Note that liquid scintillation counting measures only newly incorporated [^{3}H]biotin but not those histones that contain unlabelled biotin. Thus, the true content of biotinylated histones per cell exceeds the one reported here for liquid scintillation counting.

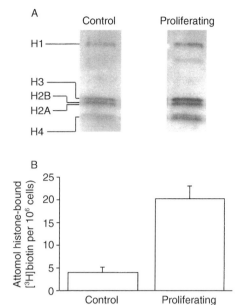

Fig. 17.4. Human PBMCs respond to proliferation with increased biotinylation of histones. (A) PBMCs were cultured with concanavalin A to induce cell proliferation; non-proliferating controls were cultured without concanavalin A. Histones were isolated from an equal number of proliferating and non-proliferating PBMCs (5.1 × 10^6 PBMCs) and analysed by SDS–PAGE and chemiluminescence. (B) Liquid scintillation counting was used to quantitate binding of [^{3}H]biotin to histones in proliferating and non-proliferating PBMCs. Histones were isolated from an equal number of proliferating and non-proliferating (control) PBMCs (20.48 × 10^6 PBMCs); aliquots of the PBMCs described in (A) were used. $P < 0.05$; $n = 3$ (Stanley et al., 2001).

Biotinylation of histones during the cell cycle

Theoretically, the increased biotinylation of histones in proliferating PBMCs might be restricted to a specific phase of the cell cycle. Cell cycle-arrested PBMCs were used to determine biotinylation of histones at various phases of the cycle (Stanley et al., 2001). PBMCs were cultured with 475 pmol l^{-1} [^{3}H]biotin and 20 μg ml^{-1} concanavalin at 37°C for 30 h to induce cell proliferation; quiescent controls were incubated without concanavalin A. Then, one of the following chemicals was added to the proliferating PBMCs and incubation was continued for 16 h in order to arrest the cells at specific phases of the cell cycle: 100 nmol l^{-1} (final concentration) wortmannin to cause arrest in G_1 phase (Yano et al., 1993); 118 μmol l^{-1} aphidicolin to cause arrest in S phase (Dasso and Newport, 1990); 1 μmol l^{-1} okadaic acid to cause arrest in G_2 phase (Cohen, 1989; Cohen et al., 1989; Haavik et al., 1989); or 5 μmol l^{-1} colchicine to cause arrest in M phase (Chalifour and Dakshinamurti, 1982). Histones were extracted from PBMC nuclei and ^{3}H-biotinylated histones were determined by liquid scintillation counting. ^{3}H-Biotinylation of histones increased early in the cell cycle (G_1 phase) and remained increased during later phases (S, G_2 and M phase) compared with quiescent controls (G_0); the increase was greater than fourfold if normalized per 10^6 PBMCs (Fig. 17.5A).

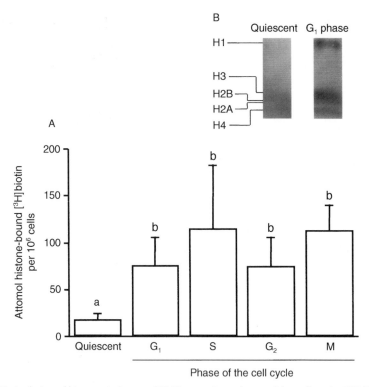

Fig. 17.5. Biotinylation of histones in human PBMCs at various phases of the cell cycle. PBMCs were cultured with [^{3}H]biotin as described in the text; cells were arrested chemically at various phases of the cell cycle. (A) ^{3}H-Biotinylation of histones was tracked by liquid scintillation counting. Quiescent PBMCs were used as controls. Bars not sharing a common letter were significantly different; $P < 0.01$; $n = 4$. (B) Biotinylation of histones was tracked by SDS–PAGE and chemiluminescence. PBMCs from phases G_0 and G_1 are depicted; G_1-arrested PBMCs show a biotinylation pattern that is representative for cells from G_1, S, G_2 and M phase (Stanley et al., 2001).

Is biotinylation increased for all classes of histones at the various phases of the cell cycle or is biotinylation increased for only some of the histones? Analysis by SDS–PAGE and chemiluminescence suggested that biotinylation increased for all classes of histones in G_1, S, G_2 and M phase of the cell cycle compared with G_0 phase (Fig. 17.5B). This might suggest that biotinylation of histones is part of a general mechanism that involves all histones to a similar extent; the nature of this mechanism remains uncertain.

In summary, these data are consistent with the hypothesis that biotinylation of all classes of histones is increased during all phases of the cell cycle (G_1, S, G_2 and M phase) compared with quiescent PBMCs (G_0 phase). It remains uncertain why biotinylation of histones is increased throughout the cell cycle rather than at a specific phase of the cycle.

Expression of the biotinidase gene

Studies by Hymes et al. provided evidence that biotinylation of histones is catalysed by biotinidase (Hymes et al., 1995; Hymes and Wolf, 1999). In our laboratory, we quantitated expression of the biotinidase gene in proliferating and non-proliferating PBMCs in order to determine whether increased biotinylation of histones in proliferating PBMCs is paralleled by increased biotinidase activity (Stanley et al., 2001). The cellular level of mRNA encoding biotinidase was similar in proliferating and non-proliferating cells (Fig. 17.6). The slight increase in mRNA encoding biotinidase (~0.6-fold) was much less pronounced than increases seen for mRNA encoding other proteins involved in biotin metabolism; presumably, this increase in biotinidase mRNA is not physiologically significant. For example, mRNA

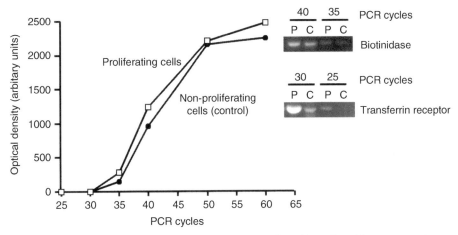

Fig. 17.6. Expression of the biotinidase gene in PBMCs is not affected by cell proliferation. mRNA encoding biotinidase was measured by quantitative polymerase chain reaction (PCR) in proliferating PBMCs (P) and in non-proliferating controls (C). Inserts depict agarose gels of mRNA encoding biotinidase and transferrin receptor (Stanley et al., 2001).

encoding the sodium-dependent multivitamin transporter (which mediates biotin transport) increased about tenfold in proliferating PBMCs (Zempleni et al., 2001). Likewise, mRNA encoding the transferrin receptor (control) increased 5.5-fold (Fig. 17.6).

Studies of the phenotype (i.e. biotinidase activity) in PBMCs also suggest that the enzyme is not affected by cell proliferation (Stanley et al., 2001). Biotinidase activity was measured in proliferating and in quiescent PBMCs; in both, biotinidase activity was barely above background (<30 U 10^{-9} PBMCs) and did not provide quantifiable data (1 U = nmol of p-aminobenzoic acid (PABA) released from N-D-biotinyl-p-aminobenzoic acid per min). This is consistent with an earlier report by Pispa (1965), suggesting that human leukocytes have low biotinidase activity. For comparison, biotinidase activity in plasma is much greater; in our laboratory, biotinidase activity in plasma from five healthy adults was 7980 ± 1324 U l^{-1}. Taken together, biotinidase activity in PBMCs is very low, and expression of the biotinidase gene is not greatly increased in proliferating cells. Notwithstanding the well-established capability of biotinidase to catalyse biotinylation of histones (Hymes et al., 1995), we propose that either other enzymes are also capable of biotinylating histones, or that levels of histone biotinylation are regulated by rates of degradation, or both.

Debiotinylation of histones

Recent studies are consistent with the hypothesis that biotinidase may catalyse both biotinylation and debiotinylation of histones (Ballard et al., 2002). If biotinylation of histones were an important mechanism to regulate cellular processes, how would cells regulate the biotinylation status of histones despite the fact that both biotinylation and debiotinylation of histones are catalysed by the same enzyme? The following explanations have been offered.

1. Enzymes other than biotinidase might also catalyse biotinylation or debiotinylation of histones. For example, holocarboxylase synthetase (EC 6.3.4.10) catalyses covalent binding of biotin to ε-amino groups of lysine residues in four mammalian carboxylases (Zempleni, 2001). The possibility that holocarboxylase synthetase catalyses binding of biotin to histones is an untested hypothesis.

2. Covalent modification of biotinidase might be a mechanism to favour either biotinylation or debiotinylation of histones. Currently, glycosylation is the only post-translational modification of biotinidase that has been identified (Cole et al., 1994).

3. The presence of cofactors might favour either biotinylation or debiotinylation of histones. For example, high concentrations of the substrate biocytin may increase the rate of histone

biotinylation. Similarly, biotinylated peptides might inhibit debiotinylation of histones by competing for binding to biotinidase. In contrast, the pH optimum is similar (pH 8) for both the biotinylating activity (Hymes *et al.*, 1995) and the debiotinylating activity of biotinidase. Thus, changes of the pH in the microenvironment of histones should not affect the biotinylation status.

Acknowledgements

Supported by the National Institutes of Health (DK 60447) and USDA/CSREES project award 2001-35200-10187.

References

Allfrey, V., Faulkner, R.M. and Mirsky, A.E. (1964) Acetylation and methylation of histones and their possible role in the regulation of RNA synthesis. *Proceedings of the National Academy of Sciences USA* 51, 786–794.

Althaus, F. (1992) Poly ADP-ribosylation: a histone shuttle mechanism in DNA excision repair. *Journal of Cell Science* 102, 663–670.

Ausio, J. and van Holde, K.E. (1986) Histone hyperacetylation: its effect on nucleosome conformation and stability. *Biochemistry* 25, 1421–1428.

Ballard, T.D., Wolffe, J., Griffin, J.B., Stanley, J. S., van Calcar, S. and Zempleni, J. (2002) Biotinidase catalyzes debiotinylation of histones. *European Journal of Biochemistry* 41, 78–84.

Boulikas, T. (1988) At least 60 ADP-ribosylated variant histones are present in nuclei from dimethylsulfate-treated and untreated cells. *EMBO Journal* 7, 57–67.

Boulikas, T., Bastin, B., Boulikas, P. and Dupuis, G. (1990) Increase in histone poly(ADP-ribosylation) in mitogen-activated lymphoid cells. *Experimental Cell Research* 187, 77–84.

Brownell, J.E., Zhou, J., Ranalli, T., Kobayashi, R., Edmondson, D.G., Roth, S.Y. and Allis, C.D. (1996) *Tetrahymena* histone acetyltransferase A: a homolog to yeast Gcn5p linking histone acetylation to gene activation. *Cell* 84, 843–851.

Chalifour, L.E. and Dakshinamurti, K. (1982) The biotin requirement of human fibroblasts in culture. *Biochemical and Biophysical Research Communications* 104, 1047–1053.

Chambon, P., Weill, J.D., Doly, J., Strosser, M.T. and Mandel, P. (1966) On the formation of a novel adenylic compound by enzymatic extracts of liver nuclei. *Biochemical and Biophysical Research Communications* 25, 638–643.

Cohen, P. (1989) The structure and regulation of protein phosphatases. *Annual Review of Biochemistry* 58, 453–508.

Cohen, P., Klumpp, S. and Schelling, D.L. (1989) An improved procedure for identifying and quantitating protein phosphatases in mammalian tissues. *FEBS Letters* 250, 596–600.

Cole, H., Reynolds, T.R., Lockyer, J.M., Buck, G.A., Denson, T., Spence, J.E., Hymes, J. and Wolf, B. (1994) Human serum biotinidase cDNA cloning, sequence, and characterization. *Journal of Biological Chemistry* 269, 6566–6570.

Dasso, M. and Newport, J.W. (1990) Completion of DNA replication is monitored by a feedback system that controls the initiation of mitosis *in vitro*: studies in *Xenopus. Cell* 61, 811–823.

Durkacz, B.W., Omidiji, O., Gray, D.A. and Shall, S. (1980) (ADP-ribose)$_n$ participates in DNA excision repair. *Nature* 283, 593–596.

Gorovsky, M.A. (1973) Macro- and micronuclei of *Tetrahymena pyriformis*: a model system for studying the structure of eukaryotic nuclei. *Journal of Protozoology* 20, 19–25.

Haavik, J., Schelling, D.L., Campbell, D.G., Andersson, K.K., Flatmark, T. and Cohen, P. (1989) Identification of protein phosphatase 2A as the major tyrosine hydroxylase phosphatase in adrenal medulla and corpus striatum: evidence from the effects of okadaic acid. *FEBS Letters* 251, 36–42.

Hebbes, T.R., Thorne, A.W. and Crane-Robinson, C. (1988) A direct link between core histone acetylation and transcriptionally active chromatin. *EMBO Journal* 7, 1395–1402.

Hymes, J. and Wolf, B. (1999) Human biotinidase isn't just for recycling biotin. *Journal of Nutrition* 129, 485S-489S.

Hymes, J., Fleischhauer, K. and Wolf, B. (1995) Biotinylation of histones by human serum biotinidase: assessment of biotinyl-transferase activity in sera from normal individuals and children with biotinidase deficiency. *Biochemical and Molecular Medicine* 56, 76–83.

Juarez-Salinas, H., Sims, J.L. and Jacobson, M.K. (1979) Poly(ADP-ribose) levels in carcinogen-treated cells. *Nature* 282, 740–741.

Lee, D.Y., Hayes, J.J., Pruss, D. and Wolffe, A.P. (1993) A positive role for histone acetylation in transcription factor access to nucelosomal DNA. *Cell* 72, 73–84.

Mathis, D.J., Oudet, P., Waslyk, B. and Chambon, P. (1978) Effect of histone acetylation on structure and *in vitro* transcription of chromatin. *Nucleic Acids Research* 5, 3523–3547.

Megee, P.C., Morgan, B.A. and Smith, M.M. (1995) Histone H4 and the maintenance of genome integrity. *Genes and Development* 9, 1716–1727.

Nishizuka, Y., Ueda, K. and Hayaishi, O. (1971) Adenosine diphosphoribosyltransferase in chromatin. In: McCormick, D.B. and Wright, L.D. (eds) *Vitamins and Coenzymes*. Academic Press, New York, pp. 230–233.

Pham, A.-D. and Sauer, F. (2000) Ubiquitin-activating/conjugating activity of TAF$_{II}$250, a mediator of activation of gene expression in *Drosophila*. *Science* 289, 2357–2360.

Pispa, J. (1965) Animal biotinidase. *Annals of Medicine and Experimental Biology Fenniae* 43, 4–39.

Rawling, J.M., Jackson, T.M., Driscoll, E.R. and Kirkland, J.B. (1994) Dietary niacin deficiency lowers tissue poly(ADP-ribose) and NAD$^+$ concentrations in Fischer-344 rats. *Journal of Nutrition* 124, 1597–1603.

Sims, J.L., Sikorski, G.W., Catino, D.M., Berger, S.J. and Berger, N.A. (1982) Poly(adenosinediphosphoribose) polymerase inhibitors stimulate unscheduled deoxyribonucleic acid synthesis in normal human lymphocytes. *Biochemistry* 21, 1813–1821.

Sommerville, J., Baird, J. and Turner, B.M. (1993) Histone H4 acetylation and transcription in amphibian chromatin. *Journal of Cell Biology* 120, 277–290.

Stanley, J.S., Griffin, J.B. and Zempleni, J. (2001) Biotinylation of histones in human cells: effects of cell proliferation. *European Journal of Biochemistry* 268, 5424–5429.

Taunton, J., Hassig, C.A. and Schreiber, S.L. (1996) A mammalian histone deacetylase related to a yeast transcriptional regulator Rpd3. *Science* 272, 408–411.

Wolffe, A. (1998) *Chromatin*. Academic Press, San Diego, California.

Yano, H., Nakanishi, S., Kimura, K., Hanai, N., Saitoh, Y., Fukui, Y., Nonomura, Y. and Matsuda, Y. (1993) Inhibition of histamine secretion by wortmannin through the blockade of phosphatidylinositol 3-kinase in RBL-2H3 cells. *Journal of Biological Chemistry* 268, 25846–25856.

Yoon, Y.S., Kim, J.W., Kang, K.W., Kim, Y.S., Choi, K.H. and Joe, C.O. (1996) Poly(ADP-ribosyl)ation of histone H1 correlates with internucleosomal DNA fragmentation during apoptosis. *Journal of Biological Chemistry* 271, 9129–9134.

Zempleni, J. (2001) Biotin. In: Bowman, B.A. and Russell, R.M. (eds) *Present Knowledge in Nutrition*, 8th edn. International Life Sciences Institute, Washington, DC, pp. 241–252.

Zempleni, J. and Mock, D.M. (1999) Chemical synthesis of biotinylated histones and analysis by sodium dodecyl sulfate–polyacrylamide gel electrophoresis/streptavidin-peroxidase. *Archives of Biochemistry and Biophysics* 371, 83–88.

Zempleni, J., Stanley, J.S. and Mock, D.M. (2001) Proliferation of peripheral blood mononuclear cells causes increased expression of the sodium-dependent multivitamin transporter gene and increased uptake of pantothenic acid. *FASEB Journal* 15, A964 [abstract].

18 Niacin Status, Poly(ADP-ribose) Metabolism and Genomic Instability

Jennifer C. Spronck and James B. Kirkland
Department of Human Biology and Nutritional Sciences, University of Guelph, Ontario, Canada

Introduction

Cell cycle checkpoints and DNA repair pathways have evolved to handle the detection and repair of DNA defects and prevent their incorporation into permanent mutations during cell division (Kaufmann and Paules, 1996). The pathways that recognize and repair DNA lesions are essential for the accurate transmission of genetic information. If DNA damage is extensive, apoptosis aids in the elimination of genetically altered cells. The coordinated efforts of these processes are required to maintain genomic stability in actively dividing cells throughout the body. Caretaker genes are viewed as guardians of the genome that ensure accurate genomic composition and replication. These include genes involved in mismatch repair, nucleotide and base excision repair, and double strand break repair. Gatekeeper genes encode proteins that determine cellular fates, such as survival or apoptosis. Increased cancer incidence in families with polymorphisms in these genes demonstrates the importance of a wide variety of cellular responses in the maintenance of genomic stability.

In addition to genotype, these cellular processes may be influenced by environmental factors. Epidemiological, laboratory and clinical investigations indicate a relationship between various nutrients and genomic instability or cancer. Niacin deficiency in humans causes the disease pellagra, which is characterized by the '4 Ds', i.e. diarrhoea, dermatitis, dementia and death. The dermatitis is limited to areas of skin that are exposed to sunlight, suggesting that it is caused by problems in DNA repair pathways. Historically, outbreaks of this disease were associated with the introduction of maize as a dietary staple, although maize use by native Americans in North and South America was not associated with pellagra. Native Americans used fireplace ashes and lime water in the preparation of maize products, and alkali is now known to increase the bioavailability of niacin in maize (Carpenter, 1981). Sources of niacin in industrialized countries include meats such as fish, poultry, pork and beef. Cereal grains, seeds and legumes also contain niacin, mainly in the form of nicotinic acid, but it may be bound to complex carbohydrates and proteins, limiting absorption. For this reason, grain products are supplemented with niacin in most developed countries. NAD may also be synthesized from tryptophan in the liver, where 1 mg of excess dietary tryptophan may yield 1/60 mg of niacin (Kirkland and Rawling, 2001). Frank pellagra is seldom reported in developed countries, but suboptimal niacin status may be difficult to diagnose, due to mild symptoms. Despite access to foodstuffs fortified with niacin, there are fractions of the population with surprisingly low niacin status. A study of women in the Malmo Diet and Cancer Study in Sweden revealed that 15% of the females had subclinical niacin deficiency (Jacobson, 1993), while a similar finding was observed in the geriatric population in the USA, with 22% of this group having low niacin status (Knebl and Jacobson, 1992).

NAD Functions and Metabolism

NAD and NADP are known mainly for their redox functions in metabolism. In the oxidized form, they are designated as NAD^+ and $NADP^+$, and in the reduced form as NADH and NADPH, respectively. The major redox role of NADH is to transfer electrons from metabolic intermediates to the electron transport chain for the production of ATP, while NADPH acts as a reducing agent in fatty acid synthesis and other biosynthetic pathways, as well as in oxidant defence (Kirkland and Rawling, 2001).

Poly(ADP-ribose)

In more recent years, non-redox functions of NAD^+ have been discovered that include the use of NAD^+ in the synthesis of poly(ADP-ribose) (pADPr), mono ADP-ribose and cyclic ADP-ribose. The majority of cellular pADPr is made by the nuclear enzyme poly(ADP-ribose) polymerase (PARP). PARP binds to, and is specifically activated by, DNA strand breaks, representing one of the earliest responses to DNA damage in the cell. On activation, PARP uses NAD^+ to synthesize pADPr, on itself (automodification) and on a number of other acceptor proteins (post-translational modification) involved in the maintenance of chromatin architecture and DNA metabolism (D'Amours et al., 1999) (Fig. 18.1). pADPr can vary in length and complexity, containing both linear and branched portions of >200 residues (D'Amours et al., 1999). PARP protein and associated poly(ADP-ribosylation) reactions are thought to play a role in a number of different biological processes such as DNA repair, recombination, apoptosis and the maintenance of genomic stability (Le Rhun et al., 1998; D'Amours et al., 1999).

PARP is a highly conserved, abundant protein found in all higher eukaryotes (D'Amours et al., 1999). The zinc fingers act as a nick sensor, recognizing and binding to DNA strand breaks, independently of DNA sequence (Ikejima et al., 1990). In addition to DNA binding, the N-terminal domain serves as an interface for protein–protein interactions with several partners including histones, DNA polymerase α, X-ray cross-complementing factor-1 (XRCC1) and various transcription factors (Niedergang et al., 2000). This domain also contains a nuclear localization signal and a cysteine protease cleavage site (DEVD) that is targeted during apoptosis (D'Amours et al., 1999).

The automodification of PARP with negatively charged pADPr causes it to lose its affinity for the strand break on DNA, thereby losing its catalytic activity. The automodification domain (middle region of the protein) also contains a BRCT (BRCA1 C-terminus) motif that has also been identified in a number of cell cycle checkpoint and DNA repair proteins (Niedergang et al., 2000). This module enables specific protein–protein interaction between PARP and XRCC1 and DNA ligase III (D'Amours et al., 1999).

The C-terminal domain contains the NAD^+-binding site, and catalytic activities include NAD^+ hydrolysis and initiation, elongation, branching and termination of pADPr synthesis (de Murcia and Menissier, 1994). The synthesis of pADPr is transient and reversible and the half-life of pADPr following DNA damage is <1 min (Alvarez-Gonzalez and Althaus, 1989). Poly(ADP-ribose) glycohydrolase degrades pADPr, releasing free ADP-ribose units and short chains of free pADPr (D'Amours et al., 1999).

PARP and DNA Repair

DNA excision repair pathways include base excision repair for removal of a single, damaged base (e.g. alkylation adduct, oxidatively damaged base, etc.), and nucleotide excision repair for the removal of bulky distortive DNA lesions (e.g. UV light-induced dimers). The initial evidence for a role for PARP in excision repair was the observation that pADPr was synthesized during this process (Durkacz et al., 1980; Jacobson et al., 1980), and inhibition of PARP, nutritional depletion of NAD or overexpression of a mutant PARP DNA-binding domain all potentiated genotoxicity (Nduka et al., 1980; Chatterjee et al., 1991; Molinete et al., 1993) or delayed strand break repair/rejoining (Durkacz et al., 1980; Nduka et al., 1980). Complementary studies demonstrated that lymphocytes from human subjects receiving niacin supplements had less DNA damage

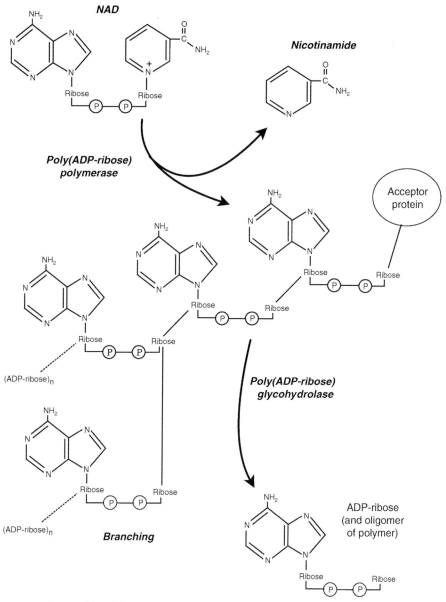

Fig. 18.1. Synthesis and breakdown of poly(ADP-ribose) (pADPr).

following exposure to oxygen radicals (Weitberg, 1989), and addition of niacin to human lymphocytes reduced the level of DNA strand breaks induced by an adenosine deaminase inhibitor (Weitberg and Corvese, 1990). Experiments conducted in one model of PARP-null mice concluded that PARP protein is required for optimal repair (Dantzer et al., 1999). Recent findings have shown a physical interaction between PARP and various components of the excision repair machinery (Masson et al., 1998). PARP interacts via its BRCT motif with XRCC1, an adaptor protein that choreographs the interaction of two important excision repair proteins, DNA ligase III and DNA polymerase β. It has been suggested that PARP may facilitate DNA repair *in vivo* by directly recruiting the DNA repair machinery to the site of DNA damage (D'Amours et al., 1999). If the ability to synthesize pADPr is compromised, unmodified, DNA-bound PARP interferes with the repair process, and DNA strand breaks persist (Satoh and Lindahl, 1992).

Conversely, when PARP was removed from a cell-free system, DNA repair lost its dependence on NAD and proceeded at the same rate as controls (Lindahl et al., 1995). This implies that PARP, per se, is not required for excision repair or the rejoining of DNA strand breaks (D'Amours et al., 1999), but that PARP inhibits repair when it is catalytically inactive. In support of this conclusion, cells from an alternative PARP-null mouse exhibited efficient excision repair following treatment with an alkylating agent or γ-irradiation (Vodenicharov et al., 2000). Reconstitution of these repair pathways in vitro with purified human proteins, in the absence of PARP, has also demonstrated that PARP is not required directly for excision repair (Vodenicharov et al., 2000), but these systems do not have the complexity of chromatic structure found in vivo. Despite these conflicting views on the effect of PARP removal, it is clear that catalytically inactive PARP disrupts excision repair, and this is the situation that may occur during niacin deficiency.

PARP and Apoptosis

Apoptosis is an essential physiological process that regulates cell number during embryogenesis, tissue homeostasis and specialization of the immune system. Suppression of apoptosis and subsequent failure to eliminate cells with damaged DNA is commonly associated with cancer (Schmitt and Lowe, 1999). Some of the biochemical hallmarks of apoptosis include an early, transient burst of pADPr synthesis (Simbulan-Rosenthal et al., 1998), activation of caspases and specific proteolytic cleavage of PARP into 24 and 89 kDa polypeptide fragments (Kaufmann et al., 1993). There are several proposed mechanisms by which the cleavage of PARP may affect apoptotic signalling. Inactivation of PARP via proteolytic cleavage may ensure normal apoptosis by preventing excessive NAD^+ consumption, leading to ATP depletion, and unwanted necrosis. It has also been proposed that PARP fragments play a direct role during apoptosis, or that intact PARP inhibits apoptosis. This is supported by work showing that expression of mutant caspase-uncleavable PARP protects cells from DNA damage-induced apoptosis and actually allows long-term survival and cell division (Halappanavar et al., 1999). Survival of damaged cells can lead to increased evidence of genomic instability and, eventually, carcinogenesis.

Despite the abundance of studies, the exact role PARP plays in the apoptotic pathway remains controversial. Studies using various PARP-null mice have failed to clarify the role of PARP in apoptosis (de Murcia et al., 1997; Leist et al., 1997; Wang et al., 1997; Simbulan-Rosenthal et al., 1998; Masutani et al., 1999). Much of the controversy may stem from the diversity of apoptotic inducers utilized, the cell types studied and the method used for disrupting PARP function.

PARP and p53

p53 is another multifunctional protein that plays a central role in regulating cellular processes such as cell cycle arrest and apoptosis in response to DNA damage (Gottlieb and Oren, 1998). Following genotoxic stress, such as that produced by chemotherapy drugs, UV and ionizing irradiation, p53 protein is stabilized and activated by a number of different post-translational modifications including phosphorylation, dephosphorylation, acetylation and poly(ADP-ribosylation) (Steegenga et al., 1996; Colman et al., 2000). Functioning as a transcription factor, p53 binds to DNA in a sequence-specific manner and mediates transactivation of target genes involved in the execution of cell cycle arrest (e.g. $p21^{WAF1}$, GADD45 and B99) (el Deiry, 1998) and apoptosis (e.g. Bax, Fas/APO1, Killer/DR5) (el Deiry, 1998; Gottlieb and Oren, 1998). p53 can also repress the transcription of many genes including the anti-apoptotic gene Bcl-2 via non-sequence-specific DNA binding (el Deiry, 1998).

Both PARP and p53 are activated in response to DNA damage. The response of PARP to DNA damage and pADPr synthesis is stimulated within seconds to minutes (Berger, 1985), suggesting that poly(ADP-ribosylation) may act as a DNA damage signal generator for downstream effectors. It has been shown that competitive inhibition of PARP results in decreased basal p53 protein levels and impaired p53 stabilization following DNA damage. Cell lines defective in pADPr synthesis, or selected for growth in the absence of NAD^+, displayed reduced basal p53 levels that were insensitive to etoposide-induced up-regulation (Whitacre et al., 1995). In addition, chemical PARP inhibition with 3-aminobenzamide in human glioblastoma cells was shown to suppress radiation-induced accumulation of p53, suggesting that poly (ADP-ribosylation) of p53 is required for protein stabilization and accumulation (Wang et al., 1998). Basal p53 protein was decreased by approximately 50% in PARP-null mouse cells, and p53 up-regulation was also defective (Agarwal et al., 1997). However, PARP-null cells from a different source had low basal levels of p53, but normal up-regulation in response to DNA damage (Wesierska-Gadek et al., 1999). Interestingly, an alternatively spliced form of p53 (p53AS) was found in PARP-null cells. Functionally, the p53AS differs from regular p53 in that it constituitively possesses sequence-specific DNA-binding activity (Bayle et al., 1995) and attenuates apoptosis (Almog et al., 1997). The results from this study demonstrated that PARP-null cells constituitively express the p53AS, and the regular form is extremely unstable in these cells (Wesierska-Gadek et al., 1999). The induction of the two p53 splice forms in response to DNA damage differs between PARP-null and wild-type PARP cells. Induction of regularly spliced p53 occurred solely in the wild-type PARP cells, whereas induction of p53AS was kinetically different between the PARP-null and wild-type PARP cells.

There are several mechanisms by which PARP may regulate p53 expression or function. There are pADPr-binding motifs in the p53 DNA-binding domain and oligomerization domain that enable strong non-covalent interaction between p53 and free pADPr or pADPr-bound PARP (Malanga et al., 1998). In addition, earlier studies have shown that p53 is modified post-translationally with pADPr *in vitro* (Wesierska-Gadek et al., 1996a; Kumari et al., 1998) as well as during the early stages of apoptosis *in vivo* (Simbulan-Rosenthal et al., 1999a; Smulson et al., 2000). p53–PARP complexes have been isolated from whole-cell and *in vitro* environments (Wesierska-Gadek et al., 1996a; Vaziri et al., 1997; Kumari et al., 1998). It has been proposed that this type of modification of p53 with covalent and non-covalent association with pADPr regulates p53 expression (basal and induced) as well as the DNA-binding properties of p53 (Wesierska-Gadek et al., 1996b; Wang et al., 1998; Malanga et al.,

1998). Functionally, inactivation of PARP with chemical inhibitors prevented the accumulation of p53-responsive gene products at both the protein and mRNA level following DNA damage (Vaziri et al., 1997; Wang et al., 1998).

The biological relevance of impaired p53 function has been evaluated in terms of the ability of cells to undergo apoptosis and/or cell cycle arrest. Again, the literature is divergent, with some investigators reporting impaired apoptosis in response to DNA damage (Whitacre et al., 1995) and cell cycle arrest (Nozaki et al., 1994; Masutani et al., 1995), while others conclude that p53 activation and function is PARP independent (Agarwal et al., 1997).

PARP Homologues

For many years, it was assumed that PARP activity was associated with a single protein. Recently, the discovery that PARP-null cells synthesize small quantities of pADPr (Shieh et al., 1998) led to the identification of multiple proteins that exhibited PARP-like activity. Like PARP (sometimes referred to now as PARP-1), PARP-2 is activated by genotoxic stress, although the DNA-binding domains differ in structure (Ame et al., 1999). PARP-2 is localized in the nucleus and may provide a distinctive catalytic activity or represent a redundant back-up for PARP (Oliver et al., 1999). Another homologue, PARP-3, has been identified, but its function is unknown (Johansson, 1999).

Tankyrase, a protein with homology to ankyrin and to the catalytic domain of PARP, was identified and localized to human telomeres (Smith et al., 1998). *In vitro*, tankyrase binds to telomere repeat-binding factor-1 (TRF-1), a protein involved in negatively regulating telomere length (Smith et al., 1998). Tankyrase will synthesize pADPr on TRF-1, removing it from the telomere and encouraging telomerase activity. In addition, tankyrase has also been found to relocate to nuclear pore complexes and centrosomes during different phases of the cell cycle (Smith and de Lange, 1999).

A fourth PARP homologue, vault-PARP (vPARP), does not require DNA for activity and is found in the cytoplasm associated with vault particles. It is also associated with the mitotic spindle and telomerase-associated proteins (Kickhoefer et al., 1999a,b). Although the function of vPARP is unclear, the identification and localization of PARP homologues to mitotic apparatus and DNA suggests a role for vPARP activity in the maintenance of genomic integrity (Jacobson and Jacobson, 1999).

Mono ADP-ribosylation and Cyclic ADP-ribose Formation

There are a multitude of mono ADP-ribosyltransferases in most cells, with poorly defined roles. Many of these post-translational modifications occur on G proteins, and may play a role in regulating signal transduction associated with cell division (Okazaki and Moss, 1999). Niacin is also required for the synthesis of cyclic ADP-ribose and the newly discovered molecule nicotinic acid adenine dinucleotide phosphate (NAADP), both of which are involved in the regulation of intracellular calcium signalling pathways (Lee, 2001). Altered calcium signalling could lead to changes in the regulation of apoptosis and the cell cycle.

Non-ADP-ribosylation Roles for Niacin in Genomic Stability

Oxidant defence/stress

NADPH is synthesized in the pentose-phosphate pathway, and is the engine that drives detoxification of oxygen radicals by glutathione peroxidase. This enzyme converts hydrogen peroxide to water, preventing the formation of highly damaging hydroxyl radicals, which can react with all cellular macromolecules, including DNA. The formation of mutagenic DNA lesions, such as 8-hydroxy-deoxy-guanosine (8-OHdG), may be increased during niacin deficiency, but this has not been investigated.

SIR2

An additional role for NAD in genomic stability may be through the action of SIR2, an NAD-dependent histone deacetylase. Deacetylation leads to a more compact chromatin structure and gene silencing. It also appears to protect sensitive areas of chromatin, such as telomeres, against translocation events, and to play a role in extended life span associated with caloric restriction (Lin et al., 2000). SIR2 is also active in the regulation of p53 through deacetylation (Vaziri et al., 2001).

PARP, Niacin Status and Genomic Stability

There is not a large body of direct work on niacin status and genomic instability or cancer. In this section, we have combined the topics of niacin deficiency and PARP disruption, due to the likely overlap in mechanisms. We will start with PARP, and then summarize the current knowledge on niacin deficiency.

Recently, PARP has been described as a guardian of the genome, and both caretaker and gatekeeper functions have been ascribed to this highly conserved enzyme (Tsutsumi et al., 2001). PARP-null mice are the latest in a number of different models used to study PARP function, each with specific advantages and limitations. Often, the findings from the early competitive inhibitor studies and expression of catalytically inactive PARP fragments are in agreement with the newer PARP-null models. However, there are also contradictions, but there are good mechanistic reasons for these. For example, chemical inhibitors and PARP fragment models create catalytically inactive forms of PARP that bind strongly to DNA strand breaks, due to the lack of automodification. This leads to an inhibition of subsequent repair steps. In contrast, PARP-null models have no PARP protein present to influence events at the site of damage. PARP-null models do remove the protein–protein interactions initiated by PARP at the damage site, and both types of models have nullified PARP-1-dependent synthesis of pADPr. A better understanding of the activity and function of PARP homologues (PARP-2, PARP-3, tankyrase and vPARP) may also help to clarify the discrepancies in PARP-null models.

In comparing these models with the possible effects of niacin deficiency, it must be noted that niacin deficiency may decrease PARP activity, but it does not affect the level of PARP expression

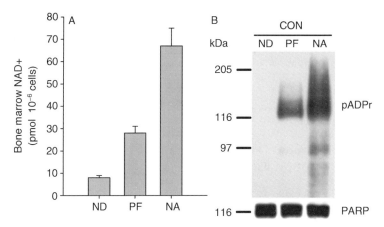

Fig. 18.2. (A) The high sensitivity of bone marrow cells to niacin deficiency (ND) and pharmacological intakes of nicotinic acid (NA, 4 g kg^{-1} diet), versus pair-fed controls (PF) receiving the required intake of 0.03 g kg^{-1} diet. Bone marrow NAD$^+$ varies about tenfold in this model. (B) Western blot of protein-bound pADPr (top) and PARP protein (bottom). The total protein and distribution pattern were identical in each lane by fast green staining, but the pADPr modification pattern was affected dramatically by niacin deficiency (ND) and pharmacological supplementation (NA), relative to pair-fed controls (PF). The differences in pADPr levels are not due to changes in PARP protein expression (for more extensive data from these models, see Boyonoski et al. (2002a,b)).

in any of the tissues that we have examined (unpublished data; Fig. 18.2). As discussed above, catalytically inactive PARP molecules are known to stay bound to DNA strand breaks and inhibit repair (Satoh and Lindahl, 1992; Satoh et al., 1994). In addition, the activity of alternative PARPs, such as PARP-2 and 3, tankyrase 1 and 2, and vPARP, may be affected directly by niacin status and the availability of NAD$^+$. Mono ADP-ribosylation reactions and the formation of the calcium signalling molecules cADPr and NAADP may also be affected by niacin deficiency, as well as NAD-dependant deacetylation of histones and p53. Thus, niacin deficiency is more complex than the specific models designed to study PARP function.

There are a variety of methods used to quantify genomic instability. Structural and numerical chromosomal aberrations can be assessed using cytogenetic methods (e.g. classical chromosomal aberration analysis or micronuclei formation), while DNA damage (adducts, strand breaks, alkali-labile sites, cross-linking) can be measured using a wide variety of assays, including the comet assay (Albertini et al., 2000). Sister chromatid exchange (SCE) is thought to be induced by DNA double strand breaks and homologous recombination (Tucker and Preston, 1996). The exact mechanisms underlying the formation of SCEs may be poorly understood, but they are often included as a genotoxic end point to reflect DNA damage (Tucker and Preston, 1996; Albertini et al., 2000).

Inhibition of PARP, either with chemical inhibitors, with substrate limitation or through molecular genetic approaches, invariably increases SCE frequency. Competitive inhibitors of PARP increased the basal frequency of SCE three- to tenfold (Oikawa et al., 1980; Hori, 1981; Morgan and Cleaver, 1982) and potentiated the increase in SCE caused by alkylating agents (Morgan and Cleaver, 1982; Park et al., 1983). Cells expressing catalytically inactive PARP DNA-binding fragments were extremely sensitive to alkylating agents and had significantly higher levels of spontaneous SCEs (Schreiber et al., 1995).

In addition, all PARP-null mice are extremely sensitive to genotoxic stressors such as ionizing radiation and alkylating drugs (Wang et al., 1997; Trucco et al., 1999; Masutani et al., 2000). PARP-null mice displayed a two- to threefold increase in spontaneous SCEs and enhanced SCE and micronuclei formation following genotoxic stress (de Murcia et al., 1997; Wang et al., 1997). Alkylating agent- or ionizing radiation-induced damage resulted in a 33- to 36-fold increase in chromatid breaks in one model of PARP-null mice (de Murcia et al., 1997). Exciting additional support for the role of PARP in genomic stability and carcinogenesis comes from a recent study in which another model of PARP-null mice exhibited an enhanced susceptibility to nitrosamine-induced tumorigenesis (Tsutsumi et al., 2001).

Repair of DNA double strand breaks appears to occur by both homologous and non-homologous recombination (Alberts et al., 1994). SCEs occur by homologous recombination and it has been proposed that PARP decreases inappropriate homologous recombination by acting as an anti-recombinogenic factor at the site of breakage (Satoh and Lindahl, 1992). Two potential mechanisms have been proposed for the prevention of unwanted recombination events: (i) PARP rapidly and tightly binds to DNA strand breaks thus blocking the access of exonucleases to the DNA; and (ii) the cloud of negatively charged pADPr at the site of strand breakage repels free DNA ends by electrostatic repulsion, preventing unwanted chromosomal translocation (Satoh and Lindahl, 1992; Lindahl et al., 1995).

Other markers of genomic instability were assessed in different experimental systems. Analysis of telomere length from wild-type PARP, heterozygous PARP (PARP +/−) and PARP-null mice revealed an inverse effect of PARP activity on telomere length (d'Adda et al., 1999). Telomeres are thought to prevent chromosomal fusions and are therefore considered to be involved in the regulation of genomic stability (McClintock, 1941). In one model of PARP-null mice, there was a dramatic increase in the number of end-to-end chromosome fusions, aneuploid cells and chromosome fragments compared with wild-type PARP mice, with heterozygous PARP mice exhibiting an intermediate level of genomic instability (d'Adda et al., 1999). In contrast, an alternative model demonstrated that telomere lengths in PARP-null primary cells were not different from those in wild-type PARP cells, and that the frequencies of end-to-end fusions were elevated, but to a lower extent than previously reported (Samper et al., 2001).

Immortalized fibroblasts from PARP-null mice were found to have unstable tetraploid populations, which were eliminated by the reintroduction of PARP cDNA (Simbulan-Rosenthal et al., 1999b). The genetic alterations were characterized by comparative genomic hybridization analysis and it was determined that PARP-null mice and the immortalized PARP-null fibroblasts had gains in regions of chromosome 4, 5 and 14 as well as a deletion in chromosome 14 (Simbulan-Rosenthal et al., 1999b). As both aneuploidy and polyploidy have been associated with carcinogenesis (Halloway et al., 1999), these findings provide another end point implicating PARP in the maintenance of genomic stability and prevention of cancer.

To provide further insight into the mechanisms by which PARP might be affecting genomic stability, gene expression was assessed by oligonucleotide microarray analysis. Primary fibroblasts from PARP-null mice down-regulated several genes involved in the regulation of cell cycle progression and mitosis, while up-regulating genes that have been implicated in cancer initiation or progression (Simbulan-Rosenthal et al., 2000).

Thus, decreased PARP activity may enhance genomic instability by affecting DNA repair, recombination, apoptosis and regulation of p53 expression/function, and therefore increase the risk of cancer development.

Niacin, Genomic Instability and Cancer

We have discussed the mechanisms by which niacin status might impact on genomic instability. This last section is a summary of studies that have addressed this question more directly. The largest group of studies in this area have focused on some aspect of carcinogenesis as an end point, which is an indirect but sensitive indicator of genomic instability.

Epidemiological studies report an association between oesophageal cancer and marginal niacin intake in populations consuming low-protein, maize-based diets in South Africa and specific provinces in China (Hageman and Stierum, 2001; Kirkland and Rawling, 2001). In Italy, the risk of oesophageal cancer associated with maize consumption is elevated further when combined with heavy alcohol consumption (Franceschi et al., 1990), and low niacin intake in the USA is associated with an increased oral cancer risk (Marshall et al., 1992). In the Linxian region in China, marginal niacin intake has been associated with oesophageal cancer; however, in a subsequent intervention study, supplementation with a combination of niacin and riboflavin did not decrease the incidence of oesophageal cancer nor reduce cancer mortality (Blot et al., 1995). In this study, 5 years of supplementation may not have addressed the role of niacin status early enough in the cancer process. There are very few data regarding the effects of large doses of nicotinic acid or nicotinamide in humans on carcinogenesis. However, people receiving pharmacological doses of nicotinic acid for the treatment of hyperlipidaemia do not exhibit lower overall cancer incidence (Kirkland and Rawling, 2001).

The relationship between niacin nutrition in animals and other forms of cancer is complex (Bryan, 1986). An animal model of malnutrition and oesophageal cancer induced by an alkylating agent in maize-fed rats showed that addition of nicotinic acid reduced tumour incidence, size and progression (Van Rensburg et al., 1986). In contrast, the results from animal models of renal and/or liver cancer have not fully supported a role for niacin in carcinogenesis. Niacin deficiency or pharmacological supplementation with either nicotinic acid or nicotinamide did not affect diethylnitrosamine-induced altered hepatic foci in rats (Jackson et al., 1995; Rawling et al., 1995), while niacin deficiency enhanced N-nitrosodimethylamine-induced renal carcinogenesis in rats (Miller and Burns, 1984).

The differences in susceptibility to various carcinogens and the development of tissue-specific cancers are influenced by the sensitivity of these target tissues to changes in dietary niacin status. Our results have demonstrated that there is a large variation among tissues with respect to NAD depletion during niacin deficiency. Tissues such as the liver, lung and kidney have very high initial NAD levels and they are relatively resistant to depletion during deficiency (Rawling et al., 1994, 1995, 1996). Conversely, bone marrow and skin are very sensitive to NAD depletion, and both of these tissues display an increased susceptibility to carcinogenesis when challenged by alkylating agents or UV light, respectively (Boyonoski et al., 2002a; Shah et al., 2002). The common thread between these two tissues is a very high rate of cell division followed by loss (skin) or export (bone marrow). It is logical that such tissues struggle to maintain niacin status as they constantly produce new cells.

Using a rat model, we have found that bone marrow is the tissue most sensitive to changes in dietary niacin levels (Fig. 18.2), and we have shown that niacin deficiency enhances the rate of appearance and number of ethylnitrosourea-induced cancers, most of which were leukaemias that originated in the bone marrow (Boyonoski et al., 2002a). We have also shown that pharmacological supplementation with nicotinamide or nicotinic acid decreases the progression of ethylnitrosourea-induced carcinogenesis (Boyonoski et al., 2002b). In these experiments, niacin status, varying from deficient through pharmacological intake, caused dramatic changes in bone marrow NAD^+ (Fig. 18.2A), and in basal and DNA damage-induced pADPr levels (Fig. 18.2B) (Boyonoski et al., 2002a,b).

The skin is another tissue that is very sensitive to niacin status, which is interesting, given the sun-sensitive lesions that occur in humans during niacin deficiency. Like the bone marrow models, experiments with skin have been conducted in both deficient and pharmacological ranges of niacin intake. Niacin deficiency in rats has been shown to increase the development of UV-induced skin cancers (Shah et al., 2002), even though the deficiency state was relatively mild, due to the efficient conversion of tryptophan to niacin in mice. Larger quantities of niacin, provided to mice that were already receiving the prescribed amount of dietary niacin, by oral or topical administration, also prevented UV-induced photocarcinogenesis (Gensler, 1997; Gensler et al., 1999), further supporting a role for niacin nutrition in the prevention of carcinogenesis. These experiments showed that niacin supplementation exerted its effects, at least in part, through immune enhancement (Gensler et al., 1999).

We recently have conducted studies to examine genomic instability more directly. Niacin deficiency, alone, caused a sixfold increase in micronuclei frequency in rat bone marrow polychromatic erythrocytes (Spronck and Kirkland, 2002). Figure 18.3 shows a fluorescent image of acridine orange-stained bone marrow cells from a niacin-deficient rat, with micronucleated polychromatic erythrocytes. Micronuclei represent whole chromosomes that did not sort correctly into daughter nuclei (aneugenic) or fragments of chromosomes that lack centromeres (clastogenic). These two types represent different mechanisms of formation, with aneugenic micronuclei resulting from disruption of spindle fibres or centromere function, and clastogenic micronuclei originating from sites of DNA damage. We have not yet determined the type of micronuclei that are present during niacin deficiency, but we suspect that a significant proportion of the micronuclei are aneugenic, as we did not find any effect of niacin deficiency

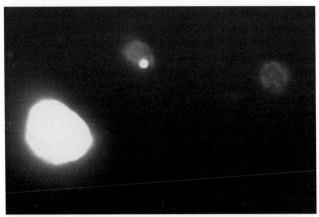

Fig. 18.3. Rat bone marrow cells stained with acridine orange. Polychromatic erythrocytes (PCEs) are young red blood cells that have lost their nucleus, but still contain RNA, which provides a red fluorescence. Micronuclei, which stain bright green, are scored only in the PCE fraction, as their presence in nucleated cells is difficult to detect, and older red cells with micronuclei tend to be removed by the spleen. A lymphocyte with a large complete nucleus is visible to the left.

alone on DNA strand breaks (Boyonoski et al., 2002a), suggesting that a dramatic level of clastogenic events is not the underlying cause. Niacin deficiency also appears to increase the additional micronuclei induced by genotoxic drugs and the level of strand breaks induced by DNA-damaging agents (unpublished data), suggesting a sensitivity to clastogenic events during niacin deficiency.

Figure 18.4 shows a metaphase spread from niacin-deficient rat bone marrow, differentially stained with bromodeoxyuridine to allow the identification of homologous recombination events between chromatids (arrows), referred to as SCE. Although the mechanisms associated with SCE are not well understood, they are known to reflect various types of genomic instability and to increase in response to almost all types of genotoxic insult. As discussed earlier, all models of PARP disruption cause an increase in SCE frequency. We found that niacin deficiency alone caused a threefold increase in SCE frequency in rat bone marrow cells (Spronck and Kirkland, 2002).

We have also found that niacin deficiency appears to increase the frequency of chromosomal aberrations in bone marrow cells (Fig. 18.5), and appears to change the structure of the chromosomes in metaphase spreads (unpublished data). We also have preliminary data showing that niacin deficiency causes abnormal expression of p53 and increases DNA strand breaks following etoposide exposure (unpublished data).

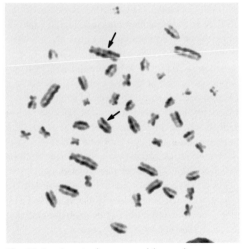

Fig. 18.4. A metaphase spread from a bone marrow cell of a niacin-deficient rat. Pre-treatment of rats with bromodeoxyuridine allows differential staining of newly synthesized chromatids. Exchanges between the old and new are referred to as sister chromatid exchanges and they reflect homologous recombination events, which are increased with DNA damage or genomic instability.

The developing picture is that niacin deficiency alone causes increases in gross chromosomal instability. The mechanisms by which niacin deficiency induces genomic instability in our model remain largely unknown. The large changes in

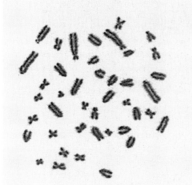 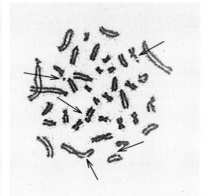

Fig. 18.5. Metaphase spreads of bone marrow cells from control and niacin-deficient rats. The niacin-deficient spread shows a variety of aberrations and the chromosomes have a rough appearance.

pADPr accumulation implicate pADPr synthesis as an underlying causative factor. However, there are a number of PARP homologues capable of pADPr synthesis as well as a number of different mechanisms by which pADPr may modulate the genomic stability. Altered p53 signalling can impair cell cycle arrest, decrease DNA repair efficiency and allow the survival of heavily damaged cells through disruption in apoptotic signalling. A comprehensive understanding of these mechanisms would further our knowledge of niacin metabolism, PARP function and carcinogenesis.

The implications of niacin deficiency-induced genomic instability are manyfold. Although frank niacin deficiency is rare in the Western world, there exist groups of individuals (women and the elderly) in the human population that exhibit subclinical niacin deficiency. In addition, 40% of cancer patients receiving chemotherapy are niacin deficient (Inculet et al., 1987). We propose that deficiency may increase the risk of developing secondary, treatment-related malignancies. Most leukaemias are characterized by chromosomal translocations, and impaired pADPr synthesis may encourage these types of illegitimate recombination events. Skin cancer is another important health problem in human populations. In animal models, this cancer is highly influenced by niacin status. Due to the limited research focus on niacin in the past, significant relationships between niacin status and cancer incidence may be escaping our attention. The basic research reviewed in this article illustrates the potential of niacin status to impact on genomic stability and cancer incidence in human populations.

References

Agarwal, M.L., Agarwal, A., Taylor, W.R., Wang, Z.Q., Wagner, E.F. and Stark, G.R. (1997) Defective induction but normal activation and function of p53 in mouse cells lacking poly-ADP-ribose polymerase. *Oncogene* 15, 1035–1041.

Albertini, R.J., Anderson, D., Douglas, G.R., Hagmar, L., Hemminki, K., Merlo, F., Natarajan, A.T., Norppa, H., Shuker, D.E., Tice, R., Waters, M.D. and Aitio, A. (2000) IPCS guidelines for the monitoring of genotoxic effects of carcinogens in humans. International Programme on Chemical Safety. *Mutation Research* 463, 111–172.

Alberts, B., Bray, D., Lewis, J., Raff, M., Roberts, K. and Watson, J.D. (1994) Cancer. In: Robertson, M., Adams, R., Cobert, S.M. and Goertzen, D. (eds) *Molecular Biology of the Cell*. Garland Publishing, New York, pp. 1255–1294.

Almog, N., Li, R., Peled, A., Schwartz, D., Wolkowicz, R., Goldfinger, N., Pei, H. and Rotter, V. (1997) The murine C′-terminally alternatively spliced form of p53 induces attenuated apoptosis in myeloid cells. *Molecular and Cellular Biology* 17, 713–722.

Alvarez-Gonzalez, R. and Althaus, F.R. (1989) Poly(ADP-ribose) catabolism in mammalian cells exposed to DNA-damaging agents. *Mutation Research* 218, 67–74.

Ame, J.C., Rolli, V., Schreiber, V., Niedergang, C., Apiou, F., Decker, P., Muller, S., Hoger, T., Menissier-de Murcia, J. and de Murcia, G. (1999)

PARP-2, a novel mammalian DNA damage-dependent poly(ADP-ribose) polymerase. *Journal of Biological Chemistry* 274, 17860–17868.

Bayle, J.H., Elenbaas, B. and Levine, A.J. (1995) The carboxyl-terminal domain of the p53 protein regulates sequence-specific DNA binding through its nonspecific nucleic acid-binding activity. *Proceedings of the National Academy of Sciences USA* 92, 5729–5733.

Berger, N.A. (1985) Poly(ADP-ribose) in the cellular response to DNA damage. *Radiation Research* 101, 4–15.

Blot, W.J., Li, J.Y., Taylor, P.R., Guo, W., Dawsey, S.M. and Li, B. (1995) The Linxian trials: mortality rates by vitamin–mineral intervention group. *American Journal of Clinical Nutrition* 62 (Supplement), 1424S–1426S.

Boyonoski, A.C., Spronck, J.C., Gallacher, L.M., Jacobs, R.M., Shah, G.M., Poirier, G.G. and Kirkland, J.B. (2002a) Niacin deficiency decreases bone marrow poly(ADP-ribose) and the latency of ethylnitrosourea-induced carcinogenesis in rats. *Journal of Nutrition* 132, 108–114.

Boyonoski, A.C., Spronck, J.C., Jacobs, R.M., Shah, G.M., Poirier G.G. and Kirkland, J.B. (2002b) Pharmacological intakes of niacin increase bone marrow poly(ADP-ribose) and the latency of ethylnitrosourea-induced carcinogenesis in rats. *Journal of Nutrition* 132, 115–120.

Bryan, G.T. (1986) The influence of niacin and nicotinamide on *in vivo* carcinogenesis. *Advances in Experimental Medicine and Biology* 206, 331–338.

Carpenter, K.J. (1981) *Pellagra*. Hutchinson Ross, Stroudsburg.

Chatterjee, S., Cheng, M.F., Berger, S.J. and Berger, N.A. (1991) Alkylating agent hypersensitivity in poly(adenosine diphosphate-ribose) polymerase deficient cell lines. *Cancer Communications* 3, 71–75.

Colman, M.S., Afshari, C.A. and Baffett, J.C. (2000) Regulation of p53 stability and activity in response to genotoxic stress. *Mutation Research* 462, 179–188.

d'Adda, D.F., Hande, M.P., Tong, W.M., Lansdorp, P.M., Wang, Z.Q. and Jackson, S.P. (1999) Functions of poly(ADP-ribose) polymerase in controlling telomere length and chromosomal stability. *Nature Genetics* 23, 76–80.

D'Amours, D., Desnoyers, S., D'Silva, I. and Poirier, G.G. (1999) Poly(ADP-ribosyl)ation reactions in the regulation of nuclear functions. *Biochemical Journal* 342, 249–268.

Dantzer, F., Schreiber, V., Niedergang, C., Trucco, C., Flatter, E., De La Rubia, G., Oliver, J., Rolli, V., Menissier-de Murcia, J. and de Murcia, G. (1999) Involvement of poly(ADP-ribose) polymerase in base excision repair. *Biochimie* 81, 69–75.

de Murcia, G. and Menissier, D.M. (1994) Poly(ADP-ribose) polymerase: a molecular nick-sensor. *Trends in Biochemical Sciences* 19, 172–176.

de Murcia, J.M., Niedergang, C., Trucco, C., Ricoul, M., Dutrillaux, B., Mark, M., Oliver, F.J., Masson, M., Dierich, A., LeMeur, M., Walztinger, C., Chambon, P. and de Murcia, G. (1997) Requirement of poly(ADP-ribose) polymerase in recovery from DNA damage in mice and in cells. *Proceedings of the National Academy of Sciences USA* 94, 7303–7307.

Durkacz, B.W., Omidiji, O., Gray, D.A. and Shall, S. (1980) (ADP-ribose)$_n$ participates in DNA excision repair. *Nature* 283, 593–596.

el Deiry, W.S. (1998) Regulation of p53 downstream genes. *Seminars in Cancer Biology* 8, 345–357.

Franceschi, S., Bidoli, E., Baron, A.E. and La Vecchia, C. (1990) Maize and risk of cancers of the oral cavity, pharynx, and esophagus in northeastern Italy. *Journal of the National Cancer Institute* 82, 1407–1411.

Gensler, H.L. (1997) Prevention of photoimmunosuppression and photocarcinogenesis by topical nicotinamide. *Nutrition and Cancer* 29, 157–162.

Gensler, H.L., Williams, T., Huang, A.C. and Jacobson, E.L. (1999) Oral niacin prevents photocarcinogenesis and photoimmunosuppression in mice. *Nutrition and Cancer* 34, 36–41.

Gottlieb, T.M. and Oren, M. (1998) p53 and apoptosis. *Seminars in Cancer Biology* 8, 359–368.

Hageman, G.J. and Stierum, R.H. (2001) Niacin, poly (ADP-ribose) polymerase-1 and genomic stability. *Mutation Research* 475, 45–56.

Halappanavar, S.S., Rhun, Y.L., Mounir, S., Martins, L.M., Huot, J., Earnshaw, W.C. and Shah, G.M. (1999) Survival and proliferation of cells expressing caspase-uncleavable poly(ADP-ribose) polymerase in response to death-inducing DNA damage by an alkylating agent. *Journal of Biological Chemistry* 274, 37097–37104.

Halloway, S.L., Poruthu, J. and Scata, K. (1999) Chromosome segregation and cancer. *Experimental Cell Research* 253, 308–314.

Hori, T. (1981) High incidence of sister chromatid exchanges and chromatid interchanges in the conditions of lowered activity of poly(ADP-ribose) polymerase. *Biochemical and Biophysical Research Communications* 102, 38–45.

Ikejima, M., Noguchi, S., Yamashita, R., Ogura, T., Sugimura, T., Gill, D.M. and Miwa, M. (1990) The zinc fingers of human poly(ADP-ribose) polymerase are differentially required for the recognition of DNA breaks and nicks and the consequent enzyme activation. *Journal of Biological Chemistry* 265, 21907–21913.

Inculet, R.I., Norton, J.A., Nichoalds, G.E., Maher, M.M., White, D.E. and Brennan, M.F. (1987) Water-soluble vitamins in cancer patients on parenteral nutrition: a prospective study. *Journal of Parenteral and Enteral Nutrition* 11, 243–249.

Jackson, T.M., Rawling, J.M., Roebuck, B.D. and Kirkland, J.B. (1995) Large supplements of nicotinic acid and nicotinamide increase tissue NAD$^+$ and poly(ADP-ribose) levels but do not affect diethylnitrosamine-induced altered hepatic foci in Fischer-344 rats. *Journal of Nutrition* 125, 1455–1461.

Jacobson, E.L. (1993) Niacin deficiency and cancer in women. *Journal of the American College of Nutrition* 12, 412–416.

Jacobson, M.K. and Jacobson, E.L. (1999) Discovering new ADP-ribose polymer cycles: protecting the genome and more. *Trends in Biochemical Sciences* 24, 415–417.

Jacobson, M.K., Levi, V., Juarez-Salinas, H., Barton, R.A. and Jacobson, E.L. (1980) Effect of carcinogenic *N*-alkyl-*N*-nitroso compounds on nicotinamide adenine dinucleotide metabolism. *Cancer Research* 40, 1797–1802.

Johansson, M. (1999) A human poly(ADP-ribose) polymerase gene family (ADPRTL): cDNA cloning of two novel poly(ADP-ribose) polymerase homologues. *Genomics* 57, 442–445.

Kaufmann, W.K. and Paules, R.S. (1996) DNA damage and cell cycle checkpoints. *FASEB Journal* 10, 238–247.

Kaufmann, S.H., Desnoyers, S., Ottaviano, Y., Davidson, N.E. and Poirier, G.G. (1993) Specific proteolytic cleavage of poly(ADP-ribose) polymerase: an early marker of chemotherapy-induced apoptosis. *Cancer Research* 53, 3976–3985.

Kickhoefer, V.A., Siva, A.C., Kedersha, N.L., Inman, E.M., Ruland, C., Streuli, M. and Rome, L.H. (1999a) The 193-kD vault protein, VPARP, is a novel poly(ADP-ribose) polymerase. *Journal of Cell Biology* 146, 917–928.

Kickhoefer, V.A., Stephen, A.G., Harrington, L., Robinson, M.O. and Rome, L.H. (1999b) Vaults and telomerase share a common subunit, TEP1. *Journal of Biological Chemistry* 274, 32712–32717.

Kirkland, J.B. and Rawling, J.M. (2001) Niacin. In: Rucker R, Suttie J.W., McCormick D.M. and Machlin L.J. (eds) *Handbook of Vitamins*. Marcel Dekker, New York, pp. 211–252.

Knebl, J.A. and Jacobson, E.L. (1992) Assessment of niacin status in an elderly population. *Gerontology* 32, 247A.

Kumari, S.R., Mendoza-Alvarez, H. and Alvarez-Gonzalez, R. (1998) Functional interactions of p53 with poly(ADP-ribose) polymerase (PARP) during apoptosis following DNA damage: covalent poly (ADP-ribosyl)ation of p53 by exogenous PARP and noncovalent binding of p53 to the M(r) 85,000 proteolytic fragment. *Cancer Research* 58, 5075–5078.

Le Rhun, Y., Kirkland, J.B. and Shah, G.M. (1998) Cellular responses to DNA damage in the absence of poly(ADP-ribose) polymerase. *Biochemical and Biophysical Research Communications* 245, 1–10.

Lee, H.C. (2001) Physiological functions of cyclic ADP-ribose and NAADP as calcium messengers. *Annual Reviews of Pharmacology and Toxicology* 41, 317–345.

Leist, M., Single, B., Kunstle, G., Volbracht, C., Hentze, H. and Nicotera, P. (1997) Apoptosis in the absence of poly-(ADP-ribose) polymerase. *Biochemical and Biophysical Research Communications* 233, 518–522.

Lin, S.J., Defossez, P.A. and Guarente, L. (2000) Requirement of NAD and SIR2 for life-span extension by calorie restriction in *Saccharomyces cerevisiae*. *Science* 289, 2126–2128.

Lindahl, T., Satoh, M.S., Poirier, G.G. and Klungland, A. (1995) Post-translational modification of poly (ADP-ribose) polymerase induced by DNA strand breaks. *Trends in Biochemical Sciences* 20, 405–411.

Malanga, M., Pleschke, J.M., Kleczkowska, H.E. and Althaus, F.R. (1998) Poly(ADP-ribose) binds to specific domains of p53 and alters its DNA binding functions. *Journal of Biological Chemistry* 273, 11839–11843.

Marshall, J.R., Graham, S., Haughey, B.P., Shedd, D., O'Shea, R., Brasure, J., Wilkinson, G.S. and West, D. (1992) Smoking, alcohol, dentition and diet in the epidemiology of oral cancer. *European Journal of Cancer B Oral Oncology* 28B, 9–15.

Masson, M., Niedergang, C., Schreiber, V., Muller, S., Menissier-de Murcia, J. and de Murcia, G. (1998) XRCC1 is specifically associated with poly(ADP-ribose) polymerase and negatively regulates its activity following DNA damage. *Molecular and Cellular Biology* 18, 3563–3571.

Masutani, M., Nozaki, T., Wakabayashi, K. and Sugimura, T. (1995) Role of poly(ADP-ribose) polymerase in cell-cycle checkpoint mechanisms following gamma-irradiation. *Biochimie* 77, 462–465.

Masutani, M., Suzuki, H., Kamada, N., Watanabe, M., Ueda, O., Nozaki, T., Jishage, K., Watanabe, T., Sugimoto, T., Nakagama, H., Ochiya, T. and Sugimura, T. (1999) Poly(ADP-ribose) polymerase gene disruption conferred mice resistant to streptozotocin-induced diabetes. *Proceedings of the National Academy of Sciences USA* 96, 2301–2304.

Masutani, M., Nozaki, T., Nakamoto, K., Nakagama, H., Suzuki, H., Kusuoka, O., Tsutsumi, M. and Sugimura, T. (2000) The response of Parp knockout mice against DNA damaging agents. *Mutation Research* 462, 159–166.

McClintock, B. (1941) The stability of broken ends of chromosomes in *Zea mays*. *Genetics* 25, 234–282.

Miller, E.G. and Burns, H. Jr (1984) *N*-Nitrosodimethylamine carcinogenesis in nicotinamide-deficient rats. *Cancer Research* 44, 1478–1482.

Molinete, M., Vermeulen, W., Burkle, A., Menissier-de Murcia, J., Kupper, J.H., Hoeijmakers, J.H. and de Murcia, G. (1993) Overproduction of the poly (ADP-ribose) polymerase DNA-binding domain blocks alkylation-induced DNA repair synthesis in mammalian cells. *EMBO Journal* 12, 2109–2117.

Morgan, W.F. and Cleaver, J.E. (1982) 3-Aminobenzamide synergistically increases sister-chromatid exchanges in cells exposed to methyl methanesulfonate but not to ultraviolet light. *Mutation Research* 104, 361–366.

Nduka, N., Skidmore, C.J. and Shall, S. (1980) The enhancement of cytotoxicity of *N*-methyl-*N*-nitrosourea and of gamma-radiation by inhibitors of poly(ADP-ribose) polymerase. *European Journal of Biochemistry* 105, 525–530.

Niedergang, C., Oliver, F.J., Menissier-de-Murcia, J. and de Murcia, G. (2000) Involvement of poly(ADP-ribose) polymerase in the cellular response to DNA damage. In: Szabo, C. (ed.) *Cell Death: the Role of PARP*. CRC Press, Boca Raton, Florida, pp. 183–207.

Nozaki, T., Masutani, M., Akagawa, T., Sugimura, T. and Esumi, H. (1994) Suppression of G1 arrest and enhancement of G2 arrest by inhibitors of poly(ADP-ribose) polymerase: possible involvement of poly(ADP-ribosyl)ation in cell cycle arrest following gamma-irradiation. *Japanese Journal of Cancer Research* 85, 1094–1098.

Oikawa, A., Tohda, H., Kanai, M., Miwa, M. and Sugimura, T. (1980) Inhibitors of poly(adenosine diphosphate ribose) polymerase induce sister chromatid exchanges. *Biochemical and Biophysical Research Communications* 97, 1311–1316.

Okazaki, I.J. and Moss, J. (1999) Characterization of glycosylphosphatidylinositol-anchored, secreted, and intracellular vertebrate mono-ADP-ribosyltransferases. *Annual Review of Nutrition* 19, 485–509.

Oliver, F.J., Menissier-de Murcia, J. and de Murcia, G. (1999) Poly(ADP-ribose) polymerase in the cellular response to DNA damage, apoptosis, and disease. *American Journal of Human Genetics* 64, 1282–1288.

Park, S.D., Kim, C.G. and Kim, M.G. (1983) Inhibitors of poly(ADP-ribose) polymerase enhance DNA strand breaks, excision repair, and sister chromatid exchanges induced by alkylating agents. *Environmental Mutagenesis* 5, 515–525.

Rawling, J.M., Jackson, T.M., Driscoll, E.R. and Kirkland, J.B. (1994) Dietary niacin deficiency lowers tissue poly(ADP-ribose) and NAD$^+$ concentrations in Fischer-344 rats. *Journal of Nutrition* 124, 1597–1603.

Rawling, J.M., Jackson, T.M., Roebuck, B.D., Poirier, G.G. and Kirkland, J.B. (1995) The effect of niacin deficiency on diethylnitrosamine-induced hepatic poly(ADP-ribose) levels and altered hepatic foci in the Fischer-344 rat. *Nutrition and Cancer* 24, 111–119.

Rawling, J.M., ApSimon, M.M. and Kirkland, J.B. (1996) Lung poly(ADP-ribose) and NAD$^+$ concentrations during hyperoxia and niacin deficiency in the Fischer-344 rat. *Free Radicals in Biology and Medicine* 20, 865–871.

Samper, E., Goytisolo, F.A., Menissier-de Murcia, J., Gonzalez-Suarez, E., Cigudosa, J.C., de Murcia, G. and Blasco, M.A. (2001) Normal telomere length and chromosomal end capping in poly(ADP-ribose) polymerase-deficient mice and primary cells despite increased chromosomal instability. *Journal of Cell Biology* 154, 49–60.

Satoh, M.S. and Lindahl, T. (1992) Role of poly (ADP-ribose) formation in DNA repair. *Nature* 356, 356–358.

Satoh, M.S., Poirier, G.G. and Lindahl, T. (1994) Dual function for poly(ADP-ribose) synthesis in response to DNA strand breakage. *Biochemistry* 33, 7099–7106.

Schmitt, C.A. and Lowe, S.W. (1999) Apoptosis and therapy. *Journal of Pathology* 187, 127–137.

Schreiber, V., Hunting, D., Trucco, C., Gowans, B., Grunwald, D., de Murcia, G. and de Murcia, J.M. (1995) A dominant-negative mutant of human poly(ADP-ribose) polymerase affects cell recovery, apoptosis, and sister chromatid exchange following DNA damage. *Proceedings of the National Academy of Sciences USA* 92, 4753–4757.

Shah, G.M., Le Rhun, Y., Sutarjono, I. and Kirkland, J.B. (2002) Niacin deficient SKH-1 mice are more susceptible to ultraviolet B radiation-induced skin carcinogenesis. *Cancer Research* 131, 3150S.

Shieh, W.M., Ame, J.C., Wilson, M.V., Wang, Z.Q., Koh, D.W., Jacobson, M.K. and Jacobson, E.L. (1998) Poly(ADP-ribose) polymerase null mouse cells synthesize ADP-ribose polymers. *Journal of Biological Chemistry* 273, 30069–30072.

Simbulan-Rosenthal, C.M., Rosenthal, D.S., Iyer, S., Boulares, A.H. and Smulson, M.E. (1998) Transient poly(ADP-ribosyl)ation of nuclear proteins and role of poly(ADP-ribose) polymerase in the early stages of apoptosis. *Journal of Biological Chemistry* 273, 13703–13712.

Simbulan-Rosenthal, C.M., Rosenthal, D.S., Luo, R. and Smulson, M.E. (1999a) Poly(ADP-ribosyl)ation of p53 during apoptosis in human osteosarcoma cells. *Cancer Research* 59, 2190–2194.

Simbulan-Rosenthal, C.M., Haddad, B.R., Rosenthal, D.S., Weaver, Z., Coleman, A., Luo, R., Young, H.M., Wang, Z.Q., Ried, T. and Smulson, M.E. (1999b) Chromosomal aberrations in PARP(−/−) mice: genome stabilization in immortalized cells by reintroduction of poly(ADP-ribose) polymerase cDNA. *Proceedings of the National Academy of Sciences USA* 96, 13191–13196.

Simbulan-Rosenthal, C.M., Ly, D.H., Rosenthal, D.S., Konopka, G., Luo, R., Wang, Z.Q., Schultz, P.G. and Smulson, M.E. (2000) Misregulation of gene expression in primary fibroblasts lacking poly

(ADP-ribose) polymerase. *Proceedings of the National Academy of Sciences USA* 97, 11274–11279.

Smith, S. and de Lange, T. (1999) Cell cycle dependent localization of the telomeric PARP, tankyrase, to nuclear pore complexes and centrosomes. *Journal of Cell Science* 112, 3649–3656.

Smith, S., Giriat, I., Schmitt, A. and de Lange, T. (1998) Tankyrase, a poly(ADP-ribose) polymerase at human telomeres. *Science* 282, 1484–1487.

Smulson, M.E., Simbulan-Rosenthal, C.M., Boulares, A.H., Yakovlev, A., Stoica, B., Iyer, S., Luo, R., Haddad, B., Wang, Z.Q., Pang, T., Jung, M., Dritschilo, A. and Rosenthal, D.S. (2000) Roles of poly(ADP-ribosyl)ation and PARP in apoptosis, DNA repair, genomic stability and functions of p53 and E2F-1. *Advances in Enzyme Regulation* 40, 183–215.

Spronck, J.C. and Kirkland, J.B. (2002) Niacin deficiency increases spontaneous and etoposide-induced chromosomal instability in rat bone marrow cells *in vivo*. *Mutation Research* 508, 83–97.

Steegenga, W.T., van der Eb, A.J. and Jochemsen, A.G. (1996) How phosphorylation regulates the activity of p53. *Journal of Molecular Biology* 263, 103–113.

Trucco, C., Rolli, V., Oliver, F.J., Flatter, E., Masson, M., Dantzer, F., Niedergang, C., Dutrillaux, B., Menissier-de Murcia, J. and de Murcia, G. (1999) A dual approach in the study of poly (ADP-ribose) polymerase: *in vitro* random mutagenesis and generation of deficient mice. *Molecular and Cellular Biochemistry* 193, 53–60.

Tsutsumi, M., Masutani, M., Nozaki, T., Kusuoka, O., Tsujiuchi, T., Nakagama, H., Suzuki, H., Konishi, Y. and Sugimura, T. (2001) Increased susceptibility of poly(ADP-ribose) polymerase-1 knockout mice to nitrosamine carcinogenicity. *Carcinogenesis* 22, 1–3.

Tucker, J.D. and Preston, R.J. (1996) Chromosome aberrations, micronuclei, aneuploidy, sister chromatid exchanges, and cancer risk assessment. *Mutation Research* 365, 147–159.

Van Rensburg, S.J., Hall, J.M. and Gathercole, P.S. (1986) Inhibition of esophageal carcinogenesis in corn-fed rats by riboflavin, nicotinic acid, selenium, molybdenum, zinc, and magnesium. *Nutrition and Cancer* 8, 163–170.

Vaziri, H., West, M.D., Allsopp, R.C., Davison, T.S., Wu, Y.S., Arrowsmith, C.H., Poirier, G.G. and Benchimol, S. (1997) ATM-dependent telomere loss in aging human diploid fibroblasts and DNA damage lead to the post-translational activation of p53 protein involving poly(ADP-ribose) polymerase. *EMBO Journal* 16, 6018–6033.

Vaziri, H., Dessain, S.K., Eaton, E.N., Imai, S.I., Frye, R.A., Pandita, T.K., Guarente, L. and Weinberg, R.A. (2001) hSir2(SIRT1) functions as an NAD-dependent p53 deacetylase. *Cell* 107, 149–159.

Vodenicharov, M.D., Sallmann, F.R., Satoh, M.S. and Poirier, G.G. (2000) Base excision repair is efficient in cells lacking poly(ADP-ribose) polymerase 1. *Nucleic Acids Research* 28, 3887–3896.

Wang, X., Ohnishi, K., Takahashi, A. and Ohnishi, T. (1998) Poly(ADP-ribosyl)ation is required for p53-dependent signal transduction induced by radiation. *Oncogene* 17, 2819–2825.

Wang, Z.Q., Stingl, L., Morrison, C., Jantsch, M., Los, M., Schulze-Osthoff, K. and Wagner, E.F. (1997) PARP is important for genomic stability but dispensable in apoptosis. *Genes and Development* 11, 2347–2358.

Weitberg, A.B. (1989) Effect of nicotinic acid supplementation *in vivo* on oxygen radical-induced genetic damage in human lymphocytes. *Mutation Research* 216, 197–201.

Weitberg, A.B. and Corvese, D. (1990) Niacin prevents DNA strand breakage by adenosine deaminase inhibitors. *Biochemical and Biophysical Research Communications* 167, 514–519.

Wesierska-Gadek, J., Bugajska-Schretter, A. and Cerni, C. (1996a) ADP-ribosylation of p53 tumor suppressor protein: mutant but not wild-type p53 is modified. *Journal of Cellular Biochemistry* 62, 90–101.

Wesierska-Gadek, J., Schmid, G. and Cerni, C. (1996b) ADP-ribosylation of wild-type p53 *in vitro*: binding of p53 protein to specific p53 consensus sequence prevents its modification. *Biochemical and Biophysical Research Communications* 224, 96–102.

Wesierska-Gadek, J., Wang, Z.Q. and Schmid, G. (1999) Reduced stability of regularly spliced but not alternatively spliced p53 protein in PARP-deficient mouse fibroblasts. *Cancer Research* 59, 28–34.

Whitacre, C.M., Hashimoto, H., Tsai, M.L., Chatterjee, S., Berger, S.J. and Berger, N.A. (1995) Involvement of NAD-poly(ADP-ribose) metabolism in p53 regulation and its consequences. *Cancer Research* 55, 3697–3701.

19 Assembly of Triglyceride-transporting Plasma Lipoproteins

Joan A. Higgins
*Department of Molecular Biology and Biotechnology,
University of Sheffield, Sheffield, UK*

Introduction

Lipids play a number of important physiological roles, e.g. triglycerides stored in adipose tissue are a major energy store; phospholipids are essential components of membranes; and cholesterol is also a membrane component and is a precursor of biological active molecules including the steroid hormones and the bile salts. The ability of lipids to carry out these roles is based on a common characteristic – low solubility in aqueous systems. Thus, the ability of phospholipids to form bilayers, the basic structure of biological membranes, is a consequence of their amphipathic character, and the physiological advantage of storing triglyceride rather than carbohydrate is because the lipids are not hydrated and thus provide a compact store adding minimal weight to the body. However, the characteristic, lack of solubility in aqueous systems, which allows lipids to carry out their physiological roles, also presents a problem in transporting lipids between organs in the plasma or lymph. This problem is solved by transport of lipids as complexes with protein, the plasma lipoproteins.

The Plasma Lipoproteins

Lipids are transported in a number of classes of plasma lipoprotein, which are defined and/or separated on the basis of differences in density, composition and electrophoretic mobility (Table 19.1). Four main classes of plasma lipoprotein have been identified: chylomicrons, very low-density lipoproteins (VLDLs), low-density lipoproteins (LDLs) and high-density lipoproteins (HDLs), and each of these major classes consists of several subclasses differing in characteristics and function. The relationships between the different classes and subclasses of lipoprotein are complex, and a detailed account of the plasma lipoproteins, their subclasses and their interrelationships is outside the scope of this chapter. A brief account is given below.

Triglyceride-transporting lipoproteins

Triglycerides are transported in chylomicrons, which carry dietary lipids from the small intestine into the plasma via the lymph, and VLDLs, which carry endogenous triglyceride from the liver into the plasma. Chylomicrons are the largest and lightest of the plasma lipoproteins. Non-polar lipid, mainly triglyceride with some cholesterol ester, accounts for >90% (w/w) of the lipid. The lipid droplets are stabilized by a shell of amphipathic phospholipid and cholesterol, and several proteins, termed apolipoproteins, which are important both for the intracellular assembly of chylomicrons in the enterocytes and for their intravascular metabolism (Tables 19.1 and 19.2). VLDLs are smaller and denser than the chylomicrons. However, they have a similar basic

Table 19.1. The plasma lipoproteins.

A. General characteristics

Lipoprotein	Diameter (nM)	Density range (g ml^{-1})	Electrophoretic mobility[1]	Flotation rate (Sf)	Major tissue source	Major function
Chylomicrons	80–500	<0.94	Origin	>400	Intestine	Transport of dietary lipids
VLDL	30–100	0.94–1.006	β	20–400	Liver	Transport of liver-derived lipids
LDL	19–25	1.006–1.063	Pre-β	0–20	Liver (formed from VLDL)	Delivery of cholesterol to tissues
HDL	4–12	1.063–1.21	α	0–9	Liver intestine	Reverse cholesterol transport from tissues to liver

B. Components

Lipoprotein	Protein (~%)	Triglyceride (~%)	Cholesterol (~%)	Phospholipid (~%)	Apolipoproteins
Chylomicrons	2	89	5	4	A1, 11 and 1V B48 C1, 11 and 111 E
VLDL	10	57	17	16	B100 C1, 11 and 111 E
LDL	23	6	47	24	B100
HDL	48	4	17	30	A1, 11 and 111 E

[1]Movement on agarose gels, where α moves farthest to the cathode.

Table 19.2. Human plasma apolipoproteins.

Apolipoprotein	Molecular weight (kDa)	Site of synthesis	Lipoprotein	Plasma levels (g l^{-1})	Major functions
Apo-A1	28.1	Liver and intestine	Chylomicron, HDL	1.3–1.8	LCAT activator; HDL structural protein
Apo-A11	17.4	Liver and intestine	Chylomicron, HDL	0.4–0.65	LCAT inhibitor; HDL structural protein
Apo-A1V	46.0	Liver and intestine	Chylomicron	0.4	
Apo-B48	260.0	Intestine	Chylomicron	0.005–0.2	Chylomicron structural protein
Apo-B100	550.0	Liver	VLDL	0.6–1.2	VLDL structural protein
Apo-C1	6.3	Liver	Chylomicron, VLDL, HDL	0.1–0.18	LCAT activator (?)
Apo-C11	8.8	Liver	Chylomicron, VLDL, HDL	0.78–2.0	Lipoprotein lipase activator
Apo-C111	8.8	Liver	Chylomicron, VLDL, HDL	0.2–0.4	Lipoprotein lipase inhibitor
Apo-E	36.5	Liver	Chylomicron, VLDL, HDL	0.78–1.5	Clearance of chylomicron remnants; role in VLDL assembly

structure, consisting of a droplet of non-polar lipid stabilized by a shell of amphipathic lipid and apolipoproteins (Fig. 19.1).

Chylomicrons and VLDLs supply respectively dietary and endogenous triglyceride fatty acids to cells and organs. This is achieved by hydrolysis of the triglyceride, through the action of lipoprotein lipase at the surface of the capillary endothelial cells of the target organ. The non-esterified fatty acid released is taken up by the cells. As the triglyceride core is depleted, the lipoprotein particle becomes smaller and some surface components are transferred to other lipoproteins. Chylomicrons are converted to chylomicron remnants, which are cleared by the liver, and the VLDLs are converted to intermediate-density lipoproteins (IDLs) and subsequently to LDLs. IDLs are also cleared by the liver, while LDLs provide a source of cholesterol for a variety of cells, and the excess LDLs are also cleared by the liver.

Cholesterol-transporting lipoproteins

LDLs are small lipoprotein particles in which the core non-polar lipids are mainly cholesterol esters stabilized by a shell of phospholipid and cholesterol and a single molecule of apolipoprotein B (apo-B). These particles are taken up by cholesterol-requiring cells by receptor-mediated-endocytosis mediated by the LDL receptor, which binds the apo-B. Uptake of cholesterol results in down-regulation of the LDL receptor and inhibition of cholesterol synthesis, thus regulating the total cellular cholesterol level. When plasma levels of LDLs are elevated, this tends to result in damage or modification, for example oxidation, to the LDL apo-B. The damaged protein is a ligand for the scavenger receptors of the macrophages in the artery wall. These receptors are not down-regulated, and cholesterol accumulates in the macrophages producing foam cells, the first step in atherosclerosis.

HDLs are the smallest densest class of lipoprotein and are produced in nascent form by the liver and the intestine. The role of HDLs appears to be transport of cholesterol from peripheral tissues, including the artery wall, either directly or via other lipoproteins, to the liver. Cholesterol is not degraded, and the main pathway for its elimination from the body is by excretion into the bile. Cholesterol transport carried out by HDLs is therefore important in reversing atherosclerosis.

Assembly of Triglyceride-transporting Lipoproteins

General features of chylomicrons and VLDLs

Chylomicrons and VLDLs share many characteristics. Both classes of lipoprotein consist of droplets of non-polar lipid, triglyceride and a small amount of cholesterol ester, with a shell of phospholipid,

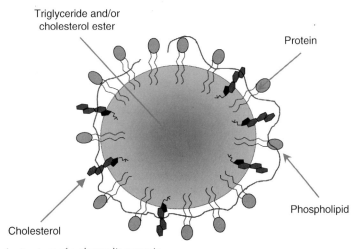

Fig. 19.1. Basic structure of a plasma lipoprotein.

cholesterol and apolipoproteins (Tables 19.1 and 19.2). The main structural apolipoprotein of chylomicrons and VLDLs is apo-B, and this is essential for their assembly and secretion. Other apolipoproteins are also present, and these have similar functional roles in the intravascular metabolism of the lipoproteins (Table 19.2).

Despite their overall similarities, there are critical structural and functional differences between chylomicrons and VLDLs. Chylomicrons are larger with proportionally more triglyceride and have a lower buoyant density than VLDLs (Table 19.1). Apo-B is an essential structural protein for both classes of lipoprotein; however, VLDLs contain apo-B100, the full-length polypeptide, while chylomicrons contain apo-B48, a truncated form consisting of the N-terminal 48% of apo-B100 (Hussain *et al.*, 1996; Davidson and Shelness, 2000; Hussain, 2000). As the LDL receptor-binding domain resides in the C-terminal half of apo-B100, this means that chylomicron remnants are not cleared from the circulation by the same pathway as LDLs (Hussain *et al.*, 1996). Chylomicron remnants are taken up by the liver through interaction of apo-E with the LDL receptor-related protein (Hussain *et al.*, 1996). Chylomicrons are processed by lipoprotein lipase more rapidly than VLDLs and, in consequence, the half-life in the circulation of chylomicrons is minutes, while that of VLDLs is days. During the post-prandial period when chylomicrons reach relatively high levels in the circulation, this can result in elevated VLDLs – leading to increased risk of atherosclerosis (Bjorkegren *et al.*, 1996; Packard *et al.*, 2000).

The small intestine is subjected to large fluctuations in the amount and composition of dietary fats, which must be absorbed efficiently and assembled into chylomicrons. In contrast, the liver is provided with fatty acid substrates for assembly into VLDLs within a fairly narrow range of concentration. Regulation of chylomicron assembly must therefore differ from that of VLDLs. This is also suggested by the different sensitivity of chylomicron assembly and VLDL assembly to agents such as pluronic acid, which inhibit chylomicron assembly but do not affect VLDL assembly (Tso *et al.*, 1981). In addition, in the genetic disorder chylomicron retention disease, there is specific inhibition of chylomicron assembly without an effect on assembly of VLDL (Anderson *et al.*, 1961; Hussain *et al.*, 2000). Thus, there is at least one, as yet unidentified protein, specifically involved in chylomicron assembly.

Editing of apo-B mRNA

Both apo-B48 and apo-B100 are a product of the same gene (Fig. 19.2). However, in the intestine,

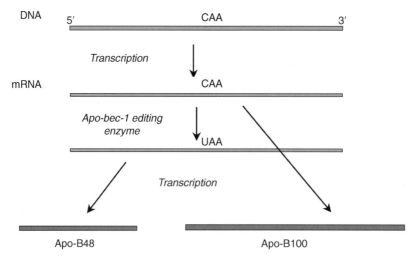

Fig. 19.2. Editing of apo-B mRNA. Apo-B100 is the product of the complete apo-B mRNA. Apo-B48 is a truncated protein formed as a result of editing of the apo-B100 mRNA to produce a stop codon at nucleotide 6666. LCAT, lecithin cholesterol acyltransferase.

the mRNA for apo-B100 undergoes post-transcriptional editing. The nucleotide 6666 in apo-B100 mRNA is altered by deamination of cytosine to uridine by the mRNA-editing zinc-dependent deaminase apo-bec-1. This results in production of a stop codon (UAA) in place of a codon for glutamine (CAA), and synthesis of apo-B is terminated after 48% is completed (Chen et al., 1987; Powell et al., 1987; Davidson and Shelness, 2000).

Apo-B mRNA editing occurs in the intestine of all mammalian species studied, suggesting that production of apo-B48 is important in chylomicron assembly and fat absorption. In humans and several animal species, including hamster and rabbit, editing is restricted to the intestine. However, in some animal species, notably the commonly used laboratory animals, rat and mouse, some post-transcriptional editing of apo-B mRNA does take place in the liver, resulting in secretion of VLDLs containing both apo-B48 and apo-B100 (Davidson and Shelness, 2000; Hussain et al., 2000).

Apo-B editing in the intestine is developmentally regulated. Fetal intestine does not express the apo-bec-1 editing enzyme, and only apo-B100 is produced. In the neonate and adult, expression of apo-bec-1 increases and apo-B48 is produced (Teng et al., 1990; Patterson et al., 1992). Thus, editing is a very specific and carefully regulated event. Recent studies have suggested that editing may result in subtle changes in the response of enterocytes to dietary change and confer an advantage under conditions in which nutrient availability is restricted (see below) (Kendrick et al., 2001).

Molecular events in the assembly of VLDLs in the liver

VLDLs are assembled by the hepatocytes in the liver and secreted into the space of Disse, from where they move into the plasma. The rate of secretion of VLDLs is one factor determining the level of LDLs in the plasma. Because elevated LDL plasma cholesterol is a risk factor for atherosclerosis, there has been considerable interest in elucidating the mechanisms involved in regulation of VLDL assembly and secretion by the liver. Morphological studies including electron microscopy, ultrastructural cytochemistry and autoradiography have shown that the assembly and secretion of VLDL utilize the classical secretory pathway. Apo-B100 apparently is synthesized by bound ribosomes in the rough endoplasmic reticulum (RER); lipid droplets (VLDLs and their precursors) are observed within the lumen of the smooth endoplasmic reticulum (SER) and in the Golgi lumen.

Experimental cell and animal models for investigation of VLDL assembly

CULTURE CELL LINES. The ultimate aim of investigations of VLDL assembly is to extrapolate the findings to humans. The choice of a suitable experimental system for investigation is therefore important. Many studies have been carried out using cultured cell lines, mainly the human hepatoma cell line Hep-G2. However, these cells are deficient in triglyceride synthesis, possibly because they have a relatively small amount of SER compared with normal liver cells, and synthesize and secrete particles of the size and density of LDLs (Gibbons, 1994). Studies of Hep-G2 cells have provided considerable information concerning the molecular details of the early stages in intracellular transit of apo-B100, but have been less useful for studies of intracellular lipid transit. McArdle cells, a rat heptoma cell line, which secretes larger, lighter apo-B-containing particles, have also been used (Boren et al., 1994; Gordon et al., 1996). However, in contrast to human liver, these cells secrete both apo-B100 and apo-B48, and assembly of the two forms of apo-B into lipoproteins differs. Cell lines have the obvious advantages that access to experimental material is relatively easy in comparison with preparation of adult hepatocytes from donor animals. However, cell lines are tumour derived and often have modified biochemistry, which raises problems in interpretation and extrapolation of results to the normal state.

ISOLATED RAT, HAMSTER AND RABBIT HEPATOCYTES. Hepatocytes freshly prepared in suspension or in primary cultures have also been used to investigate VLDL assembly (Dixon and Ginsberg, 1993; Gibbons, 1994; Cartwright and Higgins, 1995, 1996; Cartwright et al., 1997; Kendrick and Higgins, 1999). Although use of adult cells is technically demanding, this experimental approach has the advantage that the donor animals can be treated in different ways prior to isolation of the hepatocytes, e.g. subjected to short- and/or long-term dietary changes known to affect the nature of the VLDLs secreted and the rate of secretion. Hamster and rabbit hepatocytes are also similar to human liver in that they secrete VLDLs containing only apo-B100.

SUBCELLULAR FRACTIONATION. Methods have been developed for the isolation of the major components of the secretory apparatus involved in the synthesis and assembly of VLDLs (RER, SER, cis-Golgi and trans-Golgi) (Higgins and Hutson, 1984). More recently, in this laboratory, we have developed methods for the preparation of subcellular fractions of the secretory compartment in a single self-generating gradient (Plonne et al., 1999). Combined with analysis of the molecular components, the use of subcellular fractionation has provided a method for dissection of the intracellular and molecular events in VLDL assembly.

Apo-B100 and VLDL secretion are regulated post-translationally

Under conditions in which apo-B100 secretion by hepatocytes varies up to sevenfold, there is no significant change in the mRNA for apo-B (Kosykh et al., 1988; Pullinger et al., 1989). Studies on hepatoma cells and isolated hepatocytes have shown that apo-B100 is synthesized in excess of that secreted and that the excess protein is degraded intracellularly (Borchardt and Davis, 1987; Boren et al., 1990; Dixon and Ginsberg, 1991; White et al., 1992; Cartwright and Higgins, 1997; Yao et al., 1997). Although this appears to be an inefficient process, it does mean that apo-B100 is available for VLDL assembly instantly if it is required. The availability of lipids thus drives VLDL assembly, and the post-translational intracellular transit of apo-B100 is important in regulation.

A two-step model for assembly of VLDLs

Studies from a number of laboratories have suggested the following events in the assembly of VLDLs in hepatocytes (Fig. 19.3). Newly synthesized apo-B100 is incorporated into the RER membrane. Experiments using monoclonal antibodies and/or proteases as probes have shown that apo-B100 is associated with the membrane so that part of it is on the cytosolic side of the membrane and part on the lumenal side (Davis et al., 1989; Dixon et al., 1992; Furakawa et al., 1992; Wilkinson et al., 1993). This is in contrast to normal secreted proteins, which move completely into the ER lumen. The membrane-bound form of apo-B100 is degraded mainly in the RER, although, if degradation is inhibited, apo-B100 accumulates in the SER and Golgi membranes. The membrane-associated apo-B100 is degraded after ubiquitination by the proteasome pathway (Yeung et al., 1996). Part of the apo-B100 moves into the lumen of the RER in association with a small amount of lipid in VLDL precursor particles with the density of HDLs. The apo-B100 in these particles may also by degraded within the lumen, possibly if lipidation is not complete or when the apo-B100 is not folded properly (Cartwright and Higgins, 1996; Wu et al., 1997; Shelness et al., 1999; Davidson and Shelness, 2000; Fisher et al., 2001). Most of the lipids (triglyceride, cholesterol ester, cholesterol) destined for incorporation into VLDLs are transferred into the lumen of the SER, as large light droplets, that fuse with the apo-B100-containing precursor particles to form VLDLs. Thus, step 1 in VLDL assembly is formation of dense apo-B100-containing precursor particles, and step 2 is fusion of these particles with large triglyceride-rich droplets.

There has been considerable progress in understanding the molecular details of the first step in VLDL assembly, i.e. production of the dense apo-B100-containing precursor particles. However, far less is known about the second step, i.e. addition of the bulk of the lipids. The latter step is important because this determines the size, density and composition of the VLDL particles secreted.

TRANSLOCATION OF NEWLY SYNTHESIZED APO-B100 ACROSS THE MEMBRANE OF THE ROUGH ENDOPLASMIC RETICULUM. The first regulatory step in VLDL assembly is the translocation of apo-B100 across the membrane of the RER and incorporation into the precursor particles. Apo-B100 apparently undergoes translational arrest when only a small part of the protein is at the lumenal side of the membrane (Du et al., 1994; Zhou et al., 1995; Davidson and Shelness, 2000). Synthesis of the protein is completed, leaving a large part of the polypeptide at the cytosolic side of the membrane. At this stage, the protein is not associated with the membrane but is presumed to be within the translocon complex (Shelness et al., 1999; Davidson and Shelness, 2000). The cytosolic part of the

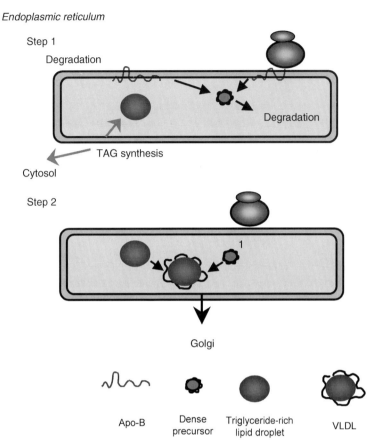

Fig. 19.3. A two-step model for the assembly of VLDLs in hepatocytes. Assembly of VLDLs occurs in two steps in the endoplasmic reticulum lumen. In step 1, apo-B100 is synthesized by bound ribosomes and is transferred into the RER lumen in particles of the density of HDLs, which contain phospholipids, cholesterol and a small amount of non-polar lipid. Triglyceride (TAG) is synthesized by enzymes in the SER and is transferred into the lumen as a lipid-rich particle lacking apo-B100. In step 2, the two particles fuse to form a VLDL, which may undergo further modification in the SER and/or the Golgi. Apo-B100 can undergo translation arrest and remains membrane associated so that part of the molecule is on the cytosolic side of the membrane. There is also evidence that apo-B100 can undergo reverse translocation from the lumenal side of the ER to the cytosolic side. The membrane-associated apo-B100 is degraded through the proteosome system in the RER, SER and Golgi. Lumenal apo-B100 may also be degraded, when triglyceride is limiting or when the dense particle is not properly lipidated or the protein folded.

protein is associated with the cytosolic heat-shock protein 70, which dissociates in the presence of ATP, and the apo-B100 is degraded by the proteosome system (Zhou *et al.*, 1995; Yeung *et al.*, 1996). In the presence of oleate, apo-B100 appears to be protected from degradation and to move into the lumen of the RER.

THE ROLE OF MICROSOMAL TRIGLYCERIDE TRANSPORT PROTEIN (MTP). MTP is a heterodimer consisting of MTP large subunit (97 kDa) non-covalently associated with protein disulphide isomerase (PDI) (Wetterau *et al.*, 1997). PDI is an ER lumen resident protein, which catalyses disulphide bond formation and isomerization, and plays an important role in folding of newly synthesized proteins. PDI is an abundant lumenal protein, much of which exists in homodimers, as well as in association with MTP large subunit. PDI has the C-terminal ER lumen retention signal KDEL, and probably serves to retain MTP in the ER lumen (Gordon, 1997). Mutations in the large subunit gene are associated with the human genetic disease, abetalipoproteinaemia, in which there is a virtual absence of the apo-B-containing plasma lipoproteins and very low secretion of VLDLs and chylomicrons (Sharp et al., 1993; Gordon, 1997). In vitro, MTP catalyses the transfer of cholesterol ester and triglyceride between plasma lipoproteins and/or membrane vesicles, suggesting that it may play a role in the assembly of the triglyceride-rich lipoproteins in the transfer of non-polar lipids to apo-B. Consistent with this, immunoprecipitation studies have shown that MTP is transiently associated with apo-B and that the association is stimulated by triglyceride synthesis (Wu et al., 1996). A variety of experiments have been carried out in different laboratories using MTP inhibitors, which have implicated MTP in the first step in assembly of VLDLs, formation of the dense apo-B100-containing particle (Gordon et al., 1996; Jamil et al., 1996). There is also evidence of an involvement of MTP in the second step, formation of the triglyceride-rich particle (Tietge et al., 1999). The mechanism of transfer of triglyceride from the RER membrane is not understood; however, models have been suggested in which, as the triglyceride content of the membrane increases, phase partition occurs and triglyceride droplets bud from the membrane. These can move either into the cytoplasm to form triglyceride storage droplets or into the lumen.

Effects of nutrients on VLDL assembly

VLDL SUBCLASSES. Two subclasses of VLDLs have been identified in human plasma by ultracentrifugation: larger, lighter triglyceride-enriched VLDL1s and smaller denser VLDL2s (Demant et al., 1993, 1996). The liver secretes both subclasses, and their formation appears to be regulated independently. In vivo insulin acutely inhibits VLDL1 production in normal subjects, but not in insulin-resistant subjects, while production of VLDL2s is unaffected (Malmstrom et al., 1997). Oestrogen stimulates production of VLDL1s without affecting VLDL2s. Both forms of VLDL have been implicated as risk factors for atherosclerosis. VLDL2s are converted to LDLs to a greater extent that VLDL1s, thus resulting in hypercholesterolaemia (Gaw et al., 1995). During the post-prandial period, plasma VLDL1 levels increase, possibly as a result of competition between chylomicrons and VLDL1s for clearance (Karpe et al., 1993; Bjorkegren et al., 1996, 1997). The increase in VLDL1s results in increased transfer of triglyceride to LDLs, and the triglyceride-rich LDLs produced are a precursor of the more atherogenic small dense LDLs.

REGULATION OF THE RATE OF VLDL PRODUCTION AND THE NATURE OF THE PARTICLES SECRETED. Studies on humans and experimental animals have shown that substrate delivery to the liver is a major regulator of the rate of VLDL secretion (Sniderman and Cianflone, 1993). Thus, a diet rich in carbohydrate results in production of larger lighter VLDL1s, while one rich in triglyceride results in production of smaller denser VLDL2s. The nature of the fatty acid provided is also a determinant of VLDL secretion. The best documented effect of dietary fatty acids in vivo is the finding that dietary fish oils (enriched with n-3 unsaturated fatty acids) inhibit VLDL assembly and secretion in vivo and in vitro (Wilkinson et al., 1998; Kendrick and Higgins, 1999). This is due to inhibition of the second step in VLDL assembly, resulting in reduced secretion of VLDLs and targeting of the apo-B100 precursor particles for degradation in the RER lumen (Wilkinson et al., 1998; Kendrick and Higgins, 1999). In the same studies, it was shown that increased dietary sunflower oil (enriched in n-6 unsaturated fatty acids) stimulates VLDL assembly and accelerates transit of the apo-B100 through the secretory pathway.

Molecular events in the assembly of chylomicrons in enterocytes

Chylomicrons are assembled and secreted by the enterocytes (absorptive epithelial cells) lining the villi at the lumenal surface of the small intestine. In contrast to the liver in which VLDL secretion is relatively continuous, chylomicron production occurs in the post-prandial phase after a fatty meal. In humans, the level of chylomicrons in plasma is highest about 2–4 h after ingestion of a fatty meal. However, there are many factors involved in the regulation of the rate of appearance and clearance of chylomicrons in the plasma. These include differences in the amount of fat and the fatty acid composition of the diet, and there are also individual and population differences related to both genetic and lifestyle factors.

Fat absorption occurs in a number of steps: (i) digestion in the intestinal lumen; (ii) transfer of the products of digestion across the brush border of the enterocytes; and (iii) resynthesis of lipids and assembly of these into chylomicrons, which are released between the lateral borders of the enterocytes from where they move into the lacteals in the lamina propria. The lacteals drain into the lymph and the chylomicrons move into the blood. The main products of digestion of triglyceride are monoglycerides and free fatty acids. These, together with bile salts and other lipids (phospholipids and cholesterol), form a fine dispersion of lipid in micelles, the physiological substrate for enterocytes. After transfer of the products of digestion into the enterocytes, the triglyceride is resynthesized by a pathway involving acylation of monoglyceride:

The monoglyceride pathway
fatty acyl CoA + monoglyceride →
diglyceride (1)
fatty acyl CoA + diglyceride → triglyceride (2)

This is in contrast to the liver in which synthesis of triglyceride is by acylation of α-glycerophosphate:

The α-glycerophosphate pathway
α-glycerophosphate + 2 fatty acyl CoA →
phosphatidic acid (1)
phosphatidic acid → diglyceride + phosphate (2)
diglyceride + fatty acyl CoA → triglyceride (3)

This monoglyceride pathway is unique to enterocytes and allows the direct use of the main products of digestion.

Experimental cell and animal models for investigation of chylomicron assembly

CACO-2 CELLS. Studies of chylomicron assembly have been fewer than those of VLDL assembly. This is due partly to a lack of appropriate cultured cell models. Caco-2 cells have been used by a number of investigators (Hussain, 2000). These cells are colonic in origin, but can be induced to differentiate into cells with enterocyte-like characteristics. Caco-2 cells secrete lipoproteins with flotation properties similar to LDLs. However, provision of oleate results in secretion of VLDL-sized particles in addition to LDL-sized particles. Hussain and co-workers have modified the culture conditions for Caco-2 cells such that they secrete small chylomicrons (Luchoomun and Hussain, 1999). However, these contain both apo-B48 and apo-B100, and thus differ from adult enterocytes, which secret only apo-B48. At present, the Caco-2 cell line cultured under appropriate conditions and provided with a physiological substrate containing taurocholate appears to be the best cultured cell model. However, the caveats indicated above for using cultured cells for studies of VLDL assembly must also be applied to Caco-2 cells; namely, cell lines are tumour derived and may have modified biochemistry, which raises problems in interpretation and extrapolation of results to the normal state. Physiological and dietary experiments are also not possible.

ISOLATED ADULT ENTEROCYTES. We recently have developed methods for the isolation of enterocytes from rabbits and mice by exposure of the lumenal surface of the small intestines to low concentrations of chelating agents – sodium citrate and EDTA (Fig. 19.4) (Cartwright and Higgins, 1999a,b). The use of oxygenated and iso-osmotically balanced solutions is important for the preparation of viable cells, which subsequently can be maintained in suspension in Dulbecco's modified Eagle's medium (DMEM) for up to 2 h.

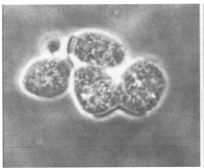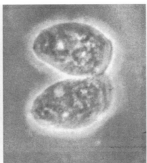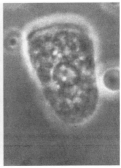

Fig. 19.4. Isolated villus rabbit enterocytes.

When provided with a physiological micellar substrate, the isolated enterocytes synthesize and secrete apo-B48 and triglyceride. More than 90% of the secreted apo-B48 and triglyceride co-precipitate in particles, which have the flotation properties of chylomicrons.

CELLS FROM THE VILLUS/CRYPT GRADIENT. The epithelial lining of the mucosa of the small intestine forms finger-like villi, which project into the lumen, and tubular invaginations, the crypts, at the base of the villi (Fig. 19.5). The cells at the tips of the villi are the most active in lipid absorption and chylomicron assembly. These cells have a half-life of about 2 days and are replaced by cells moving from the base of the crypts, the site of cell division, to the tips of the villi. As the epithelial cells move towards the villi, they acquire the functional and morphological characteristics of adult enterocytes. This unique organization and rapid rate of cell renewal allows the intestine to adapt to changes in dietary conditions by changing the characteristics of the absorptive cells.

By exposing the lumen of the small intestine to chelating agents for a range of times from 5 to 30 min, it is possible to prepare populations of epithelial cells derived from different regions of the villus/crypt gradient (Cartwright and Higgins, 1999a). Differentiated enterocytes from the tips of the villi are released first (marked by alkaline phosphatase activity), and crypt cells (marked by thymidine kinase activity) are released subsequently (Fig. 19.5). The availability of these preparations is useful for studies of the adaptation of the intestine to physiological variation including dietary changes.

SUBCELLULAR FRACTIONATION. Recently, we have developed methods for the preparation of subcellular fractions of the secretory compartment in isolated enterocytes in a single self-generating gradient (Cartwright *et al.*, 2000). Combined with analysis of the molecular components, subcellular fractionation has provided a method for dissection of the intracellular and molecular events in chylomicron assembly.

A two-step model for assembly of chylomicrons in enterocytes

Electron microscopic studies including cytochemistry and autoradiography have shown that the basic events in chylomicron assembly are similar to those involved in VLDL assembly (Fig. 19.6).

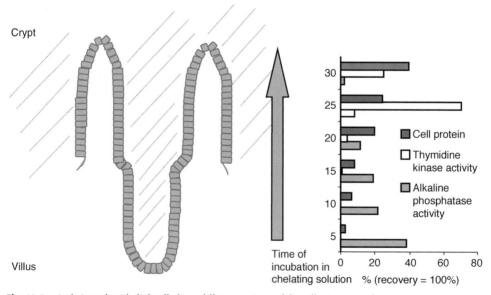

Fig. 19.5. Isolation of epithelial cells from different regions of the villus/crypt gradient. Enterocytes are isolated by exposing the lumenal surface to chelating agents. The length of time of exposure determines the population of cells released from the lumenal surface. Villus cells (alkaline phosphatase marker) are released first followed by the crypt cells (thymidine kinase marker).

Apo-B48 is synthesized by ribosomes in the RER; lipid droplets, presumed to be chylomicrons and/or their precursors, are seen in the lumen of the SER, in the lumen of the Golgi and between the lateral borders of the epithelial cells. However, apart from this basic similarity, there are differences both in the intracellular events and in the regulation of assembly of the chylomicrons and VLDLs.

Studies of isolated enterocytes and subcellular fractions suggest the following events in assembly of chylomicrons (Fig. 19.6) (Kumar and Mansbach, 1997; Mansbach and Nevin, 1998; Cartwright and Higgins, 2000, 2001; Mansbach and Dowell, 2000). Apo-B48 is synthesized by bound ribosomes in the RER but remains membrane bound and diffuses to the SER. Apo-B48 is transferred to the SER lumen together with some phospholipid to form particles of the density of HDLs (step 1). Triglyceride is synthesized in the SER and moves into the lumen. Only two forms of chylomicron precursors are isolated from the lumen of the SER: dense apo-B48-containing particles (of the density of HDLs) and light triglyceride-enriched apo-B48-containing particles (of the density of VLDLs/chylomicrons). This suggests that, in contrast to VLDL assembly in hepatocytes, triglyceride-rich particles lacking apo-B48 are not formed, or

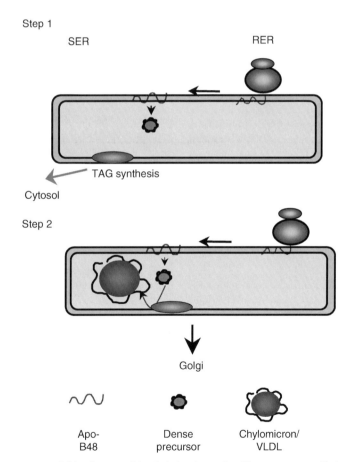

Fig. 19.6. A two-step model for the assembly of chylomicrons in villus enterocytes. Chylomicrons are assembled in two steps in the lumen of the enterocytes. In step 1, apo-B48 is synthesized by bound ribosomes, moves to the SER membrane and is incorporated into small phospholipid-rich particles of the density of HDLs. Triglyceride (TAG) is synthesized in the SER. In step 2, the apo-B48-containing particles acquire triglyceride This occurs either simultaneously or immediately after the triglyceride droplets separate from the membrane, as apo-B48-free triglyceride droplets are not present in the lumenal contents.

that these rapidly associate with the apo-B48-containing dense particles as they separate from the membrane and move into the lumen (step 2). MTP plays a role in chylomicron assembly as its inhibition reduces the transfer of triglyceride from the SER membrane into the lumen (Cartwright, *et al.*, 2000); however, other factors, for example the protein affected in chylomicron retention disease, are also involved. As in VLDL assembly, the mechanisms involved and the factors regulating whether triglyceride is transferred to cytosolic stores or into ER lumen are not understood.

Adaptation of enterocytes to fat absorption

In low-fat chow-fed rabbit enterocytes, most of the cellular triglyceride is associated with the SER membrane and the rate-limiting step in chylomicron assembly appears to be transfer of triglyceride from the membrane to the apo-B48-containing particles (Cartwright *et al.*, 2000; Cartwright and Higgins, 2001). After feeding a diet enriched with fat, triglyceride accumulates in the cytosol, and triglyceride-rich apo-B48-containing particles are found in the lumen of the SER. These observations suggest that transfer of chylomicron precursors from the SER to the Golgi is a saturatable step and becomes rate limiting when the diet is enriched with fat.

Feeding a high-fat diet for 2 weeks has a marked effect on the ability of isolated rabbit villus enterocytes to assemble and secrete chylomicrons. Sunflower oil (fat contribution = 21% of calories) compared with low-fat chow (fat contributing 7% of the calories) results in a large increase in cytosolic triglyceride stores and a 20- and 50-fold stimulation of the secretion of apo-B48 and triglyceride, respectively (Cartwright and Higgins, 1999a). Different dietary fats have different stimulatory effects, with sunflower oil > fat of the composition of a typical Western diet > fish oil. The effect is not acute, as, compared with chow, feeding fat for 18 h had only a small effect on the secretion of chylomicrons by isolated enterocytes. These observations have been interpreted to indicate that the differentiation of the enterocytes, as they move from the crypt to the villus tip, is modified by feeding fat so that the cells become 'turned on' to fat absorption, and that different fatty acids have different stimulatory effects.

Role of apo-B editing in chylomicron assembly

Editing of the mRNA for apo-B100 to yield apo-B48 is a specific and developmentally regulated step in enterocytes. However, mice in which the editing enzyme is deleted (apo-bec-1) are capable of absorbing lipid and forming apo-B100-containing chylomicrons (Hirano *et al.*, 1996; Morrison *et al.*, 1996; Nakumuta *et al.*, 1996; Kendrick *et al.*, 2001). These have a diameter about 50% larger than that of chylomicrons produced by wild-type mice under the same conditions (Kendrick *et al.*, 2001). Chylomicron assembly is thus a characteristic of enterocytes and is not a property of apo-B48. This raises questions concerning the role of apo-B mRNA editing. Why has such a specific, precise and developmentally regulated process evolved? Recent studies have shown that apo-B48 is more efficient in the assembly of chylomicrons than apo-B100 under conditions of fasting. Enterocytes from apo-bec-1 mice fed low-fat chow accumulate more cytosolic triglyceride than those from wild-type mice. An overnight fast causes the stored triglyceride to be depleted. Under these fasted conditions, chylomicron assembly *in vitro* and fat absorption *in vivo* are considerably reduced in the apo-bcc-1 mice compared with the wild-type mice. Thus, under conditions of restricted food intake and a low-fat diet, apo-B mRNA editing confers a nutritional advantage by increasing the efficiency of fat absorption.

Comparison of VLDL and chylomicron assembly

The two major triglyceride-transporting lipoproteins, chylomicrons and VLDLs, are produced by the intestine and the liver, respectively, by pathways that are similar in outline. However, there are differences in the details of the assembly process and its regulation. Thus, the intestinal enterocytes have an enormous capacity for chylomicron absorption compared with the liver and, because the intestinal epithelia has a rapid replacement rate, the population of cells involved in absorption can adapt rapidly to dietary changes. There are also differences in the intracellular events in lipoprotein assembly between hepatocytes and intestine. However, at present, the significance of these differences is not understood.

References

Anderson, C.M., Townley, R.R. and Freeman, J.P. (1961) Unusual causes of steatorrhea in infancy and childhood. *Medical Journal of Australia* 11, 617–621.

Bjorkegren, J., Packard, C.J., Hamsten, A., Bedford, D., Caslake, M., Foster, L., Shepherd, J., Stewart, P. and Karpe, F. (1996) Accumulation of large very low density lipoprotein in plasma during intravenous infusion of a chylomicron-like triglyceride emulsion reflects competition for a common lipolytic pathway. *Journal of Lipid Research* 37, 76–86.

Bjorkegren, J., Hamsten, A., Milne, R.W. and Karpe, F. (1997) Alterations of VLDL composition during alimentary lipemia. *Journal of Lipid Research* 38, 301–314.

Borchardt, R.A. and Davis, R.A. (1987) Intracellular assembly of very low density lipoproteins. *Journal of Biological Chemistry* 262, 16394–16402.

Boren, J., Wettesten, M., Sjoberg, A., Thorin, T., Bondjer, G., Wiklund, A., Carlsson, P. and Olofsson, S.O. (1990) Studies on the assembly of apo-B100 containing lipoproteins in Hep-G2 cells. *Journal of Biological Chemistry* 263, 4434–4442.

Cartwright I.J. and Higgins, J.A. (1995) Intracellular events in the assembly of very low density lipoprotein lipids with apolipoprotein B in rabbit hepatocytes. *Biochemical Journal* 310, 897–907.

Cartwright I.J. and Higgins, J.A. (1996) Intracellular degradation in the regulation of apolipoprotein B 100 by rabbit hepatocytes. *Biochemical Journal* 314, 977–984.

Cartwright, I.J. and Higgins, J.A. (1999a) Isolated enterocytes as a model cell system for investigations of chylomicron assembly and secretion. *Journal of Lipid Research* 40, 1357–1365.

Cartwright, I.J. and Higgins, J.A. (1999b) Increased dietary triacylglycerol markedly enhances the ability of isolated rabbit enterocytes to secrete chylomicrons: an effect related to dietary fatty acid composition. *Journal of Lipid Research* 40, 1858–1866.

Cartwright, I.J. and Higgins, J.A. (2001) Direct evidence for a two-step assembly of apo-B48 containing lipoproteins in the lumen of the smooth endoplasmic reticulum of rabbit enterocytes. *Journal of Biological Chemistry* 276, 48048–48057.

Cartwright, I.J., Higgins, J.A., Wilkinson, J. Bellevia, S., Kendrick, J.S, and Graham, J.M. (1997) Investigation of the role of lipids in the assembly of VLDL in rabbit hepatocytes. *Journal of Lipid Research* 38, 531–545.

Cartwright, I.J., Plonne, D. and Higgins, J.A. (2000) Intracellular events in the assembly of chylomicrons in rabbit enterocytes. *Journal of Lipid Research* 41, 1728–1739.

Chen, S.H, Habib, G., Yang, C.Y., Gu, Z.W., Lee, B.T., Weng, S.A., Siberman, S.R., Cai, S.J., Desylpere, J.P., Rossenau, M. and Chan, L. (1987) Apolipoprotein B48 is the product of a messenger RNA with an organ specific in-frame stop codon. *Science* 238, 363–366.

Davidson, N.O. and Shelness, G.S. (2000) Apolipoprotein B: mRNA editing, lipoprotein assembly, and presecretory degradation. *Annual Review of Nutrition* 20, 169–193.

Davis, R.A., Prewett, D.C., Chan, D.C.F. Thompson, J.J., Borchardt, R.A. and Gallagher, W.R. (1989) Intraheptic assembly of very low density lipoproteins; immunological characterisation of apolipoprotein B in lipoproteins and hepatic membrane fractions and its intracellular distribution. *Journal of Lipid Research* 30, 1185–1196.

Demant, T., Gaw, A., Watts, G.F., Durrington, P., Buckley, B., Imrie, C.W., Wilson, C., Packard, C.J. and Shepherd, J. (1993) Metabolism of apoB-100-containing lipoproteins in familial hyperchylomicronemia. *Journal of Lipid Research* 34, 147–156.

Demant, T., Packard, C.J., Demmelmair, H., Stewart, P., Bedynek, A., Bedford, D., Seidel, D. and Shepherd, J. (1996) Sensitive methods to study human apolipoprotein B metabolism using stable isotope-labeled amino acids. *American Journal of Physiology* 270, E1022–E1036.

Dixon J.L. and Ginsberg, H.N. (1991) Oleate stimulates secretion of apo-B containing lipoproteins from Hep-G2 cells by inhibiting early intracellular degradation of apo-B. *Journal of Biological Chemistry* 266, 5080–5086.

Dixon, J.L. and Ginsberg, H.N. (1993) Regulation of hepatic secretion of apolipoprotein B containing lipoproteins: information obtained from cultured cells. *Journal of Lipid Research* 34, 167–178.

Dixon, J.L. Chattapadhyay, R., Hulma, T., Redman, C.N. and Banjeree, D. (1992) Biosynthesis of lipoprotein: location of the nascent A1 and apo-B in the rough endoplasmic reticulum of chicken hepatocytes. *Journal of Cell Biology* 117, 1161–1169.

Du, E.Z., Kurth, J., Wang, S.L., Humiston, P. and Davis, R.A. (1994) Proteolysis-coupled secretion of the N terminus of apolipoprotein B. Characterization of a transient, translocation arrested intermediate. *Journal of Biological Chemistry* 269, 24169–24176.

Fisher, E.A., Pan, M., Chen, X., Wu, X., Wang, H., Jamil, H., Sparks, J.D. and Williams, K.J. (2001) The triple threat to nascent apolipoprotein B. *Journal of Biological Chemistry* 276, 27855–27863.

Furukawa, S., Sakata, N. and Ginsberg, H.N. (1992) Studies of the sites of intracellular degradation of apolipoprotein B in Hep-G2 cells. *Journal of Biological Chemistry* 271, 18445–18455.

Gaw, A. Packard, C.J., Lindsay, G.M., Griffin, B.A., Caslake, M.J., Lorimer, A.R. and Sheperd, J. (1995) Overproduction of small very low density lipoproteins (Sf 20–60) in moderate hypercholesterolaemia: relationships between apolipoprotein B kinetics and plasma lipoproteins. *Journal of Lipid Research* 36, 158–171.

Gibbons, G.F. (1994) A comparison of *in vitro* models to study hepatic lipid and lipoprotein metabolism. *Current Opinion in Lipidology* 23, 465–500.

Gordon, D.A. (1997) Recent advances in elucidation of the role of the microsomal triglyceride transfer protein in apolipoprotein B lipoprotein assembly. *Current Opinion in Lipidology* 8, 136–150.

Gordon, D.A., Jamil, H., Gregg, R.E., Olofsson, S.-V. and Boren, J. (1996) Inhibition of microsomal triglyceride transfer protein blocks the first step of apolipoprotein B lipoprotein assembly but not the addition of the core bulk lipids in the second step. *Journal of Biological Chemistry* 271, 33047–33053.

Higgins, J.A. and Hutson, J.L. (1984) The role of Golgi and endoplasmic reticulum in the synthesis and assembly of lipoprotein lipids in rat hepatocytes *Journal of Lipid Research* 25, 1295–1305.

Hirano, K.I., Young, S.G., Farese, R.V. Jr, Ng, J., Sande, E., Warburton, C., Powell-Braxton, L.M. and Davidson, N.O. (1996) Targeted disruption of the mouse apo-bec-gene-1 abolished apolipoprotein B mRNA editing and eliminates apolipoprotein apoB48. *Journal of Biological Chemistry* 271, 9887–9890.

Hussain, M.M. (2000) A proposed model for the assembly of chylomicrons. *Atherosclerosis* 14, 1–15.

Hussain, M.M., Kancha, R.K., Zhou, Z., Luchoomun, J., Zu, H. and Bakillah, A. (1996) Chylomicron assembly and catabolism: role of apolipoproteins and receptors. *Biochimica et Biophysica Acta* 1300, 151–170.

Jamil, H., Gordon, D.A., Eustice, D., Brooks, C.M., Dickson, J.K. Jr, Chen, Y., Ricci, B., Chu, C.H., Harrity, T.W., Ciosek, C.P. Jr, Biller, S.A., Gregg, R.E. and Wetterau, J.R. (1996) An inhibitor of the microsomal transfer protein inhbibits apo-B secretion from Hep-G2 cells. *Proceedings of the National Academy of Sciences USA* 93, 11991–11995.

Karpe, F., Steiner, G., Olivecrona, T., Carlson, L.A. and Hamsten, A. (1993) Metabolism of triglyceride-rich lipoproteins during alimentary lipemia. *Journal of Clinical Investigations* 91, 748–758.

Kendrick, J.S. and Higgins, J.A. (1999) Dietary fish oils inhibit early events in the assembly of very low density lipoproteins and target apolipoprotein B for degradation within the lumen of the rough endoplasmic reticulum of hamster hepatocytes. *Journal of Lipid Research* 40, 504–514.

Kendrick, J.S., Chan, L. and Higgins, J.A. (2001) Superior role of apolipoprotein B48 over apolipoprotein B100 in chylomicron assembly and fat absorption: an investigation of apobec-1 knock-out and wild-type mice. *Biochemical Journal* 356, 821–827.

Kosykh, V.A., Surguchow, A.P., Podres, E.A., Novikov, D.K. and Sudarickoc, N. (1988) VLDL apoprotein secretion and apo-B mRNA levels in primary cultures of cholesterol loaded rabbit hepatocytes. *FEBS Letters* 232, 103–106.

Kumar, N.S. and Mansbach, C.M. (1997) Determinants of triglyceride transport from the endoplasmic reticulum to the Golgi in intestine. *American Journal of Physiology* 273, G18–G30.

Luchoomun, J. and Hussain, M.M. (1999) Assembly and secretion of chylomicrons by differentiated Caco-2 cells. *Journal of Biological Chemistry* 274 19565–19572.

Malmstrom, R., Packard, C.J., Watson, T.G., Ranniko, S., Caslake, M., Bedford, D., Stewart, P., Yid-Jarvinen, H., Shepherd, J. and Taskinen, M.R. (1997) Metabolic basis of hypertriglyceridaemic effect of insulin in normal men. *Arteriosclerosis, Thrombosis and Vascular Biology* 17, 1454–1464.

Mansbach, C.M. and Dowell, R. (2000) Effect of increasing lipid load on the ability of the endoplasmic reticulum to tranport lipid to the Golgi. *Journal of Lipid Research* 41, 605–612.

Mansbach, C.M. and Nevin, P. (1998) Intracellular movement of triacylglycerols in the intestine. *Journal of Lipid Research* 39, 963–968.

Morrison, J.R., Pasty, Ch., Stevens, M.E., Hughes, S.D., Forte, T., Scott, J. and Rubin, E.M. (1996) Apolipoprotein B RNA editing enzyme deficient mice are viable despite alterations in lipoprotein metabolism. *Proceedings of the National Academy of Sciences USA* 93, 7154–7159.

Nakamuta, M., Chang, B.H.J., Zsigmond, E., Kobayashi, K., Lei, H., Ishida, B.Y., Oka, K., Li, E. and Chan. L. (1996) Complete phenotypic characterisation of apobec-1 knock-out mice with a wild type genetic background and restoration of mRNA editing by somatic transfer of apobec-1. *Journal of Biological Chemistry* 271, 25981–25988.

Packard, C.J., Demant, T., Stewart, J.P., Bedford, D., Caslake, M.J., Schwertfeger, G., Bedynek, A., Shepherd, J. and Seidel, D. (2000) Apolipoprotein B metabolism and the distribution of VLDL and LDL subfractions. *Journal of Lipid Research* 41, 305–317.

Patterson, A.P., Tennyson, G.E., Hoeg, J.M., Sviridov, D.D. and Brewer, H.B. (1992) Ontogenic regulation of apo-lipoprotein B mRNA editing during human and rat development *in vivo*. *Arteriosclerosis, Thrombosis and Vascular Biology* 12, 463–473.

Plonne, D., Cartwright, I.J., Graham, J.M., Dargel, R. and Higgins, J.A. (1999) Separation of the components of the secretory compartment from rat liver and isolated rat heptocytes in a single step in self generating gradients of iodixanol. *Analytical Biochemistry* 279, 88–89.

Powell, L.M., Wallis, S.C., Pease, R.J. Edwards, Y.H., Knott, T.J. and Scott, J. (1987) A novel form of tissue specific RNA processing produces apolipoprotein B48 in intestine. *Cell* 50, 831–846.

Pullinger, C.R., North, J.D., Teng, B., Rifici, V.A. Ronhild de Brito, A.E. and Scott, J. (1989) The apolipoprotein B gene is constitutively expressed in Hep-G2 cells. *Journal of Lipid Research* 38, 1065–1076

Sharp, D.L., Blinderman. K.A., Combe, R.A., Kienzie, B., Ricci, B., Wager-Smith, K., Gil, C.M., Turck, C.W., Bourma, M.E., Rader, D.J. *et al.* (1993) Cloning and gene defects in microsomal triglyceride transfer proteins associated with abetalipoproteinaemia. *Nature* 356, 65–69.

Shelness, G.S., Ingram, M., Huand, X.F. and DeLozier, J.A. (1999) Apolipoprotein B in the rough endoplasmic reticulum: translation, translocation and initiation of lipoprotein assembly. *Journal of Nutrition.* 129, 456S–462S.

Sniderman, A.D. and Cianflone, K. (1993) Substrate delivery as a determinant of hepatic apo-B secretion. *Arteriosclerosis, Thrombosis and Vascular Biology* 13, 629–636.

Teng, B., Verp, M., Salomon, J. and Davidson, N.O. (1990) Apolipoprotein B messenger RNA editing is developmentally regulated and widely expressed in human tissues. *Journal of Biological Chemistry* 265, 20616–20620.

Tietge, U.J., Bakillah, A., Maugeatis, C.M., Tsukamoto, K., Hussain, M. and Rader, D.J. (1999) Hepatic overexpression of microsomal triglyceride transfer protein (MTP) results in increased *in vivo* secretion of VLDL triglycerides and apolipoprotein B. *Journal of Lipid Research* 40, 2134–2138.

Tso, P., Balint, J.A., Bishop. M.B. and Rodgers, J.B. (1981) Acute inhibition of intestinal lipid transport by pluronic L-81 in the rat. *American Journal of Physiology* 241, G487–G497.

Wetterau, J.R., Lin, M.C. and Jamil, H. (1997) Microsomal triglyceride transfer protein. *Biochemica et Biophysica Acta* 1345, 136–150.

White A.L., Graham, D.L., LeGros, J., Pease, R.J. and Scott, J. (1992) Oleate mediated stimulation of apolipoprotein B secretion from rat hepatoma cells. *Journal of Biological Chemistry* 267, 15657–15684.

Wilkinson, J. Higgins, J.A., Groot, P.H.E., Gherardi, E. and Bowyer, D.E. (1993) Topography of apolipoprotein B in subcellular fractions of rabbit liver probed with a panel of monoclonal antibodies. *Journal of Lipid Research* 34, 815–825.

Wilkinson, J., Fitzsimmons, C., Higgins, J.A. and Bowyer, D.E (1998) Dietary fish-oils modify the assembly of very low density lipoproteins and expression of the low density lipoprotein receptor in rabbit liver. *Arteriosclerosis, Thrombosis and Vascular Biology* 18, 1490–1497.

Wu, X., Zhou, L.-S., Wetterau, J. and Ginsberg. H.N. (1996) Demonstration of a physical interaction between MTP protein and apo-B during assembly of apo-B containing lipoproteins. *Journal of Biological Chemistry* 271 10277–10281.

Wu, X., Sakata, N., Lele, K.M., Zhou, M., Jiang, H and Ginsberg, H.N. (1997) A two site model for apo-B degradation in Hep G2 cells. *Journal of Biological Chemistry* 272, 11575–11580.

Yao, Z., Tran, K. and McCleod, R.S. (1997) Intracellular degradation of newly synthesised apolipoprotein B. *Journal of Lipid Research* 38, 1937–1953.

Yeung, S.J., Chen, S.-W. and Chan, L. (1996) Ubiquitin-proteasome mediated pathway for the degradation of apolipoprotein B. *Biochemistry* 35, 13843–13848.

Zhou, M. Wu, X., Huang, L.-S. and Ginsberg, H.N. (1995) Apolipoprotein B 100 an inefficiently translocated secretory protein is bound to cytosolic chaperone heat shock protein. *Journal of Biological Chemistry* 270, 25220–25224.

20 Regulation of Cellular Cholesterol

Ji-Young Lee, Susan H. Mitmesser and Timothy P. Carr
*Department of Nutritional Science and Dietetics,
University of Nebraska, Lincoln, Nebraska, USA*

Introduction

Cholesterol was discovered over two centuries ago as an alcohol-soluble component of gallstones. It was later identified as a constituent of most animal tissues and, more specifically, as an essential component of cellular membranes. In mammals, cholesterol also serves as a precursor of bile acids and steroid hormones. While primarily of animal origin, cholesterol is found in trace amounts in fungi and green plants.

Humans require cholesterol for normal metabolic function. Virtually all cells are capable of synthesizing cholesterol when needed; dietary sources also contribute to the total body pool of cholesterol. Most cholesterol in the body is found in cell membranes, while a smaller proportion is associated with plasma lipoproteins as part of an elaborate intravascular transport system. Excess cholesterol can be eliminated from the body through biliary excretion, either as free cholesterol or as bile acids. The liver, therefore, is a critical organ in cholesterol metabolism because it represents the primary route of cholesterol removal from the body as well as the main arbitrator of plasma cholesterol concentration. Failure to maintain cholesterol homeostasis within the liver cell can lead to accumulation of cholesterol in the plasma, thus increasing the risk of coronary heart disease and stroke.

The intracellular cholesterol concentration must be maintained at a constant level to prevent the over-accumulation of cholesterol to cytotoxic levels. While dozens of regulatory proteins are involved in maintaining hepatic cholesterol homeostasis, those having received the greatest attention in the scientific literature are: (i) 3-hydroxy-3-methylglutaryl-CoA (HMG-CoA) reductase, the rate-limiting enzyme in cholesterol synthesis; (ii) low-density lipoprotein (LDL) receptors, the primary mechanism of LDL removal from plasma; and (iii) cholesterol 7α-hydroxylase, the rate-limiting enzyme in the classical bile acid synthetic pathway. Each of these proteins is under the control of transcription factors, including sterol regulatory element-binding proteins (SREBPs), liver X receptors (LXRs) and farnesoid X receptor (FXR).

This chapter focuses on the inter-relationships among the transcription factors and the major regulatory proteins as illustrated in Fig. 20.1. In the human population, genetic mutations provide the most dramatic evidence of cholesterol imbalance (Goldstein and Brown, 2001), although most people are able to influence cholesterol metabolism, and, hence, the progression of atherosclerotic disease, through external factors such as diet. Therefore, in keeping with the theme of 'molecular nutrition', the role of specific nutrients on intracellular cholesterol regulation will also be discussed.

Transcription Factors

SREBPs

SREBPs are integral proteins bound to the endoplasmic reticulum (ER) and the nuclear envelope

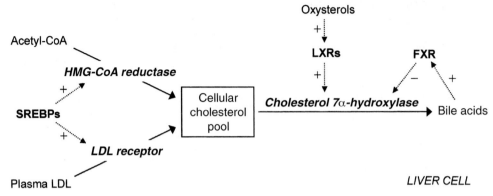

Fig. 20.1. Roles of SREBPs, LXRs and FXR in hepatic cholesterol metabolism. SREBPs increase the cellular cholesterol pool by activating the transcription of HMG-CoA reductase and LDL receptors, resulting in cholesterol synthesis and LDL uptake from the circulation, respectively. Increased concentrations of oxysterols, which reflect the increased cellular cholesterol concentration, bind to LXRs and thus activate the transcription of the *Cyp7a1* gene encoding cholesterol 7α-hydroxylase, the rate-limiting enzyme in classical (neutral) bile acid synthesis. Bile acids are ligands for FXR and exert a negative feedback on bile acid synthesis.

membrane by a hairpin domain consisting of two membrane-spanning domains separated by a short 31-amino-acid loop facing the lumen of the ER and nuclear envelope (Hua *et al.*, 1995). The N-terminal domain of approximately 480 amino acids and the C-terminal domain of approximately 590 amino acids face the cytosol. The N-terminal domain of SREBPs contains a basic helix–loop–helix leucine zipper (bHLH-ZIP) motif, which provides a DNA-binding domain. To function as a transcription factor, therefore, SREBPs need to be activated so that the N-terminal domain can translocate to the nucleus where it activates the transcription of sterol-responsive genes involved in cholesterol and fatty acid metabolism (Brown and Goldstein, 1999).

In response to signals such as cellular cholesterol depletion, two sequential proteolytic cleavages are required to release an N-terminal transcription factor domain of SREBPs. The first cleavage by site-1 protease (S1P) occurs in the lumenal loop between two membrane-spanning domains (site-1 cleavage). The second cleavage occurs within the first transmembrane domain (site-2 cleavage) by site-2 protease (S2P), releasing the N-terminal fragment of SREBP (Sakai *et al.*, 1996). The mature N-terminal fragment of SREBP now can enter the nucleus to function as a transcription factor. Figure 20.2 illustrates proteolytic cleavages of SREBPs and the translocation of mature SREBPs into the nucleus. Sterols inhibit the activation of SREBPs by selectively blocking the cleavage initiated by S1P. S2P is regulated indirectly by sterols because it can only act on SREBP after site-1 cleavage.

SREBP cleavage-activating protein (SCAP) is another factor involved in the activation of SREBPs. SCAP is an ER membrane-bound glycoprotein and was discovered by means of a mutant line of Chinese hamster ovary (CHO) cells. The cells fail to suppress cleavage of SREBPs in the presence of sterols whose mutation is in the SCAP gene (Nohturfft *et al.*, 1996). The N-terminal domain of SCAP is thought to have a sterol-sensing function. SCAP and SREBP form a complex in the ER membrane through interaction between their cytoplasmic C-terminal domains. The SREBP–SCAP complex directs the S1P to its target in the lumenal domain of SREBP, and the disruption of this complex inactivates site-1 cleavage reaction (Sakai *et al.*, 1998). The mechanism for the sterol-mediated proteolytic cleavage of the N-terminal domain of SREBP is well demonstrated (Brown and Goldstein, 1999).

SREBPs were first discovered as specific transcription factors that bind to the 10 bp sterol regulatory element (SRE) within the promoters of the genes encoding the LDL receptor and HMG-CoA synthase (Briggs *et al.*, 1993; Wang *et al.*, 1993). Ubiquitous expression of SREBPs suggests their important role in regulating cellular cholesterol and fatty acid levels by activating a

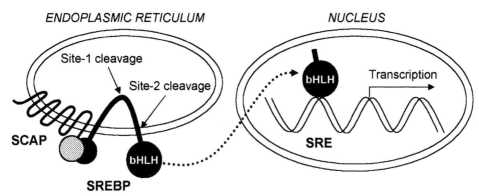

Fig. 20.2. Maturation of SREBPs and their transactivation function. Sequential proteolytic cleavages at site-1 and subsequently at site-2 by S1P and S2P, respectively, release the N-terminal domain of SREBPs from the ER membrane. This domain contains a basic helix–loop–helix leucine zipper DNA-binding motif (bHLH), which translocates to the nucleus. Binding of the mature SREBP to a sterol regulatory element (SRE) activates the transcription of sterol-responsive genes, including genes encoding HMG-CoA reductase and LDL receptors. SREBP cleavage-activating protein (SCAP) functions as a sterol sensor; interaction between SCAP and the C-terminal domain of SREBP is necessary for site-1 cleavage.

variety of sterol-responsive genes. When cellular cholesterol is depleted, SREBPs are activated, increasing the transcription of genes encoding enzymes in the cholesterol biosynthetic pathway. Transcription of the LDL receptor, which supplies cholesterol through receptor-mediated endocytosis of plasma LDL, is also regulated by SREBPs (Wang et al., 1993).

Three isoforms of SREBPs have been identified: SREBP-1a, SREBP-1c and SREBP-2. SREBP-1a and SREBP-1c are produced from a single gene through the use of alternative transcription start sites encoding alternative first exons that are spliced into a common second exon (Shimomura et al., 1997). The ratio of SREBP-1c to SREBP-1a transcripts varied markedly, showing high ratios in the liver (9:1), adrenal gland and adipose tissue, whereas low ratios were observed in the spleen (1:10). These differences indicate that SREBP-1a and SREBP-1c transcripts are controlled independently to respond differentially to organ-specific factors related to cholesterol and fatty acid metabolism (Shimomura et al., 1997). A separate single gene encodes SREBP-2, which is 47% identical to SREBP-1a (Hua et al., 1993). SREBP-1 and SREBP-2 share some amino acid sequences and have structural similarities, but their functions may not be identical. SREBP-1 plays an important role in fatty acid metabolism rather than in the regulation of cholesterol, whereas SREBP-2 plays a greater regulatory role in cholesterol metabolism. Moreover, regulation of fatty acid metabolism by SREBP-1 also involves LXRs, which is discussed in the following section.

LXRs

It has long been thought that cholesterol or its metabolites could activate their own catabolism (i.e. conversion to bile acids) in response to excess cellular cholesterol. Recent identification of LXRs and their natural ligands (Fig. 20.1) represents a breakthrough in understanding the mechanism underlying how cholesterol affects its own catabolism.

LXRs were first identified as orphan nuclear receptors. They consist of a central DNA-binding domain and a hydrophobic C-terminal domain that mediates ligand recognition, receptor dimerization and ligand-dependent activation (Peet et al., 1998a). LXRs are activated by naturally occurring oxysterols, which are oxidized derivatives of cholesterol. 22(R)-hydroxycholesterol, 24(S)-hydroxycholesterol and 24(S),25-epoxycholesterol are known to be high-affinity ligands of LXRs (Janowski et al., 1996; Lehmann et al., 1997). To function as transcription factors, LXRs have to heterodimerize with the retinoid X receptors (RXRs). The LXR–RXR heterodimer can be activated by ligands for either RXR (e.g.

9-*cis* retinoic acid) or LXR. Binding of ligands induces the conformational changes in the LXR–RXR heterodimer, resulting in enhanced transcription of a target gene.

Two LXRs have been identified in mammals: LXRα and LXRβ. LXRα is expressed mainly in the liver, whereas lower concentrations are expressed in the kidney, intestine, spleen and adrenals. LXRβ is ubiquitously expressed in most tissues (Willy *et al.*, 1995). LXRα and LXRβ have specificity similar to oxysterol ligands. However, in LXRα knockout mice, LXRβ could not compensate for the absence of LXRα with regard to bile acid synthesis and secretion (Peet *et al.*, 1998b). The LXR–RXR heterodimer is known to bind to the hormone response element of DR-4, which is a direct repeat of two similar hexanucleotides separated by four nucleotides (Lehmann *et al.*, 1997).

The pattern of expression of LXRs and the fact that their ligands are oxysterols indicate LXRs' involvement in cholesterol metabolism. Identification of the LXR response element (LXRE) in the promoter region of the *Cyp7a1* gene encoding cholesterol 7α-hydroxylase strongly supports the role of LXRs as key transcriptional regulators in bile acid metabolism (Lehmann *et al.*, 1997). Peet *et al.* (1998b) showed that under a low-cholesterol diet, LXRα knockout mice appeared normal. When fed a high-cholesterol diet, however, LXRα knockout mice failed to increase transcription of *Cyp7a1* and consequently showed a marked increase in hepatic cholesteryl ester concentrations. Conversely, wild-type mice fed a high-cholesterol diet increased transcription of *Cyp7a1*.

LXRα may be involved in fatty acid metabolism as well as in cholesterol metabolism. Genes whose products are involved in fatty acid metabolism, including SREBP-1, stearoyl-CoA desaturase-1 and fatty acid synthase, were down-regulated when LXRα knockout mice were fed chow or high-cholesterol diets compared with wild-type mice (Peet *et al.*, 1998b). The identification of the LXR–RXR-binding site in the promoter of the SREBP-1c gene suggests that lipogenesis exerted by LXRs is mediated by SREBP-1c (Repa *et al.*, 2000). These studies suggest that the metabolism of both cholesterol and fatty acids is regulated in a coordinated manner, mediated through LXRs.

FXR

Conversion of cholesterol to bile acids has a significant physiological role because it represents a major route of cholesterol elimination from the body. Catabolism of cholesterol to bile acids accounts for about 50% of the daily excretion of excess cholesterol (Vlahcevic *et al.*, 1999). However, unlike the role of LXRs in up-regulating *Cyp7a1*, FXR down-regulates bile acid synthesis by decreasing the transcription of *Cyp7a1* (Fig. 20.1). This effect is mediated by the return of bile acids to the liver via the enterohepatic circulation. The recent discovery that bile acids are ligands for FXR has added a new dimension to understanding the regulatory role of bile acids.

FXR is a nuclear hormone receptor, which is expressed exclusively in liver, intestine, kidney and the adrenal cortex (Forman *et al.*, 1995). It is necessary for FXR to form a heterodimer with RXR. The FXR–RXR heterodimer binds to an inverted hexanucleotide repeat separated by a single nucleotide known as an IR-1 motif (Forman *et al.*, 1995). *In vitro* experiments have shown that chenodeoxycholic acid (CDCA), deoxycholic acid (DCA) and litocholic acid (LCA) are able to strongly activate FXR, and the activation occurs within a physiological intracellular concentration, indicating that bile acids are naturally occurring FXR ligands (Makishima *et al.*, 1999; Parks *et al.*, 1999; Wang *et al.*, 1999).

The activity of cholesterol 7α-hydroxylase is regulated negatively at a transcriptional level by the flux of bile acids/salts within the enterohepatic circulation. In rat hepatocytes, relatively hydrophobic bile acids such as taurocholate and taurodeoxycholate repressed mRNA for cholesterol 7α-hydroxylase, whereas hydrophilic acids tauroursodeoxycholate and taurohyodeoxycholate had no effect (Stravitz *et al.*, 1993). Identification of endogenous bile acid-activated FXR prompted researchers to investigate *Cyp7a1* as a target gene of FXR.

The *Cyp7a1* gene promoter was repressed by bile acids and FXR, suggesting the involvement of bile acids and FXR in regulating *Cyp7a1* transcription (Makishima *et al.*, 1999). Two bile acid response elements – BARE-I at nucleotides −75 to −54 and BARE-II at nucleotides −149 to −118 of the *Cyp7a1* promoter – were identified (Chiang and Stroup, 1994). Deletion of BARE-II from the

Cyp7a1 promoter demonstrated that BARE-II is responsible for the bile acid-mediated down-regulation of *Cyp7a1* gene expression (Stroup *et al.*, 1997). Despite the identification of BARE-II as the FXR response element, the FXR–RXR heterodimer apparently does not bind to this sequence. The latter observation suggests that, after activation by bile acids, FXR represses the transcription of the *Cyp7a1* gene in an indirect manner (Chiang *et al.*, 2000).

The liver receptor homologue-1 (LRH-1; also known as CPF) and small heterodimer partner (SHP) also appear to play a role in the indirect repression of FXR. LRH-1 is a monomeric nuclear orphan receptor that functions as a tissue-specific transcription factor and is limitedly expressed in liver, intestine, and pancreas (Becker-Andre *et al.*, 1993). The transactivation of *Cyp7a1* by LXRα requires that LRH-1 binds to BARE-II in the *Cyp7a1* promoter (Lu *et al.*, 2000). SHP is an unusual nuclear receptor because it lacks a DNA-binding domain and forms a heterodimer with several other nuclear hormone receptors (Seol *et al.*, 1996). Generally, SHP is known to decrease transactivation of its partner. The *SHP* promoter has an IR-1, which is a binding site for the FXR–RXR heterodimer; bile acid-activated FXR increases the transcription of the *SHP* gene (Lu *et al.*, 2000). Consequently, an increase in SHP protein inactivates LRH-1 by forming a heterodimer, repressing the transcription of *Cyp7a1*. The mechanisms by which LXRs and FXR induce or repress, respectively, the transcription of the *Cyp7a1* gene are illustrated in Fig. 20.3.

Negative regulation of bile acid synthesis is not the only function exerted by FXR in bile acid metabolism. FXR is also involved in bile acid transport in the enterohepatic circulation. The ileal bile acid-binding protein (IBABP) is a cytosolic protein that binds to bile acids with a high affinity and is thought to play a role in cellular bile acid uptake and trafficking (Kramer *et al.*, 1993). In human *IBABP* promoters, IR-1 located in nucleotides −160 to −148 was found to be an FXR–RXR heterodimer-binding site, suggesting that bile acids induce the transcription of *IBABP* through the activation of FXR (Grober *et al.*, 1999; Makishima *et al.*, 1999). The bile salt export pump (BSEP) plays

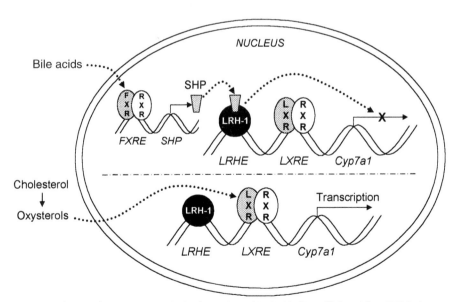

Fig. 20.3. Regulation of *Cyp7a1* transcription by LXRs and FXR. Binding of bile acids to FXR induces the expression of the *SHP* gene (upper panel). SHP protein binds to LRH-1 already attached to the LRH response element (LRHE) located in the promoter of *Cyp7a1*, thus decreasing the transcription of *Cyp7a1*. Transcription of *Cyp7a1* is increased when cholesterol is converted to oxysterols, which are ligands of LXR (lower panel). Increased cellular concentrations of oxysterols represent increased cholesterol concentrations. *LXRE*, LXR response element; *FXRE*, FXR response element.

an important role in transporting bile acids across the hepatocyte canalicular membrane. The promoter of the *BSEP* gene contains an IR-1 element, and the effect of bile acids on the transactivation of BSEP is FXR dependent (Ananthanarayanan *et al.*, 2001).

FXR is an important regulator in bile acid metabolism and, moreover, in cholesterol metabolism. It exerts its function in both a negative and a positive manner. In contrast to the indirect negative regulation of FXR in bile acid synthesis, FXR functions as a positive regulator in the expression of *IBABP* and *BSEP* by binding to the promoter of these genes. FXR knockout mice studies suggest other possible roles of FXR (Sinal *et al.*, 2000). The hepatic basolateral sodium taurocholate cotransporter protein (NTCP) and the liver fatty acid-binding protein (L-FABP) appear to be controlled by FXR. By alterations in the expression of several genes, FXR plays an essential role as a bile acid sensor in cholesterol, bile acid and fatty acid metabolism.

Regulatory Proteins and Effects of Diet

HMG-CoA reductase

Cholesterol is synthesized in a multienzyme pathway in which HMG-CoA reductase is the rate-limiting step. HMG-CoA reductase is an intrinsic membrane protein of the ER that catalyses the synthesis of mevalonate, a crucial intermediate in the formation of sterols and non-sterol isoprenoid compounds. Cholesterol biosynthesis is controlled by the regulation of genes in response to cellular cholesterol levels (Goldstein and Brown, 1990). Transcription is high in cells deprived of sterols and low in cells exposed to excess sterols. Increased intake, absorption and transport of cholesterol to cells decreases HMG-CoA reductase activity through SREBP-mediated regulation of gene expression.

When cellular cholesterol is depleted, mature forms of SREBPs (via proteolytic cleavage) are transported to the nucleus. The SREBPs then bind to SREs in the promoter regions of HMG-CoA reductase, thus increasing its transcription. In contrast, when sufficient cholesterol is present, the metabolites of cholesterol, such as hydroxy-cholesterol, prevent the proteolytic cleavage of SREBPs to their active from, resulting in the down-regulation of HMG-CoA reductase (Edwards *et al.*, 2000).

Unlike dietary cholesterol, the effects of individual dietary fatty acids on HMG-CoA reductase are varied and not fully understood. In general, saturated fatty acids (SFAs) tend to increase HMG-CoA reductase activity compared with polyunsaturated fatty acids (PUFAs) and monounsaturated fatty acid (MUFAs). Dietary MUFA (i.e. oleic acid) appears to down-regulate HMG-CoA reductase indirectly by inhibiting SREBP maturation through its effects on oxysterols (Thewke *et al.*, 1998). Both MUFA and PUFA decreased HMG-CoA reductase mRNA by decreasing the mature forms of SREBPs, whereas SFA has little effect (Worgall *et al.*, 1998).

Dietary fibres – primarily water-soluble fibres – tend to lower serum cholesterol concentrations. Several mechanisms have been suggested to explain the hypocholesterolaemic effect of dietary fibres, including alterations in intestinal cholesterol absorption, lipoprotein metabolism, bile acid metabolism, and fermentation by-products and their effects on hepatic cholesterol synthesis (Kay, 1982). Soluble dietary fibres, such as pectin, guar gum and psyllium, increased HMG-CoA reductase activity in animal studies (Fernandez *et al.*, 1995; Moundras *et al.*, 1997). The effects are attributed largely to the reduced hepatic cholesterol concentrations usually observed with consumption of dietary fibres.

LDL receptor

Very low-density lipoproteins (VLDLs) are triacylglycerol-rich particles made by the liver whose function is to transport endogenous lipid to peripheral tissues. Once depleted of its triacylglycerol, the VLDLs are transformed into cholesterol-rich remnant particles called LDLs. The LDLs have no overt biological function, although excessive amounts in plasma are linked to coronary heart disease and stroke. Consequently, researchers have focused much attention on the mechanisms involved in cellular uptake of LDL.

LDL can be removed from the plasma by receptor-dependent and receptor-independent mechanisms. Receptor-dependent LDL uptake accounts for 60–80% of LDL clearance (Spady *et al.*, 1986), most of which occurs in the liver. LDL receptors are present on the cell surface and recognize apolipoprotein-B_{100}, the only protein constituent of LDL. The LDL receptor is highly sensitive to changes in dietary cholesterol, fatty acids and fibre, whereas receptor-independent uptake of LDL is not influenced by diet.

The activity of LDL receptors is regulated at a transcriptional level in response to cellular free cholesterol concentration. As described in the section on SREBPs, the promoter of the LDL receptor gene contains an SRE, which is the binding site for SREBPs. Low concentrations of cellular cholesterol induce the maturation of SREBPs, and the N-terminal transcriptional factor domain is released, resulting in the transactivation of the LDL receptor gene (Hua et al., 1995). In contrast, when sufficient cholesterol is present in cells, cholesterol and/or the metabolites of cholesterol such as oxysterols prevent the maturation of SREBPs, resulting in the down-regulation of the LDL receptor gene (Towle, 1995).

Dietary cholesterol can affect the cholesterol concentration in the liver, which in turn influences LDL receptor activity. Although several homeostatic mechanisms respond to increases in hepatic cholesterol concentration, decreased LDL receptor activity is a major compensatory mechanism. Cholesterol-rich diets can significantly increase hepatic cholesterol concentration, repressing the abundance of mRNA for LDL receptors (Boucher et al., 1998). Very high dietary cholesterol can exceed the ability of down-regulated LDL receptors to clear LDL adequately from plasma, resulting in increased plasma LDL concentrations (Applebaum-Bowden et al., 1984; Spady and Dietschy, 1988). Thus, it is generally accepted that dietary cholesterol can decrease the gene expression of LDL receptors by influencing the hepatic cholesterol pool and subsequent repression of SREBP activation by cholesterol or its metabolites.

The type of fatty acids consumed can influence the fatty acid composition of the liver, which in turn affects LDL receptor activity. In general, dietary SFAs decrease LDL receptor activity and consequently are hypercholesterolaemic with regard to plasma LDLs. Dietary MUFAs and PUFAs are comparatively hypocholesterolaemic. One exception is dietary stearic acid. Chemically, stearic acid is an SFA, but it does not decrease LDL receptor activity or mRNA abundance and does not increase plasma LDL concentration (Nicolosi, 1997). The exact mechanisms are not fully understood but are likely to include transcriptional and post-translational mechanisms.

The hypocholesterolaemic effect of soluble dietary fibre is attributed mainly to the fibres' ability to decrease cholesterol absorption and bile acid reabsorption in the small intestine. Animal studies have shown that consumption of soluble fibre increased LDL receptor activity and mRNA levels, with concomitant decreases in plasma cholesterol concentration (Fernandez et al., 1995; Fukushima et al., 2000). At the cellular level, the hypocholesterolaemic effect of soluble fibres is mediated through decreases in the hepatic cholesterol regulatory pool and the SREBP signalling pathway.

Cholesterol 7α-hydroxylase

Bile acids are important in the digestion and absorption of dietary lipids. Recent studies have indicated the role of bile acids in regulating the transcription of genes involved in cholesterol and bile acid metabolism (Makishima et al., 1999; Parks et al., 1999; Wang et al., 1999). There are two distinct pathways for bile acid synthesis: the classical (neutral) and the alternative (acidic) pathway. Microsomal cholesterol 7α-hydroxylase is the rate-limiting enzyme in classical bile acid synthesis, and mitochondrial sterol 27-hydroxylase is suggested to be the limiting enzyme in alternative bile acid synthesis. The contribution of the alternative pathway to total bile acid synthesis is not known. Under conditions that repress the activity of cholesterol 7α-hydroxylase, bile acid synthesis remains relatively high, indicating that in the face of persistent suppression of cholesterol 7α-hydroxylase, the alternative pathway becomes a main pathway of bile acid synthesis (Xu et al., 1999). Bile acids are conjugated with glycine or taurine in the liver before being secreted into the gallbladder for temporary storage. Approximately 95% of bile acids/salts are reabsorbed in the small intestine via passive or active transport mechanisms and shuttled back to the liver. The enterohepatic circulation of bile acids/salts plays an important physiological role in lipid absorption, bile acid synthesis and cholesterol homeostasis.

In response to increased dietary cholesterol, the hepatic cholesterol pool is maintained largely by an increased conversion of cholesterol to bile acids or secretion of free cholesterol into bile. The extent of bile acid synthesis responding to dietary cholesterol varies among species. Dietary cholesterol increases mRNA and activity of cholesterol 7α-hydroxylase as a compensatory response to maintain cholesterol homeostasis in rats (Pandak et al., 1991; Shefer et al., 1992) and mice (Deuland et al., 1993; Torchia et al., 1996). However, in New Zealand rabbits (Xu et al., 1995), African green monkeys (Rudel et al., 1994) and hamsters (Horton et al., 1995), mRNA and activity of cholesterol 7α-hydroxylase were repressed by cholesterol feeding. Further study in rabbits indicated that cholesterol feeding increased cholesterol 7α-hydroxylase activity in the short term, but was followed by decreased enzyme activity (Xu et al., 1999). Nevertheless, the bile acid pool size was significantly increased due to increased activity of sterol 27-hydroxylase, an enzyme in the alternative bile acid synthetic pathway. Recall that the expression of the *Cyp7a1* gene is controlled either positively by LXRs or negatively by FXR. Therefore, we can speculate that initial cholesterol feeding might activate LXRs. However, as the hepatic bile acid flux increases, FXR may become activated and repress the transactivation of the *Cyp7a1* gene. In species that are sensitive to the repression of *Cyp7a1* by FXR, the alternative bile acid synthesis pathway may predominate.

The activity of cholesterol 7α-hydroxylase can also be regulated by dietary fatty acids, although the precise mechanisms have not been identified. In general, PUFAs and MUFAs tend to increase cholesterol 7α-hydroxylase activity in animal studies, whereas SFAs have a limited effect on activity (Bravo et al., 1996). Inclusion of oleic acid in hamster diets increased hepatic cholesterol 7α-hydroxylase activity, whereas palmitic acid suppressed activity (Kurushima et al., 1995). There clearly is little information on the specific effects of individual fatty acids at the cellular and molecular level.

Hypocholesterolaemic effects of water-soluble fibres have been attributed, at least in part, to their ability to form a viscous matrix that can inhibit the absorption of bile acids/salts and sterols from the small intestine (Eastwood and Morris, 1992). The resultant increase in the excretion of bile acids and total sterols can up-regulate bile acid synthesis in order to replenish the hepatic bile acid pool. Normolipidaemic young men consuming a diet containing oat bran for 2 months had an increased synthesis and faecal excretion of bile acids (Marlett et al., 1994). Rats fed a pectin-supplemented diet showed lower serum and liver cholesterol concentrations, increased faecal bile acid excretion and increased cholesterol 7α-hydroxylase activity (Garcia-Diez et al., 1996). Psyllium intake increased hepatic cholesterol 7α-hydroxylase activity in male guinea pigs and mRNA for cholesterol 7α-hydroxylase in rats and hamsters (Horton et al., 1994; Fernandez et al., 1995; Buhman et al., 1998). Each of these observations points to a greater need to synthesize cholesterol and convert it to bile acids when intestinal sources are diverted for excretion from the body.

Summary

Three major regulatory proteins in hepatic cholesterol metabolism – HMG-CoA reductase, LDL receptor and cholesterol 7α-hydroxylase – are controlled at the transcriptional level by SREBPs, LXRs and/or FXR in a coordinated manner to maintain cholesterol homeostasis. By regulating the expression of genes encoding HMG-CoA reductase and LDL receptor, SREBPs play an important role in cholesterol synthesis and cellular uptake. LXRs and FXR regulate the conversion of cellular cholesterol to bile acids via cholesterol 7α-hydroxylase. Because bile acid excretion represents a major route of cholesterol elimination from the body, LXRs and FXR are also instrumental in regulating whole-body cholesterol turnover and balance. Furthermore, all of these processes are influenced by diet, particularly cholesterol, fatty acids and fibre. Other dietary constituents may also impact cellular cholesterol metabolism and undoubtedly will be the subject of future research endeavours.

References

Ananthanarayanan, M., Balasubramanian, N., Makishima, M., Mangelsdorf, J.J. and Suchy, F.J. (2001) Human bile salt export pump (BSEP) promoter is transactivated by the farnesoid X receptor/bile acid receptor (FXR/BAR). *Journal of Biological Chemistry* 276, 28857–28865.

Applebaum-Bowden, D., Haffner, S.M., Hartsook, E., Luk, K.H., Albers, J.J. and Hazzard, W.R. (1984) Down-regulation of the low-density lipoprotein receptor by dietary cholesterol. *American Journal of Clinical Nutrition* 39, 360–367.

Becker-Andre, M., Andre, E. and DeLamarter, J.F. (1993) Identification of nuclear receptor mRNAs by RT–PCR amplication of conserved zinc-finger motif sequences. *Biochemical and Biophysical Research Communications* 194, 1371–1379.

Boucher, P., de Lorgeril, M., Salen, P., Crozier, P., Delaye, J., Vallor, J.J., Geyssant, A. and Dante, R. (1998) Effect of dietary cholesterol on low density lipoprotein-receptor, 3-hydroxy-3-methylglutaryl-CoA reductase, and low density lipoprotein receptor-related protein mRNA expression in healthy humans. *Lipids* 33, 1177–1186.

Bravo, E., Cantafora, A., Marinelli, T., Avella, M., Mayes, P.A. and Botham, K.M. (1996) Differential effects of chylomicron remnants derived from corn oil or palm oil on bile acid synthesis and very low density lipoprotein secretion in cultured rat hepatocytes. *Life Sciences* 59, 331–337.

Briggs, M.R., Yokoyama, C., Wang, X., Brown, M.S. and Goldstein, J.L. (1993) Nuclear protein that binds sterol regulatory element of low density lipoprotein receptor promoter. *Journal of Biological Chemistry* 268, 14490–14496.

Brown, M.S. and Goldstein, J.L. (1999) A proteolytic pathway that controls the cholesterol content of membranes, cells and blood. *Proceedings of the National Academy of Sciences USA* 96, 11041–11048.

Buhman, K.K., Furumoto, E.J., Donkin, S.S. and Story, J.A. (1998) Dietary psyllium increases fecal bile acid excretion, total steroid excretion and bile acid biosynthesis in rats. *Journal of Nutrition* 128, 1199–1203.

Chiang, J.Y.L. and Stroup, D. (1994) Identification and characterization of a putative bile acid-response element in cholesterol 7α-hydroxylase gene promoter. *Journal of Biological Chemistry* 269, 17502–17507.

Chiang, J.Y.L., Kimmel, R., Weinberger, C. and Stroup, D. (2000) Farnesoid X receptor responds to bile acids and repressed cholesterol 7α-hydroxylase gene (CYP7A1) transcription. *Journal of Biological Chemistry* 275, 10918–10924.

Deuland, S., Drisko, J., Graf, L., Machleder, D., Lusis, A.J. and Davis, R.A. (1993) Effect of dietary cholesterol and taurocholate on cholesterol 7 alpha-hydroxylase and hepatic LDL receptors in inbred mice. *Journal of Lipid Research* 34, 923–931.

Eastwood, M.A. and Morris, E.R. (1992) Physical properties of dietary fibre that influence physiological function: a model for polymers along the gastrointestinal tract. *American Journal of Clinical Nutrition* 55, 436–442.

Edwards, P.A., Tabor, D., Kast, H.R., Venkateswaran, A. (2000) Regulation of gene expression by SREBP and SCAP. *Biochimica et Biophysica Acta* 1529, 103–113.

Fernandez, M.L., Ruiz, L.R., Conde, A.K., Sun, D.M., Erickson, S.K. and McNamara, D.J. (1995) Psyllium reduces plasma LDL in guinea pigs by altering hepatic cholesterol homeostasis. *Journal of Lipid Research* 36, 1128–1138.

Forman, B.M., Goode, E., Chen, J., Oro, A.E., Bradley, D.J., Perimann, T., Noonan, D.J., Burka, L.T., McMorris, T., Lamph, W.W., Evans, R.M. and Weinberger, C. (1995) Identification of a nuclear receptor that is activated by farnesol metabolites. *Cell* 81, 687–693.

Fukushima, M., Nakano, M., Morii, Y., Ohashi, T., Fujiwara, Y. and Sonoyama, K. (2000) Hepatic LDL receptor mRNA in rats is increased by dietary mushroom (*Agaricus bisporus*) fibre and sugar beet fibre. *Journal of Nutrition* 130, 2151–2156.

Garcia-Diez, F., Garcia-Mediavilla, V., Bayon, J.E. and Gonzalez-Gallego, J. (1996) Pectin feeding influences fecal bile acid excretion, hepatic bile acid and cholesterol synthesis and serum cholesterol in rats. *Journal of Nutrition* 126, 1766–1771.

Goldstein, J.L. and Brown, M.S. (1990) Regulation of the mevalonate pathway. *Nature* 343, 425–430.

Goldstein, J.L. and Brown, M.S. (2001) The cholesterol quartet. *Science* 292, 1310–1312.

Grober, J., Zaghin, I., Fujii, H., Jones, S.A., Kliewer, S.A., Willson, T.M., Ono, T. and Besnard, P. (1999) Identification of a bile acid-responsive element in the human ileal bile acid-binding protein gene. *Journal of Biological Chemistry* 274, 29749–29754.

Horton, J.D., Cutbert, J.A. and Spady, D.K. (1995) Regulation of hepatic 7α-hydroxylase expression and response to dietary cholesterol in the rat and hamster. *Journal of Biological Chemistry* 270, 5381–5387.

Hua, X., Yokoyama, C., Wu, J., Briggs, M.R., Brown, M.S., Goldstein, J.L and Wang, X. (1993) SREBP-2, a second basic-helix–loop–helix-leucine zipper protein that stimulates transcription by binding to a sterol regulatory element. *Proceedings of the National Academy of Sciences USA* 90, 11603–11607.

Hua, X., Wu, J., Goldstein, J.L., Brown, M.S. and Hobbs, H.H. (1995) Structure of the human gene encoding sterol regulatory element binding protein-1 (SREBF-1) and localization of SREBF-1 and SREBF-2 to chromosomes 17p11.2 and 22q13. *Genomics* 25, 667–673.

Janowski, B.A., Willy, P.J., Devi, T.R., Falck, J.R. and Mangelsdorf, D.J. (1996) An oxysterol signaling pathway mediated by the nuclear receptor LXRα. *Cell* 383, 728–731.

Kay, R.M. (1982) Dietary fibre. *Journal of Lipid Research* 23, 221–242.

Kramer, W., Birbig, F., Gutjahr, U., Kowalewski, S., Jouvenal, K., Muller, G., Tripier, D. and Wess, G. (1993) Intestinal bile acid absorption. Na(+)-dependent bile acid transport activity in rabbit small intestine correlates with the coexpression of an integral 93-kDa and a peripheral 14-kDa bile acid-binding membrane protein along the duodenum–ileum axis. *Journal of Biological Chemistry* 268, 18035–18046.

Kurushima, H., Hayashi, K., Shingu, T., Kuga, Y., Ohtani, H., Okura, Y., Tanaka, K., Yasunobu, Y., Nomura, K. and Kajiyama, G. (1995) Opposite effects on cholesterol metabolism and their mechanisms induced by dietary oleic acid and palmitic acid in hamsters. *Biochimica et Biophysica Acta* 1258, 251–256.

Lehmann, J.M., Kliewer, S.A., Moore, L.B., Smith-Oliver, T.A., Oliver, B.B., Su, J.L., Sundseth, S.S., Winegar, D.A., Blanchard, D.E., Spencer, T.A. and Willson, T.M. (1997) Activation of the nuclear receptor LXR by oxysterols defines a new hormone response pathway. *Journal of Biological Chemistry* 272, 3137–3140.

Lu, T.T., Makishima, M., Repa, J.J., Schoonjans, K., Kerr, T.A., Auwerx, J. and Mangelsdorf, D.D. (2000) Molecular basis for feedback regulation of bile acid synthesis by nuclear receptors. *Molecular Cell* 6, 507–515.

Makishima, M., Lkamoto, A.Y., Repa, J.J., Tu, H., Learned, M., Luk, A., Hull, M.V., Lustig, K.D., Mangelsdorf, D.J. and Shan, B. (1999) Identification of a nuclear receptor for bile acids. *Science* 284, 1362–1365.

Marlett, J.A., Hosig, K.B., Vollendorf, N.W., Shinnick, F.L., Haack, V.S. and Story, J.A. (1994) Mechanism of serum cholesterol reduction by oat bran. *Hepatology* 20, 1450–1457.

Moundras, C., Behr, S.R., Remesy, C. and Demigne, C. (1997) Fecal losses of sterols and bile acids induced by feeding rats guar gum are due to greater pool size and liver bile acid secretion. *Journal of Nutrition* 127, 1068–1076.

Nicolosi, R.J. (1997) Dietary fat saturation effects on low-density-lipoprotein concentrations and metabolism in various animal models. *American Journal of Clinical Nutrition* 65 (Supplement 5), 1617S–1627S.

Nohturfft, A., Hua, X., Brown, M.S. and Goldstein, J.L. (1996) Recurrent G-to-A substitution in a single codon of SREBP cleavage-activating protein causes sterol resistance in three mutant Chinese hamster ovary cell lines. *Proceedings of the National Academy of Sciences USA* 93, 13709–13714.

Pandak, W.M., Li, Y.C., Chiang, J.Y.L., Studer, E.J., Gurley, E.C., Heuman, D.M., Vlahcevic, Z.R. and Hylemon, P.B. (1991) Regulation of cholesterol 7α-hydroxylase mRNA and transcriptional activity by taurocholate and cholesterol in the chronic biliary diverted rat. *Journal of Biological Chemistry* 266, 3416–3421.

Parks, D.J., Blanchard, S.G., Bledsoe, R.K., Chandra, G., Consler, T.G., Kliewer, S.A., Stimmel, J.B., Wilson, T.M., Zavacke, A.M., Moore, D.D. and Lehmann,

J.M. (1999) Bile acids: natural ligand for an orphan nuclear receptor. *Science* 284, 1365–1368.

Peet, D.J., Janowski, B.A. and Mangelsdorf, D.J. (1998a) The LXRs: a new class of oxysterol receptors. *Current Opinion in Genetics and Development* 8, 571–575.

Peet, D.J., Turley, S.D., Ma, W., Janowski, B.A., Lobaccaro, J.A., Hammer, R.E. and Mangelsdorf, D.J. (1998b) Cholesterol and bile acid metabolism are impaired in mice lacking the nuclear oxysterol receptor LXRα. *Cell* 93, 693–704.

Repa, J.J., Liang, G., Ou, J., Bashmakov, Y., Lobaccaro, J.M.A., Shimomura, I., Shan, B., Brown, M.S., Goldstein, J.L. and Mangelsdorf, D.J. (2000) Regulation of mouse sterol regulatory element-binding protein-1c gene (SREBP-1c) by oxysterol receptors, LXRalpha and LXRbeta. *Genes and Development* 14, 2819–2830.

Rudel, L., Deckelman, C., Wilson, M., Scobey, M. and Anderson, R. (1994) Dietary cholesterol and downregulation of cholesterol 7α-hydroxylase and cholesterol absorption in African green monkeys. *Journal of Clinical Investigation* 93, 2463–2472.

Sakai, J., Duncan, E.A., Rawson, R.B., Hua, X., Brown, M.S. and Goldstein, J.L. (1996) Sterol-regulated release of SREBP-2 from cell membranes requires two sequential cleavages, one within a transmembrane segment. *Cell* 85, 1037–1046.

Sakai, J., Nohturfft, A., Goldstein, J.L. and Brown, M.S. (1998) Cleavage of sterol regulatory element-binding proteins (SREBPs) at site-1 requires interaction with SREBP cleavage-activating protein. Evidence from *in vivo* competition studies. *Journal of Biological Chemistry* 273, 5785–5793.

Seol, W., Choi, H.-S. and Moore, D.D. (1996) An orphan nuclear hormone receptor that lacks a DNA binding domain and heterodimerizes with other receptors. *Science* 272, 1336–1339.

Shefer, S., Nguyen, L.B., Salen, G., Ness, G.C., Chowdhary, I.R., Lerner, S., Batta, A.K. and Tint, G.S. (1992) Differing effects of cholesterol and taurocholate on steady state hepatic HMG-CoA reductase and cholesterol 7 alpha-hydroxylase activities and mRNA levels in the rat. *Journal of Lipid Research* 33, 1193–1200.

Shimomura, I., Shimano, H., Horton, J.D., Goldstein, J.L. and Brown, M.S. (1997) Differential expression of exons 1a and 1c in mRNAs for sterol regulatory element binding protein-1 in human and mouse organs and cultured cells. *Journal of Biological Chemistry* 99, 838–845.

Sinal, C.J., Tohkin, M., Miyata, M., Ward, J.M., Lambert, G. and Gonzalez, F.J. (2000) Targeted disruption of the nuclear receptor FXR/BAR impairs bile acid and lipid homeostasis. *Cell* 102, 731–744.

Spady, D.K. and Dietschy, J.M. (1988) Interaction of dietary cholesterol and triglycerides in the regulation of hepatic low density lipoprotein transport in hamster. *Journal of Clinical Investigation* 81, 300–309.

Spady, D.K., Stange, E.F., Bilhartz, L.E. and Dietschy, J.M. (1986) Bile acids regulate hepatic low density lipoprotein receptor activity in the hamster by altering cholesterol flux across the liver. *Proceedings of the National Academy of Sciences USA* 83, 1916–1920.

Stravitz, R.T., Hylemon, P.B., Heuman, D.M., Hagey, L.R., Schteingart, C.D., Ton-Nu, H.-T., Hofmann, A.F. and Vlahcevic, Z.R. (1993) Transcriptional regulation of cholesterol 7α-hydroxylase mRNA by conjugated bile acids in primary cultures of rat hepatocytes. *Journal of Biological Chemistry* 268, 13967–13993.

Stroup, D., Crestani, M. and Chiang, J. Y. L. (1997) Identification of a bile acid response element in the cholesterol 7 alpha-hydroxylase gene CYP7A. *American Journal of Physiology* 273, G508–G517.

Thewke, D.P., Panini, S.R. and Sinensky, M. (1998) Oleate potentiates oxysterols inhibition of transcription from sterol regulatory element-1-regulated promoters and maturation of sterol regulatory element-binding proteins. *Journal of Biological Chemistry* 173, 21402–21407.

Torchia, E.C., Cheema, S.K. and Agellon, L.B. (1996) Coordinate regulation of bile acid biosynthetic and recovery pathways. *Biochemical and Biophysical Research Communications* 225, 128–133.

Towle, H.C. (1995) Metabolic regulation of gene transcription in mammals. *Journal of Biological Chemistry* 270, 23235–23238.

Vlahcevic, Z.R., Pandak, W.M. and Stravitz, R.T. (1999) Regulation of bile acid biosynthesis. *Gastroenterology Clinics of North America* 28, 1–25.

Wang, X., Briggs, M.R., Hua, X., Yokoyama, C., Goldstein, J.L. and Brown, M.S. (1993) Nuclear protein that binds sterol regulatory element of low density lipoprotein promoter. II. Purification and characterization. *Journal of Biological Chemistry* 268, 14497–14504.

Wang, H., Chen, J., Hollister, K., Sowers, L.C. and Forman, B.M. (1999) Endogenous bile acids are ligands for the nuclear receptor FXR/BAR. *Molecular Cell* 3, 543–553.

Willy, P.J., Umesono, K., Ong, E.S., Evans, R.M., Heyman, R.A. and Mangelsdorf, D.J. (1995) LXR, a nuclear receptor that defines a distinct retinoid response pathway. *Genes and Development* 9, 1033–1045.

Worgall, T.S., Sturley, S.L., Seo, T., Osborne, T.F. and Deckelbaum, R.J. (1998) Polyunsaturated fatty acids decrease expression of promoters with sterol regulatory elements by decreasing levels of mature sterol regulatory element-binding protein. *Journal of Biological Chemistry* 273, 25537–25540.

Xu, G., Salen, G., Shefer, S., Ness, G.C., Nguyen, L.B., Parker, T.S., Chen, T.S., Zhao, Z., Donnelly, T.M. and Tint G.S. (1995) Unexpected inhibition of cholesterol 7 alpha-hydroxylase by cholesterol in New Zealand white and Watanabe heritable hyperlipidemic rabbits. *Journal of Clinical Investigation* 95, 1497–1504.

Xu, G., Salen, G., Shefer, S., Tint, G.S., Nguyen, L.B., Chen, T.S. and Greenblatt, D. (1999) Increasing dietary cholesterol induces different regulation of classic and alternative bile acid synthesis. *Journal of Clinical Investigation* 103, 89–95.

21 2002 Assessment of Nutritional Influences on Risk for Cataract

Allen Taylor and Mark Siegal
Laboratory for Nutrition and Vision Research, USDA Human Nutrition Research Center on Aging, Tufts University, Boston, Massachusetts, USA

Cataract as a Public Health Issue

It is clear that oxidative stress is associated with compromises to the lens. Recent literature has indicated that antioxidants may ameliorate that risk and actually decrease risk for cataract. This chapter briefly reviews the aetiology of cataract. It also reviews the epidemiological information, with an emphasis on roles for vitamins C and E and carotenoids. More thorough recent reviews are available (Taylor, 1999b,c).

Annual Cost of Age-related Prevalence

Cataract is one of the major causes of preventable blindness throughout the world (Kupfer, 1985; Schwab, 1990; World Health Organization, 1991). In the USA, the prevalence of visually significant cataract increases from approximately 5% at age 65 years to about 50% for persons older than 75 years (Leibowitz *et al.*, 1980; Klein *et al.*, 1992, 1993). In less developed countries, such as India (Chatterjee *et al.*, 1982), China (Wang *et al.*, 1990) and Kenya (Whitfield *et al.*, 1990), cataracts are more common and develop earlier in life than in more developed countries (Leibowitz *et al.*, 1980; Chatterjee *et al.*, 1982). The impact of cataract on impaired vision is much greater in less developed countries, where >90% of the cases of blindness and visual impairment are found

(Taylor, 1999a,c), and where there is a dearth of ophthalmologists to perform lens extractions.

It is estimated that a delay in cataract formation of about 10 years would reduce the prevalence of visually disabling cataract by about 45% (Kupfer, 1985). Such a delay would enhance the quality of life for much of the world's older population and substantially reduce the economic burden (US$5 billion to US$6 billion) due to cataract-related disability and cataract surgery (Young, 1993).

Age-related Damage Involves Oxidative Insult and Declines in Protective Capabilities

The primary function of the eye lens is to collect and focus light on the retina (Fig. 21.1). To do so, it must remain clear throughout life. The lens is exquisitely organized. A single layer of epithelial cells is directly under the anterior surface of the collagenous membrane in which it is encapsulated (Fig. 21.1b). The epithelial cells at the germinative region divide, migrate posteriorly and differentiate into lens fibres. As their primary gene products, the fibres elaborate the predominant proteins of the lens, called crystallins. New cells are formed throughout life, but older cells usually are not lost. Instead, they are compressed into the centre or nucleus of the lens. As the lens ages, or with stress

©CAB *International* 2003. *Molecular Nutrition*
(eds J. Zempleni and H. Daniel)

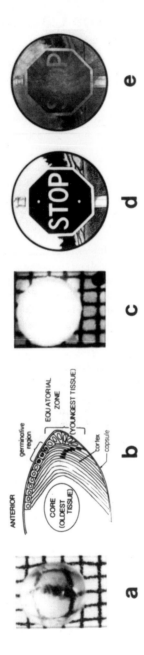

Fig. 21.1. Clear and cataractous lens. (a) Clear lens allows an unobstructed view of the wire grid placed behind it. (b) Cartoon of the structure of the lens. The anterior surface of the lens has a unicellular layer of epithelial cells (youngest tissue). Cells at the anterior equatorial region divide and migrate to the cortex as they are overlaid by less mature cells. These cells produce a majority of the crystallins. As development and maturation proceed, the cells denucleate and elongate. Tissue originally found in the embryonic lens is found in the core or nucleus (oldest tissue). (c) The cataractous lens prohibits viewing the wire grid behind it. (d) Artist's view through a clear, uncoloured young lens. The image is clear and crisp. (e) Artist's view through a lens with developing cataract. The image is partially obscured and the field is darkened due to browning of the lens that accompanies ageing (from Taylor (1999a)).

due to light exposure (McCarty and Taylor, 1999) or smoking (West, 1999), the proteins are damaged photooxidatively and aggregate. Among the oxidative insults are high-energy radiation, reactive oxygen species, sunlight and sunlight exposure, and failure of secondary defence systems (Fig. 21.2). There is a coincident dehydration of the proteins and of the lens itself. Consequently, protein concentrations rise to hundreds of milligrams per millilitre (Taylor et al., 1981). Together with other age-related modifications of the protein such as glycation, glycoxidation and modification of proteins by lipid, these changes result in a less flexible lens with limited accommodative capability. Eventually, the large aggregates of protein precipitate in lens opacities, or cataract.

The term age-related cataract distinguishes lens opacification associated with old age from opacification associated with other causes such as congenital and metabolic disorders or trauma. There are several systems for evaluating and grading cataracts. Most of these use an assessment of extent, or density, and location of the opacity (Chylack et al., 1993; Chylack, 1999). Usually evaluated are opacities in the posterior subcapsular, nuclear, cortical and multiple (mixed) locations and, possibly, coloration (Wolfe et al., 1993; Chylack et al., 1994). However, it has not been established that cataract at each location has a completely different aetiology.

Antioxidants as primary defences against lens damage

What systems are available to protect the lens? Protection against photooxidative insult can be

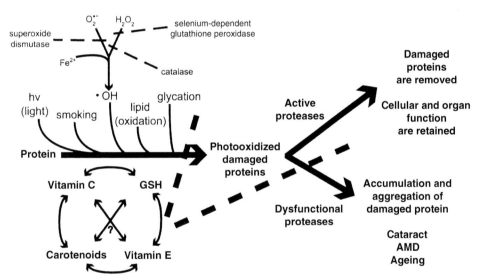

Fig. 21.2. Proposed interaction between lens proteins, oxidants, light, smoking, antioxidant enzymes and proteases. Lens proteins are extremely long lived and subject to alteration by light and various forms of oxygen. They are protected indirectly by antioxidant enzymes: superoxide dismutase, catalase and GSH reductase/peroxidase. Those enzymes convert active oxygen to less-damaging species. Direct protection is offered by antioxidants: GSH, ascorbate (vitamin C), tocopherol (vitamin E) and carotenoids. Levels of reduced and oxidized forms of some, but perhaps not all, of those molecules are determined by an interaction among the four and with the environment (Taylor and Davies, 1987; Taylor et al., 1991a; Chylack et al., 1993, 1994; Wolfe et al., 1993; Chylack, 1999). In many systems, GSH and ascorbate levels are related, but we did not find this to be the case in ascorbate-sufficient, ascorbate-requiring rats (Bunce et al., 1990; Taylor and Jacques, 1997). When the proteolytic capability is sufficient, obsolete and damaged proteins might be reduced to their constituent amino acids. With ageing, some of the eye antioxidant supplies are diminished, antioxidant enzymes inactivated and proteases less active. That action appears to be related to the accumulation, aggregation and eventual precipitation in cataractous opacities of damaged proteins. GSH, glutathione; hv, light; AMD, age-related macular degeneration.

conceived as due to two inter-related processes (Fig. 21.2). Antioxidants and antioxidant enzymes provide primary protection against insults by attenuating the insult or repairing the protein. Secondary defences include proteolytic and repair processes, which degrade and eliminate damaged proteins and other biomolecules in a timely fashion (Taylor and Davies, 1987).

The major aqueous antioxidants in the lens are ascorbate and glutathione (GSH) (Bunce et al., 1990; Reddy, 1990; Taylor et al., 1991a, 1995b, 1997; Sastre et al., 1994; Mune et al., 1995; Taylor and Jacques, 1997; Smith et al., 1999). Both are present in the lens at millimolar concentrations.

Ascorbate probably is the most effective and least toxic antioxidant identified in mammalian systems (Levine, 1986; Berger et al., 1988, 1989; Frei et al., 1988; Taylor et al., 1991a; Taylor and Jacques, 1997). Ocular levels of ascorbate are related to dietary intake in humans and animals that require ascorbate (Berger et al., 1988, 1989). Interestingly, the concentration of vitamin C in the lens was increased with dietary supplements beyond levels achieved in persons who already consumed more than twice the recommended dietary allowance (70 mg day) for vitamin C (Taylor et al., 1991a, 1997).

Feeding-elevated ascorbate delayed progress of, or prevented, galactose cataract in guinea pigs (Kosegarten and Maher, 1978; Yokoyama et al., 1994) and rats (Vinson et al., 1986), selenite-induced cataracts in rats (Devamanoharan et al., 1991) and lens opacification in GSH-depleted chick embryos (Nishigori et al., 1986), and delayed UV-induced protein and protease damage in guinea pig lenses (Blondin et al., 1986, 1987; Blondin and Taylor, 1987; Taylor et al., 1995a). Increasing lens ascorbate concentrations by only twofold has protected against cataract-like damage (Blondin et al., 1986).

Because ascorbate is a carbohydrate, it is biochemically plausible that vitamin C induces damage in the lens in vivo (Garland, 1991; Nagaraj and Monnier, 1992). However, at present, there are no data to support this as a medical concern. Although we found that glycohaemoglobin levels increase with increasing dietary ascorbate in ascorbate-requiring rats (Smith et al., 1999), mice fed 8% of the weight of their diet as ascorbate did not develop cataract (Bensch et al., 1985).

GSH levels in the eye are several-fold the levels found in whole blood, and orders of magnitude greater than the concentration observed in the plasma. As for ascorbate, GSH levels also diminish in the older and cataractous lens (Reddy, 1990). Preliminary evidence from studies with galactose-induced cataract also has indicated some advantage of maintaining elevated GSH status in rats (Sastre et al., 1994). However, it is not clear that feeding GSH is associated with higher ocular levels of this antioxidant (Sastre et al., 1994). In many systems, GSH and ascorbate levels are related. We did not find this to be the case in ascorbate-sufficient, ascorbate-requiring rats (Smith et al., 1999).

Tocopherols and carotenoids are lipid-soluble antioxidants (Schalch et al., 1999; Yeum et al., 1999) with probable roles in maintaining membrane integrity (Machlin and Bendich, 1987) and GSH recycling (Costagliola et al., 1986). Concentrations of tocopherol in the whole lens are in the micromolar range, and levels are elevated in younger tissues (Yeum et al., 1995), but lens and dietary levels of tocopherol seem unrelated (Stephens et al., 1988). Because most of the compound is found in the membranes, the concentrations in membranes may be orders of magnitude higher. Other than differential distribution within lenses (Yeum et al., 1999), age-related changes in levels of tocopherol and carotenoids have not been documented. Tocopherol has been reported to be effective in delaying a variety of induced cataracts in animals, including those induced by galactose (Bhuyan et al., 1983; Creighton et al., 1985; Jacques and Taylor, 1991), and by aminotriazole in rabbits (Bhuyan and Bhuyan, 1984).

Relationships between carotenoids and eye health have been reviewed recently (Schalch et al., 1999; Taylor, 1999b). β-Carotene levels in the human lenses are limited (Yeum et al., 1995, 1999). Instead, the major lens and macular carotenoids (present at 11–44 ng g^{-1} wet weight) are lutein and zeaxanthin (Hammond et al., 1997; Schalch et al., 1999; Snodderly and Hammond, 1999). Also present are retinol, retinyl ester (21–50 ng g^{-1} wet weight) and tocopherols (1232–2550 ng g^{-1} wet weight).

The lens also contains antioxidant enzymes: glutathione peroxidase/reductase, catalase, and superoxidase dismutase and enzymes of the glutathione redox cycle (Giblin et al., 1982; Fridovich, 1984; Varma et al., 1984; Zigler and Goosey, 1984; Rathbun et al., 1996). These interact with the forms of oxygen and with the antioxidants, i.e. GSH is a substrate for glutathione peroxidase. The activities of many antioxidant enzymes are compromised with development, ageing and cataract formation (Berman, 1991).

Proteolytic enzymes provide secondary defences against lens damage

The accumulation of damaged proteins is cytotoxic, and such toxicity has been well documented in many age-related neurodegenerative diseases including Alzheimer's and Pick's diseases. However, nowhere is it more obvious than in cataract, where the precipitation of damaged proteins causes cataract. Proteolytic systems can be considered secondary defence capabilities that remove cytotoxic damaged or obsolete proteins from lenses and other tissues (Kosegarten and Maher, 1978; Jahngen et al., 1986, 1990; Taylor and Davies, 1987; Eisenhauer et al., 1988; Taylor et al., 1991b, 1993; Jahngen-Hodge et al., 1992, 1997; Huang et al., 1993; Obin et al., 1994, 1996, 1998, 1999; Shang and Taylor, 1995; Shang et al., 1997a,b). Such proteolytic systems exist in younger lens tissue and the outer tissues in all lenses, probably protected by the primary antioxidants. However, with ageing or oxidative stress, most of these enzymatic capabilities are found in a state of reduced activity (Taylor and Davies, 1987; Taylor, 1999b). The observed accumulation of oxidized (and/or otherwise modified) proteins in older lenses is consistent with the failure of these protective systems to keep pace with the insults that damage lens proteins. This occurs in part because, like bulk proteins, enzymes that comprise some of the protective systems are damaged by photooxidation (Blondin and Taylor, 1987; Taylor et al., 1993; Shang and Taylor, 1995; Shang et al., 1995; Jahngen-Hodge et al., 1997; Taylor, 1999b).

Several studies have suggested interactions between antioxidants such that one might spare the other (Fig. 21.2). Direct sparing effects of ascorbate and GSH on (photo)oxidatively induced compromises of proteolytic function have been demonstrated (Shang and Taylor, 1995; Jahngen-Hodge et al., 1997; Obin et al., 1998). From these data, it is clear that the young lens has significant primary and secondary protection. However, age-related compromises in the activity of antioxidant enzymes, concentrations of the antioxidants and activities of secondary defences might lead to diminished protection against oxidative insults. This diminished protection leaves the long-lived proteins and other constituents vulnerable. Lens opacities develop as the damaged proteins aggregate and precipitate. Current data have predicted that elevated antioxidant intake can be exploited to extend the function of some of these proteolytic capabilities (Blondin and Taylor, 1987).

Epidemiological Studies Regarding Associations Between Antioxidants and Cataract

Because cataract is due in part to oxidative stress on lens constituents and the enzymes that normally might remove these damaging moieties (reviewed in Fig. 21.2 and Taylor, 2000), considerable effort is being dedicated to determining whether antioxidants can be used to diminish risk for cataract. The overall impression created by the data indicates that nutrient intake is related to risk for cataract and that nutrition might be exploited to diminish the risk for this debility.

The data regarding a specific nutrient are grouped together in this review. The data are subdivided to consider cataracts in the subcapsular, cortical and nuclear zones because of the assumption that the opacities in these metabolically and developmentally distinguishable areas of the lens have different aetiologies. Nutrition status was assessed with various questionnaires and/or by measuring blood nutrient levels. Within a cataract type, we discuss, in turn, the effects of supplements, diet (including supplements) and blood levels.

Studies differ in design, and comparisons are not always straightforward. Assets and limitations of various designs were discussed previously (Christen, 1999). Most of the previous studies were retrospective case–control or cross-sectional studies in which levels of cataract patients were compared with levels of individuals with clear lenses (Mohan et al., 1989; Robertson et al., 1989; Jacques and Chylack, 1991; Leske et al., 1991, 1995; The Italian–American Cataract Study, 1991; Vitale et al., 1993; Mares-Perlman et al., 1994; Luthra et al., 1997; Cumming et al., 2000; Jacques et al., 2001; Taylor et al., 2002). Our ability to interpret data from retrospective studies such as those is limited by the concurrent assessment of lens status and nutrient levels. Prior diagnosis of cataract might influence the behaviour of cases including diet and bias reporting of the usual diet.

Other studies (Hankinson et al., 1992; Knekt et al., 1992; Seddon et al., 1994; Mares-Perlman et al., 1996; Rouhiainen et al., 1996; Leske et al., 1998a,b; Brown et al., 1999; Chasan-Taber et al., 1999a,b; Lyle et al., 1999a,b; Christen, 2001) assessed levels and/or supplement use and then followed individuals with intact lenses for up to 13 years. Such prospective studies are less prone to bias because assessment of exposure is performed before the outcome is known. Some of those studies (Hankinson et al., 1992; Knekt et al., 1992; Seddon et al., 1994; Brown et al., 1999; Chasan-Taber et al., 1999a,b) used cataract extraction or reported diagnosis of cataract as a measure of cataract risk. Whereas risk ratios, which are presented in some studies, are adjusted for many potentially confounding variables, others are not. Data from retrospective and prospective studies are separated in the figures.

Duration of measurement of dietary intake of nutrients and frequency of assessment can also affect the accuracy of these analyses because cataract develops over many years, and the frequency of one measure may not provide as accurate an assessment of usual intake as multiple measures over time.

In addition to the different study designs, various studies used different lens classification schemes, different definitions of high and low levels of nutrients and different age groups of subjects.

Although there are now epidemiological data regarding many nutrients (Taylor, 1999b), in this review only data for which there is a significant body of information are displayed.

Ascorbate

The available epidemiological data regarding ascorbate intake and risk for cataract is particularly intriguing because dietary ascorbate intake is related to eye tissue ascorbate levels and because there are reports indicating potential anti-cataractogenic and cataractogenic roles for vitamins.

Most studies have investigated relationships between vitamin C intake and risk for cataract (Mohan et al., 1989; Robertson et al., 1989; Jacques and Chylack, 1991; Leske et al., 1991, 1998a; The Italian–American Cataract Study, 1991; Hankinson et al., 1992; Vitale et al., 1993; Jacques et al., 1994, 1997, 2001; Mares-Perlman et al., 1994, 1995a,c, 2000; Brown et al., 1999; Lyle et al., 1999b; Simon and Hudes, 1999; Cumming et al., 2000; Taylor et al., 2002). In many of those studies, vitamin C, in particular vitamin C supplement use, was inversely associated with at least one type of cataract (Fig. 21.3).

In our Vitamin C and Cataract Study (VCCS), age-adjusted analyses based on 165 women with high intake of vitamin C (mean = 294 mg day^{-1}) and 136 women with relatively low intake of vitamin C (mean = 77 mg day^{-1}) indicated that the women who took vitamin C supplements for at least 10 years had >70% lower prevalence of early opacities (relative risk (RR) = 0.23, 95% confidence interval (CI) = 0.09–0.60; Fig. 21.3a) and >80% lower risk of moderate opacities (RR = 0.17, CI = 0.03–0.87; Fig. 21.3b) at any site compared with women who did not use vitamin C supplements (Jacques et al., 1994). Data from the baseline phase of the 'Nutrition and Vision Project' corroborated that observation. There was a 64% decrease in RR (0.36, CI = 0.18–0.72, fully adjusted, including adjustment for other nutrients) for early nuclear opacities as assessed with the Lens Opacification Classification System III (LOCS III) for persons who took vitamin C supplements for >10 years (Jacques et al., 2001) (Fig. 21.3c). Although it is not clear that cortical and nuclear cataract have the same aetiology, it is of interest that women <60 years old who used vitamin C supplements for at least 10 years had 60% lower odds (odds ratio (OR) = 0.40, CI = 0.18–0.87; Fig. 21.3d) for cortical opacities (Taylor et al., 2002). In corroboration of our observation, Mares-Perlman et al. (2000) indicated that, compared with non-users, the 5-year incidence for any cataract was 60% lower among persons who used vitamin C supplements for >10 years. Blood level measures also indicated inverse relationships between vitamin C status and risk for nuclear opacity (Jacques et al., 2001). Those studies were consistent with work by Hankinson et al. (1992) who found that women who consumed vitamin C supplements for >10 years had a 45% reduction in rate of cataract surgery (RR = 0.55, CI = 0.32–0.96; Fig. 21.3g). However, after controlling for nine potential confounders including age, diabetes, smoking and energy intake, they found no association between vitamin C intake and rate of cataract surgery.

Moreover, after looking at the same groups, Chasan-Taber et al. (1999a) found no benefit of prolonged use of vitamin C supplement with respect to risk for cataract extraction (Fig. 21.3g).

In comparison with those data noted above, Mares-Perlman et al. (1994) associated past use of supplements containing vitamin C with a reduced prevalence of nuclear cataract (RR = 0.7, CI = 0.5–1.0; Fig. 21.3c) but an increased prevalence of cortical cataract (adjusted RR = 1.8, CI = 1.2–2.9) after controlling for age, sex, smoking and history of heavy alcohol consumption (Fig. 21.3d).

The inverse relationship reported more often has been corroborated by data from other studies. Robertson et al. (1989) compared cases (with cataracts that impaired vision) with age- and sex-matched controls who had no cataract or minimal opacities that did not impair vision. The prevalence of cataract in those who consumed >300 mg day^{-1} of vitamin C supplements was approximately one-third that in persons who did not consume vitamin C supplements (RR = 0.30, CI = 0.24–0.77; Fig. 21.3b). Elevated dietary ascorbate also was found to be beneficial with respect to cataract in some studies. Leske et al. (1991) observed that persons with vitamin C intake in the highest 20% of their population group had a 52% lower prevalence for nuclear cataract (RR = 0.48, CI = 0.24–0.99) compared with persons who had intakes among the lowest 20% after controlling for age and sex (Fig. 21.3c). Weaker inverse associations were noted for other types of cataract (Fig. 21.3f). Jacques and Chylack (1991) observed that, among persons with higher vitamin C intakes (>490 mg day^{-1}), the prevalence of cataract was 25% of the prevalence among persons with lower intakes (<125 mg day^{-1}; RR = 0.25, CI = 0.06–1.09; Fig. 21.3a). In our assessments of early opacities in the baseline phase of the 'Nutrition and Vision Project', we found a 69% decreased risk for nuclear (RR = 0.31, CI = 0.16–0.58; Fig. 21.3c) opacities in the entire cohort, and a 57% decreased risk for cortical opacities in persons younger than 60 years (RR = 0.43, CI = 0.2–0.93; Fig. 21.3d) who had elevated vitamin C intake (Taylor et al., 2002). Intakes of 200–360 mg day^{-1} of vitamin C appear to provide maximum benefit (Taylor et al., 1991a; Taylor and Jacques, 1997; Jacques et al., 2001). This is of interest because intake in the reference group was twice the recommended dietary allowance. In a recent prospective study from the Beaver Dam Group, Lyle et al. (1999a) found a 70% decreased risk for nuclear cataract for heavy smokers (RR = 0.3, CI = 0.1–0.8) and persons with hypertension, but no significant relationship between vitamin C intake and nuclear opacity in the overall group (Fig. 21.3c).

However, Vitale et al. (1993) observed no differences in cataract prevalence between persons with high (>261 mg day^{-1}) and relatively low (<115 mg day^{-1}) vitamin C intakes. The Italian–American Cataract Study Group (1991) also observed no association between prevalence of cataract and vitamin C intake. In addition, in a large prospective study, a comparison of women with high intakes (median = 705 mg day^{-1}) and low intakes (median = 70 mg day^{-1}) did not show a significant correlation with risk for cataract extraction (RR = 0.98, CI = 0.72–1.32; Fig. 21.3g) (Hankinson et al., 1992).

Attempts to corroborate those inverse associations between cataract risk and intake using plasma vitamin C levels have yielded mixed results. Jacques and Chylack (1991) observed that persons with high plasma levels of vitamin C (>90 μM) had less than one-third the prevalence of early cataract than persons with low plasma levels of vitamin C (<40 μM), although this difference was not statistically significant (RR = 0.29, CI = 0.06–1.32) after adjustment for age, sex, race and history of diabetes (Fig. 21.3a). We corroborated those results in the baseline phase of our 'Nutrition and Vision Project' (Fig. 21.3c) (Jacques et al., 2001). Mohan et al. (1989) noted an 87% (RR = 1.87, CI = 1.29–2.69) increased prevalence of mixed cataract (posterior subcapsular and nuclear involvement) for each standard deviation increase in plasma levels of vitamin C (Fig. 21.3f). Vitale et al. (1993) observed that persons with plasma levels >80 μM and <60 μM had similar prevalences of nuclear (RR = 1.31, CI = 0.61–2.39) and cortical (RR = 1.01, CI = 0.45–2.26) cataract after controlling for age, sex and diabetes (Fig. 21.3c and d).

Results from one 5-year intervention trial showed no benefits of vitamin C supplementation to the population studied (Sperduto et al., 1993). Another large (n = 4596), double-masked, randomized, placebo-controlled intervention, AREDS, in which subjects used 500 mg supplements of vitamin C along with vitamin E, β-carotene and zinc for 6.3 years, was also null (Age-Related Eye Disease Study Research Group, 2001) (Fig. 21.4).

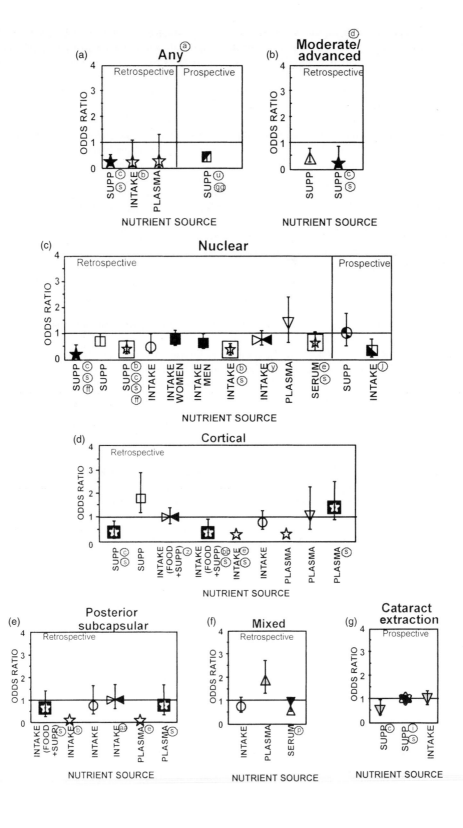

In comparison, the Roche European American Cataract Trial (REACT) ($n = 158$) used 750 mg vitamin C day^{-1}, vitamin E and β-carotene in a double-masked, randomized, placebo-controlled study and found some delay in the 3-year progress of early opacification in subjects

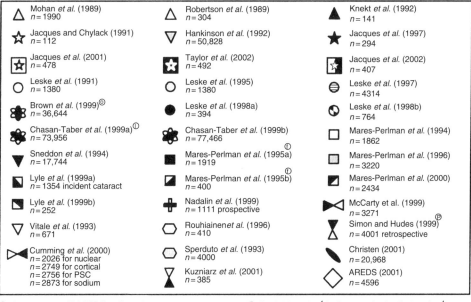

ⓐ Any = cataract with LOCS II grade ≫ 1
ⓑ $P \ll 0.05$
ⓒ Supp. use >10 years
ⓓ Adv = cataract with LOCS II grade ≫ 2
ⓔ $P \ll 0.1$
ⓕ When data were available for comparable groups of subjects, the most recent data from the largest sample were used for these displays
ⓖ Applicable for persons <70 years of age
ⓗ Lutein/zeaxanthin, particularly broccoli and kale
ⓘ Vitamins E, C, A and multivitamin use > 10 years
ⓙ Risk for incident opacities for person with ≫ 25 pack-years smoking ($n = 80$). Same OR for persons with hypertension ($n = 90$). For overall group, there was no significant relation between vitamin C intake and risk for nuclear opacity
ⓚ Risk for incident opacities for persons > 65 years of age ($n = 102$) for diet assessed 1978–1980. Not significant for diet assessed 1988–1990. Decreased OR was also observed for persons with high egg and spinach consumption
Risk for incident opacities for persons ≫ 65 years of age
ⓛ ($n = 143$). For overall group there was no significant relationship between vitamin E intake and risk for nuclear opacity
ⓜ Sum of α and γ-tocopherol Risk for incident opacities >5 years
ⓝ Risk for incident opacities in 5-year follow-up
ⓞ Particularly broccoli and spinach
ⓟ Self-reported cataract in 60–74-year-olds
Each mg dl^{-1} increase in serum ascorbate is associated with a 26% decrease in prevalence of cataract (CI 0.56–0.97)
ⓠ Multivariate risk factor including interaction with UV-B exposure

ⓡ For 5–10 mg day^{-1} intake versus intake of <5 mg day^{-1}
ⓢ Women
ⓣ Intervention trial 5 years
ⓤ Cohort study, use of supplement > 10 years
ⓥ Adjusted for age, sex, smoking, diabetes, hypertension and steroid use
ⓦ Vitamin A supplement intake
ⓧ 13.2 years supplementation with β-carotene
ⓨ $n = 2026$, intake=food + supp.
ⓩ $n = 2749$, intake=food + supp.
㉑ $n = 2756$, intake=food + supp.
㉒ Women < 60 years of age
㉓ Neversmoking women, intake=food + supp.
㉔ Adjusted for age, sex, energy, education, smoking, diabetes, hypertension and steroid use
㉕ Age and sex adjusted
㉖ Corrected for use of vitamin E and multivitamin supplements
㉗ Adjusted for age, sex, smoking, diabetes, hypertension, BMI, UV-B exposure and nutrient use during teen years
㉘ Adjusted for age, alcohol intake, smoking, BMI
㉙ Adjusted for age, sex, hypertension, diabetes, education and use of oral and inhaled steroids
㉚ For opacity of 60 pixel density units (manuscript submitted)
㉛ AREDS used daily supplements containing 500 mg vitamin C, 400 IU vitamin E, 15 mg β-carotene, 80 mg zinc oxide, 2 mg cupric oxide. Mean intervention was for 6.3 years. Changes from baseline in photographic grade: nuclear opacity (a 1.5 U increase on a scale of 0.9–6.1 U); cortical opacity (10% absolute increase in the area of opacity within a standard central 5 mm circle); and PSC (5% absolute increase in the area of opacity within a standard central 5 mm circle)

Fig. 21.3 (opposite and above). Cataract risk ratio, high vs. low intake (with or without supplements), and plasma levels of vitamin C. Types of cataract are any, moderate/advanced, nuclear, cortical, posterior subcapsular, mixed and cataract extraction. Data for retrospective and prospective studies are presented independently. Adapted from Taylor et al., *Vitamin C in Health and Disease* (New York: Marcel Dekker, 1997). SUPP, supplement.

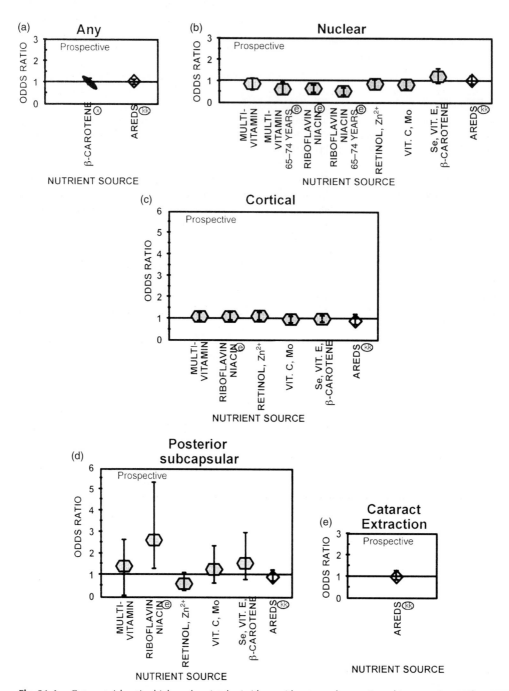

Fig. 21.4. Cataract risk ratio, high vs. low intake (with or without supplements), and intervention studies. Types of cataract are any, nuclear, cortical, posterior subcapsular and cataract extraction. Adapted from Taylor *et al.*, *Vitamin C in Health and Disease* (New York: Marcel Dekker, 1997).

>40 years old in the USA, but not in more poorly nourished British subjects (Chylack et al., 2002) (Table 21.1).

Vitamin E

Vitamin E supplements frequently provide significantly higher levels of this nutrient than can be obtained in the diet. A variety of studies examined relationships between vitamin E supplement use, dietary intake and plasma levels, and risk for various forms of cataract. The results were mixed. Consumption of vitamin E supplements was inversely correlated with cataract risk in several retrospective studies (Fig. 21.5). Robertson et al. (1989) found among age- and sex-matched cases and controls that the prevalence of advanced cataract was 56% lower (RR = 0.44, CI = 0.24–0.77; Fig. 21.5b) in persons who consumed vitamin E supplements (>400 IU day^{-1}) than in those not consuming supplements. P.F. Jacques and L.T. Chylack (unpublished) observed a 67% (RR = 0.33, CI = 0.12–0.96) reduction in the prevalence of cataract for users of vitamin E supplements after adjusting for age, sex, race and diabetes (Fig. 21.5a). In the baseline 'Nutrition and Vision Project', we found that risk for nuclear cataract was decreased by 55% (RR = 0.45, CI = 0.23–0.86) in women with intakes of >90 vs. <6.7 mg day^{-1}, and 51% (RR = 0.49, CI = 0.22–1.09) in those using supplements for >10 years (Fig. 21.5c) (Jacques et al., 2001). Correction of the data for use of vitamin C supplements resulted in loss of significance of the trend. The Beaver Dam Group published several observations regarding use of vitamin E supplements. Mares-Perlman et al. (1994) observed only weak, non-significant associations between vitamin E supplement use and nuclear cataract (RR = 0.9, CI = 0.6–1.5; Fig. 21.5c); in the same group, Lyle et al. (1999a) in a prospective study found a non-significant inverse relationship between vitamin E intake and nuclear cataract (RR = 0.5, CI = 0.3–1.1; Fig. 21.5c). However, their most recent work indicated a 60% decreased risk (RR = 0.4, CI = 0.3–0.6) for any cataract in users of vitamin E supplements (Fig. 21.5a) (Mares-Perlman et al., 2000). Vitamin E supplementation was also related to a lower risk for progress of nuclear opacity in the Longitudinal Study of Cataract (RR = 0.43, CI = 0.19–0.99; Fig. 21.5c) (Leske et al., 1998a).

In contrast to their data on nuclear cataract, cortical (RR = 1.2, CI = 0.6–2.3; Fig. 21.5d) cataract was positively (although non-significantly) related to vitamin E supplement intake by Mares-Perlman et al. (1994). While working in the Blue Mountain Group, McCarty et al. (1999) found no effect of supplemental intake of vitamin E on risk for nuclear cataract (Fig. 21.5c) and a positive relationship between vitamin E supplementation and posterior subcapsular cataract (RR = 1.47, CI = 1.04–2.09; Fig. 21.5e). Nadalin et al. (1999)

Table 21.1. Results from studies that do not use fixed endpoints.

Study	Design	Type of change	Change			
			Not supp.	Supp.	Diff. (not supp. − supp.)	P
REACT[a] Chylack et al. (2001) n = 158	Placebo-controlled intervention trial using daily supp.: 18 mg β-carotene, 750 mg vitamin C, 600 mg vitamin E	% pixels opaque over 3 years	3.3% pixels opaque in placebo	1.7% pixels opaque in supp. users	1.6%	0.05
NVP Jacques et al. (2002) n = 407	Vitamin E supp. use ≥ 10 years	Change in pixels opaque over 5 years	17	12	5	0.004

[a]Double masked, randomized, placebo-controlled; subjects >40 years, had some early cataract at baseline in one eye, cataract was more severe in UK than US; UK 2/3 lower plasma vitamin C, vitamin E, lycopene, 1/3 lower cryptoxanthin; UK 23% smokers; US 15% smokers; difference in pixels opaque in US 0.0014 after 3 years, UK not significant.

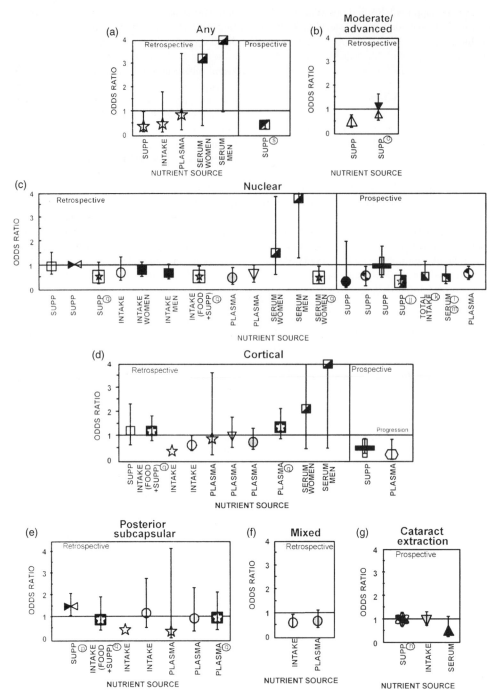

Fig. 21.5. Cataract risk ratio, high vs. low intake (with or without supplements), and plasma levels of vitamin E (α-tocopherol). Types of cataract are any, moderate/advanced, nuclear, cortical, posterior subcapsular, mixed and cataract extraction. Data for retrospective and prospective studies are presented independently. Adapted from Taylor et al., *Vitamin C in Health and Disease* (New York: Marcel Dekker, 1997). SUPP, supplement.

associated supplement intake with a 53% decreased risk for incident cortical cataract in a prospective study (RR = 0.47, CI = 0.28– 0.83; Fig. 21.5d) but not with a reduced risk for nuclear cataract (Fig. 21.5c).

Simon and Hudes (1999), in the National Health and Nutrition Examination Survey (NHANES), found no significant relationship between use of vitamin E supplements and self-reported cataract (RR = 0.93, CI = 0.52–1.67; Fig. 21.5b), and Chasan-Taber et al. (1999a) did not find cataract extraction to be related to intake of vitamin E supplements (RR = 0.99, CI = 0.74–1.32), even when intake was for 10 years (Fig. 21.5g).

Leske et al. (1991) observed that, after controlling for age and sex, persons with vitamin E intakes among the highest 20% had an approximately 40% lower prevalence of cortical (RR = 0.59, CI = 0.36–0.97; Fig. 21.5d) and mixed (RR = 0.58, CI = 0.37–0.93; Fig. 21.5f) cataract relative to persons with intakes among the lowest 20%. Jacques and Chylack (1991) observed a non-significant inverse association when they related total vitamin E intake (combined dietary and supplemental intake) to cataract prevalence. Persons with vitamin E intake >35.7 mg day^{-1} had a 55% lower prevalence of early cataract (RR = 0.45, CI = 0.12–1.79) than did those with intakes of <8.4 mg day^{-1} (Fig. 21.5a) (Jacques and Chylack, 1991). They also showed a significant reduction in risk for cortical cataract (Fig. 21.5d; RR = 0.37; $P < 0.1$) for persons with higher vitamin E intakes.

However, Hankinson et al. (1992) found no association between vitamin E intake and cataract surgery (Fig. 21.5g). Women with high intakes of vitamin E (median = 210 mg day^{-1}) had similar rates of cataract surgery (RR = 0.96, CI = 0.72–1.29) to women with low intakes (median = 3.3 mg day^{-1}). This finding is consistent with data regarding supplement intake in the same cohort (Chasan-Taber et al., 1999a). Mares-Perlman et al. (1995a) associated dietary vitamin E (non-significantly) with diminished risk for nuclear cataract in men but not in women (Fig. 21.5c), although this information contrasts with their positive correlations between serum γ-tocopherol levels and cataract (Fig. 21.5a, c and d).

Lyle et al. (1999a,b) in prospective studies associated higher serum levels of vitamin E with diminished risk for incident cataract (Fig. 21.5c; RR = 0.4, CI = 0.2–0.9). Several other studies assessing plasma levels of vitamin E also reported significant inverse associations with cataract (Fig. 21.5). We observed a 52% decreased risk (RR = 0.48, CI = 0.25–0.95) for women in the baseline phase of the 'Nutrition and Vision Project', with serum vitamin E levels of 41 μM versus women with levels of 23 μM (Fig. 21.5c) (Jacques et al., 2001). Knekt et al. (1992) followed a cohort of 1419 Finns for 15 years and identified 47 patients admitted to ophthalmological wards for mature cataract. They selected two controls per patient matched for age, sex and municipality. Those investigators reported that persons with serum vitamin E concentrations above approximately 20 μM had about one-half the rate of subsequent cataract surgery (RR = 0.53, CI = 0.24–1.1; Fig. 21.5 g) compared with persons with vitamin E concentrations below that concentration. In a retrospective study, Vitale et al. (1993) observed the age-, sex- and diabetes-adjusted prevalence of nuclear cataract to be about 50% less (RR = 0.52, CI = 0.27–0.99; Fig. 21.5c) among persons with plasma concentrations of vitamin E >29.7 μM compared with those with levels <18.6 μM. A similar comparison showed that the prevalence of cortical cataract did not differ between those with high and low plasma levels of vitamin E (RR = 0.96, CI = 0.52–1.78; Fig. 21.5d). Jacques and Chylack (1991) observed the prevalence of posterior subcapsular cataract to be 67% (RR = 0.33, CI = 0.03–4.13; Fig. 21.5e) lower among persons with plasma vitamin E levels >35 μM compared with those with levels <21 μM after adjustment for age, sex, race and diabetes; however, the effect was not statistically significant. Prevalence of any early (RR = 0.83, CI = 0.20–3.40; Fig. 21.5a) or cortical (RR = 0.84, CI = 0.20–3.60; Fig. 21.5d) cataract did not differ between those with high and low plasma levels. In a prospective study examining risk for cortical cataract progress among individuals with higher plasma levels of vitamin E, Rouhiainen et al. (1996) found a 73% reduction in risk (RR = 0.27, CI = 0.08–0.83; Fig. 21.5d). Plasma vitamin E was also inversely associated with prevalence of cataract in a large Italian study after adjusting for age and sex, but the relationship was no longer statistically significant after adjusting for other factors such as education, sunlight exposure and family history of cataract (The Italian–American Cataract Study, 1991). Leske et al. (1995) also demonstrated that individuals with high plasma levels of vitamin E had a significantly lower prevalence of nuclear cataract (RR = 0.44, CI = 0.21–0.90), but vitamin E was not associated with cataracts at other lens sites (Fig. 21.5c). In a prospective study, Leske et al. (1998a) reported a 42% reduction in risk for nuclear cataract progression among persons with higher plasma levels of vitamin E (RR = 0.58, CI = 0.36–0.94; Fig. 21.5c).

Mares-Perlman et al. (1995b), working with similar cohorts as Lyle et al. (1999b), noted a significant, elevated prevalence of nuclear cataract in men with high serum levels of vitamin E (RR = 3.74, CI = 1.25–11.2; Fig. 21.5c) and in women (RR = 1.47, CI = 0.57–3.82; Fig. 21.5c). Another study did not associate cataract and plasma levels of vitamin E (Mohan et al., 1989).

Since the recognition of the antioxidant potential of γ-tocopherol, studies have investigated whether this form of tocopherol is associated with altered risk for cataract. γ-Tocopherol has lower biological activity of vitamin E than α-tocopherol. Mares-Perlman et al. (1995b) observed an inverse (non-significant) relationship (RR = 0.61, CI = 0.32–1.19) between serum γ-tocopherol and severity of nuclear sclerosis, but a significant, positive relationship between elevated serum levels of γ-tocopherol and severity of cortical cataract for men and women.

We recently monitored progress of opacities in the 'Nutrition and Vision Project'. Progress of nuclear cataract was delayed in persons who used vitamin E supplements for >10 years (Table 21.1) (P.F. Jacques et al., unpublished observations).

Results from two masked interventions yielded discrepant results with respect to value for supplements that contain high levels, 600 mg (Chylack et al., 2002) and 400 IU (Age-Related Eye Disease Study Research Group, 2001). Whereas a delay in progress of early cataract was seen in some of the REACT cohort members, the prevalence results were null for users of the multivitamin for 6.3 years for all forms of opacities in the AREDS study (Age-Related Eye Disease Study Research Group, 2001).

Carotenoids, vitamin A and other nutrients

The carotenoids, like vitamin E, are also natural lipid-soluble antioxidants (Machlin and Bendich, 1987; Shang et al., 1997a; Curran Celentano et al., 2002; Krinsky; 2002). β-Carotene is the best known carotenoid because of its importance as a vitamin A precursor. Levels of β-carotene in lens are vanishingly low (Yeum et al., 1999). In addition to β-carotene, α-carotene, lutein, zeaxanthin and lycopene are important carotenoid components of the human diet, the latter being found in the lens in concentrations nearly equal to 10 ng^{-1} g of wet weight (Daicker et al., 1987; Yeum et al., 1995, 1999). Data that relate carotenoid supplement use to risk for cataracts are appearing, and carotenoid supplements already are being included in several 'eye vitamin' preparations.

Jacques and Chylack (1991) were the first to observe that persons with carotene intakes >18 700 IU day^{-1} had the same prevalence of cataract as those with intakes below 5677 IU day^{-1} (RR = 0.91, CI = 0.23–3.78; Fig. 21.6a). Hankinson et al. (1992) followed that report with a study that reported that the multivariate-adjusted rate of cataract surgery was about 30% lower (RR = 0.73, CI = 0.55–0.97) for women with high carotene intakes (median = 14,558 IU day^{-1}) than for those with low intakes of this nutrient (median = 2935 IU day^{-1}; Fig. 21.6e). Using members of the same population in the 'Nutrition and Vision Project', we observed that women with daily intakes of β-carotene of 6.6 versus 3.0 mg day^{-1} had 48% decreased odds (RR = 0.52, CI = 0.28–0.97) for nuclear cataract (Fig. 21.7b) (Jacques et al., 2001). In that study, we also observed that intakes of folate (RR = 0.44, CI = 0.24–0.81) and riboflavin (RR = 0.37, CI = 0.19–0.73; Fig. 21.8a) were associated with decreased RR for nuclear cataract. For women who were never smokers, we noted decreased RR for posterior subcapsular cataract: 81% decreased odds (RR = 0.19, CI = 0.08–0.68) in persons with higher (24 vs. 12 mg day^{-1}) total carotenoid intake (Fig. 21.6d), 71% decreased odds (RR = 0.29, CI = 0.08–1.05, non-significant) with α-carotene intakes of 1.2 vs. 0.44 mg day^{-1} (Fig. 21.9d) and 72% decreased odds (RR = 0.28, CI 0.08–0.96) with β-carotene intakes of 6.6 vs. 3.0 mg day^{-1} (Fig. 21.7d) (Taylor et al., 2002). Elevated folate intake also was associated with benefit: 74% decreased odds (RR = 0.26, CI = 0.09–0.77) for intakes of >548 μg day^{-1} vs. <284 μg day^{-1}. When the data were corrected for folate and the carotenoids, only the relationship with total carotenoids remained significant (Taylor et al., 2002). Mares-Perlman et al. (1996) also found decreased RR for nuclear cataract in users of β-carotene supplements.

In another group of women, that risk for cataract extraction was slightly reduced in users of supplements (Fig. 21.6e) Chasan-Taber *et al.* (1999a) and in women with higher total carotenoid intake (RR = 0.85, CI = 0.7–1.03) (Chasan-Taber *et al.*, 1999b). Similarly, Brown *et al.* (1999) showed

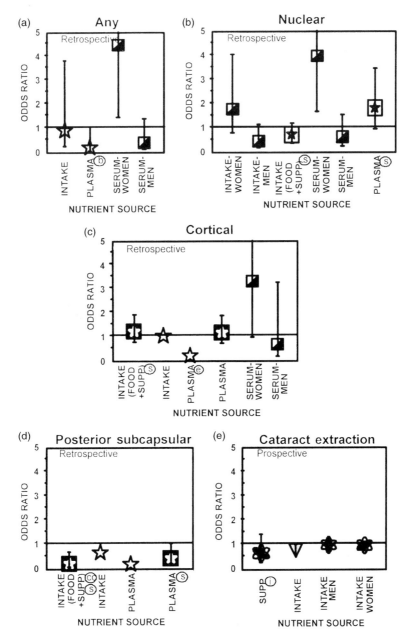

Fig. 21.6. Cataract risk ratio, high vs. low intake (with or without supplements), and plasma levels of carotenoids. Types of cataract are any, moderate/advanced, nuclear, cortical, posterior subcapsular, mixed and cataract extraction. Data for retrospective and prospective studies are presented independently. Adapted from Taylor *et al.*, *Vitamin C in Health and Disease* (New York: Marcel Dekker, 1997). SUPP, supplement.

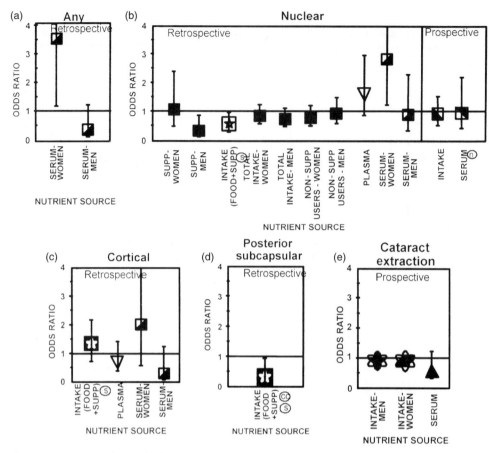

Fig. 21.7. Cataract risk ratio, high vs. low intake (with or without supplements), and plasma levels of β-carotene. Types of cataract are any, moderate/advanced, nuclear, cortical, posterior subcapsular, mixed and cataract extraction. Data for retrospective and prospective studies are presented independently. Adapted from Taylor et al., *Vitamin C in Health and Disease* (New York: Marcel Dekker, 1997). SUPP, supplement.

that men with higher carotenoid intakes had reduced risk for cataract extraction (RR = 0.85, CI = 0.68–1.07; Fig. 21.6e). However, whereas cataract surgery was inversely associated with total carotene intake in those studies, it was not strongly associated with consumption of carotene-rich foods such as carrots. Rather, cataract surgery was associated with lower intakes of foods such as spinach that are rich in lutein and xanthin carotenoids, rather than β-carotene. This finding seems consistent with our observation that the human lens contains lutein and zeaxanthin but no β-carotene, and with Christen's (2001) observation that use of β-carotene supplements was without effect (see below). Unfortunately, cataract surgery was not an end point in other studies that considered xanthaphylls (Mares-Perlman et al., 1994, 1995b). Mares-Perlman et al. (1995a) found no significant associations between specific carotenoid intake and risk for cataract.

In another prospective study, Chasan-Taber et al. (1999b) showed that intakes of α-carotene (RR = 0.95, CI = 0.7–2.1; Fig. 21.9b) and β-carotene (RR = 0.9, CI = 0.5–1.4; Fig. 21.7b) were not significantly related to incident cataract. Intakes of α-carotene, β-carotene (Fig. 21.7e), lycopene and β-cryptoxanthin were not related to risk for cataract extraction. For men, intakes of α-carotene (RR = 0.89, CI = 0.72–1.1; Fig. 21.9e), β-carotene (RR = 0.92, CI = 0.73–1.16; Fig. 21.7e), lycopene (RR = 1.10, CI = 0.88–1.36) or β-cryptoxanthin (RR = 1.09, CI = 0.87–1.37) were not significantly

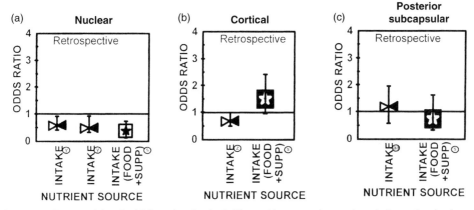

Fig. 21.8. Cataract risk ratio, high vs. low intake (with or without supplements), and plasma levels of riboflavin. Types of cataract are nuclear, cortical, and posterior subcapsular. Adapted from Taylor et al., *Vitamin C in Health and Disease* (New York: Marcel Dekker, 1997). SUPP, supplement.

related to risk for cataract extraction (Brown et al., 1999), and Mares-Perlman et al. (1994, 1995a) also did not detect significantly altered risk for cataract among consumers of those nutrients.

Plasma levels of carotenoids also were related to risk for cataract. Jacques and Chylack (1991) noted that persons with high plasma total carotenoid concentrations (>3.3 μM) had less than one-fifth the prevalence of cataract than did persons with low plasma carotenoid levels (<1.7 μM; RR = 0.18, CI = 0.03–1.03) after adjustment for age, sex, race and diabetes (Fig. 21.6a). However, they did not observe an association between carotene intake and cataract prevalence. Knekt et al. (1992) reported that, among age- and sex-matched cases and controls, persons with serum β-carotene concentrations above approximately 0.1 μM had a 40% reduction in the rate of cataract surgery compared with persons with concentrations below that level (RR = 0.59, CI = 0.26–1.25; Fig. 21.7e).

Mares-Perlman et al. (1995b) correlated serum carotenoids and severity of nuclear and cortical opacities and found that higher levels of individual or total carotenoids in the serum were not associated with less severe nuclear or cortical cataract overall (Figs 21.6, 21.7 and 21.9). Associations between risk for some forms of cataract and nutriture differed between men and women, e.g. nuclear cataract and α-carotene intake (Mares-Perlman et al., 1995a). Other nutrients for which cataract risk in women versus men showed opposing relationships include serum β-carotene (Fig. 21.7a and b) and serum lycopene. A marginally significant trend for lower risk ratio for cortical opacity with increasing serum levels of β-carotene was observed in men, but not in women. Higher serum levels of α-carotene, β-cryptoxanthin and lutein related significantly to lower risk for nuclear sclerosis only in men who smoked. In contrast, higher levels of some carotenoids often were associated directly with elevated risk for nuclear sclerosis and cortical cataract (Figs 21.6b and c, 21.7b and c, and 21.9b), particularly in women. Following up on these relationships, Lyle et al. (1999b) found little effect of serum levels of α-carotene (RR = 0.9, CI = 0.4–2.2; Fig. 21.9b), β-carotene (RR = 0.9, CI = 0.4–2.2; Fig. 21.7b), lycopene (RR = 1.1, CI = 0.5–2.6), β-cryptoxanthin (RR = 0.7, CI = 0.3–1.6) or lutein (RR = 0.7, CI = 0.3–1.6; Fig. 21.10b) on incident cataract.

Vitale et al. (1993) also examined the relationship between plasma levels of β-carotene and age-, sex- and diabetes-adjusted prevalence of cortical and nuclear cataract (Fig. 21.7b and c). Although the data suggested a weak inverse association between plasma β-carotene and cortical cataract and a weak positive association between this and nuclear cataract, neither association was statistically significant. Persons with plasma β-carotene concentrations >0.88 μM had a 28% lower prevalence of cortical cataract (RR = 0.72, CI = 0.37–1.42; Fig. 21.7c) and a 57% higher prevalence of nuclear cataract (RR = 1.57, CI = 0.84–2.93; Fig. 21.7b) compared with persons with levels <0.33 mM.

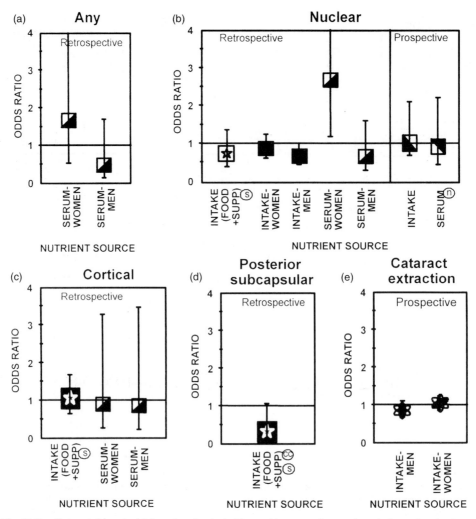

Fig. 21.9. Cataract risk ratio, high vs. low intake (with or without supplements), and plasma levels of α-carotene. Types of cataract are any, nuclear, cortical, posterior subcapsular and cataract extraction. Data for retrospective and prospective studies are presented independently. Adapted from Taylor et al., *Vitamin C in Health and Disease* (New York: Marcel Dekker, 1997). SUPP, supplement.

Lutein intake has been related to risk for age-related maculopathy, and it recently has been given more careful examination with respect to risk for cataract. As indicated above, several studies did not find significant inverse relationships with respect to lutein intake and risk for cataract. However, other studies found indications that elevated intake of foods rich in lutein is related to decreased risk for cataract. In keeping with food analyses, intakes of lutein and zeaxanthin were inversely related to risk for incident nuclear cataract (RR = 0.4, CI = 0.2–0.8; Fig. 21.10b) (Lyle et al., 1999a) and cataract extraction in women (Fig. 21.10e) (Chasan-Taber et al., 1999b) (RR = 0.78, CI = 0.63–0.95) and men (RR = 0.81; CI = 0.65–1.01) (Brown et al., 1999). The foods most clearly associated with decreased risk for cataract extraction were broccoli and cooked spinach. These are foods which are rich in lutein and zeaxanthin (Brown et al., 1999). Recently, we found that risk for nuclear cataract was 51% lower (RR = 0.49, CI = 0.25–0.94) in persons who had intakes of at least 5.6 versus at most 2.4 mg day^{-1} of lutein and zeaxanthin (Fig. 21.10b) (Jacques et al., 2001).

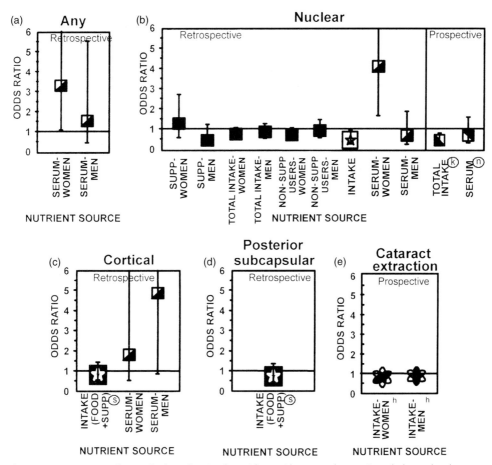

Fig. 21.10. Cataract risk ratio, high vs. low intake (with or without supplements), and plasma levels of lutein. Types of cataract are any, moderate/advanced, nuclear, cortical, posterior subcapsular, mixed and cataract extraction. Data for retrospective and prospective studies are presented independently. Adapted from Taylor et al., *Vitamin C in Health and Disease* (New York: Marcel Dekker, 1997). SUPP, supplement.

In the Blue Mountain Study, Cumming et al. (2000) found a 50% decreased risk (RR = 0.5, CI = 0.3–0.8) for nuclear cataract in persons with vitamin A intakes of 3.2 versus 0.3 mg day^{-1}.

However, intervention studies yielded mixed results. The 6.3-year AREDS intervention administered 15 mg of β-carotene daily, along with vitamins C and E and zinc, and found no benefit with respect to prevalence or progress of opacities (Age-Related Eye Disease Study Research Group, 2001) (Fig. 21.4). In comparison, the REACT study, which used 18 mg of β-carotene daily, along with vitamins C and E supplements, did indicate that progress of cataract could be slowed – at least in the American members of their cohort (Chylack et al., 2002).

Antioxidant combinations

To approximate combined effects on cataract risk of the multiple antioxidants contained in food, we adopted 'antioxidant indices.' However, single nutrients appear to have strong influences on the indices, and we now question the usefulness of the indices. There are relatively few new data regarding the usefulness of multivitamin supplements on risk for cataract. Thus, readers are referred to previous reviews (Taylor, 1999b). Several studies have associated intakes of various foods with risk for cataract, and some of that material was reviewed (Taylor, 2000).

Multivitamins may also provide the functional equivalent of antioxidant combinations. These have been reviewed previously. A new addition to the literature is data from Kuzniarz et al. (2001). In the Blue Mountain population ($n = 385$), they found that odds for nuclear and cortical opacities were 0.1 (CI = 0.0–1.0) and 0.9 (CI = 0.3–1.1), respectively, for users of multivitamins for >10 years (Fig. 21.11b and c). Although suggestive of benefit, these results fail to reach standards for statistical significance.

Intervention studies

There have been several intervention trials designed to assess the effect of vitamin supplements on cataract risk (Sperduto et al., 1993; Age-Related Eye Disease Study Research Group, 2001; Christen, 2001; Chylack et al., 2002). Sperduto et al. (1993) took advantage of two ongoing, randomized, double-blinded vitamin and cancer trials to assess the impact of vitamin supplements on cataract prevalence. The trials were conducted among almost 4000 poorly nourished participants aged 45–74 years from rural communes in Linxian, China. Participants in one trial received a multivitamin supplement or placebo. In the second trial, a more complex factorial design evaluated the effects of four different combinations of vitamins and minerals: retinol (5000 IU) and zinc (22 mg); riboflavin (3 mg) and niacin (40 mg); vitamin C (120 mg) and molybdenum (30 µg); and vitamin E (30 mg), β-carotene (15 mg) and selenium (50 µg). At the end of the 5–6 year follow-up, the investigators conducted eye examinations to measure the prevalence of cataract.

In the first trial, there was a significant 43% reduction in the prevalence of nuclear cataract for persons aged 65–74 years receiving the multivitamin supplement (RR = 0.57, CI = 0.36–0.90; Fig. 21.4a). The second trial showed a significantly reduced prevalence of nuclear cataract in persons receiving the riboflavin/niacin supplement compared with those not receiving this supplement (RR = 0.59, CI = 0.45–0.79; Fig. 21.4a). The effect was strongest in those aged 65–74 years (RR = 0.45, CI = 0.31–0.64; Fig. 21.4a). However, the riboflavin/niacin supplement appeared to increase the risk of posterior subcapsular cataract (RR = 2.64, CI = 1.31–5.35; Fig. 21.4d). Decreased risks for cortical (RR = 0.7, CI = 0.5–1.0) and nuclear (RR = 0.6; 0.4–0.9) cataract were also indicated in the Blue Mountain Study for persons with niacin intakes of 44 vs. 11.7 mg day^{-1} in analyses that had been adjusted for energy (Cumming et al., 2000). The results also suggested a protective effect of the retinol/zinc (RR = 0.77, CI = 0.58–1.02) and the vitamin C/molybdenum (RR = 0.78, CI = 0.59–1.04) supplements on the prevalence of nuclear cataract (Fig. 21.4d). Suggestions of benefit for zinc intake of 14.3 vs. 7.8 mg day^{-1} (RR = 0.7, CI = 0.5–1.1) for nuclear cataract (Cumming et al., 2000) remains a possibility that needs further testing.

In an intervention using β-carotene (50 mg every other day) that lasted >13 years, Christen (2001) found no effect of such supplementation on the risk for self-reported cataract. A large ($n = 4596$) intervention trial was completed recently. A daily combination of 500 mg of vitamin C, 400 IU of vitamin E, 15 mg of β-carotene and 80 mg of zinc used for 6.3 years was without effect on all forms of lens opacities (Age-Related Eye Disease Study Research Group, 2001). In comparison, the REACT intervention trial ($n = 158$) found benefit with respect to delayed progress of opacities in Americans but not British subjects who used daily supplements that contained 750 mg of vitamin C, 600 mg of vitamin E and 18 mg of β-carotene for 3 years (Chylack et al., 2002) (Table 21.1).

Iron metabolism has been implicated in cataractogenesis because of association with free radical reactions. However, Cumming et al. (2000) associated elevated iron intakes (15.4 vs. 8.5 mg day^{-1}) with a 40% decreased risk for nuclear cataract (RR = 0.5, CI = 0.4–0.9).

Conclusion

Light and oxygen appear to be a boon and a bane. Although necessary for physiological function, when present in excess or in uncontrolled circumstances, they appear to be causally related to cataractogenesis. With ageing, compromised function of the lens is exacerbated by depleted

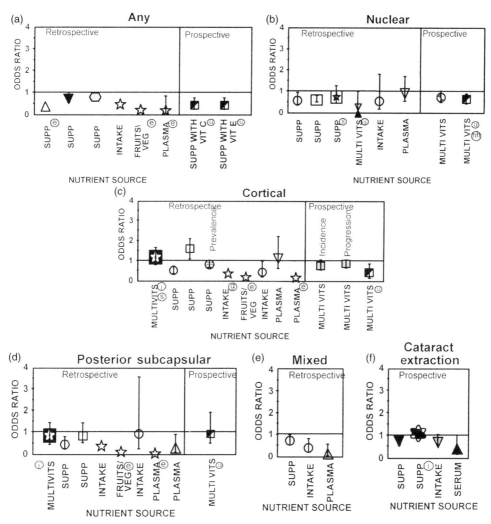

Fig. 21.11. Cataract risk ratio, high vs. low intake (with or without supplements), and antioxidant index. Types of cataract are any, nuclear, cortical, posterior subcapsular, mixed and cataract extraction. Data for retrospective and prospective studies are presented independently. Adapted from Taylor *et al.*, *Vitamin C in Health and Disease* (New York: Marcel Dekker, 1997). SUPP, supplement.

or diminished primary antioxidant reserves, antioxidant enzyme capabilities and diminished secondary defences such as proteases. Smoking (West, 1999) and light exposure (McCarty and Taylor, 1999) appear to provide oxidative challenges and are associated with an elevated risk of cataract. Poor education and lower socio-economic status also markedly increase risk for cataract (McLaren, 1980; Harding and van Heyningen, 1987; Mohan *et al.*, 1989; Leske *et al.*, 1991). These are related to poor nutrition. The impression created by the literature is that there is some benefit to enhanced antioxidant intake with respect to diminished risk for cataract. Assuming that other studies corroborate our observations that vitamins C and E and lutein provide benefit with respect to risk for nuclear cataract, it may be useful to obtain intakes of approximately 250, 90 and 3 mg day^{-1}, respectively, of vitamins C and E and lutein from diet and/or supplements (Taylor and Jacques, 1997; Jacques *et al.*, 2001). It probably is possible to adjust normal dietary practice to obtain those intake levels in well-fed populations (Jacob *et al.*, 1988; Taylor *et al.*, 1991a; Hankinson *et al.*, 1992;

Jacques et al., 1997). Because the bioavailability of ascorbate may decrease with age, slightly higher intakes might be required in the elderly. Defining useful levels of the other nutrients should await further studies. It is difficult to compare the various studies. That the correlations were not always with the same form of cataract may indicate, in addition to the conclusions reached, that the cataracts were graded differently and/or that there are common aetiological features of each of the forms of cataract described. Most of the studies reviewed used case–control designs, and most assessed status only once. Because intake or status measures are highly variable and the effects of diet are likely to be cumulative, studies should be performed on populations for which long-term dietary records are available. Intake studies seem to be preferable to the use of plasma measures, if a single measure of status must be chosen. Optimization of nutriture can be achieved through better diets and supplement use, especially among the poor and at-risk populations. Thus, the overall impression created by these data suggests that nutrition, possibly including antioxidant vitamin supplement use, provides the least costly and most practicable means to delay cataract.

Acknowledgements

The authors acknowledge the assistance of Tom Nowell and Marisa Hobbs in the preparation of figures, and Paul Jacques for invaluable assistance in evaluating the epidemiological data.

References

Age-Related Eye Disease Study Research Group (2001) A randomized, placebo-controlled, clinical trial of high-dose supplementation with vitamins C and E and beta carotene for age-related cataract and vision loss: AREDS report no. 9. *Archives of Ophthalmology* 119, 1439–1452.

Bensch, K.G., Fleming, J.E. and Lohmann, W. (1985) The role of ascorbic acid in senile cataract. *Proceedings of the National Academy of Sciences USA* 82, 7193–7196.

Berger, J., Shephard, D., Morrow, F., Sadowski, J., Haire, T. and Taylor, A. (1988) Reduced and total ascorbate in guinea pig eye tissues in response to dietary intake. *Current Eye Research* 7, 681–686.

Berger, J., Shepard, D., Morrow, F. and Taylor, A. (1989) Relationship between dietary intake and tissue levels of reduced and total vitamin C in the nonscorbutic guinea pig. *Journal of Nutrition* 119, 734–740.

Berman, E.R. (1991) *Biochemistry of the Eye*. Plenum Press, New York.

Bhuyan, K.C. and Bhuyan, D.K. (1984) Molecular mechanism of cataractogenesis: III. Toxic metabolites of oxygen as initiators of lipid peroxidation and cataract. *Current Eye Research* 3, 67–81.

Bhuyan, D.K., Podos, S.M., Machlin, L.T., Bhagavan, H.N., Chondhury, D.N., Soja, W.S. and Bhuyan, K.C. (1983) Antioxidant in therapy of cataract II: effect of all-roc-alpha-tocopherol (vitamin E) in sugar-induced cataract in rabbits (ARVO abstracts). *Investigative Ophthalmology and Visual Science* 24, 74.

Blondin, J. and Taylor, A. (1987) Measures of leucine aminopeptidase can be used to anticipate UV-induced age-related damage to lens proteins: ascorbate can delay this damage. *Mechanisms of Ageing and Development* 41, 39–46.

Blondin, J., Baragi, V., Schwartz, E., Sadowski, J.A. and Taylor, A. (1986) Delay of UV-induced eye lens protein damage in guinea pigs by dietary ascorbate. *Journal of Free Radicals in Biology and Medicine* 2, 275–281.

Blondin, J., Baragi, V.J., Schwartz, E., Sadowski, J. and Taylor, A. (1987) Dietary vitamin C delays UV-induced age-related eye lens protein damage. *Annals of the New York Academy of Sciences* 498, 460–463.

Brown, L., Rimm, E.B., Seddon, J.M., Giovannucci, E.L., Chasan-Taber, L., Spiegelman, D., Willett, W.C. and Hankinson, S.E. (1999) A prospective study of carotenoid intake and risk of cataract extraction in US men. *American Journal of Clinical Nutrition* 70, 517–524.

Bunce, G.E., Kinoshita, J. and Horwitz, J. (1990) Nutritional factors in cataract. *Annual Review of Nutrition* 10, 233–254.

Chasan-Taber, L., Willett, W.C., Seddon, J.M., Stampfer, M.J., Rosner, B., Colditz, G.A. and Hankinson, S.E. (1999a) A prospective study of vitamin supplement intake and cataract extraction among U.S. women. *Epidemiology* 10, 679–684.

Chasan-Taber, L., Willett, W.C., Seddon, J.M., Stampfer, M.J., Rosner, B., Colditz, G.A., Speizer, F.E. and Hankinson, S.E. (1999b) A prospective study of carotenoid and vitamin A intakes and risk of cataract extraction in US women. *American Journal of Clinical Nutrition* 70, 509–516.

Chatterjee, A., Milton, R.C. and Thyle, S. (1982) Prevalence and aetiology of cataract in Punjab. *British Journal of Ophthalmology* 66, 35–42.

Christen, W.G. (1999) Evaluation of epidemiologic studies of nutrition and cataract. In: Taylor, A. (ed.)

Nutritional and Environmental Influences on the Eye. CRC Press, Boca Raton, Florida, pp. 95–104.

Christen, W.G. (2001) Beta-carotene and age-related cataract in a randomized trial of U.S. physicians. *Investigative Ophthalmology and Visual Science* 42, S518.

Chylack, L.T. Jr (1999) Function of the lens and methods of quantifying cataract. In: Taylor, A. (ed.) *Nutritional and Environmental Influences on the Eye*. CRC Press, Boca Raton, Florida, pp. 25–52.

Chylack, L.T. Jr, Wolfe, J.K., Singer, D.M., Leske, M.C., Bullimore, M.A., Bailey, I.L., Friend, J., McCarthy, D. and Wu, S.Y. (1993) The Lens Opacities Classification System III. The Longitudinal Study of Cataract Study Group. *Archives of Ophthalmology* 111, 831–836.

Chylack, L.T. Jr, Wolfe, J.K., Judith, F., Singer, D.M., Wu, S.Y. and Leske, M.C. (1994) Nuclear cataract: relative contributions to vision loss of opalescence and brunescence. *Investigative Ophthalmology and Visual Science* 35, 42632 [abstract].

Chylack, L.T. Jr, Brown, N.P., Bron, A., Hurst, M., Kopcke, W., Thien, U. and Schalch, W. (2002) The Roche European American Cataract Trial (REACT): a randomized clinical trial to investigate the efficacy of an oral antioxidant micronutrient mixture to slow progression of age-related cataract. *Ophthalmic Epidemiology* 9, 49–80.

Costagliola, C., Iuliano, G., Menzione, M., Rinaldi, E., Vito, P. and Auricchio, G. (1986) Effect of vitamin E on glutathione content in red blood cells, aqueous humor and lens of humans and other species. *Experimental Eye Research* 43, 905–914.

Creighton, M.O., Ross, W.M., Stewart-DeHaan, P.J., Sanwal, M. and Trevithick, J.R. (1985) Modelling cortical cataractogenesis VII: effects of vitamin E treatment on galactose-induced cataracts. *Experimental Eye Research* 40, 213–222.

Cumming, R.G., Mitchell, P. and Smith, W. (2000) Diet and cataract: the Blue Mountains Eye Study. *Ophthalmology* 107, 450–456.

Curran Celentano, J., Burke, J.D. and Hammond, B.R. Jr (2002) *In vivo* assessment of retinal carotenoids: macular pigment detection techniques and their impact on monitoring pigment status. *Journal of Nutrition* 132, 535S–539S.

Daicker, B., Schiedt, K., Adnet, J.J. and Bermond, P. (1987) Canthaxantin retinopathy. An investigation by light and electron microscopy and physicochemical analysis. *Graefe's Archive for Clinical and Experimental Ophthalmology* 225, 189–197.

Devamanoharan, P.S., Henein, M., Morris, S., Ramachandran, S., Richards, R.D. and Varma, S.D. (1991) Prevention of selenite cataract by vitamin C. *Experimental Eye Research* 52, 563–568.

Eisenhauer, D.A., Berger, J.J., Peltier, C.Z. and Taylor, A. (1988) Protease activities in cultured beef lens epithelial cells peak and then decline upon progressive passage. *Experimental Eye Research* 46, 579–590.

Frei, B., Stocker, R. and Ames, B.N. (1988) Antioxidant defenses and lipid peroxidation in human blood plasma. *Proceedings of the National Academy of Sciences USA* 85, 9748–9752.

Fridovich, I. (1984) Oxygen: aspects of its toxicity and elements of defense. *Current Eye Research* 3, 1–2.

Garland, D.L. (1991) Ascorbic acid and the eye. *American Journal of Clinical Nutrition* 54, 1198S–1202S.

Giblin, F.J., McCready, J.P. and Reddy, V.N. (1982) The role of glutathione metabolism in the detoxification of H_2O_2 in rabbit lens. *Investigative Ophthalmology and Visual Science* 22, 330–335.

Hammond, B.R. Jr, Wooten, B.R. and Snodderly, D.M. (1997) Density of the human crystalline lens is related to the macular pigment carotenoids, lutein and zeaxanthin. *Optometry and Vision Science* 74, 499–504.

Hankinson, S.E., Stampfer, M.J., Seddon, J.M., Colditz, G.A., Rosner, B., Speizer, F.E. and Willett, W.C. (1992) Nutrient intake and cataract extraction in women: a prospective study. *British Medical Journal* 305, 335–339.

Harding, J.J. and van Heyningen, R. (1987) Epidemiology and risk factors for cataract. *Eye* 1, 537–541.

Huang, L.L., Jahngen-Hodge, J. and Taylor, A. (1993) Bovine lens epithelial cells have a ubiquitin-dependent proteolysis system. *Biochimica et Biophysica Acta* 1175, 181–187.

Italian-American Cataract Study (1991) Risk factors for age-related cortical, nuclear, and posterior subcapsular cataracts. *American Journal of Epidemiology* 133, 541–553.

Jacob, R.A., Otradovec, C.L., Russell, R.M., Munro, H.N., Hartz, S.C., McGandy, R.B., Morrow, F.D. and Sadowski, J.A. (1988) Vitamin C status and nutrient interactions in a healthy elderly population. *American Journal of Clinical Nutrition* 48, 1436–1442.

Jacques, P.F. and Chylack, L.T. Jr (1991) Epidemiologic evidence of a role for the antioxidant vitamins and carotenoids in cataract prevention. *American Journal of Clinical Nutrition* 53, 352S–355S.

Jacques, P.F. and Taylor, A. (1991) Micronutrients and age-related cataracts. In: Bendich, A. and Butterworth, C.E. (eds) *Micronutrients in Health and in Disease Prevention*. Marcel Dekker, New York, pp. 359–379.

Jacques, P.F., Chylack, L.T. Jr and Taylor, A. (1994) Relationships between natural antioxidants and cataract formation. In: Frei, B. (ed.) *Natural Antioxidants in Human Health and Disease*. Academic Press, Orlando, Florida, pp. 513–533.

Jacques, P.F., Taylor, A., Hankinson, S.E., Willett, W.C., Mahnken, B., Lee, Y., Vaid, K. and Lahav, M. (1997) Long-term vitamin C supplement use and prevalence of early age-related lens opacities. *American Journal of Clinical Nutrition* 66, 911–916.

Jacques, P.F., Chylack, L.T. Jr, Hankinson, S.E., Khu, P.M., Rogers, G., Friend, J., Tung, W., Wolfe, J.K., Padhye, N., Willett, W.C. and Taylor, A. (2001) Long-term nutrient intake and early age-related nuclear lens opacities. *Archives of Ophthalmology* 119, 1009–1019.

Jahngen, J.H., Haas, A.L., Ciechanover, A., Blondin, J., Eisenhauer, D. and Taylor, A. (1986) The eye lens has an active ubiquitin–protein conjugation system. *Journal of Biological Chemistry* 261, 13760–13767.

Jahngen, J.H., Lipman, R.D., Eisenhauer, D.A., Jahngen, E.G. Jr and Taylor, A. (1990) Aging and cellular maturation cause changes in ubiquitin–eye lens protein conjugates. *Archives of Biochemistry and Biophysics* 276, 32–37.

Jahngen-Hodge, J., Cyr, D., Laxman, E. and Taylor, A. (1992) Ubiquitin and ubiquitin conjugates in human lens. *Experimental Eye Research* 55, 897–902.

Jahngen-Hodge, J., Obin, M.S., Gong, X., Shang, F., Nowell, T.R. Jr, Gong, J., Abasi, H., Blumberg, J. and Taylor, A. (1997) Regulation of ubiquitin-conjugating enzymes by glutathione following oxidative stress. *Journal of Biological Chemistry* 272, 28218–28226.

Klein, B.E., Klein, R. and Linton, K.L. (1992) Prevalence of age-related lens opacities in a population: the Beaver Dam Eye Study. *Ophthalmology* 99, 546–552.

Klein, R., Klein, B.E., Linton, K.L. and DeMets, D.L. (1993) The Beaver Dam Eye Study: the relation of age-related maculopathy to smoking. *American Journal of Epidemiology* 137, 190–200.

Knekt, P., Heliovaara, M., Rissanen, A., Aromaa, A. and Aaran, R.K. (1992) Serum antioxidant vitamins and risk of cataract. *British Medical Journal* 305, 1392–1394.

Kosegarten, D.C. and Maher, T.J. (1978) Use of guinea pigs as model to study galactose-induced cataract formation. *Journal of Pharmaceutical Sciences* 67, 1478–1479.

Krinsky, N.I. (2002) Possible biologic mechanisms for a protective role of xanthophylls. *Journal of Nutrition* 132, 540S–542S.

Kupfer, C. (1985) The conquest of cataract: a global challenge. *Transactions of the Ophthalmological Societies of the United Kingdom* 104, 1–10.

Kuzniarz, M., Mitchell, P., Cumming, R.G. and Flood, V.M. (2001) Use of vitamin supplements and cataract: the Blue Mountains Eye Study. *American Journal of Ophthalmology* 132, 19–26.

Leibowitz, H.M., Krueger, D.E., Maunder, L.R., Milton, R.C., Kini, M.M., Kahn, H.A., Nickerson, R.J., Pool, J., Colton, T.L., Ganley, J.P., Loewenstein, J.I. and Dawber, T.R. (1980) The Framingham Eye Study monograph: an ophthalmological and epidemiological study of cataract, glaucoma, diabetic retinopathy, macular degeneration, and visual acuity in a general population of 2631 adults, 1973–1975. *Survey of Ophthalmology* 24, 335–610.

Leske, M.C., Chylack, L.T. Jr and Wu, S.Y. (1991) The Lens Opacities Case–Control Study. Risk factors for cataract. *Archives of Ophthalmology* 109, 244–251.

Leske, M.C., Wu, S.Y., Hyman, L., Sperduto, R., Underwood, B., Chylack, L.T., Milton, R.C., Srivastava, S. and Ansari, N. (1995) Biochemical factors in the lens opacities. Case–control study. The Lens Opacities Case–Control Study Group. *Archives of Ophthalmology* 113, 1113–1119.

Leske, M.C., Wu, S.Y., Connel, A.M., Hyman, L. and Schachat, A.P. (1997) Lens opacities, demographic factors and nutritional suppplements in the Barbados Eye Study. *International Journal of Epidemiology* 26, 1314–1322.

Leske, M.C., Chylack, L.T. Jr, He, Q., Wu, S.Y., Schoenfeld, E., Friend, J. and Wolfe, J. (1998a) Antioxidant vitamins and nuclear opacities: the longitudinal study of cataract. *Ophthalmology* 105, 831–836.

Leske, M.C., Chylack, L.T. Jr, He, Q., Wu, S.Y., Schoenfeld, E., Friend, J. and Wolfe, J. (1998b) Risk factors for nuclear opalescence in a longitudinal study. LSC Group. Longitudinal Study of Cataract. *American Journal of Epidemiology* 147, 36–41.

Levine, M. (1986) New concepts in the biology and biochemistry of ascorbic acid. *New England Journal of Medicine* 314, 892–902.

Luthra, R., Wa, S. and Leske, M.C. (1997) Lens opacities and use of nutritional supplements: the Barbados Study. *Investigative Ophthalmology and Visual Science* 8, S450.

Lyle, B.J., Mares-Perlman, J.A., Klein, B.E., Klein, R. and Greger, J.L. (1999a) Antioxidant intake and risk of incident age-related nuclear cataracts in the Beaver Dam Eye Study. *American Journal of Epidemiology* 149, 801–809.

Lyle, B.J., Mares-Perlman, J.A., Klein, B.E., Klein, R., Palta, M., Bowen, P.E. and Greger, J.L. (1999b) Serum carotenoids and tocopherols and incidence of age-related nuclear cataract. *American Journal of Clinical Nutrition* 69, 272–277.

Machlin, L.J. and Bendich, A. (1987) Free radical tissue damage: protective role of antioxidant nutrients. *FASEB Journal* 1, 441–445.

Mares-Perlman, J.A., Klein, B.E., Klein, R. and Ritter, L.L. (1994) Relation between lens opacities and vitamin and mineral supplement use. *Ophthalmology* 101, 315–325.

Mares-Perlman, J.A., Brady, W.E., Klein, B.E., Klein, R., Haus, G.J., Palta, M., Ritter, L.L. and Shoff, S.M. (1995a) Diet and nuclear lens opacities. *American Journal of Epidemiology* 141, 322–334.

Mares-Perlman, J.A., Brady, W.E., Klein, B.E., Klein, R., Palta, M., Bowen, P. and Stacewicz-Sapuntzakis, M. (1995b) Serum carotenoids and tocopherols and severity of nuclear and cortical opacities. *Investigative Ophthalmology and Visual Science* 36, 276–288.

Mares-Perlman, J.A., Brady, W.E., Klein, R., Klein, B.E., Bowen, P., Stacewicz-Sapuntzakis, M. and Palta, M. (1995c) Serum antioxidants and age-related macular degeneration in a population-based case–control study. *Archives of Ophthalmology* 113, 1518–1523.

Mares-Perlman, J.A., Brady, W.E., Klein, B.E.K., Klein, R. and Palta, M. (1996) Supplement use and 5-year progression of cortical opacities. *Investigative Ophthalmology and Visual Science* 37, S237.

Mares-Perlman, J.A., Lyle, B.J., Klein, R., Fisher, A.I., Brady, W.E., VandenLangenberg, G.M., Trabulsi, J.N. and Palta, M. (2000) Vitamin supplement use and incident cataracts in a population-based study. *Archives of Ophthalmology* 118, 1556–1563.

McCarty, C. and Taylor, H.R. (1999) Light and risk for age-related diseases. In: Taylor, A. (ed.) *Nutritional and Environmental Influences on the Eye*. CRC Press, Boca Raton, Florida, pp. 135–150.

McCarty, C.A., Mukesh, B.N., Fu, C.L. and Taylor, H.R. (1999) The epidemiology of cataract in Australia. *American Journal of Ophthalmology* 128, 446–465.

McLaren, D.S. (1980) *Nutritional Ophthalmology*. Academic Press, London.

Mohan, M., Sperduto, R.D., Angra, S.K., Milton, R.C., Mathur, R.L., Underwood, B.A., Jaffery, N., Pandya, C.B., Chhabra, V.K., Vajpayee, R.B. et al. (1989) India–US case–control study of age-related cataracts. India–US Case–Control Study Group. *Archives of Ophthalmology* 107, 670–676.

Mune, M., Meydani, M., Jahngen-Hodge, J., Martin, A., Smith, D., Palmer, V., Blumberg, J.B. and Taylor, A. (1995) Effect of calorie restriction on liver and kidney glutathione in aging Emory mice. *Age* 18, 43–49.

Nadalin, G., Robman, L.D., McCarty, C.A., Garrett, S.K., McNeil, J.J. and Taylor, H.R. (1999) The role of past intake of vitamin E in early cataract changes. *Ophthalmic Epidemiology* 6, 105–112.

Nagaraj, R.H. and Monnier, V.M. (1992) Isolation and characterization of a blue fluorophore from human eye lens crystallins: *in vitro* formation from Maillard reaction with ascorbate and ribose. *Biochimica et Biophysica Acta* 1116, 34–42.

Nishigori, H., Lee, J.W., Yamauchi, Y. and Iwatsuru, M. (1986) The alteration of lipid peroxide in glucocorticoid-induced cataract of developing chick embryos and the effect of ascorbic acid. *Current Eye Research* 5, 37–40.

Obin, M., Nowell, T. and Taylor, A. (1994) The photoreceptor G-protein transducin (Gt) is a substrate for ubiquitin-dependent proteolysis. *Biochemical and Biophysical Research Communications* 200, 1169–1176.

Obin, M.S., Jahngen-Hodge, J., Nowell, T. and Taylor, A. (1996) Ubiquitinylation and ubiquitin-dependent proteolysis in vertebrate photoreceptors (rod outer segments). Evidence for ubiquitinylation of Gt and rhodopsin. *Journal of Biological Chemistry* 271, 14473–14484.

Obin, M., Shang, F., Gong, X., Handelman, G., Blumberg, J. and Taylor, A. (1998) Redox regulation of ubiquitin-conjugating enzymes: mechanistic insights using the thiol-specific oxidant diamide. *FASEB Journal* 12, 561–569.

Obin, M., Mesco, E., Gong, X., Haas, A.L., Joseph, J. and Taylor, A. (1999) Neurite outgrowth in PC12 cells. Distinguishing the roles of ubiquitylation and ubiquitin-dependent proteolysis. *Journal of Biological Chemistry* 274, 11789–11795.

Rathbun, W.B., Killen, C.E., Holleschau, A.M. and Nagasawa, H.T. (1996) Maintenance of hepatic glutathione homeostasis and prevention of acetaminophen-induced cataract in mice by L-cysteine prodrugs. *Biochemical Pharmacology* 51, 1111–1116.

Reddy, V.N. (1990) Glutathione and its function in the lens – an overview. *Experimental Eye Research* 50, 771–778.

Robertson, J.M., Donner, A.P. and Trevithick, J.R. (1989) Vitamin E intake and risk of cataracts in humans. *Annals of the New York Academy of Sciences* 570, 372–382.

Rouhiainen, P., Rouhiainen, H. and Salonen, J.T. (1996) Association between low plasma vitamin E concentration and progression of early cortical lens opacities. *American Journal of Epidemiology* 144, 496–500.

Sastre, J., Meydani, M., Martin, A., Biddle, L., Taylor, A. and Blumberg, J. (1994) Effect of glutathione monoethyl ester administration on galactose-induced cataract in the rat. *Life Chemistry Reports* 12, 89–95.

Schalch, W., Dayhaw-Barker, P. and Barker, F.M. II (1999) The carotenoids of the human retina. In: Taylor, A. (ed.) *Nutritional and Environmental Influences on the Eye*. CRC Press, Boca Raton, Florida, pp. 215–250.

Schwab, L. (1990) Cataract blindness in developing nations. *International Ophthalmology Clinics* 30, 16–18.

Seddon, J.M., Christen, W.G., Manson, J.E., LaMotte, F.S., Glynn, R.J., Buring, J.E. and Hennekens, C.H. (1994) The use of vitamin supplements and the risk of cataract among US male physicians. *American Journal of Public Health* 84, 788–792.

Shang, F. and Taylor, A. (1995) Oxidative stress and recovery from oxidative stress are associated with altered ubiquitin conjugating and proteolytic activities in bovine lens epithelial cells. *Biochemical Journal* 307, 297–303.

Shang, F., Gong, X. and Taylor, A. (1995) Changes in ubiquitin conjugation activities in young and old lenses in response to oxidative stress. *Investigative Ophthalmology and Visual Science* 36, S528.

Shang, F., Gong, X., Palmer, H.J., Nowell, T.R. Jr and Taylor, A. (1997a) Age-related decline in ubiquitin conjugation in response to oxidative stress in the lens. *Experimental Eye Research* 64, 21–30.

Shang, F., Gong, X. and Taylor, A. (1997b) Activity of ubiquitin-dependent pathway in response to oxidative stress. Ubiquitin-activating enzyme is transiently up-regulated. *Journal of Biological Chemistry* 272, 23086–23093.

Simon, J.A. and Hudes, E.S. (1999) Serum ascorbic acid and other correlates of self-reported cataract among older Americans. *Journal of Clinical Epidemiology* 52, 1207–1211.

Smith, D., Shang, F., Nowell, T.R., Asmundsson, G., Perrone, G., Dallal, G., Scott, L., Kelliher, M., Gindelsky, B. and Taylor, A. (1999) Decreasing ascorbate intake does not affect the levels of glutathione, tocopherol or retinol in the ascorbate-requiring osteogenic disorder shionogi rats. *Journal of Nutrition* 129, 1229–1232.

Snodderly, D.M. and Hammond, B.R. Jr (1999) *In vivo* psychophysical assessment of nutritional and environmental influences on human ocular tissues: lens and macular pigment. In: Taylor, A. (ed.) *Nutritional and Environmental Influences on the Eye*. CRC Press, Boca Raton, Florida, pp. 251–285.

Sperduto, R.D., Hu, T.S., Milton, R.C., Zhao, J.L., Everett, D.F., Cheng, Q.F., Blot, W.J., Bing, L., Taylor, P.R., Li, J.Y., Dawsey, S. and Guo, W. (1993) The Linxian cataract studies. Two nutrition intervention trials. *Archives of Ophthalmology* 111, 1246–1253.

Stephens, R.J., Negi, D.S., Short, S.M., van Kuijk, F.J., Dratz, E.A. and Thomas, D.W. (1988) Vitamin E distribution in ocular tissues following long-term dietary depletion and supplementation as determined by microdissection and gas chromatography–mass spectrometry. *Experimental Eye Research* 47, 237–245.

Taylor, A. (1999a) Lens and retina function: introduction and challenge. In: Taylor, A. (ed.) *Nutrition and Environmental Influences on the Eye*. CRC Press, Boca Raton, Florida, pp. 1–4.

Taylor, A. (1999b) Nutritional and environmental influences on risk for cataract. In: Taylor, A. (eds) *Nutritional and Environmental Influences on the Eye*. CRC Press, Boca Raton, Florida, pp. 53–93.

Taylor, A. (ed.) (1999c) *Nutritional and Environmental Influences on the Eye*. CRC Press, Boca Raton, Florida.

Taylor, A. (2000) Nutritional influences on risk for cataract. In: Fuchs, J. and Packer, L. (eds) *Environmental Stressors: Effects on Lung, Skin, Eye and Immune System Function*. Marcel Dekker, New York, pp. 457–487.

Taylor, A. and Davies, K.J. (1987) Protein oxidation and loss of protease activity may lead to cataract formation in the aged lens. *Free Radical Biology and Medicine* 3, 371–377.

Taylor, A. and Jacques, P. (1997) Antioxidant status and risk for cataract. In: Bendich, A. and Deckelbaum, R.J. (eds) *Preventive Nutrition: the Guide for Health Professionals*. Humana Press, Totowa, New Jersey.

Taylor, A., Tisdell, F.E. and Carpenter, F.H. (1981) Leucine aminopeptidase (bovine lens): synthesis and kinetic properties of ortho-, meta-, and para-substituted leucyl-anilides. *Archives of Biochemistry and Biophysics* 210, 90–97.

Taylor, A., Jacques, P.F., Nadler, D., Morrow, F., Sulsky, S.I. and Shepard, D. (1991a) Relationship in humans between ascorbic acid consumption and levels of total and reduced ascorbic acid in lens, aqueous humor, and plasma. *Current Eye Research* 10, 751–759.

Taylor, A., Jahngen-Hodge, J., Huang, L. and Jacques, P. (1991b) Aging in the eye lens: roles for proteolysis and nutrition in formation of cataract. *Age* 14, 65–71.

Taylor, A., Jacques, P.F. and Dorey, C.K. (1993) Oxidation and aging: impact on vision. *Toxicology and Industrial Health* 9, 349–371.

Taylor, A., Jacques, P.F. and Epstein, E.M. (1995a) Relations among aging, antioxidant status, and cataract. *American Journal of Clinical Nutrition* 62, 1439S–1447S.

Taylor, A., Jahngen-Hodge, J., Smith, D.E., Palmer, V.J., Dallal, G.E., Lipman, R.D., Padhye, N. and Frei, B. (1995b) Dietary restriction delays cataract and reduces ascorbate levels in Emory mice. *Experimental Eye Research* 61, 55–62.

Taylor, A., Lipman, R.D., Jahngen-Hodge, J., Palmer, V., Smith, D., Padhye, N., Dallal, G.E., Cyr, D.E., Laxman, E., Shepard, D., Morrow, F., Salomon, R., Perrone, G., Asmundsson, G., Meydani, M., Blumberg, J., Mune, M., Harrison, D.E., Archer, J.R. and Shigenaga, M. (1995c) Dietary calorie restriction in the Emory mouse: effects on lifespan, eye lens cataract prevalence and progression, levels of ascorbate, glutathione, glucose, and glycohemoglobin, tail collagen breaktime, DNA and RNA oxidation, skin integrity, fecundity, and cancer. *Mechanisms of Ageing and Development* 79, 33–57.

Taylor, A., Jacques, P.F., Nowell, T., Perrone, G., Blumberg, J., Handelman, G., Jozwiak, B. and Nadler, D. (1997) Vitamin C in human and guinea pig aqueous, lens and plasma in relation to intake. *Current Eye Research* 16, 857–864.

Taylor, A., Jacques, P.F., Chylack, L.T. Jr, Hankinson, S.E., Khu, P.M., Rogers, G., Friend, J., Tung, W., Wolfe, J.K., Padhye, N. and Willett, W.C. (2002)

Long-term intake of vitamins and carotenoids and odds of early age-related cortical and posterior subcapsular lens opacities. *American Journal of Clinical Nutrition* 75, 540–549.

Varma, S.D., Chand, D., Sharma, Y.R., Kuck, J.F. Jr and Richards, R.D. (1984) Oxidative stress on lens and cataract formation: role of light and oxygen. *Current Eye Research* 3, 35–57.

Vinson, J.A., Possanza, C.J. and Drack, A.V. (1986) The effect of ascorbic acid on galactose-induced cataracts. *Nutrition Reports International* 33, 665–668.

Vitale, S., West, S., Hallfrisch, J., Alston, C., Wang, F., Moorman, C., Muller, D., Singh, V. and Taylor, H.R. (1993) Plasma antioxidants and risk of cortical and nuclear cataract. *Epidemiology* 4, 195–203.

Wang, G.M., Spector, A., Luo, C.Q., Tang, L.Q., Xu, L.H., Guo, W.Y. and Huang, Y.Q. (1990) Prevalence of age-related cataract in Ganzi and Shanghai. The Epidemiological Study Group. *Chinese Medical Journal (England)* 103, 945–951.

West, S.K. (1999) Smoking and the risk of eye disease. In: Taylor, A. (ed.) *Nutritional and Environmental Influences on the Eye.* CRC Press, Boca Raton, Florida, pp. 151–164.

Whitfield, R., Schwab, L., Ross-Degnan, D., Steinkuller, P. and Swartwood, J. (1990) Blindness and eye disease in Kenya: ocular status survey results from the Kenya Rural Blindness Prevention Project. *British Journal of Ophthalmology* 74, 333–340.

Wolfe, J.K., Chylack, L.T., Leske, M.C. and Wu, S.Y. (1993) Lens nuclear color and visual function. *Investigative Ophthalmology and Visual Science* 34 (Supplement), 2550.

World Health Organization (1991) Use of intraocular lenses in cataract surgery in developing countries: memorandum from a WHO meeting. *Bulletin of the World Health Organization* 69, 657–666.

Yeum, K.J., Taylor, A., Tang, G. and Russell, R.M. (1995) Measurement of carotenoids, retinoids, and tocopherols in human lenses. *Investigative Ophthalmology and Visual Science* 36, 2756–2761.

Yeum, K.J., Shang, F.M., Schalch, W.M., Russell, R.M. and Taylor, A. (1999) Fat-soluble nutrient concentrations in different layers of human cataractous lens. *Current Eye Research* 19, 502–505.

Yokoyama, T., Sasaki, H., Giblin, F.J. and Reddy, V.N. (1994) A physiological level of ascorbate inhibits galactose cataract in guinea pigs by decreasing polyol accumulation in the lens epithelium: a dehydroascorbate-linked mechanism. *Experimental Eye Research* 58, 207–218.

Young, R.W. (1993) The Charles F. Prentice Medal Award Lecture 1992: optometry and the preservation of visual health. *Optometry and Vision Science* 70, 255–262.

Zigler, J.S. Jr and Goosey, J.D. (1984) Singlet oxygen as a possible factor in human senile nuclear cataract development. *Current Eye Research* 3, 59–65.

22 Nutrition and Immune Function

Parveen Yaqoob[1] and Philip C. Calder[2]
[1]School of Food Biosciences, University of Reading, Whiteknights, Reading, UK;
[2]Institute of Human Nutrition, School of Medicine,
University of Southampton, Southampton, UK

Introduction

Associations between famine and epidemics of infectious disease have been noted throughout history – Hippocrates recognized that poorly nourished people are more susceptible to infectious disease as early as 370 BC. In general, undernutrition impairs the immune system, suppressing immune functions that are fundamental to host protection against pathogenic organisms. Undernutrition leading to impairment of immune function can be due to insufficient intake of energy and macronutrients and/or due to deficiencies in specific micronutrients (vitamins and minerals). Often these occur in combination – this is particularly notable for protein energy malnutrition and deficiencies in micronutrients such as vitamin A, iron, zinc and iodine. Clearly the impact of undernutrition is greatest in developing countries, but it is also important in developed countries, especially amongst the elderly, individuals with eating disorders, alcoholics, patients with certain diseases and premature and small-for-gestational-age babies. The precise effects of individual nutrients on different aspects of immune function have been notoriously difficult to study. However, it is becoming clear that many nutrients have defined roles in the immune response and that each nutrient has a distinct range of intakes over which it supports optimal immune function. Lowering the level of the nutrient below this range or increasing it in excess of the range can impair immune function. Thus, the functioning of the immune system is influenced by nutrients consumed as normal components of the diet, and appropriate nutrition is required in order for the host to maintain adequate immune defences towards bacteria, viruses, fungi and parasites. The cellular and molecular mechanisms by which nutrients affect immune function are only just beginning to be understood. This chapter begins with an overview of the key components of the immune system. The main body of the chapter is devoted to an evaluation of the influence of individual macro- and micronutrients on immune function and their proposed mechanisms of action.

The Immune System

Innate immunity

The immune system acts to protect the host from infectious agents that exist in the environment (bacteria, viruses, fungi, parasites) and from other noxious insults. The immune system has two functional divisions: the innate (or natural) immune system and the acquired (also termed specific or adaptive) immune system. Innate immunity consists of physical barriers, soluble factors and phagocytic cells, which include granulocytes (neutrophils, basophils, eosinophils), monocytes and macrophages. Innate immunity has no memory and is therefore not influenced by prior exposure to an organism. Phagocytic cells, the main effectors of innate immunity, express

©CAB International 2003. Molecular Nutrition
(eds J. Zempleni and H. Daniel)

surface receptors specific for bacterial surface antigens. Binding of antigen to the receptors triggers phagocytosis and subsequent destruction of the pathogenic microorganism by complement or by toxic chemicals, such as superoxide radicals and hydrogen peroxide. Natural killer (NK) cells also possess surface receptors and destroy pathogens by release of cytotoxic proteins. In this way, innate immunity provides a first line of defence against invading pathogens. However, an immune response often requires the coordinated actions of both innate immunity and the more powerful and flexible acquired immunity.

Acquired immunity

Acquired immunity involves the specific recognition of molecules (antigens) on an invading pathogen that distinguish it as being foreign to the host. Lymphocytes, which are subdivided into T and B lymphocytes, effect this form of immunity. All lymphocytes (indeed all cells of the immune system) originate in the bone marrow. B lymphocytes undergo further development and maturation in the bone marrow before being released into the circulation, whilst T lymphocytes mature in the thymus. From the bloodstream, lymphocytes can enter peripheral lymphoid organs, which include lymph nodes, the spleen, mucosal lymphoid tissue, tonsils and gut-associated lymphoid tissue. Immune responses occur largely in these lymphoid organs, which are highly organized to promote the interaction of cells and invading pathogens.

The acquired immune system is highly specific, since each lymphocyte carries surface receptors for a single antigen. However, acquired immunity is extremely diverse; the lymphocyte repertoire in humans has been estimated at recognition of approximately 10^{11} antigens. The high degree of specificity, combined with the huge lymphocyte repertoire, means that only a relatively small number of lymphocytes will be able to recognize any given antigen. The acquired immune system has developed the ability for clonal expansion to deal with this. Clonal expansion involves the proliferation of a lymphocyte once an interaction with its specific antigen has occurred, so that a single lymphocyte gives rise to a clone of lymphocytes, each of which has the ability to recognize and destroy the antigen causing the initial response. This feature of acquired immunity has often been likened to building up an army in order to fight a foreign invasion. The acquired immune response becomes effective over several days after the initial activation, but it also persists for some time after the removal of the initiating antigen. This persistence gives rise to immunological memory, which is also a characteristic feature of acquired immunity. It is the basis for the stronger, more effective immune response to re-exposure to an antigen (i.e. re-infection with the same pathogen) and is the basis for vaccination. Eventually, the immune system will re-establish homeostasis using self-regulatory mechanisms that involve communication between cells.

B and T lymphocytes

B lymphocytes are characterized by their ability to produce antibodies, or immunoglobulins (Igs), which confer antigen specificity to the acquired immune system (i.e. the antibodies produced by B lymphocytes are specific for individual antigens). This form of protection against infections is termed humoral immunity and is conducted exclusively by B lymphocytes. B lymphocytes carry immunoglobulins, which are capable of binding an antigen, on their cell surfaces. Binding of immunoglobulin with antigen causes proliferation of the B lymphocyte and subsequent transformation into plasma cells, which secrete large amounts of antibody with the same specificity as the parent cell.

Immunoglobulins are proteins consisting of two identical heavy chains and two identical light chains. Five different types of heavy chain give rise to five major classes of immunoglobulin (IgA, IgD, IgG, IgM and IgE), each of which elicits different components of the humoral immune response. Antibodies work in several ways to combat invading pathogens. They can 'neutralize' toxins or microorganisms by binding to them and preventing their attachment to host cells, and they can activate complement proteins in plasma, which in turn promote the destruction of bacteria by phagocytes. Since they have binding sites both for an antigen and for receptors on phagocytic cells, antibodies can also promote the interaction of the two components by forming physical 'bridges', a process known as opsonization. The type of phagocytic cell bound by the antibody will be determined by the antibody class; macrophages and neutrophils are specific for IgM and IgG, whilst eosinophils are specific for IgE. In this way, antibodies are a form of communication between the acquired and the innate immune response; they are elicited through highly specific mechanisms, but ultimately are translated to a form that can be interpreted by the innate immune system, enabling it to destroy the pathogen.

Humoral immunity deals with extracellular pathogens. However, some pathogens, particularly viruses, but also some bacteria, infect individuals by entering cells. These pathogens will escape humoral immunity and instead are dealt with by cell-mediated immunity, which is conferred by T lymphocytes. T lymphocytes express antigen-specific T-cell receptors (TCRs) on their surface, which have an enormous antigen repertoire. However, unlike B lymphocytes, they are only able to recognize antigens that are 'presented' to them on a cell surface; this is the distinguishing feature between humoral and cell-mediated immunity. Therefore, infection of a cell by an intracellular pathogen is signalled to T lymphocytes by cell surface expression of peptide fragments derived from the pathogen. These fragments are transported to the surface of the infected cell and expressed there in conjunction with proteins termed major histocompatibility complex (MHC); in humans, MHC is termed human leukocyte antigen (HLA). It is the combination of the pathogen-derived peptide fragment bound to MHC that is recognized by T lymphocytes. There are two classes of MHC, MHC I and MHC II, and the source of the peptide bound to each differs. MHC I binds peptides that originate from pathogen proteins synthesized within the host cell cytosol; typically these are from viruses or certain bacteria. The peptides bound to MHC II are derived from pathogens that have been phagocytosed by macrophages or endocytosed by antigen-presenting cells (macrophages, dendritic cells, B lymphocytes). The MHC–peptide complex is recognized by the TCR on T lymphocytes. T lymphocytes expressing CD8 (cytotoxic T cells) recognize MHC I, while T lymphocytes expressing CD4 (helper T cells) recognize MHC II. Thus, intracellular pathogens stimulate cytotoxic T lymphocytes to destroy the infected cell, while extracellular pathogens stimulate a helper T cell-mediated response. In delayed-type hypersensitivity (DTH), antigen-activated $CD4^+$ T lymphocytes (helper T cells) secrete cytokines, which have several effects, including recruitment of neutrophils and monocytes from the blood to the site of antigen challenge and activation of monocytes in order to effect elimination of the antigen. The DTH reaction can be induced in humans by contact sensitization with chemicals and environmental antigens or by intradermal injection of microbial antigens, and as such has been widely used as a rapid *in vivo* marker of cell-mediated immunity. The degree of the DTH reaction can be assessed by measuring the thickening of the skin, normally 48 h after exposure to the antigen.

Communication within the immune system: cytokines

Communication within the acquired immune system and between the innate and acquired systems is brought about by direct cell–cell contact involving adhesion molecules and by the production of chemical messengers, which send signals from one cell to another (Fig. 22.1). Chief among these chemical messengers are proteins called cytokines, which can act to regulate the activity of the cell that produced the cytokine, or of other cells. Each cytokine can have multiple activities on different cell types. Cytokines act by binding to specific receptors on the cell surface, and thereby induce changes in growth, development or activity of the target cell.

Tumour necrosis factor (TNF)-α, interleukin (IL)-1 and IL-6 are among the most important cytokines produced by monocytes and macrophages. These cytokines activate neutrophils, monocytes and macrophages to initiate bacterial and tumour cell killing, increase adhesion molecule expression on the surface of neutrophils and endothelial cells, stimulate T- and B-lymphocyte proliferation and initiate the production of other pro-inflammatory cytokines (e.g. TNF induces production of IL-1 and IL-6, and IL-1 induces production of IL-6). Thus, TNF, IL-1 and IL-6 are mediators of both natural and acquired immunity and are an important link between them (Fig. 22.1). In addition, these cytokines mediate the systemic effects of inflammation such as fever, weight loss and acute phase protein synthesis in the liver. Inflammation is the body's immediate response

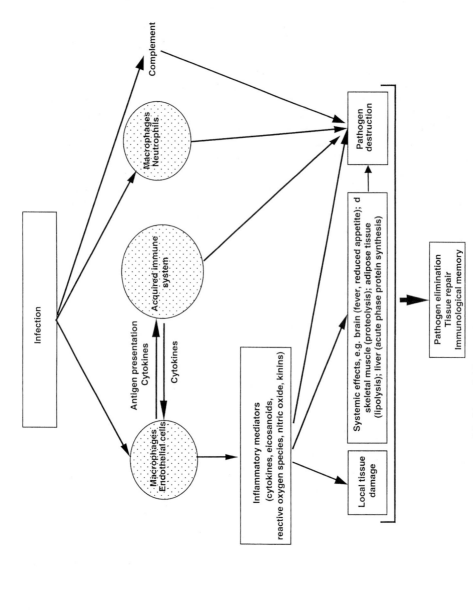

Fig. 22.1. Overview of the immune system. Modified from Calder (2002).

to infection or injury. It is typified by redness, swelling, heat and pain. These occur as a result of increased blood flow, increased permeability across blood capillaries, which permits large molecules (e.g. complement, antibodies, cytokines) to leave the bloodstream and cross the endothelial wall, and increased movement of leukocytes from the bloodstream into the surrounding tissue. Thus, inflammation is an integral part of the innate immune response, and production of appropriate amounts of TNF, IL-1 and IL-6 is clearly important in response to infection. However, inappropriate production or overproduction can be dangerous, and these cytokines, particularly TNF, are implicated in causing some of the pathological responses that occur in chronic inflammatory conditions (e.g. rheumatoid arthritis, psoriasis).

Helper T lymphocytes can be subdivided into two broad categories according to the pattern of cytokines they produce (Fig. 22.2). It is believed that helper T cells that have not encountered antigen previously produce mainly IL-2 upon initial encounter with antigen. These cells may differentiate into a population sometimes referred to as Th0 cells, which differentiate further into either Th1 or Th2 cells (Fig. 22.2). This differentiation is regulated by cytokines: IL-12 and interferon-γ (IFN-γ) promote the development of Th1 cells, while IL-4 promotes the development of Th2 cells (Fig. 22.2). Th1 and Th2 cells themselves have relatively restricted profiles of cytokine production: Th1 cells produce IL-2 and IFN-γ, which activate macrophages, NK cells and cytotoxic T lymphocytes and are the principal effectors of cell-mediated immunity. Th2 cells produce IL-4, which stimulates immunoglobulin E production, IL-5, an eosinophil-activating factor, and IL-10, which together with IL-4 suppresses cell-mediated immunity (Fig. 22.2). Th2 cells are responsible for defence against helminthic parasites, which is due to IgE-mediated activation of mast cells and basophils. The patterns of cytokine secretion by Th1 and Th2 lymphocytes were demonstrated first in mice, and it has been demonstrated subsequently that, while human helper T lymphocytes do show

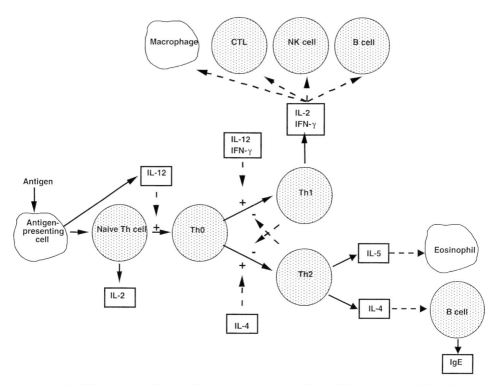

Fig. 22.2. The differentiation of helper T lymphocytes along the Th1 and Th2 pathways and the roles of Th1 and Th2 cytokines in eliciting effector cell function.

differences in cytokine profile, the divisions are not clear as in mice.

Why Should Nutrients Affect Immune Function?

Although the immune system is functioning at all times, specific immunity becomes activated when the host is challenged by pathogens. This activation is associated with a marked increase in the demand of the immune system for substrates and nutrients to provide a ready source of energy, which can be supplied from exogenous sources (i.e. from the diet) and/or from endogenous pools. The cells of the immune system are metabolically active and are able to utilize glucose, amino acids and fatty acids as fuels. Energy generation involves electron carriers, which are nucleotide derivatives (e.g. nicotinamide adenine dinucleotide (NAD), flavin adenine dinucleotide (FAD)) and a range of coenzymes. The electron carriers and coenzymes are usually derivatives of vitamins: thiamine pyrophosphate is derived from thiamine (vitamin B_1), FAD and flavin mononucleotide from riboflavin (vitamin B_2), NAD from nicotinate (niacin), pyridoxal phosphate from pyridoxine (vitamin B_6), coenzyme A from pantothenate, tetrahydrofolate from folate and cobamide from cobalamin (vitamin B_{12}). In addition, biotin is required by some enzymes for activity. The final component of the pathway for energy generation (the mitochondrial electron transfer chain) includes electron carriers that have iron or copper at their active site.

Activation of the immune response gives rise to production of proteins (immunoglobulins, cytokines, cytokine receptors, adhesion molecules, acute phase proteins) and lipid-derived mediators (prostaglandins, leukotrienes). To respond optimally, it is clear that there must be the appropriate enzymatic machinery in place (for RNA synthesis and protein synthesis and their regulation) and ample substrate available (nucleotides for RNA synthesis, the correct mix of amino acids for protein synthesis, polyunsaturated fatty acids for eicosanoid synthesis).

An important component of the immune response is oxidative burst, during which superoxide anion radicals are produced from oxygen in a reaction linked to the oxidation of NADPH. The reactive oxygen species produced can be damaging to host tissues, and thus antioxidant protective mechanisms are necessary. Among these are the classic antioxidant vitamins, α-tocopherol (vitamin E) and ascorbic acid (vitamin C), glutathione, a tripeptide composed of glutamate, cysteine and glycine, the antioxidant enzymes superoxide dismutase and catalase, and the glutathione recycling enzyme glutathione peroxidase. Superoxide dismutase has two forms, a mitochondrial form and a cytosolic form; the mitochondrial form includes manganese at its active site, while the cytosolic form includes copper and zinc. Catalase contains iron at its active site, while glutathione peroxidase contains selenium.

Cellular proliferation is a key component of the immune response providing amplification and memory: prior to division, there must be replication of DNA and then of all cellular components (proteins, membranes, intracellular organelles, etc.). In addition to energy, this clearly needs a supply of nucleotides (for DNA and RNA synthesis), of amino acids (for protein synthesis), of fatty acids, bases and phosphate (for phospholipid synthesis), and of other lipids (e.g. cholesterol) and cellular components. Although nucleotides are synthesized mainly from amino acids, some of the cellular building blocks cannot be synthesized in mammalian cells and must come from the diet (e.g. essential fatty acids, essential amino acids, minerals). Amino acids (e.g. arginine) are precursors for synthesis of polyamines, which have roles in the regulation of DNA replication and cell division. Various micronutrients (e.g. iron, folic acid, zinc, magnesium) are also involved in nucleotide and nucleic acid synthesis.

Thus, the roles for nutrients in immune function are many and varied, and it is easy to appreciate that an adequate and balanced supply of these is essential if an appropriate immune response is to be mounted. However, our understanding of the molecular basis for nutrient action on immune function is incomplete, although there is evidence to suggest that it involves effects on surface receptor/protein regulation, production of mediators and redox status.

Protein Energy Malnutrition and Immune Function

Protein energy malnutrition, although often considered a problem solely of developing countries, has been described in even the most affluent of countries. Moderate malnutrition in the developed world is encountered amongst the elderly, anorexics, bulimics, premature babies, hospitalized patients and patients with various diseases (e.g. cystic fibrosis, acquired immunodeficiency syndrome, some cancers). A large number of studies in animals have demonstrated the adverse effects of protein deficiency on immunity (see Woodward, 1998), and these effects have been confirmed in various human settings (see Gross and Newberne, 1980; Chandra, 1991). It is not surprising that protein deficiency diminishes immune responses and increases susceptibility to infection, because immune defences are dependent upon cell replication and the production of proteins with biological activities (e.g. antibodies, cytokines, acute phase proteins). It is important, however, to recognize that protein energy malnutrition often is characterized not simply by insufficient intake of total energy and macronutrients but by micronutrient deficiencies as well.

Practically all forms of immunity may be affected by protein energy malnutrition, but non-specific defences and cell-mediated immunity are more severely affected than humoral (antibody) responses (Chandra, 1991; Woodward, 1998). Protein energy malnutrition causes atrophy of the lymphoid organs (thymus, spleen, lymph nodes, tonsils) in laboratory animals and humans. The circulating white blood cell count can be increased, but this is due to increased numbers of neutrophils; the absolute and relative numbers of monocytes, lymphocytes, $CD4^+$ cells and $CD8^+$ cells are decreased, as is the CD4:CD8 ratio. The decline in the number of circulating T lymphocytes is proportional to the extent of malnutrition (Rivera *et al.*, 1986). The proliferative responses of T lymphocytes to mitogens and antigens is decreased by malnutrition, as is the synthesis of the cytokines IL-2 and IFN-γ, and the activity of NK cells. Production of cytokines by monocytes, including TNF-α, IL-1 and IL-6, is also decreased by malnutrition. The *in vivo* DTH response to challenge with specific recall antigens is reduced by malnutrition (Rivera *et al.*, 1986). Bactericidal activity and respiratory burst of neutrophils are decreased by malnutrition, but the phagocytic capacity of neutrophils and monocytes appears to be unaffected. Numbers of circulating B lymphocytes and circulating immunoglobulin levels do not seem to be affected, or may even be increased, by malnutrition (Chandra, 1991); this increase may represent underlying infection. Although the level of secretory IgA in tears, saliva and intestinal washings is decreased by malnutrition, this may relate to decreased expression of the polymeric Ig receptor, which is responsible for transepithelial transport of IgA, rather than to reduced IgA synthesis (for references, see Woodward, 1998).

Micronutrients and Immune Function

Vitamin A

Vitamin A deficiency is accompanied by increased incidence and severity of infectious diseases and some cancers, and this has been linked to an impaired immune response (see Semba, 1998, 1999). Almost all immune responses studied have been shown to be impaired by vitamin A deficiency (see Semba, 1998, 1999). Furthermore, vitamin A is essential for maintaining epidermal and mucosal integrity, and vitamin A-deficient mice have histopathological changes in the gut mucosa consistent with a breakdown in gut barrier integrity and impaired mucus secretion, both of which would facilitate entry of pathogens. Mice fed a very low protein diet displayed a decline in mucosal IgA and cytokine (IL-4, IL-5, IL-6) concentrations, which was prevented by inclusion of an increased level of vitamin A in the diet (Nikawa *et al.*, 1999). Vitamin A regulates keratinocyte differentiation, and vitamin A deficiency induces changes in skin keratinization, which may explain the observed increased incidence of skin infections (Semba, 1998).

The impact of vitamin A deficiency on infectious disease has been studied widely in the developing world. Vitamin A deficiency is associated with increased morbidity and mortality in children and appears to predispose to respiratory infections, diarrhoea and severe measles (Chandra, 1991; Scrimshaw and SanGiovanni, 1997; Semba, 1999; Calder and Jackson, 2000). Replenishment of vitamin A in depleted or deficient animals or subjects leads to restoration of lymphoid organ development, circulating immune cell populations, immune cell functions and DTH response, and improved resistance to infection by pathogens (Semba, 1998). There have been a number of studies of vitamin A administration and morbidity and mortality in children. Most, though not all, of these studies show reduced morbidity and a drop in mortality related to measles and diarrhoeal disease, but not respiratory tract infections (Chandra, 1991; Scrimshaw and SanGiovanni, 1997; Semba, 1999; Calder and Jackson, 2000). There are suggestions that high levels of vitamin A in the diet might enhance immune responsiveness above 'normal'. However, opposing effects have been observed, with high levels of dietary vitamin A causing increased, unchanged or decreased immune responses (Friedman and Sklan, 1993). Animal experiments indicate that an excess vitamin A intake can decrease immune functions and resistance to infection to a similar extent as vitamin A deficiency (Friedman and Sklan, 1993). Vitamin A exerts its effects via nuclear retinoic acid and retinoid receptors, which act as transcriptional activators for many specific target genes (see Chapter 10).

Folic acid and B-group vitamins

Folic acid deficiency in laboratory animals causes thymus and spleen atrophy, and decreases circulating T-cell numbers, the activity of cytotoxic T lymphocytes and spleen lymphocyte proliferation, but does not alter phagocytosis or bactericidal capacity of neutrophils (Gross and Newberne, 1980). In contrast, vitamin B_{12} deficiency decreases phagocytosis and the bactericidal capacity of neutrophils (Gross and Newberne, 1980). Vitamin B_6 deficiency in laboratory animals causes thymus and spleen atrophy, decreases lymphocyte proliferation and the DTH response, and increases allograft survival (Gross and Newberne, 1980). In a study in healthy elderly humans, a vitamin B_6-deficient diet (3 µg kg^{-1} body weight day^{-1} or ~0.17 and 0.1 mg day^{-1} for men and women, respectively) for 21 days resulted in a decreased percentage and total number of circulating lymphocytes, decreased T- and B-cell proliferation in response to mitogens and decreased IL-2 production (Meydani et al., 1991). Repletion at 15 or 22.5 µg kg^{-1} body weight day^{-1} for 21 days did not return the immune functions to starting values. However, repletion at 33.75 µg kg^{-1} body weight day^{-1} (~1.9 and 1.1 mg day^{-1} for men and women, respectively) returned immune parameters to starting values. Providing 41 mg of vitamin B_6 day^{-1} for 4 days caused a further increase in lymphocyte proliferation and IL-2 production. These data indicate that vitamin B_6 deficiency impairs human immune function (at least in the elderly), that this impairment is reversible by repletion, and that lymphocyte functions are enhanced at levels of vitamin B_6 above those typical of habitual consumption.

Vitamin C

Vitamin C is a water-soluble antioxidant found in high concentrations in circulating leukocytes, and it appears to be utilized during infections. The effects of vitamin C on immune function are very much dependent on dosage, and confusion has arisen as a result of attempts to compare studies where different doses have been used (see Siegel, 1993). Vitamin C deficiency in humans does not impair lymphocyte proliferation, but it may increase circulating Ig concentrations (although some studies fail to show this; Siegel, 1993). Jacob et al. (1991) studied the effect of vitamin C at different levels in the diet on immune function in a group of young healthy non-smokers: a vitamin C-deficient diet decreased mononuclear cell vitamin C content by 50% and decreased the DTH response to seven recall antigens, but did not alter lymphocyte proliferation. Adding back vitamin C at 10, 20, 60 or 250 mg day^{-1} (each for 28 days) did not induce recovery of the DTH response even though the mononuclear cell vitamin C content returned to baseline.

Vitamin D

Vitamin D receptors have been identified in most cells of the immune system, suggesting that it has immunoregulatory properties (see Lemire, 1992). Its effects on immune function include inhibition of T-lymphocyte proliferation, of IL-2 and IFN-γ production, of monocyte differentiation, of NK cell activity and of antibody production by B lymphocytes (see Lemire, 1992; Overbergh et al., 2000). The active metabolite, 1,25(OH$_2$)D$_3$ appears to have specific effects on cytokine production, driving the immune response towards a Th2 profile, partly by inhibition of macrophage/dendritic cell IL-12 production (the major cytokine that induces the Th1 response), but also by direct inhibition of IL-2 and IFN-γ production (Th1 cytokines) and stimulation of IL-4 production (Alroy et al., 1995; D'Ambrosio et al., 1998; Takeuchi et al., 1998; Singh et al., 1999). This might explain the observation that vitamin D appears to prolong graft survival (Lemire, 1992; Overbergh et al., 2000) and is of benefit in animal models of autoimmune disease. For example, vitamin D deficiency resulted in increased susceptibility of mice to the autoimmune disease experimental allergic encephalomyelitis (a model of multiple sclerosis) (Cantorna et al., 1996). Administration of vitamin D prior to and after the induction of this disease in mice and rats prevented its development and abrogated the rise in antibodies to myelin basic protein and the histological changes (see Lemire, 1992; see also Cantorna et al., 1996). Intraperitoneal injections of vitamin D prevented the skin lesions, proteinuria and production of autoantibodies that accompany the development of lupus in a genetically susceptible mouse strain (see Lemire, 1992), reduced the incidence of insulitis in genetically susceptible mice (Mathieu et al., 1992) and prevented the progress of collagen-induced arthritis in mice (Cantorna et al., 1998). These observations suggest that vitamin D suppresses the activity of Th1 cells.

According to the classic genomic action of vitamin D, $1,25(OH_2)D_3$ is released from its serum-binding protein, diffuses through the cell membrane and binds to a classic zinc finger-containing receptor (the vitamin D receptor; VDR) in the nucleus (Macdonald et al., 1994). There is some interaction between the nuclear co-receptors of vitamins A (retinoid X receptor; RXR) and D, as well as thyroxin; the receptors form heterodimers in different combinations, which can affect target gene transcription (Haag, 1999). Over the last decade, it has become clear that in addition to the slow genomic mode of action, vitamin D can also act via rapid non-genomic actions, which involve a number of second messenger pathways.

The IFN-γ promoter is negatively regulated by vitamin D, and this effect is enhanced by RXR (Cippitelli and Santoni, 1998). Inhibition of IL-2 gene expression by $1,25(OH_2)D_3$ through direct interference with promoter activation has been described and appears to be due to the VDR–RXR heterodimer blocking the formation of a nuclear factor of activated T cells/activator protein-1 (NFAT–AP-1) complex (Alroy et al., 1995). A repressive effect of $1,25(OH_2)D_3$ on nuclear factor-kappa B (NF-κB) activity in T cells has also been described, suggesting a regulatory mechanism whereby excessive lymphocyte activation might be prevented (Yu et al., 1995). These effects probably underlie the beneficial effects of vitamin D in the treatment of animal models of autoimmune diseases. Furthermore, the immunosuppressive agent, cyclosporin, has been shown to increase the expression of the enzyme responsible for the final activation of $1,25(OH_2)D_3$, and a defect in the expression of this enzyme has been detected in autoimmune mice (Overbergh et al., 2000). Although the defect is not necessarily responsible for the condition, it is possible that it contributes and that up-regulation (resulting in synthesis of $1,25(OH_2)D_3$) might provide a negative feedback loop in inflammation (Casteels et al., 1998a,b).

Vitamin E

Vitamin E is the major lipid-soluble antioxidant in the body and is required for protection of membrane lipids from peroxidation. Since free radicals and lipid peroxidation are immunosuppressive, it is considered that vitamin E should act to optimize and even enhance the immune response (see Bendich, 1993; Meydani and Beharka, 1998). Except in premature infants and the elderly, clinical vitamin E deficiency is rare in humans, although many individuals have vitamin E intakes below the recommended daily intake in many countries.

In laboratory animals, vitamin E deficiency decreases spleen lymphocyte proliferation, NK cell activity, specific antibody production following vaccination, and phagocytosis by neutrophils, and increases the susceptibility of animals to infectious pathogens (Bendich, 1993; Meydani and Beharka, 1998; Han and Meydani, 1999). The effects of vitamin E depletion are more marked if animals are fed a diet containing a high level of polyunsaturated fatty acids. Vitamin E supplementation of the diet of laboratory animals enhances antibody production, lymphocyte proliferation, IL-2 production, NK cell activity, macrophage phagocytosis and the DTH response (Meydani and Beharka, 1998). Dietary vitamin E promotes resistance to pathogens in chickens, turkeys, mice, pigs, sheep and cattle (see Meydani and Beharka, 1998; Han and

Meydani, 1999). Feeding extra vitamin E to mice prevented the decrease in production of IL-2 and IFN-γ by spleen lymphocytes and in NK cell activity induced by retrovirus (Wang et al., 1994), and decreased lung titres of influenza virus (Hayek et al., 1997).

A positive association exists between plasma vitamin E levels and DTH responses and a negative association between plasma vitamin E levels and incidence of infections in healthy adults aged over 60 years (see Bendich, 1993; Meydani and Beharka, 1998; Han and Meydani, 1999). There appears to be a particular benefit of vitamin E supplementation in the elderly; a comprehensive study demonstrated increased DTH responses in elderly subjects supplemented with 60, 200 and 800 mg of vitamin E day^{-1}, with a maximal effect at a dose of 200 mg day^{-1} (Meydani et al., 1997). This dose also significantly increased the antibody responses to hepatitis B, tetanus toxoid and *Pneumococcus* vaccinations. Thus, it appears that adding vitamin E to the diet at levels beyond those normally achievable in the diet enhances some immune responses. However, as with many other micronutrients, doses that are greatly in excess of normal requirements may suppress the immune response; e.g. 800 mg of vitamin E day^{-1} decreased some of the antibody responses to below those of the placebo group (Meydani et al., 1997).

Zinc

Zinc deficiency in animals is associated with a wide range of immune impairments (see Fraker et al., 1993; Shankar and Prasad, 1998). Zinc deficiency has a marked impact on bone marrow, decreasing the number of nucleated cells and the number and proportion of cells that are lymphoid precursors (see Fraker et al., 1993; Fraker and King, 1998). In patients with zinc deficiency related to sickle cell disease, NK cell activity is decreased, but can be returned to normal by zinc supplementation. In acrodermatitis enteropathica, which is characterized by reduced intestinal zinc absorption, thymic atrophy, impaired lymphocyte development and reduced lymphocyte responsiveness and DTH are observed. Moderate or mild zinc deficiency or experimental zinc deficiency in man (induced by consumption of <3.5 mg of zinc day^{-1}; habitual intakes among adults in the UK are 9–12 mg day^{-1}) decrease thymulin activity, NK cell activity, lymphocyte proliferation, IL-2, IFN-γ and TNF-α production and the DTH response; all can be corrected by zinc repletion (Shankar and Prasad, 1998).

Low plasma zinc levels can be used to predict the subsequent development of lower respiratory tract infections and diarrhoea in malnourished populations (Shankar and Prasad, 1998; Calder and Jackson, 2000). Indeed, diarrhoea is considered a symptom of zinc deficiency. Zinc administration (2 mg kg^{-1} body weight daily) to malnourished children has been demonstrated to decrease by >50% the incidence of diarrhoea, and of respiratory and skin infections, and results in increased growth compared with children given low-dose zinc (3.5 mg daily) (see Calder and Jackson, 2000). A number of studies show that zinc supplementation decreases the incidence of childhood diarrhoea and respiratory illness, although some studies fail to show benefit of zinc supplementation in respiratory disease (see Chandra, 1991; Scrimshaw and SanGiovanni, 1997; Shankar and Prasad, 1998; Calder and Jackson, 2000). Zinc administration to pre-term low body weight infants (1 mg kg^{-1} day^{-1} for 30 days) increased the number of circulating T lymphocytes and lymphocyte proliferation (Chandra, 1991). Providing 5 mg of zinc day^{-1} to low birth weight, small-for-gestational-age infants for 6 months increased measures of cell-mediated immune function and decreased the incidence of gastrointestinal and upper respiratory tract infections (see Calder and Jackson, 2000); a zinc dose of 1 mg day^{-1} was without effect. Although increasing zinc intake enhances immune function, excessive zinc intakes impair immune responses. For example, giving 300 mg zinc day^{-1} for 6 weeks to young adult humans decreased lymphocyte and phagocyte function (Chandra, 1984). High zinc intakes can also result in copper depletion, and copper deficiency impairs immune function (see below).

Zinc is a cofactor for many enzymes, including Cu–Zn superoxide dismutase, the cytoplasmic antioxidant, and is therefore involved in protecting host cells from the cytotoxic effects of free radical species produced during immune responses. Zinc is required for the activity of thymidine kinase, DNA polymerase, RNA polymerase and aminoacyl-tRNA synthetase. Thus, zinc has a key role in supporting DNA, RNA and protein synthesis and progress through the cell cycle. Furthermore, the

DNA-binding domains of many transcription factors, including NF-κB, contain 'zinc fingers', further highlighting the importance of zinc in gene transcription. A recent study points to a key role for zinc in regulating cytokine production through its effects on the NF-κB pathway (Prasad et al., 2001).

Copper

Although overt copper deficiency is believed to be rare in humans, modest deficiency is likely to be present among some populations. Zinc and iron impair copper uptake so that taking high doses of these might induce mild copper deficiency. Copper deficiency has been described in premature infants and in patients receiving total parenteral nutrition. The classic example of copper deficiency is Menkes syndrome, a rare congenital disease that results in the complete absence of ceruloplasmin, the copper-carrying protein in the blood. Children with Menkes syndrome have increased bacterial infections, diarrhoea and pneumonia. Copper deficiency in experimental and farm animals impairs a range of immune functions and increases susceptibility to bacterial and parasitic challenges (see Prohaska and Failla, 1993; Failla and Hopkins, 1998). Human studies show that subjects on a low-copper diet have decreased lymphocyte proliferation and IL-2 production, while copper administration reverses these effects (Kelley et al., 1995). As with many other micronutrients, excess copper can be immunosuppressive.

Iron

Iron deficiency has multiple effects on immune function in laboratory animals and humans (see Sherman and Spear, 1993). Iron-deficient individuals have normal phagocytic function but there is impaired ability to kill bacteria by neutrophils, probably as a result of an alteration in respiratory burst. Iron deficiency is associated with gastrointestinal and respiratory infections (see Calder and Jackson, 2000). Despite the suppressing effects of iron deficiency on immune responses, diseases of iron overload and excessive iron supplementation are also associated with increased risk of infection (see Calder and Jackson, 2000). The effects of iron administration might be, in part, because microorganisms require iron and so providing it may favour the pathogen. Indeed, it has been argued that the decline in circulating iron concentrations that accompanies infection is an attempt by the host to 'starve' the infectious agent of iron. However, iron overload also decreases the number of T lymphocytes, lymphocyte proliferation, IL-2 production, cytotoxic T cell and NK cell activity, and phagocytic activity of neutrophils (Sherman and Spear, 1993). The mechanism for the effect of iron overload is not clear, but it may relate to deposition of iron in lymphoid tissues affecting cell movement and function or to increased peroxidative damage, which is favoured by increased levels of free iron.

The situation regarding iron status and malaria is particularly complicated, since red blood cells host the parasite, so that host stores of iron and the invading pathogen coincide. This might explain the observations that malaria is more common in iron-replete than in iron-deficient individuals, and that the levels of malaria infection and the severity of the disease are increased by iron supplementation (see Calder and Jackson, 2000). Indeed, it could be argued that lowering the iron status might improve malaria outcome: this is borne out by the observation that iron chelation therapy enhances the clearance of parasites and accelerates the effect of anti-malarials (see Calder and Jackson, 2000).

Selenium

Selenium is found in high concentrations in the liver, spleen and lymph nodes. Selenium deficiency in laboratory animals decreases a range of immune functions and increases susceptibility to bacterial, viral, fungal and parasitic challenges (see Stabel and Spears, 1993; McKenzie et al., 1998). Selenium deficiency does not affect the ability of neutrophils or macrophages to engage in phagocytosis, but it does diminish the ability of the cells to kill microorganisms once they have been ingested (Stabel and Spears, 1993). Selenium supplementation studies in animals demonstrate that selenium increases antibody titres in response to immunization and antigen challenges, lymphocyte proliferation, IFN-γ production, NK cell activity,

DTH response and rejection of skin allografts, and decreases susceptibility to infections (see Stabel and Spears, 1993; McKenzie et al., 1998). In man, selenium deficiency results in decreased circulating IgG and IgM concentrations.

A fascinating relationship has been shown to exist between impaired host selenium status and susceptibility to viral infection. A mouse model of coxsackievirus B3-induced myocarditis was employed to show that selenium-deficient mice were more suceptible to the virus than selenium-supplemented mice (see Beck, 1999). Furthermore, a normally benign strain of the virus became virulent in selenium-deficient mice, and it was demonstrated that the increased virulence occurred as a result of specific alterations of the viral genome (see Beck, 1999).

The selenium-dependent glutathione peroxidases and thioredoxin reductases are important examples of antioxidant enzymes that are expressed to protect cells from the cytotoxic effects of free radical species released to destroy microorganisms. These enzymes are up-regulated during monocyte differentiation and are under the influence of vitamin D. Thioredoxin reductases reduce sulphydryl groups and are involved in mediating protein–protein interactions and protein–DNA interactions (e.g. DNA binding of transcription factors), representing a potentially important molecular link for the action of selenium in the immune system.

Micronutrient combinations

Several studies have examined the effects of micronutrient combinations on immune outcome. Penn et al. (1991) investigated the effects of supplementing healthy elderly individuals with vitamins A, C and E, and reported increased numbers of circulating T cells, increased numbers of helper cells, an increased helper to cytotoxic cell ratio and increased lymphocyte responses to mitogens. In another micronutrient combination study, healthy elderly subjects received placebo or a multivitamin/trace element supplement for 12 months. The multivitamin supplement increased T-cell numbers, NK cell activity and IL-2 production, and improved antibody responses (Chandra, 1992). There was also less infection-related illness than in the placebo group (Chandra, 1992).

Dietary Fat and Immune Function

Amount of dietary fat and immune function

Animal studies indicate that high-fat diets diminish lymphocyte proliferation and NK cell activity compared with low-fat diets (see Calder, 1998, 2001a,b). However, the precise effect depends upon the exact level of fat used in the high-fat diet and its source. A reduction in total dietary fat intake (from 40 to 25% of total energy) resulted in greatly enhanced human blood lymphocyte proliferation in response to mitogens (see Calder, 1998, 2001a,b). Human NK cell activity was significantly increased by a reduction in fat intake to <30% energy (see Calder, 1998, 2001a,b). IL-1 production by lipopolysaccharide-stimulated monocytes from elderly subjects is increased when fat intake is reduced from 36 to 27% of energy (see Calder, 1998). Taken together, these data suggest that a high-fat diet suppresses the activity of cellular components of both natural and cell-mediated immunity in humans.

Essential fatty acids and immune function

There are two main families of essential fatty acids, the n-6 (or ω-6) and the n-3 (or ω-3) families, which cannot be interconverted in animals. The simplest members of these families are linoleic acid (18:2n-6) and α-linolenic acid (18:3n-3). Plant tissues and plant oils tend to be rich sources of linoleic and α-linolenic acids. Once consumed in the diet, linoleic acid can be converted to arachidonic acid (20:4n-6) (Fig. 22.3). Using the same pathway, dietary α-linolenic acid can be converted into eicosapentaenoic acid (20:5n-3) and docosahexaenoic acid (22:6n-3) (Fig. 22.3). Eicosapentaenoic acid and docosahexaenoic acid are found in relatively high proportions in the tissues of oily fish and in fish oil preparations. In the absence of significant consumption of oily fish, α-linolenic acid is the major dietary n-3 fatty acid.

Feeding rats or mice diets deficient in both linoleic and α-linolenic acids results in decreased thymus and spleen weight, lymphocyte proliferation, neutrophil chemotaxis, macrophage-mediated cytotoxicity and DTH response compared with animals fed diets containing adequate amounts of

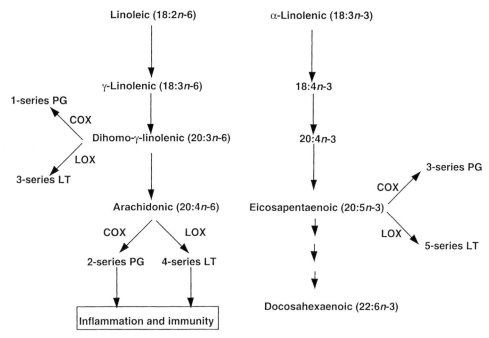

Fig. 22.3. The pathways of metabolism of n-6 and n-3 polyunsaturated fatty acids. COX, cyclooxygenase; LOX, lipoxygenase; LT, leukotriene; PG, prostaglandin.

these fatty acids (Kelley and Daudu, 1993). This is most probably because immune cells require polyunsaturated fatty acids for membrane synthesis and as precursors for synthesis of eicosanoids.

Eicosanoids: a link between fatty acids and the immune system

The key link between fatty acids and immune function is that the family of mediators termed eicosanoids are synthesized from fatty acids, in particular dihomo-γ-linolenic acid, arachidonic acid and eicosapentaenoic acid (Fig. 22.3). Since the membranes of most cells contain large amounts of arachidonic acid, compared with dihomo-γ-linolenic and eicosapentaenoic acids, arachidonic acid is usually the principal precursor for eicosanoid synthesis. Arachidonic acid in cell membranes can be released by various phospholipase enzymes, most notably phospholipase A_2, and the free arachidonic acid subsequently can act as a substrate for cyclooxygenase, forming prostaglandins and related compounds, or for one of the lipoxygenase enzymes, forming leukotrienes and related compounds. These compounds are involved in regulating inflammation and the functions of neutrophils, monocyte/macrophages, T cells and B cells. The capacity of inflammatory cells to produce arachidonic acid-derived eicosanoids can be decreased by increased availability of n-3 fatty acids, especially eicosapentaenoic acid from fish oil (Calder, 2001a,b,c). The reduction in generation of arachidonic acid-derived mediators that accompanies fish oil consumption has led to the idea that fish oil is anti-inflammatory. There is a significant amount of evidence supporting this notion; for example n-3 fatty acids influence the pattern of production of cytokines by monocyte/macrophages (Calder, 2001a,b,c). However, the effects on susceptibility to infection remain controversial, with some animal studies showing increased susceptibility to infection in animals that have been fed large amounts of fish oil and other studies showing decreased susceptibility (Calder, 2001a,b). Nevertheless, it is clear that fish oil protects against the deleterious effects of endotoxin (Calder, 2001a,b), and is a useful adjunct to existing therapies for inflammatory diseases, such as rheumatoid arthritis (Calder, 2001a,b,c).

Mechanisms of action

In addition to effects directly induced by decreasing production of eicosanoids from arachidonic acid, recent studies show that *n*-3 fatty acids can affect the expression of cytokine and other immune-related genes (e.g. Renier *et al.*, 1993; Curtis *et al.*, 2000; Miles *et al.*, 2000; Wallace *et al.*, 2001). This might be due to altered eicosanoid production or might be independent of this (see Miles and Calder, 1998; Yaqoob, 1998). For example, different fatty acids might interact with various transcription factors. A number of transcription factors have been identified as having pro-inflammatory functions. Some of these are cell specific, while others, such as NF-κB, are ubiquitous. NF-κB is important because it governs the expression of genes that encode cytokines, cell adhesion molecules, growth factors and some acute phase proteins, both in health and in chronic disease (Barnes and Karin, 1997). It is possible that fatty acids may influence the activation and DNA binding of NF-κB, either directly or indirectly. There is recent evidence for this mode of action of certain *n*-3 fatty acids (Lo *et al.*, 1999; Xi *et al.*, 2001).

Peroxisome proliferator-activated receptors (PPARs) are a more recently discovered class of transcription factors, and have been shown to be bound and activated by fatty acids and some of their metabolic derivatives. Two of the isoforms, PPARα and PPARγ, have been demonstrated to be present in cells of the immune system and suggested to have anti-inflammatory actions (see Gelman *et al.*, 1999). However, many of these effects have been demonstrated using synthetic activators of PPARs, and it appears that for PPARγ at least, some of the anti-inflammatory effects of these compounds are independent of the transcription factor itself (Thieringer *et al.*, 2000). Furthermore, the nature of the physiological ligands of PPAR is still unclear. Thus, while the idea of PPARs acting as molecular mediators for the anti-inflammatory effects of some fatty acids is attractive, the effects of PPARs in immune and inflammatory responses require clarification and the true molecular targets for fatty acids remain to be demonstrated. It is also pertinent to note that the effects of fatty acids on gene expression do not necessarily involve a direct action on a single transcription factor. Many transcription factors engage in cross-talk, resulting in transactivation or trans-repression, and this has been clearly demonstrated for PPARs, which engage in cross-talk with NF-κB and AP-1 (Delerive *et al.*, 1999).

Dietary Amino Acids and Related Compounds and Immune Function

Sulphur amino acids and glutathione

Sulphur amino acids are essential in humans. Deficiency in methionine and cysteine results in atrophy of the thymus, spleen and lymph nodes, and prevents recovery from protein energy malnutrition (see Gross and Newberne, 1980). When combined with a deficiency of isoleucine and valine, also essential amino acids, sulphur amino acid deficiency results in severe depletion of gut lymphoid tissue, very similar to the effect of protein deprivation (Gross and Newberne, 1980).

Glutathione is an antioxidant tripeptide that consists of glycine, cysteine and glutamate. Glutathione concentrations in the liver, lung, small intestine and immune cells fall in response to inflammatory stimuli, and this fall can be prevented in some organs by provision of cysteine in the diet (Hunter and Grimble, 1997). Glutathione itself can enhance the activity of human cytotoxic T cells, and depletion of intracellular glutathione diminishes lymphocyte proliferation and the generation of cytotoxic T lymphocytes (Droge *et al.*, 1994; Kinscherf *et al.*, 1994). Glutathione depletion is associated with diminished IFN-γ, but not IL-2 or IL-4, production by antigen-stimulated murine lymph node lymphocytes (Peterson *et al.*, 1998); this effect was mediated by antigen-presenting cells, and the authors suggest that glutathione acts via inducing IL-12 production by these cells to alter the Th1/Th2 balance in favour of a Th1 response.

Arginine

Arginine is a non-essential amino acid in humans and is involved in protein, urea and nucleotide synthesis and ATP generation. It is also is the precursor of nitric oxide, a potent immunoregulatory mediator, which is cytotoxic to tumour cells and to some microorganisms. Arginine is the precursor for synthesis of polyamines, which have a key role in DNA replication, regulation of the cell cycle and cell division. In laboratory animals, arginine was found to decrease the thymus involution associated with trauma, to promote thymus repopulation and cellularity, to increase lymphocyte proliferation, NK cell activity and macrophage cytotoxicity, to improve DTH, to increase resistance to bacterial infections and survival to sepsis and burns, and to promote wound healing and the rejection of skin allografts (see Redmond and Daly, 1993; Evoy et al., 1998). In healthy human subjects, arginine supplementation (30 g day^{-1}; consumption as part of a habitual Western diet is ~4 g day^{-1}) has been shown to increase blood lymphocyte proliferation in response to mitogens and to promote wound healing (Redmond and Daly, 1993; Evoy et al., 1998). There is particular interest in the inclusion of arginine in enteral formulae given to patients hospitalized for surgery, trauma and burns, since it appears to reduce the severity of infectious complications and the length of hospital stay (Evoy et al., 1998). However, in many of the clinical studies carried out in these patients, the enteral formulae have contained a variety of nutrients with potential immunomodulatory actions, so it has been difficult to ascribe observed effects to specific nutrients.

Glutamine

Glutamine is the most abundant amino acid in the blood and in the free amino acid pool in the body. Skeletal muscle is considered to be the most important glutamine producer in the body, and it supplies glutamine to act as an interorgan nitrogen transporter. The immune system is considered to be an important user of glutamine. Intramuscular and plasma glutamine concentrations are decreased in stress situations such as in sepsis and cancer cachexia and following burn injury (see Wilmore and Shabert, 1998; Calder and Yaqoob, 1999). These observations indicate that a significant depletion of the skeletal muscle glutamine pool is characteristic of trauma. The lowered plasma glutamine concentrations that occur are most probably the result of demand for glutamine (by the liver, kidney, gut and immune system) exceeding the supply, and it is proposed that glutamine be considered a conditionally essential amino acid during stress. It has been suggested that the lowered plasma glutamine contributes, at least in part, to the impaired immune function that accompanies such situations.

Animal studies have reported that enrichment of the diet with glutamine increases T-lymphocyte proliferation and IL-2 production, and increases the ability of rodents to survive infectious challenges (Wilmore and Shabert, 1998; Calder and Yaqoob, 1999). Studies in various patient groups (bone marrow transplant patients, very low birthweight babies, patients in intensive care) have revealed significant clinical benefits from glutamine, including decreased rates of infection and shorter hospital stay (Wilmore and Shabert, 1998; Calder and Yaqoob, 1999); in some of these studies, improved outcome was associated with improved immune function (others did not monitor immune function). In addition to this direct immunological effect, glutamine, even provided parenterally, improves gut barrier function in patients at risk of infection. This would have the benefit of decreasing the translocation of bacteria from the gut and eliminating a key source of infection.

Nucleotides

Nucleotides have a wide range of cellular functions, including roles in DNA and RNA structure, energy metabolism, signal transduction, biosynthesis of phospholipids and regulation of enzyme activity. Nucleotides can be synthesized *de novo*, largely from amino acids; in addition, all diets contain nucleic acids and nucleotides. Activation of lymphocytes causes a rapid increase in the synthesis of nucleotides, which are required immediately for the increase in metabolism to provide energy, and later for synthesis of nucleic acids (RNA for protein synthesis and DNA for cell division). Thus, exogenous nucleotides (or their precursors) supplied by the diet potentially could increase immunity by contributing to the pool of nucleotides available to lymphocytes, and other leukocytes, and so sparing the need for complete *de novo* synthesis.

A nucleotide-free diet impairs murine immune responses, whereas addition of RNA or nucleotides to the diet of mice enhances T-cell-dependent antibody responses, lymphocyte proliferation, IL-2 production, DTH and survival to bacterial and fungal challenges (see Boza, 1998). Nucleotides have often been included in enteral and parenteral feeds for hospitalized patients, but, as with the amino acids, it is difficult to ascribe specific effects to single components in these preparations. There has been considerable interest in supplementation of infant formulae with nucleotides, since human breast milk is very rich in them. However, there is little evidence that such fortification produces benefits in terms of resistance to infection in newborn infants (Boza, 1998).

Concluding Comments

Deficiencies of total energy or of one or more essential nutrients, including vitamins A, B_6, B_{12}, C, E, folic acid, zinc, iron, copper, selenium, essential amino acids and essential fatty acids, impair immune function and increase susceptibility to infectious pathogens. This is most probably because these nutrients are involved in the molecular and cellular responses to challenge of the immune system. Providing these nutrients to deficient individuals restores immune function and improves resistance to infection. For several nutrients, the dietary intakes that result in greatest enhancement of immune function are greater than recommended intakes. However, excess intake of some nutrients also impairs immune responses. Thus, four potential general relationships between the intake of a nutrient and immune function appear to exist (Fig. 22.4). These different types of relationship might, in part, reflect interactions between nutrients such that an excess of one nutrient negatively affects the status of a second nutrient (e.g. zinc and copper). It is often assumed when defining the relationship between nutrient intake and immune function that all components of the immune system will respond in the same dose-dependent fashion to a given nutrient. This is not correct, at least as far as some nutrients are concerned, and it appears likely that different components of the immune system show an individual dose–response relationship to the availability of a given nutrient.

One aspect not touched upon in this overview is the role of hormones in regulating immune function. An inadequate supply of nutrients to the body may cause physiological stress leading to elevations in the circulating concentrations

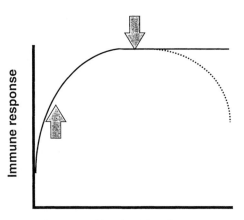

Fig. 22.4. The potential relationships between intake of a nutrient and the immune response. It is assumed that low or deficient intake of a particular nutrient impairs the immune response, so that increasing intake of that nutrient increases the response. Two potential relationships result: intake significantly above that at which the maximal immune response occurs may not result in further change in the response, or intake significantly above that at which the maximal immune response occurs may impair the response. The habitual (or recommended) intake of the nutrient is indicated by the arrows. Thus, the greatest immune response could occur at an intake above the habitual (or recommended) intake or may occur at an intake that coincides with the habitual (or recommended) intake.

of glucocorticoids and catecholamines. Both these classes of hormones have an inhibitory effect on immune function, and so may be important factors when considering the relationship between nutrient supply and immunological outcome.

Finally, while it is clear that the roles for nutrients in immune function are many and varied and it is easy to appreciate that an adequate and balanced supply of these is essential if an appropriate immune response is to be mounted, our understanding of the molecular basis for the action of most nutrients on immune function is incomplete.

References

Alroy, I., Towers, T.L. and Freedman, L.P. (1995) Transcriptional repression of the interleukin-2 gene by vitamin D3: direct inhibition of NFAT/AP-1 complex formation by a nuclear hormone receptor. *Molecular and Cellular Biology* 15, 5789–5799.

Barnes, P.J. and Karin, M. (1997) Nuclear factor-κB: pivotal transcription factor in chronic inflammatory diseases. *New England Journal of Medicine* 336, 1066–1071.

Beck, M.A. (1999) Selenium and host defence towards viruses. *Proceedings of the Nutrition Society* 58, 707–711.

Bendich, A. (1993) Vitamin E and human immune functions. In: Klurfeld, D.M. (ed.) *Nutrition and Immunology*. Plenum Press, New York, pp. 217–228.

Boza, J. (1998) Nucleotides in infant nutrition. *Monatsschrift für Kinderheilkunde* 98, S39–S48.

Calder, P.C. (1998) Dietary fatty acids and the immune system. *Nutrition Reviews* 56, S70–S83.

Calder, P.C. (2001a) The effect of dietary fatty acids on the immune response and susceptibility to infection. In: Suskind, R. and Tontisirin, K (eds) *Nutrition, Immunity and Infectious Diseases in Infants and Children*. Karger, Basel, pp. 137–172.

Calder, P.C. (2001b) Polyunsaturated fatty acids, inflammation and immunity. *Lipids* 36, 1007–1024.

Calder, P.C. (2001c) *N*-3 polyunsaturated fatty acids, inflammation and immunity: pouring oil on troubled waters or another fishy tale? *Nutrition Research* 21, 309–341.

Calder, P.C. (2002) Dietary modification of inflamation with lipids. *Proceedings of the Nutrition Society* 61, 345–358.

Calder, P.C. and Jackson, A.A. (2000) Undernutrition, infection and immune function. *Nutrition Research Reviews* 13, 3–29.

Calder, P.C. and Yaqoob, P. (1999) Glutamine and the immune system. *Amino Acids* 17, 227–241.

Cantorna, M.T., Hayes, C.E. and de Luca, H.F. (1996) 1,25-Dihydroxyvitamin D3 reversibly blocks the progression of relapsing encephalomyelitis, a model of multiple sclerosis. *Proceedings of the National Academy of Sciences USA* 93, 7861–7864.

Cantorna, M.T., Hayes, C.E. and de Luca, H.F. (1998) 1,25-Dihydrocholecalciferol inhibits the progression of arthritis in murine models of human arthritis. *Journal of Nutrition* 128, 68–72.

Casteels, K., Waer, M., Laureys, J., Valckx, D., Depovere, J., Bouillon, R. and Mathieu, C. (1998a) Prevention of autoimmune destruction of syngeneic islet grafts in spontaneously diabetic nonobese mice by a combination of a vitamin D3 analog and cyclosporin. *Transplantation* 65, 1225–1232.

Casteels, K., Bouillo, R., Waer, M., Valckx, D., Overbergh, L., Laureys, J. and Mathieu C. (1998b) Prevention of type I diabetes by late intervention with non-hypercalcaemic analogues of vitamin D3 in combination with cyclosporin A. *Endocrinology* 139, 95–102.

Chandra, R.K. (1984) Excessive intake of zinc impairs immune responses. *Journal of the American Medical Association* 252, 1443–1446.

Chandra, R.K. (1991) 1990 McCollum Award Lecture. Nutrition and immunity: lessons from the past and new insights into the future. *American Journal of Clinical Nutrition* 53, 1087–1101.

Chandra, R.K. (1992) Effect of vitamin and trace-element supplementation on immune responses and infection in elderly patients. *Lancet* 340, 1124–1127.

Cippitelli, M. and Santoni, A. (1998) Vitamin D3: a transcriptional modulator of the interferon-γ gene. *European Journal of Immunology* 28, 3017–3030.

Curtis, C.L., Hughes, C.E., Flannery, C.R., Little, C.B., Harwood, J.L. and Caterson, B. (2000) *n*-3 Fatty acids specifically modulate catabolic factors involved in articular cartilage degradation. *Journal of Biological Chemistry* 275, 721–724.

D'Ambrosio, D., Cippitelli, M., Cocciolo, M.G., Mazzeo, D., Di Lucia, P., Lang, R., Sinigaglia, F. and Panina-Bordignon, P. (1998) Inhibition of IL-12 production by 1,25-dihydroxyvitamin D3. Involvement of NFκB downregulation in transcriptional repression of the p40 gene. *Journal of Clinical Investigation* 101, 252–262.

Delerive, P., De Bosscher, K., Besnard, S., Vanden Berghe, W., Peters, J.M., Gonzalez, F.J., Fruchart, J.-C., Tedgui, Alain, Haegman, G. and Staels, B. (1999) Peroxisome proliferator-activated receptor α negatively regulates the vascular inflammatory gene response by negative cross talk with transcription factors NFκB and AP-1. *Journal of Biological Chemistry* 274, 32048–32054.

Droge, W., Schulzeosthoff, K., Mihm, S., Galter, D., Schenk, H., Eck, H.P., Roth, S. and Gmunder, H. (1994) Functions of glutathione and glutathione disulfide immunology and immunopathology. *FASEB Journal* 8, 1131–1138.

Evoy, D., Lieberman, M.D., Fahey, T.J. and Daly, J.M. (1998) Immunonutrition: the role of arginine. *Nutrition* 14, 611–617.

Failla, M.L. and Hopkins, R.G. (1998) Is low copper status immunosuppressive? *Nutrition Reviews* 56, S59–S64.

Fraker, P. and King, L. (1998) Changes in regulation of lymphopoiesis and myelopoiesis in the zinc-deficient mouse. *Nutrition Reviews* 56, S65–S69.

Fraker, P.J., King, L.E., Garvy, B.A. and Medina, C.A. (1993) The immunopathology of zinc deficiency in humans and rodents: a possible role for programmed cell death. In: Klurfeld, D.M. (ed.) *Nutrition and Immunology*. Plenum Press, New York, pp. 267–283.

Friedman, A. and Sklan, D. (1993) Vitamin A and immunity. In: Klurfeld, D.M. (ed.) *Nutrition and Immunology*. Plenum Press, New York, pp. 197–216.

Gelman, L., Fruchart, J.-C. and Auwerx, J. (1999) An update on the mechanisms of action of the peroxisome proliferator-activated receptors (PPARs) and their roles in inflammation and cancer. *Cellular and Molecular Life Science* 55, 932–943.

Gross, R.L. and Newberne, P.M. (1980) Role of nutrition in immunologic function. *Physiological Reviews* 60, 188–302.

Haag, M. (1999) Vitamin D – new action mechanisms and effects. *South African Medical Journal* 89, 1195–1199.

Han, S.N. and Meydani, S.N. (1999) Vitamin E and infectious disease in the aged. *Proceedings of the Nutrition Society* 58, 697–705.

Hayek, M.G., Taylor, S.F., Bender, B.S., Han, S.N., Meydani, M., Smith, D.E., Eghtesada, S. and Meydani, S.N. (1997) Vitamin E supplementation decreases lung virus titers in mice infected with influenza. *Journal of Infectious Diseases* 176, 273–276.

Hunter, E.A.L. and Grimble, R.F. (1997) Dietary sulphur amino acid adequacy influences glutathione synthesis and glutathione-dependent enzymes during the inflammatory response to endotoxin and tumour necrosis factor-α in rats. *Clinical Science* 92, 297–305.

Jacob, R.A., Kelley, D.S., Pianalto, F.S., Swendseid, M.E., Henning, S.M., Zhang, J.Z., Ames, B.N., Fraga, C.G. and Peters, J.H. (1991) Immunocompetence and oxidant defense during ascorbate depletion in healthy men. *American Journal of Clinical Nutrition* 54, 1302S–1309S.

Kelley, D.S. and Daudu, P.A. (1993) Fat intake and immune response. *Progress in Food and Nutritional Science* 17, 41–63.

Kelley, D.S., Daudu, P.A., Taylor, P.C., Mackey, B.E. and Turnlund, J.R. (1995) Effects of low-copper diets on human immune response. *American Journal of Clinical Nutrition* 62, 412–416.

Kinscherf, R., Fischbach, T., Mihm, S., Roth, S., Hohenhaus-Sievert, E., Weiss, C., Edler, L., Bartsch, P. and Droge, W. (1994) Effect of glutathione depletion and oral N-acetyl-cysteine treatment on CD4$^+$ and CD8$^+$ cells. *FASEB Journal* 8, 448–451.

Lemire, J.M. (1992) Immunomodulatory role of 1.25-dihydroxyvitamin D$_3$. *Journal of Cellular Biochemistry* 49, 26–31.

Lo, C.J., Chiu, K.C., Fu, M., Lo, R. and Helton, S. (1999) Fish oil decreases macrophage tumour necrosis factor gene transcription by altering the NFkappaB activity. *Journal of Surgical Research* 82, 216–221.

MacDonald, P.N., Dowd, D.R. and Hayssler, M.R. (1994) New insight into the structure and functions of the vitamin D receptor. *Seminars in Nephrology* 14, 101–118.

Mathieu, C., Laureys, J., Sobis, H., van de Putte, M., Waer, M. and Bouillon, R. (1992) 1,25 Dihydroxyvitamin D3 prevents insulitis in NOD mice. *Diabetes* 41, 1491–1495.

McKenzie, R.C., Rafferty, T.S. and Beckett, G.J. (1998) Selenium: an essential element for immune function. *Immunology Today* 19, 342–345.

Meydani, S.N. and Beharka, AA. (1998) Recent developments in vitamin E and immune response. *Nutrition Reviews* 56, S49–S58.

Meydani, S.N., Ribaya-Mercado, J.D., Russell, R.M., Sahyoun, N., Morrow, F.D. and Gershoff, S.N. (1991) Vitamin B6 deficiency impairs interleukin-2 production and lymphocyte proliferation in elderly adults. *American Journal of Clinical Nutrition* 53, 1275–1280.

Meydani, S.N., Meydani, M., Blumberg, J.B., Leka, L.S., Siber, G., Loszewski, R., Thompson, C., Pedrosa, M.C., Diamond, R.D. and Stollar, B.D. (1997) Vitamin E supplememtation and *in vivo* immune response in healthy subjects. *Journal of the American Medical Association* 277, 1380–1386.

Miles, E.A. and Calder, P.C. (1998) Modulation of immune function by dietary fatty acids. *Proceedings of the Nutrition Society* 57, 277–292.

Miles, E.A., Wallace, F.A. and Calder, P.C. (2000) Dietary fish oil reduces intercellular adhesion molecule 1 and scavenger receptor expression on murine macrophages. *Atherosclerosis* 152, 43–50.

Nikawa, T., Odahara, K., Koizumi, H., Kido, Y., Teshima, S., Rokutan, K. and Kishi, K. (1999) Vitamin A prevents the decline in immunoglobulin A and Th2 cytokine levels in small intestine mucosa of protein-malnourished mice. *Journal of Nutrition* 129, 934–941.

Overbergh, L., Decallone, B., Valckx, D., Verstuyf, A., Depovere, J., Laureys, J., Rutgeerts, O., Saint-Arnaud, R., Bouillon, R. and Mathieu, C. (2000) Identification and immune regulation of 25-hydroxyvitamin D-1-α-hydroxylase in murine macrophages. *Clinical and Experimental Immunology* 120, 139–146.

Penn, N.D., Purkins, I. and Kelleher, J. (1991) The effects of dietary supplementation with vitamins A, C and E on cell-mediated immune function in elderly long-stay patients: a randomised controlled study. *Age and Ageing* 20, 169–174.

Peterson, J.D., Herzenberg, L.A., Vasquez, K. and Waltenbaugh, C. (1998) Glutathione levels in antigen-presenting cells modulate Th1 versus Th2 response patterns. *Proceedings of the National Academy of Sciences USA* 95, 3071–3076.

Prasad, A.S., Bao, B., Beck, F.W.J. and Sarkar, F.H. (2001) Zinc activates NF-κB in HUT-78 cells. *Journal of Laboratory and Clinical Medicine* 138, 250–256.

Prohaska, J.R. and Failla, M.L. (1993) Copper and immunity. In: Klurfeld, D.M. (ed.) *Nutrition and Immunology*. Plenum Press, New York, pp. 309–332.

Redmond, H.P. and Daly, J.M. (1993) Arginine. In: Klurfeld, D.M. (ed.) *Nutrition and Immunology*. Plenum Press, New York, pp. 157–166.

Renier, G., Skamene, E., de Sanctis, J. and Radzioch, D. (1993) Dietary n-3 polyunsaturated fatty acids prevent the development of atherosclerotic lesions in mice: modulation of macrophage secretory activities. *Arteriosclerosis and Thombosis* 13, 1515–1524.

Rivera, J., Habicht, J.-P., Torres, N., Cossio, T., Utermohlen, V., Tovar, A., Robson, D.S. and Bourges, H. (1986) Decreased cellular immune response in wasted but not in stunted children. *Nutrition Research* 6, 1161–1170.

Scrimshaw, N.S. and SanGiovanni, J.P. (1997) Synergism of nutrition, infection and immunity: an overview. *American Journal of Clinical Nutrition* 66, 464S–477S.

Semba, R.D. (1998) The role of vitamin A and related retinoids in immmune function. *Nutrition Reviews* 56, S38–S48.

Semba, R.D. (1999) Vitamin A and immunity to viral, bacterial and protozoan infections. *Proceedings of the Nutrition Society* 58, 719–727.

Shankar, A.H. and Prasad, A.S. (1998) Zinc and immune function: the biological basis of altered resistance to infection. *American Journal of Clinical Nutrition* 68, 447S–463S.

Sherman, A.R. and Spear, A.T. (1993) Iron and immunity. In: Klurfeld, D.M. (ed.) *Nutrition and Immunology*. Plenum Press, New York, pp. 285–307.

Siegel, B.V. (1993) Vitamin C and the immune response in health and disease. In: Klurfeld, D.M. (ed.) *Nutrition and Immunology*. Plenum Press, New York, pp. 167–196.

Singh, S., Aiba, S., Manome, H. and Tagami, H. (1999) The effects of dexamethasone, cyclosporine, and vitamin D3 on the activation of dendritic cells stimulated by haptens. *Archives of Dermatological Disease* 291, 548–554.

Stabel, J.R. and Spears, J.W. (1993) Role of selenium in immune responsiveness and disease resistance. In: Klurfeld, D.M. (ed.) *Nutrition and Immunology*. Plenum Press, New York, pp. 333–356.

Takeuchi, A., Reddy, G.S., Kobayashi, T., Okani, T., Park, J. and Sharma, S. (1998) Nuclear factor of activated T cells (NFAT) as a molecular target for 1,25-dihydroxyvitamin D3-mediated effects. *Journal of Immunology* 160, 209–218.

Thieringer, R., Fenyk-Melody, J.E., Le Grand C.B., Shelton, B.A., Detmers, P.A., Somers, E.P., Carbin, L., Moller, D.E., Wright, S.D. and Berger, J. (2000) Activation of peroxisome proliferator-activated receptor γ does not inhibit IL-6 or TNF-α responses of macrophages to lipopolysaccharide *in vitro* or *in vivo*. *Journal of Immunology* 164, 1046–1054.

Wallace, F.A., Miles, E.A., Evans, C., Stock, T.E., Yaqoob, P. and Calder, P.C. (2001) Dietary fatty acids influence the production of Th1- but not Th2-type cytokines. *Journal of Leukocyte Biology* 69, 449–457.

Wang, Y., Huang, D.S., Eskelson, C.D. and Watson, R.R (1994) Long-term dietary vitamin E retards development of retrovirus-induced dysregulation in cytokine production. *Clinical Immunology and Immunopathology* 72, 70–75.

Wilmore, D.W. and Shabert, J.K. (1998) Role of glutamine in immunologic responses. *Nutrition* 14, 618–626.

Woodward, B. (1998) Protein, calories and immune defences. *Nutrition Reviews* 56, S84–S92.

Xi, S., Cohen, D., Barve, S. and Chen, L.H. (2001) Fish oil suppressed cytokines and nuclear factor kappaB induced by murine AIDS virus infection. *Nutrition Research* 21, 865–878.

Yaqoob, P. (1998) Lipids and the immune response. *Current Opinion in Clinical Nutrition and Metabolic Care* 1, 153–161.

Yu, X., Bellido, T. and Manolagas, S.C. (1995) Down-regulation of NFκB protein levels in activated human lymphocytes by 1,25-dihydroxyvitamin D3. *Proceedings of the National Academy of Sciences USA* 92, 10990–10994.

23 Molecular Mechanisms of Food Allergy

J. Steven Stanley[1] and Gary A. Bannon[2]*
[1]Departments of Pediatrics, and [2]Biochemistry Molecular Biology, Arkansas Children's Hospital Research Institute and the University of Arkansas for Medical Sciences, Little Rock, Arkansas, USA

Prevalence

A consumer survey regarding the prevalence of adverse reactions to foods indicated that 30% of the people interviewed felt that they or some family member had an allergy to a food product (Anderson and Sogn, 1984). This survey also found that 22% avoided particular foods on the mere possibility that the food may contain an allergen. In reality, food allergic reactions affect only 6–8% of children and 1–2% of the adult population (Burks and Sampson, 1993; Jansen et al., 1994) and are elicited by a small subset of the food we consume. The incidence of all allergic diseases appears to be on the increase in industrialized societies (Hanson and Telemo, 1997). The most common food allergies known to affect children are immunoglobulin E (IgE)-mediated reactions to cow's milk, eggs, groundnuts, soybeans, wheat, fish and tree nuts. Approximately 80% of all reported food allergies in children are due to groundnuts, milk or eggs. While most childhood food allergies are outgrown, allergies to groundnuts, tree nuts and fish are rarely resolved in adulthood. In adults, the most common food allergies are to groundnuts, tree nuts, fish and shellfish. The incidence of IgE-mediated reactions to specific food crops is increasing, particularly in developed countries, probably due to increased levels of protein consumption. Allergic reactions typically are elicited by a defined subset of proteins that are found in abundance in the food.

Immunological Basis of Allergic Sensitization and Elicitation

The human gastrointestinal (GI) tract processes a wide variety of foodstuffs into nutrients that can then be transferred efficiently across the epithelium. Mucus membranes of the GI tract also provide a route of entry for a number of harmful organisms. The gut-associated lymphoid tissues are the specialized branch of the immune system responsible for distinguishing between potential pathogens and innocuous dietary nutrients, self-proteins and gut flora. The majority of food proteins trigger oral tolerance, while pathogens are targeted for attack and elimination. Immune defences against GI pathogens include humoral and cell-mediated immune responses, increased mucus secretions, IgA production, vomiting and diarrhoea. Oral tolerance, on the other hand, is the systemic, immunological unresponsiveness to an antigen produced when the antigen is ingested. It is important to note that immune tolerance is an active process that begins with immune activation and ends with either active suppression or immune anergy/deletion. Thus food allergy can be viewed as a breakdown of immune tolerance.

Most allergic reactions to food proteins are mediated by IgE. The immune process that results in the production of IgE is dependent on CD4$^+$ T cells. The delineation of CD4$^+$ T cells into the Th1 and Th2 subgroups has become widely accepted.

---

* Present address: Product Safety Center, Monsanto Company, St Louis, Missouri, USA.

Although originating from the same progenitor cells, the Th1 and Th2 cell types represent opposing ends of a spectrum. The secretion of interleukin (IL)-4, IL-5, IL-10, IL-13 and their responsiveness to inteferon (IFN)-γ and unresponsiveness to IL-12 distinguishes Th2 cells. Th1 cells can be distinguished by their secretion of IFN-γ, IL-2, tumour necrosis factor-β (TNF-β), their responsiveness to IL-12 and unresponsiveness to IFN-γ. Th1 cells regulate cell-mediated immune responses. Th1 and Th2 cells not only promote very different types of immune responses but actively oppose the immune response of the other. The process of immune deviation, in which the immune system responds to a pathogen with either a Th1 response or a Th2 response, but not both, is a result of this opposition. Th2 cells promote humoral immune responses to extracellular pathogens, and under certain conditions will drive the immune system into an atopic response.

The end result of the atopic immune response is the production of allergen-specific IgE antibodies. IgE antibodies mediate food allergy by binding to high-affinity IgE receptors (FcεRI) on mast cells and basophils. Cross-linking of these FcεRI receptors by the allergen results in the release of inflammatory mediators. The release of these mediators triggers the pruritis, urticaria, vomiting, diarrhoea and anaphylaxis that are symptomatic of an allergic reaction. It is also the specific IgE antibody that is required for the common diagnostic allergy tests, such as the skin prick test and the *in vitro* assays for identifying the allergen source. The high affinity of the FcεRI receptor for IgE ensures that the vast majority of FcεRI receptors on mast cells and basophils are occupied throughout the body, even with relatively low titres of circulating IgE.

The requirement that IgE be cross-linked on effector cells in order to release the mediators of allergic symptoms dictates that there be at least two high affinity IgE-binding epitopes on a single allergen. Since food allergens encounter a variety of proteases as they traverse the GI tract, a peptide size below which clinical symptoms cannot be elicited could theoretically be calculated. In order to evaluate the minimum peptide size that might effectively cross-link receptors on mast cells, various assumptions must be made regarding epitope size and peptide conformation. The first assumption regards the size of a typical IgE-binding epitope observed in a food allergen. Most food allergen IgE-binding epitopes are reported to range in size from six to 15 amino acids in length (Stanley and Bannon, 1999). Therefore, the absolute minimum size of a peptide would have to be 12–30 amino acids long and contain two IgE-binding epitopes. However, this does not take into account the data of Kane *et al.* (1986, 1988) that show that the IgE-binding epitopes must be at least 80–240 Å apart to provide optimum degranulation. Assuming the two IgE-binding epitopes are separated by the minimum length of 80 Å and that the diameter size for an amino acid such as alanine is 5 Å, the minimum size for a peptide that would be expected to elicit the clinical symptoms of an allergic reaction would be 29 amino acids long or a peptide of about 3190 Da (29 (amino acids) × 110 (average molecular weight of amino acids)). These calculations do not take into account the secondary structure of the peptide. For example, the peptide could be in an α-helical arrangement, a β-pleated sheet or a random coil, dependent on its amino acid sequence. Dependent on the secondary structure of the peptide, mast cell degranulation would only be possible if each end of the fragment represents a strong IgE-binding epitope and if the peptide is in a β-strand conformation. Based on this rationale, it appears improbable that the presence of a protease-resistant fragment of <3 kDa would have the ability to degranulate mast cells.

Clinical Manifestations

Allergic reactions to a food usually will become evident after a few minutes, but may occur any time up to 2 h after ingesting the food. The symptoms will depend on which part of the body is affected (Sampson, 1997). An itching or tingling of the lips, tongue, palate or throat may be the first symptom. These sensations can develop very quickly and may be followed by swelling of the lips and tongue, a feeling of tightness in the throat, and a dry cough and hoarseness. Nausea, abdominal cramps, vomiting and/or diarrhoea can develop as the allergen moves into the stomach and intestines. Itching, tearing and nasal congestion associated with rhinoconjunctivitis can occur in those cases where the allergen is carried to the eyes and nose. Involvement of the lungs can trigger asthma, resulting in tightness of the chest, wheezing and shortness of breath (Quirce *et al.*, 1997). The skin is affected commonly, resulting in urticaria, angio-oedema, eczema and atopic dermatitis (Burks *et al.*, 1998). The most dangerous outcome of an allergic reaction is systemic anaphylaxis, in which mast cells and basophils release histamine, prostaglandins and leukotrienes that produce vasodilation, vascular permeability and smooth muscle contraction. A typical anaphylactic reaction could include any of the symptoms described above in addition to hypotension, tachycardia, cyanosis, difficulty in breathing and shock. Systemic anaphylaxis can develop rapidly, starting a few minutes to a few hours after eating the offending food. Groundnuts, tree nuts, fish and shellfish are the foods most likely to cause systemic anaphylaxis. Fifty milligrams of food is generally required to produce an allergic reaction, but in some extremely allergic individuals as little as 2 mg can produce symptoms (Hourihane *et al.*, 1997).

Diagnosis

Double-blind placebo-controlled food challenges, in which patients are actually given the suspected allergen or a placebo until clinical symptoms are evident, are the only reliable method for determining whether or not an individual is allergic to a specific food (Sicherer *et al.*, 2000). However, because of the time, expense and danger to the patient they are not practical in testing large numbers of suspected food allergens. Skin prick tests are the most common method for testing an individual's reactivity to a suspected allergen. Skin prick testing is an excellent negative predictor of allergenicity, but is only 50–60% predictive if a positive result is obtained (Hill *et al.*, 2001). Cross-reactivity between similar proteins from other foods is responsible for most of the false-positive results. Radioallergosorbent tests (RASTs) and other similar *in vitro* assays for identifying allergen-specific IgE antibodies are similar to prick skin tests in their accuracy. Therefore, they are useful in eliminating suspected foods if a negative result is obtained. However, their predictive value for identifying foods that a patient will actually develop clinical symptoms to is very limited.

General Characteristics of Food Allergens

A number of allergens have been identified that stimulate IgE production and cause allergic disease in man. The information gained from the study of food allergens from a wide variety of sources has revealed some general traits. Most allergens are glycoproteins with acidic isoelectric points that are highly abundant in the allergen source. These molecules usually are resistant to proteases, heat and denaturants, allowing these proteins to resist degradation during food preparation and digestion (Metcalf, 1985). These properties may be important in allowing rapid penetration of the allergen at mucosal membranes, facilitating the initial sensitization and later the rapid production of symptoms observed in allergic patients. Despite our increasing knowledge of the primary sequences and structure of allergens, features that can be identified as required for allergic sensitization have not yet been determined (Anderson and Sogn, 1984).

Two categories of IgE-binding epitopes, linear and conformational, are generally accepted to occur in food allergens. Conformational epitopes occur when both secondary and tertiary structure of the allergen are required before IgE will bind. In contrast, linear epitopes only require the primary amino acid sequence of the allergen for IgE to bind. While conformational IgE-binding epitopes are prevalent and important to the aetiology of aeroallergen-mediated allergic reactions, linear epitopes are important to food allergens mainly because the immune system will encounter them only after they have been partially denatured and digested by the human GI tract. Therefore, the linear IgE-binding epitopes of food allergens have garnered more attention than the less prevalent conformational epitopes.

Identification of Allergens from the Most Common Allergenic Foods

Groundnut allergens

The groundnut (*Arachis hypogaea* L.) is an annual plant native to South America that belongs to the family Leguminosae (legumes). The Virginia, Spanish and runner are the most common and important commercial groundnut varieties. Various studies over the last several years have examined the nature and location of the allergens in groundnuts (Gillespie *et al.*, 1976; Bush *et al.*, 1989). These studies verified that the allergenic proteins of the groundnut are stored in the cotyledon and do not differ significantly between the different varieties of groundnuts (Taylor et al., 1981).

Numerous groundnut proteins have been identified as allergens by their ability to bind IgE from groundnut allergic patients. The first major groundnut allergen identified in this manner was Ara h 1 (Burks et al., 1991). The Ara h 1 protein has a mean molecular weight of 63.5 kDa on SDS–polyacrylamide gels and an isoelectric point of 4.55. The gene for this allergen recently has been cloned and sequenced and found to have significant sequence homology with the plant 7S globulin proteins (Burks et al., 1995). Ara h 1 is known to be glycosylated and has one asparagine-linked consensus carbohydrate addition site (Burks et al., 1995; van Rhee et al., 2000). More than 90% of patients with positive challenges to groundnut have specific IgE to Ara h 1 and based on IgE recognition, it is considered one of the major allergens of groundnut. The second allergen isolated and cloned was Ara h 2 (Burks et al., 1992). Ara h 2 has a mean molecular weight of 17.5 kDa and a pI of 5.2, and has significant sequence homology with the plant conglutins (Stanley et al., 1997; Viquez et al., 2001). A third groundnut allergen (Ara h 3) was identified by using soy-adsorbed serum IgE from groundnut allergic patients (Eigenman et al., 1996), an approach used to identify proteins specific to groundnut allergy. The deduced amino acid sequence of Ara h 3 was found to be homologous to the 11S family of seed storage proteins. The recombinant form of this allergen was expressed in a bacterial system and was recognized by serum IgE from approximately 50% of a groundnut allergic patient population (Rabjohn et al., 1999). Four other proteins have been identified as groundnut allergens and designated Ara h 4–7 (Kleber-Janke et al., 1999). With the exception of Ara h 5, they all share significant homology with either Ara h 1, 2 or 3 (Kleber-Janke et al., 1999). Ara h 5 is a member of the profilin family, but is only recognized by IgE from a small fraction (13%) of the groundnut allergic population (Kleber-Janke et al., 1999).

Egg allergens

In the US paediatric population, allergy to chicken egg is one of the most common food allergies (Bock et al., 1988; Bock and Atkins, 1989; Sampson et al., 1992). In a study of 470 children evaluated for possible food allergies by double-blind placebo-controlled food challenges, egg proteins were responsible for as many as two-thirds of the positive food challenges (Sampson, 1997). Ovomucoid (Gal d 1), ovalbumin (Gal d 2), ovotransferrin (Gal d 3) and lysozyme (Gal d 4) are the most important allergens in egg white protein (Langeland, 1982; Holen and Elsyed, 1990; Hansen et al., 1997). All four allergens are glycoproteins, and have been cloned and sequenced. Ovomucoid makes up approximately 10% of the total egg white protein and is the major allergen (Bernhisel-Broadbent et al., 1994).

Ovomucoid is 186 amino acids in length, has a molecular weight of 28 kDa, an isoelectric point of 4.1, and has significant homology to pancreatic secretory trypsin inhibitor. The protein is arranged into three domains of about 60 amino acids. Two interdomain disulphide bridges secure the tertiary structure of each domain. The first two domains contain two glycosylation sites each, and the third domain contains a single site. Carbohydrate represents about 25% of the total glycoprotein's mass (Egge et al., 1983).

Ovotransferrin (Gal d 3) and lysozyme (Gal d 4) have not been studied to the same degree as ovomucoid and ovalbumin. Ovotransferrin is an approximately 70 kDa iron-binding glycoprotein with bactericidal activity in egg white and an isoelectric point between pH 5.6 and 6.2. Lysozyme is an approximately 14.3 kDa glycoprotein with an isoelectric point at pH 11 (Holen and Elsayed, 1990). The four allergens ovomucoid, ovalbumin, ovotransferrin and lysozyme account for approximately 80% of the protein in egg white.

Milk allergens

In addition to being one of the most allergenic foods, cow's milk is one of the first foods added to an infant's diet. Serum IgE antibodies have been used to identify α-casein (~32.4 kDa), β-casein (~26.6 kDa), α-lactalbumin (~14.2 kDa), β-lactoglobulin (~18 kDa), bovine serum albumin (~66 kDa) and bovine γ-globulin (~150 kDa) as milk allergens. However, α-casein (~32.4 kDa) and β-lactoglobulin (~26.6 kDa) appear to be the most common allergens for milk allergic individuals (Docena et al., 1996).

Fish and shellfish allergens

Allergic reactions to fish and shellfish are common causes of food allergy in both children and adults. Fish and shellfish allergies resemble groundnut allergy in that they are less likely to resolve naturally with age. Two allergens that have been studied extensively are Gad c 1 and Pen a 1. Gad c 1 is the parvalbumin homologue present in codfish. It is a glycoprotein approximately 12.5 kDa in size. Parvalbumin is present in most fish species examined and is generally the most IgE-reactive protein present (Hansen et al., 1997; James et al., 1997). As yet, the IgE-binding epitopes of Gad c 1 have not been mapped.

Tropomyosin (~36 kDa) is a major allergen identified in shrimp. Pen a 1 is the brown shrimp (*Penaeus aztecus*) homologue of tropomyosin, and Met e 1 is the tropomyosin homologue isolated from the greasyback shrimp (*Metapenaeus ensis*). Recently, tropomyosin was implicated as an allergen in lobster (Leung et al., 1998). Pan s 1 from the spiny lobster (*Panulirus stimpsoni*) and Hom a 1 from the American lobster (*Homarus americanus*) are both tropomyosin homologues. Both allergens have been cloned and sequenced and have significant homology to the shrimp allergen Pen a 1. Inhibition studies show that the two lobster allergens will completely inhibit each other's binding of IgE from lobster allergic individuals. In addition, the recombinant shrimp allergen Met e 1 is able to block IgE binding to lobster. This suggests that tropomyosin is also the major allergen in lobster and that many of the IgE-binding epitopes are cross-reactive (Leung et al., 1998).

Molecular Characterization of IgE-Binding Epitopes on Common Food Allergens

Groundnut allergens

The linear IgE-binding epitopes for the major groundnut allergens, Ara h 1, Ara h 2 and Ara h 3, were mapped using overlapping peptides and serum IgE from patients with documented groundnut hypersensitivity. Twenty-one different linear IgE-binding epitopes were identified throughout the length of the Ara h 1 molecule (Burks et al., 1997). Ten IgE-binding epitopes were identified in Ara h 2 and Ara h 3 using the same methods (Stanley et al., 1997; Rabjohn et al., 1999). The epitopes ranged in length from six to 15 amino acids, but there was no obvious sequence motif shared by all peptides. Four of the Ara h 1 epitopes appeared to be immunodominant IgE-binding peptides in that they were recognized by serum from >80% of the patients tested and bound more IgE than any of the other Ara h 1 IgE-binding epitopes. Similarly, three of the Ara h 2 IgE-binding epitopes and one of the Ara h 3 epitopes were determined to be immunodominant.

Each of the IgE-binding epitopes for the three major groundnut allergens was subjected to site-directed mutational analysis. The analysis revealed that single amino acid changes within these peptides had dramatic effects on IgE-binding characteristics. One or more amino acids within each epitope were found to be critical for IgE binding. Substitution of one of these critical amino acids led to the loss of a majority of IgE binding (Burks et al., 1997; Stanley et al., 1997; Shin et al., 1998). Analysis of the type and position of amino acids within the IgE-binding epitopes that had this effect indicated that substitution of hydrophobic residues in the centre of the epitopes were more likely to lead to loss of IgE binding (Shin et al., 1998). These results have been used to develop recombinant forms of these allergens for use in immunotherapy.

Ara h 1 forms a highly stable homotrimer that is held together by hydrophobic interactions. We found that the hydrophobic amino acids that contribute to trimer formation are located at the terminal ends of the three-dimensional structure where monomer–monomer contacts occur. Coincidentally, the IgE-binding epitopes are also located in this region. The stable nature of the Ara h 1 trimer and the location of the IgE-binding epitopes at sites of monomer–monomer contact indicate that these features may play a role in the overall allergenicity of this major groundnut allergen (Shin et al., 1998).

Egg allergens

IgE and IgG binding to ovomucoid is resistant to extensive heating, proteolytic digestion, deglycosylation and chemical denaturation (Matsuda et al., 1982; Honma et al., 1996; Cooke and Sampson, 1997). These results suggest that linear epitopes are important in antibody recognition of ovomucoid. The importance of linear and conformation epitopes for both IgE and IgG antibodies varies considerably from person to person. In general, conformational epitopes play a more significant role in individuals with newly acquired egg hypersensitivity, while linear epitopes were more important in individuals who had had their allergies for longer. Overlapping synthetic dodecapeptides representing the full-length protein were used to map the linear IgE- and IgG-binding epitopes of ovomucoid. Pooled serum IgE from seven egg allergic individuals recognized five regions of the ovomucoid protein, while pooled serum IgG bound to seven regions. Most of the IgE antibody binding sites also bound IgG.

The allergen ovalbumin (Gal d 2) is a 42.8 kDa glycoprotein that represents >50% of the protein in egg white. Much of the IgE binding to ovalbumin is resistant to enzymatic or chemical degradation and denaturation, indicating the importance of linear IgE epitopes (Kahlert et al., 1992; Honma et al., 1996). Using chemical and enzymatic digestion of the full-length protein and synthetic peptides based on the known protein sequence, at least seven IgE-binding regions have been identified by several different laboratories. The IgE epitopes identified so far cluster at the N- and C-terminal ends of the allergen (Honma et al., 1994).

Milk allergens

Bovine α_{S1}-casein is an acidic phosphoprotein (pI 4.1–4.5). Three aspects of its primary amino acid sequence dictate its tertiary structure. Nearly half of its 199 amino acids are hydrophobic, eight of its 16 serine residues may be phosphorylated and there are 17 proline residues located throughout its length. These factors, plus a lack of disulphide bridges, cause the protein to exhibit a linear conformation with reduced tertiary interactions (Kumosinski et al., 1991). This lack of higher-order structure in turn increases the likelihood that important IgE-binding epitopes are linear and not conformational. Two independent research groups recently have finished mapping the linear IgE- and IgG-binding epitopes of α_{S1}-casein and produced differing results. Spuergin et al. (1996) identified seven IgE-binding epitopes using overlapping synthetic peptides and pooled serum from 15 cow's milk allergic individuals. Three of these epitopes appeared to be immunodominant, being recognized by IgE from the sera of all 15 individuals. These three immunodominant epitopes were distinguished by a high content of non-polar and aromatic amino acids. The IgG-binding epitopes were at essentially the same positions in α_{S1}-casein as the IgE-binding epitopes. Nakajima-Adachi et al. (1998) mapped the IgE, IgG$_4$ and T-cell epitopes of α_{S1}-casein using overlapping synthetic peptides in combination with digestion fragments. A single immunodominant IgE-binding peptide was identified at the C-terminal end of the allergen. The peptide was mapped to the amino acids at positions 181–199 of α_{S1}-casein and was recognized by serum IgE from all nine milk allergic patients used in this study. A single individual recognized one other peptide, at amino acid position 106–125. These results are very different from those of Spuergin et al. (1996) who identified only two minor epitopes in this area (position 172–183 and 189–199) and major epitopes at positions where Nakajima-Adachi et al. (1998) saw no IgE binding. The difference in results may be due to differences in the way the peptides were made or the population of milk allergic patients used in the two studies. Spuergin et al. (1996) used 188 peptides that were ten residues long and overlapped by nine residues. Nakajima-Adachi et al. (1998) used 13 peptides that were 20 residues long and overlapped by five residues. Both groups measured IgE binding to the synthetic peptides by enzyme-linked immunosorbent assay (ELISA) and used different formulae to decide what level of binding was significant. The formula used by Nakajima-Adachi et al. (1998) appeared to be more stringent and may explain why they mapped fewer IgE-binding sites. In addition, it is possible that, because one study was carried out in Germany and the other in Japan, there is a difference in the two allergic populations. Nakajima-Adachi et al. (1998) reported no difference in the peptides bound by IgE and IgG$_4$, which agrees with the findings of Spuergin et al. (1996).

Using T-cell lines isolated from two individuals Nakajima-Adachi et al. (1998) identified six different T-cell epitopes distributed throughout the length of α_{S1}-casein. When the T-cell epitopes recognized by each individual were compared, two different binding motifs were observed. The T-cell lines isolated from the first individual recognized peptides with a glutamic acid anchor residue at the N-terminal end and a leucine anchor residue at the C-terminal end separated by seven amino acids (-E-(X)$_7$-L-). The T-cell lines from the second individual recognized peptides with a glutamic acid anchor residue at the N-terminal end and a lysine anchor residue at the C-terminal end separated by six amino acids (-E-(X)$_6$-K-). Interestingly, a T-cell activating motif with the arrangement of -E-(X)$_5$-K- has been identified as a cross-reacting T-cell epitope present in a number of different inhaled allergens. This motif is very similar to the -E-(X)$_6$-K- motif found by Nakajima-Adachi et al. (1998). In addition, α$_{S1}$-casein contains the -E-(X)$_5$-K- at two other positions in the sequence. Although none of the seven T-cell lines from the two individuals used in this study recognized this cross-reactive epitope, it might very well play a role in vivo in these individuals or in others.

Fish and shellfish allergens

Tropomyosins are also present in vertebrate sources of food (e.g. bony fish, beef, pork and chicken), yet are rarely allergenic in these foods. This has generated interest in the molecular characteristics that might explain this difference in allergenicity. Reese et al. (1997) mapped IgE-reactive peptides from Pen a 1 using a peptide expression library and serum IgE from shrimp allergic patients. Four peptides (13–21 amino acids long) were identified and sequenced. All four peptides were located in the C-terminal end of the allergen. The peptides aligned with both conserved and non-conserved regions of tropomyosin. However, none of the peptides was identical with the vertebrate homologues. This observation suggests that the amino acid differences between the vertebrate and shrimp tropomyosin may be responsible for the differences in allergenicity. However, experiments involving site-directed mutagenesis to determine which of the amino acid residues are critical for IgE binding have not yet been performed. Ayuso et al. (2002) mapped the linear IgE-binding epitopes of shrimp using synthetic overlapping peptides and human sera from shrimp allergic patients. In contrast to the work by Reese et al. (1997), Ayuso et al. (2002) identified five major IgE-binding regions that were positioned at regular intervals of about 42 amino acids along the length of the molecule. This observation suggested a relationship between the repetitive coiled-coil structure of the troposmyosin protein and the allergenicity of this molecule.

Strategies to Alter Food Allergens

Conventional allergen immunotherapy involves the injection of increasing doses of allergen extracts until a maintenance level is achieved. The relatively high level of allergen administered appears to be key to the effectiveness of the therapy. Unfortunately, conventional allergen immunotherapy has not proved very beneficial for food allergy. Eliminating the allergen from the diet presently is the only proven therapy for food allergies. Elimination diets for foods such as wheat, milk, egg, soy and groundnut can be very difficult to adhere to without accidental ingestion of the allergen. In addition, with important foods such as milk, there is the risk of negative effects on the nutritional state of the allergic patient. Madsen and Henderson (1997) assessed calcium intake in a population of 58 patients diagnosed as allergic to cow's milk. They found calcium intake was below the recommended daily allowance (RDA) for 53% of the patients. Twenty-one per cent of the patients taking calcium supplements still did not reach their RDA. The ability to alter allergens to make them non-allergenic or less allergenic holds the promise of making the food supply safer and lowering the incidence of food allergy.

Physiochemical and enzymatic processing

An alternative approach to treating the clinical symptoms of allergy is to modify the allergenic proteins so that they are no longer capable of eliciting an allergic response. Some labile allergens do not survive standard processing. For example, many fruit allergens are heat sensitive and are destroyed in the canning process. Not surprisingly, many food allergens are resistant to extremes of heat, pH and enzymatic degradation, since this resistance allows the allergen to survive the digestive system long enough to interact with the immune system and either sensitize the individual or trigger an allergic reaction. The most effective and common method for reducing the danger posed by food allergens is enzymatic hydrolysis of food proteins. Milk was hydrolysed successfully to a non-allergenic state >50 years ago. An allergy to cow's milk is not only one of the most common but also one of the most problematic food allergies because there are so few alternatives for infants. In addition, switching from a cow's milk-based formula to a soybean-based formula can result in allergy to soybeans (Bishop et al., 1990). Because of this, hydrolysed cow's milk is a common substitute in infant formulae for treating this allergy. Hydrolysed formulae can be classified as either partially hydrolysed or extensively hydrolysed. The partially hydrolysed cow's milk still contains some full-length allergen and large allergen fragments capable of binding IgE and triggering allergic reactions, and for this reason is not suitable for use by cow's milk allergic individuals (Halken et al., 1993). Studies of extensively hydrolysed formulae have detected allergen-derived peptides still capable of binding IgE and/or triggering positive skin prick tests. However, it was rare for the extensively hydrolysed formulae to trigger allergic reactions in patient challenges (Sampson and McCaskill, 1985; Rugo et al., 1992; Halken et al., 1993; Isolauri et al., 1995). Enzymatic hydrolysis does not harm the nutritional value of the food, since it specifically degrades proteins and has no effect on the lipid, carbohydrate or mineral content of the food source. As a bonus, the peptides and amino acids that are produced are taken up and utilized more easily by the digestive tract. Unfortunately, extensive hydrolysis produces a bitter taste, and the increase in free amino acids produces an increase in osmotic pressure. Some formula manufacturers eliminate the possibility of intact allergens in the final product by using only free amino acids to begin with (Sampson and McCaskill, 1985; Isolauri et al., 1995). Enzymatic hydrolysis has also been used on rice (Watanabe et al., 1990) and wheat (Tanabe et al., 1996).

Traditional plant breeding methods and genetic engineering

An economical and desirable alternative to chemical or enzymatic hydrolysis of foods would be to remove the allergenic protein from the food source through breeding or genetic engineering. Coeliac disease is a non-IgE-mediated food hypersensitivity to wheat gliadins. Strains of wheat lacking different gliadin genes have been bred to produce wheat strains that are missing most of the gliadin proteins. The same strategy could be used to breed naturally hypoallergenic strains of allergenic fruits and vegetables where such strains exist. Apples are an example of an allergenic fruit in which the concentration of the allergenic protein, Mal d 1, varies greatly between different strains (Vieths et al., 1994). Therefore, hypoallergenic strains of apple might be developed and substituted for the more allergenic strains in processed foods. The concentrations of the major allergens in groundnut strains, however, are more uniform. Groundnut allergens are storage proteins laid down by the plant to supply amino acids for the growth and development of the groundnut embryo. Developing hypoallergenic groundnuts by traditional breeding methods will be difficult because these allergenic proteins are both abundant and required for seed germination. Genetic engineering may offer a way to deal with allergens such as those found in the groundnut. The major IgE-binding epitopes of the three major groundnut allergens have been identified (Egge et al., 1983; Burks et al., 1997; Rabjohn et al., 1999). These epitopes have been mutated to a non-IgE-binding form in the recombinant allergen using site-directed mutagenesis. Care has been taken in the mutagenesis to avoid disturbing folding of the proteins (Rabjohn et al., 2002). This process could be used in groundnut or other allergenic foods to produce hypoallergenic protein genes that could be substituted for the endogenous genes. The potential advantages of this approach are numerous. Allergen avoidance currently is the only effective therapy for food allergy. However, the long-term nature of the groundnuts hypersensitivity and the increasing use of groundnuts and their by-products as additives in other foods make avoidance very difficult. Bock and Atkins (1989) found that of 32 groundnut allergic individuals trying to eliminate groundnuts from their diets, 24 had accidentally consumed groundnuts within the last 5 years, 16 within the last year alone. Not only will the hypoallergenic version of the groundnut be safer if accidentally ingested, but it may also reduce the number of new individuals becoming sensitized to groundnut.

A great deal of thought and research has been directed at attempts to define the mechanisms that determine which proteins will be allergenic and which benign. Resistance to pH, heat and proteolytic enzymes has been suggested as a possible requirement for food allergens (Astwood et al., 1996). However, it is doubtful that these properties alone are sufficient to make a protein allergenic. The possibility exists that the allergen is simply the target of the immune system and that a second protein or some non-protein compound is acting as an adjuvant and triggering the atopic immune response. If the allergen is not responsible for triggering the atopic response itself, then substituting a non-IgE-binding mutant for the original allergen may not reduce the number of people who become sensitized to groundnuts in the future. However, this problem could be addressed by introducing pepsin cleavage sites at exposed positions in the folded protein, thus creating a labile protein without affecting the protein's function. If resistance to pepsin is required for food allergens, then making the allergens labile to enzymatic cleavage should disrupt the sensitization process.

Summary

The proteins within foods that are allergenic represent a very small percentage of consumed proteins with functions that vary from storage to enzymatic activity. General characteristics of food allergens include an acidic isoelectric point, abundance in the food, stability to digestion and the presence of multiple, linear IgE-binding epitopes. However, there are exceptions even to these very general characteristics, making it difficult to predict whether a protein is a potential allergen. The development of an IgE response to an allergen involves a series of steps from ingestion to interaction with the immune system. Because of the complexity of the sensitization process, it is important to continue to characterize known allergens at the molecular level with respect to their ability to survive digestion and stimulate an allergic immune response. With this information, it will be possible to improve diagnosis, develop new immunotherapeutic strategies and possibly develop hypoallergenic plants for a safer food supply.

References

Anderson, J.A. and Sogn, D.D. (eds) (1984) *Adverse Reactions to Food.* American Academy of Allergy and Immunology Committee on Adverse Reactions to Foods, US National Institute of Allergy and Infectious Diseases, NIH publication no. 84-2442.

Astwood, J.D., Leach, J.N. and Fuchs, R.L. (1996) Stability of food allergens to digestion *in vitro. Nature Biotechnology* 14, 1269–1273.

Ayuso, R., Lehrer, S.B. and Reese, G. (2002) Identification of continuous, allergenic regions of the major shrimp allergen Pen a 1 (tropomyosin). *International Archives of Allergy and Immunology* 127, 27–37.

Bernhisel-Broadbent, J., Dintzis, H.M., Dintzis, R.Z. and Sampson, H.A. (1994) Allergenicity and antigenicity of chicken egg ovomucoid (Gal d III) compared with ovalbumin (Gal d I) in children with egg allergy and in mice. *Journal of Allergy and Clinical Immunology* 93, 1047–1059.

Bishop, J.M., Hill, D.J. and Hosking, C.S. (1990) Natural history of cow milk allergy: clinical outcome. *Journal of Pediatrics* 116, 862–867.

Bock, S.A. and Atkins, F.M. (1989) The natural history of peanut allergy. *Journal of Allergy and Clinical Immunology* 83, 900–904.

Bock, S.A., Sampson, H.A., Atkins, F.M., Zeiger, R.S., Lehrer, S., Sachs, M., Bush, R.K. and Metcalfe, D.D. (1988) Double-blind, placebo-controlled food challenge (DBPCFC) as an office procedure: a manual. *Journal of Allergy and Clinical Immunology* 82, 986–997.

Burks, A.W., and Sampson, H.A. (1993) Food allergies in children. *Current Problems in Pediatrics* 23, 230–252.

Burks, A.W., Williams, L.W., Helm, R.M., Connaughton, C., Cockrell, G. and O'Brien, T. (1991) Identification of a major peanut allergen, Ara h I, in patients with atopic dermatitis and positive peanut challenges. *Journal of Allergy and Clinical Immunology* 88, 172–179.

Burks, A.W., Williams, L.W., Connaughton, C., Cockrell, G., O'Brien, T.J. and Helm, R.M. (1992) Identification and characterization of a second major peanut allergen, Ara h II, with use of the sera of patients with atopic dermatitis and positive peanut challenge. *Journal of Allergy and Clinical Immunology* 90, 962–969.

Burks, A.W., Cockrell, G., Stanley, J.S., Helm, R.M. and Bannon, G.A. (1995) Recombinant peanut allergen Ara h I expression and IgE binding in patients with peanut hypersensitivity. *Journal of Clinical Investigation* 96, 1715–1721.

Burks, A.W., Shin, D., Cockrell, G., Stanley, J.S., Helm, R.M. and Bannon, G.A. (1997) Mapping and mutational analysis of the IgE-binding epitopes on Ara h 1, a legume vicilin protein and a major allergen in peanut hypersensitivity. *European Journal of Biochemistry* 245, 334–339.

Burks, A.W., James, J.M., Hiegel, A., Wilson, G., Wheeler, J.G., Jones, S.M. and Zuerlein, N. (1998) Atopic dermatitis and food hypersensitivity reactions. *Journal of Pediatrics* 132, 132–136.

Bush, R.K., Taylor, S.L. and Nordlee, J.A. (1989) Peanut sensitivity. *Allergy Proceedings* 10, 261–264.

Cooke, S.K. and Sampson, H.A. (1997) Allergenic properties of ovomucoid in man. *Journal of Immunology* 159, 2026–2032.

Docena, G.H., Fernandez, R., Chirdo, F.G. and Fossati, C.A. (1996) Identification of casein as the major allergenic and antigenic protein of cow's milk. *Allergy* 51, 412–416.

Egge, H., Peter-Katalinic, J., Paz-Parente, J., Strecker, G., Montreuil, J. and Fournet, B. (1983) Carbohydrate structures of hen ovomucoid. A mass spectrometric analysis. *FEBS Letters* 156, 357–362.

Eigenmann, P.A., Burks, A.W., Bannon, G.A. and Sampson, H.A. (1996) Identification of unique peanut and soy allergens in sera adsorbed with cross-reacting antibodies. *Journal of Allergy and Clinical Immunology* 98, 969–978.

Gillespie, D.N., Nakajima, S. and Gleich, G.J. (1976) Detection of allergy to nuts by the radioallergosorbent test. *Journal of Allergy and Clinical Immunology* 57, 302–309.

Halken, S., Host, A., Hansen, L.G. and Osterballe, O. (1993) Safety of a new, ultrafiltrated whey hydrolysate formula in children with cow milk allergy: a clinical investigation. *Pediatric Allergy and Immunology* 4, 53–59.

Hansen, T.K., Bindslev-Jensen, C., Skov, P.S. and Poulsen, L.K. (1997) Codfish allergy in adults: IgE cross-reactivity among fish species. *Annals of Allergy, Asthma and Immunology* 78, 187–194.

Hanson, L. and Telemo, E. (1997) The growing allergy problem. *Acta Paediatrica* 86, 916–918.

Hill, D.J., Hosking, C.S. and Reyes-Benito, L.V. (2001) Reducing the need for food allergen challenges in young children: a comparison of *in vitro* with *in vivo* tests. *Clinical and Experimental Allergy* 31, 1031–1035.

Holen, E. and Elsayed, S. (1990) Characterization of four major allergens of hen egg-white by IEF/SDS–PAGE combined with electrophoretic transfer and IgE-immunoautoradiography. *International Archives of Allergy and Applied Immunology* 91, 136–141.

Honma, K., Kohno, Y., Saito, K., Shimojo, N., Tsunoo, H. and Niimi, H. (1994) Specificities of IgE, IgG and IgA antibodies to ovalbumin. Comparison of binding activities to denatured ovalbumin or ovalbumin fragments of IgE antibodies with those of IgG or IgA antibodies. *International Archives of Allergy and Immunology* 103, 28–35.

Honma, K., Kohno, Y., Saito, K., Shimojo, N., Horiuchi, T., Hayashi, H., Suzuki, N., Hosoya, T., Tsunoo, H. and Niimi, H. (1996) Allergenic epitopes of ovalbumin (OVA) in patients with hen's egg allergy: inhibition of basophil histamine release by haptenic ovalbumin peptide. *Clinical and Experimental Immunology* 103, 446–453.

Hourihane, J.O., Kilburn, S.A., Nordlee, J.A., Hefle, S.L., Taylor, S.L. and Warner, J.O. (1997) An evaluation of the sensitivity of subjects with peanut allergy to very low doses of peanut protein: a randomized, double-blind, placebo-controlled food challenge study. *Journal of Allergy and Clinical Immunology* 100, 596–600.

Isolauri, E., Sutas, Y., Makinen-Kiljunen, S., Oja, S.S., Isosomppi, R. and Turjanmaa, K. (1995) Efficacy and safety of hydrolyzed cow milk and amino acid-derived formulas in infants with cow milk allergy. *Journal of Pediatrics* 127, 550–557.

James, J.M., Helm, R.M., Burks, A.W. and Lehrer, S.B. (1997) Comparison of pediatric and adult IgE antibody binding to fish proteins. *Annals of Allergy, Asthma and Immunology* 79, 131–137.

Jansen, J., Kardinaal, A., Huijber, G., Vleig-Boestra, B. and Ockhuizen, T. (1994) Prevalence of food allergy and intolerance in the adult Dutch population. *Journal of Allergy and Clinical Immunology* 93, 446–456.

Kahlert, H., Petersen, A., Becker, W.M. and Schlaak, M. (1992) Epitope analysis of the allergen ovalbumin (Gal d II) with monoclonal antibodies and patients' IgE. *Molecular Immunology* 29, 1191–1201.

Kane, P., Erickson, J., Fewtrell, C., Baird, B. and Holowka, D. (1986) Cross-linking of IgE–receptor complexes at the cell surface: synthesis and characterization of a long bivalent hapten that is capable of triggering mast cells and rat basophilic leukemia cells. *Molecular Immunology* 23, 783–790.

Kane, P.M., Holowka, D. and Baird, B. (1988) Cross-linking of IgE–receptor complexes by rigid bivalent antigens greater than 200 Å in length triggers cellular degranulation. *Journal of Cell Biology* 107, 969–980.

Kleber-Janke, T., Crameri, R., Appenzeller, U., Schlaak, M. and Becker, W.M. (1999) Selective cloning of peanut allergens, including profilin and 2S albumins, by phage display technology. *International Archives of Allergy and Immunology* 119, 265–274.

Kumosinski, T.F., Brown, E.M. and Farrell, H.M. Jr (1991) Three-dimensional molecular modeling of bovine caseins: kappa-casein. *Journal of Dairy Science* 74, 2879–2887.

Langeland, T. (1982) A clinical and immunological study of allergy to hen's egg white. III. Allergens in hen's egg white studied by crossed radioimmunoelectrophoresis (CRIE). *Allergy* 37, 521–530.

Leung, P.S., Chen, Y.C., Mykles, D.L., Chow, W.K., Li, C.P. and Chu, K.H. (1998) Molecular identification of the lobster muscle protein tropomyosin as a seafood allergen. *Molecular Marine Biology and Biotechnology* 7, 12–20.

Madsen, C.D. and Henderson, R.C. (1997) Calcium intake in children with positive IgG RAST to cow's milk. *Journal of Paediatrics and Child Health* 33, 209–212.

Matsuda, T., Watanabe, K. and Nakamura, R. (1982) Immunochemical studies on thermal denaturation of ovomucoid. *Biochimica et Biophysica Acta* 707, 121–128.

Metcalfe, D.D. (1985) Food allergens. *Clinical Reviews in Allergy* 3, 331–349.

Nakajima-Adachi, H., Hachimura, S., Ise, W., Honma, K., Nishiwaki, S., Hirota, M., Shimojo, N., Katsuki, T., Ametani, A., Kohno, Y. and Kaminogawa, S. (1998) Determinant analysis of IgE and IgG4 antibodies and T cells specific for bovine alpha(s)1-casein from the same patients allergic to cow's milk: existence of alpha(s)1-casein-specific B cells and T cells characteristic in cow's-milk allergy. *Journal of Allergy and Clinical Immunology* 101, 660–671.

Quirce, S., Blanco, R., Diez-Gomez, M.L., Cuevas, M., Eiras, P. and Losada, E. (1997) Carrot-induced asthma: immunodetection of allergens. *Journal of Allergy and Clinical Immunology* 99, 718–719.

Rabjohn, P., Helm, E.M., Stanley, J.S., West, C.M., Sampson, H.A., Burks, A.W. and Bannon, G.A. (1999) Molecular cloning and epitope analysis of the peanut allergen Ara h 3. *Journal of Clinical Investigation* 103, 535–542.

Rabjohn, P., West, C.M., Connaughton, C., Sampson, H.A., Helm, R.M., Burks, A.W. and Bannon, G.A. (2002) Modification of peanut allergen Ara h 3: effects on IgE binding and T cell stimulation. *International Archives of Allergy and Immunology* 128, 15–23.

Reese, G., Jeoung, B.J., Daul, C.B. and Lehrer, S.B. (1997) Characterization of recombinant shrimp allergen Pen a 1 (tropomyosin). *International Archives of Allergy and Immunology* 113, 240–242.

Rugo, E., Wahl, R. and Wahn, U. (1992) How allergenic are hypoallergenic infant formulae? *Clinical and Experimental Allergy* 22, 635–639.

Sampson, H.A. (1997) Food allergy. *Journal of the American Medical Association* 278, 1888–1894.

Sampson, H.A. and McCaskill, C.C. (1985) Food hypersensitivity and atopic dermatitis: evaluation of 113 patients. *Journal of Pediatrics* 107, 669–675.

Sampson, H.A., James, J.M. and Bernhisel-Broadbent, J. (1992) Safety of an amino acid-derived infant formula in children allergic to cow milk. *Pediatrics* 90, 463–465.

Shin, D.S., Compadre, C.M., Maleki, S.J., Kopper, R.A., Sampson, H., Huang, S.K., Burks, A.W. and Bannon, G.A. (1998) Biochemical and structural analysis of the IgE binding sites on ara h1, an abundant and highly allergenic peanut protein. *Journal of Biological Chemistry* 273, 13753–13759.

Sicherer, S.H., Morrow, E.H. and Sampson, H.A. (2000) Dose-response in double-blind, placebo-controlled oral food challenges in children with atopic dermatitis. *Journal of Allergy and Clinical Immunology* 105, 582–586.

Spuergin, P., Mueller, H., Walter, M., Schiltz, E. and Forster, J. (1996) Allergenic epitopes of bovine alpha S1-casein recognized by human IgE and IgG. *Allergy* 51, 306–312.

Stanley, J.S. and Bannon, G.A. (1999) Biochemical aspects of food allergens. *Immunology and Allergy Clinics of North America* 19, 605–617.

Stanley, J.S., King, N., Burks, A.W., Huang, S.K., Sampson, H., Cockrell, G., Helm, R.M., West, C.M. and Bannon, G.A. (1997) Identification and mutational analysis of the immunodominant IgE binding epitopes of the major peanut allergen Ara h 2. *Archives of Biochemistry and Biophysics* 342, 244–253.

Tanabe, S., Arai, S. and Watanabe, M. (1996) Modification of wheat flour with bromelain and baking hypoallergenic bread with added ingredients. *Bioscience Biotechnology and Biochemistry* 60, 1269–1272.

Taylor, S.L., Busse, W.W., Sachs, M.I., Parker, J.L. and Yunginger, J.W. (1981) Peanut oil is not allergenic to peanut-sensitive individuals. *Journal of Allergy and Clinical Immunology* 68, 372–375.

van Rhee, R., Cabanes-Macheteau, M., Akkerdaas, J., Milazzo, J.P., Loutelier-Bourhis, C., Rayon, C., Villalba, M., Koppelman, S., Aalberse, R., Rodriguez, R., Faye, L. and Lerouge, P. (2000) Beta(1,2)-xylose and alpha(1,3)-fucose residues have a strong contribution in IgE binding to plant glycoallergens. *Journal of Biological Chemistry* 275, 11451–11458.

Vieths, S., Jankiewicz, A., Schoning, B. and Aulepp, H. (1994) Apple allergy: the IgE-binding potency of apple strains is related to the occurrence of the 18-kDa allergen. *Allergy* 49, 262–271.

Viquez, O.M., Summer, C.G., and Dodo, H.W. (2001) Isolation and molecular characterization of the first genomic clone of a major peanut allergen, Ara h 2. *Journal of Allergy and Clinical Immunology* 107, 713–717.

Watanabe, M., Miyakawa, J., Ikezawa, Z., Suzuki, Y., Hirao, T., Yashizawa, T. and Arai, S. (1990) Production of hypoallergenic rice by enzymatic decomposition of constituent proteins. *Journal of Food Science* 55, 781.

24 Safety Assessment of Genetically Modified Foods

Steve L. Taylor
Department of Food Science and Technology, University of Nebraska-Lincoln, Lincoln, Nebraska, USA

Introduction

Foods produced through agricultural biotechnology are beginning to appear in the consumer marketplace, especially in North America. While these products have been accepted and successful in some locales, some opposition to the presence of foods produced through agricultural biotechnology has surfaced in several countries. Opposition groups have raised a number of arguments over the entry of these products into the consumer marketplace, including environmental, ethical, economic and safety concerns. This chapter will deal solely with the issues surrounding the safety of genetically modified foods and the nature and adequacy of the safety assessment approach.

Impact of Agricultural Biotechnology on Production Agriculture

Agricultural biotechnology has already had considerable impact on production agriculture, especially with respect to certain crop plants. Thus far, the number of crops that have been commercially developed through agricultural biotechnology is relatively small, including maize, potatoes, canola, soybeans, cotton, squash and papaya (James, 2001). Agricultural biotechnology has had its greatest impact on two staple crops, soybeans and maize, where, on a worldwide basis, 25.8 million ha were planted to transgenic soybeans and 10.3 million ha were planted to transgenic maize in 2000 (James, 2001). These two crops have been especially successful in North America where planting of transgenic soybeans and maize has been increasing steadily to a level of 30–50% of the total acreage planted to these crops. Transgenic cotton and canola have also been introduced successfully in several countries, although the oil fractions of these crops are the only products likely to appear in the food supplies. Transgenic potatoes have also been developed commercially, although the acreage planted to transgenic potatoes remains small. Transgenic squash and papaya have also been introduced in small amounts to the market. The Hawaiian papaya industry that was virtually destroyed by a viral infection has been revitalized through the development of a virus-resistant variety through agricultural biotechnology. Transgenic crops have been planted in 15 countries, although the USA, Argentina and Canada were responsible for 97% of all plantings of transgenic crops in 2000 (James, 2001).

Clearly, farmers see advantages to genetically modified crops; otherwise, these varieties would be less successful. Only a limited number of beneficial traits have been introduced into crops commercially developed through agricultural biotechnology (IFT, 2000a). Maize, soybeans, potatoes, canola and cotton with either improved insect resistance or enhanced herbicide tolerance have been introduced commercially. Very recently, crops with a combination of improved insect resistance and enhanced herbicide tolerance have been developed. The beneficial trait introduced into squash and papaya is virus resistance. All of the existing traits in this current generation of genetically modified crops provide primarily agronomic benefits.

Thus, the benefits accrue primarily to farmers. The commercial success of crops with enhanced agronomic traits will probably lead to the development and introduction of additional transgenic crops with these and similar benefits. Additional crops that protect themselves from diseases and pests and that prosper under adverse climatic conditions such as heat, cold and drought are likely in the future.

The introduction of genetically modified animal products has not yet occurred on a widespread basis, with the exception of milk from cows treated with recombinant bovine growth hormone. However, a genetically modified salmon with enhanced growth characteristics is poised to enter the US marketplace if regulatory approvals can be obtained.

With the growing population worldwide, agricultural production will experience growing pressure to provide sufficient food to feed the masses. Conventional, modern farming practices simply will not suffice with the current constraints on arable land. Much of the current, potential harvest is lost each year to insect pests, competition from weeds, drought conditions and other agronomic factors. The application of conventional agricultural chemicals can eliminate insects and weeds, but many consumers are concerned about the safety of pesticide residues in their foods and the potential adverse environmental impacts associated with these practices. Agricultural biotechnology already offers crops with improved insect resistance and herbicide tolerance that allow the control of losses due to insects and weeds with limited agrichemical application. As noted earlier, transgenic crops with improved tolerance to climatic stresses and poor soil conditions will probably be developed in the foreseeable future. Also, this improved technology is contained in the seed itself so that the benefits potentially can be enjoyed by farmers who do not have access to modern and expensive agricultural machinery. For example, virus-resistant squash varieties are especially well suited to Africa, where great crop losses occur associated with viruses. Hopefully, with the development of virus-resistant squash, this crop will become a more reliable food source and improve the food security in some African countries.

Impact of Agricultural Biotechnology on Consumers

Currently, most consumers probably perceive that few benefits accrue from agricultural biotechnology since the major existing benefits are associated with improved agronomic traits. Few foods from agricultural biotechnology with direct consumer benefits have reached the marketplace. Tomatoes with improved ripening characteristics that provided enhanced flavour attributes were developed (Redenbaugh, 1992), but these tomatoes were not introduced successfully into the marketplace. Canola oil enriched with high levels of oleic acid, a monounsaturated fatty acid that is thought to be beneficial to cardiovascular health, has been developed and may soon be available (FAO/WHO, 2000). The recent development of the so-called golden rice with enhanced levels of β-carotene has received considerable publicity (Ye et al., 2000). Although golden rice is not yet available commercially, it holds some promise to help eradicate widespread vitamin A deficiency and night blindness in certain Asian populations (Underwood, 1994). Agricultural biotechnology offers the potential for development of many more traits that provide direct benefits to consumers. The consumer benefits that could be developed through agricultural biotechnology include enhanced nutritional and nutraceutical composition, prolonged shelf-life, resistance to spoilage, improved flavour and appearance, and the elimination of naturally occurring toxicants including allergens.

Consumers, particularly those in North America, have been widely exposed to foods and food ingredients derived from genetically modified crops. Maize and soybeans are particularly important foods that are used to produce many popular food ingredients including corn oil, corn starch, corn syrup, soybean oil, soy protein isolate, soy protein concentrate, soy sauce and soybean lecithin. In addition, canola oil and cottonseed oil are common food ingredients that would be derived from those particular genetically modified crops.

Although exposure to foods and food ingredients derived from crops produced through agricultural biotechnology has been high, the impact of agricultural biotechnology on consumers arguably has been invisible due to the lack of readily apparent consumer benefits. Most consumers probably are not aware that they are ingesting foods and food ingredients derived from genetically modified crops. However, it is also important to indicate that the introduction of genetically modified foods into the marketplace has not been accompanied by any known adverse health consequences. Obviously, the safety evaluation of foods produced through agricultural biotechnology is critically important to ensure the continuation of this safety record.

Primer on Agricultural Biotechnology

Agricultural biotechnology provides a precise method for enhancing the beneficial traits of foods produced from plants, animals and microorganisms. With modern recombinant DNA methodologies, genes can be selected from virtually any biological source and be moved into the genome of some other species (IFT, 2000a). In this approach, genes would be selected for specific beneficial traits ranging from improved agronomic characteristics to improved nutritional profiles. In recombinant DNA approaches, only a small number of genes are inserted into the genome of the recipient species to provide the selected beneficial trait. In the case of insect-resistant maize, a single gene is needed that leads to production of the selected insect toxicant (Sanders et al., 1998). A few other genes are also added as markers or promoters (Sanders et al., 1998). In the case of golden rice, the introduction of four genes was necessary to provide the biological capability for enhanced β-carotene synthesis (Ye et al., 2000). With more complex traits such as improved nutritional composition, more genes may need to be added to accomplish the intended effect. The same is true with more complex agronomic benefits including varieties with combined insect resistance and herbicide tolerance where several selected genes must be introduced simultaneously.

In contrast, with conventional breeding, hundreds of genes are moved from one biological source into another. The identity of all of these genes often is unknown. In the process of selecting for the beneficial trait, other unwanted genes may also appear in the new variety. With conventional breeding, genes may only be introduced from closely related biological sources. In contrast, with recombinant DNA approaches, the genes can be selected from a very wide variety of biological sources. Historically, only a few episodes have occurred with conventional breeding where the new variety was found to have potentially hazardous components (Zitnak and Johnson, 1970; IFT, 2000b). However, a thorough safety assessment is not performed typically with new varieties produced by conventional breeding.

Whether by traditional plant breeding or by modern recombinant DNA agricultural biotechnology, novel proteins can be introduced into the edible portion of the new variety. As noted, with traditional plant breeding, the safety of these novel proteins has not been assessed historically. With conventional breeding, safety assessment of the novel proteins would be extremely difficult since the number and precise identity of the novel proteins typically are unknown. However, with foods produced through modern agricultural biotechnology, only a limited number of well-defined novel proteins produced by specific inserted genes are introduced into the new plant variety. The opportunity thus exists for the assessment of the safety of these specific novel proteins. The assessment of the safety of the novel proteins is just one part of the overall safety assessment process for a food produced through agricultural biotechnology.

Safety Assessment Process for Foods Produced Through Agricultural Biotechnology

The products of agricultural biotechnology should be subjected to a careful and complete safety assessment prior to commercialization. Regulatory agencies around the world require agricultural biotechnology companies to conduct thorough safety assessments that are reviewed before the approval of the crop for commercialization. Some concern has arisen because, in the USA, the safety assessment process has been voluntary. However, the US Food and Drug Administration has announced its intent to create a mandatory safety assessment for foods produced through agricultural biotechnology. It should be recognized that the genetically modified foods allowed on the market in the USA thus far have been subjected to a *de facto* mandatory safety assessment system since governmental regulatory agencies have reviewed the available safety assessment information prior to commercialization of these products.

Thus far, all of the genetically modified products brought into the marketplace have been subjected to an intensive safety assessment. The data from the safety assessments have been reviewed by regulatory authorities around the world. The current generation of genetically modified products are deemed safe for human and feed animal consumption by one or more regulatory agencies in various countries.

Safety assessment must be a consideration at all stages in the development of a commercial genetically modified food. The processes involved in the commercialization of genetically modified foods can be categorized into three phases: gene discovery, line selection and product advancement to commercialization. Safety assessment plays a role in all three of these stages.

As noted above, agricultural biotechnology involves the introduction of novel genes that confer desirable traits of various kinds. For genetically modified products, the safety evaluation usually focuses on the introduced trait or gene product.

Gene discovery phase

In the gene discovery phase, a product concept is strategically chosen, and scientists then screen and select genes that might fulfil that concept. The safety assessment process ideally should begin at the earliest stages of the initial gene discovery phase. If any questions arise, these questions must be answered either immediately or later in the developmental process. The careful selection of genes from sources with a history of safe use by humans often will allow developers to avoid many of the more difficult safety assessment questions. The selection of suitable genes should take into account various factors including the source of the gene, any previous consumer exposure to this source material, and the history of safe use of the source material, the gene and the product of the gene. Ethical issues usually are also considered at this early stage. Although they are beyond the scope of this review, environmental and ecological concerns are also addressed initially during the gene discovery phase. If unacceptable or highly questionable food safety risks are identified during this early phase, the concept probably should be abandoned. This is especially true if the identified risks cannot be evaluated completely and clearly before commercialization.

This initial safety assessment during the gene discovery phase is important because it serves to highlight concerns and questions that must be addressed effectively later in the safety assessment process unless the concept is abandoned at this point. Among other issues, concerns might be raised during the gene discovery phase regarding such issues as the allergenicity of the source material from which the gene was obtained, known naturally occurring toxicants in this source material, or various environmental/ecological issues associated with the source material or the gene product. For example, if a gene is selected from a source with a known history of allergy such as groundnuts, tree nuts, fish, etc., or even ragweed, birch trees or grass (common sources of pollen allergens), then assurance must be sought that the gene product is not one of the allergens from that source. If the source of the gene is commonly allergenic, the concerns are relatively obvious. However, in some cases, the concerns are not always so obvious. For example, chitinase genes might be selected as a potential means to prevent various fungal diseases common to some crop plants, because mould hyphae are composed largely of chitin. However, chitinases from some sources are allergens (Breiteneder and Ebner, 2000), so the possible cross-reactivity with known allergenic chitinases needs to be assessed.

Insect-resistant crops can serve as a good illustration of the considerations involved at the gene discovery phase. Insect-resistant crops were developed by selecting genes from *Bacillus thuringiensis* that code for the production of various insecticidal Bt proteins. Several different, but related, genes have been cloned into several agricultural crops by commercial seed companies to produce insect-resistant varieties. Bt proteins occur naturally in *B. thuringiensis*; many different, but related, forms of Bt are known to exist (Schnepf *et al.*, 1998). Bt proteins have been included in commercial agricultural sprays as one option for insect control for several decades. The Bt sprays have been widely used, especially by organic farmers and home gardeners. The microbial products used as Bt sprays contain the Bt proteins as the active insecticidal component. Bt proteins exhibit selective toxicity to specific insects, but are essentially non-toxic to mammalian species. The Bt proteins present in these commercial agricultural sprays have been subjected to a rather comprehensive toxicological assessment including acute, subchronic and chronic toxicity testing in experimental animals and even oral gavage studies in humans (McClintock et al., 1995). Regulatory assessments of the data arising from these various tests have led to the conclusion that the Bt proteins in these commercial products are safe. Additional studies conducted with Bt crops produced through agricultural biotechnology also confirm their safety (Sanders, 1998). Various Bt crops including maize, cotton and potatoes have been approved by regulatory authorities in the USA and other countries, and have been on the market since the mid-1990s (James, 2001).

Line selection phase

Once an appropriate and promising gene has been identified for its potential to produce the desired beneficial trait, efforts begin to produce the desired transformation in the recipient plant and develop the genetically modified crop. At the outset, a large number of transformants are produced in the laboratory. By the end of the line selection phase, one or a few of these lines will be identified as the most promising for entry into the commercialization phase. During the line selection stage, the transgenic lines progress from the laboratory through a variety of greenhouse trials. Eventually, if the results are promising, field trials will be conducted.

Regulatory approvals are usually needed to proceed to greenhouse and field trials (Taylor, 2001). In the USA, several regulatory hurdles must be cleared during the line selection phase. First, the USDA must review and approve facility plans for greenhouses and other facilities where the plants will be developed and tested. Next, the developer must seek and obtain USDA approval in order to conduct field trials. Additionally, the USDA must provide approval to the developer for the shipment of seeds from the greenhouse to the field trial site. After the completion of greenhouse and field trials, the commercial developer must submit a full data package generated from the trials to the USDA and request that the genetically modified crop be moved to non-regulated status. At this juncture in the approval process, the USDA invites public comment. Once USDA approves the crop for non-regulated status, then the crop can be grown, tested or used for traditional crop breeding without seeking further USDA permission.

Although the primary focus of the line selection phase is to identify the best line from amongst the many choices, the safety of the genetically modified crop should also be considered while making these choices. Numerous agronomic traits are taken into consideration during the growth trials that occur in laboratories, greenhouses and fields in the line selection phase. These traits include such attributes as plant height, leaf orientation, leaf colour, early plant vigour, root strength and yield. During the line selection phase, the biological and agronomic equivalence of the genetically modified crop is compared with its traditional counterpart. Although not done specifically for the purposes of safety assessment, potential products with unusual characteristics are eliminated during this phase of the developmental process. The elimination of such products with unusual agronomic or biological characteristics probably enhances the prospect of selecting a safe product. As noted earlier, most of the prospective novel varieties are discarded from development during the line selection phase.

In the USA, during the line selection phase, additional governmental overseeing occurs in cases where the novel plant variety contains pesticidal properties such as Bt proteins. In crops possessing these traits, the Environmental Protection Agency (EPA) is also involved in the regulatory approval process. As part of the EPA regulatory process, the developer must obtain an experimental use permit before planting >10 acres (0.405 ha) of any crop that possesses a pesticidal protein or other component. Public comments are invited on requests for this type of EPA approval. The EPA continues to be involved in the regulatory process for genetically modified crops containing pesticidal components beyond providing this initial field permit. Prior to commercialization, the EPA establishes limits (tolerances) on the amount of the pesticidal component that can be permitted in food derived from the genetically modified plant. To reach this decision, the EPA reviews and relies upon data on the human, animal and environmental safety of the pesticidal component. If considerable data should already exist on the safety of the pesticidal component and if there is a history of safe use, as might be the case with some of the Bt proteins, an exemption from the requirement for a tolerance can be requested. However, in those cases, the EPA must rule on whether such an exemption should be granted. Typically, an extensive review of the safety of the crop and its pesticidal component are conducted by the EPA at this stage, and public comments are sought on the establishment of a tolerance.

Product advancement to commercialization phase

Following the line selection phase and before commercialization, the selected variety of the genetically modified crop should go through a detailed and more formal safety assessment process. This safety assessment should focus on the gene products associated with the introduced gene and any other likely toxicological or anti-nutrient factors associated with the source of the novel gene and the crop to which it was introduced. All of the potential uses of the genetically modified product, for both food and feed uses, must be considered during the safety assessment process. In the USA, when a prospective genetically modified crop moves from greenhouse and field trials toward commercialization, the Food and Drug Administration (FDA) becomes involved in the regulatory approval process. The FDA reviews all of the available data relating to the safety of any food or food ingredient produced from the genetically modified crops. In the existing voluntary safety assessment approach used by the FDA, the agency typically would meet with the developer early in the developmental process and offer guidance on the studies that should be performed, in the FDA's opinion, to provide assurance that the genetically modified food will be safe for its intended food and animal feed uses. The nature of the pending mandatory safety assessment approach in the USA is not known at this time. The specifics of the safety assessment process are likely to vary dependent upon the nature of the genetically modified food. The FDA did publish a list of questions that it considers appropriate in the evaluation of the safety of genetically modified foods and feeds (FDA, 1992). Despite the existing voluntary nature of the regulatory safety approach in the USA, all current genetically modified foods in the US marketplace have been subjected to this FDA scrutiny.

Of course, the safety of genetically modified foods must also be considered on a worldwide basis. Internationally, several worldwide organizations including the Food and Agricultural Organization of the United Nations (FAO), the World Health Organization (WHO) and the Organization for Economic Cooperation and Development (OECD) have established the background for the safety assessment of foods produced through agricultural biotechnology (FAO/WHO, 2000; OECD, 2000). Regulatory agencies in individual countries have, in general, based their approaches to the approval of genetically modified foods on this background. The general conclusion of these reports is that the products of plant biotechnology are not inherently less safe than those developed by traditional breeding (FAO/WHO, 1996). These organizations have also concluded that the food safety considerations are basically of the same nature as those arising from the products of conventional breeding. Therefore, traditional approaches to safety assessment are appropriate for the assessment of foods and food ingredients produced through agricultural biotechnology. The accepted standard for genetically modified foods is identical to that expressed in the US and most international food laws for all food products – a reasonable certainty that no harm will result from intended uses under anticipated conditions of consumption. Under these conditions, genetically modified foods should provide equal or greater assurance of safety by comparison with foods derived through conventional breeding practices.

Concept of substantial equivalence

The concept of substantial equivalence has been a key, though sometimes misunderstood, feature of the safety assessment of genetically modified foods from the outset of international deliberations (OECD, 1992; FAO/WHO, 1996; IFT, 2000b). The concept of substantial equivalence arose from the goal that the genetically modified food should be as safe as its traditional counterpart. In practice, the currently approved genetically modified varieties of traditional crops, such as maize and soybeans, are altered very little from their traditional counterparts. In the concept of substantial equivalence, the safety evaluation would then focus on those differences while assuming that the unaltered components are just as safe as the same components in the traditional counterpart varieties.

In the concept of substantial equivalence, the genetically modified food (or food ingredient) is compared with its traditional counterpart for such attributes as the origin of the novel genes, agronomic parameters, composition including key nutrients, anti-nutrients and allergens, and consumption patterns. Three possible outcomes can arise from the comparisons made through the concept of substantial equivalence: (i) the genetically modified food would be considered substantially equivalent to its traditional counterpart; (ii) the genetically modified food would be considered substantially equivalent to its traditional counterpart except for one or more defined differences; or (iii) the genetically modified food would not be considered to be substantially equivalent to its traditional counterpart (FAO/WHO, 1996).

In the first scenario, the genetically modified food could be judged to be substantially equivalent to its conventional counterpart. In this case, no further safety testing would be required. In reality, this possibility would occur only on rare occasions. This situation might occur if a gene is removed or silenced but no novel genes are introduced. This situation might also occur with ingredients such as cottonseed oil where the plant would have some defined differences from the expression of specific novel proteins but the oil fraction would not contain these novel proteins and would be equivalent in composition to cottonseed oil derived from traditional varieties of cottonseed. Some of the confusion surrounding the adequacy of the safety assessment of genetically modified foods is based on the incorrect assumption that most genetically modified foods are considered to be substantially equivalent to their traditional counterparts and, therefore, that no safety studies are conducted. In fact, the vast majority of genetically modified foods are not considered to be completely equivalent to their traditional counterparts, and regulatory agencies have required safety studies on those features of the genetically modified food that are distinct and different from the traditional counterpart.

With most of the current genetically modified crops, the foods or food ingredients derived from these crops could be judged to be substantially equivalent to their conventional counterpart except for the defined differences associated with the introduced traits. In this situation, the safety testing would focus upon the safety of the introduced trait or gene product, usually a novel protein. Even in the cases where several genes are introduced, the product could be considered to be substantially equivalent to its conventional counterpart except for the various gene products derived from the specific introduced and novel genes.

Finally, the genetically modified food could be judged to be not substantially equivalent to the conventional food or food ingredient. More extensive safety assessments would probably be required for such products. Since no products have yet been released to the commercial marketplace or approved by worldwide regulatory agencies, the nature of the requirements for safety assessment of such products has not been delineated specifically. Certainly, the safety assessments would need to be conducted in a flexible manner depending upon the nature of such novel food products. More rigorous nutritional and toxicological evaluation would be likely to be desired with any products of this type that are developed. Although no such products have yet been released into the commercial marketplace, agricultural biotechnology offers the promise of many such products in the future.

Extensive compositional analyses are conducted on genetically modified foods to compare them with their conventional counterparts. Of course, the composition of the conventional counterpart can vary significantly as a result of varietal differences, climatic conditions and agronomic conditions. The choice of the conventional counterpart for comparative purposes is extremely important. In such comparisons, the crops and the foods or food components produced therefrom are compared for protein, carbohydrate, fat, fatty acid composition, starch, amino acid composition, fibre, ash, minerals, vitamins and other factors. If known anti-nutrients are present in either the source material for the novel gene or the host plant, the comparative levels of these anti-nutrients are determined. Similarly, if the biological sources of the novel gene or the host plant are known to be allergenic, the presence and levels of the allergens are determined in the transgenic variety and compared with the traditional counterpart. The allergenicity of the host plant is a lesser concern because, for example, consumers with soybean allergy will probably avoid all soybeans whether genetically modified or not.

Safety of DNA

As already noted, the vast majority of the current genetically modified foods on the market are considered to be substantially equivalent to their traditional counterparts except for certain defined differences. Thus, the safety evaluation focuses on the introduced gene and the traits or products that it produces. The DNA associated with the inserted genes in the genetically modified food is not a safety issue of great concern. DNA in the diet, regardless of source, is not considered to be toxicologically significant (Beever and Kemp, 2000). Virtually all foods contain DNA, and DNA is ingested in significant quantities. Human dietary intakes of RNA and DNA are estimated to range from 0.1 to 1.0 g per day (Doerfler and Schubbert, 1997). Ingestion of novel DNA from genetically modified foods is insignificant by comparison and would probably represent <1/250,000 of the total amount of DNA consumed (FAO/WHO, 2000). DNA is also highly digestible (FAO/WHO, 2000). Thus, the novel DNA contained in genetically modified foods is extremely unlikely to pose any safety concerns.

The possibility of the transfer of genes from genetically modified foods to mammalian cells has also been examined, and the probability of such occurrence is considered very low. The transfer of DNA from genetically modified plants into mammalian cells under normal conditions of dietary exposure would require that all of the following conditions be met (FAO/WHO, 2000):

- The relevant gene in the plant DNA would have to be released probably as a linear fragment.
- The gene would have to survive nucleases in the plant and in the gastrointestinal tract.
- The gene would have to compete for uptake with dietary DNA from traditional sources.
- The recipient mammalian cells would have to be competent for transformation and the gene would have to survive their restriction enzymes.
- The gene would have to be inserted into the host DNA by rare repair and recombination events.

Experiments have been conducted aimed at evaluating the possibility that the transfer of plant DNA to microbial or mammalian cells might occur. In one such experiment, mice were orally administered high doses of bacterially derived DNA, and some indications were obtained for the apparent incorporation of the test DNA into mouse cells (Schubbert et al., 1998). However, this observation has been seriously questioned (Beever and Kemp, 2000). FAO/WHO (2000) concluded that no data exist to demonstrate that plant DNA can be transferred to and stably maintained in mammalian cells. Additionally, FAO/WHO (2000) concluded that no evidence exists that intact genes from plants can be transferred to and be expressed in mammalian cells.

The possibility of the transfer of genes from genetically modified foods to gastrointestinal bacteria has also been examined. No evidence exists to suggest that such transfer has occurred in humans ingesting genetically modified foods. The likelihood of such transfer is considered to be quite low (FAO/WHO, 2000). The transfer of genes from plant cells into microbial cells is considered very unlikely except in circumstances where the gene from the genetically modified plant shows homology with prokaryotic genes (FAO/WHO, 2000). Gene transfer has been observed under laboratory conditions, but only in situations where homologous recombination is possible (Nielsen et al., 1998). Gene transfer has not been demonstrated to occur from plant cells to microbes in experiments conducted in the gastrointestinal tracts of animals.

Safety of antibiotic resistance markers

Antibiotic resistance marker genes are often inserted along with the novel gene of interest into genetically modified foods. The purpose of the antibiotic resistance marker is to aid in tracking the successful incorporation of the desired genetic material into the host genome. The use of antibiotic resistance markers has raised safety concerns in the minds of some. However, no evidence exists that the antibiotic resistance markers currently in use pose any health risk to humans (FAO/WHO, 2000). The likelihood of transfer of the antibiotic resistance marker genes to mammalian cells or the cells of gastrointestinal bacteria is quite low for the reasons cited in the previous section. However, even if gene transfer were to occur, the consequences to human health would be insignificant (FAO/WHO, 2000). The most common antibiotic resistance marker gene encodes neomycin phosphotransferase II. The safety of the common antibiotic resistance markers has been well established (WHO, 1993; FAO/WHO, 2000). Neomycin has limited therapeutic value. However, this part of the safety assessment will become more important if novel markers are used, because the safety of these alternative strategies must be documented.

Safety of novel protein

Since the safety of DNA is not considered to be an important consideration, the safety assessment appropriately focuses on the novel protein produced from the inserted gene and any components that might be created in the genetically modified food from the action of that protein if it is an enzyme. Typically, the safety assessment of the novel protein initially would involve structural comparisons with known toxins or allergens based upon homologies in amino acid sequences. The digestive fate or resistance to pepsin hydrolysis of the novel protein is also usually assessed, because, if the novel protein is hydrolysed rapidly by pepsin or in simulated gastric digestibility tests, then the protein is unlikely to exert untoward responses such as toxicity or allergenicity. Another approach in the safety assessment for the novel protein would be to test the purified protein for toxicity in an acute oral toxicity screen in mice or another suitable animal model. When purified proteins are used in toxicological tests, assurance must be sought that the purified protein is virtually identical to the protein as expressed in the genetically modified food. Proteins in plants are often glycosylated. Glycosylation and other post-translational modifications potentially can influence the toxicity of a novel protein. Another approach used in the safety assessment of the novel protein is to assess its heat stability, although the relevance of this attribute to the safety of the protein is debatable. The application of these various approaches should be flexible based upon the known information about the protein and its possible toxicity. However, of course, it must be sufficient to convince worldwide regulatory agencies that the novel protein is not inherently toxic or likely to be or become allergenic.

The assessment of the allergenicity of the novel protein is another important part of the safety evaluation process. All food allergens are proteins. Thus, some possibility exists that any newly introduced protein will be an allergen or become a novel allergen. However, foods contain many thousands of proteins, and only a few hundred are known to be allergens under typical circumstances of exposure. Thus, the likelihood that any specific novel protein will be an allergen on a probabilistic basis alone is slight.

Despite the low probability, generally it is acknowledged that the potential allergenicity of a novel protein is perhaps the most relevant safety concern to be evaluated as part of the safety assessment of genetically modified foods. Several strategies have been promoted for the assessment of the potential allergenicity of the novel proteins in genetically modified foods (Metcalfe et al., 1996; FAO/WHO, 2000, 2001). All of these strategies share several common elements including considerations based upon the source of the novel gene(s) and the allergenicity of the source material for the novel gene, the degree of amino acid sequence homology of the novel protein to known allergens, the immunoreactivity of the novel protein with serum immunoglobulin E (IgE) from humans with known allergies to the source material, and the resistance of the novel protein to pepsin, which would be an indication of its digestive stability.

The allergenicity assessment usually begins by focusing on the source of DNA to be introduced into the host organism. If the gene source is known to be allergic, then the allergenicity of the gene product must be determined. Both environmental allergens, e.g. pollens, and food allergens should be considered because some known food allergens are cross-reactive with pollen allergens (Calkoven et al., 1987). If the gene source is known to be allergenic, then the potential allergenicity of the novel gene product can be determined with a reasonable degree of certainty using a specific serum screening test that employs blood serum from human subjects known to be allergic to the biological source of the gene. The assumption should be that the source gene in such circumstances encodes an allergen unless data are generated to disprove that assumption. In agricultural biotechnology, the greatest concerns arise when the gene is obtained from a commonly allergenic source. Commonly allergenic foods include groundnuts, soybeans, tree nuts and wheat from the plant kingdom, and milk, eggs, fish and crustacea from the animal kingdom (FAO, 1995). These few foods or food groups probably account for >90% of all food allergies on a worldwide basis (FAO, 1995). In addition to these eight foods or food groups, >160 foods and food-related substances have been associated with allergic reactions in individuals on at least some occasions (Hefle *et al.*, 1996). Alternatively, if the gene is obtained from a source with no history of allergenicity, then no human sera can exist, and the specific serum screening test obviously is not possible. In the vast majority of cases in agricultural biotechnology, the gene is obtained from a source with no history of allergenicity. In such situations, the potential allergenicity of the novel proteins in the transgenic product must be assessed by a combination of less reliable approaches, such as the amino acid sequence homology with known allergens and the comparative resistance of the novel protein to pepsin.

A comparison of the degree of amino acid sequence homology between the novel introduced protein and known food and environmental allergens is a key strategy that has been advocated in all of the approaches devised thus far (Metcalfe *et al.*, 1996; FAO/WHO, 2000, 2001). The amino acid sequences of many food and environmental allergens are known (Metcalfe *et al.*, 1996; Gendel, 1998). A comparison of the amino acid sequence of the novel protein with those of known allergens is a useful approach in the determination of the allergenic potential of the novel protein (Metcalfe *et al.*, 1996; FAO/WHO, 2001). If sufficient homology exists, then the novel protein might cross-react with the known allergen and provoke symptoms when ingested by individuals with that particular allergy.

In IgE-mediated food allergy, only certain, specific proteins are known to induce IgE sensitization under typical circumstances of exposure. The immune system recognizes only a portion of the overall structure of the allergen rather than the entire structure. These smaller sections are called allergenic determinants or epitopes. Epitopes can be either continuous (a linear sequence of amino acids) or discontinuous (dependent upon the three-dimensional conformational structure of the protein) (Taylor and Lehrer, 1996). Linear epitopes may be more important with food allergens because food allergens often are stable to heat processing and digestion, conditions that often would lead to unfolding of the three-dimensional structure of a protein (Taylor *et al.*, 1987). However, recent evidence obtained with Ara h 1, one of the major groundnut allergens, suggests that discontinuous conformational epitopes may be important in the case of this particular food allergen (Shin *et al.*, 1998). Despite this recent evidence, linear epitopes probably remain critical in assessment of the allergenic potential of the novel proteins introduced into foods through agricultural biotechnology. IgE-mediated food allergies occur through a two-phase process of sensitization followed by elicitation (Mekori, 1996). In the sensitization process, dietary proteins are processed by proteolysis in an antigen-presenting cell. The resulting peptides then react with T cells to provoke B cells to switch to production of allergen-specific IgE antibodies. Once the allergen-specific IgE antibodies are attached to mast cells and basophils, the interaction of IgE-binding epitopes on these same proteins with the cell-bound IgE antibodies elicits the elicitation phase with the release of histamine and the other mediators of the allergic reactions from the mast cells and basophils. The allergen-specific IgE antibodies can bind to either linear or conformational epitopes depending upon the particular allergen involved. If a novel protein introduced through agricultural biotechnology is identical to an existing allergen or is cross-reactive with an existing allergen, then any linear or conformational epitopes recognized by the cell-bound IgE could be important. However, for truly novel proteins, prior sensitization would not exist. In these situations, sensitization through the T-cells would be critical. T-cell epitopes are most likely to be exclusively linear.

Several different criteria have been proposed to determine a significant degree of amino acid sequence homology between the novel protein and known allergens. In some of the approaches (Metcalfe *et al.*, 1996; FAO/WHO, 2000), a criterion of eight contiguous identical amino acids was used to define a positive match. In the most recent FAO/WHO consultation, a criterion of six contiguous identical amino acids was used to define a positive match (FAO/WHO, 2001). The use of six contiguous, identical amino acids as a match was predicted upon observations that the minimal IgE-binding epitopes of Ara h 1 and Ara h 2 involve six contiguous amino acids (Burks *et al.*, 1997; Stanley *et al.*, 1997). The minimum peptide length for a T-cell-binding epitope is probably eight contiguous amino acids (Metcalfe *et al.*, 1996). The use of the criterion of six contiguous identical amino acids would probably lead to identification of a large number of false-positive matches. Therefore, the use of either seven or eight contiguous identical amino acids as the criterion for a match may be more reasonable. Since such approaches assess the entire protein sequence, they are based upon the identity of amino acid sequences, and not only on known T- and B-cell-binding epitopes of known allergens. Thus, such approaches may (and probably will) identify matching sequences that are unrelated to the allergenic potential of the novel proteins. However, specific serum screening could be used to eliminate clinically insignificant matches.

FAO/WHO (2001) also proposed that a criterion of 35% overall structural homology between the amino acid sequence of the novel protein and the sequences of known food and environmental allergens be used to identify proteins that share similar functions and which thus might be cross-reactive allergens. Many common plant allergens fall within a few functional categories (Breiteneder and Ebner, 2000). The pathogenesis-related proteins of several different types are prominently involved (Breiteneder and Ebner, 2000). If the novel proteins introduced into foods developed through agricultural biotechnology fall into functional categories that contain known food allergens, the criterion for 35% overall structural homology will probably allow their identification. The potential allergenicity of these particular proteins should be carefully assessed.

If the gene is obtained from a known allergenic source or if the search for sequence homology identifies a match with a known food or environmental allergen, an assessment of the immunoreactivity of the novel protein with IgE antibodies from the sera of individuals allergic to the source material or allergen should then be conducted (FAO/WHO, 2001). Since the structures of all of the allergens from all allergenic sources are not yet known, specific serum screening has been advocated in every case where the gene is obtained from a known allergenic source (FAO/WHO, 2001). In serum screening, blood serum from individuals known to be allergic to the specific allergenic source is tested to determine if the allergen-specific IgE in the serum reacts with epitopes on the novel protein. The availability of sera from well-characterized patients with that particular allergy is an important and sometimes challenging issue. Well-characterized human blood serum for some types of allergies may be rare and difficult to obtain. A positive serum screening test certainly raises concerns about the possible allergenicity of the novel protein. A positive result with serum screening will lead to the conclusion that the novel protein is likely to be allergenic, and further commercial development will probably cease.

While serum screening using sera from individuals allergic to the source of the novel gene or to determine if a homologous sequence from structural comparisons is relevant is a well-accepted approach in the allergenicity assessment (Metcalfe et al., 1996; FAO/WHO, 2000, 2001), the use of targeted serum screening represents a more controversial approach. The recent FAO/WHO consultation (FAO/WHO, 2001) advocated the use of targeted serum screening where the immunoreactivity of the novel protein is assessed with serum IgE from humans with known allergies to materials that are broadly related to the source material. Examples would include serum from grass pollen allergic individuals if the gene is obtained from a monocot source, or serum from individuals with cockroach allergy if the gene is obtained from an insect source. The FAO/WHO approach suggests several broad categories for targeted serum screening: monocots, dicots, invertebrates, vertebrates and moulds (FAO/WHO, 2001). If the gene source is bacterial, then targeted serum screening is not conducted, because bacterial proteins are rarely allergenic due to the low exposure levels and lack of allergic sensitization to these proteins. Further research will be likely to be needed to determine whether targeted serum screening will be useful in the assessment of the potential allergenicity of novel proteins in foods produced through agricultural biotechnology. Certainly, false-positive results would probably occur, and the acquisition of well-characterized blood serum would be a critical issue.

The proteolytic stability of the novel protein is another useful criterion in the assessment of the allergenic potential of novel proteins. An allergenic protein must reach the intestinal tract in a form that is sufficiently intact to provoke the immune system. Rapidly digestible proteins would be unlikely to reach the intestinal tract in an immunologically intact form. Known food allergens exhibited greater proteolytic stability than known non-allergenic food proteins in simulated gastric and intestinal digestive models (Astwood et al., 1996). Certain novel proteins introduced into foods produced through agricultural biotechnology were also digested rapidly in these same model systems (Astwood et al., 1996). For example, the enzyme transferred into soybeans to make them tolerant to the herbicide, glyphosate, is digested rapidly in vitro (Harrison et al., 1996). This novel protein is therefore unlikely to induce allergic sensitization. Digestive stability and/or resistance to pepsin have been advocated as approaches to use in the assessment of the potential allergenicity of the novel proteins introduced into genetically modified foods (Metcalfe et al., 1996; FAO/WHO, 2000, 2001). Since human digestion is individually variable, digestive stability tests cannot be construed to be predictive of the digestive stability of a novel protein in all humans. However, the comparative resistance to pepsin proteolysis is likely to be a reasonable comparative measure in the allergenicity assessment. Pepsin resistance is not a perfect indicator of allergenic potential. Some allergens in fresh fruits and vegetables are known to be sensitive to proteolysis (Moneret-Vautrin et al., 1997). These particular allergens tend to be ones that are cross-reactive with known pollen allergens (Calkoven et al., 1987) and would thus probably be discovered in the sequence homology tests.

The use of animal models in the assessment of the allergencity of novel proteins would be premature. No well-validated animal models exist for such purposes. However, some progress has been made in the development of suitable animal models (Taylor, 2002).

The suitability of these approaches to the assessment of the potential allergenicity of novel proteins introduced through agricultural biotechnology was demonstrated when a high-methionine protein from Brazil nuts was introduced into soybeans to correct their inherent methionine deficiency. At the time, Brazil nuts were known to be allergenic (Arshad et al., 1991) but the identity of the allergens in Brazil nuts was not known. The novel high-methionine protein from Brazil nuts was demonstrated to be the previously unidentified, major allergen from Brazil nuts on the basis of serum screening with blood serum from Brazil nut allergic individuals (Nordlee et al., 1996). Accordingly, commercialization of the high-methionine soybeans was not pursued.

Safety for animal feeding purposes

Safety assessments should also take into consideration possible animal feeding uses for the genetically modified crops. Maize, soybeans, canola and cottonseed are among the genetically modified crops that are important in the feeding of domestic animals. Feed safety assessment typically involves feeding studies with the appropriate target animal species and comparisons for typical performance parameters such as growth rate (Hammond et al., 1996).

Conclusions

In the USA and other countries, genetically modified foods pass through numerous regulatory hurdles between the early gene discovery phase and product commercialization. Accordingly, the products of agricultural biotechnology currently sold for food purposes around the world have been subjected to intensive safety evaluations. The safety assessment of genetically modified foods usually begins with a comparison of the novel food with its traditional counterpart. In most cases currently, the novel food is comparable with its conventional counterpart except for a few defined differences resulting from the introduction of the particular gene(s) of interest. As a result, the safety assessment is then focused upon the safety of the introduced gene and especially the novel protein produced from the gene. Current genetically modified foods are well documented to be safe for their intended uses under the anticipated conditions of consumption because they have been subjected to such safety evaluations. With the future introduction of genetically modified foods that are not comparable with any traditional food, the safety assessment process will become more difficult. However, no such genetically modified crops have yet been introduced into the marketplace. A regulatory system exists on a worldwide basis to ensure that suitable testing will be conducted to ensure the safety of current and future novel foods for consumers before their introduction into the marketplace. The major safety concern involves the potential allergenicity of the novel protein. However, the likelihood that a novel protein will become a new and unique allergen is quite low unless the novel gene is obtained from a known allergenic source or the novel protein is expressed in large amounts in the genetically modified foods. Approaches have been developed to assess the potential allergenicity of novel proteins introduced through agricultural biotechnology. Research continues on the development of improved approaches that will increase the assurance that these novel proteins will not become allergens.

References

Arshad, S.H., Malmberg, E., Krapf, K. and Hide, D.W. (1991) Clinical and immunological characteristics of Brazil nut allergy. *Clinical and Experimental Allergy* 21, 373–376.

Astwood, J.D., Leach, J.N. and Fuchs, R.L. (1996) Stability of food allergens to digestion *in vitro*. *Nature Biotechnology* 14, 1269–1273.

Beever, D.E. and Kemp, C.F. (2000) Safety issues associated with the DNA in animal feed derived from genetically modified crops. A review of scientific and regulatory procedures. *Nutrition Abstracts and Reviews Series B: Livestock Feeds and Feeding* 70, 175–182.

Breiteneder, H. and Ebner, C. (2000) Molecular and biochemical classification of plant-derived food allergens. *Journal of Allergy and Clinical Immunology* 106, 27–36.

Burks, A.W., Shin, D., Cockrell, G., Stanley, J.S., Helm, R.M. and Bannon, G.A. (1997) Mapping and mutational analysis of the IgE-binding epitopes of Ara h 1, a legume vicilin protein and a major allergen in peanut hypersensitivity. *European Journal of Biochemistry* 245, 334–339.

Calkoven, P.G., Aalbers, M., Koshte, V.L., Pos, O., Oei, H.D. and Aalberse, R.C. (1987) Cross-reactivity among birch pollen, vegetables and fruits as detected by IgE antibodies is due to at least three distinct cross-reactive structures. *Allergy* 42, 382–390.

Doerfler, W. and Schubert, R. (1997) Fremde DNA im Saugersystem. *Deutsches Arzteblatt* 94, 51–52.

FAO (1995) *Report of the FAO Technical Consultation on Food Allergies*. Food and Agriculture Organization of the United Nations. FAO, Rome, Italy.

FAO/WHO (1996) Biotechnology and food safety. *Report of a Joint FAO/WHO Expert Consultation*. Food and Agriculture Organization of the United Nations and World Health Organization. FAO, Rome, Italy.

FAO/WHO (2000) Safety aspects of genetically modified foods of plant origin. *Report of a Joint FAO/WHO Expert Consultation*. Food and Agriculture Organization of the United Nations and World Health Organization. WHO, Geneva, Switzerland.

FAO/WHO (2001) Evaluation of the allergenicity of genetically modified foods. *Report of a Joint FAO/WHO Expert Consultation*. Food and Agriculture Organization of the United Nations and World Health Organization. FAO, Rome, Italy.

FDA (1992) Statement of policy: foods derived from new plant varieties – Food and Drug Administration. *Federal Register* 57, 22984–23005.

Gendel, S.M. (1998) The use of amino acid sequence alignments to assess potential allergenicity of proteins used in genetically modified foods. *Advances in Food and Nutrition Research* 42, 45–62.

Hammond, B.G., Vicini, J.L., Hartnell, G.F., Naylor, M.W., Knight, C.D., Robinson, E.H., Fuchs, R.L. and Padgette, S.R. (1996) The feeding value of soybeans fed to rats, chickens, catfish and dairy cattle is not altered by genetic incorporation of glyphosate tolerance. *Journal of Nutrition* 126, 717–727.

Harrison, L.A., Bailey, M.R., Naylor, M.W., Ream, J.E., Hammond, B.G., Nida, D.L., Burnette, B.L., Nickson, T.E., Mitsky, T.A., Taylor, M.L., Fuchs, R.L. and Padgette, S.R. (1996) The expressed protein in glyphosate-tolerant soybean, 5-enolpyruvylshikimate-3-phosphate synthase from *Agrobacterium* sp. strain CP4, is rapidly digested *in vitro* and is not toxic to acutely gavaged mice. *Journal of Nutrition* 126, 728–740.

Hefle, S.L., Nordlee, J.A. and Taylor, S.L. (1996) Allergenic foods. *Critical Reviews in Food Science and Nutrition* 36, S69–S89.

IFT (2000a) IFT expert report on biotechnology and foods. Introduction. *Food Technology* 54(8), 124–136.

IFT (2000b) IFT expert report on biotechnology and foods. Human food safety of rDNA biotechnology-derived foods. *Food Technology* 54(9), 53–61.

James, C. (2001) *Global Review of Commercialized Transgenic Crops, 2000. ISAAA Briefs No. 23*. International Service for the Acquisition of Agri-biotech Applications, Ithaca, New York.

McClintock, J.T., Schaffer, C.R. and Sjoblad, R.D. (1995) A comparative review of the mammalian toxicity of *Bacillus thuringiensis*-based pesticides. *Pesticide Science* 45, 95–105.

Mekori, Y.A. (1996) Introduction to allergic diseases. *Critical Reviews in Food Science and Nutrition* 36, S1–S18.

Metcalfe, D.D., Astwood, J.D., Townsend, R., Sampson, H.A., Taylor, S.L. and Fuchs, R.L. (1996) Assessment of the allergenic potential of foods derived from genetically engineered crop plants. *Critical Reviews in Food Science and Nutrition* 36S, 165–186.

Moneret-Vautrin, D.A., Kanny, G., Rance, F. and Lemerdy, P. (1997) Les allergènes végétaux alimentaires. Allergies associées et réactions croisées. *Revue Francaise d'Allergologie et d'Immunologie Clinique* 37, 316–324.

Neilsen, K.M., Bones, A.M., Smalla, K. and van Elsas, J.D. (1998) Horizontal gene transfer from transgenic plants to terrestrial bacteria – a rare event? *FEMS Microbiology Reviews* 22, 79–103.

Nordlee, J.A., Taylor, S.L., Townsend, J.A., Thomas, L.A. and Bush, R.K. (1996) Identification of Brazil nut allergen in transgenic soybeans. *New England Journal of Medicine* 334, 688–692.

OECD (1992) *Safety Evaluation of Foods Derived by Modern Biotechnology – Concepts and Principles*. Organisation for Economic Cooperation and Development, Paris.

OECD (2000) *Report of the Task Force for the Safety of Novel Foods and Feeds*. Organisation for Economic Cooperation and Development, Paris, 86/ADDI, May 17.

Redenbaugh, K., Hiatt, W., Martineau, B., Kramer, M., Sheehy, R., Sanders, R., Houck, C. and Emlay, D. (1992) *Safety Assessment of Genetically Engineered Fruits and Vegetables. A Case Study of the Flavr Savr Tomato*. CRC Press, Boca Raton, Florida.

Sanders, P.R., Lee, T.C., Groth, M.E., Astwood, J.D. and Fuchs, R.L. (1998) Safety assessment of insect-protected corn. In: Thomas, J.A. (ed.) *Biotechnology and Safety Assessment*, 2nd edn. Hemisphere Publishing, New York, pp. 241–256.

Schnepf, E., Crickmore, N., Van Rie, J., Lereclus, D., Baum, J., Feitelson, J., Zeigler, D.R. and Dean, D.H. (1998) *Bacillus thuringiensis* and its pesticidal crystal proteins. *Microbiology and Molecular Biology Reviews* 62, 775–806.

Schubert, R., Hohlweg, U., Renz, D. and Doerfler, W. (1998) On the fate of orally ingested foreign DNA in mice: chromosomal association and placental transfer to the fetus. *Molecular and General Genetics* 259, 569–576.

Shin, D.S., Compadre, C.M., Maleki, S.J., Kopper, R.A., Sampson, H., Huang, S.K., Burks, A.W. and Bannon, G.A. (1998) Biochemical and structural analysis of the IgE binding sites on Ara h 1, an abundant and highly allergenic peanut protein. *Journal of Biological Chemistry* 273, 13753–13759.

Stanley, J.S., King, N., Burks, A.W., Huang, S.K., Sampson, H., Cockrell, G., Helm, R.M., West, C.M. and Bannon, G.A. (1997) Identification and mutational analysis of the immunodominant IgE binding epitopes of the major peanut allergen Ara h 2. *Archives of Biochemistry and Biophysics* 342, 244–253.

Taylor, S.L. (2001) Safety assessment of genetically modified foods. *Journal of Nematology* 33, 178–182.

Taylor, S.L. (2002) Protein allergenicity assessment of foods produced through agricultural biotechnology. *Annual Reviews of Pharmacology and Toxicology* 42, 99–112.

Taylor, S.L., and Lehrer, S.B. (1996) Principles and characteristics of food allergens. *Critical Reviews in Food Science and Nutrition* 36, S91–S118.

Taylor, S.L., Lemanske, R.F. Jr, Bush, R.K. and Busse, W.W. (1987) Food allergens: structure and immunologic properties. *Annals of Allergy* 59, 93–99.

Underwood, B.A. (1994) Vitamin A in human nutrition: public health considerations. In: Sporn, M.B., Roberts, A.B. and Goodman, D.S. (eds) *The Retinoids: Biology, Chemistry and Medicine*, 2nd edn. Raven Press, New York, pp. 217–227.

WHO (1993) *Health Aspects of Marker Genes in Genetically Modified Plants. Report of a WHO Workshop*. World Health Organization, Geneva, Switzerland.

Ye, X., Al-Babili, S., Kloti, A., Zhang, J., Lucca, P., Beyer, P. and Potrykus, I. (2000) Engineering the provitamin A (β-carotene) biosynthetic pathway into (carotenoid-free) rice endosperm. *Science* 287, 303–305.

Zitnak, A. and Johnston, G.R. (1970) Glycoalkaloid content of B5141-6 potatoes. *American Potato Journal* 47, 256–260.

Index

Note: page numbers in *italics* refer to figures and tables

aberrant crypt foci (ACF) in colon cancer 190, 191, 192
ACC gene 96
acetyl-CoA carboxylase 154
N-acetylsphingosine 190
acute phase proteins 179–180
S-adenosylhomocysteine (AdoHcy) 238, 241
    DNA methylation loss 245
    folate deficiency 251
    intestinal cancer 252
    methyl deficiency 249
S-adenosylmethionine (AdoMet) 238, 241
    DNA methylation loss 245
    folate deficiency 251
    intestinal cancer 252
    methyl deficiency 249
    synthesis 239, 241
    universal methyl donor 241–242
adipose tissue, gene regulation 127–128
ADP-ribose 50
aglycones 201–202
agricultural biotechnology
    animal feeding 392
    antibiotic resistance marker safety 388–389
    DNA safety 388
    gene discovery phase 384–385
    impact
        on consumers 382–383
        on production agriculture 381–382
    line selection phase 385–386
    novel protein safety 389–392
    product advancement to commercialization phase 386
    safety assessment process 383–392
    substantial equivalence concept 386–387
    transfer to gastrointestinal bacteria 388
AhR nuclear translocator (ARNT) 212
AIDS 79
albumin gene
    vitamin B_6-induced expression 159
    vitamin C deficiency 161
alcohol consumption
    dietary folate and cancer risk 247, 248
    folate deficiency 239
allergen immunotherapy 375
allergenic proteins
    modification 375–376
    see also food allergens/allergy
Alzheimer's disease 79, 325
amino acid response elements (AARE) 108–110, 115
amino acid response (AAR) pathways 108, 110, 115
amino acids
    anionic 32–33
    asparagine synthetase regulation 110–111
    availability 106–107
    cationic 32
    dietary and immune function 362–364
    gene expression regulation in yeast 105–106
    homology with novel protein 390–391
    N transporters 32
    signal molecules 105–106
    transcriptional control 107–115
    transmembrane transport 28–30, *31*, 32–35
    zwitterionic 29–30, *31*, 32
aminophospholipids 73
AMP-activated protein kinase (AMPK) 94–95
anaphylaxis 371
angiotensin converting enzyme (ACE) inhibitors 79
animal feeding 392

animal products, genetically modified 382
anti-nutrients 387
antibiotic resistance marker safety 388–389
antibodies 350
antigens 350
antioxidant defence system 169
    lens protection 323–325
antioxidant index, cataract 339, *341*
antioxidants and cataract risk 325–342
    association 325–326
    combinations 339–340
    defences 323–325
    vitamin A 339, 340
    see also ascorbate; carotenoids; vitamin E
APC gene 80, 193–194
    mutation 190, 191
apo-B 296
    editing in chylomicron assembly 304
    mRNA editing 296–297
    non-polar lipid transfer 300
apo-B48 296–297, 302, 303
apo-B100 296–297
    apo-B48 yielding 304
    post-translational regulation of secretion 298
    precursor particles 298, 300
    translational arrest 298
    translocation across RER membrane 298–299
    VLDL assembly 298
apo-bec-1 297
apocarboxylases 67–68, 69
apolipoprotein(s) 293, *294*, 296
apolipoprotein A-1 161
apoptosis 73–85
    cancer 79–84, 129
    death signals 75–77, 79
    DNA damage 277
        response 74–75
    effector molecules 78
    intestinal disorders 78–79
    mitochondrial pathway 77
    modulation in disease prevention and treatment 78–80
    p53AS 281
    PARP 280
    plant secondary metabolites 82–84
    process 75–78
    roles 73–74
    selenium 178
    sphingolipids 188, 193
    survival signals 75, *77*
apoptotic bodies 73
apples 376
aquaporins 36
Ara h proteins 372, 373
    amino acid sequence homology 390–391
arachidonate 12-lipoxygenase (ALOX 12) 179
arachidonic acid 361, 362

arginine 362–363
aryl hydrocarbon receptor (AhR) 212
arylamine-induced carcinogenesis 162
ASC transporter family 33
ascorbate 324, 325
    cataract 326–327, *328*, 329, *330*, 331
ascorbic acid *see* vitamin C
L-ascorbic acid 54
ascorbic acid-transporting proteins 37
asialoglycoprotein receptor (ASGR) 156, 157
asparaginase, leukaemia therapy 111–112
asparagine synthetase 109, 110–115
    cell cycle control 111
    mRNA expression 110–111
    promoter 112
asparagine synthetase (AS) gene
    NSREs 113–115
    nutrient sensing 112–113
aspartate aminotransferase, cytosolic (cAST) 159
ataxia with vitamin E deficiency (AVED) 170
atherosclerosis, VLDL risk factor 300
ATM gene 74
ATRX domain 254
5-azacytidine 251

B lymphocytes 350–351
Bacillus thuringiensis (Bt) 384
rBAT 30, *31*
Bcl-2 protein family 77–78, 79
1,4-benzoquin 230
best-matching unit (BMU) 8
BH domains 77–78
bile acid response elements (BAREs) 312–313
bile acids 309
    metabolism 313–314
bile salt export pump (BSEP) 313–314
biocytin 63, 156
    cleavage 268, *269*
biomarkers of nutrient status 18
biotin 52, 62–64
    brain regions 155
    carboxylases 155
    cell culture requirements 153
    cell differentiation 153–154
    cell proliferation 70, 152
        effect on uptake 64–67, 70
    cellular transport 63
    deficiency 154
    egg-yolk 155
    embryological development 154
    functions 152
    gene expression regulation 152–158
    glucokinase induction 157
    cGMP 157
    HDL 153
    holocarboxylase synthetase 155

immune system 155–156
insulin secretion 158
lymphocytes 156
neuronal plasticity 155
optically clear nuclei (OCN) 152–153
protein synthesis regulation 156–158
transporter synthesis 66–67
uptake rate 64–65
    cell cycle phase 65, *66*
biotin-binding protein 155
    nuclear 152
biotin-dependent carboxylases 67–69
biotinidase 64, 155
    biocytin cleavage 268, *269*
    covalent modification 273
    gene expression 272–273
bone marrow, niacin dietary levels 285, *286*, *287*
brain-derived neurotrophic factor (BDNF) 154–155
Brazil nut allergens 392
breast cancer
    dietary folate 248
    folate levels 248
    oestrogens 210
    ω3-PUFA 129
    soy intake 208, 210
breast milk, isoflavone levels 207
BSEP gene 314
Bt proteins 384–385
butyrate 27, 84

calbindin 45–46
calciferols 44–46
calcium signalling, intracellular 282, 283
calpains 219, 230
cancer
    apoptosis 79–84, 129
    cachexia 230
    DNA methylation 237–239
    fatty acids 129–130
    folate in diet 246–249
    isoflavones 208–210
    malignant genome mapping 3–4
    methyl dietary deficiency 254–255
    methyl donor dietary deficiency link 246–249
    niacin 284–287
    phyto-oestrogens 209
    plant secondary metabolites 82–84
    selenium 178–179
    sphingolipid-rich foods 189
    sphingolipids 189–196
        suppression mechanisms 193–196
    vitamin B_6 159
    vitamin C 162
canola
    animal feeding 392
    oil 382

carbohydrate availability in diet 92
carbohydrate response element (ChoRE) 95–96, 97
carbohydrate response element-binding protein (ChREBP) 97
carboxylases, biotin-dependent 63
α-carotene, cataract 334, 336–337, *338*
β-carotene 81, 82
    cataract 327, 329, 334, 336–337, 339, 340
    lens 324
carotenoids 324
    cataract 334, *335*, 336–339
cartilage 160
$α_{S1}$-casein 374, 375
α-casein 372
caspase(s) 78, 79
    skeletal muscle 219
caspase-8 76, 78
cataract
    age-related damage 321, *322*, 323–325
    age-related prevalence 321
    antioxidant index 339, *341*
    antioxidants 325–342
        association 325–326
        combinations 339–340
        defences 323–325
    ascorbate 326–327, *328*, 329, *330*, 331
    α-carotene 334, 336–337, *338*
    β-carotene 327, 329, 334, 336–337, 339, 340
    carotenoids 334, *335*, 336–339
    β-cryptoxanthin 336, 337
    economic burden 321
    folate 334
    iron intake 340
    lutein 334, 336, 338
    lycopene 334, 336, 337
    molybdenum 340
    multivitamin supplements 339–340
    niacin 340
    nutritional influences on risk 321–342
    oxidative insult 321, *322*, 323–325
    public health issues 321
    retinol 340
    riboflavin 334, *337*, 340
    vitamin A 339, 340
    vitamin C 339
    vitamin E 327, 329, 331, *332*, 333–334, 339, 340
    zeaxanthin 334, 336, 338
    zinc 339, 340
β-catenin 193–195
cathepsin L 230
cathepsins 219
cationic amino acid transporter (CAT) 32, 107
$CD4^+$ cells *see* helper T cells
$CD8^+$ cells *see* cytotoxic T cells
CD36 deficiency 28
cDNA pools 21

cell cycle
    arrest 70, 74
        p53 altered signalling 287
    biotin uptake 65, *66*
    control and asparagine synthetase 111
    histone biotinylation 271–272
    mammalian 61–62
    selenium 178–179
cell differentiation, vitamin C 160
cell migration 195
cell proliferation
    biotin 64–67, 70, 152
    colon cancer 193
    gene expression 61
    histones 268
        biotinylation 69–70, 270–271
    immune response 354
    linoleic acid 70
    metabolic flux 61–62
    nutrient homeostasis 62–64
    nutrient status 70
    PBMCs 271
        cell model 64–70
    vitamin B_6 158
    vitamin C 160
ceramide 188, 191, 194–195
    lipophilic derivatives 193
cervical carcinoma 162, 247
chemotherapy 79
    niacin deficiency 287
    sphingolipid metabolites 192–193
cholesterol
    biliary excretion 309
    cell membrane 309
    dietary 315
    intracellular concentration 309
    metabolism 309, *310*
    plasma lipoprotein 309
    regulation of cellular 309–316
    regulatory proteins 314–316
    transcription factors 309–314
    transport 295
cholesterol 7α-hydroxylase 309, *310*, 312, 315–316
chondrogenesis, vitamin C 160
CHOP gene 107–110
    AAR pathways 108, 110, 115
    AARE 108–110, 115
chromatin-associated proteins 253–254
chromosome *in situ* hybridization 2–3
chromosome microdissection 4
chylomicron retention disease 296
chylomicrons 122, 126, 293, *294*, 295
    assembly 295–296
        apo-B editing 304
        in enterocytes 300–304
        models 301–302
        two-step model 302–304

    characteristics 295–296
    precursors 303
    SER 303
    vitamin E transport 170
CoA 51, 52
cobalamins 53–54
coeliac disease 79, 376
coenzyme compartmentalization 54
collagen synthesis 160
colon/colorectal cancer 80
    aberrant crypt foci 190, 191, 192
    apoptosis target in chemoprevention 80–84
    colonic proliferation 193
    dietary folate 246–247, 248–249, 252
    DNA methylation 249
    fat-soluble vitamins 81–82
    MTHFR function 248–249
    soy intake 208
    sphingolipids 189
    sphingomyelin 190
colonocytes 27
comparative genome hybridization 4
concentrative nucleoside transporter (CNT) proteins 25
congenital malformations, biotin deficiency 154
copper, immune function 359
coronary heart disease (CHD) 179
cottonseed, animal feeding 392
cow's milk
    allergy 372, 375, 376
    hydrolysed 376
    infant formula 207
CpG sites, methylation status 255, 256
crop plants 382–383
cruciferous vegetables 82
β-cryptoxanthin 336, 337
crystallins 321
curcumin 82–83
cyclooxygenase (COX) 81, 84
    vitamin E regulation 171–172
Cyp7a1 gene promoter 312–313, 316
cystathionase 159
cystathionine-β-synthase (CBS) *240*, 241
cysteine scanning 22
cytochrome c 77, 78
cytochrome P450
    dietary factors 211
    expression alteration 212
    vitamin C deficiency 161
cytokines 351, *352*, 353–354
    malnutrition 355
cytotoxic T cells 351, 353, 355

1,25-$(OH_2)D_3$ 45, 46, 357
daidzein 201, 202, *203*, 205
death receptor pathway 75–77
death signals 75–77, 79

delayed-type hypersensitivity (DHT) 351, 355, 356
    essential fatty acids 360
    nucleotide role 364
    vitamin E 358
deubiquitinating enzymes (DUBs) 222, 223
deubiquitination 222–223
    rate in skeletal muscle 227–228
diabetes insipidus 35
diabetes mellitus, non-insulin-dependent (NIDDM) 23
diet, cellular methylation status 239, *240*, 241–246
dietary fibres, hypocholesterolaemic effect 314, 315, 316
dihomo-γ-linolenic acid 361
1α,25-dihydroxy-D$_3$ 45, 46, 357
5'DIs 177
disease gene isolation 3
DNA
    cancer malignant genome mapping 3, 4
    damage
        apoptosis 74–75, 277
            PARP activation 280
    human dietary intake 388
    mismatch repair pathway 75
    molecule 1, 2
    mutagenic lesions 282
    nucleosides 25
    nucleotide role 363
    p53 activation 280
    repair 277
        nutritional status 268
            PARP 278–280, 284
    replication *246*, 354
    safety 388
    sequencing 3, 4, 10
    synthesis 358
        purine metabolism 239, 241
    transfer of plant to animal cells 388
DNA-binding proteins, methylated 253
DNA cytosine (C5) methyltransferases 242–246
DNA methylation
    bisulphite sequencing 255
    cancer 237–239
    colon cancer 249
    dietary methyl deficiency 254–255
    dietary sources of methyl groups 238
    folate deficiency 239, 241, 251–252
    global loss 255
    hypomethylation 256
    liver tumours 249–251
    loss 245–246
    methyl donor deficiency 242
    remethylation 256
    SNF2 protein 253–254
DNA methyltransferases (DNMTs) 238, 242–245, *246*
    chromatin-associated proteins 253–254
    chromatin remodelling 253–254
    functions 252–254

    structure *243*
    see also *individual DNMTs*
DNA microarrays 4–5, *6*, 14
    clustering algorithms 8
    data analysis 5–6, *7*, 8–10
    similarity measurements 5–6, 8
    technology 1
DNMT1 243–244, 245, 254–255
    functions 253
    intestinal cancer 252
    maintenance 256
    structure *243*, 244
DNMT3a *243*, 244–245
    functions 253
DNMT3b *243*, 244–245
    functions 253
docosahexaenoic acid (DHA) 81, 121, 360

egg allergens 372, 373–374
Ehlers–Danlos syndrome 161
eicosanoids 361–362
eicosapentaenoic acid (EPA) 80–81, 121, 360, *361*
    proteasomes 230
elastin synthesis 160
electro-spray ionization mass spectrometry (ESI-MS) 16, 17
elimination diets 375
endoplasmic reticulum, SREBPs 309–310
endoplasmic reticulum stress response (ERSR)
    nutrient-sensing pathways 107, 108, 109
    AS gene 112–113
enterocytes
    chylomicron assembly 300–304
    fat absorption adaptation 304
    isolated adult 301–302
    subcellular fractions 302
environmental allergens 389
Environmental Protection Agency (EPA) 385–386
epidermal growth factor 8 (EGF-8) 73
epoxide hydrolase 212
equilibrative nucleoside transporter (ENT) protein 25
L-ethionine 251, 254
Euclidean distance 5, *7*
excitatory amino acid transporters (EAAT) 32–33
expressed sequence tags (ESTs) 2
expression profiling 13, 19
extracellular matrix, vitamin C in production 160, 161

F-box proteins 222
4F2hc 30, *31*
familial adenomatous polyposis (FAP) 190, 191, 193–194
    DNA methylation 238
    see also APC gene

famine 349
Fanconi–Bickel syndrome 23
farnesoid X receptor (FXR) 309, 312–314, 316
    hepatic cholesterol metabolism *310*
FAS 93, 94
    inhibition 95
Fas-associated death domain (FADD) 76, 77
Fas–FasL apoptosis pathway 76–77, 195
fat, dietary, immune function 360–362
fat absorption 301, 304
fatty acid(s) 121–130
    cancer 129–130
    derivatives 121–122
    dietary 315, 316
    essential and immune function 360–361
    gene expression regulation 124, 126–130
    immune system 361–362
    monounsaturated (MUFAs) 314
    PPAR system ligand actions 123–124
    short chain 27, 315
    synthesis 51
    transcription factors 123
    transporters 27–28, 126–127
    tumor cell effects 129–130
    see also long chain fatty acids (LCFAs);
        polyunsaturated fatty acids (PUFAs)
fatty acid-binding proteins (FABPs) 122
    plasma membrane (FABPpm) 28
fatty acid translocase (FAT) 28
fatty acid transporter proteins (FATP) 28
fish allergens 373, 375
fish oils, dietary 300, 361
flavin adenine dinucleotide (FAD) 354
fluorescent *in situ* hybridization (FISH) 3
folate-binding protein 52
folate/folic acid
    immune function 356
    intake and cataract 334
    transporters 37, 52–53
folate/folic acid deficiency 52–53, 79
    cancer risk 246–249, 252
    colon cancer 246–247, 248–249, 252
    DNA methylation 239, 241
        inhibition in liver and colon 251–252
    homocysteine 241
food allergens/allergy 375–376
    characteristics 371
    clinical manifestations 370–371
    diagnosis 371
    identification 371–373
    IgE-binding epitope characterization 373–375
    IgE-mediated 390
    immunological basis 369–370
    novel proteins 389–390
    plant breeding 376–377
    prevalence 369
    strategies to alter 375–377

Food and Agriculture Organization of the United
    Nations (FAO) 386
Food and Drug Administration (FDA, USA) 383, 386
foods, functional 189
Fourier transform infrared spectroscopy 16
functional genomics 4–6, *7*, 8–10, 14
    gene-/phenotype-driven 14, *15*

Gad c 1 protein 373
Gal d proteins 372
α-galactosylceramide 195
gangliosides
    anti-tumour 196
    cell surface 195
gas chromatographs 18
gas chromatography–mass spectrometry (GC–MS)
    techniques 17, 18
gatekeeper genes 277
    hypothesis 80
gene clustering 8
    self-organizing map 9–10
general control non-repressible (GCN) pathway
    105–106
genetic distances 1–2
genetic engineering 376–377
genetic maps 1–2
    comparative 3
    self-organizing 9–10
genetically modified foods 381–393
    agricultural biotechnology
        impact 381–383
        safety assessment 384–393
    animal feeding 392
    antibiotic resistance marker safety 388–389
    gene transfer to gastrointestinal bacteria 388
    novel protein safety 389–392
    substantial equivalence concept 386–387
genistein 201, 202, *203*, 205
    oestrogenic effects 208
genome
    human 1, 4
    mapping 3, 4
genome-wide differential display *7*, 9–10
genomics 1–10
    see also functional genomics
GI-GPx 177, 178
gliadins 376
glucocorticoid receptors 159
glucocorticoid-responsive elements 159
glucokinase
    biotin-induced induction 157
    insulin secretion 158
gluconeogenic genes 92
glucose
    cellular cascade to transcription machinery
        94–95

cis-acting DNA sequences in response 95–96
    gene expression regulation 91–99
    gluconeogenic enzyme gene expression effect 97, 99
    hepatic gene regulation *98–99*
    membrane sensors 94
    metabolic hypothesis 93–94
    non-metabolic hypothesis 94
    renal reabsorption 25
    trans-acting factors 96–97
glucose-6-phosphatase 97, 99
glucose-6-phosphate 93
glucose galactose malabsorption syndrome 25
glucose response elements (GlREs) 95–96, 97
glucosinolates 82
GLUT transporters 22–23, 37
    malfunctions 23
GLUT2 23, 93, 94
GLUT4 23, 93
glutamine 32, 363
glutathione 324, 325
    immune function 362
glutathione peroxidase 282, 324
    selenium-dependent 360
glutathione transferases 212
gluten 78–79
α-glycerophosphate pathway 54, 301
glycine 33
    metabolism 53
glycitein 201, *203*
glycogen 91, 93
glycogen storage disease, type I 23
glycolytic genes 92–93
GLYT transporters 33
cGMP 157
GPx 177–178
groundnut allergens 371–372, 373, 376

heat-shock protein 70 299
HECT domain enzymes 221
helper T cells 351, 353, 355
    food allergy 369–370
hepatic gene regulation 128–129
    by glucose 91–92
hepatic nuclear factor 4α (HNF4α) 128, 129
herbicide tolerance 382
hierarchical clustering 8
high density lipoprotein (HDL) 126, 293, *294*, 295
    biotin 153
high-performance liquid chromatography (HPLC) 18
high-performance liquid chromatography–mass spectrometry (HPLC–MS) techniques 18
histone(s)
    cell proliferation 268
    debiotinylation 273–274
    DNA repair 267–268
    DNA transcription 267
    poly(ADP-ribosylation) 267–268
    post-translational modification 267, 268
    spermatogenesis 268
    ubiquitination 267
histone biotinylation 63–64, 69–70, 156, 267–274
    biotinidase 272–273
    cell cycle 271–272
    cell proliferation 270–271
    cofactors 273–274
    enzymatic 268, *269*
    human cells 270–274
    identification 270
    PBMCs 69–70, 270–272
histone deacetylases (HDACs) 253
HIV protease inhibitors 23, 79
holocarbonase synthetase 273
holocarboxylase synthetase 67–69, *70*
    biotin binding 155
homocysteine 241
    folate status 247
human genome 1, 4
human leukocyte antigens (HLA) 351
human papilloma virus (HPV) 162
human–rodent hybrid panels 2
3-hydroxy-3-methylglutaryl-CoA (HMG-CoA) reductase 309, 314
3-hydroxy-3-methylglutaryl-CoA (HMG-CoA) synthetase 310
15S-hydroxyeicosatetraenoicin (15S-HETE) 130
hypoxia, MCT expression effects 27

IBABP gene 313, 314
ileal bile acid-binding protein (IBABP) 313, 314
immune function 349–364
    copper 359
    dietary amino acids 362–364
    dietary fat 360–362
    folic acid 356
    impaired 74
    iron intake 359
    micronutrients 355–360
    multivitamin supplements 360
    nucleotides 363–364
    nutrient effects 354
    protein energy malnutrition 354–355
    selenium 359–360
    vitamin A 355–356
    vitamin B group 355–356
    vitamin B_6 356
    vitamin B_{12} 356
    vitamin C 356
    vitamin D 356–357
    vitamin E 357–358
    zinc 358–359
immune response 354

immune system 349–351, *352*, 353–354
    biotin 155–156
    communication 351, 353–354
    fatty acids 361–362
immunity
    acquired 350
    humoral 351
    innate 349–351
immunoglobulin(s) 350
immunoglobulin A (IgA) 355
immunoglobulin E (IgE) 353
    antibodies 370, 390
        assays 371
    food allergy 369, 370
    sensitization 390
immunoglobulin E (IgE)-binding epitopes 371, 373–375, 376, 390
immunoglobulin G (IgG) 373–374
immunoproteasomes 225
infant growth/development, soy proteins 211
infectious disease 349
    selenium 360
    vitamin A 355–356
inflammation
    PPARs 127
    selenium deficiency 180
insect resistance 382, 383
insulin 92, 93
    hepatic gene regulation *98–99*
    proteasomes 229
    secretion with biotin 158
    VLDL effects 300
insulin-like growth factor-binding protein 1 (IGFBP-1) 107, 109, 110
insulin-like growth factor I (IGF-I) 109, 110
insulin receptor (IR) 156, 157
interferon γ (IFN-γ) 353
    vitamin D effects 356, 357
interleukin 1 (IL-1) 351, 353, 355
interleukin 2 (IL-2) 74
    glutamine role 363
    nucleotide role 364
    vitamin D effects 356, 357
    vitamin E effects 357–358
interleukin 4 (IL-4) 353
interleukin 5 (IL-5) 353
interleukin 6 (IL-6) 351, 353, 355
interleukin 10 (IL-10) 353
interleukin 12 (IL-12) 353
    vitamin D effects 356
intermediate density lipoprotein (IDL) 295
intestine
    apo-B mRNA editing 296–297
    cancer and DNMT1 252
    coeliac disease 79, 376
    dietary fats 296

    disorders and apoptosis 78–79
    villus/crypt gradient cells 302
    see also colon/colorectal cancer; familial adenomatous polyposis (FAP)
intracellular adhesion molecule 1 (ICAM-1) 172, 173
iron intake
    cataract 340
    deficiency 359
    immune system 359
isoflavone glycosides 201, *203*
isoflavones 201–212
    biological activity 206
    cancer 208–210
    exposure profiles 207
    fetal exposure 207
    gene transcriptional repression triggering 209
    health effects 206
    human exposure 206–207
    identification 205–206
    infant growth/development 211
    intracellular conjugated 202, 204
    metabolites *203*, *204*
    oestrogen-mediated effects 207–211
    phase I enzyme effects 212
    reproductive effects 210–211
    serum levels after soy ingestion 205
    structure *202*
isothiocyanates 82

K-mean clustering 7, 8
kidneys, glucose filtering 25

L-PK gene 93, 94
    inhibition 95
lactacystin 228
β-lactoglobulin 372
lens
    defences against damage 323–325
    functions 321, *322*, 323
    proteolytic enzymes 325
leukaemia, asparaginase therapy 111–112
light exposure, cataract 323
linoleic acid 70, 360, *361*
    cell proliferation 70
    conjugated 81
α-linolenic acid 360, *361*
linolenic acids, conjugated 122
lipids, dietary 122
lipogenic genes 92
    glucose effect on expression 92–93
lipoprotein lipase (LPL) 296
    gene 126
lipoproteins 45
    cholesterol 309

cholesterol-transporting 295
   plasma 293, *294*
      structure *295*
   triglyceride-transporting 293, *294*, 295
      assembly 295–304
lipotrope deficiency 251
liver
   cholesterol metabolism 309, *310*
   fatty acids 296, 315
   VLDL assembly 297–300
liver cancer, DNA methylation 249–251
liver receptor homologue 1 (LRH-1) 313
liver X receptors (LXR) 128–129, 309, 311–312, 316
   hepatic cholesterol metabolism *310*
long chain fatty acids (LCFAs) 27–28, 121, 122
   colorectal cancer 80–81
low density lipoprotein (LDL) 293, *294*, 295
   oxidized 179
   tocopherol transfer 170
low density lipoprotein (LDL) receptors 295, 309, 310, 311, 314–315
lung cancer, folate levels in smokers 247–248
lungs and food allergy 370
lutein, cataract 334, 336, 338
LXR response element (LXRE) 312
lycopene, cataract 334, 336, 337
lymphocyte-specific helicase (LSH) 254
lymphocytes 350–351
   biotin 156
lysinuric protein intolerance 32
lysozyme 372
lysyl hydroxylase 161

macrophages 353
maize
   agricultural biotechnology impact 381–382
   animal feeding 392
   staple diet 277, 284
major histocompatibility complex (MHC) 351
Mal d 1 protein 376
malaria, iron status 359
malate–aspartate shuttle 54
malnutrition 230, 354–355
   marker 74
mass spectrometers 18
matrix-assisted laser desorption/ionization time-of-flight mass spectrometry (MALDI-TOFMS) 17
membrane protein
   structure 21–22
   topology analysis 22
menadione 47
menaquinone-7 47
Menkes syndrome 359
menstrual cycle, soy protein consumption 210

metabolism
   molecular descriptors 13–14
   steady-state determination 14
metabolite analysis 16
   NMR techniques 18
metabolomics 14, 16, 17–18
metalloproteinases, skeletal muscle 219
metastasis, sphingolipids 195
methionine adenosyl transferase (MAT) 239, *240*
methyl dietary deficiency in cancer 254–255
methyl donors
   dietary deficiency 241
      cancer link 246–249
      liver cancer 249–251
   universal 241–242
methyl trap hypothesis 53–54
3-methylcrotonyl-CoA carboxylase (MCC) 67, *68*, 155
methylenetetrahydrofolate reductase (MTHFR) 239, 248
3-methylhistidine 228
methylsphingosine 190
MG132 228
microsomal triglyceride transport protein (MTP) 299–300, 304
milk, sphingolipids 189, 190
milk allergens/allergy 372–373, 374–375
   calcium intake 375
   hydrolysed protein 376
mitogen-activated protein kinase (MAPK) 161–162
mitosis 61, *62*
molybdenum, cataract 340
mono ADP-ribosylation 282, 283
monocarboxylate transporters (MCTs) 25–27, 63, 67
monocyte chemotactic protein-1 (MCP-1) 172, 173, 174
monoglyceride pathway 301
mRNA molecules 13, 14
   biomarkers of changes 18, 19
   expression 16
multivitamin supplements 360
   cataract 339–340

N-CoR (nuclear receptor co-repressor) 141
Na^+/H^+ antiporters 34
NAD 50, 278, 354
   depletion in niacin deficiency 285
NADP 50, 278
NADPH, oxygen radical detoxification 282
natural killer (NK) cells 350, 353
   immune function 357–358
neomycin phosphotransferase II 389
niacin 49–50, 279
   bone marrow effects 285, *286*, *287*
   cancer 284–287
   cataract 340

niacin *continued*
    deficiency 277–278, 282
        chemotherapy 287
        chromosomal instability 286
        genomic instability 285–287
        NAD depletion 285
        oesophageal cancer 284
        PARP 282–284
        sister chromatid exchange (SCE) 286
        skin cancer 285, 287
    genomic stability role 282, 284–287
    oxidant defence role 282
    skin effects 285
nicotinamide adenine dinucleotide *see* NAD
nicotinic acid 49, 277
nicotinic acid adenine dinucleotide phosphate (NAADP) 282
nicotinomide 49
nitric oxide 174
nitric oxide synthase, endothelial (eNOS) 174
nitrogen *see* reactive nitrogen species (RNS)
nuclear factor-κB (NF-κB) 362
    transcription 229
    vitamin C 161–162
    vitamin E 172, *173*
    zinc 359
nuclear magnetic resonance (NMR) spectroscopy 16, 18
nucleoside transport 25
nucleotides, immune function 363–364
Nur77 74
nutrient deprivation 105, 107–108
nutrient homeostasis
    cell proliferation 62–64
    proliferating cells 61–70
nutrient-sensing response elements (NSRE) 113–115
nutrient-sensing response unit (NSRU) 114
nutrient transporters 21–38
    cloning 22–23
    coenzyme compartmentalization 54
    folic acid 37, 52–53
    pantothenic acid 51–52
    thyroid hormones 33
    vitamins 36–38
    water 35–36
    see also amino acids, transmembrane transport; fatty acid(s), transporters; monocarboxylate transporters (MCTs); sodium-dependent glucose transporter (SGLT) family

oesophageal cancer, niacin deficiency 284
oestrogen, VLDL effects 300
oestrogen receptors (ER) 207, 208–209
    activation 209–210
    kinase cascade 210
    repression 209–210
oestrogens 207–208
    breast cancer 210
    soy-fed infants 208
opsonization 351
optically clear nuclei (OCN) 152–153
Organization for Economic Cooperation and Development (OECD) 386
osteoarthritis 79
osteocalcin 48
osteoporosis 79
ovalbumin 372, 374
ovomucoid 372, 373–374
ovotransferrin 372
oxidative burst 354
oxygen *see* reactive oxygen species (ROS)

p53 74–75, 79–80
    cell cycle arrest impairment 287
    expression loss in liver tumours 251
    impaired function 281
    PARP 280–281
    SIR2 in regulation 282
p53AS 281
p53–PARP complexes 281
PA28 *224*, 225–226
PA200 226
PA700 *224*, 225, 228
    subunit expression 229–230
pantothenic acid 51–52
paraptosis 73
Parkinson's disease, ascorbic acid 161
PARP-2 and -3 281, 282, 283
pellagra 277
Pen a 1 protein 373, 375
pepsin 377, 392
PEPT1 and 2 33–35
peptide mapping 16–17
peptide mass analysis 17
peptide transport 33–35
peptidomimetic drugs 35
peripheral blood mononuclear cells (PBMCs)
    biotin uptake 64–70
    biotinidase gene expression 272–273
    carboxylase activity 67
    cell proliferation 271
    histone biotinylation 69–70, 270–272
peroxisome proliferator-activated receptor(s) (PPARs) 123
    agonists 126, 128
        tumour cell effects 129–130
    fatty acid ligands 123–124, *125*, 362
    gene regulation 127
    inflammation 127
    tumour cell effects 129–130

peroxisome proliferator-activated receptor α (PPARα) 362
peroxisome proliferator-activated receptor γ (PPARγ) 81, 82
    gene regulation 127
    immune system 362
    lipid metabolism 127
pesticide residues 382
phagocytic cells 349–350
pharmacogenomics 19
phase I enzymes 211, 212
phase II enzymes 211–212
phenomics 16
phenylbutyrate 84
PHGPx 177, 178, 179
phosphoenolpyruvate carboxykinase (PEPCK)
    transcription inhibition 97, *98*, 157–158
phylloquinone 47
physical mapping 2
    chromosome *in situ* hybridization 2–3
    disease gene isolation 3
    DNA sequence 10
    human–rodent hybrid panels 2
phytanic acid 122, 124
phyto-oestrogens 207, 208–209
    cancer risk 209
Pick's disease 325
plant breeding, traditional 376–377
plant secondary metabolites 82–84
plasma membrane
    amino acid transport 28–30, *31*, 32–33
    GLUT transporters 22–23
    long chain fatty acids 27–28
    monocarboxylate transporters 25–27
    nucleoside transport 25
    nutrient transport 21
    SGLT transporters 23–25
    short chain peptide transport 33–35
    transporters 21–38
platelet aggregation 174
pluronic acid 296
pollen allergens 389
poly(ADP-ribose) 278
    p53 expression regulation 281
    synthesis 278, *279*
        during apoptosis 280
        impaired 287
poly(ADP-ribosylation) 267–268, 280
poly(ADP-ribose) glycohydrolase 278
polymorphic DNA markers 2
poly(ADP-ribose) polymerase (PARP) 278
    apoptosis 280
    carcinogenesis 284
    DNA repair 278–280
    genomic stability 282–287
    homologues 281–282, 283
    inhibition 283

niacin deficiency 282–284
null mouse models 283–284
p53 280–281
p53–PARP complexes 281
polyubiquitination 220–222
polyunsaturated fatty acids (PUFAs) 121–122, 124
    free radical attack 168–169
    HMG-CoA reductase activity 314
    PPAR agonists 128
ω3-polyunsaturated fatty acids (PUFAs) 81, 360, 361, 362
    breast cancer 129
    VLDL assembly inhibition 300
ω6-polyunsaturated fatty acids (PUFAs) 300, 360, *361*
post-genomic nutrition 13–19
PPAR-response element (PPARE) 123, 124
pregnancy, vitamin C 161
premature rupture of the membranes (PROM) 161
pristanic acid 124
promyelocytic leukaemia gene (PML) 253
propionyl-CoA carboxylase 67, *68, 69*
    biotin deficiency 155
prostate cancer, PPARs 130
proteasome(s) 223–236
    activators 225
    eicosapentaenoic acid 230
    hybrid 226
    insulin effects 229
    skeletal muscle 228–230
19S proteasome 226
20S proteasome 224–225, 228, 229, 230
26S proteasome 223–225, 226, 228, 230
protein(s)
    cytotoxicity of damaged 325
    enzymatic hydrolysis 376
    introduction of novel in plant breeding 383
    modification of allergenic 375–376
    safety of novel 389–392
        allergenicity 389–390
        amino acid homology 390–391
        proteolytic stability 391–392
    synthesis regulation 156–158
protein disulphide isomerase (PDI) 299
protein kinase C (PKC) 167, 174
    sphingosine inhibition 187
    α-tocopherol inhibition 171
protein phosphatase 2A 171
protein spot identification 16–17
protein-energy malnutrition 354–355
proteome analysis 17
proteomics 13, 14, 16–17
PTR family transport 33
purine metabolism 239, 241
pyridoxal 50
pyridoxal-5-phosphate (PLP) 158–159
pyridoxal lipase 51

pyridoxamine 50
pyridoxine 50

quinone reductase 212

reactive nitrogen species (RNS) 77, 167
reactive oxygen species (ROS) 74, 77, 167–168
    oxidative burst 354
    selenium 178
recombinant DNA technologies 383
reproduction, isoflavones 210–211
restriction sites 2
retina 321, *322*
retinitis pigmentosa 79
retinoic acid receptors (RARs) 44, 135
    co-repressors/-activators 141–142
    domain structure 136–137
    expression pattern 142
    heterodimerization 140–141
    homodimerization 140–141
    identification 136
    inactivation in F9 teratocarcinoma cells 144
    isoforms 138–139
    ligand-binding domain structure 139–140
    ligand specificity 137–138
    PML fusion protein 253
    retinoid receptor knockout mutant mice 142–144
    subtype roles 142–144
    transcriptional regulation mechanism 140–142
retinoic acid response elements (RAREs) 140, 141
retinoic acids 44
retinoid binding proteins (RBPs) 44
retinoid receptor knockout mutant mice 142–144
retinoid X receptor response elements (RXREs) 140, 141
retinoid X receptors (RXRs) 44, 74, 123, 135
    co-repressors/-activators 141–142
    domain structure 136–137
    expression pattern 142
    FXR heterodimerization 312
    heterodimerization 140–141
    homodimerization 140–141
    identification 136
    isoforms 138–139
    ligand-binding domain structure 139–140
    ligand specificity 137–138
    LXR heterodimerization 311–212
    PPAR system 124
    retinoid receptor knockout mutant mice 142–144
    subtype roles 142–144
    transcriptional regulation mechanism 140–142
    vitamin D interaction 357
retinol, cataract 340
retinol-binding protein (RBP) 133
retinyl esters 43–44

RFC-1 37
riboflavin 49
    cataract 334, *337*, 340
rice, golden 382, 383
RING finger proteins 221–222
RNA
    human dietary intake 388
    nucleosides 25
    nucleotide role 363
    synthesis 354, 358
rough endoplasmic reticulum (RER)
    apo-B48 synthesis 303
    apo-B100 298
    triglyceride transfer 300

S14 gene 92, 93, 94
    inhibition 95
safingol 190, 192–193
salicylates 83–84
selective oestrogen receptor modulators (SERMs) 207, 209
selenium 174–180
    apoptosis prevention 178
    cancer 178–179
    cell cycle 178–179
    coronary heart disease 179
    deficiency and gene expression 179–180
    immune function 359–360
    reactive oxygen species 178
selenocysteine
    incorporation into eukaryotes 174–175
    mammalian proteins 176
selenocysteine insertion sequence (SECIS) 175, 176, 177
selenocysteine-specific elongation factor (SELB) 174
selenocysteine synthase (SELA) 174
selenophosphate synthetase (SELD) 174
selenoproteins
    biosynthesis 174–176, 177
    expression regulation 176–177
    importance 177–179
self-organizing map (SOM) 7, 8–10
sequence-tagged sites (STSs) 2
serum screening, specific/targeted 391
shellfish allergens 373, 375
short chain peptides, transmembrane transport 33–35
shuttle mechanisms 54
silencing mediator for RAR and TR (SMRT) 141
single nucleotide polymorphisms (SNPs) 13, 19
SIR2 282
sister chromatid exchange (SCE) 283, 284, 286
skeletal muscle
    myofibrillar protein breakdown 230, *231*
    proteasome-dependent pathways 219
    proteasomes 228–230
    proteolysis rate 228–229

ubiquitin 226
ubiquitin-activating enzyme 227
ubiquitin-conjugating enzyme 227
ubiquitin–protein ligases 227
ubiquitination mechanisms 219–230, *231*
ubiquitin–proteasome-dependent proteolysis regulation 226–230
skin
    food allergy 370–371
    niacin status 285
skin cancer
    niacin deficiency 285, 287
    sphingolipids 190
SLC2A group 22
SLC16A1–SLC16A9 genes 26
SLC27A family 28
small heterodimer partner (SHP) 313
small intestine, villus/crypt gradient cells 302
smoking
    cataract 323, 327
    folate levels and lung cancer 247–248
smooth endoplasmic reticulum (SER) 303
    apo-B100 298
SMRT (silencing mediator for RAR and TR) 141
SN1 transporter 32
SN2 transporter 32
SNPs (single nucleotide polymorphisms) 13, 19
sodium-dependent glucose transporter 1 (SGLT-1) 23, 24
    expression 25
sodium-dependent glucose transporter (SGLT) family 23–25
    malfunction 25
sodium-dependent multivitamin transporter (SMVT) 37, 63
    expression 66, *67*
soy beans
    agricultural biotechnology impact 381–382
    animal feeding 392
    Brazil nut high methionine protein introduction 392
soy proteins 201, 205
    biological activities 210
    infant growth/development 211
    infant nutrition 206–207, 208, 376
    reproductive effects 210–211
    see also daidzein; genistein; glycitein
spectrum karyotyping (SKY) 3–4
spermatogenesis
    biotin deficiency 154
    histones 268
sphingolipids
    apoptosis 188, 193
    cancer 189–196
        suppression mechanisms 193–196
    β-catenin modulation 193–195
    colon cancer 189

    dietary and tumour formation 190–191, 194–195
    foods 189
    intracellular pathways *196*
    metabolism 187, *188*
    metabolites in chemotherapy 192–193
    metastasis 195
    second messengers 187–189
    skin cancer 190
    structure 187, 192
    tumour growth inhibition 193
sphingomyelin, dietary 190
sphingosine
    skin cancer 190
    tumour growth inhibition 193, 194–195
sphingosine-1-phosphate 195
sphingosine kinase 188
squash, virus-resistant 381, 382
SREBP cleavage-activating protein (SCAP) 310
starvation
    protein catabolism 230
    see also malnutrition
stearic acid 315
steroid hormones 158–159
sterol regulatory element (SRE) 310, 314
sterol regulatory element-binding protein(s) (SREBPs) 309–311, 314
    isoforms 311
sterol regulatory element-binding protein type 1 (SREBP1) 128–129, 311, 312
sterol regulatory element-binding protein type 2 (SREBP2) 311
sterols 310
substantial equivalence concept 386–387
sucrose non-fermenter gene (SNF2) 253–254
sulphotransferases 212
sulphur amino acids 362
sunflower oil 300
survival signals 75, *77*
SVCT transporter proteins 37

T-cell receptors 351
T lymphocytes 350–351
    proliferation 363
    vitamin D effects 356
    see also cytotoxic T cells; helper T cells
T3 and T4 transport 33
tankyrase 281, 283
telomere repeat-binding factor-1 (TRF-1) 281
telomeres 282, 284
testosterone, biotin deficiency 154
thiamine 48–49
thiamine-responsive megaloblastic anaemia (TRMA) syndrome 37
thiamine transporter (ThTr1) 37
thioredoxin 178
thioredoxin reductase, selenium-dependent 360

thioredoxin/thioredoxin reductase system 48
thymus involution 74
thyroid hormone transport 33
tocopherol(s) 45
    intracellular transport 171
    lens 324
    vitamin E absorption/transport 170
α-tocopherol 169
    anti-atherogenic effect 174
    TAP binding 171
γ-tocopherol 334
tocopherol-associated protein (TAP) 167, 171
tocopherol-binding protein, membrane (TBPpms) 170
α-tocopherol transfer protein (TTP) 167, 170
tocotrienol(s) 45
α-tocotrienol 169
tomatoes, genetically modified 382
TP53 gene 80
transcription factors 122–123
transcriptomics 13, 19
transgenic crops 381–382
    see also agricultural biotechnology
transgenics 14
    see also genetically modified foods
tributyrin 84
triglycerides 92
    transfer from RER membrane 300
tripeptidyl-peptidase II (TPPII) 230
tRNA^{Sec} 174, 176
tropomyosin 373, 375
α-tropomyosin 173
TrxR 177, 178
tumour cells
    gangliosides 195
    gene regulation 129–130
tumour necrosis factor (TNF) 75–76, 77, 351, 353
    malnutrition 355
two-dimensional polyacrylamide gel electrophoresis (2D-PAGE) 16

U-matrix (unified distance matrix) 9
ubiquitin
    conjugates 227–228
    conjugation regulation 226–228
    skeletal muscle 226
ubiquitin-activating enzyme 221, 227
ubiquitin carboxy-terminal hydrolases (UCH) 222, 223
ubiquitin-conjugating enzyme 221, 227
ubiquitin–protein ligases 221–222, 227
ubiquitin-specific processing proteases (UBPs) 222, 223, 228
ubiquitin-specific protease (USP) 228
ubiquitination 220–222
    histones 267
    molecular basis 223–226
    rate in skeletal muscle 227–228

ubiquitin–proteasome pathway 219
    proteolysis 223–226
        mechanisms 220–223
        regulation 226–230
UDP-glucuronosyltransferases 212
uncoupling proteins (UCPs), mitochondrial 77
United States Department of Agriculture (USDA) 385
upstream stimulatory factor (USF) 96–97
UV-mediated cytotoxicity 162

vascular cell adhesion molecule 1 (VCAM-1) 172, 173
vasopressin 35
vault-PARP 281–282, 283
very low density lipoprotein (VLDL) 45, 92, 126, 293, *294*, 295
    assembly 295–296
        liver 297–300
        nutrient effects 300
        secretory apparatus 298
        two-step model 298–300
    characteristics 295–296
    post-translational regulation of secretion 298
    production rate regulation 300
    subclasses 300
    transformation to LDL 314
    vitamin E transport 170
viral infection
    resistance 381, 382
    selenium 360
vitamin(s)
    deficiency 241
    fat-soluble 81–82
    intracellular trafficking 43–54
    multivitamin supplements 339–340, 360
    supplements in cataract risk 340
    transport 36–38
vitamin A 135–144
    cataract 339, 340
    deficiency 382
    immune function 355–356
    intracellular trafficking 43–44
    signal transduction pathway *136*
vitamin B_1 *see* thiamine
vitamin B_2 *see* riboflavin
vitamin B_6 50–51, 158–159
    albumin gene expression 159
    cancer 159
    cell proliferation 158
    deficiency 158, 241
    immune function 356
vitamin B_{12}
    deficiency 79, 241
    immune function 356
    lung cancer in smokers 248
    transport 53–54

vitamin C 159–162
    antioxidant activity 159
    cancer 162
    cataract 326–327, *328*, 329, *330*, 331, 339, 340
    cell proliferation/differentiation 160
    chondrogenesis 160
    collagen gene expression 160–161
    cytotoxicity 160
    deficiency 161
    extracellular matrix production 160, 161
    immune function 356
    pregnancy 161
    transport 54
vitamin D 44–46
    immune function 356–357
vitamin E 46, 82, 167–180
    absorption 170–171
    antioxidant properties 167–170
    cataract 327, 329, 331, *332*, 333–334, 339, 340
    cell adhesion proteins 173–174
    chemokines 173–174
    COX regulation 171–172
    deficiency and gene expression 179–180
    immune function 357–358
    NF-κB 172, *173*
    nitric oxide activity 174
    platelet aggregation 174
    regeneration 170
    scavenger receptors 173–174
    transport 170–171
vitamin K 47–48
VLDL1 300
VLDL2 300

water transport pathways 35–36
wheat gliadins 376
World Health Organization (WHO) 386

XaG transport system 32
Xc transport system 32
xylulose-5-phosphate 93–94

zeaxanthin, cataract 334, 336, 338
zinc
    cataract 339, 340
    deficiency 358
    immune function 358–359

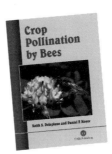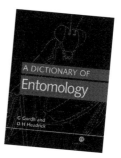